Subject	Symbol	Meaning	Page
Counting and Probability	$N(A)$	the number of elements in set A	405
	$P(A)$	the probability of a set A	405
	$P(n, r)$	the number of r-permutations of a set of n elements	418
	$\binom{n}{r}$	n choose r, the number of r-combinations of a set of n elements, the number of r-element subsets of a set of n elements	182, 447
	ϵ	the null string	414
Functions	$f : X \to Y$	f is a function from X to Y	294
	$f(x)$	the value of f at x	294
	$x \xrightarrow{f} y$	f sends x to y	294
	$f(A)$	the image of A	305
	$f^{-1}(C)$	the inverse image of C	305
	I_x	the identity function on X	297
	b^x	b raised to the power x	312
	$\exp_b(x)$	b raised to the power x	312
	$\log_b(x)$	logarithm with base b of x	299
	F^{-1}	the inverse function of F	317
	$f \circ g$	the composition of g and f	322
Relations	$x\, R\, y$	x is related to y by R	14
	R^{-1}	the inverse relation of R	347
	$m \equiv n \pmod{d}$	m is congruent to n modulo d	363
	$[a]$	the equivalence class of a	364
	Z_n	the set of equivalence classes of integers modulo n	381
Graphs and Trees	$V(G)$	the set of vertices of a graph G	477
	$E(G)$	the set of edges of a graph G	477
	$\{v, w\}$	the edge joining v and w in a simple graph	483, 484
	K_n	complete graph on n vertices	484
	$K_{m,n}$	complete bipartite graph on (m, n) vertices	484
	$deg(v)$	degree of vertex v	486
	$v_0 e_1 v_1 e_2 \cdots e_n v_n$	a walk from v_0 to v_n	495

DISCRETE MATHEMATICS

DISCRETE MATHEMATICS

AN INTRODUCTION TO MATHEMATICAL REASONING

SUSANNA S. EPP
DePaul University

BROOKS/COLE
CENGAGE Learning™

Australia • Brazil • Japan • Korea • Mexico • Singapore • Spain • United Kingdom • United States

Cover Photo: *The stones are discrete objects placed one on top of another like a chain of careful reasoning. A person who decides to build such a tower aspires to the heights and enjoys playing with a challenging problem. Choosing the stones takes both a scientific and an aesthetic sense. Getting them to balance requires patient effort and careful thought. And the tower that results is beautiful. A perfect metaphor for discrete mathematics!*

Discrete Mathematics: An Introduction to Mathematical Reasoning
Susanna S. Epp

Publisher: Richard Stratton

Senior Sponsoring Editor: Molly Taylor

Assistant Editor: Shaylin Walsh

Editorial Assistant: Alexander Gontar

Associate Media Editor: Andrew Coppola

Senior Marketing Manager:
Jennifer Pursley Jones

Marketing Communications Manager:
Mary Anne Payumo

Marketing Coordinator: Michael Ledesma

Content Project Manager: Alison Eigel Zade

Senior Art Director: Jill Ort

Senior Print Buyer: Diane Gibbons

Right Acquisition Specialists:
Timothy Sisler and Don Schlotman

Production Management and Composition:
Integra

Photo Manager: Chris Althof,
Bill Smith Group

Cover Designer: Hanh Luu

Cover Image: GettyImages.com

For product information and technology assistance, contact us at
Cengage Learning Customer & Sales Support, 1-800-354-9706.

For permission to use material from this text or product,
submit all requests online at **www.cengage.com/permissions**.
Further permissions questions can be emailed to
permissionrequest@cengage.com.

Library of Congress Control Number: 2010940881

Student Edition:
ISBN-13: 978-0-495-82617-0
ISBN-10: 0-495-82617-0

Brooks/Cole
20 Channel Center Street
Boston, MA 02210
USA

Cengage Learning is a leading provider of customized learning solutions with office locations around the globe, including Singapore, the United Kingdom, Australia, Mexico, Brazil, and Japan. Locate your local office at: **international.cengage.com/region.**

Cengage Learning products are represented in Canada by
Nelson Education, Ltd.

For your course and learning solutions, visit
www.cengage.com

Purchase any of our products at your local college store or at our preferred online store **www.cengagebrain.com.**

Printed in the United States of America
5 6 7 18 17 16 15

To my students, in appreciation for all they have taught me

CONTENTS

PREFACE

My purpose in developing this book was to provide a clear, accessible treatment of the essential aspects of discrete mathematics with a special focus on introducing students to mathematical proof. The book is based on the fourth edition of my Discrete Mathematics with applications, which has been used successfully by students at hundreds of institutions in North and South America, Europe, the Middle East, Asia, and Australia.

This version was written in response to requests by its users interested in a shorter, more streamlined treatment of the subject. Like its predecessor, however, its goal is to lay a mathematical foundation for upper-level courses in mathematics and computer science. The book may be used by students either before or after a course in calculus; a good background in algebra is the only prerequisite.

Recent curricular recommendations from the Institute for Electrical and Electronic Engineers Computer Society (IEEE-CS) and the Association for Computing Machinery (ACM) include discrete mathematics as the largest portion of "core knowledge" for computer science students and state that students should take at least a one-semester course in the subject as part of their first-year studies, with a two-semester course preferred when possible. This book includes most of the topics recommended by those organizations; the ones that it omits are often covered elsewhere in computer science programs. Coverage of sets, relations, and functions is foundational for all mathematics and computer science courses; inclusion of basic number theory provides background for students' future study of abstract algebra, certain computer algorithms, and cryptography; extensive work with quantifiers is especially useful for students who will take a course in real analysis or go on to study artificial intelligence, database theory, or techniques for establishing program correctness; and discussion of counting principles, graph theory, and the calculation of the likelihood of events prepares the way for further study of combinatorics and probability.

At one time, most of the topics in discrete mathematics were taught only to upper-level undergraduates. Discovering how to present these topics in ways that can be understood by first- and second-year students was the major and most interesting challenge of the work I have done. The presentation has been developed over a long period of experimentation during which my students were in many ways my teachers. Their questions, comments, and written work continue to show me what concepts and techniques cause them difficulty, and their reaction to my exposition shows me what works to build their understanding and to encourage their interest.

Themes of a Discrete Mathematics Course

Discrete mathematics describes processes that consist of a sequence of individual steps. This contrasts with calculus, which describes processes that change in a continuous fashion. Whereas the ideas of calculus were fundamental to the science and technology of the industrial revolution, the ideas of discrete mathematics underlie the science and technology of the computer age. Important themes of a first course in discrete mathematics are logic and proof, induction and recursion, discrete structures, and combinatorics and discrete probability.

Logic and Proof Probably the most important goal of a first course in discrete mathematics is to help students develop the ability to think abstractly. This means learning to use logically valid forms of argument and avoid common logical errors, appreciating what it means to reason from definitions, knowing how to use both direct and indirect argument to derive new results from those already known to be true, and being able to work with symbolic representations as if they were concrete objects.

Induction and Recursion An exciting development of recent years has been the increased appreciation for the power and beauty of "recursive thinking." To think recursively means to address a problem by assuming that similar problems of a smaller nature have already been solved and figuring out how to put those solutions together to solve the larger problem. Such thinking is used in modeling biological and financial systems, developing data management algorithms, and analyzing algorithms. Recurrence relations that result from recursive thinking often give rise to formulas that are verified by mathematical induction.

Discrete Structures Discrete mathematical structures are the abstract structures that describe, categorize, and reveal the underlying relationships among discrete mathematical objects. Those studied in this book are the sets of integers and rational numbers, general sets, functions, relations, and graphs and trees. In addition, the book includes brief introductions to Boolean algebras and commutative rings and fields.

Combinatorics and Discrete Probability Combinatorics is the mathematics of counting and arranging objects, and probability is the study of laws concerning the measurement of random or chance events. Discrete probability focuses on situations involving discrete sets of objects, such as finding the likelihood of obtaining a certain number of heads when an unbiased coin is tossed a certain number of times. Skill in using combinatorics and probability is needed in almost every discipline where mathematics is applied, from economics to biology, to computer science, to chemistry and physics, to business management.

Special Features of This Book

Mathematical Reasoning The feature that most distinguishes this book from other discrete mathematics texts is that it teaches—explicitly but in a way that is accessible to first- and second-year college and university students—the unspoken logic and reasoning that underlie mathematical thought. For many years I taught an intensively interactive transition-to-abstract-mathematics course to mathematics and computer science majors. This experience showed me that while it is possible to teach the majority of students to understand and construct straightforward mathematical arguments, the obstacles to doing so cannot be passed over lightly. To be successful, a text for such a course must address students' difficulties with logic and language directly and at some length. It must also include enough concrete examples and exercises to enable students to develop the mental models needed to conceptualize more abstract problems. The treatment of logic and proof in this book blends common sense and rigor in a way that explains the essentials, yet avoids overloading students with formal detail.

Spiral Approach to Concept Development A number of concepts in this book appear in increasingly more sophisticated forms in successive chapters to help students develop the ability to deal effectively with increasing levels of abstraction. For example, by the time students encounter the theory behind modular arithmetic and the solutions for linear Diophantine equations in Sections 8.4 and 8.5, they have been introduced to the logic of mathematical discourse in Chapters 1, 2, and 3, learned the basic methods of proof and the concepts of *mod* and *div* in Chapter 4, explored *mod* and *div* as functions in Chapter 7, and become familiar with equivalence relations in Sections 8.2 and 8.3. This approach builds in useful review and develops mathematical maturity in natural stages.

Support for the Student   Students at colleges and universities inevitably have to learn a great deal on their own. Though it is often frustrating, learning to learn through self-study is a crucial step toward eventual success in a professional career. This book has a number of features to facilitate students' transition to independent learning.

Worked Examples

The book contains over 300 worked examples, which are written using a problem-solution format and are keyed in type and in difficulty to the exercises. Many solutions for the proof problems are developed in two stages: first a discussion of how one might come to think of the proof or disproof and then a summary of the solution, which is enclosed in a box. This format allows students to read the problem and skip immediately to the summary, if they wish, only going back to the discussion if they have trouble understanding the summary. The format also saves time for students who are rereading the text in preparation for an examination.

Marginal Notes and Test Yourself Questions

Notes about issues of particular importance and cautionary comments to help students avoid common mistakes are included in the margins throughout the book. Questions designed to focus attention on the main ideas of each section are located between the text and the exercises. For convenience, the questions use a fill-in-the-blank format, and the answers are found immediately after the exercises.

Exercises

The book contains almost 2000 exercises. The sets at the end of each section have been designed so that students with widely varying backgrounds and ability levels will find some exercises they can be sure to do successfully and also some exercises that will challenge them.

Solutions for Exercises

To provide adequate feedback for students between class sessions, Appendix B contains a large number of complete solutions to exercises. Students are strongly urged not to consult solutions until they have tried their best to answer the questions on their own. Once they have done so, however, comparing their answers with those given can lead to significantly improved understanding. In addition, many problems, including some of the most challenging, have partial solutions or hints so that students can determine whether they are on the right track and make adjustments if necessary. There are also plenty of exercises without solutions to help students learn to grapple with mathematical problems in a realistic environment.

Reference Features

My rationale for screening statements of definitions and theorems, for putting titles on exercises, and for giving the meanings of symbols and a list of reference formulas in the endpapers is to make it easier for students to use this book for review in a current course and as a reference in later ones. Figures and tables are included where doing so would help readers to a better understanding. In most, a second color is used to highlight meaning.

Support for the Instructor   I have received a great deal of valuable feedback from instructors who have used editions of *Discrete Mathematics with Applications*, on which this book is based. Many aspects of this book have been improved through their suggestions. In addition to the following items, there is additional instructor support on the book's website, described later in the preface.

Exercises

The large variety of exercises at all levels of difficulty allows instructors great freedom to tailor a course to the abilities of their students. Exercises with solutions in the back of the book have numbers in blue, and those whose solutions are given in a separate *Student Solutions Manual and Study Guide* have numbers that are a multiple of three. There are exercises of every type represented in this book that have no answer in either location to enable instructors to assign whatever mixture they prefer of exercises with and without answers. The ample number of exercises of all kinds gives instructors a significant choice of problems to use for review assignments and exams. Instructors are invited to use the many exercises stated as questions rather than in "prove that" form to stimulate class discussion on the role of proof and counterexample in problem solving.

Flexible Sections

Most sections are divided into subsections so that an instructor who is pressed for time can choose to cover certain subsections only and either omit the rest or leave them for the students to study on their own. The division into subsections also makes it easier for instructors to break up sections if they wish to spend more then one day on them.

Presentation of Proof Methods

It is inevitable that the proofs and disproofs in this book will seem easy to instructors. Many students, however, find them difficult. In showing students how to discover and construct proofs and disproofs, I have tried to describe the kinds of approaches that mathematicians use when confronting challenging problems in their own research.

Instructor Solutions

Complete instructor solutions to all exercises are available to anyone teaching a course from this book via Cengage's Solution Builder service. Instructors can sign up for access at www.cengage.com/solutionbuilder.

Companion Website
www.cengage.com/math/epp

A website has been developed for this book that contains information and materials for both students and instructors. It includes:

- descriptions and links to many sites on the Internet with accessible information about discrete mathematical topics,
- links to applets that illustrate or provide practice in the concepts of discrete mathematics,
- additional examples and exercises with solutions,
- review guides for the chapters of the book.

A special section for instructors contains:

- suggestions about how to approach the material of each chapter,
- solutions for all exercises not fully solved in Appendix B,
- ideas for projects and writing assignments,
- PowerPoint slides,
- review sheets and additional exercises for quizzes and exams.

Student Solutions Manual and Study Guide
(ISBN-10: 0-495-82618-9; ISBN-13: 978-0-495-82618-7)

In writing this book, I strove to give sufficient help to students through the exposition in the text, the worked examples, and the exercise solutions, so that the book itself would provide all that a student would need to successfully master the material of the course. I believe that students who finish the study of this book with the ability to solve, on their own, all the exercises with full solutions in Appendix B will have developed an excellent command of the subject. Nonetheless, I have become aware that some students want the opportunity to obtain additional helpful materials. In response, I developed a Student Solutions Manual and Study Guide, available separately from this book, which contains complete solutions to every exercise that is not completely answered in Appendix B and whose number is divisible by 3. The guide also includes alternative explanations for some of the concepts and review questions for each chapter.

Organization

The following tree diagram shows, approximately, how the chapters of this book depend on each other. Chapters on different branches of the tree are sufficiently independent that instructors need to make at most minor adjustments if they skip chapters but follow paths along branches of the tree.

In most cases, covering only the core material in each chapter is adequate preparation for moving down the tree.

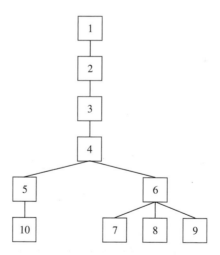

Acknowledgments

I owe a debt of gratitude to many people at DePaul University for their support and encouragement throughout the years I worked on *Discrete Mathematics with Applications*, on which this book is based. A number of my colleagues used early versions and various editions and provided many excellent suggestions for improvement. For this, I am thankful to Louis Aquila, J. Marshall Ash, Allan Berele, Jeffrey Bergen, William Chin, Barbara Cortzen, Constantine Georgakis, Sigrun Goes, Jerry Goldman, Lawrence Gluck, Leonid Krop, Carolyn Narasimhan, Walter Pranger, Eric Rieders, Ayse Sahin, Yuen-Fat Wong, and, most especially, Jeanne LaDuke. The thousands of students to whom I have taught discrete mathematics have had a profound influence on the book's form.

By sharing their thoughts and thought processes with me, they taught me how to teach them better. I am very grateful for their help. I owe the DePaul University administration, especially my dean, Charles Suchar, and my former deans, Michael Mezey and Richard Meister, a special word of thanks for considering the writing of this book a worthwhile scholarly endeavor.

My thanks to the reviewers for their valuable suggestions for editions of *Discrete Mathematics with Applications*: David Addis, Texas Christian University; Itshak Borosh, Texas A & M University; Douglas M. Campbell, Brigham Young University; David G. Cantor, University of California at Los Angeles; C. Patrick Collier, University of Wisconsin-Oshkosh; Kevan H. Croteau, Francis Marion University; Irinel Drogan, University of Texas at Arlington; Pablo Echeverria, Camden County College; Rachel Esselstein, California State University-Monterrey Bay; Henry A. Etlinger, Rochester Institute of Technology; Melvin J. Friske, Wisconsin Lutheran College; William Gasarch, University of Maryland; Ladnor Geissinger, University of North Carolina; Jerrold R. Griggs, University of South Carolina; Nancy Baxter Hastings, Dickinson College; Lillian Hupert, Loyola University Chicago; Joseph Kolibal, University of Southern Mississippi; Benny Lo, International Technological University; George Luger, University of New Mexico; Leonard T. Malinowski, Finger Lakes Community College; William Marion, Valparaiso University; Michael McClendon, University of Central Oklahoma; Steven Miller, Brown University; John F. Morrison, Towson State Unviersity; Paul Pederson, University of Denver; George Peck, Arizona State University; Roxy Peck, California Polytechnic State University, San Luis Obispo; Dix Pettey, University of Missouri; Anthony Ralston, State University of New York at Buffalo; Norman Richert, University of Houston–Clear Lake; John Roberts, University of Louisville; and George Schultz, St. Petersburg Junior College, Clearwater. Special thanks are due John Carroll, San Diego State University; Dr. Joseph S. Fulda; and Porter G. Webster, University of Southern Mississippi; Peter Williams, California State University at San Bernardino; and Jay Zimmerman, Towson University for their unusual thoroughness and their encouragement.

I have also benefitted greatly from the suggestions of the many instructors who have generously offered me their ideas for improvement based on their experiences with *Discrete Mathematics with Applications*, especially Jonathan Goldstine, Pennsylvania State University; David Hecker, St. Joseph's University; Edward Huff, Northern Virginia Community College; Robert Messer, Albion College; Sophie Quigley, Ryerson University; Piotr Rudnicki, University of Alberta; Anwar Shiek, Diné College; Norton Starr, Amherst College; and Eng Wee, National University of Singapore. Production of the third edition received valuable assistance from Christopher Novak, University of Michigan, Dearborn, and Ian Crewe, Ascension Collegiate School. I am especially grateful for the many excellent suggestions for improvement made by Tom Jenkyns, Brock University.

I owe many thanks to the Brooks/Cole staff, especially my editors, Dan Seibert and Shaylin Walsh, for their advice and direction during the production process, and my previous editors, Stacy Green, Robert Pirtle, Barbara Holland, and Heather Bennett, for their encouragement and enthusiasm.

The older I get the more I realize the profound debt I owe my own mathematics teachers for shaping the way I perceive the subject. My first thanks must go to my husband, Helmut Epp, who, on a high school date (!), introduced me to the power and beauty of the field axioms and the view that mathematics is a subject with ideas as well as formulas and techniques. In my formal education, I am most grateful to Daniel Zelinsky and Ky Fan at Northwestern University and Izaak Wirszup, I. N. Herstein, and Irving Kaplansky at the University of Chicago, all of whom, in their own ways, helped lead me to appreciate the elegance, rigor, and excitement of mathematics.

To my family, I owe thanks beyond measure. I am grateful to my mother, whose keen interest in the workings of the human intellect started me many years ago on the track that led ultimately to this book, and to my late father, whose devotion to the written word has been a constant source of inspiration. I thank my children and grandchildren for their affection and cheerful acceptance of the demands this book has placed on my life. And, most of all, I am grateful to my husband, who for many years has encouraged me with his faith in the value of this project and supported me with his love and his wise advice.

Susanna Epp

CHAPTER 1

SPEAKING MATHEMATICALLY

Therefore O students study mathematics and do not build without foundations. —Leonardo da Vinci (1452–1519)

The aim of this book is to introduce you to a mathematical way of thinking that can serve you in a wide variety of situations. Often when you start work on a mathematical problem, you may have only a vague sense of how to proceed. You may begin by looking at examples, drawing pictures, playing around with notation, rereading the problem to focus on more of its details, and so forth. The closer you get to a solution, however, the more your thinking has to crystallize. And the more you need to understand, the more you need language that expresses mathematical ideas clearly, precisely, and unambiguously.

This chapter will introduce you to some of the special language that is a foundation for much mathematical thought, the language of variables, sets, relations, and functions. Think of the chapter like the exercises you would do before an important sporting event. Its goal is to warm up your mental muscles so that you can do your best.

1.1 Variables

A variable is sometimes thought of as a mathematical "John Doe" because you can use it as a placeholder when you want to talk about something but either (1) you imagine that it has one or more values but you don't know what they are, or (2) you want whatever you say about it to be equally true for all elements in a given set, and so you don't want to be restricted to considering only a particular, concrete value for it. To illustrate the first use, consider asking

> Is there a number with the following property: doubling it and adding 3 gives the same result as squaring it?

In this sentence you can introduce a variable to replace the potentially ambiguous word "it":

> Is there a number x with the property that $2x + 3 = x^2$?

The advantage of using a variable is that it allows you to give a temporary name to what you are seeking so that you can perform concrete computations with it to help discover its possible values. To emphasize the role of the variable as a placeholder, you might write the following:

> Is there a number $\square$ with the property that $2 \cdot \square + 3 = \square^2$?

The emptiness of the box can help you imagine filling it in with a variety of different values, some of which might make the two sides equal and others of which might not.

To illustrate the second use of variables, consider the statement:

> No matter what number might be chosen, if it is greater than 2,
> then its square is greater than 4.

In this case introducing a variable to give a temporary name to the (arbitrary) number you might choose enables you to maintain the generality of the statement, and replacing all instances of the word "it" by the name of the variable ensures that possible ambiguity is avoided:

> No matter what number n might be chosen, if n is greater than 2,
> then n^2 is greater than 4.

Example 1.1.1 Writing Sentences Using Variables

Use variables to rewrite the following sentences more formally.

a. Are there numbers with the property that the sum of their squares equals the square of their sum?

b. Given any real number, its square is nonnegative.

Solution

Note In part (a) the answer is yes. For instance, $a = 1$ and $b = 0$ would work. Can you think of other numbers that would also work?

a. Are there numbers a and b with the property that $a^2 + b^2 = (a + b)^2$?
   Or: Are there numbers a and b such that $a^2 + b^2 = (a + b)^2$?
   Or: Do there exist any numbers a and b such that $a^2 + b^2 = (a + b)^2$?

b. Given any real number r, r^2 is nonnegative.
   Or: For any real number r, $r^2 \geq 0$.
   Or: For all real numbers r, $r^2 \geq 0$. ∎

Some Important Kinds of Mathematical Statements

Three of the most important kinds of sentences in mathematics are universal statements, conditional statements, and existential statements:

> A **universal statement** says that a certain property is true for all elements in a set. (For example: *All positive numbers are greater than zero.*)
>
> A **conditional statement** says that if one thing is true then some other thing also has to be true. (For example: *If 378 is divisible by 18, then 378 is divisible by 6.*)
>
> Given a property that may or may not be true, an **existential statement** says that there is at least one thing for which the property is true. (For example: *There is a prime number that is even.*)

In later sections we will define each kind of statement carefully and discuss all of them in detail. The aim here is for you to realize that combinations of these statements can be expressed in a variety of different ways. One way uses ordinary, everyday language and another expresses the statement using one or more variables. The exercises are designed to help you start becoming comfortable in translating from one way to another.

Universal Conditional Statements

Universal statements contain some variation of the words "for all" and conditional statements contain versions of the words "if-then." A ***universal conditional statement*** is a statement that is both universal and conditional. Here is an example:

> For all animals a, if a is a dog, then a is a mammal.

One of the most important facts about universal conditional statements is that they can be rewritten in ways that make them appear to be purely universal or purely conditional. For example, the previous statement can be rewritten in a way that makes its conditional nature explicit but its universal nature implicit:

> If a is a dog, then a is a mammal.
> *Or*: If an animal is a dog, then the animal is a mammal.

The statement can also be expressed so as to make its universal nature explicit and its conditional nature implicit:

> For all dogs a, a is a mammal.
> *Or*: All dogs are mammals.

The crucial point is that the ability to translate among various ways of expressing universal conditional statements is enormously useful for doing mathematics and many parts of computer science.

Example 1.1.2 Rewriting a Universal Conditional Statement

Fill in the blanks to rewrite the following statement:

> For all real numbers x, if x is nonzero then x^2 is positive.

a. If a real number is nonzero, then its square ____.

Note If you introduce x in the first part of the sentence, be sure to include it in the second part of the sentence.

b. For all nonzero real numbers x, ____.

c. If x ____, then ____.

d. The square of any nonzero real number is ____.

e. All nonzero real numbers have ____.

Solution

a. is positive

b. x^2 is positive

c. is a nonzero real number; x^2 is positive

d. positive

e. positive squares (*or*: squares that are positive) ∎

Universal Existential Statements

A ***universal existential statement*** is a statement that is universal because its first part says that a certain property is true for all objects of a given type, and it is existential because its second part asserts the existence of something. For example:

Note For a number b to be an additive inverse for a number a means that $a + b = 0$.

> Every real number has an additive inverse.

In this statement the property "has an additive inverse" applies universally to all real numbers. "Has an additive inverse" asserts the existence of something—an additive inverse—for each real number. However, the nature of the additive inverse depends on the real number; different real numbers have different additive inverses. Knowing that an additive inverse is a real number, you can rewrite this statement in several ways, some less formal and some more formal*:

> All real numbers have additive inverses.
> *Or*: For all real numbers r, there is an additive inverse for r.
> *Or*: For all real numbers r, there is a real number s such that s is an additive inverse for r.

Introducing names for the variables simplifies references in further discussion. For instance, after the third version of the statement you might go on to write: When r is positive, s is negative, when r is negative, s is positive, and when r is zero, s is also zero.

One of the most important reasons for using variables in mathematics is that it gives you the ability to refer to quantities unambiguously throughout a lengthy mathematical argument, while not restricting you to consider only specific values for them.

Example 1.1.3 Rewriting a Universal Existential Statement

Fill in the blanks to rewrite the following statement: Every pot has a lid.

a. All pots _____.

b. For all pots P, there is _____.

c. For all pots P, there is a lid L such that _____.

Solution

a. have lids

b. a lid for P

c. L is a lid for P ∎

Existential Universal Statements

An ***existential universal statement*** is a statement that is existential because its first part asserts that a certain object exists and is universal because its second part says that the object satisfies a certain property for all things of a certain kind. For example:

> There is a positive integer that is less than or equal to every positive integer:

This statement is true because the number one is a positive integer, and it satisfies the property of being less than or equal to every positive integer. We can rewrite the statement in several ways, some less formal and some more formal:

> Some positive integer is less than or equal to every positive integer.
> *Or*: There is a positive integer m that is less than or equal to every positive integer.
> *Or*: There is a positive integer m such that every positive integer is greater than or equal to m.
> *Or*: There is a positive integer m with the property that for all positive integers $n, m \leq n$.

---

*A conditional could be used to help express this statement, but we postpone the additional complexity to a later chapter.

Example 1.1.4 Rewriting an Existential Universal Statement

Fill in the blanks to rewrite the following statement in three different ways:

There is a person in my class who is at least as old as every person in my class.

a. Some _____ is at least as old as _____.

b. There is a person p in my class such that p is _____.

c. There is a person p in my class with the property that for every person q in my class, p is _____.

Solution

a. person in my class; every person in my class

b. at least as old as every person in my class

c. at least as old as q ∎

Some of the most important mathematical concepts, such as the definition of limit of a sequence, can only be defined using phrases that are universal, existential, and conditional, and they require the use of all three phrases "for all," "there is," and "if-then." For example, if $a_1, a_2, a_3, \ldots$ is a sequence of real numbers, saying that

the limit of u_n as n approaches infinity is L

means that

for all positive real numbers ε, **there is** an integer N such that
for all integers n, **if** $n > N$ **then** $-\varepsilon < a_n - L < \varepsilon$.

Test Yourself

Answers to Test Yourself questions are located at the end of each section.

1. A universal statement asserts that a certain property is _____ for _____.

2. A conditional statement asserts that if one thing _____ then some other thing _____.

3. Given a property that may or may not be true, an existential statement asserts that _____ for which the property is true.

Exercise Set 1.1

Appendix B contains either full or partial solutions to all exercises with blue numbers. When the solution is not complete, the exercise number has an **H** next to it. A ✶ next to an exercise number signals that the exercise is more challenging than usual. Be careful not to get into the habit of turning to the solutions too quickly. Make every effort to work exercises on your own before checking your answers. See the Preface for additional sources of assistance and further study.

In each of 1–6, fill in the blanks using a variable or variables to rewrite the given statement.

1. Is there a real number whose square is -1?
  a. Is there a real number x such that _____?
  b. Does there exist _____ such that $x^2 = -1$?

2. Is there an integer that has a remainder of 2 when it is divided by 5 and a remainder of 3 when it is divided by 6?
  a. Is there an integer n such that n has _____?
  b. Does there exist _____ such that if n is divided by 5 the remainder is 2 and if _____?
  Note: There are integers with this property. Can you think of one?

3. Given any two real numbers, there is a real number in between.
   a. Given any two real numbers a and b, there is a real number c such that c is ____.
   b. For any two ____, ____ such that c is between a and b.

4. Given any real number, there is a real number that is greater.
   a. Given any real number r, there is ____ s such that s is ____.
   b. For any ____, ____ such that $s > r$.

5. The reciprocal of any positive real number is positive.
   a. Given any positive real number r, the reciprocal of ____.
   b. For any real number r, if r is ____, then ____.
   c. If a real number r ____, then ____.

6. The cube root of any negative real number is negative.
   a. Given any negative real number s, the cube root of ____.
   b. For any real number s, if s is ____, then ____.
   c. If a real number s ____, then ____.

7. Rewrite the following statements less formally, without using variables. Determine, as best as you can, whether the statements are true or false.
   a. There are real numbers u and v with the property that $u + v < u - v$.
   b. There is a real number x such that $x^2 < x$.
   c. For all positive integers n, $n^2 \geq n$.
   d. For all real numbers a and b, $|a + b| \leq |a| + |b|$.

In each of 8–13, fill in the blanks to rewrite the given statement.

8. For all objects J, if J is a square then J has four sides.
   a. All squares ____.
   b. Every square ____.
   c. If an object is a square, then it ____.
   d. If J ____, then J ____.
   e. For all squares J, ____.

9. For all equations E, if E is quadratic then E has at most two real solutions.
   a. All quadratic equations ____.
   b. Every quadratic equation ____.
   c. If an equation is quadratic, then it ____.
   d. If E ____, then E ____.
   e. For all quadratic equations E, ____.

10. Every nonzero real number has a reciprocal.
    a. All nonzero real numbers ____.
    b. For all nonzero real numbers r, there is ____ for r.
    c. For all nonzero real numbers r, there is a real number s such that ____.

11. Every positive number has a positive square root.
    a. All positive numbers ____.
    b. For any positive number e, there is ____ for e.
    c. For all positive numbers e, there is a positive number r such that ____.

12. There is a real number whose product with every number leaves the number unchanged.
    a. Some ____ has the property that its ____.
    b. There is a real number r such that the product of r ____.
    c. There is a real number r with the property that for every real number s, ____.

13. There is a real number whose product with every real number equals zero.
    a. Some ____ has the property that its ____.
    b. There is a real number a such that the product of a ____.
    c. There is a real number a with the property that for every real number b, ____.

Answers for Test Yourself

1. true; all elements of a set   2. is true; also has to be true   3. there is at least one thing

1.2 The Language of Sets

... when we attempt to express in mathematical symbols a condition proposed in words. First, we must understand thoroughly the condition. Second, we must be familiar with the forms of mathematical expression. —George Polyá (1887–1985)

Use of the word *set* as a formal mathematical term was introduced in 1879 by Georg Cantor (1845–1918). For most mathematical purposes we can think of a set intuitively, as

Cantor did, simply as a collection of elements. For instance, if C is the set of all countries that are currently in the United Nations, then the United States is an element of C, and if I is the set of all integers from 1 to 100, then the number 57 is an element of I.

• Notation

If S is a set, the notation $x \in S$ means that x is an element of S. The notation $x \notin S$ means that x is not an element of S. A set may be specified using the **set-roster notation** by writing all of its elements between braces. For example, $\{1, 2, 3\}$ denotes the set whose elements are 1, 2, and 3. A variation of the notation is sometimes used to describe a very large set, as when we write $\{1, 2, 3, \ldots, 100\}$ to refer to the set of all integers from 1 to 100. A similar notation can also describe an infinite set, as when we write $\{1, 2, 3, \ldots\}$ to refer to the set of all positive integers. (The symbol $\ldots$ is called an **ellipsis** and is read "and so forth.")

The **axiom of extension** says that a set is completely determined by what its elements are—not the order in which they might be listed or the fact that some elements might be listed more than once.

Example 1.2.1 Using the Set-Roster Notation

a. Let $A = \{1, 2, 3\}$, $B = \{3, 1, 2\}$, and $C = \{1, 1, 2, 3, 3, 3\}$. What are the elements of A, B, and C? How are A, B, and C related?

b. Is $\{0\} = 0$?

c. How many elements are in the set $\{1, \{1\}\}$?

d. For each nonnegative integer n, let $U_n = \{n, -n\}$. Find U_1, U_2, and U_0.

Solution

a. A, B, and C have exactly the same three elements: 1, 2, and 3. Therefore, A, B, and C are simply different ways to represent the same set.

b. $\{0\} \neq 0$ because $\{0\}$ is a set with one element, namely 0, whereas 0 is just the symbol that represents the number zero.

c. The set $\{1, \{1\}\}$ has two elements: 1 and the set whose only element is 1.

d. $U_1 = \{1, -1\}$, $U_2 = \{2, -2\}$, $U_0 = \{0, -0\} = \{0, 0\} = \{0\}$.

Certain sets of numbers are so frequently referred to that they are given special symbolic names. These are summarized in the table on the next page.

Symbol	Set
R	set of all real numbers
Z	set of all integers
Q	set of all rational numbers, or quotients of integers

Note The **Z** is the first letter of the German word for integers, *Zahlen*. It stands for the *set* of all integers and should not be used as a shorthand for the word *integer*.

Addition of a superscript $+$ or $-$ or the letters *nonneg* indicates that only the positive or negative or nonnegative elements of the set, respectively, are to be included. Thus $\mathbf{R}^+$ denotes the set of positive real numbers, and $\mathbf{Z}^{nonneg}$ refers to the set of nonnegative integers: 0, 1, 2, 3, 4, and so forth. Some authors refer to the set of nonnegative integers as the set of **natural numbers** and denote it as **N**. Other authors call only the positive integers natural numbers. To prevent confusion, we simply avoid using the phrase *natural numbers* in this book.

The set of real numbers is usually pictured as the set of all points on a line, as shown below. The number 0 corresponds to a middle point, called the *origin*. A unit of distance is marked off, and each point to the right of the origin corresponds to a positive real number found by computing its distance from the origin. Each point to the left of the origin corresponds to a negative real number, which is denoted by computing its distance from the origin and putting a minus sign in front of the resulting number. The set of real numbers is therefore divided into three parts: the set of positive real numbers, the set of negative real numbers, and the number 0. *Note that 0 is neither positive nor negative* Labels are given for a few real numbers corresponding to points on the line shown below.

The real number line is called *continuous* because it is imagined to have no holes. The set of integers corresponds to a collection of points located at fixed intervals along the real number line. Thus every integer is a real number, and because the integers are all separated from each other, the set of integers is called *discrete*. The name *discrete mathematics* comes from the distinction between continuous and discrete mathematical objects.

Another way to specify a set uses what is called the *set-builder notation*.

Note We read the left-hand brace as "the set of all" and the vertical line as "such that." In all other mathematical contexts, however, we do not use a vertical line to denote the words "such that"; we abbreviate "such that" as "s. t." or "s. th." or "· ∋ ·."

• Set-Builder Notation

Let S denote a set and let $P(x)$ be a property that elements of S may or may not satisfy. We may define a new set to be **the set of all elements x in S such that $P(x)$ is true**. We denote this set as follows:

$$\{x \in S \mid P(x)\}$$

the set of all   such that

Occasionally we will write $\{x \mid P(x)\}$ without being specific about where the element x comes from. It turns out that unrestricted use of this notation can lead to genuine contradictions in set theory. We will discuss one of these in Section 6.4 and will be careful to use this notation purely as a convenience in cases where the set S could be specified if necessary.

Example 1.2.2 Using the Set-Builder Notation

Given that $\mathbf{R}$ denotes the set of all real numbers, $\mathbf{Z}$ the set of all integers, and $\mathbf{Z}^+$ the set of all positive integers, describe each of the following sets.

a. $\{x \in \mathbf{R} \mid -2 < x < 5\}$

b. $\{x \in \mathbf{Z} \mid -2 < x < 5\}$

c. $\{x \in \mathbf{Z}^+ \mid -2 < x < 5\}$

Solution

a. $\{x \in \mathbf{R} \mid -2 < x < 5\}$ is the open interval of real numbers (strictly) between -2 and 5. It is pictured as follows:

b. $\{x \in \mathbf{Z} \mid -2 < x < 5\}$ is the set of all integers (strictly) between -2 and 5. It is equal to the set $\{-1, 0, 1, 2, 3, 4\}$.

c. Since all the integers in $\mathbf{Z}^+$ are positive, $\{x \in \mathbf{Z}^+ \mid -2 < x < 5\} = \{1, 2, 3, 4\}$. ■

Subsets

A basic relation between sets is that of subset.

• Definition

If A and B are sets, then A is called a **subset** of B, written $A \subseteq B$, if, and only if, every element of A is also an element of B.
    Symbolically:

$$A \subseteq B \quad \text{means that} \quad \text{For all elements } x, \text{ if } x \in A \text{ then } x \in B.$$

The phrases A *is contained in* B and B *contains* A are alternative ways of saying that A is a subset of B.

It follows from the definition of subset that for a set A not to be a subset of a set B means that there is at least one element of A that is not an element of B. Symbolically:

$$A \nsubseteq B \quad \text{means that} \quad \text{There is at least one element } x \text{ such that } x \in A \text{ and } x \notin B.$$

• Definition

Let A and B be sets. A is a **proper subset** of B if, and only if, every element of A is in B but there is at least one element of B that is not in A.

Example 1.2.3 Subsets

Let $A = \mathbf{Z}^+$, $B = \{n \in \mathbf{Z} \mid 0 \leq n \leq 100\}$, and $C = \{100, 200, 300, 400, 500\}$. Evaluate the truth and falsity of each of the following statements.

a. $B \subseteq A$
b. C is a proper subset of A
c. C and B have at least one element in common
d. $C \subseteq B$ e. $C \subseteq C$

Solution

a. False. Zero is not a positive integer. Thus zero is in B but zero is not in A, and so $B \nsubseteq A$.

b. True. Each element in C is a positive integer and, hence, is in A, but there are elements in A that are not in C. For instance, 1 is in A and not in C.

c. True. For example, 100 is in both C and B.

d. False. For example, 200 is in C but not in B.

e. True. Every element in C is in C. In general, the definition of subset implies that all sets are subsets of themselves. ■

Example 1.2.4 Distinction between $\in$ and $\subseteq$

Which of the following are true statements?

a. $2 \in \{1, 2, 3\}$ b. $\{2\} \in \{1, 2, 3\}$ c. $2 \subseteq \{1, 2, 3\}$
d. $\{2\} \subseteq \{1, 2, 3\}$ e. $\{2\} \subseteq \{\{1\}, \{2\}\}$ f. $\{2\} \in \{\{1\}, \{2\}\}$

Solution   Only (a), (d), and (f) are true.

For (b) to be true, the set $\{1, 2, 3\}$ would have to contain the element $\{2\}$. But the only elements of $\{1, 2, 3\}$ are 1, 2, and 3, and 2 is not equal to $\{2\}$. Hence (b) is false.

For (c) to be true, the number 2 would have to be a set and every element in the set 2 would have to be an element of $\{1, 2, 3\}$. This is not the case, so (c) is false.

For (e) to be true, every element in the set containing only the number 2 would have to be an element of the set whose elements are $\{1\}$ and $\{2\}$. But 2 is not equal to either $\{1\}$ or $\{2\}$, and so (e) is false. ■

Cartesian Products

*Kazimierz Kuratowski
(1896–1980)*

Problemy monthly, July 1959

With the introduction of Georg Cantor's set theory in the late nineteenth century, it began to seem possible to put mathematics on a firm logical foundation by developing all of its various branches from set theory and logic alone. A major stumbling block was how to use sets to define an ordered pair because the definition of a set is unaffected by the order in which its elements are listed. For example, $\{a, b\}$ and $\{b, a\}$ represent the same set, whereas in an ordered pair we want to be able to indicate which element comes first.

In 1914 crucial breakthroughs were made by Norbert Wiener (1894–1964), a young American who had recently received his Ph.D. from Harvard and the German mathematician Felix Hausdorff (1868–1942). Both gave definitions showing that an ordered pair can be defined as a certain type of set, but both definitions were somewhat awkward. Finally, in 1921, the Polish mathematician Kazimierz Kuratowski (1896–1980) published the following definition, which has since become standard. It says that an ordered pair is a set of the form

$$\{\{a\}, \{a, b\}\}.$$

This set has elements, $\{a\}$ and $\{a, b\}$. If $a \neq b$, then the two sets are distinct and a is in both sets whereas b is not. This allows us to distinguish between a and b and say that a is the first element of the ordered pair and b is the second element of the pair. If $a = b$, then we can simply say that a is both the first and the second element of the pair. In this case the set that defines the ordered pair becomes $\{\{a\}, \{a, a\}\}$, which equals $\{\{a\}\}$.

However, it was only long after ordered pairs had been used extensively in mathematics that mathematicians realized that it was possible to define them entirely in terms of sets, and, in any case, the set notation would be cumbersome to use on a regular basis. The usual notation for ordered pairs refers to $\{\{a\}, \{a, b\}\}$ more simply as (a, b).

• Notation

Given elements a and b, the symbol (a, b) denotes the **ordered pair** consisting of a and b together with the specification that a is the first element of the pair and b is the second element. Two ordered pairs (a, b) and (c, d) are equal if, and only if, $a = c$ and $b = d$. Symbolically:

$$(a, b) = (c, d) \quad \text{means that} \quad a = c \text{ and } b = d.$$

Example 1.2.5 Ordered Pairs

a. Is $(1, 2) = (2, 1)$?

b. Is $\left(3, \frac{5}{10}\right) = \left(\sqrt{9}, \frac{1}{2}\right)$?

c. What is the first element of $(1, 1)$?

Solution

a. No. By definition of equality of ordered pairs,

$$(1, 2) = (2, 1) \text{ if, and only if, } 1 = 2 \text{ and } 2 = 1.$$

But $1 \neq 2$, and so the ordered pairs are not equal.

b. Yes. By definition of equality of ordered pairs,

$$\left(3, \frac{5}{10}\right) = \left(\sqrt{9}, \frac{1}{2}\right) \text{ if, and only if, } 3 = \sqrt{9} \text{ and } \frac{5}{10} = \frac{1}{2}.$$

Because these equations are both true, the ordered pairs are equal.

c. In the ordered pair $(1, 1)$, the first and the second elements are both 1. ■

• Definition

Given sets A and B, the **Cartesian product of A and B**, denoted $A \times B$ and read "A cross B," is the set of all ordered pairs (a, b), where a is in A and b is in B. Symbolically:

$$A \times B = \{(a, b) \mid a \in A \text{ and } b \in B\}.$$

Example 1.2.6 Cartesian Products

Let $A = \{1, 2, 3\}$ and $B = \{u, v\}$.

a. Find $A \times B$

b. Find $B \times A$

c. Find $B \times B$

d. How many elements are in $A \times B$, $B \times A$, and $B \times B$?

e. Let **R** denote the set of all real numbers. Describe $\mathbf{R} \times \mathbf{R}$.

Solution

a. $A \times B = \{(1, u), (2, u), (3, u), (1, v), (2, v), (3, v)\}$

b. $B \times A = \{(u, 1), (u, 2), (u, 3), (v, 1), (v, 2), (v, 3)\}$

c. $B \times B = \{(u, u), (u, v), (v, u), (v, v)\}$

Note This is why it makes sense to call a Cartesian product a product!

d. $A \times B$ has six elements. Note that this is the number of elements in A times the number of elements in B. $B \times A$ has six elements, the number of elements in B times the number of elements in A. $B \times B$ has four elements, the number of elements in B times the number of elements in B.

e. $\mathbf{R} \times \mathbf{R}$ is the set of all ordered pairs (x, y) where both x and y are real numbers. If horizontal and vertical axes are drawn on a plane and a unit length is marked off, then each ordered pair in $\mathbf{R} \times \mathbf{R}$ corresponds to a unique point in the plane, with the first and second elements of the pair indicating, respectively, the horizontal and vertical positions of the point. The term **Cartesian plane** is often used to refer to a plane with this coordinate system, as illustrated in Figure 1.2.1.

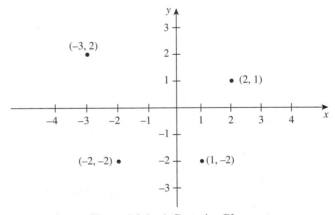

Figure 1.2.1: A Cartesian Plane ■

Test Yourself

1. When the elements of a set are given using the set-roster notation, the order in which they are listed ____.

2. The symbol **R** denotes ____.

3. The symbol **Z** denotes ____.

4. The symbol **Q** denotes ____.

5. The notation $\{x \mid P(x)\}$ is read ____.

6. For a set A to be a subset of a set B means that, ____.

7. Given sets A and B, the Cartesian product $A \times B$ is ____.

Exercise Set 1.2

1. Which of the following sets are equal?

$A = \{a, b, c, d\}$ $B = \{d, e, a, c\}$

$C = \{d, b, a, c\}$ $D = \{a, a, d, e, c, e\}$

2. Write in words how to read each of the following out loud.
  a. $\{x \in \mathbf{R}^+ \mid 0 < x < 1\}$
  b. $\{x \in \mathbf{R} \mid x \le 0 \text{ or } x \ge 1\}$
  c. $\{n \in \mathbf{Z} \mid n \text{ is a factor of } 6\}$
  d. $\{n \in \mathbf{Z}^+ \mid n \text{ is a factor of } 6\}$

3. **a.** Is $4 = \{4\}$?
  b. How many elements are in the set $\{3, 4, 3, 5\}$?
  c. How many elements are in the set $\{1, \{1\}, \{1, \{1\}\}\}$?

4. **a.** Is $2 \in \{2\}$?
  b. How many elements are in the set $\{2, 2, 2, 2\}$?
  c. How many elements are in the set $\{0, \{0\}\}$?
  d. Is $\{0\} \in \{\{0\}, \{1\}\}$?
  e. Is $0 \in \{\{0\}, \{1\}\}$?

H 5. Which of the following sets are equal?

$A = \{0, 1, 2\}$

$B = \{x \in \mathbf{R} \mid -1 \le x < 3\}$

$C = \{x \in \mathbf{R} \mid -1 < x < 3\}$

$D = \{x \in \mathbf{Z} \mid -1 < x < 3\}$

$E = \{x \in \mathbf{Z}^+ \mid -1 < x < 3\}$

H 6. For each integer n, let $T_n = \{n, n^2\}$. How many elements are in each of T_2, T_{-3}, T_1 and T_0? Justify your answers.

7. Use the set-roster notation to indicate the elements in each of the following sets.
  a. $S = \{n \in \mathbf{Z} \mid n = (-1)^k, \text{ for some integer } k\}$.
  b. $T = \{m \in \mathbf{Z} \mid m = 1 + (-1)^i, \text{ for some integer } i\}$.

c. $U = \{r \in \mathbf{Z} \mid 2 \le r \le -2\}$
d. $V = \{s \in \mathbf{Z} \mid s > 2 \text{ or } s < 3\}$
e. $W = \{t \in \mathbf{Z} \mid 1 < t < -3\}$
f. $X = \{u \in \mathbf{Z} \mid u \le 4 \text{ or } u \ge 1\}$

8. Let $A = \{c, d, f, g\}$, $B = \{f, j\}$, and $C = \{d, g\}$. Answer each of the following questions. Give reasons for your answers.
  a. Is $B \subseteq A$?   **b.** Is $C \subseteq A$?
  c. Is $C \subseteq C$?   **d.** Is C a proper subset of A?

9. **a.** Is $3 \in \{1, 2, 3\}$?   **b.** Is $1 \subseteq \{1\}$?
  c. Is $\{2\} \in \{1, 2\}$?   **d.** Is $\{3\} \in \{1, \{2\}, \{3\}\}$?
  e. Is $1 \in \{1\}$?   **f.** Is $\{2\} \subseteq \{1, \{2\}, \{3\}\}$?
  g. Is $\{1\} \subseteq \{1, 2\}$?   **h.** Is $1 \in \{\{1\}, 2\}$?
  i. Is $\{1\} \subseteq \{1, \{2\}\}$?   **j.** Is $\{1\} \subseteq \{1\}$?

10. **a.** Is $((-2)^2, -2^2) = (-2^2, (-2)^2)$?
  b. Is $(5, -5) = (-5, 5)$?
  c. Is $\left(8 - 9, \sqrt[3]{-1}\right) = (-1, -1)$?
  d. Is $\left(\frac{-2}{-4}, (-2)^3\right) = \left(\frac{3}{6}, -8\right)$?

11. Let $A = \{w, x, y, z\}$ and $B = \{a, b\}$. Use the set-roster notation to write each of the following sets, and indicate the number of elements that are in each set:
  a. $A \times B$ b. $B \times A$
  c. $A \times A$ d. $B \times B$

12. Let $S = \{2, 4, 6\}$ and $T = \{1, 3, 5\}$. Use the set-roster notation to write each of the following sets, and indicate the number of elements that are in each set:
  a. $S \times T$ b. $T \times S$
  c. $S \times S$ d. $T \times T$

Answers for Test Yourself

1. does not matter   2. the set of all real numbers   3. the set of all integers   4. the set of all rational numbers   5. the set of all x such that $P(x)$ 6. every element in A is an element in B 7. the set of all ordered pairs (a, b) where a is in A and b is in B

1.3 The Language of Relations and Functions

Mathematics is a language. — Josiah Willard Gibbs (1839–1903)

There are many kinds of relationships in the world. For instance, we say that two people are related by blood if they share a common ancestor and that they are related by marriage if one shares a common ancestor with the spouse of the other. We also speak of the relationship between student and teacher, between people who work for the same employer, and between people who share a common ethnic background.

Similarly, the objects of mathematics may be related in various ways. A set A may be said to be related to a set B if A is a subset of B, or if A is not a subset of B, or if A and B have at least one element in common. A number x may be said to be related to a number y if $x < y$, or if x is a factor of y, or if $x^2 + y^2 = 1$. Two identifiers in a computer program may be said to be related if they have the same first eight characters, or if the same memory location is used to store their values when the program is executed. And the list could go on!

Let $A = \{0, 1, 2\}$ and $B = \{1, 2, 3\}$ and let us say that an element x in A is related to an element y in B if, and only if, x is less than y. Let us use the notation $x\,R\,y$ as a shorthand for the sentence "x is related to y." Then

$$
\begin{array}{llll}
0\ R\ 1 & \text{since} & 0 < 1, \\
0\ R\ 2 & \text{since} & 0 < 2, \\
0\ R\ 3 & \text{since} & 0 < 3, \\
1\ R\ 2 & \text{since} & 1 < 2, \\
1\ R\ 3 & \text{since} & 1 < 3, & \text{and} \\
2\ R\ 3 & \text{since} & 2 < 3.
\end{array}
$$

On the other hand, if the notation $x\,\not\!R\,y$ represents the sentence "x is not related to y," then

$$
\begin{array}{llll}
1\ \not\!R\ 1 & \text{since} & 1 \not< 1, \\
2\ \not\!R\ 1 & \text{since} & 2 \not< 1, & \text{and} \\
2\ \not\!R\ 2 & \text{since} & 2 \not< 2.
\end{array}
$$

Recall that the Cartesian product of A and B, $A \times B$, consists of all ordered pairs whose first element is in A and whose second element is in B:

$$A \times B = \big\{(x, y) \mid x \in A \text{ and } y \in B\big\}.$$

In this case,

$$A \times B = \big\{(0, 1), (0, 2), (0, 3), (1, 1), (1, 2), (1, 3), (2, 1), (2, 2), (2, 3)\big\}.$$

The elements of some ordered pairs in $A \times B$ are related, whereas the elements of other ordered pairs are not. Consider the set of all ordered pairs in $A \times B$ whose elements are related

$$\big\{(0, 1), (0, 2), (0, 3), (1, 2), (1, 3), (2, 3)\big\}.$$

Observe that knowing which ordered pairs lie in this set is equivalent to knowing which elements are related to which. The relation itself can therefore be thought of as the totality of ordered pairs whose elements are related by the given condition. The formal mathematical definition of relation, based on this idea, was introduced by the American mathematician and logician C. S. Peirce in the nineteenth century.

• Definition

Let A and B be sets. A **relation R from A to B** is a subset of $A \times B$. Given an ordered pair (x, y) in $A \times B$, **x is related to y by R**, written $x\,R\,y$, if, and only if, (x, y) is in R. The set A is called the **domain** of R and the set B is called its **co-domain**.

The notation for a relation R may be written symbolically as follows:

$$x\,\boldsymbol{R}\,y \quad \text{means that} \quad (x, y) \in \boldsymbol{R}.$$

The notation $x\,\not\!\boldsymbol{R}\,y$ means that x is not related to y by R:

$$x\,\not\!\boldsymbol{R}\,y \quad \text{means that} \quad (x, y) \notin \boldsymbol{R}.$$

Example 1.3.1 A Relation as a Subset

Let $A = \{1, 2\}$ and $B = \{1, 2, 3\}$ and define a relation R from A to B as follows: Given any $(x, y) \in A \times B$,

$$(x, y) \in R \quad \text{means that} \quad \frac{x - y}{2} \text{ is an integer.}$$

a. State explicitly which ordered pairs are in $A \times B$ and which are in R.

b. Is $1\ R\ 3$? Is $2\ R\ 3$? Is $2\ R\ 2$?

c. What are the domain and co-domain of R?

Solution

a. $A \times B = \{(1, 1), (1, 2), (1, 3), (2, 1), (2, 2), (2, 3)\}$. To determine explicitly the composition of R, examine each ordered pair in $A \times B$ to see whether its elements satisfy the defining condition for R.

$(1, 1) \in R$ because $\frac{1-1}{2} = \frac{0}{2} = 0$, which is an integer.

$(1, 2) \notin R$ because $\frac{1-2}{2} = \frac{-1}{2}$, which is not an integer.

$(1, 3) \in R$ because $\frac{1-3}{2} = \frac{-2}{2} = -1$, which is an integer.

$(2, 1) \notin R$ because $\frac{2-1}{2} = \frac{1}{2}$, which is not an integer.

$(2, 2) \in R$ because $\frac{2-2}{2} = \frac{0}{2} = 0$, which is an integer.

$(2, 3) \notin R$ because $\frac{2-3}{2} = \frac{-1}{2}$, which is not an integer.

Thus

$$R = \{(1, 1), (1, 3), (2, 2)\}$$

b. Yes, $1\ R\ 3$ because $(1, 3) \in R$.
No, $2\ \rlap{/}{R}\ 3$ because $(2, 3) \notin R$.
Yes, $2\ R\ 2$ because $(2, 2) \in R$.

c. The domain of R is $\{1, 2\}$ and the co-domain is $\{1, 2, 3\}$. ■

Example 1.3.2 The Circle Relation

Define a relation C from $\mathbf{R}$ to $\mathbf{R}$ as follows: For any $(x, y) \in \mathbf{R} \times \mathbf{R}$,

$$(x, y) \in C \quad \text{means that} \quad x^2 + y^2 = 1.$$

a. Is $(1, 0) \in C$? Is $(0, 0) \in C$? Is $\left(-\frac{1}{2}, \frac{\sqrt{3}}{2}\right) \in C$? Is $-2\ C\ 0$? Is $0\ C\ (-1)$? Is $1\ C\ 1$?

b. What are the domain and co-domain of C?

c. Draw a graph for C by plotting the points of C in the Cartesian plane.

Solution

a. Yes, $(1, 0) \in C$ because $1^2 + 0^2 = 1$.
No, $(0, 0) \notin C$ because $0^2 + 0^2 = 0 \neq 1$.
Yes, $\left(-\frac{1}{2}, \frac{\sqrt{3}}{2}\right) \in C$ because $\left(-\frac{1}{2}\right)^2 + \left(\frac{\sqrt{3}}{2}\right)^2 = \frac{1}{4} + \frac{3}{4} = 1$.
No, $-2\ \rlap{/}{C}\ 0$ because $(-2)^2 + 0^2 = 4 \neq 1$.
Yes, $0\ C\ (-1)$ because $0^2 + (-1)^2 = 1$.
No, $1\ \rlap{/}{C}\ 1$ because $1^2 + 1^2 = 2 \neq 1$.

b. The domain and co-domain of C are both $\mathbf{R}$, the set of all real numbers.

c.

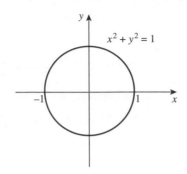

Arrow Diagram of a Relation

Suppose R is a relation from a set A to a set B. The **arrow diagram for R** is obtained as follows:

1. Represent the elements of A as points in one region and the elements of B as points in another region.

2. For each x in A and y in B, draw an arrow from x to y if, and only if, x is related to y by R. Symbolically:

Draw an arrow from x to y

if, and only if, $x \ R \ y$

if, and only if, $(x, y) \in R.$

Example 1.3.3 Arrow Diagrams of Relations

Let $A = \{1, 2, 3\}$ and $B = \{1, 3, 5\}$ and define relations S and T from A to B as follows: For all $(x, y) \in A \times B$,

$$(x, y) \in S \quad \text{means that} \quad x < y \quad \text{S is a "less than" relation.}$$

$$T = \{(2, 1), (2, 5)\}.$$

Draw arrow diagrams for S and T.

Solution

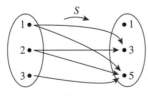 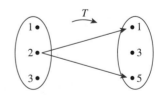

These example relations illustrate that it is possible to have several arrows coming out of the same element of A pointing in different directions. Also, it is quite possible to have an element of A that does not have an arrow coming out of it. ■

Functions

In Section 1.2 we showed that ordered pairs can be defined in terms of sets and we defined Cartesian products in terms of ordered pairs. In this section we introduced relations as subsets of Cartesian products. Thus we can now define functions in a way that depends only on the concept of set. Although this definition is not obviously related to the way we usually work with functions in mathematics, it is satisfying from a theoretical point of view and computer scientists like it because it is particularly well suited for operating with functions on a computer.

• Definition

A **function F from a set A to a set B** is a relation with domain A and co-domain B that satisfies the following two properties:

1. For every element x in A, there is an element y in B such that $(x, y) \in F$.

2. For all elements x in A and y and z in B,

$$\text{if} \quad (x, y) \in F \text{ and } (x, z) \in F, \quad \text{then} \quad y = z.$$

Properties (1) and (2) can be stated less formally as follows: A relation F from A to B is a function if, and only if:

1. Every element of A is the first element of an ordered pair of F.

2. No two distinct ordered pairs in F have the same first element.

In most mathematical situations we think of a function as sending elements from one set, the domain, to elements of another set, the co-domain. Because of the definition of function, each element in the domain corresponds to one and only one element of the co-domain.

More precisely, if F is a function from a set A to a set B, then given any element x in A, property (1) from the function definition guarantees that there is at least one element of B that is related to x by F and property (2) guarantees that there is at most one such element. This makes it possible to give the element that corresponds to x a special name.

• Notation

If A and B are sets and F is a function from A to B, then given any element x in A, the unique element in B that is related to x by F is denoted $F(x)$, which is read "F of x."

Example 1.3.4 Functions and Relations on Finite Sets

Let $A = \{2, 4, 6\}$ and $B = \{1, 3, 5\}$. Which of the relations R, S, and T defined below are functions from A to B?

a. $R = \{(2, 5), \ (4, 1), \ (4, 3), \ (6, 5)\}$

b. For all $(x, y) \in A \times B$, $(x, y) \in S$ means that $y = x + 1$.

c. T is defined by the arrow diagram

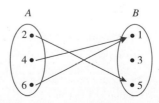

Solution

a. R is not a function because it does not satisfy property (2). The ordered pairs (4, 1) and (4, 3) have the same first element but different second elements. You can see this graphically if you draw the arrow diagram for R. There are two arrows coming out of 4: One points to 1 and the other points to 3.

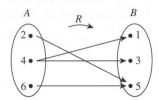

b. S is not a function because it does not satisfy property (1). It is not true that every element of A is the first element of an ordered pair in S. For example, $6 \in A$ but there is no y in B such that $y = 6 + 1 = 7$. You can also see this graphically by drawing the arrow diagram for S.

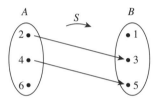

Note In part (c), $T(4) = T(6)$. This illustrates the fact that although no element of the domain of a function can be related to more than one element of the co-domain, several elements in the domain can be related to the same element in the co-domain.

c. T is a function: Each element in $\{2, 4, 6\}$ is related to some element in $\{1, 3, 5\}$ and no element in $\{2, 4, 6\}$ is related to more than one element in $\{1, 3, 5\}$. When these properties are stated in terms of the arrow diagram, they become (1) there is an arrow coming out of each element of the domain, and (2) no element of the domain has more than one arrow coming out of it. So you can write $T(2) = 5$, $T(4) = 1$, and $T(6) = 1$. ■

Example 1.3.5 Functions and Relations on Sets of Real Numbers

a. In Example 1.3.2 the circle relation C was defined as follows:

$$\text{For all } (x, y) \in \mathbf{R} \times \mathbf{R}, \quad (x, y) \in C \quad \text{means that} \quad x^2 + y^2 = 1.$$

Is C a function? If it is, find $C(0)$ and $C(1)$.

b. Define a relation from $\mathbf{R}$ to $\mathbf{R}$ as follows:

$$\text{For all } (x, y) \in \mathbf{R} \times \mathbf{R}, \quad (x, y) \in L \quad \text{means that} \quad y = x - 1.$$

Is L a function? If it is, find $L(0)$ and $L(1)$.

Solution

a. The graph of C, shown on the next page, indicates that C does not satisfy either function property. To see why C does not satisfy property (1), observe that there are many real numbers x such that $(x, y) \notin C$ for any y.

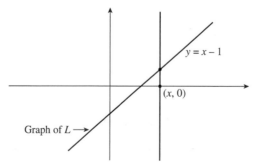

For instance, when $x = 2$, there is no real number y so that

$$x^2 + y^2 = 2^2 + y^2 = 4 + y^2 = 1$$

because if there were, then it would have to be true that

$$y^2 = -3.$$

which is not the case for any real number y.

To see why C does not satisfy property (2), note that for some values of x there are two distinct values of y so that $(x, y) \in C$. One way to see this graphically is to observe that there are vertical lines, such as $x = \frac{1}{2}$, that intersect the graph of C at two separate points: $\left(\frac{1}{2}, \frac{\sqrt{3}}{2} \right)$ and $\left(\frac{1}{2}, -\frac{\sqrt{3}}{2} \right)$.

b. L is a function. For each real number x, $y = x - 1$ is a real number, and so there is a real number y with $(x, y) \in L$. Also if $(x, y) \in L$ and $(x, z) \in L$, then $y = x - 1$ and $z = x - 1$, and so $y = z$. In particular, $L(0) = 0 - 1 = -1$ and $L(1) = 1 - 1 = 0$.

You can also check these results by inspecting the graph of L, shown below. Note that for every real number x, the vertical line through $(x, 0)$ passes through the graph of L exactly once. This indicates both that every real number x is the first element of an ordered pair in L and also that no two distinct ordered pairs in L have the same first element.

Function Machines

Another useful way to think of a function is as a machine. Suppose f is a function from X to Y and an input x of X is given. Imagine f to be a machine that processes x in a certain way to produce the output $f(x)$. This is illustrated in Figure 1.3.1 on the next page.

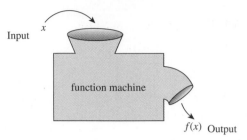

Input x

function machine

$f(x)$ Output

Figure 1.3.1

Example 1.3.6 Functions Defined by Formulas

The **squaring function** f from $\mathbf{R}$ to $\mathbf{R}$ is defined by the formula $f(x) = x^2$ for all real numbers x. This means that no matter what real number input is substituted for x, the output of f will be the square of that number. This idea can be represented by writing $f(\square) = \square^2$. In other words, f sends each real number x to x^2, or, symbolically, $f: x \to x^2$. Note that the variable x is a dummy variable; any other symbol could replace it, as long as the replacement is made everywhere the x appears.

The **successor function** g from $\mathbf{Z}$ to $\mathbf{Z}$ is defined by the formula $g(n) = n + 1$. Thus, no matter what integer is substituted for n, the output of g will be that number plus one: $g(\square) = \square + 1$. In other words, g sends each integer n to $n + 1$, or, symbolically, $g: n \to n + 1$.

An example of a **constant function** is the function h from $\mathbf{Q}$ to $\mathbf{Z}$ defined by the formula $h(r) = 2$ for all rational numbers r. This function sends each rational number r to 2. In other words, no matter what the input, the output is always 2: $h(\square) = 2$ or $h: r \to 2$.

The functions f, g, and h are represented by the function machines in Figure 1.3.2.

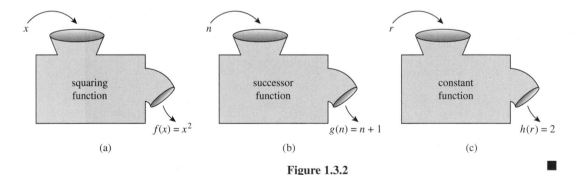

x squaring function $f(x) = x^2$ (a)

n successor function $g(n) = n + 1$ (b)

r constant function $h(r) = 2$ (c)

Figure 1.3.2

A function is an entity in its own right. It can be thought of as a certain relationship between sets or as an input/output machine that operates according to a certain rule. This is the reason why a function is generally denoted by a single symbol or string of symbols, such as f, G, of log, or sin.

A relation is a subset of a Cartesian product and a function is a special kind of relation. Specifically, if f and g are functions from a set A to a set B, then

$$f = \{(x, y) \in A \times B \mid y = f(x)\} \quad \text{and} \quad g = \{(x, y) \in A \times B \mid y = g(x)\}.$$

It follows that

f **equals** g, written $f = g$, if, and only if, $f(x) = g(x)$ for all x in A.

Example 1.3.7 Equality of Functions

Define $f\colon \mathbf{R} \to \mathbf{R}$ and $g\colon \mathbf{R} \to \mathbf{R}$ by the following formulas:

$$f(x) = |x| \quad \text{for all } x \in \mathbf{R}.$$
$$g(x) = \sqrt{x^2} \quad \text{for all } x \in \mathbf{R}.$$

Does $f = g$?

Solution

Yes. Because the absolute value of any real number equals the square root of its square, $|x| = \sqrt{x^2}$ for all $x \in \mathbf{R}$. Hence $f = g$. ∎

Test Yourself

1. Given sets A and B, a relation from A to B is ____.

2. A function F from A to B is a relation from A to B that satisfies the following two properties:

   a. for every element x of A, there is ____.

   b. for all elements x in A and y and z in B, if ____ then ____.

3. If F is a function from A to B and x is an element of A, then $F(x)$ is ____.

Exercise Set 1.3

1. Let $A = \{2, 3, 4\}$ and $B = \{6, 8, 10\}$ and define a relation R from A to B as follows: For all $(x, y) \in A \times B$,

$$(x, y) \in R \quad \text{means that} \quad \frac{y}{x} \text{ is an integer.}$$

a. Is $4\,R\,6$? Is $4\,R\,8$? Is $(3, 8) \in R$? Is $(2, 10) \in R$?
b. Write R as a set of ordered pairs.
c. Write the domain and co-domain of R.
d. Draw an arrow diagram for R.

2. Let $C = D = \{-3, -2, -1, 1, 2, 3\}$ and define a relation S from C to D as follows: For all $(x, y) \in C \times D$,

$$(x, y) \in S \quad \text{means that} \quad \frac{1}{x} - \frac{1}{y} \text{ is an integer.}$$

a. Is $2\,S\,2$? Is $-1\,S - 1$? Is $(2, 2) \in S$? Is $(2, -2) \in S$?
b. Write S as a set of ordered pairs.
c. Write the domain and co-domain of S.
d. Draw an arrow diagram for S.

3. Let $E = \{1, 2, 3\}$ and $F = \{-2, -1, 0\}$ and define a relation T from E to F as follows: For all $(x, y) \in E \times F$,

$$(x, y) \in T \quad \text{means that} \quad \frac{x - y}{3} \text{ is an integer.}$$

a. Is $3\,T\,0$? Is $1\,T(-1)$? Is $(2, -1) \in T$? Is $(3, -2) \in T$?
b. Write T as a set of ordered pairs.
c. Write the domain and co-domain of T.
d. Draw an arrow diagram for T.

4. Let $G = \{-2, 0, 2\}$ and $H = \{4, 6, 8\}$ and define a relation V from G to H as follows: For all $(x, y) \in G \times H$,

$$(x, y) \in V \quad \text{means that} \quad \frac{x - y}{4} \text{ is an integer.}$$

a. Is $2\,V\,6$? Is $(-2)V(8)$? Is $(0, 6) \in V$? Is $(2, 4) \in V$?

b. Write V as a set of ordered pairs.
c. Write the domain and co-domain of V.
d. Draw an arrow diagram for V.

5. Define a relation S from $\mathbf{R}$ to $\mathbf{R}$ as follows: For all $(x, y) \in \mathbf{R} \times \mathbf{R}$,

$$(x, y) \in S \quad \text{means that} \quad x \geq y.$$

a. Is $(2, 1) \in S$? Is $(2, 2) \in S$? Is $2\,S\,3$? Is $(-1)\,S\,(-2)$?
b. Draw the graph of S in the Cartesian plane.

6. Define a relation R from $\mathbf{R}$ to $\mathbf{R}$ as follows: For all $(x, y) \in \mathbf{R} \times \mathbf{R}$,

$$(x, y) \in R \quad \text{means that} \quad y = x^2.$$

a. Is $(2, 4) \in R$? Is $(4, 2) \in R$? Is $(-3)\,R\,9$? Is $9\,R\,(-3)$?
b. Draw the graph of R in the Cartesian plane.

7. Let $A = \{4, 5, 6\}$ and $B = \{5, 6, 7\}$ and define relations R, S, and T from A to B as follows: For all $(x, y) \in A \times B$,

$$(x, y) \in R \quad \text{means that} \quad x \geq y.$$
$$(x, y) \in S \quad \text{means that} \quad \frac{x - y}{2} \text{ is an integer.}$$
$$T = \{(4, 7), (6, 5), (6, 7)\}.$$

a. Draw arrow diagrams for R, S, and T.
b. Indicate whether any of the relations R, S, and T are functions.

8. Let $A = \{2, 4\}$ and $B = \{1, 3, 5\}$ and define relations U, V, and W from A to B as follows: For all $(x, y) \in A \times B$,

$$(x, y) \in U \quad \text{means that} \quad y - x > 2.$$
$$(x, y) \in V \quad \text{means that} \quad y - 1 = \frac{x}{2}.$$
$$W = \{(2, 5), (4, 1), (2, 3)\}.$$

a. Draw arrow diagrams for U, V, and W.
b. Indicate whether any of the relations U, V, and W are functions.

9. a. Find all relations from $\{0,1\}$ to $\{1\}$.
   b. Find all functions from $\{0,1\}$ to $\{1\}$.
   c. What fraction of the relations from $\{0,1\}$ to $\{1\}$ are functions?

10. Find four relations from $\{a, b\}$ to $\{x, y\}$ that are not functions from $\{a, b\}$ to $\{x, y\}$.

11. Define a relation P from $\mathbf{R}^+$ to $\mathbf{R}$ as follows: For all real numbers x and y with $x > 0$,

$$(x, y) \in P \quad \text{means that} \quad x = y^2.$$

Is P a function? Explain.

12. Define a relation T from $\mathbf{R}$ to $\mathbf{R}$ as follows: For all real numbers x and y,

$$(x, y) \in T \quad \text{means that} \quad y^2 - x^2 = 1.$$

Is T a function? Explain.

13. Let $A = \{-1, 0, 1\}$ and $B = \{t, u, v, w\}$. Define a function $F: A \to B$ by the following arrow diagram:

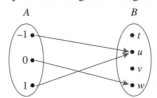

a. Write the domain and co-domain of F.
b. Find $F(-1)$, $F(0)$, and $F(1)$.

14. Let $C = \{1, 2, 3, 4\}$ and $D = \{a, b, c, d\}$. Define a function $G: C \to D$ by the following arrow diagram:

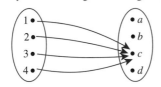

a. Write the domain and co-domain of G.
b. Find $G(1)$, $G(2)$, $G(3)$, and $G(4)$.

15. Let $X = \{2, 4, 5\}$ and $Y = \{1, 2, 4, 6\}$. Which of the following arrow diagrams determine functions from X to Y?

a.

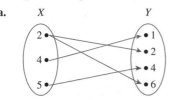

b.

X   Y

c.

X   Y

d.

X   Y

e.

X   Y

16. Let f be the squaring function defined in Example 1.3.6. Find $f(-1)$, $f(0)$, and $f\left(\frac{1}{2}\right)$.

17. Let g be the successor function defined in Example 1.3.6. Find $g(-1000)$, $g(0)$, and $g(999)$.

18. Let h be the constant function defined in Example 1.3.6. Find $h\left(-\frac{12}{5}\right)$, $h\left(\frac{0}{1}\right)$, and $h\left(\frac{9}{17}\right)$.

19. Define functions f and g from $\mathbf{R}$ to $\mathbf{R}$ by the following formulas: For all $x \in \mathbf{R}$,

$$f(x) = 2x \quad \text{and} \quad g(x) = \frac{2x^3 + 2x}{x^2 + 1}.$$

Does $f = g$? Explain.

20. Define functions H and K from $\mathbf{R}$ to $\mathbf{R}$ by the following formulas: For all $x \in \mathbf{R}$,

$$H(x) = (x - 2)^2 \quad \text{and} \quad K(x) = (x - 1)(x - 3) + 1.$$

Does $H = K$? Explain.

Answers for Test Yourself

1. a subset of the Cartesian product $A \times B$ 2. a. an element y of B such that $(x, y) \in F$ (i.e., such that x is related to y by F)   b. $(x, y) \in F$ and $(x, z) \in F$; $y = z$ 3. the unique element of B that is related to x by F

CHAPTER 2

THE LOGIC OF COMPOUND STATEMENTS

Aristotle
(384 B.C.–322 B.C.)

The first great treatises on logic were written by the Greek philosopher Aristotle. They were a collection of rules for deductive reasoning that were intended to serve as a basis for the study of every branch of knowledge. In the seventeenth century, the German philosopher and mathematician Gottfried Leibniz conceived the idea of using symbols to mechanize the process of deductive reasoning in much the same way that algebraic notation had mechanized the process of reasoning about numbers and their relationships. Leibniz's idea was realized in the nineteenth century by the English mathematicians George Boole and Augustus De Morgan, who founded the modern subject of symbolic logic. With research continuing to the present day, symbolic logic has provided, among other things, the theoretical basis for many areas of computer science such as digital logic circuit design, relational database theory, automata theory and computability, and artificial intelligence.

2.1 Logical Form and Logical Equivalence

Logic is a science of the necessary laws of thought, without which no employment of the understanding and the reason takes place. —Immanuel Kant, 1785

The central concept of deductive logic is the concept of argument form. An argument is a sequence of statements aimed at demonstrating the truth of an assertion. The assertion at the end of the sequence is called the *conclusion,* and the preceding statements are called *premises.* To have confidence in the conclusion that you draw from an argument, you must be sure that the premises are acceptable on their own merits or follow from other statements that are known to be true.

In logic, the form of an argument is distinguished from its content. Logical analysis won't help you determine the intrinsic merit of an argument's content, but it will help you analyze an argument's form to determine whether the truth of the conclusion follows *necessarily* from the truth of the premises. For this reason logic is sometimes defined as the science of necessary inference or the science of reasoning.

Consider the following two arguments, for example. Although their content is very different, their logical form is the same. Both arguments are *valid* in the sense that if their premises are true, then their conclusions must also be true. (In Section 2.3 you will learn how to test whether an argument is valid.)

Argument 1   If the program syntax is faulty or if program execution results in division by zero, then the computer will generate an error message. Therefore, if the computer does not generate an error message, then the program syntax is correct and program execution does not result in division by zero.

Argument 2 If x is a real number such that $x < -2$ or $x > 2$, then $x^2 > 4$. Therefore, if $x^2 \not> 4$, then $x \not< -2$ and $x \not> 2$.

To illustrate the logical form of these arguments, we use letters of the alphabet (such as p, q, and r) to represent the component sentences and the expression "not p" to refer to the sentence "It is not the case that p." Then the *common logical form* of both the previous arguments is as follows:

If p or q, then r.

Therefore, if not r, then not p and not q.

Example 2.1.1 Identifying Logical Form

Fill in the blanks below so that argument (b) has the same form as argument (a). Then represent the common form of the arguments using letters to stand for component sentences.

a. If Jane is a math major or Jane is a computer science major, then Jane will take Math 150.
Jane is a computer science major.
Therefore, Jane will take Math 150.

b. If logic is easy or __(1)__ , then __(2)__ .
I will study hard.
Therefore, I will get an A in this course.

Solution

1. I (will) study hard.

2. I will get an A in this course.

   Common form: If p or q, then r.

   q.

   Therefore, r.　　　　　　　　　　　　　　　　　　　　■

Statements

Most of the definitions of formal logic have been developed so that they agree with the natural or intuitive logic used by people who have been educated to think clearly and use language carefully. The differences that exist between formal and intuitive logic are necessary to avoid ambiguity and obtain consistency.

In any mathematical theory, new terms are defined by using those that have been previously defined. However, this process has to start somewhere. A few initial terms necessarily remain undefined. In logic, the words *sentence, true,* and *false* are the initial undefined terms.

• Definition
A **statement** (or **proposition**) is a sentence that is true or false but not both.

For example, "Two plus two equals four" and "Two plus two equals five" are both statements, the first because it is true and the second because it is false. On the other hand, the truth or falsity of "He is a college student" depends on the reference for the pronoun *he*. For some values of *he* the sentence is true; for others it is false. If the sentence were preceded by other sentences that made the pronoun's reference clear, then the sentence would be a statement. Considered on its own, however, the sentence is neither true nor false, and so it is not a statement. We will discuss ways of transforming sentences of this form into statements in Section 3.1.

Similarly, "$x + y > 0$" is not a statement because for some values of x and y the sentence is true, whereas for others it is false. For instance, if $x = 1$ and $y = 2$, the sentence is true; if $x = -1$ and $y = 0$, the sentence is false.

Compound Statements

We now introduce three symbols that are used to build more complicated logical expressions out of simpler ones. The symbol $\sim$ denotes *not,* $\wedge$ denotes *and,* and $\vee$ denotes *or.* Given a statement p, the sentence "$\sim p$" is read "not p" or "It is not the case that p" and is called the **negation of p**. In some computer languages the symbol $\neg$ is used in place of $\sim$. Given another statement q, the sentence "$p \wedge q$" is read "p and q" and is called the **conjunction of p and q**. The sentence "$p \vee q$" is read "p or q" and is called the **disjunction of p and q**.

In expressions that include the symbol $\sim$ as well as $\wedge$ or $\vee$, the **order of operations** specifies that $\sim$ is performed first. For instance, $\sim p \wedge q = (\sim p) \wedge q$. In logical expressions, as in ordinary algebraic expressions, the order of operations can be overridden through the use of parentheses. Thus $\sim(p \wedge q)$ represents the negation of the conjunction of p and q. In this, as in most treatments of logic, the symbols $\wedge$ and $\vee$ are considered coequal in order of operation, and an expression such as $p \wedge q \vee r$ is considered ambiguous. This expression must be written as either $(p \wedge q) \vee r$ or $p \wedge (q \vee r)$ to have meaning.

A variety of English words translate into logic as $\wedge$, $\vee$, or $\sim$. For instance, the word *but* translates the same as *and* when it links two independent clauses, as in "Jim is tall but he is not heavy." Generally, the word *but* is used in place of *and* when the part of the sentence that follows is, in some way, unexpected. Another example involves the words *neither-nor.* When Shakespeare wrote, "Neither a borrower nor a lender be," he meant, "Do not be a borrower and do not be a lender." So if p and q are statements, then

p but q	means	p and q
neither p nor q	means	$\sim p$ and $\sim q$.

Example 2.1.2 Translating from English to Symbols: *But* and *Neither-Nor*

Write each of the following sentences symbolically, letting $h =$ "It is hot" and $s =$ "It is sunny."

a. It is not hot but it is sunny.

b. It is neither hot nor sunny.

Solution

a. The given sentence is equivalent to "It is not hot and it is sunny," which can be written symbolically as $\sim h \wedge s$.

b. To say it is neither hot nor sunny means that it is not hot and it is not sunny. Therefore, the given sentence can be written symbolically as $\sim h \wedge \sim s$. ∎

The notation for inequalities involves *and* and *or* statements. For instance, if x, a, and b are particular real numbers, then

$x \leq a$	means	$x < a$	or	$x = a$
$a \leq x \leq b$	means	$a \leq x$	and	$x \leq b$.

Note that the inequality $2 \leq x \leq 1$ is not satisfied by any real numbers because

$$2 \leq x \leq 1 \quad \text{means} \quad 2 \leq x \quad \text{and} \quad x \leq 1,$$

and this is false no matter what number x happens to be. By the way, the point of specifying x, a, and b to be *particular* real numbers is to ensure that sentences such as "$x < a$" and "$x \geq b$" are either true or false and hence that they are statements.

Example 2.1.3 *And, Or,* and **Inequalities**

Suppose x is a particular real number. Let p, q, and r symbolize "$0 < x$," "$x < 3$," and "$x = 3$," respectively. Write the following inequalities symbolically:

a. $x \leq 3$ b. $0 < x < 3$ c. $0 < x \leq 3$

Solution

a. $q \vee r$ b. $p \wedge q$ c. $p \wedge (q \vee r)$ ■

Truth Values

In Examples 2.1.2 and 2.1.3 we built compound sentences out of component statements and the terms *not, and,* and *or.* If such sentences are to be statements, however, they must have well-defined **truth values**—they must be either true or false. We now define such compound sentences as statements by specifying their truth values in terms of the statements that compose them.

The negation of a statement is a statement that exactly expresses what it would mean for the statement to be false.

• Definition

If p is a statement variable, the **negation** of p is "not p" or "It is not the case that p" and is denoted $\sim p$. It has opposite truth value from p: if p is true, $\sim p$ is false; if p is false, $\sim p$ is true.

The truth values for negation are summarized in a *truth table*.

Truth Table for $\sim p$

p	$\sim p$
T	F
F	T

In ordinary language the sentence "It is hot and it is sunny" is understood to be true when both conditions—being hot and being sunny—are satisfied. If it is hot but not sunny, or sunny but not hot, or neither hot nor sunny, the sentence is understood to be false. The formal definition of truth values for an *and* statement agrees with this general understanding.

• Definition

If p and q are statement variables, the **conjunction** of p and q is "p and q," denoted $p \wedge q$. It is true when, and only when, both p and q are true. If either p or q is false, or if both are false, $p \wedge q$ is false.

The truth values for conjunction can also be summarized in a truth table. The table is obtained by considering the four possible combinations of truth values for p and q. Each combination is displayed in one row of the table; the corresponding truth value for the whole statement is placed in the right-most column of that row. Note that the only row containing a T is the first one since the only way for an *and* statement to be true is for both component statements to be true.

Truth Table for $p \wedge q$

p	q	$p \wedge q$
T	T	T
T	F	F
F	T	F
F	F	F

By the way, the order of truth values for p and q in the table above is TT, TF, FT, FF. It is not absolutely necessary to write the truth values in this order, although it is customary to do so. We will use this order for all truth tables involving two statement variables. In Example 2.1.5 we will show the standard order for truth tables that involve three statement variables.

In the case of disjunction—statements of the form "p or q"—intuitive logic offers two alternative interpretations. In ordinary language *or* is sometimes used in an exclusive sense (p or q but not both) and sometimes in an inclusive sense (p or q or both). A waiter who says you may have "coffee, tea, or milk" uses the word *or* in an exclusive sense: Extra payment is generally required if you want more than one beverage. On the other hand, a waiter who offers "cream or sugar" uses the word *or* in an inclusive sense: You are entitled to both cream and sugar if you wish to have them.

Mathematicians and logicians avoid possible ambiguity about the meaning of the word *or* by understanding it to mean the inclusive "and/or." The symbol $\vee$ comes from the Latin word *vel,* which means *or* in its inclusive sense. To express the exclusive *or,* the phrase p or q *but not both* is used.

• Definition

If p and q are statement variables, the **disjunction** of p and q is "p or q," denoted $p \lor q$. It is true when either p is true, or q is true, or both p and q are true; it is false only when both p and q are false.

Here is the truth table for disjunction:

Note The statement "$2 \leq 2$" means that 2 is less than 2 or 2 equals 2. It is true because $2 = 2$.

Truth Table for $p \lor q$

p	q	$p \lor q$
T	T	T
T	F	T
F	T	T
F	F	F

Evaluating the Truth of More General Compound Statements

Now that truth values have been assigned to $\sim p$, $p \land q$, and $p \lor q$, consider the question of assigning truth values to more complicated expressions such as $\sim p \lor q$, $(p \lor q) \land \sim(p \land q)$, and $(p \land q) \lor r$. Such expressions are called *statement forms* (or *propositional forms*).

• Definition

A **statement form** (or **propositional form**) is an expression made up of statement variables (such as p, q, and r) and logical connectives (such as $\sim$, $\land$, and $\lor$) that becomes a statement when actual statements are substituted for the component statement variables. The **truth table** for a given statement form displays the truth values that correspond to all possible combinations of truth values for its component statement variables.

To compute the truth values for a statement form, follow rules similar to those used to evaluate algebraic expressions. For each combination of truth values for the statement variables, first evaluate the expressions within the innermost parentheses, then evaluate the expressions within the next innermost set of parentheses, and so forth until you have the truth values for the complete expression.

Example 2.1.4 Truth Table for *Exclusive Or*

Construct the truth table for the statement form $(p \lor q) \land \sim(p \land q)$. Note that when *or* is used in its exclusive sense, the statement "p or q" means "p or q but not both" or "p or q and not both p and q," which translates into symbols as $(p \lor q) \land \sim(p \land q)$. This is sometimes abbreviated $p \oplus q$ or p XOR q.

Solution   Set up columns labeled $p, q, p \lor q, p \land q, \sim(p \land q)$, and $(p \lor q) \land \sim(p \land q)$. Fill in the p and q columns with all the logically possible combinations of T's and F's. Then use the truth tables for $\lor$ and $\land$ to fill in the $p \lor q$ and $p \land q$ columns with the appropriate truth values. Next fill in the $\sim(p \land q)$ column by taking the opposites of the truth values for $p \land q$. For example, the entry for $\sim(p \land q)$ in the first row is F because in the first row the truth value of $p \land q$ is T. Finally, fill in the $(p \lor q) \land \sim(p \land q)$ column by considering the truth table for an *and* statement together with the computed truth values for $p \lor q$ and $\sim(p \land q)$. For example, the entry in the first row is F because the entry for $p \lor q$ is T, the entry for $\sim(p \land q)$ is F, and an *and* statement is false unless both components are true. The entry in the second row is T because both components are true in this row.

Truth Table for *Exclusive Or:* $(p \lor q) \land \sim(p \land q)$

p	q	$p \lor q$	$p \land q$	$\sim(p \land q)$	$(p \lor q) \land \sim(p \land q)$
T	T	T	T	F	F
T	F	T	F	T	T
F	T	T	F	T	T
F	F	F	F	T	F

Example 2.1.5 Truth Table for $(p \land q) \lor \sim r$

Construct a truth table for the statement form $(p \land q) \lor \sim r$.

Solution   Make columns headed $p, q, r, p \land q, \sim r$, and $(p \land q) \lor \sim r$. Enter the eight logically possible combinations of truth values for p, q, and r in the three left-most columns. Then fill in the truth values for $p \land q$ and for $\sim r$. Complete the table by considering the truth values for $(p \land q)$ and for $\sim r$ and the definition of an *or* statement. Since an *or* statement is false only when both components are false, the only rows in which the entry is F are the third, fifth, and seventh rows because those are the only rows in which the expressions $p \land q$ and $\sim r$ are both false. The entry for all the other rows is T.

p	q	r	$p \land q$	$\sim r$	$(p \land q) \lor \sim r$
T	T	T	T	F	T
T	T	F	T	T	T
T	F	T	F	F	F
T	F	F	F	T	T
F	T	T	F	F	F
F	T	F	F	T	T
F	F	T	F	F	F
F	F	F	F	T	T

The essential point about assigning truth values to compound statements is that it allows you—using logic alone—to judge the truth of a compound statement on the basis of your knowledge of the truth of its component parts. Logic does not help you determine the truth or falsity of the component statements. Rather, logic helps link these separate pieces of information together into a coherent whole.

Logical Equivalence

The statements

6 is greater than 2   and   2 is less than 6

are two different ways of saying the same thing. Why? Because of the definition of the phrases *greater than* and *less than*. By contrast, although the statements

(1) Dogs bark and cats meow   and   (2) Cats meow and dogs bark

are also two different ways of saying the same thing, the reason has nothing to do with the definition of the words. It has to do with the logical form of the statements. Any two statements whose logical forms are related in the same way as (1) and (2) would either both be true or both be false. You can see this by examining the following truth table, where the statement variables p and q are substituted for the component statements "Dogs bark" and "Cats meow," respectively. The table shows that for each combination of truth values for p and q, $p \wedge q$ is true when, and only when, $q \wedge p$ is true. In such a case, the statement forms are called *logically equivalent,* and we say that (1) and (2) are *logically equivalent statements.*

p	q	$p \wedge q$	$q \wedge p$
T	T	T	T
T	F	F	F
F	T	F	F
F	F	F	F

↑   ↑

$p \wedge q$ and $q \wedge p$ always
have the same truth
values, so they are
logically equivalent

• Definition

Two *statement forms* are called **logically equivalent** if, and only if, they have identical truth values for each possible substitution of statements for their statement variables. The logical equivalence of statement forms P and Q is denoted by writing $P \equiv Q$.

    Two *statements* are called **logically equivalent** if, and only if, they have logically equivalent forms when identical component statement variables are used to replace identical component statements.

Testing Whether Two Statement Forms P and Q Are Logically Equivalent

1. Construct a truth table with one column for the truth values of P and another column for the truth values of Q.

2. Check each combination of truth values of the statement variables to see whether the truth value of P is the same as the truth value of Q.

    a. If in each row the truth value of P is the same as the truth value of Q, then P and Q are logically equivalent.

    b. If in some row P has a different truth value from Q, then P and Q are not logically equivalent.

Example 2.1.6 Double Negative Property: $\sim(\sim p) \equiv p$

Construct a truth table to show that the negation of the negation of a statement is logically equivalent to the statement, annotating the table with a sentence of explanation.

Solution

p	$\sim p$	$\sim(\sim p)$
T	F	T
F	T	F

p and $\sim(\sim p)$ always have the same truth values, so they are logically equivalent

There are two ways to show that statement forms P and Q are *not* logically equivalent. As indicated previously, one is to use a truth table to find rows for which their truth values differ. The other way is to find concrete statements for each of the two forms, one of which is true and the other of which is false. The next example illustrates both of these ways.

Example 2.1.7 Showing Nonequivalence

Show that the statement forms $\sim(p \wedge q)$ and $\sim p \wedge \sim q$ are not logically equivalent.

Solution

a. This method uses a truth table annotated with a sentence of explanation.

p	q	$\sim p$	$\sim q$	$p \wedge q$	$\sim(p \wedge q)$		$\sim p \wedge \sim q$
T	T	F	F	T	F		F
T	F	F	T	F	T	$\neq$	F
F	T	T	F	F	T	$\neq$	F
F	F	T	T	F	T		T

$\sim(p \wedge q)$ and $\sim p \wedge \sim q$ have different truth values in rows 2 and 3, so they are not logically equivalent

b. This method uses an example to show that $\sim(p \wedge q)$ and $\sim p \wedge \sim q$ are not logically equivalent. Let p be the statement "$0 < 1$" and let q be the statement "$1 < 0$." Then

$$\sim(p \wedge q) \quad \text{is} \quad \text{"It is not the case that both } 0 < 1 \text{ and } 1 < 0, \text{"}$$

which is true. On the other hand,

$$\sim p \wedge \sim q \quad \text{is} \quad \text{"}0 \not< 1 \quad \text{and} \quad 1 \not< 0,\text{"}$$

which is false. This example shows that there are concrete statements you can substitute for p and q to make one of the statement forms true and the other false. Therefore, the statement forms are not logically equivalent.

Example 2.1.8 Negations of *And* and *Or:* De Morgan's Laws

For the statement "John is tall and Jim is redheaded" to be true, both components must be true. So for the statement to be false, one or both components must be false. Thus the negation can be written as "John is not tall or Jim is not redheaded." In general, the negation of the conjunction of two statements is logically equivalent to the disjunction of their negations. That is, statements of the forms $\sim(p \wedge q)$ and $\sim p \vee \sim q$ are logically equivalent. Check this using truth tables.

Solution

p	q	$\sim p$	$\sim q$	$p \wedge q$	$\sim(p \wedge q)$	$\sim p \vee \sim q$
T	T	F	F	T	F	F
T	F	F	T	F	T	T
F	T	T	F	F	T	T
F	F	T	T	F	T	T

↑ ↑

$\sim(p \wedge q)$ and $\sim p \vee \sim q$ always have the same truth values, so they are logically equivalent

*Augustus De Morgan
(1806–1871)*

Symbolically,

$$\sim(p \wedge q) \equiv \sim p \vee \sim q.$$

In the exercises at the end of this section you are asked to show the analogous law that the negation of the disjunction of two statements is logically equivalent to the conjunction of their negations:

$$\sim(p \vee q) \equiv \sim p \wedge \sim q.$$

■

The two logical equivalences of Example 2.1.8 are known as **De Morgan's laws** of logic in honor of Augustus De Morgan, who was the first to state them in formal mathematical terms.

De Morgan's Laws

The negation of an *and* statement is logically equivalent to the *or* statement in which each component is negated.

The negation of an *or* statement is logically equivalent to the *and* statement in which each component is negated.

Example 2.1.9 Applying De Morgan's Laws

Write negations for each of the following statements:

a. John is 6 feet tall and he weighs at least 200 pounds.

b. The bus was late or Tom's watch was slow.

Solution

a. John is not 6 feet tall or he weighs less than 200 pounds.

b. The bus was not late and Tom's watch was not slow.

Since the statement "neither p nor q" means the same as "$\sim p$ and $\sim q$," an alternative answer for (b) is "Neither was the bus late nor was Tom's watch slow." ∎

If x is a particular real number, saying that x is not less than 2 ($x \not< 2$) means that x does not lie to the left of 2 on the number line. This is equivalent to saying that either $x = 2$ or x lies to the right of 2 on the number line ($x = 2$ or $x > 2$). Hence,

$$x \not< 2 \quad \text{is equivalent to} \quad x \geq 2.$$

Pictorially,

$$
\begin{array}{ccccccccc}
-2 & -1 & 0 & 1 & 2 & 3 & 4 & 5 \\
\end{array}
$$

If $x \not< 2$, then x lies in here.

Similarly,

$$x \not> 2 \quad \text{is equivalent to} \quad x \leq 2,$$
$$x \not\leq 2 \quad \text{is equivalent to} \quad x > 2, \text{ and}$$
$$x \not\geq 2 \quad \text{is equivalent to} \quad x < 2.$$

Example 2.1.10 Inequalities and De Morgan's Laws

Use De Morgan's laws to write the negation of $-1 < x \leq 4$.

Solution   The given statement is equivalent to

$$-1 < x \quad \text{and} \quad x \leq 4.$$

By De Morgan's laws, the negation is

Caution! The negation of $-1 < x \leq 4$ is *not* $-1 \not< x \not\leq 4$. It is also not $-1 \geq x > 4$.

$$-1 \not< x \quad \text{or} \quad x \not\leq 4,$$

which is equivalent to

$$-1 \geq x \quad \text{or} \quad x > 4.$$

Pictorially, if $-1 \geq x$ or $x > 4$, then x lies in the shaded region of the number line, as shown below.

$$
\begin{array}{cccccccccc}
-2 & -1 & 0 & 1 & 2 & 3 & 4 & 5 & 6 \\
\end{array}
$$

∎

De Morgan's laws are frequently used in writing computer programs. For instance, suppose you want your program to delete all files modified outside a certain range of dates, say from date 1 through date 2 inclusive. You would use the fact that

$$\sim(date1 \leq file\_modification\_date \leq date2)$$

is equivalent to

$$(\textit{file\_modification\_date} < \textit{date}1) \quad \text{or} \quad (\textit{date}2 < \textit{file\_modification\_date}).$$

Example 2.1.11 A Cautionary Example

According to De Morgan's laws, the negation of

$$p\text{: Jim is tall and Jim is thin}$$

is

$$\sim p\text{: Jim is not tall or Jim is not thin}$$

because the negation of an *and* statement is the *or* statement in which the two components are negated.

Unfortunately, a potentially confusing aspect of the English language can arise when you are taking negations of this kind. Note that statement p can be written more compactly as

$$p'\text{: Jim is tall and thin.}$$

When it is so written, another way to negate it is

$$\sim(p')\text{: Jim is not tall and thin.}$$

Caution! Although the laws of logic are extremely useful, they should be used as an *aid* to thinking, not as a mechanical substitute for it.

But in this form the negation looks like an *and* statement. Doesn't that violate De Morgan's laws?

Actually no violation occurs. The reason is that in formal logic the words *and* and *or* are allowed only between complete statements, not between sentence fragments.

One lesson to be learned from this example is that when you apply De Morgan's laws, you must have complete statements on either side of each *and* and on either side of each *or*. ∎

Tautologies and Contradictions

It has been said that all of mathematics reduces to tautologies. Although this is formally true, most working mathematicians think of their subject as having substance as well as form. Nonetheless, an intuitive grasp of basic logical tautologies is part of the equipment of anyone who reasons with mathematics.

• Definition

A **tautology** is a statement form that is always true regardless of the truth values of the individual statements substituted for its statement variables. A statement whose form is a tautology is a **tautological statement.**

A **contradiction** is a statement form that is always false regardless of the truth values of the individual statements substituted for its statement variables. A statement whose form is a contradiction is a **contradictory statement.**

According to this definition, the truth of a tautological statement and the falsity of a contradictory statement are due to the logical structure of the statements themselves and are independent of the meanings of the statements.

Example 2.1.12 Tautologies and Contradictions

Show that the statement form $p \vee \sim p$ is a tautology and that the statement form $p \wedge \sim p$ is a contradiction.

Solution

p	$\sim p$	$p \vee \sim p$	$p \wedge \sim p$
T	F	T	F
F	T	T	F

$\qquad\qquad\quad\uparrow\qquad\qquad\uparrow$

all T's so   all F's so
$p \vee \sim p$ is $p \wedge \sim p$ is a
a tautology   contradiction

■

Example 2.1.13 Logical Equivalence Involving Tautologies and Contradictions

If **t** is a tautology and **c** is a contradiction, show that $p \wedge \mathbf{t} \equiv p$ and $p \wedge \mathbf{c} \equiv \mathbf{c}$.

Solution

p	t	$p \wedge \mathbf{t}$	p	c	$p \wedge \mathbf{c}$
T	T	T	T	F	F
F	T	F	F	F	F

$\qquad\qquad\;\uparrow\!\!\underline{\qquad}\!\!\uparrow\qquad\uparrow\!\!\underline{\qquad}\!\!\uparrow$

same truth   same truth
values, so   values, so
$p \wedge \mathbf{t} \equiv p$ $p \wedge \mathbf{c} \equiv \mathbf{c}$

■

Summary of Logical Equivalences

Knowledge of logically equivalent statements is very useful for constructing arguments. It often happens that it is difficult to see how a conclusion follows from one form of a statement, whereas it is easy to see how it follows from a logically equivalent form of the statement. A number of logical equivalences are summarized in Theorem 2.1.1 for future reference.

Theorem 2.1.1 Logical Equivalences

Given any statement variables p, q, and r, a tautology **t** and a contradiction **c**, the following logical equivalences hold.

1. *Commutative laws:* $p \wedge q \equiv q \wedge p$ $p \vee q \equiv q \vee p$

2. *Associative laws:* $(p \wedge q) \wedge r \equiv p \wedge (q \wedge r)$ $(p \vee q) \vee r \equiv p \vee (q \vee r)$

3. *Distributive laws:* $p \wedge (q \vee r) \equiv (p \wedge q) \vee (p \wedge r)$ $p \vee (q \wedge r) \equiv (p \vee q) \wedge (p \vee r)$

4. *Identity laws:* $p \wedge \mathbf{t} \equiv p$ $p \vee \mathbf{c} \equiv p$

5. *Negation laws:* $p \vee \sim p \equiv \mathbf{t}$ $p \wedge \sim p \equiv \mathbf{c}$

6. *Double negative law:* $\sim(\sim p) \equiv p$

7. *Idempotent laws:* $p \wedge p \equiv p$ $p \vee p \equiv p$

8. *Universal bound laws:* $p \vee \mathbf{t} \equiv \mathbf{t}$ $p \wedge \mathbf{c} \equiv \mathbf{c}$

9. *De Morgan's laws:* $\sim(p \wedge q) \equiv \sim p \vee \sim q$ $\sim(p \vee q) \equiv \sim p \wedge \sim q$

10. *Absorption laws:* $p \vee (p \wedge q) \equiv p$ $p \wedge (p \vee q) \equiv p$

11. *Negations of* **t** *and* **c**: $\sim \mathbf{t} \equiv \mathbf{c}$ $\sim \mathbf{c} \equiv \mathbf{t}$

The proofs of laws 4 and 6, the first parts of laws 1 and 5, and the second part of law 9 have already been given as examples in the text. Proofs of the other parts of the theorem are left as exercises. In fact, it can be shown that the first five laws of Theorem 2.1.1 form a core from which the other laws can be derived. The first five laws are the axioms for a mathematical structure known as a Boolean algebra, which is discussed in Section 6.4.

The equivalences of Theorem 2.1.1 are general laws of thought that occur in all areas of human endeavor.

Test Yourself

Answers to Test Yourself questions are located at the end of each section.

1. An *and* statement is true if, and only if, both components are ____.

2. An *or* statement is false if, and only if, both components are ____.

3. Two statement forms are logically equivalent if, and only if, they always have ____.

4. De Morgan's laws say (1) that the negation of an *and* statement is logically equivalent to the ____ statement in which each component is ____, and (2) that the negation of an *or* statement is logically equivalent to the ____ statement in which each component is ____.

5. A tautology is a statement that is always ____.

6. A contradiction is a statement that is always ____.

Exercise Set 2.1 *

In each of 1–4 represent the common form of each argument using letters to stand for component sentences, and fill in the blanks so that the argument in part (b) has the same logical form as the argument in part (a).

1. a. If all integers are rational, then the number 1 is rational.
   All integers are rational.
   Therefore, the number 1 is rational.
   b. If all algebraic expressions can be written in prefix notation, then _____.
   _____.
   Therefore, $(a + 2b)(a^2 - b)$ can be written in prefix notation.

2. a. If all computer programs contain errors, then this program contains an error.
   This program does not contain an error.
   Therefore, it is not the case that all computer programs contain errors.
   b. If _____, then _____.
   2 is not odd.
   Therefore, it is not the case that all prime numbers are odd.

3. a. This number is even or this number is odd.
   This number is not even.
   Therefore, this number is odd.
   b. _____ or logic is confusing.
   My mind is not shot.
   Therefore, _____.

4. a. If n is divisible by 6, then n is divisible by 3.
   If n is divisible by 3, then the sum of the digits of n is divisible by 3.
   Therefore, if n is divisible by 6, then the sum of the digits of n is divisible by 3.
   (Assume that n is a particular, fixed integer.)
   b. If this function is ____ then this function is differentiable.
   If this function is ____ then this function is continuous.
   Therefore, if this function is a polynomial, then this function ____.

5. Indicate which of the following sentences are statements.
   a. 1,024 is the smallest four-digit number that is a perfect square.
   b. She is a mathematics major.
   c. $128 = 2^6$ d. $x = 2^6$

Write the statements in 6–9 in symbolic form using the symbols $\sim$, $\vee$, and $\wedge$ and the indicated letters to represent component statements.

6. Let s = "stocks are increasing" and i = "interest rates are steady."
   a. Stocks are increasing but interest rates are steady.
   b. Neither are stocks increasing nor are interest rates steady.

---

*For exercises with blue numbers or letters, solutions are given in Appendix B. The symbol **H** indicates that only a hint or a partial solution is given. The symbol ∗ signals that an exercise is more challenging than usual.

7. Juan is a math major but not a computer science major. (m = "Juan is a math major," c = "Juan is a computer science major")

8. Let h = "John is healthy," w = "John is wealthy," and s = "John is wise."
   a. John is healthy and wealthy but not wise.
   b. John is not wealthy but he is healthy and wise.
   c. John is neither healthy, wealthy, nor wise.
   d. John is neither wealthy nor wise, but he is healthy.
   e. John is wealthy, but he is not both healthy and wise.

9. Either this polynomial has degree 2 or it has degree 3 but not both. (n = "This polynomial has degree 2," k = "This polynomial has degree 3")

10. Let p be the statement "DATAENDFLAG is off," q the statement "ERROR equals 0," and r the statement "SUM is less than 1,000." Express the following sentences in symbolic notation.
    a. DATAENDFLAG is off, ERROR equals 0, and SUM is less than 1,000.
    b. DATAENDFLAG is off but ERROR is not equal to 0.
    c. DATAENDFLAG is off; however, ERROR is not 0 or SUM is greater than or equal to 1,000.
    d. DATAENDFLAG is on and ERROR equals 0 but SUM is greater than or equal to 1,000.
    e. Either DATAENDFLAG is on or it is the case that both ERROR equals 0 and SUM is less than 1,000.

11. In the following sentence, is the word *or* used in its inclusive or exclusive sense? A team wins the playoffs if it wins two games in a row or a total of three games.

Write truth tables for the statement forms in 12–15.

12. $\sim p \wedge q$

13. $\sim (p \wedge q) \vee (p \vee q)$

14. $p \wedge (q \wedge r)$

15. $p \wedge (\sim q \vee r)$

Determine whether the statement forms in 16–24 are logically equivalent. In each case, construct a truth table and include a sentence justifying your answer. Your sentence should show that you understand the meaning of logical equivalence.

16. $p \vee (p \wedge q)$ and p

17. $\sim (p \wedge q)$ and $\sim p \wedge \sim q$

18. $p \vee \mathbf{t}$ and $\mathbf{t}$

19. $p \wedge \mathbf{t}$ and p

20. $p \wedge \mathbf{c}$ and $p \vee \mathbf{c}$

21. $(p \wedge q) \wedge r$ and $p \wedge (q \wedge r)$

22. $p \wedge (q \vee r)$ and $(p \wedge q) \vee (p \wedge r)$

23. $(p \wedge q) \vee r$ and $p \wedge (q \vee r)$

24. $(p \vee q) \vee (p \wedge r)$ and $(p \vee q) \wedge r$

Use De Morgan's laws to write negations for the statements in 25–31.

25. Hal is a math major and Hal's sister is a computer science major.

26. Sam is an orange belt and Kate is a red belt.

27. The connector is loose or the machine is unplugged.

28. The units digit of 4^{67} is 4 or it is 6.

29. This computer program has a logical error in the first ten lines or it is being run with an incomplete data set.

30. The dollar is at an all-time high and the stock market is at a record low.

31. The train is late or my watch is fast.

Assume x is a particular real number and use De Morgan's laws to write negations for the statements in 32–37.

32. $-2 < x < 7$

33. $-10 < x < 2$

34. $x < 2$ or $x > 5$

35. $x \leq -1$ or $x > 1$

36. $1 > x \geq -3$

37. $0 > x \geq -7$

In 38 and 39, imagine that *num_orders* and *num_instock* are particular values, such as might occur during execution of a computer program. Write negations for the following statements.

38. (*num_orders* > 100 and *num_instock* ≤ 500) or *num_instock* < 200

39. (*num_orders* < 50 and *num_instock* > 300) or ($50 \leq$ *num_orders* < 75 and *num_instock* > 500)

Use truth tables to establish which of the statement forms in 40–43 are tautologies and which are contradictions.

40. $(p \wedge q) \vee (\sim p \vee (p \wedge \sim q))$

41. $(p \wedge \sim q) \wedge (\sim p \vee q)$

42. $((\sim p \wedge q) \wedge (q \wedge r)) \wedge \sim q$

43. $(\sim p \vee q) \vee (p \wedge \sim q)$

In 44 and 45, determine whether the statements in (a) and (b) are logically equivalent.

44. Assume x is a particular real number.

    a. $x < 2$ or it is not the case that $1 < x < 3$.

    b. $x \leq 1$ or either $x < 2$ or $x \geq 3$.

45. a. Bob is majoring in both math and computer science, and Ann is majoring in math but Ann is not majoring in both math and computer science.
    b. It is not the case that both Bob and Ann are majoring in both math and computer science, but it is the case that Ann is majoring in math and Bob is majoring in both math and computer science.

✳ 46. In Example 2.1.4, the symbol ⊕ was introduced to denote *exclusive or,* so $p \oplus q \equiv (p \vee q) \wedge \sim (p \wedge q)$. Hence the truth table for *exclusive or* is as follows:

p	q	$p \oplus q$
T	T	F
T	F	T
F	T	T
F	F	F

a. Find simpler statement forms that are logically equivalent to $p \oplus p$ and $(p \oplus p) \oplus p$.
b. Is $(p \oplus q) \oplus r \equiv p \oplus (q \oplus r)$? Justify your answer.
c. Is $(p \oplus q) \wedge r \equiv (p \wedge r) \oplus (q \wedge r)$? Justify your answer.

✳ 47. In logic and in standard English, a double negative is equivalent to a positive. There is one fairly common English usage in which a "double positive" is equivalent to a negative. What is it?

Answers for Test Yourself

1. true   2. false   3. the same truth values   4. *or*; negated; *and*; negated   5. true   6. false

2.2 Conditional Statements

. . . hypothetical reasoning implies the subordination of the real to the realm of the possible . . . — Jean Piaget, 1972

When you make a logical inference or deduction, you reason *from* a hypothesis *to* a conclusion. Your aim is to be able to say, "*If* such and such is known, *then* something or other must be the case."

Let p and q be statements. A sentence of the form "If p then q" is denoted symbolically by "$p \rightarrow q$"; p is called the *hypothesis* and q is called the *conclusion*. For instance, consider the following statement:

$$\text{If } \underbrace{4{,}686 \text{ is divisible by } 6}_{\text{hypothesis}}, \text{ then } \underbrace{4{,}686 \text{ is divisible by } 3}_{\text{conclusion}}$$

Such a sentence is called *conditional* because the truth of statement q is conditioned on the truth of statement p.

The notation $p \rightarrow q$ indicates that $\rightarrow$ is a connective, like $\wedge$ or $\vee$, that can be used to join statements to create new statements. To define $p \rightarrow q$ as a statement, therefore, we must specify the truth values for $p \rightarrow q$ as we specified truth values for $p \wedge q$ and for $p \vee q$. As is the case with the other connectives, the formal definition of truth values for $\rightarrow$ (if-then) is based on its everyday, intuitive meaning. Consider an example.

Suppose you go to interview for a job at a store and the owner of the store makes you the following promise:

If you show up for work Monday morning, then you will get the job.

Under what circumstances are you justified in saying the owner spoke falsely? That is, under what circumstances is the above sentence false? The answer is: You *do* show up for work Monday morning and you do *not* get the job.

After all, the owner's promise only says you will get the job *if* a certain condition (showing up for work Monday morning) is met; it says nothing about what will happen if the condition is *not* met. So if the condition is not met, you cannot in fairness say the promise is false regardless of whether or not you get the job.

The above example was intended to convince you that *the only combination of circumstances in which you would call a conditional sentence false occurs when the hypothesis is true and the conclusion is false. In all other cases, you would not call the sentence* false. This implies that the only row of the truth table for $p \rightarrow q$ that should be filled in with an F is the row where p is T and q is F. No other row should contain an F. But each row of a truth table must be filled in with either a T or an F. Thus all other rows of the truth table for $p \rightarrow q$ must be filled in with T's.

Truth Table for $p \rightarrow q$

p	q	$p \rightarrow q$
T	T	T
T	F	F
F	T	T
F	F	T

• Definition

If p and q are statement variables, the **conditional** of q by p is "If p then q" or "p implies q" and is denoted $p \rightarrow q$. It is false when p is true and q is false; otherwise it is true. We call p the **hypothesis** (or **antecedent**) of the conditional and q the **conclusion** (or **consequent**).

A conditional statement that is true by virtue of the fact that its hypothesis is false is often called **vacuously true** or **true by default.** Thus the statement "If you show up for work Monday morning, then you will get the job" is vacuously true if you do not show up for work Monday morning. In general, when the "if" part of an if-then statement is false, the statement as a whole is said to be true, regardless of whether the conclusion is true or false.

Example 2.2.1 A Conditional Statement with a False Hypothesis

Consider the statement:

$$\text{If } 0 = 1 \text{ then } 1 = 2.$$

As strange as it may seem, since the hypothesis of this statement is false, the statement as a whole is true. ■

The philosopher Willard Van Orman Quine advises against using the phrase "p implies q" to mean "$p \rightarrow q$" because the word *implies* suggests that q can be logically deduced from p and this is often not the case. Nonetheless, the phrase is used by many people, probably because it is a convenient replacement for the $\rightarrow$ symbol. And, of course, in many cases a conclusion can be deduced from a hypothesis, even when the hypothesis is false.

Note For example, if $0 = 1$, then, by adding 1 to both sides of the equation, you can deduce that $1 = 2$.

In expressions that include $\rightarrow$ as well as other logical operators such as $\wedge$, $\vee$, and $\sim$, the **order of operations** is that $\rightarrow$ is performed last. Thus, according to the specification of order of operations in Section 2.1, $\sim$ is performed first, then $\wedge$ and $\vee$, and finally $\rightarrow$.

Example 2.2.2 Truth Table for $p \vee \sim q \rightarrow \sim p$

Construct a truth table for the statement form $p \vee \sim q \rightarrow \sim p$.

Solution   By the order of operations given above, the following two expressions are equivalent: $p \vee \sim q \rightarrow \sim p$ and $(p \vee (\sim q)) \rightarrow (\sim p)$, and this order governs the construction of the truth table. First fill in the four possible combinations of truth values for p and q, and then enter the truth values for $\sim p$ and $\sim q$ using the definition of negation. Next fill in the $p \vee \sim q$ column using the definition of $\vee$. Finally, fill in the $p \vee \sim q \to \sim p$ column using the definition of $\to$. The only rows in which the hypothesis $p \vee \sim q$ is true and the conclusion $\sim p$ is false are the first and second rows. So you put F's in those two rows and T's in the other two rows.

				conclusion	hypothesis	
p	q	$\sim p$	$\sim q$	$p \vee \sim q$	$p \vee \sim q \to \sim p$
T	T	F	F	T	F
T	F	F	T	T	F
F	T	T	F	F	T
F	F	T	T	T	T

Logical Equivalences Involving $\to$

Imagine that you are trying to solve a problem involving three statements: p, q, and r. Suppose you know that the truth of r follows from the truth of p *and also* that the truth of r follows from the truth of q. Then no matter whether p or q is the case, the truth of r must follow. The division-into-cases method of analysis is based on this idea.

Example 2.2.3 Division into Cases: Showing that $p \vee q \to r \equiv (p \to r) \wedge (q \to r)$

Use truth tables to show the logical equivalence of the statement forms $p \vee q \to r$ and $(p \to r) \wedge (q \to r)$. Annotate the table with a sentence of explanation.

Solution   First fill in the eight possible combinations of truth values for p, q, and r. Then fill in the columns for $p \vee q$, $p \to r$, and $q \to r$ using the definitions of *or* and *if-then*. For instance, the $p \to r$ column has F's in the second and fourth rows because these are the rows in which p is true and r is false. Next fill in the $p \vee q \to r$ column using the definition of *if-then*. The rows in which the hypothesis $p \vee q$ is true and the conclusion r is false are the second, fourth, and sixth. So F's go in these rows and T's in all the others. The complete table shows that $p \vee q \to r$ and $(p \to r) \wedge (q \to r)$ have the same truth values for each combination of truth values of p, q, and r. Hence the two statement forms are logically equivalent.

p	q	r	$p \vee q$	$p \to r$	$q \to r$	$p \vee q \to r$	$(p \to r) \wedge (q \to r)$
T	T	T	T	T	T	T	T
T	T	F	T	F	F	F	F
T	F	T	T	T	T	T	T
T	F	F	T	F	T	F	F
F	T	T	T	T	T	T	T
F	T	F	T	T	F	F	F
F	F	T	F	T	T	T	T
F	F	F	F	T	T	T	T

$p \vee q \to r$ and $(p \to r) \wedge (q \to r)$ always have the same truth values, so they are logically equivalent

Representation of If-Then As Or

In exercise 13(a) at the end of this section you are asked to use truth tables to show that

$$p \rightarrow q \equiv \sim p \vee q.$$

The logical equivalence of "if p then q" and "not p or q" is occasionally used in everyday speech. Here is one instance.

Example 2.2.4 Application of the Equivalence between $\sim p \vee q$ and $p \rightarrow q$

Rewrite the following statement in if-then form.

Either you get to work on time or you are fired.

Solution   Let $\sim p$ be

You get to work on time.

and q be

You are fired.

Then the given statement is $\sim p \vee q$. Also p is

You do not get to work on time.

So the equivalent if-then version, $p \rightarrow q$, is

If you do not get to work on time, then you are fired.   ■

The Negation of a Conditional Statement

By definition, $p \rightarrow q$ is false if, and only if, its hypothesis, p, is true and its conclusion, q, is false. It follows that

> The negation of "if p then q" is logically equivalent to "p and not q."

This can be restated symbolically as follows:

$$\boxed{\sim(p \rightarrow q) \equiv p \wedge \sim q}$$

You can also obtain this result by starting from the logical equivalence $p \rightarrow q \equiv \sim p \vee q$.

Take the negation of both sides to obtain

$$
\begin{aligned}
\sim(p \rightarrow q) &\equiv \sim(\sim p \vee q) \\
&\equiv \sim(\sim p) \wedge (\sim q) \quad \text{by De Morgan's laws} \\
&\equiv p \wedge \sim q \quad \text{by the double negative law.}
\end{aligned}
$$

Yet another way to derive this result is to construct truth tables for $\sim(p \rightarrow q)$ and for $p \wedge \sim q$ and to check that they have the same truth values. (See exercise 13(b) at the end of this section.)

Example 2.2.5 Negations of *If-Then* Statements

Write negations for each of the following statements:

a. If my car is in the repair shop, then I cannot get to class.

b. If Sara lives in Athens, then she lives in Greece.

Solution

a. My car is in the repair shop and I can get to class.

b. Sara lives in Athens and she does not live in Greece. (Sara might live in Athens, Georgia; Athens, Ohio; or Athens, Wisconsin.) ■

It is tempting to write the negation of an if-then statement as another if-then statement. Please resist that temptation!

Caution! Remember that **the negation of an if-then statement does not start with the word** *if*.

The Contrapositive of a Conditional Statement

One of the most fundamental laws of logic is the equivalence between a conditional statement and its contrapositive.

> **• Definition**
>
> The **contrapositive** of a conditional statement of the form "If p then q" is
>
> $$\text{If } \sim q \text{ then } \sim p.$$
>
> Symbolically,
>
> $$\text{The contrapositive of } p \rightarrow q \text{ is } \sim q \rightarrow \sim p.$$

The fact is that

> A conditional statement is logically equivalent to its contrapositive.

You are asked to establish this equivalence in exercise 26 at the end of this section.

Example 2.2.6 Writing the Contrapositive

Write each of the following statements in its equivalent contrapositive form:

a. If Howard can swim across the lake, then Howard can swim to the island.

b. If today is Easter, then tomorrow is Monday.

Solution

a. If Howard cannot swim to the island, then Howard cannot swim across the lake.

b. If tomorrow is not Monday, then today is not Easter. ■

When you are trying to solve certain problems, you may find that the contrapositive form of a conditional statement is easier to work with than the original statement. Replacing a statement by its contrapositive may give the extra push that helps you over the top in your search for a solution. This logical equivalence is also the basis for one of the most important laws of deduction, modus tollens (to be explained in Section 2.3), and for the contrapositive method of proof (to be explained in Section 4.5).

The Converse and Inverse of a Conditional Statement

The fact that a conditional statement and its contrapositive are logically equivalent is very important and has wide application. Two other variants of a conditional statement are *not* logically equivalent to the statement.

> **• Definition**
>
> Suppose a conditional statement of the form "If p then q" is given.
>
> 1. The **converse** is "If q then p."
> 2. The **inverse** is "If $\sim p$ then $\sim q$."
>
> Symbolically,
>
> $$\text{The converse of } p \to q \text{ is } q \to p,$$
>
> and
>
> $$\text{The inverse of } p \to q \text{ is } \sim p \to \sim q.$$

Example 2.2.7 Writing the Converse and the Inverse

Write the converse and inverse of each of the following statements:

a. If Howard can swim across the lake, then Howard can swim to the island.

b. If today is Easter, then tomorrow is Monday.

Solution

a. *Converse:* If Howard can swim to the island, then Howard can swim across the lake.

   Inverse:　If Howard cannot swim across the lake, then Howard cannot swim to the island.

b. *Converse:* If tomorrow is Monday, then today is Easter.

   Inverse:　If today is not Easter, then tomorrow is not Monday.　■

Note that while the statement "If today is Easter, then tomorrow is Monday" is always true, both its converse and inverse are false on every Sunday except Easter.

> 1. A conditional statement and its converse are *not* logically equivalent.
>
> 2. A conditional statement and its inverse are *not* logically equivalent.
>
> 3. The converse and the inverse of a conditional statement are logically equivalent to each other.

⚠ Caution!　Many people believe that if a conditional statement is true, then its converse and inverse must also be true. This is not correct!

In exercises 24, 25, and 27 at the end of this section, you are asked to use truth tables to verify the statements in the box above. Note that the truth of statement 3 also follows from the observation that the inverse of a conditional statement is the contrapositive of its converse.

Only If and the Biconditional

To say "p only if q" means that p can take place *only* if q takes place also. That is, if q does not take place, then p cannot take place. Another way to say this is that if p occurs, then q must also occur (by the logical equivalence between a statement and its contrapositive).

• Definition

If p and q are statements,

$$p \text{ \textbf{only if} } q \quad \text{means} \quad \text{``if not } q \text{ then not } p\text{,''}$$

or, equivalently,

$$\text{``}if \ p \text{ then } q\text{.''}$$

Example 2.2.8 Converting *Only If* to *If-Then*

Rewrite the following statement in if-then form in two ways, one of which is the contra-positive of the other.

> John will break the world's record for the mile run only if
> he runs the mile in under four minutes.

Solution   *Version 1:*   If John does not run the mile in under four minutes, then he will not break the world's record.

           Version 2:   If John breaks the world's record, then he will have run the mile in under four minutes. ■

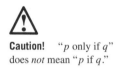

Caution!   "p only if q" does *not* mean "p if q."

Note that it is possible for "p only if q" to be true at the same time that "p if q" is false. For instance, to say that John will break the world's record only if he runs the mile in under four minutes does not mean that John will break the world's record if he runs the mile in under four minutes. His time could be under four minutes but still not be fast enough to break the record.

• Definition

Given statement variables p and q, the **biconditional of p and q** is "p if, and only if, q" and is denoted $p \leftrightarrow q$. It is true if both p and q have the same truth values and is false if p and q have opposite truth values. The words *if and only if* are sometimes abbreviated **iff.**

The biconditional has the following truth table:

Truth Table for $p \leftrightarrow q$

p	q	$p \leftrightarrow q$
T	T	T
T	F	F
F	T	F
F	F	T

In order of operations $\leftrightarrow$ is coequal with $\rightarrow$. As with $\wedge$ and $\vee$, the only way to indicate precedence between them is to use parentheses. The full hierarchy of operations for the five logical operators is on the next page.

	Order of Operations for Logical Operators
1. ~	Evaluate negations first.
2. ∧, ∨	Evaluate ∧ and ∨ second. When both are present, parentheses may be needed.
3. →, ↔	Evaluate → and ↔ third. When both are present, parentheses may be needed.

According to the separate definitions of *if* and *only if,* saying "*p* if, and only if, *q*" should mean the same as saying both "*p* if *q*" and "*p* only if *q*." The following annotated truth table shows that this is the case:

Truth Table Showing that $p \leftrightarrow q \equiv (p \rightarrow q) \wedge (q \rightarrow p)$

p	q	$p \rightarrow q$	$q \rightarrow p$	$p \leftrightarrow q$	$(p \rightarrow q) \wedge (q \rightarrow p)$
T	T	T	T	T	T
T	F	F	T	F	F
F	T	T	F	F	F
F	F	T	T	T	T

$p \leftrightarrow q$ and $(p \rightarrow q) \wedge (q \rightarrow p)$ always have the same truth values, so they are logically equivalent

Example 2.2.9 *If* and *Only If*

Rewrite the following statement as a conjunction of two if-then statements:

This computer program is correct if, and only if, it produces correct answers for all possible sets of input data.

Solution   If this program is correct, then it produces the correct answers for all possible sets of input data; and if this program produces the correct answers for all possible sets of input data, then it is correct.   ■

Necessary and Sufficient Conditions

The phrases *necessary condition* and *sufficient condition,* as used in formal English, correspond exactly to their definitions in logic.

> **• Definition**
>
> If *r* and *s* are statements:
>
> *r* is a **sufficient condition** for *s*   means   "if *r* then *s*."
>
> *r* is a **necessary condition** for *s*   means   "if not *r* then not *s*."

In other words, to say "*r* is a sufficient condition for *s*" means that the occurrence of *r* is *sufficient* to guarantee the occurrence of *s*. On the other hand, to say "*r* is a necessary condition for *s*" means that if *r* does not occur, then *s* cannot occur either:

The occurrence of r is *necessary* to obtain the occurrence of s. Note that because of the equivalence between a statement and its contrapositive,

> r is a necessary condition for s also means   "if s then r."

Consequently,

> r is a necessary and sufficient condition for s means   "r if, and only if, s."

Example 2.2.10 Interpreting Necessary and Sufficient Conditions

Consider the statement "If John is eligible to vote, then he is at least 18 years old." The truth of the condition "John is eligible to vote" is *sufficient* to ensure the truth of the condition "John is at least 18 years old." In addition, the condition "John is at least 18 years old" is *necessary* for the condition "John is eligible to vote" to be true. If John were younger than 18, then he would not be eligible to vote. ∎

Example 2.2.11 Converting a Sufficient Condition to If-Then Form

Rewrite the following statement in the form "If A then B":

> Pia's birth on U.S soil is a sufficient condition
> for her to be a U.S. citizen.

Solution   If Pia was born on U.S. soil, then she is a U.S. citizen. ∎

Example 2.2.12 Converting a Necessary Condition to If-Then Form

Use the contrapositive to rewrite the following statement in two ways:

> George's attaining age 35 is a necessary condition
> for his being president of the United States.

Solution   *Version 1:* If George has not attained the age of 35, then he cannot be president of the United States.

Version 2: If George can be president of the United States, then he has attained the age of 35. ∎

Remarks

1. *In logic, a hypothesis and conclusion are not required to have related subject matters.*

   In ordinary speech we never say things like "If computers are machines, then Babe Ruth was a baseball player" or "If $2 + 2 = 5$, then Mickey Mouse is president of the United States." We formulate a sentence like "If p then q" only if there is some connection of content between p and q.

   In logic, however, the two parts of a conditional statement need not have related meanings. The reason? If there were such a requirement, who would enforce it? What one person perceives as two unrelated clauses may seem related to someone else. There would have to be a central arbiter to check each conditional sentence before anyone could use it, to be sure its clauses were in proper relation. This is impractical, to say the least!

Thus a statement like "if computers are machines, then Babe Ruth was a baseball player" is allowed, and it is even called true because both its hypothesis and its conclusion are true. Similarly, the statement "If $2 + 2 = 5$, then Mickey Mouse is president of the United States" is allowed and is called true because its hypothesis is false, even though doing so may seem ridiculous.

In mathematics it often happens that a carefully formulated definition that successfully covers the situations for which it was primarily intended is later seen to be satisfied by some extreme cases that the formulator did not have in mind. But those are the breaks, and it is important to get into the habit of exploring definitions fully to seek out and understand *all* their instances, even the unusual ones.

2. *In informal language, simple conditionals are often used to mean biconditionals.*

The formal statement "p if, and only if, q" is seldom used in ordinary language. Frequently, when people intend the biconditional they leave out either the *and only if* or the *if and.* That is, they say either "p if q" or "p only if q" when they really mean "p if, and only if, q." For example, consider the statement "You will get dessert if, and only if, you eat your dinner." Logically, this is equivalent to the conjunction of the following two statements.

Statement 1: If you eat your dinner, then you will get dessert.

Statement 2: You will get dessert only if you eat your dinner.

> or

> If you do not eat your dinner, then you will not get dessert.

Now how many parents in the history of the world have said to their children "You will get dessert if, and only if, you eat your dinner"? Not many! Most say either "If you eat your dinner, you will get dessert" (these take the positive approach—they emphasize the reward) or "You will get dessert only if you eat your dinner" (these take the negative approach—they emphasize the punishment). Yet the parents who promise the reward intend to suggest the punishment as well, and those who threaten the punishment will certainly give the reward if it is earned. Both sets of parents expect that their conditional statements will be interpreted as biconditionals.

Since we often (correctly) interpret conditional statements as biconditionals, it is not surprising that we may come to believe (mistakenly) that conditional statements are always logically equivalent to their inverses and converses. In formal settings, however, statements must have unambiguous interpretations. If-then statements can't sometimes mean "if-then" and other times mean "if and only if." When using language in mathematics, science, or other situations where precision is important, it is essential to interpret if-then statements according to the formal definition and not to confuse them with their converses and inverses.

Test Yourself

1. An *if-then* statement is false if, and only if, the hypothesis is ____ and the conclusion is ____.

2. The negation of "if p then q" is ____.

3. The converse of "if p then q" is ____.

4. The contrapositive of "if p then q" is ____.

5. The inverse of "if p then q" is ____.

6. A conditional statement and its contrapositive are ____.

7. A conditional statement and its converse are not ____.

8. "R is a sufficient condition for S" means "if ____ then ____."

9. "R is a necessary condition for S" means "if ____ then ____."

10. "R only if S" means "if ____ then ____."

Exercise Set 2.2

Rewrite the statements in 1–4 in if-then form.

1. This loop will repeat exactly N times if it does not contain a **stop** or a **go to.**

2. I am on time for work if I catch the 8:05 bus.

3. Freeze or I'll shoot.

4. Fix my ceiling or I won't pay my rent.

Construct truth tables for the statement forms in 5–11.

5. $\sim p \vee q \rightarrow \sim q$ **6.** $(p \vee q) \vee (\sim p \wedge q) \rightarrow q$

7. $p \wedge \sim q \rightarrow r$ **8.** $\sim p \vee q \rightarrow r$

9. $p \wedge \sim r \leftrightarrow q \vee r$ **10.** $(p \rightarrow r) \leftrightarrow (q \rightarrow r)$

11. $(p \rightarrow (q \rightarrow r)) \leftrightarrow ((p \wedge q) \rightarrow r)$

12. Use the logical equivalence established in Example 2.2.3, $p \vee q \rightarrow r \equiv (p \rightarrow r) \wedge (q \rightarrow r)$, to rewrite the following statement. (Assume that x represents a fixed real number.)

$$\text{If } x > 2 \text{ or } x < -2, \text{ then } x^2 > 4.$$

13. Use truth tables to verify the following logical equivalences. Include a few words of explanation with your answers.

a. $p \rightarrow q \equiv \sim p \vee q$ **b.** $\sim(p \rightarrow q) \equiv p \wedge \sim q$.

H 14. **a.** Show that the following statement forms are all logically equivalent.

$$p \rightarrow q \vee r, \quad p \wedge \sim q \rightarrow r, \quad \text{and} \quad p \wedge \sim r \rightarrow q$$

b. Use the logical equivalences established in part (a) to rewrite the following sentence in two different ways. (Assume that n represents a fixed integer.)

$$\text{If } n \text{ is prime, then } n \text{ is odd or } n \text{ is 2.}$$

15. Determine whether the following statement forms are logically equivalent:

$$p \rightarrow (q \rightarrow r) \quad \text{and} \quad (p \rightarrow q) \rightarrow r$$

In 16 and 17, write each of the two statements in symbolic form and determine whether they are logically equivalent. Include a truth table and a few words of explanation.

16. If you paid full price, you didn't buy it at Crown Books. You didn't buy it at Crown Books or you paid full price.

17. If 2 is a factor of n and 3 is a factor of n, then 6 is a factor of n. 2 is not a factor of n or 3 is not a factor of n or 6 is a factor of n.

18. Write each of the following three statements in symbolic form and determine which pairs are logically equivalent. Include truth tables and a few words of explanation.

If it walks like a duck and it talks like a duck, then it is a duck.

Either it does not walk like a duck or it does not talk like a duck, or it is a duck.

If it does not walk like a duck and it does not talk like a duck, then it is not a duck.

19. True or false? The negation of "If Sue is Luiz's mother, then Ali is his cousin" is "If Sue is Luiz's mother, then Ali is not his cousin."

20. Write negations for each of the following statements. (Assume that all variables represent fixed quantities or entities, as appropriate.)

   a. If P is a square, then P is a rectangle.
   b. If today is New Year's Eve, then tomorrow is January.
   c. If the decimal expansion of r is terminating, then r is rational.
   d. If n is prime, then n is odd or n is 2.
   e. If x is nonnegative, then x is positive or x is 0.
   f. If Tom is Ann's father, then Jim is her uncle and Sue is her aunt.
   g. If n is divisible by 6, then n is divisible by 2 and n is divisible by 3.

21. Suppose that p and q are statements so that $p \rightarrow q$ is false. Find the truth values of each of the following:

   a. $\sim p \rightarrow q$ b. $p \vee q$ c. $q \rightarrow p$

H **22.** Write contrapositives for the statements of exercise 20.

H **23.** Write the converse and inverse for each statement of exercise 20.

Use truth tables to establish the truth of each statement in 24–27.

24. A conditional statement is not logically equivalent to its converse.

25. A conditional statement is not logically equivalent to its inverse.

26. A conditional statement and its contrapositive are logically equivalent to each other.

27. The converse and inverse of a conditional statement are logically equivalent to each other.

H **28.** "Do you mean that you think you can find out the answer to it?" said the March Hare.

   "Exactly so," said Alice.
   "Then you should say what you mean," the March Hare went on.
   "I do," Alice hastily replied; "at least—at least I mean what I say—that's the same thing, you know."

"Not the same thing a bit!" said the Hatter. "Why, you might just as well say that 'I see what I eat' is the same thing as 'I eat what I see'!"

—from "A Mad Tea-Party" in *Alice in Wonderland,* by Lewis Carroll

The Hatter is right. "I say what I mean" is not the same thing as "I mean what I say." Rewrite each of these two sentences in if-then form and explain the logical relation between them. (This exercise is referred to in the introduction to Chapter 4.)

If statement forms P and Q are logically equivalent, then $P \leftrightarrow Q$ is a tautology. Conversely, if $P \leftrightarrow Q$ is a tautology, then P and Q are logically equivalent. Use $\leftrightarrow$ to convert each of the logical equivalences in 29–31 to a tautology. Then use a truth table to verify each tautology.

29. $p \rightarrow (q \vee r) \equiv (p \wedge \sim q) \rightarrow r$

30. $p \wedge (q \vee r) \equiv (p \wedge q) \vee (p \wedge r)$

31. $p \rightarrow (q \rightarrow r) \equiv (p \wedge q) \rightarrow r$

Rewrite each of the statements in 32 and 33 as a conjunction of two if-then statements.

32. This quadratic equation has two distinct real roots if, and only if, its discriminant is greater than zero.

33. This integer is even if, and only if, it equals twice some integer.

Rewrite the statements in 34 and 35 in if-then form in two ways, one of which is the contrapositive of the other.

34. The Cubs will win the pennant only if they win tomorrow's game.

35. Sam will be allowed on Signe's racing boat only if he is an expert sailor.

36. Taking the long view on your education, you go to the Prestige Corporation and ask what you should do in college to be hired when you graduate. The personnel director replies that you will be hired *only if* you major in mathematics or computer science, get a B average or better, and take accounting. You do, in fact, become a math major, get a B+ average, and take accounting. You return to Prestige Corporation, make a formal application, and are turned down. Did the personnel director lie to you?

Some programming languages use statements of the form "*r* unless *s*"" to mean that as long as *s* does not happen, then *r* will happen. More formally:

> **Definition:** If *r* and *s* are statements,
>
> **r unless s**   means   if $\sim s$ then *r*.

In 37–39, rewrite the statements in if-then form.

37. Payment will be made on the fifth unless a new hearing is granted.

38. Ann will go unless it rains.

39. This door will not open unless a security code is entered.

Rewrite the statements in 40 and 41 in if-then form.

40. Catching the 8:05 bus is a sufficient condition for my being on time for work.

41. Having two 45° angles is a sufficient condition for this triangle to be a right triangle.

Use the contrapositive to rewrite the statements in 42 and 43 in if-then form in two ways.

42. Being divisible by 3 is a necessary condition for this number to be divisible by 9.

43. Doing homework regularly is a necessary condition for Jim to pass the course.

Note that "a sufficient condition for *s* is *r*" means *r* is a sufficient condition for *s* and that "a necessary condition for *s* is *r*" means *r* is a necessary condition for *s*. Rewrite the statements in 44 and 45 in if-then form.

44. A sufficient condition for Jon's team to win the championship is that it win the rest of its games.

45. A necessary condition for this computer program to be correct is that it not produce error messages during translation.

46. "If compound X is boiling, then its temperature must be at least 150°C." Assuming that this statement is true, which of the following must also be true?
   a. If the temperature of compound X is at least 150°C, then compound X is boiling.
   b. If the temperature of compound X is less than 150°C, then compound X is not boiling.
   c. Compound X will boil only if its temperature is at least 150°C.
   d. If compound X is not boiling, then its temperature is less than 150°C.
   e. A necessary condition for compound X to boil is that its temperature be at least 150°C.
   f. A sufficient condition for compound X to boil is that its temperature be at least 150°C.

Answers for Test Yourself

1. true; false   2. $p \wedge \sim q$ 3. if q then p 4. if $\sim q$ then $\sim p$ 5. if $\sim p$ then $\sim q$ 6. logically equivalent   7. logically equivalent   8. *R; S*   9. *S; R*   10. *R; S*

2.3 *Valid and Invalid Arguments*

"Contrariwise," continued Tweedledee, *"if it was so, it might be; and if it were so, it would be; but as it isn't, it ain't. That's logic."* — Lewis Carroll, *Through the Looking Glass*

In mathematics and logic an argument is not a dispute. It is a sequence of statements ending in a conclusion. In this section we show how to determine whether an argument is valid—that is, whether the conclusion follows *necessarily* from the preceding statements. We will show that this determination depends only on the form of an argument, not on its content.

It was shown in Section 2.1 that the logical form of an argument can be abstracted from its content. For example, the argument

> If Socrates is a man, then Socrates is mortal.
>
> Socrates is a man.
>
> ∴ Socrates is mortal.

has the abstract form

> If p then q
>
> p
>
> ∴ q

When considering the abstract form of an argument, think of p and q as variables for which statements may be substituted. An argument form is called *valid* if, and only if, whenever statements are substituted that make all the premises true, the conclusion is also true.

• Definition

An **argument** is a sequence of statements, and an **argument form** is a sequence of statement forms. All statements in an argument and all statement forms in an argument form, except for the final one, are called **premises** (or **assumptions** or **hypotheses**). The final statement or statement form is called the **conclusion.** The symbol ∴, which is read "therefore," is normally placed just before the conclusion.

To say that an *argument form* is **valid** means that no matter what particular statements are substituted for the statement variables in its premises, if the resulting premises are all true, then the conclusion is also true. To say that an *argument* is **valid** means that its form is valid.

The crucial fact about a valid argument is that the truth of its conclusion follows *necessarily* or *inescapably* or *by logical form alone* from the truth of its premises. It is impossible to have a valid argument with true premises and a false conclusion. When an argument is valid and its premises are true, the truth of the conclusion is said to be *inferred* or *deduced* from the truth of the premises. If a conclusion "ain't necessarily so," then it isn't a valid deduction.

Testing an Argument Form for Validity

1. Identify the premises and conclusion of the argument form.

2. Construct a truth table showing the truth values of all the premises and the conclusion.

3. A row of the truth table in which all the premises are true is called a **critical row**. If there is a critical row in which the conclusion is false, then it is possible for an argument of the given form to have true premises and a false conclusion, and so the argument form is invalid. If the conclusion in *every* critical row is true, then the argument form is valid.

Example 2.3.1 Determining Validity or Invalidity

Determine whether the following argument form is valid or invalid by drawing a truth table, indicating which columns represent the premises and which represent the conclusion, and annotating the table with a sentence of explanation. When you fill in the table, you only need to indicate the truth values for the conclusion in the rows where all the premises are true (the critical rows) because the truth values of the conclusion in the other rows are irrelevant to the validity or invalidity of the argument.

$$p \to q \lor \sim r$$
$$q \to p \land r$$
$$\therefore p \to r$$

Solution   The truth table shows that even though there are several situations in which the premises and the conclusion are all true (rows 1, 7, and 8), there is one situation (row 4) where the premises are true and the conclusion is false.

						premises		conclusion
p	q	r	$\sim r$	$q \lor \sim r$	$p \land r$	$p \to q \lor \sim r$	$q \to p \land r$	$p \to r$
T	T	T	F	T	T	T	T	T
T	T	F	T	T	F	T	F	
T	F	T	F	F	T	F	T	
T	F	F	T	T	F	T	T	F
F	T	T	F	T	F	T	F	
F	T	F	T	T	F	T	F	
F	F	T	F	F	F	T	T	T
F	F	F	T	T	F	T	T	T

This row shows that an argument of this form can have true premises and a false conclusion. Hence this form of argument is invalid.

■

Modus Ponens and Modus Tollens

An argument form consisting of two premises and a conclusion is called a **syllogism.** The first and second premises are called the **major premise** and **minor premise**, respectively.

The most famous form of syllogism in logic is called **modus ponens.** It has the following form:

$$\text{If } p \text{ then } q.$$
$$p$$
$$\therefore q$$

Here is an argument of this form:

If the sum of the digits of 371,487 is divisible by 3, then 371,487 is divisible by 3.

The sum of the digits of 371,487 is divisible by 3.

∴ 371,487 is divisible by 3.

The term *modus ponens* is Latin meaning "method of affirming" (the conclusion is an affirmation). Long before you saw your first truth table, you were undoubtedly being convinced by arguments of this form. Nevertheless, it is instructive to prove that modus ponens is a valid form of argument, if for no other reason than to confirm the agreement between the formal definition of validity and the intuitive concept. To do so, we construct a truth table for the premises and conclusion.

		premises		conclusion	
p	q	$p \rightarrow q$	p	q	
T	T	T	T	T	← critical row
T	F	F	T		
F	T	T	F		
F	F	T	F		

The first row is the only one in which both premises are true, and the conclusion in that row is also true. Hence the argument form is valid.

Now consider another valid argument form called **modus tollens.** It has the following form:

$$\text{If } p \text{ then } q.$$
$$\sim q$$
$$\therefore \sim p$$

Here is an example of modus tollens:

If Zeus is human, then Zeus is mortal.

Zeus is not mortal.

∴ Zeus is not human.

An intuitive explanation for the validity of modus tollens uses proof by contradiction. It goes like this:

Suppose

(1) If Zeus is human, then Zeus is mortal; and

(2) Zeus is not mortal.

Must Zeus necessarily be nonhuman?

Yes!

Because, if Zeus were human, then by (1) he would be mortal.

But by (2) he is not mortal.

Hence, Zeus cannot be human.

Modus tollens is Latin meaning "method of denying" (the conclusion is a denial). The validity of modus tollens can be shown to follow from modus ponens together with the fact that a conditional statement is logically equivalent to its contrapositive. Or it can be established formally by using a truth table. (See exercise 13.)

Studies by cognitive psychologists have shown that although nearly 100% of college students have a solid, intuitive understanding of modus ponens, less than 60% are able to apply modus tollens correctly.* Yet in mathematical reasoning, modus tollens is used almost as often as modus ponens. Thus it is important to study the form of modus tollens carefully to learn to use it effectively.

Example 2.3.2 Recognizing Modus Ponens and Modus Tollens

Use modus ponens or modus tollens to fill in the blanks of the following arguments so that they become valid inferences.

a. If there are more pigeons than there are pigeonholes, then at least two pigeons roost in the same hole.
There are more pigeons than there are pigeonholes.
∴ _____ .

b. If 870,232 is divisible by 6, then it is divisible by 3.
870,232 is not divisible by 3.
∴ _____ .

Solution

a. At least two pigeons roost in the same hole.   by modus ponens

b. 870,232 is not divisible by 6.   by modus tollens   ■

Additional Valid Argument Forms: Rules of Inference

A **rule of inference** is a form of argument that is valid. Thus modus ponens and modus tollens are both rules of inference. The following are additional examples of rules of inference that are frequently used in deductive reasoning.

Example 2.3.3 Generalization

The following argument forms are valid:

$$
\begin{array}{ll}
\text{a.} \quad p & \qquad\qquad \text{b.} \quad q \\
\therefore p \vee q & \qquad\qquad \therefore p \vee q
\end{array}
$$

These argument forms are used for making generalizations. For instance, according to the first, if p is true, then, more generally, "p or q" is true for *any* other statement q. As an example, suppose you are given the job of counting the upperclassmen at your school. You ask what class Anton is in and are told he is a junior.

Cognitive Psychology and Its Implications, 3d ed. by John R. Anderson (New York: Freeman, 1990), pp. 292–297.

You reason as follows:

Anton is a junior.

∴ (more generally) Anton is a junior or Anton is a senior.

Knowing that upperclassman means junior or senior, you add Anton to your list. ■

Example 2.3.4 Specialization

The following argument forms are valid:

a. $p \wedge q$ b. $p \wedge q$
     $\therefore p$ $\therefore q$

These argument forms are used for specializing. When classifying objects according to some property, you often know much more about them than whether they do or do not have that property. When this happens, you discard extraneous information as you concentrate on the particular property of interest.

For instance, suppose you are looking for a person who knows graph algorithms to work with you on a project. You discover that Ana knows both numerical analysis and graph algorithms. You reason as follows:

Ana knows numerical analysis and Ana knows graph algorithms.

∴ (in particular) Ana knows graph algorithms.

Accordingly, you invite her to work with you on your project. ■

Both generalization and specialization are used frequently in mathematics to tailor facts to fit into hypotheses of known theorems in order to draw further conclusions. Elimination, transitivity, and proof by division into cases are also widely used tools.

Example 2.3.5 Elimination

The following argument forms are valid:

a. $p \vee q$ b. $p \vee q$
     $\sim q$ $\sim p$
     $\therefore p$ $\therefore q$

These argument forms say that when you have only two possibilities and you can rule one out, the other must be the case. For instance, suppose you know that for a particular number x,

$$x - 3 = 0 \quad \text{or} \quad x + 2 = 0.$$

If you also know that x is not negative, then $x \neq -2$, so

$$x + 2 \neq 0.$$

By elimination, you can then conclude that

$$\therefore x - 3 = 0.$$ ■

Example 2.3.6 Transitivity

The following argument form is valid:

$$p \rightarrow q$$
$$q \rightarrow r$$
$$\therefore p \rightarrow r$$

Many arguments in mathematics contain chains of if-then statements. From the fact that one statement implies a second and the second implies a third, you can conclude that the first statement implies the third. Here is an example:

If 18,486 is divisible by 18, then 18,486 is divisible by 9.

If 18,486 is divisible by 9, then the sum of the digits of 18,486 is divisible by 9.

∴ If 18,486 is divisible by 18, then the sum of the digits of 18,486 is divisible by 9. ∎

Example 2.3.7 Proof by Division into Cases

The following argument form is valid:

$$p \lor q$$
$$p \to r$$
$$q \to r$$
$$\therefore r$$

It often happens that you know one thing or another is true. If you can show that in either case a certain conclusion follows, then this conclusion must also be true. For instance, suppose you know that x is a particular nonzero real number. The trichotomy property of the real numbers says that any number is positive, negative, or zero. Thus (by elimination) you know that x is positive or x is negative. You can deduce that $x^2 > 0$ by arguing as follows:

x is positive or x is negative.

If x is positive, then $x^2 > 0$.

If x is negative, then $x^2 > 0$.

∴ $x^2 > 0$. ∎

The rules of valid inference are used constantly in problem solving. Here is an example from everyday life.

Example 2.3.8 Application: A More Complex Deduction

You are about to leave for school in the morning and discover that you don't have your glasses. You know the following statements are true:

a. If I was reading the newspaper in the kitchen, then my glasses are on the kitchen table.

b. If my glasses are on the kitchen table, then I saw them at breakfast.

c. I did not see my glasses at breakfast.

d. I was reading the newspaper in the living room or I was reading the newspaper in the kitchen.

e. If I was reading the newspaper in the living room then my glasses are on the coffee table.

Where are the glasses?

Solution   Let RK = I was reading the newspaper in the kitchen.

GK = My glasses are on the kitchen table.

SB = I saw my glasses at breakfast.

RL = I was reading the newspaper in the living room.

GC = My glasses are on the coffee table.

Here is a sequence of steps you might use to reach the answer, together with the rules of inference that allow you to draw the conclusion of each step:

1. $RK \rightarrow GK$ by (a)

    $GK \rightarrow SB$ by (b)

    $\therefore RK \rightarrow SB$ by transitivity

2. $RK \rightarrow SB$ by the conclusion of (1)

    $\sim SB$ by (c)

    $\therefore \sim RK$ by modus tollens

3. $RL \vee RK$ by (d)

    $\sim RK$ by the conclusion of (2)

    $\therefore RL$ by elimination

4. $RL \rightarrow GC$ by (e)

    RL by the conclusion of (3)

    $\therefore GC$ by modus ponens

Thus the glasses are on the coffee table.   ■

Fallacies

A **fallacy** is an error in reasoning that results in an invalid argument. Three common fallacies are **using ambiguous premises,** and treating them as if they were unambiguous, **assuming what is to be proved** (without having derived it from the premises), and **jumping to a conclusion** (without adequate grounds). In this section we discuss two other fallacies, called *converse error* and *inverse error,* which give rise to arguments that superficially resemble those that are valid by modus ponens and modus tollens but are not, in fact, valid.

As in previous examples, you can show that an argument is invalid by constructing a truth table for the argument form and finding at least one critical row in which all the premises are true but the conclusion is false. Another way is to find an argument of the same form with true premises and a false conclusion.

> For an argument to be valid, every argument of the same form whose premises are all true must have a true conclusion. It follows that for an argument to be invalid means that there is an argument of that form whose premises are all true and whose conclusion is false.

Example 2.3.9 Converse Error

Show that the following argument is invalid:

> If Zeke is a cheater, then Zeke sits in the back row.
>
> Zeke sits in the back row.
>
> $\therefore$ Zeke is a cheater.

Solution   Many people recognize the invalidity of the above argument intuitively, reasoning something like this: The first premise gives information about Zeke *if* it is known he is a cheater. It doesn't give any information about him if it is not already known that he is a cheater. One can certainly imagine a person who is not a cheater but happens to sit in the back row. Then if that person's name is substituted for Zeke, the first premise is true by default and the second premise is also true but the conclusion is false.

The general form of the previous argument is as follows:

$$p \rightarrow q$$
$$q$$
$$\therefore p$$

In exercise 12(a) at the end of this section you are asked to use a truth table to show that this form of argument is invalid. ∎

The fallacy underlying this invalid argument form is called the **converse error** because the conclusion of the argument would follow from the premises if the premise $p \rightarrow q$ were replaced by its converse. Such a replacement is not allowed, however, because a conditional statement is not logically equivalent to its converse. Converse error is also known as the *fallacy of affirming the consequent.*

Another common error in reasoning is called the *inverse error.*

Example 2.3.10 Inverse Error

Consider the following argument:

If interest rates are going up, stock market prices will go down.

Interest rates are not going up.

∴ Stock market prices will not go down.

Note that this argument has the following form:

$$p \rightarrow q$$
$$\sim p$$
$$\therefore \sim q$$

Caution! In logic, the words *true* and *valid* have very different meanings. A valid argument may have a false conclusion, and an invalid argument may have a true conclusion.

You are asked to give a truth table verification of the invalidity of this argument form in exercise 12(b) at the end of this section.

The fallacy underlying this invalid argument form is called the **inverse error** because the conclusion of the argument would follow from the premises if the premise $p \rightarrow q$ were replaced by its inverse. Such a replacement is not allowed, however, because a conditional statement is not logically equivalent to its inverse. Inverse error is also known as the *fallacy of denying the antecedent.* ∎

Sometimes people lump together the ideas of validity and truth. If an argument seems valid, they accept the conclusion as true. And if an argument seems fishy (really a slang expression for invalid), they think the conclusion must be false. This is not correct!

Example 2.3.11 A Valid Argument with a False Premise and a False Conclusion

The argument on the next page is valid by modus ponens. But its major premise is false, and so is its conclusion.

> If John Lennon was a rock star, then John Lennon had red hair.
>
> John Lennon was a rock star.
>
> ∴ John Lennon had red hair. ■

Example 2.3.12 An Invalid Argument with True Premises and a True Conclusion

The argument below is invalid by the converse error, but it has a true conclusion.

> If New York is a big city, then New York has tall buildings.
>
> New York has tall buildings.
>
> ∴ New York is a big city. ■

• **Definition**

An argument is called **sound** if, and only if, it is valid *and* all its premises are true. An argument that is not sound is called **unsound**.

The important thing to note is that validity is a property of argument *forms:* If an argument is valid, then so is every other argument that has the same form. Similarly, if an argument is invalid, then so is every other argument that has the same form. What characterizes a valid argument is that no argument whose form is valid can have all true premises and a false conclusion. For each valid argument, there are arguments of that form with all true premises and a true conclusion, with at least one false premise and a true conclusion, and with at least one false premise and a false conclusion. On the other hand, for each invalid argument, there are arguments of that form with every combination of truth values for the premises and conclusion, including all true premises and a false conclusion. The bottom line is that we can only be sure that the conclusion of an argument is true when we know that the argument is sound, that is, when we know both that the argument is valid and that it has all true premises.

Contradictions and Valid Arguments

The concept of logical contradiction can be used to make inferences through a technique of reasoning called the *contradiction rule.* Suppose p is some statement whose truth you wish to deduce.

Contradiction Rule

If you can show that the supposition that statement p is false leads logically to a contradiction, then you can conclude that p is true.

Example 2.3.13 Contradiction Rule

Show that the following argument form is valid:

$$\sim p \to \mathbf{c}, \text{ where } \mathbf{c} \text{ is a contradiction}$$
$$\therefore p$$

Solution   Construct a truth table for the premise and the conclusion of this argument.

			premises	conclusion
p	$\sim p$	c	$\sim p \rightarrow c$	p
T	F	F	T	T
F	T	F	F	

There is only one critical row in which the premise is true, and in this row the conclusion is also true. Hence this form of argument is valid. ■

The contradiction rule is the logical heart of the method of proof by contradiction. A slight variation also provides the basis for solving many logical puzzles by eliminating contradictory answers: *If an assumption leads to a contradiction, then that assumption must be false.*

Example 2.3.14 Knights and Knaves

The logician Raymond Smullyan describes an island containing two types of people: knights who always tell the truth and knaves who always lie.* You visit the island and are approached by two natives who speak to you as follows:

> A says: B is a knight.
>
> B says: A and I are of opposite type.

What are A and B?

Raymond Smullyan
(born 1919)

Solution A and B are both knaves. To see this, reason as follows:
Suppose A is a knight.

∴ What A says is true.   by definition of *knight*

∴ B is also a knight.   That's what A said.

∴ What B says is true.   by definition of *knight*

∴ A and B are of opposite types.   That's what B said.

∴ We have arrived at the following contradiction: A and B are both knights and A and B are of opposite type.

∴ The supposition is false.   by the contradiction rule

∴ A is not a knight.   negation of supposition

∴ A is a knave.   by elimination: It's given that all inhabitants are knights or knaves, so since A is not a knight, A is a knave.

∴ What A says is false.

∴ B is not a knight.

∴ B is also a knave.   by elimination

This reasoning shows that if the problem has a solution at all, then A and B must both be knaves. It is conceivable, however, that the problem has no solution. The problem statement could be inherently contradictory. If you look back at the solution, though, you can see that it does work out for both A and B to be knaves. ■

*Raymond Smullyan has written a delightful series of whimsical yet profound books of logical puzzles starting with *What Is the Name of This Book?* (Englewood Cliffs, New Jersey: Prentice-Hall, 1978). Other good sources of logical puzzles are the many excellent books of Martin Gardner, such as *Aha! Insight* and *Aha! Gotcha* (New York: W. H. Freeman, 1978, 1982).

Summary of Rules of Inference

Table 2.3.1 summarizes some of the most important rules of inference.

Table 2.3.1 Valid Argument Forms

Modus Ponens	$p \rightarrow q$ p $\therefore q$	Elimination	**a.** $\;p \vee q$ $\sim q$ $\therefore p$	**b.** $\;p \vee q$ $\sim p$ $\therefore q$
Modus Tollens	$p \rightarrow q$ $\sim q$ $\therefore \sim p$	Transitivity	$p \rightarrow q$ $q \rightarrow r$ $\therefore p \rightarrow r$	
Generalization	**a.** $\;p$ $\therefore p \vee q$ $\qquad$ **b.** $\;q$ $\therefore p \vee q$	Proof by Division into Cases	$p \vee q$ $p \rightarrow r$ $q \rightarrow r$ $\therefore r$	
Specialization	**a.** $\;p \wedge q$ $\therefore p$ $\qquad$ **b.** $\;p \wedge q$ $\therefore q$			
Conjunction	p q $\therefore p \wedge q$	Contradiction Rule	$\sim p \rightarrow c$ $\therefore p$	

Test Yourself

1. For an argument to be valid means that every argument of the same form whose premises ____ has a ____ conclusion.

2. For an argument to be invalid means that there is an argument of the same form whose premises ____ and whose conclusion ____.

3. For an argument to be sound means that it is ____ and its premises ____. In this case we can be sure that its conclusion ____.

Exercise Set 2.3

Use modus ponens or modus tollens to fill in the blanks in the arguments of 1–5 so as to produce valid inferences.

1. If $\sqrt{2}$ is rational, then $\sqrt{2} = a/b$ for some integers a and b.
It is not true that $\sqrt{2} = a/b$ for some integers a and b.
$\therefore$ _____

2. If $1 - 0.99999\ldots$ is less than every positive real number, then it equals zero.
_____
$\therefore$ The number $1 - 0.99999\ldots$ equals zero.

3. If logic is easy, then I am a monkey's uncle.
I am not a monkey's uncle.
$\therefore$ _____

4. If this figure is a quadrilateral, then the sum of its interior angles is 360°.
The sum of the interior angles of this figure is not 360°.
$\therefore$ _____

5. If they were unsure of the address, then they would have telephoned.
_____
$\therefore$ They were sure of the address.

Use truth tables to determine whether the argument forms in 6–11 are valid. Indicate which columns represent the premises and which represent the conclusion, and include a sentence explaining how the truth table supports your answer. Your explanation should show that you understand what it means for a form of argument to be valid or invalid.

6. $p \rightarrow q$
$q \rightarrow p$
$\therefore p \vee q$

7. p
$p \rightarrow q$
$\sim q \vee r$
$\therefore r$

8. $p \vee q$
$p \rightarrow \sim q$
$p \rightarrow r$
$\therefore r$

9. $p \wedge q \rightarrow \sim r$
$p \vee \sim q$
$\sim q \rightarrow p$
$\therefore \sim r$

10. $p \to r$
    $q \to r$
    $\therefore\ p \vee q \to r$

11. $p \to q \vee r$
    $\sim q \vee \sim r$
    $\therefore\ \sim p \vee \sim r$

12. Use truth tables to show that the following forms of argument are invalid.

  a. $p \to q$
       q
       $\therefore\ p$
   (converse error)

  b. $p \to q$
       $\sim p$
      $\therefore\ \sim q$
   (inverse error)

Use truth tables to show that the argument forms referred to in 13–21 are valid. Indicate which columns represent the premises and which represent the conclusion, and include a sentence explaining how the truth table supports your answer. Your explanation should show that you understand what it means for a form of argument to be valid.

13. Modus tollens:

$$p \to q$$
$$\sim q$$
$$\therefore\ \sim p$$

14. Example 2.3.3(a)   15. Example 2.3.3(b)

16. Example 2.3.4(a)   17. Example 2.3.4(b)

18. Example 2.3.5(a)   19. Example 2.3.5(b)

20. Example 2.3.6   21. Example 2.3.7

Use symbols to write the logical form of each argument in 22 and 23, and then use a truth table to test the argument for validity. Indicate which columns represent the premises and which represent the conclusion, and include a few words of explanation showing that you understand the meaning of validity.

22.   If Tom is not on team A, then Hua is on team B.
     If Hua is not on team B, then Tom is on team A.
   $\therefore$ Tom is not on team A or Hua is not on team B.

23.   Oleg is a math major or Oleg is an economics major.
    If Oleg is a math major, then Oleg is required to take Math 362.
   $\therefore$ Oleg is an economics major or Oleg is not required to take Math 362.

Some of the arguments in 24–32 are valid, whereas others exhibit the converse or the inverse error. Use symbols to write the logical form of each argument. If the argument is valid, identify the rule of inference that guarantees its validity. Otherwise, state whether the converse or the inverse error is made.

24.   If Jules solved this problem correctly, then Jules obtained the answer 2.
     Jules obtained the answer 2.
   $\therefore$ Jules solved this problem correctly.

25.   This real number is rational or it is irrational.
     This real number is not rational.
   $\therefore$ This real number is irrational.

26.   If I go to the movies, I won't finish my homework. If I don't finish my homework, I won't do well on the exam tomorrow.
   $\therefore$ If I go to the movies, I won't do well on the exam tomorrow.

27.   If this number is larger than 2, then its square is larger than 4.
     This number is not larger than 2.
   $\therefore$ The square of this number is not larger than 4.

28.   If there are as many rational numbers as there are irrational numbers, then the set of all irrational numbers is infinite.
     The set of all irrational numbers is infinite.
   $\therefore$ There are as many rational numbers as there are irrational numbers.

29.   If at least one of these two numbers is divisible by 6, then the product of these two numbers is divisible by 6.
     Neither of these two numbers is divisible by 6.
   $\therefore$ The product of these two numbers is not divisible by 6.

30.   If this computer program is correct, then it produces the correct output when run with the test data my teacher gave me.
     This computer program produces the correct output when run with the test data my teacher gave me.
   $\therefore$ This computer program is correct.

31.   Sandra knows Java and Sandra knows C++.
   $\therefore$ Sandra knows C++.

32.   If I get a Christmas bonus, I'll buy a stereo.
     If I sell my motorcycle, I'll buy a stereo.
   $\therefore$ If I get a Christmas bonus or I sell my motorcycle, then I'll buy a stereo.

33. Give an example (other than Example 2.3.11) of a valid argument with a false conclusion.

34. Give an example (other than Example 2.3.12) of an invalid argument with a true conclusion.

35. Explain in your own words what distinguishes a valid form of argument from an invalid one.

36. Given the following information about a computer program, find the mistake in the program.
  a. There is an undeclared variable or there is a syntax error in the first five lines.
  b. If there is a syntax error in the first five lines, then there is a missing semicolon or a variable name is misspelled.
  c. There is not a missing semicolon.
  d. There is not a misspelled variable name.

37. In the back of an old cupboard you discover a note signed by a pirate famous for his bizarre sense of humor and love of logical puzzles. In the note he wrote that he had hidden treasure somewhere on the property. He listed five true statements (a–e below) and challenged the reader to use them to figure out the location of the treasure.

a. If this house is next to a lake, then the treasure is not in the kitchen.

b. If the tree in the front yard is an elm, then the treasure is in the kitchen.

c. This house is next to a lake.

d. The tree in the front yard is an elm or the treasure is buried under the flagpole.

e. If the tree in the back yard is an oak, then the treasure is in the garage.

Where is the treasure hidden?

38. You are visiting the island described in Example 2.3.14 and have the following encounters with natives.

a. Two natives A and B address you as follows:

   A says: Both of us are knights.

   B says: A is a knave.

   What are A and B?

b. Another two natives C and D approach you but only C speaks.

   C says: Both of us are knaves.

   What are C and D?

c. You then encounter natives E and F.

   E says: F is a knave.

   F says: E is a knave.

   How many knaves are there?

H d. Finally, you meet a group of six natives, U, V, W, X, Y, and Z, who speak to you as follows:

   U says: None of us is a knight.

   V says: At least three of us are knights.

   W says: At most three of us are knights.

   X says: Exactly five of us are knights.

   Y says: Exactly two of us are knights.

   Z says: Exactly one of us is a knight.

   Which are knights and which are knaves?

39. The famous detective Percule Hoirot was called in to solve a baffling murder mystery. He determined the following facts:

a. Lord Hazelton, the murdered man, was killed by a blow on the head with a brass candlestick.

b. Either Lady Hazelton or a maid, Sara, was in the dining room at the time of the murder.

c. If the cook was in the kitchen at the time of the murder, then the butler killed Lord Hazelton with a fatal dose of strychnine.

d. If Lady Hazelton was in the dining room at the time of the murder, then the chauffeur killed Lord Hazelton.

e. If the cook was not in the kitchen at the time of the murder, then Sara was not in the dining room when the murder was committed.

f. If Sara was in the dining room at the time the murder was committed, then the wine steward killed Lord Hazelton.

Is it possible for the detective to deduce the identity of the murderer from these facts? If so, who did murder Lord Hazelton? (Assume there was only one cause of death.)

40. Sharky, a leader of the underworld, was killed by one of his own band of four henchmen. Detective Sharp interviewed the men and determined that all were lying except for one. He deduced who killed Sharky on the basis of the following statements:

a. Socko: Lefty killed Sharky.

b. Fats: Muscles didn't kill Sharky.

c. Lefty: Muscles was shooting craps with Socko when Sharky was knocked off.

d. Muscles: Lefty didn't kill Sharky.

Who did kill Sharky?

In 41–44 a set of premises and a conclusion are given. Use the valid argument forms listed in Table 2.3.1 to deduce the conclusion from the premises, giving a reason for each step as in Example 2.3.8. Assume all variables are statement variables.

41. a. $\sim p \vee q \to r$

   b. $s \vee \sim q$

   c. $\sim t$

   d. $p \to t$

   e. $\sim p \wedge r \to \sim s$

   f. $\therefore \sim q$

42. a. $p \vee q$

   b. $q \to r$

   c. $p \wedge s \to t$

   d. $\sim r$

   e. $\sim q \to u \wedge s$

   f. $\therefore t$

43. a. $\sim p \to r \wedge \sim s$

   b. $t \to s$

   c. $u \to \sim p$

   d. $\sim w$

   e. $u \vee w$

   f. $\therefore \sim t$ —

44. a. $p \to q$

   b. $r \vee s$

   c. $\sim s \to \sim t$

   d. $\sim q \vee s$

   e. $\sim s$

   f. $\sim p \wedge r \to u$

   g. $w \vee t$

   h. $\therefore u \wedge w$

Answers for Test Yourself

1. are all true; true   2. are all true; is false   3. valid; are all true; is true

CHAPTER 3

THE LOGIC OF QUANTIFIED STATEMENTS

In Chapter 2 we discussed the logical analysis of compound statements—those made of simple statements joined by the connectives $\sim, \wedge, \vee, \rightarrow$, and $\leftrightarrow$. Such analysis casts light on many aspects of human reasoning, but it cannot be used to determine validity in the majority of everyday and mathematical situations. For example, the argument

> All men are mortal.
>
> Socrates is a man.
>
> ∴ Socrates is mortal.

is intuitively perceived as correct. Yet its validity cannot be derived using the methods outlined in Section 2.3. To determine validity in examples like this, it is necessary to separate the statements into parts in much the same way that you separate declarative sentences into subjects and predicates. And you must analyze and understand the special role played by words that denote quantities such as "all" or "some." The symbolic analysis of predicates and quantified statements is called the **predicate calculus.** The symbolic analysis of ordinary compound statements (as outlined in Sections 2.1–2.3) is called the **statement calculus** (or the **propositional calculus**).

3.1 Predicates and Quantified Statements I

. . . it was not till within the last few years that it has been realized how fundamental any and some are to the very nature of mathematics. — A. N. Whitehead (1861–1947)

As noted in Section 2.1, the sentence "He is a college student" is not a statement because it may be either true or false depending on the value of the pronoun *he.* Similarly, the sentence "$x + y$ is greater than 0" is not a statement because its truth value depends on the values of the variables x and y.

In grammar, the word *predicate* refers to the part of a sentence that gives information about the subject. In the sentence "James is a student at Bedford College," the word *James* is the subject and the phrase *is a student at Bedford College* is the predicate. The predicate is the part of the sentence from which the subject has been removed.

In logic, predicates can be obtained by removing some or all of the nouns from a statement. For instance, let P stand for "is a student at Bedford College" and let Q stand for "is a student at." Then both P and Q are *predicate symbols*. The sentences "x is a student at Bedford College" and "x is a student at y" are symbolized as $P(x)$ and as $Q(x, y)$ respectively, where x and y are *predicate variables* that take values in appropriate sets. When concrete values are substituted in place of predicate variables, a statement results. For simplicity, we define a *predicate* to be a predicate symbol together with suitable predicate variables. In some other treatments of logic, such objects are referred to as **propositional functions** or **open sentences.**

> **• Definition**
>
> A **predicate** is a sentence that contains a finite number of variables and becomes a statement when specific values are substituted for the variables. The **domain** of a predicate variable is the set of all values that may be substituted in place of the variable.

Example 3.1.1 Finding Truth Values of a Predicate

Let $P(x)$ be the predicate "$x^2 > x$" with domain the set **R** of all real numbers. Write $P(2)$, $P(\frac{1}{2})$, and $P(-\frac{1}{2})$, and indicate which of these statements are true and which are false.

Solution

$$P(2): \quad 2^2 > 2, \quad \text{or} \quad 4 > 2. \quad \text{True.}$$

$$P\left(\frac{1}{2}\right): \quad \left(\frac{1}{2}\right)^2 > \frac{1}{2}, \quad \text{or} \quad \frac{1}{4} > \frac{1}{2}. \quad \text{False.}$$

$$P\left(-\frac{1}{2}\right): \quad \left(-\frac{1}{2}\right)^2 > -\frac{1}{2}, \quad \text{or} \quad \frac{1}{4} > -\frac{1}{2}. \quad \text{True.} \qquad \blacksquare$$

When an element in the domain of the variable of a one-variable predicate is substituted for the variable, the resulting statement is either true or false. The set of all such elements that make the predicate true is called the *truth set* of the predicate.

Note Recall that we read these symbols as "the set of all x in D such that $P(x)$."

> **• Definition**
>
> If $P(x)$ is a predicate and x has domain D, the **truth set** of $P(x)$ is the set of all elements of D that make $P(x)$ true when they are substituted for x. The truth set of $P(x)$ is denoted
>
> $$\{x \in D \mid P(x)\}.$$

Example 3.1.2 Finding the Truth Set of a Predicate

Let $Q(n)$ be the predicate "n is a factor of 8." Find the truth set of $Q(n)$ if

a. the domain of n is the set $\mathbf{Z}^+$ of all positive integers

b. the domain of n is the set $\mathbf{Z}$ of all integers.

Solution

a. The truth set is $\{1, 2, 4, 8\}$ because these are exactly the positive integers that divide 8 evenly.

b. The truth set is $\{1, 2, 4, 8, -1, -2, -4, -8\}$ because the negative integers $-1, -2, -4,$ and -8 also divide into 8 without leaving a remainder. $\qquad \blacksquare$

The Universal Quantifier: ∀

One sure way to change predicates into statements is to assign specific values to all their variables. For example, if x represents the number 35, the sentence "x is (evenly) divisible by 5" is a true statement since $35 = 5 \cdot 7$. Another way to obtain statements from predicates is to add **quantifiers.** Quantifiers are words that refer to quantities such as "some" or "all" and tell for how many elements a given predicate is true. The formal concept of quantifier was introduced into symbolic logic in the late nineteenth century by

Charles Sanders Peirce (1839–1914)

Culver Pictures

Note Think "for all" when you see the symbol ∀.

Gottlob Frege (1848–1925)

Friedrich Schiller, Universität Jena

the American philosopher, logician, and engineer Charles Sanders Peirce and, independently, by the German logician Gottlob Frege.

The symbol ∀ denotes "for all" and is called the **universal quantifier.** For example, another way to express the sentence "All human beings are mortal" is to write

$$\forall \text{ human beings } x, \ x \text{ is mortal.}$$

When the symbol x is introduced into the phrase "∀ human beings x," you are supposed to think of x as an individual, but generic, object—with all the properties shared by every human being but no other properties. Thus you should say "x is mortal" rather than "x are mortal." In other words, use the singular "is" rather than the plural verb "are" when describing the property satisfied by x. If you let H be the set of all human beings, then you can symbolize the statement more formally by writing

$$\forall x \in H, \ x \text{ is mortal,}$$

which is read as "For all x in the set of all human beings, x is mortal."

The domain of the predicate variable is generally indicated between the ∀ symbol and the variable name (as in ∀ human beings x) or immediately following the variable name (as in $\forall x \in H$). Some other expressions that can be used instead of *for all* are *for every, for arbitrary, for any, for each,* and *given any.* In a sentence such as "∀ real numbers x and y, $x + y = y + x$," the ∀ symbol is understood to refer to both x and y.*

Sentences that are quantified universally are defined as statements by giving them the truth values specified in the following definition:

• Definition

Let $Q(x)$ be a predicate and D the domain of x. A **universal statement** is a statement of the form "$\forall x \in D, Q(x)$." It is defined to be true if, and only if, $Q(x)$ is true for every x in D. It is defined to be false if, and only if, $Q(x)$ is false for at least one x in D. A value for x for which $Q(x)$ is false is called a **counterexample** to the universal statement.

Example 3.1.3 Truth and Falsity of Universal Statements

    a. Let $D = \{1, 2, 3, 4, 5\}$, and consider the statement

$$\forall x \in D, x^2 \geq x.$$

Show that this statement is true.

    b. Consider the statement

$$\forall x \in \mathbf{R}, x^2 \geq x.$$

Find a counterexample to show that this statement is false.

Solution

    a. Check that "$x^2 \geq x$" is true for each individual x in D.

$$1^2 \geq 1, \qquad 2^2 \geq 2, \qquad 3^2 \geq 3, \qquad 4^2 \geq 4, \qquad 5^2 \geq 5.$$

Hence "$\forall x \in D, x^2 \geq x$" is true.

---

*More formal versions of symbolic logic would require writing a separate ∀ for each variable: "$\forall x \in \mathbf{R}(\forall y \in \mathbf{R}(x + y = y + x))$."

b. *Counterexample:* Take $x = \frac{1}{2}$. Then x is in $\mathbf{R}$ (since $\frac{1}{2}$ is a real number) and

$$\left(\frac{1}{2}\right)^2 = \frac{1}{4} \not\geq \frac{1}{2}.$$

Hence "$\forall x \in \mathbf{R}, x^2 \geq x$" is false. ∎

The technique used to show the truth of the universal statement in Example 3.1.3(a) is called the **method of exhaustion.** It consists of showing the truth of the predicate separately for each individual element of the domain. (The idea is to exhaust the possibilities before you exhaust yourself!) This method can, in theory, be used whenever the domain of the predicate variable is finite. In recent years the prevalence of digital computers has greatly increased the convenience of using the method of exhaustion. Computer expert systems, or knowledge-based systems, use this method to arrive at answers to many of the questions posed to them. Because most mathematical sets are infinite, however, the method of exhaustion can rarely be used to derive general mathematical results.

The Existential Quantifier: ∃

The symbol ∃ denotes "there exists" and is called the **existential quantifier.** For example, the sentence "There is a student in Math 140" can be written as

∃ a person p such that p is a student in Math 140,

or, more formally,

$\exists p \in P$ such that p is a student in Math 140,

Note Think "there exists" when you see the symbol ∃.

where P is the set of all people. The domain of the predicate variable is generally indicated either between the ∃ symbol and the variable name or immediately following the variable name. The words *such that* are inserted just before the predicate. Some other expressions that can be used in place of *there exists* are *there is a, we can find a, there is at least one, for some,* and *for at least one.* In a sentence such as "∃ integers m and n such that $m + n = m \cdot n$," the ∃ symbol is understood to refer to both m and n.[*]

Sentences that are quantified existentially are defined as statements by giving them the truth values specified in the following definition.

• Definition

Let $Q(x)$ be a predicate and D the domain of x. An **existential statement** is a statement of the form "$\exists x \in D$ such that $Q(x)$." It is defined to be true if, and only if, $Q(x)$ is true for at least one x in D. It is false if, and only if, $Q(x)$ is false for all x in D.

Example 3.1.4 Truth and Falsity of Existential Statements

a. Consider the statement

$$\exists m \in \mathbf{Z}^+ \text{ such that } m^2 = m.$$

Show that this statement is true.

---

[*]In more formal versions of symbolic logic, the words *such that* are not written out (although they are understood) and a separate ∃ symbol is used for each variable: "$\exists m \in \mathbf{Z}(\exists n \in \mathbf{Z}(m + n = m \cdot n))$."

b. Let $E = \{5, 6, 7, 8\}$ and consider the statement

$$\exists m \in E \text{ such that } m^2 = m.$$

Show that this statement is false.

Solution

a. Observe that $1^2 = 1$. Thus "$m^2 = m$" is true for at least one integer m. Hence "$\exists m \in \mathbf{Z}$ such that $m^2 = m$" is true.

b. Note that $m^2 = m$ is not true for any integers m from 5 through 8:

$$5^2 = 25 \neq 5, \qquad 6^2 = 36 \neq 6, \qquad 7^2 = 49 \neq 7, \qquad 8^2 = 64 \neq 8.$$

Thus "$\exists m \in E$ such that $m^2 = m$" is false. ∎

Formal Versus Informal Language

It is important to be able to translate from formal to informal language when trying to make sense of mathematical concepts that are new to you. It is equally important to be able to translate from informal to formal language when thinking out a complicated problem.

Example 3.1.5 Translating from Formal to Informal Language

Rewrite the following formal statements in a variety of equivalent but more informal ways. Do not use the symbol $\forall$ or $\exists$.

a. $\forall x \in \mathbf{R}, x^2 \geq 0$.

b. $\forall x \in \mathbf{R}, x^2 \neq -1$.

c. $\exists m \in \mathbf{Z}^+$ such that $m^2 = m$.

Solution

Note The singular noun is used to refer to the domain when the $\forall$ symbol is translated as *every, any,* or *each.*

a. All real numbers have nonnegative squares.
   Or: Every real number has a nonnegative square.
   Or: Any real number has a nonnegative square.
   Or: The square of each real number is nonnegative.

b. All real numbers have squares that are not equal to -1.
   Or: No real numbers have squares equal to -1.
   (The words *none are* or *no ... are* are equivalent to the words *all are not*.)

Note In ordinary English, the statement in part (c) might be taken to be true only if there are at least two positive integers equal to their own squares. In mathematics, we understand the last two statements in part (c) to mean the same thing.

c. There is a positive integer whose square is equal to itself.
   Or: We can find at least one positive integer equal to its own square.
   Or: Some positive integer equals its own square.
   Or: Some positive integers equal their own squares. ∎

Another way to restate universal and existential statements informally is to place the quantification at the end of the sentence. For instance, instead of saying "For any real number x, x^2 is nonnegative," you could say "x^2 is nonnegative for any real number x." In such a case the quantifier is said to "trail" the rest of the sentence.

Example 3.1.6 Trailing Quantifiers

Rewrite the following statements so that the quantifier trails the rest of the sentence.

a. For any integer n, $2n$ is even.

b. There exists at least one real number x such that $x^2 \leq 0$.

Solution

a. $2n$ is even for any integer n.

b. $x^2 \leq 0$ for some real number x.
Or: $x^2 \leq 0$ for at least one real number x. ∎

Example 3.1.7 Translating from Informal to Formal Language

Rewrite each of the following statements formally. Use quantifiers and variables.

a. All triangles have three sides.

b. No dogs have wings.

c. Some programs are structured.

Solution

a. $\forall$ triangles t, t has three sides.
Or: $\forall t \in T$, t has three sides (where T is the set of all triangles).

b. $\forall$ dogs d, d does not have wings.
Or: $\forall d \in D$, d does not have wings (where D is the set of all dogs).

c. $\exists$ a program p such that p is structured.
Or: $\exists p \in P$ such that p is structured (where P is the set of all programs). ∎

Universal Conditional Statements

A reasonable argument can be made that the most important form of statement in mathematics is the **universal conditional statement:**

$$\forall x, \text{ if } P(x) \text{ then } Q(x).$$

Familiarity with statements of this form is essential if you are to learn to speak mathematics.

Example 3.1.8 Writing Universal Conditional Statements Informally

Rewrite the following statement informally, without quantifiers or variables.

$$\forall x \in \mathbf{R}, \text{ if } x > 2 \text{ then } x^2 > 4.$$

Solution   If a real number is greater than 2 then its square is greater than 4.
Or: Whenever a real number is greater than 2, its square is greater than 4.
Or: The square of any real number greater than 2 is greater than 4.
Or: The squares of all real numbers greater than 2 are greater than 4. ∎

Example 3.1.9 Writing Universal Conditional Statements Formally

Rewrite each of the following statements in the form

$$\forall \underline{\hspace{1cm}}, \text{ if } \underline{\hspace{1cm}} \text{ then } \underline{\hspace{1cm}}.$$

a. If a real number is an integer, then it is a rational number.

b. All bytes have eight bits.

c. No fire trucks are green.

Solution

a. ∀ real numbers x, if x is an integer, then x is a rational number.
   Or: $\forall x \in \mathbf{R}$, if $x \in \mathbf{Z}$ then $x \in \mathbf{Q}$.

b. $\forall x$, if x is a byte, then x has eight bits.

c. $\forall x$, if x is a fire truck, then x is not green.

It is common, as in (b) and (c) above, to omit explicit identification of the domain of predicate variables in universal conditional statements. ■

Careful thought about the meaning of universal conditional statements leads to another level of understanding for why the truth table for an if-then statement must be defined as it is. Consider again the statement

$$\forall \text{ real numbers } x, \text{ if } x > 2 \text{ then } x^2 > 4.$$

Your experience and intuition tell you that this statement is true. But that means that

$$\text{If } x > 2 \text{ then } x^2 > 4$$

must be true for every single real number x. Consequently, it must be true even for values of x that make its hypothesis "$x > 2$" false. In particular, both statements

$$\text{If } 1 > 2 \text{ then } 1^2 > 4 \quad \text{and} \quad \text{If } -3 > 2 \text{ then } (-3)^2 > 4$$

must be true. In both cases the hypothesis is false, but in the first case the conclusion "$1^2 > 4$" is false, and in the second case the conclusion "$(-3)^2 > 4$" is true. Hence, regardless of whether its conclusion is true or false, an if-then statement with a false hypothesis must be true.

Note also that the definition of valid argument is a universal conditional statement:

∀ combinations of truth values for the component statements,
if the premises are all true then the conclusion is also true.

Equivalent Forms of Universal and Existential Statements

Observe that the two statements "∀ real numbers x, if x is an integer then x is rational" and "∀ integers x, x is rational" mean the same thing. Both have informal translations "All integers are rational." In fact, a statement of the form

$$\forall x \in U, \text{ if } P(x) \text{ then } Q(x)$$

can always be rewritten in the form

$$\forall x \in D, Q(x)$$

by narrowing U to be the domain D consisting of all values of the variable x that make $P(x)$ true. Conversely, a statement of the form

$$\forall x \in D, Q(x)$$

can be rewritten as

$$\forall x, \text{ if } x \text{ is in } D \text{ then } Q(x).$$

Example 3.1.10 Equivalent Forms for Universal Statements

Rewrite the following statement in the two forms "$\forall x$, if _____ then _____" and "$\forall$ _____ x, _____": All squares are rectangles.

Solution $\qquad$ $\forall x$, if x is a square then x is a rectangle.

$\forall$ squares x, x is a rectangle. ■

Similarly, a statement of the form "$\exists x$ such that $p(x)$ and $Q(x)$" can be rewritten as "$\exists x \varepsilon D$ such that $Q(x)$," where D is the set of all x for which $P(x)$ is true.

Example 3.1.11 Equivalent Forms for Existential Statements

A **prime number** is an integer greater than 1 whose only positive integer factors are itself and 1. Consider the statement "There is an integer that is both prime and even." Let Prime(n) be "n is prime" and Even(n) be "n is even." Use the notation Prime(n) and Even(n) to rewrite this statement in the following two forms:

a. $\exists n$ such that _____ $\wedge$ _____.

b. $\exists$ _____ n such that _____.

Solution

a. $\exists n$ such that Prime(n) $\wedge$ Even(n).

b. Two answers: $\exists$ a prime number n such that Even(n).
   $\qquad\qquad\qquad$ $\exists$ an even number n such that Prime(n). ■

Implicit Quantification

Consider the statement

If a number is an integer, then it is a rational number.

As shown earlier, this statement is equivalent to a universal statement. However, it does not contain the telltale word *all* or *every* or *any* or *each*. The only clue to indicate its universal quantification comes from the presence of the indefinite article a. This is an example of *implicit* universal quantification.

Existential quantification can also be implicit. For instance, the statement "The number 24 can be written as a sum of two even integers" can be expressed formally as "$\exists$ even integers m and n such that $24 = m + n$."

Mathematical writing contains many examples of implicitly quantified statements. Some occur, as in the first example above, through the presence of the word a or an. Others occur in cases where the general context of a sentence supplies part of its meaning. For example, in an algebra course in which the letter x is always used to indicate a real number, the predicate

If $x > 2$ then $x^2 > 4$

is interpreted to mean the same as the statement

$\forall$ real numbers x, if $x > 2$ then $x^2 > 4$.

Mathematicians often use a double arrow to indicate implicit quantification symbolically. For instance, they might express the above statement as

$$x > 2 \quad \Rightarrow \quad x^2 > 4.$$

> **• Notation**
>
> Let $P(x)$ and $Q(x)$ be predicates and suppose the common domain of x is D.
>
> - The notation $P(x) \Rightarrow Q(x)$ means that every element in the truth set of $P(x)$ is in the truth set of $Q(x)$, or, equivalently, $\forall x,\ P(x) \rightarrow Q(x)$.
> - The notation $P(x) \Leftrightarrow Q(x)$ means that $P(x)$ and $Q(x)$ have identical truth sets, or, equivalently, $\forall x,\ P(x) \leftrightarrow Q(x)$.

Example 3.1.12 Using $\Rightarrow$ and $\Leftrightarrow$

Let

$$Q(n) \text{ be “}n \text{ is a factor of 8,”}$$
$$R(n) \text{ be “}n \text{ is a factor of 4,”}$$
$$S(n) \text{ be “}n < 5 \text{ and } n \neq 3,”$$

and suppose the domain of n is $\mathbf{Z}^+$, the set of positive integers. Use the $\Rightarrow$ and $\Leftrightarrow$ symbols to indicate true relationships among $Q(n)$, $R(n)$, and $S(n)$.

Solution

1. As noted in Example 3.1.2, the truth set of $Q(n)$ is $\{1, 2, 4, 8\}$ when the domain of n is $\mathbf{Z}^+$. By similar reasoning the truth set of $R(n)$ is $\{1, 2, 4\}$. Thus it is true that every element in the truth set of $R(n)$ is in the truth set of $Q(n)$, or, equivalently, $\forall n$ in $\mathbf{Z}^+$, $R(n) \rightarrow Q(n)$. So $R(n) \Rightarrow Q(n)$, or, equivalently

$$n \text{ is a factor of 4} \quad \Rightarrow \quad n \text{ is a factor of 8.}$$

2. The truth set of $S(n)$ is $\{1, 2, 4\}$, which is identical to the truth set of $R(n)$, or, equivalently, $\forall n$ in $\mathbf{Z}^+$, $R(n) \leftrightarrow S(n)$. So $R(n) \Leftrightarrow S(n)$, or, equivalently,

$$n \text{ is a factor of 4} \quad \Leftrightarrow \quad n < 5 \text{ and } n \neq 3.$$

Moreover, since every element in the truth set of $S(n)$ is in the truth set of $Q(n)$, or, equivalently, $\forall n$ in $\mathbf{Z}^+$, $S(n) \rightarrow Q(n)$, then $S(n) \Rightarrow Q(n)$, or, equivalently,

$$n < 5 \text{ and } n \neq 3 \quad \Rightarrow \quad n \text{ is a factor of 8.} \quad \blacksquare$$

Some questions of quantification can be quite subtle. For instance, a mathematics text might contain the following:

a. $(x + 1)^2 = x^2 + 2x + 1.$ b. Solve $3x - 4 = 5$.

Although neither (a) nor (b) contains explicit quantification, the reader is supposed to understand that the x in (a) is universally quantified whereas the x in (b) is existentially quantified. When the quantification is made explicit, (a) and (b) become

a. $\forall$ real numbers x, $(x + 1)^2 = x^2 + 2x + 1$.

b. Show (by finding a value) that $\exists$ a real number x such that $3x - 4 = 5$.

The quantification of a statement—whether universal or existential—crucially determines both how the statement can be applied and what method must be used to establish its truth. Thus it is important to be alert to the presence of hidden quantifiers when you read mathematics so that you will interpret statements in a logically correct way.

Tarski's World

Tarski's World is a computer program developed by information scientists Jon Barwise and John Etchemendy to help teach the principles of logic. It is described in their book *The Language of First-Order Logic,* which is accompanied by a CD-Rom containing the program Tarski's World, named after the great logician Alfred Tarski.

Example 3.1.13 Investigating Tarski's World

Alfred Tarski
(1902–1983)

The program for Tarski's World provides pictures of blocks of various sizes, shapes, and colors, which are located on a grid. Shown in Figure 3.1.1 is a picture of an arrangement of objects in a two-dimensional Tarski world. The configuration can be described using logical operators and—for the two-dimensional version—notation such as Triangle(x), meaning "x is a triangle," Blue(y), meaning "y is blue," and RightOf(x, y), meaning "x is to the right of y (but possibly in a different row)." Individual objects can be given names such as a, b, or c.

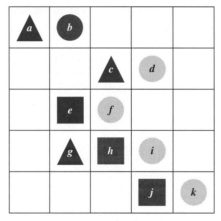

Figure 3.1.1

Determine the truth or falsity of each of the following statements. The domain for all variables is the set of objects in the Tarski world shown above.

a. $\forall t$, Triangle(t) $\rightarrow$ Blue(t).

b. $\forall x$, Blue(x) $\rightarrow$ Triangle(x).

c. $\exists y$ such that Square(y) $\wedge$ RightOf(d, y).

d. $\exists z$ such that Square(z) $\wedge$ Gray(z).

Solution

a. This statement is true: All the triangles are blue.

b. This statement is false. As a counterexample, note that e is blue and it is not a triangle.

c. This statement is true because e and h are both square and d is to their right.

d. This statement is false: All the squares are either blue or black. ∎

Test Yourself

Answers to Test Yourself questions are located at the end of each section.

1. If $P(x)$ is a predicate with domain D, the truth set of $P(x)$ is denoted ____. We read these symbols out loud as ____.

2. Some ways to express the symbol $\forall$ in words are ____.

3. Some ways to express the symbol $\exists$ in words are ____.

4. A statement of the form $\forall x \in D$, $Q(x)$ is true if, and only if, $Q(x)$ is ____ for ____.

5. A statement of the form $\exists x \in D$ such that $Q(x)$ is true if, and only if, $Q(x)$ is ____ for ____.

Exercise Set 3.1*

1. A menagerie consists of seven brown dogs, two black dogs, six gray cats, ten black cats, five blue birds, six yellow birds, and one black bird. Determine which of the following statements are true and which are false.
   a. There is an animal in the menagerie that is red.
   b. Every animal in the menagerie is a bird or a mammal.
   c. Every animal in the menagerie is brown or gray or black.
   d. There is an animal in the menagerie that is neither a cat nor a dog.
   e. No animal in the menagerie is blue.
   f. There are in the menagerie a dog, a cat, and a bird that all have the same color.

2. Indicate which of the following statements are true and which are false. Justify your answers as best as you can.
   a. Every integer is a real number.
   b. 0 is a positive real number.
   c. For all real numbers r, $-r$ is a negative real number.
   d. Every real number is an integer.

3. Let $P(x)$ be the predicate "$x > 1/x$."
   a. Write $P(2)$, $P(\frac{1}{2})$, $P(-1)$, $P(-\frac{1}{2})$, and $P(-8)$, and indicate which of these statements are true and which are false.
   b. Find the truth set of $P(x)$ if the domain of x is $\mathbf{R}$, the set of all real numbers.
   c. If the domain is the set $\mathbf{R}^+$ of all positive real numbers, what is the truth set of $P(x)$?

4. Let $Q(n)$ be the predicate "$n^2 \leq 30$."
   a. Write $Q(2)$, $Q(-2)$, $Q(7)$, and $Q(-7)$, and indicate which of these statements are true and which are false.
   b. Find the truth set of $Q(n)$ if the domain of n is $\mathbf{Z}$, the set of all integers.
   c. If the domain is the set $\mathbf{Z}^+$ of all positive integers, what is the truth set of $Q(n)$?

5. Let $Q(x, y)$ be the predicate "If $x < y$ then $x^2 < y^2$" with domain for both x and y being the set $\mathbf{R}$ of real numbers.
   a. Explain why $Q(x, y)$ is false if $x = -2$ and $y = 1$.
   b. Give values different from those in part (a) for which $Q(x, y)$ is false.
   c. Explain why $Q(x, y)$ is true if $x = 3$ and $y = 8$.
   d. Give values different from those in part (c) for which $Q(x, y)$ is true.

6. Let $R(m, n)$ be the predicate "If m is a factor of n^2 then m is a factor of n," with domain for both m and n being the set $\mathbf{Z}$ of integers.
   a. Explain why $R(m, n)$ is false if $m = 25$ and $n = 10$.
   b. Give values different from those in part (a) for which $R(m, n)$ is false.
   c. Explain why $R(m, n)$ is true if $m = 5$ and $n = 10$.
   d. Give values different from those in part (c) for which $R(m, n)$ is true.

7. Find the truth set of each predicate.
   a. predicate: $6/d$ is an integer, domain: $\mathbf{Z}$
   b. predicate: $6/d$ is an integer, domain: $\mathbf{Z}^+$
   c. predicate: $1 \leq x^2 \leq 4$, domain: $\mathbf{R}$
   d. predicate: $1 \leq x^2 \leq 4$, domain: $\mathbf{Z}$

8. Let $B(x)$ be "$-10 < x < 10$." Find the truth set of $B(x)$ for each of the following domains.
   a. $\mathbf{Z}$ b. $\mathbf{Z}^+$ c. The set of all even integers

Find counterexamples to show that the statements in 9–12 are false.

9. $\forall x \in \mathbf{R}, x > 1/x$.

10. $\forall a \in \mathbf{Z}, (a - 1)/a$ is not an integer.

11. $\forall$ positive integers m and n, $m \cdot n \geq m + n$.

12. $\forall$ real numbers x and y, $\sqrt{x + y} = \sqrt{x} + \sqrt{y}$.

13. Consider the following statement:

$$\forall \text{ basketball players } x, x \text{ is tall.}$$

   Which of the following are equivalent ways of expressing this statement?
   a. Every basketball player is tall.
   b. Among all the basketball players, some are tall. ✗
   c. Some of all the tall people are basketball players. ✗
   d. Anyone who is tall is a basketball player. ✗
   e. All people who are basketball players are tall.
   f. Anyone who is a basketball player is a tall person.

---

*For exercises with blue numbers or letters, solutions are given in Appendix B. The symbol **H** indicates that only a hint or a partial solution is given. The symbol ✷ signals that an exercise is more challenging than usual.

14. Consider the following statement:

$$\exists x \in \mathbf{R} \text{ such that } x^2 = 2.$$

Which of the following are equivalent ways of expressing this statement?
a. The square of each real number is 2. ✕
b. Some real numbers have square 2.
c. The number x has square 2, for some real number x.
d. If x is a real number, then $x^2 = 2$. ✕
e. Some real number has square 2.
f. There is at least one real number whose square is 2.

H **15.** Rewrite the following statements informally in at least two different ways without using variables or quantifiers.
a. ∀ rectangles x, x is a quadrilateral.
b. ∃ a set A such that A has 16 subsets.

16. Rewrite each of the following statements in the form "∀ _____ x, _____."
a. All dinosaurs are extinct.
b. Every real number is positive, negative, or zero.
c. No irrational numbers are integers.
d. No logicians are lazy.
e. The number 2,147,581,953 is not equal to the square of any integer.
f. The number -1 is not equal to the square of any real number.

17. Rewrite each of the following in the form "∃ _____ x such that _____."
a. Some exercises have answers.
b. Some real numbers are rational.

18. Let D be the set of all students at your school, and let $M(s)$ be "s is a math major," let $C(s)$ be "s is a computer science student," and let $E(s)$ be "s is an engineering student." Express each of the following statements using quantifiers, variables, and the predicates $M(s)$, $C(s)$, and $E(s)$.
a. There is an engineering student who is a math major.
b. Every computer science student is an engineering student.
c. No computer science students are engineering students.
d. Some computer science students are also math majors.
e. Some computer science students are engineering students and some are not.

19. Consider the following statement:

$$\forall \text{ integers } n, \text{ if } n^2 \text{ is even then } n \text{ is even.}$$

Which of the following are equivalent ways of expressing this statement?
a. All integers have even squares and are even. ✕
b. Given any integer whose square is even, that integer is itself even.
c. For all integers, there are some whose square is even. ✕
d. Any integer with an even square is even.
e. If the square of an integer is even, then that integer is even.
f. All even integers have even squares. ✕

H **20.** Rewrite the following statement informally in at least two different ways without using variables or the symbol ∀ or the words "for all."

$$\forall \text{ real numbers } x, \text{ if } x \text{ is positive, then}$$
$$\text{the square root of } x \text{ is positive.}$$

21. Rewrite the following statements so that the quantifier trails the rest of the sentence.
a. For any graph G, the total degree of G is even.
b. For any isosceles triangle T, the base angles of T are equal.
c. There exists a prime number p such that p is even.
d. There exists a continuous function f such that f is not differentiable.

22. Rewrite each of the following statements in the form "∀ _____ x, if _____ then _____."
a. All Java programs have at least 5 lines.
b. Any valid argument with true premises has a true conclusion.

23. Rewrite each of the following statements in the two forms "∀x, if _____ then _____" and "∀ _____ x, _____" (without an if-then).
a. All equilateral triangles are isosceles.
b. Every computer science student needs to take data structures.

24. Rewrite the following statements in the two forms "∃ _____ x such that _____" and "∃x such that _____ and _____."
a. Some hatters are mad.   b. Some questions are easy.

25. The statement "The square of any rational number is rational" can be rewritten formally as "For all rational numbers x, x^2 is rational" or as "For all x, if x is rational then x^2 is rational." Rewrite each of the following statements in the two forms "∀ _____ x, _____" and "∀x, if _____, then _____" or in the two forms "∀ _____ x and y, _____" and "∀x and y, if _____, then _____."
a. The reciprocal of any nonzero fraction is a fraction.
b. The derivative of any polynomial function is a polynomial function.
c. The sum of the angles of any triangle is 180°.
d. The negative of any irrational number is irrational.
e. The sum of any two even integers is even.
f. The product of any two fractions is a fraction.

26. Consider the statement "All integers are rational numbers but some rational numbers are not integers."
a. Write this statement in the form "∀x, if _____ then _____, but ∃ _____ x such that _____."
b. Let Ratl(x) be "x is a rational number" and Int(x) be "x is an integer." Write the given statement formally using only the symbols Ratl(x), Int(x), ∀, ∃, ∧, ∨, ∼, and →.

27. Refer to the picture of Tarski's world given in Example 3.1.13. Let Above(x, y) mean that x is above y (but possibly in a different column). Determine the truth or falsity of each of the following statements. Give reasons for your answers.

a. $\forall u$, Circle(u) → Gray(u).

b. $\forall u$, Gray(u) → Circle(u).

c. $\exists y$ such that Square(y) ∧ Above(y, d).

d. $\exists z$ such that Triangle(z) ∧ Above(f, z).

In 28–30, rewrite each statement without using quantifiers or variables. Indicate which are true and which are false, and justify your answers as best as you can.

28. Let the domain of x be the set D of objects discussed in mathematics courses, and let Real(x) be "x is a real number," Pos(x) be "x is a positive real number," Neg(x) be "x is a negative real number," and Int(x) be "x is an integer."

a. Pos(0)

b. $\forall x$, Real(x) ∧ Neg(x) → Pos($-x$).

c. $\forall x$, Int(x) → Real(x).

d. $\exists x$ such that Real(x) ∧ ~Int(x).

29. Let the domain of x be the set of geometric figures in the plane, and let Square(x) be "x is a square" and Rect(x) be "x is a rectangle."

a. $\exists x$ such that Rect(x) ∧ Square(x).

b. $\exists x$ such that Rect(x) ∧ ~Square(x).

c. $\forall x$, Square(x) → Rect(x).

30. Let the domain of x be the set **Z** of integers, and let Odd(x) be "x is odd," Prime(x) be "x is prime," and Square(x) be

"x is a perfect square." (An integer n is said to be a **perfect square** if, and only if, it equals the square of some integer. For example, 25 is a perfect square because $25 = 5^2$.)

a. $\exists x$ such that Prime(x) ∧ ~Odd(x).

b. $\forall x$, Prime(x) → ~Square(x).

c. $\exists x$ such that Odd(x) ∧ Square(x).

H 31. In any mathematics or computer science text other than this book, find an example of a statement that is universal but is implicitly quantified. Copy the statement as it appears and rewrite it making the quantification explicit. Give a complete citation for your example, including title, author, publisher, year, and page number.

32. Let **R** be the domain of the predicate variable x. Which of the following are true and which are false? Give counter examples for the statements that are false.

a. $x > 2 \Rightarrow x > 1$

b. $x > 2 \Rightarrow x^2 > 4$

c. $x^2 > 4 \Rightarrow x > 2$

d. $x^2 > 4 \Leftrightarrow |x| > 2$

33. Let **R** be the domain of the predicate variables a, b, c, and d. Which of the following are true and which are false? Give counterexamples for the statements that are false.

a. $a > 0$ and $b > 0 \Rightarrow ab > 0$

b. $a < 0$ and $b < 0 \Rightarrow ab < 0$

c. $ab = 0 \Rightarrow a = 0$ or $b = 0$

d. $a < b$ and $c < d \Rightarrow ac < bd$

Answers for Test Yourself

1. $\{x \in D \mid P(x)\}$; the set of all x in D such that $P(x)$ 2. *Possible answers:* for all, for every, for any, for each, for arbitrary, given any   3. *Possible answers:* there exists, there exist, there exists at least one, for some, for at least one, we can find a   4. true; every x in D (*Alternative answers:* all x in D; each x in D)   5. true; at least one x in D (*Alternative answer:* some x in D)

3.2 Predicates and Quantified Statements II

TOUCHSTONE: *Stand you both forth now: stroke your chins, and swear by your beards that I am a knave.*

CELIA: *By our beards—if we had them—thou art.*

TOUCHSTONE: *By my knavery—if I had it—then I were; but if you swear by that that is not, you are not forsworn.* — William Shakespeare, *As You Like It*

This section continues the discussion of predicates and quantified statements begun in Section 3.1. It contains the rules for negating quantified statements; an exploration of the relation among $\forall, \exists, \wedge$, and $\vee$; an introduction to the concept of vacuous truth of universal statements; examples of variants of universal conditional statements; and an extension of the meaning of *necessary, sufficient,* and *only if* to quantified statements.

Negations of Quantified Statements

Consider the statement "All mathematicians wear glasses." Many people would say that its negation is "No mathematicians wear glasses," but if even one mathematician does not wear glasses, then the sweeping statement that *all* mathematicians wear glasses is false. So a correct negation is "There is at least one mathematician who does not wear glasses."

The general form of the negation of a universal statement follows immediately from the definitions of negation and of the truth values for universal and existential statements.

Theorem 3.2.1 Negation of a Universal Statement

The negation of a statement of the form

$$\forall x \text{ in } D, Q(x)$$

is logically equivalent to a statement of the form

$$\exists x \text{ in } D \text{ such that } \sim Q(x).$$

Symbolically, $\sim(\forall x \in D, Q(x)) \equiv \exists x \in D \text{ such that } \sim Q(x).$

Thus

The negation of a universal statement ("all are") is logically equivalent to an existential statement ("some are not" or "there is at least one that is not").

Note that when we speak of **logical equivalence for quantified statements,** we mean that the statements always have identical truth values no matter what predicates are substituted for the predicate symbols and no matter what sets are used for the domains of the predicate variables.

Now consider the statement "Some snowflakes are the same." What is its negation? For this statement to be false means that not a single snowflake is the same as any other. In other words, "No snowflakes are the same," or "All snowflakes are different."

The general form for the negation of an existential statement follows immediately from the definitions of negation and of the truth values for existential and universal statements.

Theorem 3.2.2 Negation of an Existential Statement

The negation of a statement of the form

$$\exists x \text{ in } D \text{ such that } Q(x)$$

is logically equivalent to a statement of the form

$$\forall x \text{ in } D, \sim Q(x).$$

Symbolically, $\sim(\exists x \in D \text{ such that } Q(x)) \equiv \forall x \in D, \sim Q(x).$

Thus

The negation of an existential statement ("some are") is logically equivalent to a universal statement ("none are" or "all are not").

Example 3.2.1   **Negating Quantified Statements**

Write formal negations for the following statements:

a. ∀ primes p, p is odd.

b. ∃ a triangle T such that the sum of the angles of T equals $200°$.

Solution

a. By applying the rule for the negation of a ∀ statement, you can see that the answer is

∃ a prime p such that p is not odd.

b. By applying the rule for the negation of a ∃ statement, you can see that the answer is

∀ triangles T, the sum of the angles of T does not equal $200°$.   ■

You need to exercise special care to avoid mistakes when writing negations of statements that are given informally. One way to avoid error is to rewrite the statement formally and take the negation using the formal rule.

Example 3.2.2   **More Negations**

Rewrite the following statement formally. Then write formal and informal negations.

No politicians are honest.

Solution   *Formal version:* ∀ politicians x, x is not honest.

Formal negation: ∃ a politician x such that x is honest.

Informal negation: Some politicians are honest.   ■

Another way to avoid error when taking negations of statements that are given in informal language is to ask yourself, "What *exactly* would it mean for the given statement to be false? What statement, if true, would be equivalent to saying that the given statement is false?"

Example 3.2.3   **Still More Negations**

Write informal negations for the following statements:

a. All computer programs are finite.

b. Some computer hackers are over 40.

c. The number 1,357 is divisible by some integer between 1 and 37.

Solution

a. What exactly would it mean for this statement to be false? The statement asserts that all computer programs satisfy a certain property. So for it to be false, there would have to be at least one computer program that does not satisfy the property. Thus the answer is

There is a computer program that is not finite.

Or: Some computer programs are infinite.

b. This statement is equivalent to saying that there is at least one computer hacker with a certain property. So for it to be false, not a single computer hacker can have that property. Thus the negation is

No computer hackers are over 40.

Or: All computer hackers are 40 or under.

Note   Which is true: the statement in part (c) or its negation? Is 1,357 divisible by some integer between 1 and 37? Or is 1,357 not divisible by any integer between 1 and 37?

Caution!   Just inserting the word *not* to negate a quantified statement can result in a statement that is ambiguous.

c. This statement has a trailing quantifier. Written formally it becomes:

> ∃ an integer *n* between 1 and 37 such that 1,357 is divisible by *n*.

Its negation is therefore

> ∀ integers *n* between 1 and 37; 1,357 is not divisible by *n*.

An informal version of the negation is

> The number 1,357 is not divisible by any integer between 1 and 37.   ■

Informal negations of many universal statements can be constructed simply by inserting the word *not* or the words *do not* at an appropriate place. However, the resulting statements may be ambiguous. For example, a possible negation of "All mathematicians wear glasses" is "All mathematicians do not wear glasses." The problem is that this sentence has two meanings. With the proper verbal stress on the word *not*, it could be interpreted as the logical negation. (What! You say that all mathematicians wear glasses? Nonsense! All mathematicians do *not* wear glasses.) On the other hand, stated in a flat tone of voice (try it!), it would mean that all mathematicians are nonwearers of glasses; that is, not a single mathematician wears glasses. This is a much stronger statement than the logical negation: It implies the negation but is not equivalent to it.

Negations of Universal Conditional Statements

Negations of universal conditional statements are of special importance in mathematics. The form of such negations can be derived from facts that have already been established.

By definition of the negation of a *for all* statement,

$$\sim(\forall x, P(x) \rightarrow Q(x)) \equiv \exists x \text{ such that } \sim(P(x) \rightarrow Q(x)). \qquad 3.2.1$$

But the negation of an if-then statement is logically equivalent to an *and* statement. More precisely,

$$\sim(P(x) \rightarrow Q(x)) \equiv P(x) \wedge \sim Q(x). \qquad 3.2.2$$

Substituting (3.2.2) into (3.2.1) gives

$$\sim(\forall x, P(x) \rightarrow Q(x)) \equiv \exists x \text{ such that } (P(x) \wedge \sim Q(x)).$$

Written less symbolically, this becomes

Negation of a Universal Conditional Statement

$$\sim(\forall x, \text{ if } P(x) \text{ then } Q(x)) \equiv \exists x \text{ such that } P(x) \text{ and } \sim Q(x).$$

Example 3.2.4 Negating Universal Conditional Statements

Write a formal negation for statement (a) and an informal negation for statement (b).

a. ∀ people *p*, if *p* is blond then *p* has blue eyes.

b. If a computer program has more than 100,000 lines, then it contains a bug.

Solution

a. ∃ a person *p* such that *p* is blond and *p* does not have blue eyes.

b. There is at least one computer program that has more than 100,000 lines and does not contain a bug.   ■

The Relation among ∀, ∃, ∧, and ∨

The negation of a *for all* statement is a *there exists* statement, and the negation of a *there exists* statement is a *for all* statement. These facts are analogous to De Morgan's laws, which state that the negation of an *and* statement is an *or* statement and that the negation of an *or* statement is an *and* statement. This similarity is not accidental. In a sense, universal statements are generalizations of *and* statements, and existential statements are generalizations of *or* statements.

If $Q(x)$ is a predicate and the domain D of x is the set $\{x_1, x_2, \ldots, x_n\}$, then the statements

$$\forall x \in D, Q(x)$$

and
$$Q(x_1) \wedge Q(x_2) \wedge \cdots \wedge Q(x_n)$$

are logically equivalent. For example, let $Q(x)$ be "$x \cdot x = x$" and suppose $D = \{0, 1\}$. Then

$$\forall x \in D, Q(x)$$

can be rewritten as
$$\forall \text{ binary digits } x, x \cdot x = x.$$

This is equivalent to

$$0 \cdot 0 = 0 \quad \text{and} \quad 1 \cdot 1 = 1,$$

which can be rewritten in symbols as

$$Q(0) \wedge Q(1).$$

Similarly, if $Q(x)$ is a predicate and $D = \{x_1, x_2, \ldots, x_n\}$, then the statements

$$\exists x \in D \text{ such that } Q(x)$$

and
$$Q(x_1) \vee Q(x_2) \vee \cdots \vee Q(x_n)$$

are logically equivalent. For example, let $Q(x)$ be "$x + x = x$" and suppose $D = \{0, 1\}$. Then

$$\exists x \in D \text{ such that } Q(x)$$

can be rewritten as
$$\exists \text{ a binary digit } x \text{ such that } x + x = x.$$

This is equivalent to

$$0 + 0 = 0 \quad \text{or} \quad 1 + 1 = 1,$$

which can be rewritten in symbols as

$$Q(0) \vee Q(1).$$

Vacuous Truth of Universal Statements

Suppose a bowl sits on a table and next to the bowl is a pile of five blue and five gray balls, any of which may be placed in the bowl. If three blue balls and one gray ball are placed in the bowl, as shown in Figure 3.2.1(a), the statement "All the balls in the bowl are blue" would be false (since one of the balls in the bowl is gray).

Now suppose that no balls at all are placed in the bowl, as shown in Figure 3.2.1(b). Consider the statement

All the balls in the bowl are blue.

Is this statement true or false? The statement is false if, and only if, its negation is true. And its negation is

There exists a ball in the bowl that is not blue.

But the only way this negation can be true is for there actually to be a nonblue ball in the bowl. And there is not! Hence the negation is false, and so the statement is true "by default."

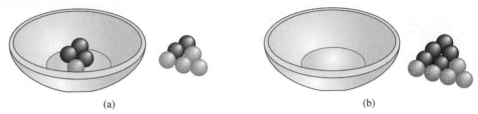

(a)   (b)

Figure 3.2.1

In general, a statement of the form

$$\forall x \text{ in } D, \text{ if } P(x) \text{ then } Q(x)$$

is called **vacuously true** or **true by default** if, and only if, $P(x)$ is false for every x in D.

By the way, in ordinary language the words *in general* mean that something is usually, but not always, the case. (In general, I take the bus home, but today I walked.) In mathematics, the words *in general* are used quite differently. When they occur just after discussion of a particular example (as in the preceding paragraph), they are a signal that what is to follow is a generalization of some aspect of the example that always holds true.

Variants of Universal Conditional Statements

Recall from Section 2.2 that a conditional statement has a contrapositive, a converse, and an inverse. The definitions of these terms can be extended to universal conditional statements.

• **Definition**

Consider a statement of the form: $\forall x \in D,$ if $P(x)$ then $Q(x)$.

1. Its **contrapositive** is the statement: $\forall x \in D,$ if $\sim Q(x)$ then $\sim P(x)$.

2. Its **converse** is the statement: $\forall x \in D,$ if $Q(x)$ then $P(x)$.

3. Its **inverse** is the statement: $\forall x \in D,$ if $\sim P(x)$ then $\sim Q(x)$.

Example 3.2.5 Contrapositive, Converse, and Inverse of a Universal Conditional Statement

Write a formal and an informal contrapositive, converse, and inverse for the following statement:

If a real number is greater than 2, then its square is greater than 4.

Solution   The formal version of this statement is $\forall x \in \mathbf{R},$ if $x > 2$ then $x^2 > 4$.

Contrapositive: $\forall x \in \mathbf{R},$ if $x^2 \le 4$ then $x \le 2$.
                  Or: If the square of a real number is less than or equal to 4, then the number is less than or equal to 2.

Converse: $\forall x \in \mathbf{R},$ if $x^2 > 4$ then $x > 2$.
                  Or: If the square of a real number is greater than 4, then the number is greater than 2.

Inverse: $\forall x \in \mathbf{R},$ if $x \le 2$ then $x^2 \le 4$.
                  Or: If a real number is less than or equal to 2, then the square of the number is less than or equal to 4.

Note that in solving this example, we have used the equivalence of "$x \not> a$" and "$x \le a$" for all real numbers x and a. (See page 33.) ∎

In Section 2.2 we showed that a conditional statement is logically equivalent to its contrapositive and that it is not logically equivalent to either its converse or its inverse. The following discussion shows that these facts generalize to the case of universal conditional statements and their contrapositives, converses, and inverses.

Let $P(x)$ and $Q(x)$ be any predicates, let D be the domain of x, and consider the statement

$$\forall x \in D, \text{if } P(x) \text{ then } Q(x)$$

and its contrapositive

$$\forall x \in D, \text{if } \sim Q(x) \text{ then } \sim P(x).$$

Any particular x in D that makes "if $P(x)$ then $Q(x)$" true also makes "if $\sim Q(x)$ then $\sim P(x)$" true (by the logical equivalence between $p \to q$ and $\sim q \to \sim p$). It follows that the sentence "If $P(x)$ then $Q(x)$" is true for all x in D if, and only if, the sentence "If $\sim Q(x)$ then $\sim P(x)$" is true for all x in D.

Thus we write the following and say that a universal conditional statement is logically equivalent to its contrapositive:

$$\forall x \in D, \text{if } P(x) \text{ then } Q(x) \equiv \forall x \in D, \text{if } \sim Q(x) \text{ then } \sim P(x)$$

In Example 3.2.5 we noted that the statement

$$\forall x \in \mathbf{R}, \text{if } x > 2 \text{ then } x^2 > 4$$

has the converse $\qquad \forall x \in \mathbf{R}, \text{if } x^2 > 4 \text{ then } x > 2.$

Observe that the statement is true whereas its converse is false (since, for instance, $(-3)^2 = 9 > 4$ but $-3 \not> 2$). This shows that a universal conditional statement may have a different truth value from its converse. Hence a universal conditional statement is not logically equivalent to its converse. This is written in symbols as follows:

$$\forall x \in D, \text{if } P(x) \text{ then } Q(x) \not\equiv \forall x \in D, \text{if } Q(x) \text{ then } P(x).$$

In the exercises at the end of this section, you are asked to show similarly that a universal conditional statement is not logically equivalent to its inverse.

$$\forall x \in D, \text{if } P(x) \text{ then } Q(x) \not\equiv \forall x \in D, \text{if } \sim P(x) \text{ then } \sim Q(x).$$

Necessary and Sufficient Conditions, Only If

The definitions of *necessary, sufficient,* and *only if* can also be extended to apply to universal conditional statements.

• Definition

- "$\forall x, r(x)$ is a **sufficient condition** for $s(x)$" means "$\forall x, \text{if } r(x) \text{ then } s(x)$."
- "$\forall x, r(x)$ is a **necessary condition** for $s(x)$" means "$\forall x, \text{if } \sim r(x) \text{ then } \sim s(x)$" or, equivalently, "$\forall x, \text{if } s(x) \text{ then } r(x)$."
- "$\forall x, r(x)$ **only if** $s(x)$" means "$\forall x, \text{if } \sim s(x) \text{ then } \sim r(x)$" or, equivalently, "$\forall x, \text{if } r(x) \text{ then } s(x)$."

Example 3.2.6 Necessary and Sufficient Conditions

Rewrite the following statements as quantified conditional statements. Do not use the word *necessary* or *sufficient*.

a. Squareness is a sufficient condition for rectangularity.

b. Being at least 35 years old is a necessary condition for being President of the United States.

Solution

a. A formal version of the statement is

$$\forall x, \text{ if } x \text{ is a square, then } x \text{ is a rectangle.}$$

Or, in informal language:

$$\text{If a figure is a square, then it is a rectangle.}$$

b. Using formal language, you could write the answer as

$$\forall \text{ people } x, \text{ if } x \text{ is younger than 35, then } x$$
$$\text{cannot be President of the United States.}$$

Or, by the equivalence between a statement and its contrapositive:

$$\forall \text{ people } x, \text{ if } x \text{ is President of the United States,}$$
$$\text{then } x \text{ is at least 35 years old.} \qquad \blacksquare$$

Example 3.2.7 Only If

Rewrite the following as a universal conditional statement:

$$\text{A product of two numbers is 0 only if one of the numbers is 0.}$$

Solution   Using informal language, you could write the answer as

$$\text{If neither of two numbers is 0, then the product of the numbers is not 0.}$$

Or, by the equivalence between a statement and its contrapositive,

$$\text{If a product of two numbers is 0, then one of the numbers is 0.} \qquad \blacksquare$$

Test Yourself

1. A negation for "All R have property S" is "There is ____ R that ____."

2. A negation for "Some R have property S" is "____."

3. A negation for "For all x, if x has property P then x has property Q" is "____."

4. The converse of "For all x, if x has property P then x has property Q" is "____."

5. The contrapositive of "For all x, if x has property P then x has property Q" is "____."

6. The inverse of "For all x, if x has property P then x has property Q" is "____."

Exercise Set 3.2

1. Which of the following is a negation for "All discrete mathematics students are athletic"? More than one answer may be correct.
   a. There is a discrete mathematics student who is nonathletic.
   b. All discrete mathematics students are nonathletic.
   c. There is an athletic person who is a discrete mathematics student.
   d. No discrete mathematics students are athletic.
   e. Some discrete mathematics students are nonathletic.
   f. No athletic people are discrete mathematics students.

2. Which of the following is a negation for "All dogs are loyal"? More than one answer may be correct.
   a. All dogs are disloyal.   b. No dogs are loyal.
   c. Some dogs are disloyal.   d. Some dogs are loyal.
   e. There is a disloyal animal that is not a dog.
   f. There is a dog that is disloyal.
   g. No animals that are not dogs are loyal.
   h. Some animals that are not dogs are loyal.

3. Write a formal negation for each of the following statements:
   a. $\forall$ fish x, x has gills.
   b. $\forall$ computers c, c has a CPU.
   c. $\exists$ a movie m such that m is over 6 hours long.
   d. $\exists$ a band b such that b has won at least 10 Grammy awards.

4. Write an informal negation for each of the following statements. Be careful to avoid negations that are ambiguous.
   a. All dogs are friendly.
   b. All people are happy.
   c. Some suspicions were substantiated.
   d. Some estimates are accurate.

5. Write a negation for each of the following statements.
   a. Any valid argument has a true conclusion.
   b. Every real number is positive, negative, or zero.

6. Write a negation for each of the following statements.
   a. Sets A and B do not have any points in common.
   b. Towns P and Q are not connected by any road on the map.

7. Informal language is actually more complex than formal language. For instance, the sentence "There are no orders from store A for item B" contains the words *there are*. Is the statement existential? Write an informal negation for the statement, and then write the statement formally using quantifiers and variables.

8. Consider the statement "There are no simple solutions to life's problems." Write an informal negation for the statement, and then write the statement formally using quantifiers and variables.

Write a negation for each statement in 9 and 10.

9. $\forall$ real numbers x, if $x > 3$ then $x^2 > 9$.

10. $\forall$ computer programs P, if P compiles without error messages, then P is correct.

In each of 11–14 determine whether the proposed negation is correct. If it is not, write a correct negation.

11.   *Statement:* The sum of any two irrational numbers is irrational.
   Proposed negation: The sum of any two irrational numbers is rational.

12.   *Statement:* The product of any irrational number and any rational number is irrational.

Proposed negation: The product of any irrational number and any rational number is rational.

13.   *Statement:* For all integers n, if n^2 is even then n is even.
   Proposed negation: For all integers n, if n^2 is even then n is not even.

14.   *Statement:* For all real numbers x_1 and x_2, if $x_1^2 = x_2^2$ then $x_1 = x_2$.
   Proposed negation: For all real numbers x_1 and x_2, if $x_1^2 = x_2^2$ then $x_1 \neq x_2$.

15. Let $D = \{-48, -14, -8, 0, 1, 3, 16, 23, 26, 32, 36\}$. Determine which of the following statements are true and which are false. Provide counterexamples for those statements that are false.
   a. $\forall x \in D$, if x is odd then $x > 0$.
   b. $\forall x \in D$, if x is less than 0 then x is even.
   c. $\forall x \in D$, if x is even then $x \leq 0$.
   d. $\forall x \in D$, if the ones digit of x is 2, then the tens digit is 3 or 4.
   e. $\forall x \in D$, if the ones digit of x is 6, then the tens digit is 1 or 2.

In 16–23, write a negation for each statement.

16. $\forall$ real numbers x, if $x^2 \geq 1$ then $x > 0$.

17. $\forall$ integers d, if $6/d$ is an integer then $d = 3$.

18. $\forall x \in \mathbf{R}$, if $x(x + 1) > 0$ then $x > 0$ or $x < -1$.

19. $\forall n \in \mathbf{Z}$, if n is prime then n is odd or $n = 2$.

20. $\forall$ integers a, b and c, if $a - b$ is even and $b - c$ is even, then $a - c$ is even.

21. $\forall$ integers n, if n is divisible by 6, then n is divisible by 2 and n is divisible by 3.

22. If the square of an integer is odd, then the integer is odd.

23. If a function is differentiable then it is continuous.

24. Rewrite the statements in each pair in if-then form and indicate the logical relationship between them.
   a. All the children in Tom's family are female.
      All the females in Tom's family are children.
   b. All the integers greater than 5 that end in 1, 3, 7, or 9 are prime.
      All the integers greater than 5 that are prime end in 1, 3, 7, or 9.

25. Each of the following statements is true. In each case write the converse of the statement, and give a counterexample showing that the converse is false.
   a. If n is any prime number that is greater than 2, then $n + 1$ is even.
   b. If m is any odd integer, then $2m$ is even.
   c. If two circles intersect in exactly two points, then they do not have a common center.

In 26–33, for each statement in the referenced exercise write the converse, inverse, and contrapositive. Indicate as best as you can which among the statement, its converse, its inverse, and its contrapositive are true and which are false. Give a counterexample for each that is false.

26. Exercise 16
27. Exercise 17
28. Exercise 18
29. Exercise 19
30. Exercise 20
31. Exercise 21
32. Exercise 22
33. Exercise 23

34. Write the contrapositive for each of the following statements.
    a. If n is prime, then n is not divisible by any prime number between 1 and $\sqrt{n}$ inclusive. (Assume that n is a fixed integer that is greater than 1.)
    b. If A and B do not have any elements in common, then they are disjoint. (Assume that A and B are fixed sets.)

35. Give an example to show that a universal conditional statement is not logically equivalent to its inverse.

★ 36. If $P(x)$ is a predicate and the domain of x is the set of all real numbers, let R be "$\forall x \in \mathbf{Z}, P(x)$," let S be "$\forall x \in \mathbf{Q}, P(x)$," and let T be "$\forall x \in \mathbf{R}, P(x)$."
    a. Find a definition for $P(x)$ (but do not use "$x \in \mathbf{Z}$") so that R is true and both S and T are false.
    b. Find a definition for $P(x)$ (but do not use "$x \in \mathbf{Q}$") so that both R and S are true and T is false.

37. Consider the following sequence of digits: 0204. A person claims that all the 1's in the sequence are to the left of all the 0's in the sequence. Is this true? Justify your answer. (*Hint:* Write the claim formally and write a formal negation for it. Is the negation true or false?)

38. True or false? All occurrences of the letter u in *Discrete Mathematics* are lowercase. Justify your answer.

Rewrite each statement of 39–42 in if-then form.

39. Earning a grade of C− in this course is a sufficient condition for it to count toward graduation.

40. Being divisible by 8 is a sufficient condition for being divisible by 4.

41. Being on time each day is a necessary condition for keeping this job.

42. Passing a comprehensive exam is a necessary condition for obtaining a master's degree.

Use the facts that the negation of a $\forall$ statement is a $\exists$ statement and that the negation of an if-then statement is an *and* statement to rewrite each of the statements 43–46 without using the word *necessary* or *sufficient*.

43. Being divisible by 8 is not a necessary condition for being divisible by 4.

44. Having a large income is not a necessary condition for a person to be happy.

45. Having a large income is not a sufficient condition for a person to be happy.

46. Being a polynomial is not a sufficient condition for a function to have a real root.

47. The computer scientists Richard Conway and David Gries once wrote:

    > The absence of error messages during translation of a computer program is only a necessary and not a sufficient condition for reasonable [program] correctness.

    Rewrite this statement without using the words *necessary* or *sufficient*.

48. A frequent-flyer club brochure states, "You may select among carriers only if they offer the same lowest fare." Assuming that "only if" has its formal, logical meaning, does this statement guarantee that if two carriers offer the same lowest fare, the customer will be free to choose between them? Explain.

Answers for Test Yourself

1. some (*Alternative answers:* at least one; an); does not have property S.   2. No R have property S.   3. There is an x such that x has property P and x does not have property Q.   4. For all x, if x has property Q then x has property P.   5. For all x, if x does not have property Q then x does not have property P.   6. For all x, if x does not have property P then x does not have property Q.

3.3 Statements with Multiple Quantifiers

It is not enough to have a good mind. The main thing is to use it well. — René Descartes

Imagine you are visiting a factory that manufactures computer microchips. The factory guide tells you,

There is a person supervising every detail of the production process.

Note that this statement contains informal versions of both the existential quantifier *there is* and the universal quantifier *every*. Which of the following best describes its meaning?

- There is one single person who supervises all the details of the production process.
- For any particular production detail, there is a person who supervises that detail, but there might be different supervisors for different details.

As it happens, either interpretation could be what the guide meant. (Reread the sentence to be sure you agree!) Taken by itself, his statement is genuinely ambiguous, although other things he may have said (the context for his statement) might have clarified it. In our ordinary lives, we deal with this kind of ambiguity all the time. Usually context helps resolve it, but sometimes we simply misunderstand each other.

In mathematics, formal logic, and computer science, by contrast, it is essential that we all interpret statements in exactly the same way. For instance, the initial stage of software development typically involves careful discussion between a programmer analyst and a client to turn vague descriptions of what the client wants into unambiguous program specifications that client and programmer can mutually agree on.

Because many important technical statements contain both $\exists$ and $\forall$, a convention has developed for interpreting them uniformly. When a statement contains more than one quantifier, we imagine the actions suggested by the quantifiers as being performed in the order in which the quantifiers occur. For instance, consider a statement of the form

$$\forall x \text{ in set } D, \exists y \text{ in set } E \text{ such that } x \text{ and } y \text{ satisfy property } P(x, y).$$

To show that such a statement is true, you must be able to meet the following challenge:

- Imagine that someone is allowed to choose any element whatsoever from the set D, and imagine that the person gives you that element. Call it x.
- The challenge for you is to find an element y in E so that the person's x and your y, taken together, satisfy property $P(x, y)$.

Note that *because you do not have to specify the y until after the other person has specified the x, you are allowed to find a different value of y for each different x you are given.*

Example 3.3.1   Truth of a $\forall\exists$ Statement in a Tarski World

Consider the Tarski world shown in Figure 3.3.1.

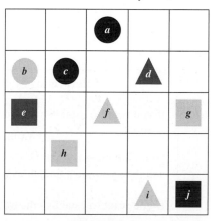

Figure 3.3.1

Show that the following statement is true in this world:

For all triangles x, there is a square y such that x and y have the same color.

Solution   The statement says that no matter which triangle someone gives you, you will be able to find a square of the same color. There are only three triangles, d, f, and i. The following table shows that for each of these triangles a square of the same color can be found.

Given $x =$	choose $y =$	and check that y is the same color as x.
d	e	yes ✓
f or i	h or g	yes ✓

Now consider a statement containing both ∀ and ∃, where the ∃ comes before the ∀:

$$\exists \text{ an } x \text{ in } D \text{ such that } \forall y \text{ in } E, x \text{ and } y \text{ satisfy property } P(x, y).$$

To show that a statement of this form is true:

You must find one single element (call it x) in D with the following property:

- After you have found your x, someone is allowed to choose any element whatsoever from E. The person challenges you by giving you that element. Call it y.
- Your job is to show that your x together with the person's y satisfy property $P(x, y)$.

Note that your x has to work for *any* y the person gives you; ***you are not allowed to change your x once you have specified it initially.***

Example 3.3.2 Truth of a ∃∀ Statement in a Tarski World

Consider again the Tarski world in Figure 3.3.1. Show that the following statement is true: There is a triangle x such that for all circles y, x is to the right of y.

Solution   The statement says that you can find a triangle that is to the right of all the circles. Actually, either d or i would work for all of the three circles, a, b, and c, as you can see in the following table.

Choose $x =$	Then, given $y =$	check that x is to the right of y.
d or i	a	yes ✓
	b	yes ✓
	c	yes ✓

Here is a summary of the convention for interpreting statements with two different quantifiers:

Interpreting Statements with Two Different Quantifiers

If you want to establish the truth of a statement of the form

$$\forall x \text{ in } D, \exists y \text{ in } E \text{ such that } P(x, y)$$

your challenge is to allow someone else to pick whatever element x in D they wish and then you must find an element y in E that "works" for that particular x.

If you want to establish the truth of a statement of the form

$$\exists x \text{ in } D \text{ such that } \forall y \text{ in } E, P(x, y)$$

your job is to find one particular x in D that will "work" no matter what y in E anyone might choose to challenge you with.

Example 3.3.3 Interpreting Multiply-Quantified* Statements

A college cafeteria line has four stations: salads, main courses, desserts, and beverages. The salad station offers a choice of green salad or fruit salad; the main course station offers spaghetti or fish; the dessert station offers pie or cake; and the beverage station offers milk, soda, or coffee. Three students, Uta, Tim, and Yuen, go through the line and make the following choices:

Uta: green salad, spaghetti, pie, milk

Tim: fruit salad, fish, pie, cake, milk, coffee

Yuen: spaghetti, fish, pie, soda

These choices are illustrated in Figure 3.3.2.

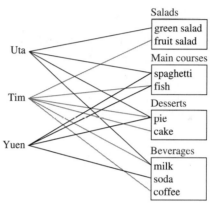

Figure 3.3.2

Write each of following statements informally and find its truth value.

a. ∃ an item I such that ∀ students S, S chose I.

b. ∃ a student S such that ∀ items I, S chose I.

c. ∃ a student S such that ∀ stations Z, ∃ an item I in Z such that S chose I.

d. ∀ students S and ∀ stations Z, ∃ an item I in Z such that S chose I.

Solution

a. There is an item that was chosen by every student. This is true; every student chose pie.

b. There is a student who chose every available item. This is false; no student chose all nine items.

c. There is a student who chose at least one item from every station. This is true; both Uta and Tim chose at least one item from every station.

d. Every student chose at least one item from every station. This is false; Yuen did not choose a salad. ■

---

*The term "multiply-quantified" is pronounced MUL-ti-plee QUAN-ti-fied. A multiply-quantified statement is a statement that contains more than one quantifier.

Translating from Informal to Formal Language

Most problems are stated in informal language, but solving them often requires translating them into more formal terms.

Example 3.3.4 Translating Multiply-Quantified Statements from Informal to Formal Language

The **reciprocal** of a real number a is a real number b such that $ab = 1$. The following two statements are true. Rewrite them formally using quantifiers and variables:

a. Every nonzero real number has a reciprocal.

b. There is a real number with no reciprocal.   The number 0 has no reciprocal.

Solution

a. $\forall$ nonzero real numbers u, $\exists$ a real number v such that $uv = 1$.

b. $\exists$ a real number c such that $\forall$ real numbers d, $cd \neq 1$.   ∎

Example 3.3.5 There Is a Smallest Positive Integer

Recall that every integer is a real number and that real numbers are of three types: positive, negative, and zero (zero being neither positive nor negative). Consider the statement "There is a smallest positive integer." Write this statement formally using both symbols $\exists$ and $\forall$.

Solution   To say that there is a smallest positive integer means that there is a positive integer m with the property that no matter what positive integer n a person might pick, m will be less than or equal to n:

$$\exists \text{ a positive integer } m \text{ such that } \forall \text{ positive integers } n, m \leq n.$$

Note that this statement is true because 1 is a positive integer that is less than or equal to every positive integer.

Example 3.3.6 There Is No Smallest Positive Real Number

Imagine any positive real number x on the real number line. These numbers correspond to all the points to the right of 0. Observe that no matter how small x is, the number $x/2$ will be both positive and less than x.*

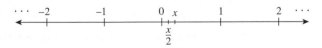

---

*This can be deduced from the properties of the real numbers given in Appendix A. Because x is positive, $0 < x$. Add x to both sides to obtain $x < 2x$. Then $0 < x < 2x$. Now multiply all parts of the inequality by the positive number $1/2$. This does not change the direction of the inequality, so $0 < x/2 < x$.

Thus the following statement is true: "There is no smallest positive real number." Write this statement formally using both symbols ∀ and ∃.

Solution   ∀ positive real numbers x, ∃ a positive real number y such that $y < x$.   ∎

Example 3.3.7 The Definition of Limit of a Sequence

The definition of limit of a sequence, studied in calculus, uses both quantifiers ∀ and ∃ and also if-then. We say that the limit of the sequence a_n as n goes to infinity equals L and write

$$\lim_{n \to \infty} a_n = L$$

if, and only if, the values of a_n become *arbitrarily* close to L as n gets larger and larger without bound. More precisely, this means that given any positive number ε, we can find an integer N such that whenever n is larger than N, the number a_n sits between $L - \varepsilon$ and $L + \varepsilon$ on the number line.

$$\begin{array}{ccc} L - \varepsilon & L & L + \varepsilon \end{array}$$

a_n must lie in here when $n > N$

Symbolically:

$\forall \varepsilon > 0$, ∃ an integer N such that ∀ integers n,
if $n > N$ then $L - \varepsilon < a_n < L + \varepsilon$.

Considering the logical complexity of this definition, it is no wonder that many students find it hard to understand.   ∎

Ambiguous Language

The drawing in Figure 3.3.3 is a famous example of visual ambiguity. When you look at it for a while, you will probably see either a silhouette of a young woman wearing a large hat or an elderly woman with a large nose. Whichever image first pops into your mind,

Figure 3.3.3

try to see how the drawing can be interpreted in the other way. (*Hint:* The mouth of the elderly woman is the necklace on the young woman.)

Once most people see one of the images, it is difficult for them to perceive the other. So it is with ambiguous language. Once you interpreted the sentence at the beginning of this section in one way, it may have been hard for you to see that it could be understood in the other way. Perhaps you had difficulty even though the two possible meanings were explained, just as many people have difficulty seeing the second interpretation for the drawing even when they are told what to look for.

Although statements written informally may be open to multiple interpretations, we cannot determine their truth or falsity without interpreting them one way or another. Therefore, we have to use context to try to ascertain their meaning as best we can.

Negations of Multiply-Quantified Statements

You can use the same rules to negate multiply-quantified statements that you used to negate simpler quantified statements. Recall that

$$\sim(\forall x \text{ in } D, P(x)) \equiv \exists x \text{ in } D \text{ such that } \sim P(x).$$

and

$$\sim(\exists x \text{ in } D \text{ such that } P(x)) \equiv \forall x \text{ in } D, \sim P(x).$$

We apply these laws to find

$$\sim(\forall x \text{ in } D, \exists y \text{ in } E \text{ such that } P(x, y))$$

by moving in stages from left to right along the sentence.

> *First version of negation:* $\exists x$ in D such that $\sim(\exists y$ in E such that $P(x, y))$.
> *Final version of negation:* $\exists x$ in D such that $\forall y$ in $E, \sim P(x, y)$.

Similarly, to find

$$\sim(\exists x \text{ in } D \text{ such that } \forall y \text{ in } E, P(x, y)),$$

we have

> *First version of negation:* $\forall x$ in $D, \sim(\forall y$ in $E, P(x, y))$.
> *Final version of negation:* $\forall x$ in $D, \exists y$ in E such that $\sim P(x, y)$.

These facts can be summarized as follows:

Negations of Multiply-Quantified Statements

$\sim(\forall x \text{ in } D, \exists y \text{ in } E \text{ such that } P(x, y)) \equiv \exists x \text{ in } D \text{ such that } \forall y \text{ in } E, \sim P(x, y).$

$\sim(\exists x \text{ in } D \text{ such that } \forall y \text{ in } E, P(x, y)) \equiv \forall x \text{ in } D, \exists y \text{ in } E \text{ such that } \sim P(x, y).$

Example 3.3.8 Negating Statements in a Tarski World

Refer to the Tarski world of Figure 3.3.1, which is reprinted here for reference.

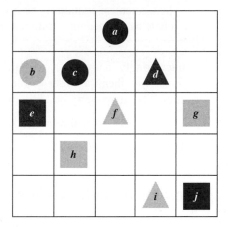

Write a negation for each of the following statements, and determine which is true, the given statement or its negation.

a. For all squares x, there is a circle y such that x and y have the same color.

b. There is a triangle x such that for all squares y, x is to the right of y.

Solution

a.   *First version of negation:* $\exists$ a square x such that $\sim(\exists$ a circle y such that x and y have the same color).

           Final version of negation: $\exists$ a square x such that $\forall$ circles y, x and y do not have the same color.

The negation is true. Square e is black and no circle is black, so there is a square that does not have the same color as any circle.

b.   *First version of negation:* $\forall$ triangles x, $\sim(\forall$ squares y, x is to the right of y).

           Final version of negation: $\forall$ triangles x, $\exists$ a square y such that x is not to the right of y.

The negation is true because no matter what triangle is chosen, it is not to the right of square g (or square j). ■

Order of Quantifiers

Consider the following two statements:

$$\forall \text{ people } x, \exists \text{ a person } y \text{ such that } x \text{ loves } y.$$
$$\exists \text{ a person } y \text{ such that } \forall \text{ people } x, x \text{ loves } y.$$

Note that except for the order of the quantifiers, these statements are identical. However, the first means that given any person, it is possible to find someone whom that person loves, whereas the second means that there is one amazing individual who is loved by all people. (Reread the statements carefully to verify these interpretations!)

The two sentences illustrate an extremely important property about multiply-quantified statements:

> In a statement containing both ∀ and ∃, changing the order of the quantifiers usually changes the meaning of the statement.

Caution! If a statement contains two different quantifiers, reversing their order can change the truth value of the statement to its opposite.

Interestingly, however, if one quantifier immediately follows another quantifier *of the same type,* then the order of the quantifiers does not affect the meaning. Consider the commutative property of addition of real numbers, for example:

$$\forall \text{ real numbers } x \text{ and } \forall \text{ real numbers } y, x + y = y + x.$$

This means the same as

$$\forall \text{ real numbers } y \text{ and } \forall \text{ real numbers } x, x + y = y + x.$$

Thus the property can be expressed more briefly as

$$\forall \text{ real numbers } x \text{ and } y, \ x + y = y + x.$$

Example 3.3.9 Quantifier Order in a Tarski World

Look again at the Tarski world of Figure 3.3.1. Do the following two statements have the same truth value?

a. For every square x there is a triangle y such that x and y have different colors.

b. There exists a triangle y such that for every square x, x and y have different colors.

Solution   Statement (a) says that if someone gives you one of the squares from the Tarski world, you can find a triangle that has a different color. This is true. If someone gives you square g or h (which are gray), you can use triangle d (which is black); if someone gives you square e (which is black), you can use either triangle f or triangle i (which are both gray); and if someone gives you square j (which is blue), you can use triangle d (which is black) or triangle f or i (which are both gray).

Statement (b) says that there is one particular triangle in the Tarski world that has a different color from every one of the squares in the world. This is false. Two of the triangles are gray, but they cannot be used to show the truth of the statement because the Tarski world contains gray squares. The only other triangle is black, but it cannot be used either because there is a black square in the Tarski world.

Thus one of the statements is true and the other is false, and so they have opposite truth values. ∎

Test Yourself

1. To establish the truth of a statement of the form "$\forall x$ in D, $\exists y$ in E such that $P(x, y)$," you imagine that someone has given you an element x from D but that you have no control over what that element is. Then you need to find ____ with the property that the x the person gave you together with the ____ you subsequently found satisfy ____.

2. To establish the truth of a statement of the form "$\exists x$ in D such that $\forall y$ in E, $P(x, y)$," you need to find ____ so that no matter what ____ a person might subsequently give you, ____ will be true.

3. Consider the statement "$\forall x, \exists y$ such that $P(x, y)$, a property involving x and y, is true." A negation for this statement is "____."

4. Consider the statement "$\exists x$ such that $\forall y$, $P(x, y)$, a property involving x and y, is true." A negation for this statement is "____."

5. Suppose $P(x, y)$ is some property involving x and y, and suppose the statement "$\forall x$ in D, $\exists y$ in E such that $P(x, y)$" is true. Then the statement "$\exists x$ in D such that $\forall y$ in E, $P(x, y)$"

a. is true. b. is false. c. may be true or may be false.

Exercise Set 3.3

1. Let C be the set of cities in the world, let N be the set of nations in the world, and let $P(c, n)$ be "c is the capital city of n." Determine the truth values of the following statements.

   a. $P(\text{Tokyo, Japan})$ **b.** $P(\text{Athens, Egypt})$
   c. $P(\text{Paris, France})$ **d.** $P(\text{Miami, Brazil})$

2. Let $G(x, y)$ be "$x^2 > y$." Indicate which of the following statements are true and which are false.

   a. $G(2, 3)$ **b.** $G(1, 1)$
   c. $G\left(\frac{1}{2}, \frac{1}{2}\right)$ **d.** $G(-2, 2)$

3. The following statement is true: "$\forall$ nonzero numbers x, $\exists$ a real number y such that $xy = 1$." For each x given below, find a y to make the predicate "$xy = 1$" true.

   a. $x = 2$ **b.** $x = -1$ **c.** $x = 3/4$

4. The following statement is true: "$\forall$ real numbers x, $\exists$ an integer n such that $n > x$."* For each x given below, find an n to make the predicate "$n > x$" true.

   a. $x = 15.83$ **b.** $x = 10^8$ **c.** $x = 10^{10^{10}}$

The statements in exercises 5–8 refer to the Tarski world given in Figure 3.3.1. Explain why each is true.

5. For all circles x there is a square y such that x and y have the same color.

6. For all squares x there is a circle y such that x and y have different colors and y is above x.

7. There is a triangle x such that for all squares y, x is above y.

8. There is a triangle x such that for all circles y, y is above x.

9. Let $D = E = \{-2, -1, 0, 1, 2\}$. Explain why the following statements are true.

   a. $\forall x$ in D, $\exists y$ in E such that $x + y = 0$.
   b. $\exists x$ in D such that $\forall y$ in E, $x + y = y$.

10. This exercise refers to Example 3.3.3. Determine whether each of the following statements is true or false.

    a. $\forall$ students S, $\exists$ a dessert D such that S chose D.
    b. $\forall$ students S, $\exists$ a salad T such that S chose T.
    c. $\exists$ a dessert D such that $\forall$ students S, S chose D.
    d. $\exists$ a beverage B such that $\forall$ students D, D chose B.
    e. $\exists$ an item I such that $\forall$ students S, S did not choose I.
    f. $\exists$ a station Z such that $\forall$ students S, $\exists$ an item I such that S chose I from Z.

11. Let S be the set of students at your school, let M be the set of movies that have ever been released, and let $V(s, m)$ be "student s has seen movie m." Rewrite each of the following statements without using the symbol $\forall$, the symbol $\exists$, or variables.

    a. $\exists s \in S$ such that $V(s, \text{Casablanca})$.
    b. $\forall s \in S$, $V(s, \text{Star Wars})$.

   c. $\forall s \in S$, $\exists m \in M$ such that $V(s, m)$.
   d. $\exists m \in M$ such that $\forall s \in S$, $V(s, m)$.
   e. $\exists s \in S$, $\exists t \in S$, and $\exists m \in M$ such that $s \neq t$ and $V(s, m) \wedge V(t, m)$.
   f. $\exists s \in S$ and $\exists t \in S$ such that $s \neq t$ and $\forall m \in M$, $V(s, m) \rightarrow V(t, m)$.

12. Let $D = E = \{-2, -1, 0, 1, 2\}$. Write negations for each of the following statements and determine which is true, the given statement or its negation.

    a. $\forall x$ in D, $\exists y$ in E such that $x + y = 1$.
    b. $\exists x$ in D such that $\forall y$ in E, $x + y = -y$.
    c. $\forall x$ in D, $\exists y$ in E such that $xy \geq y$.
    d. $\exists x$ in D such that $\forall y$ in E, $x \leq y$.

In each of 13–19, (a) rewrite the statement in English without using the symbol $\forall$ or $\exists$ or variables and expressing your answer as simply as possible, and (b) write a negation for the statement.

13. $\forall$ colors C, $\exists$ an animal A such that A is colored C.

14. $\exists$ a book b such that $\forall$ people p, p has read b.

15. $\forall$ odd integers n, $\exists$ an integer k such that $n = 2k + 1$.

16. $\exists$ a real number u such that $\forall$ real numbers v, $uv = v$.

17. $\forall r \in \mathbf{Q}$, $\exists$ integers a and b such that $r = a/b$.

18. $\forall x \in \mathbf{R}^+$, $\exists y \in \mathbf{R}^+$ such that $y > x$.

19. $\exists x \in \mathbf{R}^+$ such that $\forall y \in \mathbf{R}^+$, $x \leq y$.

20. Recall that reversing the order of the quantifiers in a statement with two different quantifiers may change the truth value of the statement—but it does not necessarily do so. All the statements in the pairs below refer to the Tarski world of Figure 3.3.1. In each pair, the order of the quantifiers is reversed but everything else is the same. For each pair, determine whether the statements have the same or opposite truth values. Justify your answers.

    a. (1) For all squares y there is a triangle x such that x and y have different color.
    (2) There is a triangle x such that for all squares y, x and y have different colors.
    b. (1) For all circles y there is a square x such that x and y have the same color.
    (2) There is a square x such that for all circles y, x and y have the same color.

21. For each of the following equations, determine which of the following statements are true:
    (1) For all real numbers x, there exists a real number y such that the equation is true.
    (2) There exists a real number x, such that for all real numbers y, the equation is true.

*This is called the Archimedean principle because it was first formulated (in geometric terms) by the great Greek mathematician Archimedes of Syracuse, who lived from about 287 to 212 B.C.E.

Note that it is possible for both statements to be true or for both to be false.

a. $2x + y = 7$
b. $y + x = x + y$
c. $x^2 - 2xy + y^2 = 0$
d. $(x - 5)(y - 1) = 0$
e. $x^2 + y^2 = -1$

In 22 and 23, rewrite each statement without using variables or the symbol $\forall$ or $\exists$. Indicate whether the statement is true or false.

22. a. $\forall$ real numbers x, $\exists$ a real number y such that $x + y = 0$.
    b. $\exists$ a real number y such that $\forall$ real numbers x, $x + y = 0$.

23. a. $\forall$ nonzero real numbers r, $\exists$ a real number s such that $rs = 1$.
    b. $\exists$ a real number r such that $\forall$ nonzero real numbers s, $rs = 1$.

24. Use the laws for negating universal and existential statements to derive the following rules:
    a. $\sim(\forall x \in D(\forall y \in E(P(x, y))))$
       $\equiv \exists x \in D(\exists y \in E(\sim P(x, y)))$
    b. $\sim(\exists x \in D(\exists y \in E(P(x, y))))$
       $\equiv \forall x \in D(\forall y \in E(\sim P(x, y)))$

Each statement in 25–28 refers to the Tarski world of Figure 3.3.1. For each, (a) determine whether the statement is true or false and justify your answer, (b) write a negation for the statement (referring, if you wish, to the result in exercise 24).

25. $\forall$ circles x and $\forall$ squares y, x is above y.

26. $\forall$ circles x and $\forall$ triangles y, x is above y.

27. $\exists$ a circle x and $\exists$ a square y such that x is above y and x and y have different colors.

28. $\exists$ a triangle x and $\exists$ a square y such that x is above y and x and y have the same color.

For each of the statements in 29 and 30, (a) write a new statement by interchanging the symbols $\forall$ and $\exists$, and (b) state which is true: the given statement, the version with interchanged quantifiers, neither, or both.

29. $\forall x \in \mathbf{R}$, $\exists y \in \mathbf{R}$ such that $x < y$.

30. $\exists x \in \mathbf{R}$ such that $\forall y \in \mathbf{R}^-$ (the set of negative real numbers), $x > y$.

31. Consider the statement "Everybody is older than somebody." Rewrite this statement in the form "$\forall$ people x, $\exists$ _____."

32. Consider the statement "Somebody is older than everybody." Rewrite this statement in the form "$\exists$ a person x such that $\forall$ _____."

In 33–39, (a) rewrite the statement formally using quantifiers and variables, and (b) write a negation for the statement.

33. Everybody loves somebody.

34. Somebody loves everybody.

35. Everybody trusts somebody.

36. Somebody trusts everybody.

37. Any even integer equals twice some integer.

38. Every action has an equal and opposite reaction.

39. There is a program that gives the correct answer to every question that is posed to it.

40. In informal speech most sentences of the form "There is _____ every _____" are intended to be understood as meaning "$\forall$ _____ $\exists$ _____," even though the existential quantifier *there is* comes before the universal quantifier *every*. Note that this interpretation applies to the following well-known sentences. Rewrite them using quantifiers and variables.
    a. There is a sucker born every minute.
    b. There is a time for every purpose under heaven.

41. Indicate which of the following statements are true and which are false. Justify your answers as best you can.
    a. $\forall x \in \mathbf{Z}^+$, $\exists y \in \mathbf{Z}^+$ such that $x = y + 1$.
    b. $\forall x \in \mathbf{Z}$, $\exists y \in \mathbf{Z}$ such that $x = y + 1$.
    c. $\exists x \in \mathbf{R}$ such that $\forall y \in \mathbf{R}$, $x = y + 1$.
    d. $\forall x \in \mathbf{R}^+$, $\exists y \in \mathbf{R}^+$ such that $xy = 1$.
    e. $\forall x \in \mathbf{R}$, $\exists y \in \mathbf{R}$ such that $xy = 1$.
    f. $\forall x \in \mathbf{Z}^+$ and $\forall y \in \mathbf{Z}^+$, $\exists z \in \mathbf{Z}^+$ such that $z = x - y$.
    g. $\forall x \in \mathbf{Z}$ and $\forall y \in \mathbf{Z}$, $\exists z \in \mathbf{Z}$ such that $z = x - y$.
    h. $\exists u \in \mathbf{R}^+$ such that $\forall v \in \mathbf{R}^+$, $uv < v$.

42. Write the negation of the definition of limit of a sequence given in Example 3.3.7.

43. The following is the definition for $\lim_{x \to a} f(x) = L$:

    For all real numbers $\varepsilon > 0$, there exists a real number $\delta > 0$ such that for all real numbers x, if $a - \delta < x < a + \delta$ and $x \neq a$ then $L - \varepsilon < f(x) < L + \varepsilon$.

    Write what it means for $\lim_{x \to a} f(x) \neq L$. In other words, write the negation of the definition.

44. The notation $\exists!$ stands for the words "there exists a unique." Thus, for instance, "$\exists!$ x such that x is prime and x is even" means that there is one and only one even prime number. Which of the following statements are true and which are false? Explain.
    a. $\exists!$ real number x such that $\forall$ real numbers y, $xy = y$.
    b. $\exists!$ integer x such that $1/x$ is an integer.
    c. $\forall$ real numbers x, $\exists!$ real number y such that $x + y = 0$.

✱ 45. Suppose that $P(x)$ is a predicate and D is the domain of x. Rewrite the statement "$\exists!$ $x \in D$ such that $P(x)$" without using the symbol $\exists!$. (See exercise 44 for the meaning of $\exists!$.)

In 46–54, refer to the Tarski world given in Figure 3.1.1, which is printed again here for reference. The domains of all variables consist of all the objects in the Tarski world. For each statement, indicate whether the statement is true or false and justify your answer.

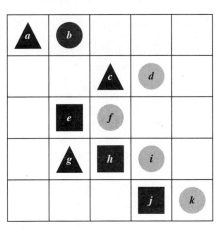

46. There is a triangle x such that for all squares y, x is above y.

47. There is a triangle x such that for all circles y, x is above y.

48. For all circles x, there is a square y such that y is to the right of x.

49. For every object x, there is an object y such that x and y have different colors.

50. For every object x, there is an object y such that if $x \neq y$ then x and y have different colors.

51. There is an object y such that for all objects x, if $x \neq y$ then x and y have different colors.

52. For all circles x and for all triangles y, x is to the right of y.

53. There is a circle x and there is a square y such that x and y have the same color.

54. There is a circle x and there is a triangle y such that x and y have the same color.

Answers for Test Yourself

1. an element y in E; y; $P(x, y)$ 2. an element x in D; y in E; $P(x, y)$ 3. $\exists x$ such that $\forall y$, the property $P(x, y)$ is false.   4. $\forall x$, $\exists y$ such that the property $P(x, y)$ is false.   5. The answer is (c): the truth or falsity of a statement in which the quantifiers are reversed depends on the nature of the property involving x and y.

3.4 Arguments with Quantified Statements

The only complete safeguard against reasoning ill, is the habit of reasoning well; familiarity with the principles of correct reasoning; and practice in applying those principles. — John Stuart Mill

The rule of *universal instantiation* (in-stan-she-AY-shun) says the following:

> If some property is true of *everything* in a set, then it is true of *any particular* thing in the set.

Use of the words *universal instantiation* indicates that the truth of a property in a particular case follows as a special instance of its more general or universal truth. The validity of this argument form follows immediately from the definition of truth values for a universal statement. One of the most famous examples of universal instantiation is the following:

All men are mortal.

Socrates is a man.

∴ Socrates is mortal.

Universal instantiation is *the* fundamental tool of deductive reasoning. Mathematical formulas, definitions, and theorems are like general templates that are used over and over in a wide variety of particular situations. A given theorem says that such and such is true for all things of a certain type. If, in a given situation, you have a particular object of

law of universal instantiation

ROSS ♡

that type, then by universal instantiation, you conclude that such and such is true for that particular object. You may repeat this process 10, 20, or more times in a single proof or problem solution.

As an example of universal instantiation, suppose you are doing a problem that requires you to simplify

$$r^{k+1} \cdot r,$$

where r is a particular real number and k is a particular integer. You know from your study of algebra that the following universal statements are true:

1. For all real numbers x and all integers m and n, $x^m \cdot x^n = x^{m+n}$.

2. For all real numbers x, $x^1 = x$.

So you proceed as follows:

$$
\begin{aligned}
r^{k+1} \cdot r &= r^{k+1} \cdot r^1 && \text{Step 1} \\
&= r^{(k+1)+1} && \text{Step 2} \\
&= r^{k+2} && \text{by basic algebra.}
\end{aligned}
$$

The reasoning behind step 1 and step 2 is outlined as follows.

Step 1:   For all real numbers x, $x^1 = x$.   universal truth
          r is a particular real number.   particular instance
          $\therefore r^1 = r$.   conclusion

Step 2:   For all real numbers x and all integers
          m and n, $x^m \cdot x^n = x^{m+n}$.   universal truth
          r is a particular real number and $k + 1$
          and 1 are particular integers.   particular instance
          $\therefore r^{k+1} \cdot r^1 = r^{(k+1)+1}$. conclusion

Both arguments are examples of universal instantiation.

Universal Modus Ponens

The rule of universal instantiation can be combined with modus ponens to obtain the valid form of argument called *universal modus ponens*.

Universal Modus Ponens	
Formal Version	*Informal Version*
$\forall x$, if $P(x)$ then $Q(x)$.	If x makes $P(x)$ true, then x makes $Q(x)$ true.
$P(a)$ for a particular a.	a makes $P(x)$ true.
$\therefore Q(a)$.	$\therefore a$ makes $Q(x)$ true.

Note that the first, or major, premise of universal modus ponens could be written "All things that make $P(x)$ true make $Q(x)$ true," in which case the conclusion would follow by universal instantiation alone. However, the if-then form is more natural to use in the majority of mathematical situations.

Example 3.4.1 Recognizing Universal Modus Ponens

Rewrite the following argument using quantifiers, variables, and predicate symbols. Is this argument valid? Why?

> If an integer is even, then its square is even.
>
> k is a particular integer that is even.
>
> $\therefore k^2$ is even.

Solution   The major premise of this argument can be rewritten as

$$\forall x, \text{if } x \text{ is an even integer then } x^2 \text{ is even.}$$

Let $E(x)$ be "x is an even integer," let $S(x)$ be "x^2 is even," and let k stand for a particular integer that is even. Then the argument has the following form:

> $\forall x, \text{if } E(x) \text{ then } S(x).$
>
> $E(k), \text{for a particular } k.$
>
> $\therefore S(k).$

This argument has the form of universal modus ponens and is therefore valid. ■

Example 3.4.2 Drawing Conclusions Using Universal Modus Ponens

Write the conclusion that can be inferred using universal modus ponens.

> If T is any right triangle with hypotenuse c and legs a and b, then $c^2 = a^2 + b^2$.
>
> The triangle shown at the right is a right triangle with both legs equal to 1 and hypotenuse c.
>
> $\therefore$ _____.

Pythagorean theorem

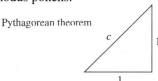

Solution $c^2 = 1^2 + 1^2 = 2$

Note that if you take the nonnegative square root of both sides of this equation, you obtain $c = \sqrt{2}$. This shows that there is a line segment whose length is $\sqrt{2}$. Section 4.7 contains a proof that $\sqrt{2}$ is not a rational number. ■

Use of Universal Modus Ponens in a Proof

In Chapter 4 we discuss methods of proving quantified statements. Here is a proof that the sum of any two even integers is even. It makes use of the definition of even integer, namely, that an integer is *even* if, and only if, it equals twice some integer. (Or, more formally: $\forall$ integers x, x is even if, and only if, $\exists$ an integer k such that $x = 2k$.)

Suppose m and n are particular but arbitrarily chosen even integers. Then $m = 2r$ for some integer r,[1] and $n = 2s$ for some integer s.[2] Hence

$$m + n = 2r + 2s \qquad \text{by substitution}$$
$$= 2(r + s)^{[3]} \qquad \text{by factoring out the 2.}$$

Now $r + s$ is an integer,[4] and so $2(r + s)$ is even.[5] Thus $m + n$ is even.

The following expansion of the proof shows how each of the numbered steps is justified by arguments that are valid by universal modus ponens.

Note The logical principle of **existential instantiation** says that if we know something exists, we may give it a name. This principle, discussed further in Section 4.1 allows us to give the integers the names r and s.

(1)   If an integer is even, then it equals twice some integer.
    m is a particular even integer.
    $\therefore$ m equals twice some integer r.

(2)   If an integer is even, then it equals twice some integer.
    n is a particular even integer.
    $\therefore$ n equals twice some integer s.

(3)   If a quantity is an integer, then it is a real number.
    r and s are particular integers.
    $\therefore$ r and s are real numbers.

   For all a, b, and c, if a, b, and c are real numbers, then $ab + ac = a(b + c)$.
   2, r, and s are particular real numbers.
   $\therefore$ $2r + 2s = 2(r + s)$.

(4)   For all u and v, if u and v are integers, then $u + v$ is an integer.
    r and s are two particular integers.
    $\therefore$ $r + s$ is an integer.

(5)   If a number equals twice some integer, then that number is even.
    $2(r + s)$ equals twice the integer $r + s$.
    $\therefore$ $2(r + s)$ is even.

Of course, the actual proof that the sum of even integers is even does not explicitly contain the sequence of arguments given above. (Heaven forbid!) And, in fact, people who are good at analytical thinking are normally not even conscious that they are reasoning in this way. But that is because they have absorbed the method so completely that it has become almost as automatic as breathing.

Universal Modus Tollens

Another crucially important rule of inference is *universal modus tollens*. Its validity results from combining universal instantiation with modus tollens. Universal modus tollens is the heart of proof of contradiction, which is one of the most important methods of mathematical argument.

Universal Modus Tollens

Formal Version	*Informal Version*
$\forall x$, if $P(x)$ then $Q(x)$.	If x makes $P(x)$ true, then x makes $Q(x)$ true.
$\sim Q(a)$, for a particular a.	a does not make $Q(x)$ true.
$\therefore \sim P(a)$.	$\therefore$ a does not make $P(x)$ true.

Example 3.4.3 Recognizing the Form of Universal Modus Tollens

Rewrite the following argument using quantifiers, variables, and predicate symbols. Write the major premise in conditional form. Is this argument valid? Why?

All human beings are mortal.

Zeus is not mortal.

$\therefore$ Zeus is not human.

Solution   The major premise can be rewritten as

$$\forall x, \text{if } x \text{ is human then } x \text{ is mortal.}$$

Let $H(x)$ be "x is human," let $M(x)$ be "x is mortal," and let Z stand for Zeus. The argument becomes

$$\forall x, \text{if } H(x) \text{ then } M(x)$$
$$\sim M(Z)$$
$$\therefore \sim H(Z).$$

This argument has the form of universal modus tollens and is therefore valid.   ■

Example 3.4.4 Drawing Conclusions Using Universal Modus Tollens

Write the conclusion that can be inferred using universal modus tollens.

> All professors are absent-minded.
>
> Tom Hutchins is not absent-minded.
>
> $\therefore$ _____.

Solution   Tom Hutchins is not a professor.   ■

Proving Validity of Arguments with Quantified Statements

The intuitive definition of validity for arguments with quantified statements is the same as for arguments with compound statements. An argument is valid if, and only if, the truth of its conclusion follows *necessarily* from the truth of its premises. The formal definition is as follows:

> **• Definition**
>
> To say that an *argument form* is **valid** means the following: No matter what particular predicates are substituted for the predicate symbols in its premises, if the resulting premise statements are all true, then the conclusion is also true. An *argument* is called **valid** if, and only if, its form is valid.

As already noted, the validity of universal instantiation follows immediately from the definition of the truth value of a universal statement. General formal proofs of validity of arguments in the predicate calculus are beyond the scope of this book. We give the proof of the validity of universal modus ponens as an example to show that such proofs are possible and to give an idea of how they look.

Universal modus ponens asserts that

$$\forall x, \text{if } P(x) \text{ then } Q(x).$$
$$P(a) \text{ for a particular } a.$$
$$\therefore Q(a).$$

To prove that this form of argument is valid, suppose the major and minor premises are both true. *[We must show that the conclusion "$Q(a)$" is also true.]* By the minor premise, $P(a)$ is true for a particular value of a. By the major premise and universal instantiation, the statement "If $P(a)$ then $Q(a)$" is true for that particular a. But by modus ponens, since the statements "If $P(a)$ then $Q(a)$" and "$P(a)$" are both true, it follows that $Q(a)$ is true also. *[This is what was to be shown.]*

The proof of validity given above is abstract and somewhat subtle. We include the proof not because we expect that you will be able to make up such proofs yourself at this stage of your study. Rather, it is intended as a glimpse of a more advanced treatment of the subject, which you can try your hand at in exercises 35 and 36 at the end of this section if you wish.

One of the paradoxes of the formal study of logic is that the laws of logic are used to prove that the laws of logic are valid!

In the next part of this section we show how you can use diagrams to analyze the validity or invalidity of arguments that contain quantified statements. Diagrams do not provide totally rigorous proofs of validity and invalidity, and in some complex settings they may even be confusing, but in many situations they are helpful and convincing.

Using Diagrams to Test for Validity

Consider the statement

All integers are rational numbers.

Or, formally,

$\forall$ integers n, n is a rational number.

Picture the set of all integers and the set of all rational numbers as disks. The truth of the given statement is represented by placing the integers disk entirely inside the rationals disk, as shown in Figure 3.4.1.

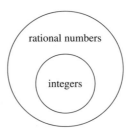

Figure 3.4.1

Because the two statements "$\forall x \in D$, $Q(x)$" and "$\forall x$, if x is in D then $Q(x)$" are logically equivalent, both can be represented by diagrams like the foregoing.

Perhaps the first person to use diagrams like these to analyze arguments was the German mathematician and philosopher Gottfried Wilhelm Leibniz. Leibniz (LIPE-nits) was far ahead of his time in anticipating modern symbolic logic. He also developed the main ideas of the differential and integral calculus at approximately the same time as (and independently of) Isaac Newton (1642–1727).

To test the validity of an argument diagrammatically, represent the truth of both premises with diagrams. Then analyze the diagrams to see whether they necessarily represent the truth of the conclusion as well.

Culver Pictures

G. W. Leibniz
(1646–1716)

Example 3.4.5 Using a Diagram to Show Validity

Use diagrams to show the validity of the following syllogism:

All human beings are mortal.

Zeus is not mortal.

∴ Zeus is not a human being.

Solution   The major premise is pictured on the left in Figure 3.4.2 by placing a disk labeled "human beings" inside a disk labeled "mortals." The minor premise is pictured on the right in Figure 3.4.2 by placing a dot labeled "Zeus" outside the disk labeled "mortals."

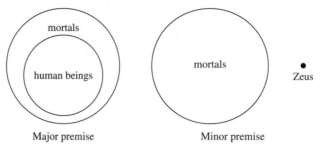

Major premise   Minor premise

Figure 3.4.2

The two diagrams fit together in only one way, as shown in Figure 3.4.3.

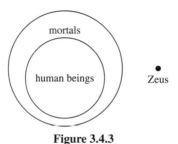

Figure 3.4.3

Since the Zeus dot is outside the mortals disk, it is necessarily outside the human beings disk. Thus the truth of the conclusion follows necessarily from the truth of the premises. It is impossible for the premises of this argument to be true and the conclusion false; hence the argument is valid.   ■

Example 3.4.6 Using Diagrams to Show *In*validity

Use a diagram to show the invalidity of the following argument:

All human beings are mortal.

Felix is mortal.

∴ Felix is a human being.

Solution   The major and minor premises are represented diagrammatically in Figure 3.4.4.

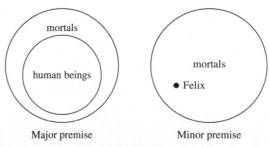

Major premise   Minor premise

Figure 3.4.4

All that is known is that the Felix dot is located *somewhere* inside the mortals disk. Where it is located with respect to the human beings disk cannot be determined. Either one of the situations shown in Figure 3.4.5 might be the case.

Caution! Be careful when using diagrams to test for validity! For instance, in this example if you put the diagrams for the premises together to obtain only Figure 3.4.5(a) and not Figure 3.4.5(b), you would conclude erroneously that the argument was valid.

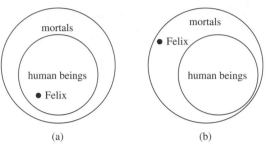

(a)   (b)

Figure 3.4.5

The conclusion "Felix is a human being" is true in the first case but not in the second (Felix might, for example, be a cat). Because the conclusion does not necessarily follow from the premises, the argument is invalid. ■

The argument of Example 3.4.6 would be valid if the major premise were replaced by its converse. But since a universal conditional statement is not logically equivalent to its converse, such a replacement cannot, in general, be made. We say that this argument exhibits the converse error.

Converse Error (Quantified Form)

Formal Version   *Informal Version*

$\forall x$, if $P(x)$ then $Q(x)$.   If x makes $P(x)$ true, then x makes $Q(x)$ true.

$Q(a)$ for a particular a. a makes $Q(x)$ true.

$\therefore P(a)$. ← invalid $\therefore a$ makes $P(x)$ true. ← invalid
        conclusion   conclusion

The following form of argument would be valid if a conditional statement were logically equivalent to its inverse. But it is not, and the argument form is invalid. We say that it exhibits the inverse error. You are asked to show the invalidity of this argument form in the exercises at the end of this section.

Inverse Error (Quantified Form)

Formal Version   *Informal Version*

$\forall x$, if $P(x)$ then $Q(x)$.   If x makes $P(x)$ true, then x makes $Q(x)$ true.

$\sim P(a)$, for a particular a. a does not make $P(x)$ true.

$\therefore \sim Q(a)$. ← invalid $\therefore a$ does not make $Q(x)$ true. ← invalid
        conclusion   conclusion

Example 3.4.7 An Argument with "No"

Use diagrams to test the following argument for validity:

No polynomial functions have horizontal asymptotes.

This function has a horizontal asymptote.

∴ This function is not a polynomial function.

Solution   A good way to represent the major premise diagrammatically is shown in Figure 3.4.6, two disks—a disk for polynomial functions and a disk for functions with horizontal asymptotes—that do not overlap at all. The minor premise is represented by placing a dot labeled "this function" inside the disk for functions with horizontal asymptotes.

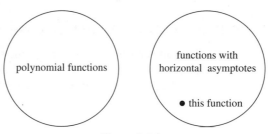

Figure 3.4.6

The diagram shows that "this function" must lie outside the polynomial functions disk, and so the truth of the conclusion necessarily follows from the truth of the premises. Hence the argument is valid.   ■

An alternative approach to this example is to transform the statement "No polynomial functions have horizontal asymptotes" into the equivalent form "$\forall x$, if x is a polynomial function, then x does not have a horizontal asymptote." If this is done, the argument can be seen to have the form

$\forall x$, if $P(x)$ then $Q(x)$.

$\sim Q(a)$, for a particular a.

∴ $\sim P(a)$.

where $P(x)$ is "x is a polynomial function" and $Q(x)$ is "x does not have a horizontal asymptote." This is valid by universal modus tollens.

Creating Additional Forms of Argument

Universal modus ponens and modus tollens were obtained by combining universal instantiation with modus ponens and modus tollens. In the same way, additional forms of arguments involving universally quantified statements can be obtained by combining universal instantiation with other of the valid argument forms given in Section 2.3. For instance, in Section 2.3 the argument form called transitivity was introduced:

$$p \rightarrow q$$
$$q \rightarrow r$$
$$\therefore p \rightarrow r$$

This argument form can be combined with universal instantiation to obtain the following valid argument form.

Universal Transitivity

Formal Version	*Informal Version*
$\forall x\, P(x) \rightarrow Q(x).$	Any x that makes $P(x)$ true makes $Q(x)$ true.
$\forall x\, Q(x) \rightarrow R(x).$	Any x that makes $Q(x)$ true makes $R(x)$ true.
$\therefore \forall x\, P(x) \rightarrow R(x).$	$\therefore$ Any x that makes $P(x)$ true makes $R(x)$ true.

Example 3.4.8 Evaluating an Argument for Tarski's World

The following argument refers to the kind of arrangement of objects of various types and colors described in Examples 3.1.13 and 3.3.1. Reorder and rewrite the premises to show that the conclusion follows as a valid consequence from the premises.

1. All the triangles are blue.

2. If an object is to the right of all the squares, then it is above all the circles.

3. If an object is not to the right of all the squares, then it is not blue.

$\therefore$ All the triangles are above all the circles.

Solution   It is helpful to begin by rewriting the premises and the conclusion in if-then form:

1. $\forall x$, if x is a triangle, then x is blue.

2. $\forall x$, if x is to the right of all the squares, then x is above all the circles.

3. $\forall x$, if x is not to the right of all the squares, then x is not blue.

$\therefore$ $\forall x$, if x is a triangle, then x is above all the circles.

The goal is to reorder the premises so that the conclusion of each is the same as the hypothesis of the next. Also, the hypothesis of the argument's conclusion should be the same as the hypothesis of the first premise, and the conclusion of the argument's conclusion should be the same as the conclusion of the last premise. To achieve this goal, it may be necessary to rewrite some of the statements in contrapositive form.

In this example you can see that the first premise should remain where it is, but the second and third premises should be interchanged. Then the hypothesis of the argument is the same as the hypothesis of the first premise, and the conclusion of the argument's conclusion is the same as the conclusion of the third premise. But the hypotheses and conclusions of the premises do not quite line up. This is remedied by rewriting the third premise in contrapositive form.

Thus the premises and conclusion of the argument can be rewritten as follows:

1. $\forall x$, if x is a triangle, then x is blue.

3. $\forall x$, if x is blue, then x is to the right of all the squares.

2. $\forall x$, if x is to the right of all the squares, then x is above all the circles.

$\therefore$ $\forall x$, if x is a triangle, then x is above all the circles.

The validity of this argument follows easily from the validity of universal transitivity. Putting 1 and 3 together and using universal transitivity gives that

4. $\forall x$, if x is a triangle, then x is to the right of all the squares.

And putting 4 together with 2 and using universal transitivity gives that

$\forall x$, if x is a triangle, then x is above all the circles,

which is the conclusion of the argument.   ■

Remark on the Converse and Inverse Errors

One reason why so many people make converse and inverse errors is that the forms of the resulting arguments would be valid if the major premise were a biconditional rather than a simple conditional. And, as we noted in Section 2.2, many people tend to conflate biconditionals and conditionals.

Consider, for example, the following argument:

> All the town criminals frequent the Den of Iniquity bar.
>
> John frequents the Den of Iniquity bar.
>
> ∴ John is one of the town criminals.

The conclusion of this argument is invalid—it results from making the converse error. Therefore, it may be false even when the premises of the argument are true. This type of argument attempts unfairly to establish guilt by association.

The closer, however, the major premise comes to being a biconditional, the more likely the conclusion is to be true. If hardly anyone but criminals frequents the bar and John also frequents the bar, then it is likely (though not certain) that John is a criminal. On the basis of the given premises, it might be sensible to be suspicious of John, but it would be wrong to convict him.

A variation of the converse error is a very useful reasoning tool, provided that it is used with caution. It is the type of reasoning that is used by doctors to make medical diagnoses and by auto mechanics to repair cars. It is the type of reasoning used to generate explanations for phenomena. It goes like this: If a statement of the form

For all x, if $P(x)$ then $Q(x)$

is true, and if

$Q(a)$ is true, for a particular a,

then check out the statement $P(a)$; it just might be true. For instance, suppose a doctor knows that

For all x, if x has pneumonia, then x has a fever and chills, coughs deeply, and feels exceptionally tired and miserable.

And suppose the doctor also knows that

John has a fever and chills, coughs deeply, and feels exceptionally tired and miserable.

On the basis of these data, the doctor concludes that a diagnosis of pneumonia is a strong possibility, though not a certainty. The doctor will probably attempt to gain further support for this diagnosis through laboratory testing that is specifically designed to detect pneumonia. Note that the closer a set of symptoms comes to being a necessary and sufficient condition for an illness, the more nearly certain the doctor can be of his or her diagnosis.

This form of reasoning has been named **abduction** by researchers working in artificial intelligence. It is used in certain computer programs, called expert systems, that attempt to duplicate the functioning of an expert in some field of knowledge.

Test Yourself

1. The rule of universal instantiation says that if some property is true for _____ in a domain, then it is true for _____.

2. If the first two premises of universal modus ponens are written as "If x makes $P(x)$ true, then x makes $Q(x)$ true" and "For a particular value of a _____," then the conclusion can be written as "_____."

3. If the first two premises of universal modus tollens are written as "If x makes $P(x)$ true, then x makes $Q(x)$ true" and

"For a particular value of a _____," then the conclusion can be written as "_____."

4. If the first two premises of universal transitivity are written as "Any x that makes $P(x)$ true makes $Q(x)$ true" and "Any x that makes $Q(x)$ true makes $R(x)$ true," then the conclusion can be written as "_____."

5. Diagrams can be helpful in testing an argument for validity. However, if some possible configurations of the premises are not drawn, a person could conclude that an argument was _____ when it was actually _____.

Exercise Set 3.4

1. Let the following law of algebra be the first statement of an argument: For all real numbers a and b,

$$(a + b)^2 = a^2 + 2ab + b^2.$$

Suppose each of the following statements is, in turn, the second statement of the argument. Use universal instantiation or universal modus ponens to write the conclusion that follows in each case.
   a. $a = x$ and $b = y$ are particular real numbers.
   b. $a = f_i$ and $b = f_j$ are particular real numbers.
   c. $a = 3u$ and $b = 5v$ are particular real numbers.
   d. $a = g(r)$ and $b = g(s)$ are particular real numbers.
   e. $a = \log(t_1)$ and $b = \log(t_2)$ are particular real numbers.

Use universal instantiation or universal modus ponens to fill in valid conclusions for the arguments in 2–4.

2. If an integer n equals $2 \cdot k$ and k is an integer, then n is even.
   0 equals $2 \cdot 0$ and 0 is an integer.
   ∴ _____

3. For all real numbers a, b, c, and d, if $b \neq 0$ and $d \neq 0$, then $a/b + c/d = (ad + bc)/bd$.
   $a = 2, \ b = 3, \ c = 4$, and $d = 5$ are particular real numbers such that $b \neq 0$ and $d \neq 0$.
   ∴ _____

4. ∀ real numbers r, a, and b, if r is positive, then $(r^a)^b = r^{ab}$.
   $r = 3, a = 1/2$, and $b = 6$ are particular real numbers such that r is positive.
   ∴ _____

Use universal modus tollens to fill in valid conclusions for the arguments in 5 and 6.

5. All irrational numbers are real numbers
   $\frac{1}{0}$ is not a real number.
   ∴ _____

6. If a computer program is correct, then compilation of the program does not produce error messages.
   Compilation of this program produces error messages.
   ∴ _____

Some of the arguments in 7–18 are valid by universal modus ponens or universal modus tollens; others are invalid and exhibit the converse or the inverse error. State which are valid and which are invalid. Justify your answers.

7. All healthy people eat an apple a day.
   Keisha eats an apple a day.
   ∴ Keisha is a healthy person.

8. All freshmen must take writing.
   Caroline is a freshman.
   ∴ Caroline must take writing.

9. All healthy people eat an apple a day.
   Herbert is not a healthy person.
   ∴ Herbert does not eat an apple a day.

10. If a product of two numbers is 0, then at least one of the numbers is 0.
    For a particular number x, neither $(2x + 1)$ nor $(x - 7)$ equals 0.
    ∴ The product $(2x + 1)(x - 7)$ is not 0.

11. All cheaters sit in the back row.
    Monty sits in the back row.
    ∴ Monty is a cheater.

12. All honest people pay their taxes.
    Darth is not honest.
    ∴ Darth does not pay his taxes.

13. For all students x, if x studies discrete mathematics, then x is good at logic.
    Tarik studies discrete mathematics.
    ∴ Tarik is good at logic.

14. If compilation of a computer program produces error messages, then the program is not correct.
    Compilation of this program does not produce error messages.
    ∴ This program is correct.

15. Any sum of two rational numbers is rational.
    The sum $r + s$ is rational.
    ∴ The numbers r and s are both rational.

16. If a number is even, then twice that number is even.
    The number $2n$ is even, for a particular number n.
    ∴ The particular number n is even.

17. If an infinite series converges, then the terms go to 0.
    The terms of the infinite series $\sum_{n=1}^{\infty} \dfrac{1}{n}$ go to 0.
    ∴ The infinite series $\sum_{n=1}^{\infty} \dfrac{1}{n}$ converges.

18. If an infinite series converges, then its terms go to 0.
    The terms of the infinite series $\sum_{n=1}^{\infty} \dfrac{n}{n+1}$ do not go to 0.
    ∴ The infinite series $\sum_{n=1}^{\infty} \dfrac{n}{n+1}$ does not converge.

19. Rewrite the statement "No good cars are cheap" in the form "$\forall x$, if $P(x)$ then $\sim Q(x)$." Indicate whether each of the following arguments is valid or invalid, and justify your answers.
    a. No good car is cheap.
    A Rimbaud is a good car.
    ∴ A Rimbaud is not cheap.
    b. No good car is cheap.
    A Simbaru is not cheap.
    ∴ A Simbaru is a good car.
    c. No good car is cheap.
    A VX Roadster is cheap.
    ∴ A VX Roadster is not good.
    d. No good car is cheap.
    An Omnex is not a good car.
    ∴ An Omnex is cheap.

20. a. Use a diagram to show that the following argument can have true premises and a false conclusion.

    All dogs are carnivorous.
    Aaron is not a dog.
    ∴ Aaron is not carnivorous.

b. What can you conclude about the validity or invalidity of the following argument form? Explain how the result from part (a) leads to this conclusion.

$$\forall x, \text{ if } P(x) \text{ then } Q(x).$$
$$\sim P(a) \text{ for a particular } a.$$
$$\therefore \sim Q(a).$$

Indicate whether the arguments in 21–27 are valid or invalid. Support your answers by drawing diagrams.

21. All people are mice.
    All mice are mortal.
    ∴ All people are mortal.

22. All discrete mathematics students can tell a valid argument from an invalid one.
    All thoughtful people can tell a valid argument from an invalid one.
    ∴ All discrete mathematics students are thoughtful.

23. All teachers occasionally make mistakes.
    No gods ever make mistakes.
    ∴ No teachers are gods.

24. No vegetarians eat meat.
    All vegans are vegetarian.
    ∴ No vegans eat meat.

25. No college cafeteria food is good.
    No good food is wasted.
    ∴ No college cafeteria food is wasted.

26. All polynomial functions are differentiable.
    All differentiable functions are continuous.
    ∴ All polynomial functions are continuous.

27. [Adapted from Lewis Carroll.]
    Nothing intelligible ever puzzles *me*.
    Logic puzzles me.
    ∴ Logic is unintelligible.

In exercises 28–32, reorder the premises in each of the arguments to show that the conclusion follows as a valid consequence from the premises. It may be helpful to rewrite the statements in if-then form and replace some statements by their contrapositives. Exercises 28–30 refer to the kinds of Tarski worlds discussed in Example 3.1.13 and 3.3.1. Exercises 31 and 32 are adapted from *Symbolic Logic* by Lewis Carroll.*

28. 1. Every object that is to the right of all the blue objects is above all the triangles.
    2. If an object is a circle, then it is to the right of all the blue objects.
    3. If an object is not a circle, then it is not gray.
    ∴ All the gray objects are above all the triangles.

*Lewis Carroll, *Symbolic Logic* (New York: Dover, 1958), pp. 118, 120, 123.

29. 1. All the objects that are to the right of all the triangles are above all the circles.
    2. If an object is not above all the black objects, then it is not a square.
    3. All the objects that are above all the black objects are to the right of all the triangles.
    ∴ All the squares are above all the circles.

30. 1. If an object is above all the triangles, then it is above all the blue objects.
    2. If an object is not above all the gray objects, then it is not a square.
    3. Every black object is a square.
    4. Every object that is above all the gray objects is above all the triangles.
    ∴ If an object is black, then it is above all the blue objects.

31. 1. I trust every animal that belongs to me.
    2. Dogs gnaw bones.
    3. I admit no animals into my study unless they will beg when told to do so.
    4. All the animals in the yard are mine.
    5. I admit every animal that I trust into my study.
    6. The only animals that are really willing to beg when told to do so are dogs.
    ∴ All the animals in the yard gnaw bones.

32. 1. When I work a logic example without grumbling, you may be sure it is one I understand.
    2. The arguments in these examples are not arranged in regular order like the ones I am used to.
    3. No easy examples make my head ache.
    4. I can't understand examples if the arguments are not arranged in regular order like the ones I am used to.
    5. I never grumble at an example unless it gives me a headache.
    ∴ These examples are not easy.

In 33 and 34 a single conclusion follows when all the given premises are taken into consideration, but it is difficult to see because the premises are jumbled up. Reorder the premises to make it clear that a conclusion follows logically, and state the valid conclusion that can be drawn. (It may be helpful to rewrite some of the statements in if-then form and to replace some statements by their contrapositives.)

33. 1. No birds except ostriches are at least 9 feet tall.
    2. There are no birds in this aviary that belong to anyone but me.
    3. No ostrich lives on mince pies.
    4. I have no birds less than 9 feet high.

34. 1. All writers who understand human nature are clever.
    2. No one is a true poet unless he can stir the human heart.
    3. Shakespeare wrote *Hamlet*.
    4. No writer who does not understand human nature can stir the human heart.
    5. None but a true poet could have written *Hamlet*.

✴ 35. Derive the validity of universal modus tollens from the validity of universal instantiation and modus tollens.

✴ 36. Derive the validity of universal form of part(a) of the elimination rule from the validity of universal instantiation and the valid argument called elimination in Section 2.3.

Answers for Test Yourself

1. all elements; any particular element in the domain (*Or:* each individual element of the domain) 2. $P(a)$ is true; $Q(a)$ is true
3. $Q(a)$ is false; $P(a)$ is false   4. Any x that makes $P(x)$ true makes $R(x)$ true.   5. valid; invalid (*Or:* invalid; valid).

CHAPTER 4

ELEMENTARY NUMBER THEORY AND METHODS OF PROOF

The underlying content of this chapter is likely to be familiar to you. It consists of properties of integers (whole numbers), rational numbers (integer fractions), and real numbers. The underlying theme of this chapter is the question of how to determine the truth or falsity of a mathematical statement.

As an example consider the following question:

Given an odd integer n, is it always the case that $n^2 - 1$ is (evenly) divisible by 8?

Take a few minutes to try to answer this question for yourself.

It turns out that the answer is yes. Is this the answer you got? If not, don't worry. In Section 4.4 you will learn the technique you need to answer it and more. If you did get the correct answer, congratulations! You have excellent mathematical intuition. Now ask yourself, "How sure am I of my answer? Was it a plausible guess or an absolute certainty? Would I have been willing to bet a large sum of money on the correctness of my answer?"

One of the best ways to think of a mathematical proof is as a carefully reasoned argument to convince a skeptical listener (often yourself) that a given statement is true. Imagine the listener challenging your reasoning every step of the way, constantly asking, "Why is that so?" If you can counter every possible challenge, then your proof as a whole will be correct.

As an example, imagine proving to someone not very familiar with mathematical notation that if x is a number with $5x + 3 = 33$, then $x = 6$. You could argue as follows:

If $5x + 3 = 33$, then $5x + 3$ minus 3 will equal $33 - 3$ since subtracting the same number from two equal quantities gives equal results. But $5x + 3$ minus 3 equals $5x$ because adding 3 to $5x$ and then subtracting 3 just leaves $5x$. Also, $33 - 3 = 30$. Hence $5x = 30$. This means that x is a number which when multiplied by 5 equals 30. But the only number with this property is 6. Therefore, if $5x + 3 = 33$ then $x = 6$.

Of course there are other ways to phrase this proof, depending on the level of mathematical sophistication of the intended reader. In practice, mathematicians often omit reasons for certain steps of an argument when they are confident that the reader can easily supply them. When you are first learning to write proofs, however, it is better to err on the side of supplying too many reasons rather than too few. All too frequently, when even the best mathematicians carefully examine some "details" in their arguments, they discover that those details are actually false. One of the most important reason's for requiring proof in mathematics is that writing a proof forces us to become aware of weaknesses in our arguments and in the unconscious assumptions we have made.

Sometimes correctness of a mathematical argument can be a matter of life or death. Suppose, for example, that a mathematician is part of a team charged with designing a new type of airplane engine, and suppose that the mathematician is given the job of determining whether the thrust delivered by various engine types is adequate. If you knew that the mathematician was only fairly sure, but not positive, of the correctness of his analysis, you would probably not want to ride in the resulting aircraft.

At a certain point in Lewis Carroll's *Alice in Wonderland* (see exercise 28 in Section 2.2), the March Hare tells Alice to "say what you mean." In other words, she should be precise in her use of language: If she means a thing, then that is exactly what she should say. In this chapter, perhaps more than in any other mathematics course you have ever taken, you will find it necessary to say what you mean. Precision of thought and language is essential to achieve the mathematical certainty that is needed if you are to have complete confidence in your solutions to mathematical problems.

4.1 Direct Proof and Counterexample I: Introduction

Mathematics, as a science, commenced when first someone, probably a Greek, proved propositions about "any" things or about "some" things without specification of definite particular things. — Alfred North Whitehead, 1861–1947

Both discovery and proof are integral parts of problem solving. When you think you have discovered that a certain statement is true, try to figure out why it is true. If you succeed, you will know that your discovery is genuine. Even if you fail, the process of trying will give you insight into the nature of the problem and may lead to the discovery that the statement is false. For complex problems, the interplay between discovery and proof is not reserved to the end of the problem-solving process but, rather, is an important part of each step.

Assumptions

- In this text we assume a familiarity with the laws of basic algebra, which are listed in Appendix A.
- We also use the three properties of equality: For all objects A, B, and C,
  (1) $A = A$, (2) if $A = B$ then $B = A$, and (3) if $A = B$ and $B = C$, then $A = C$.
- In addition, we assume that there is no integer between 0 and 1 and that the set of all integers is closed under addition, subtraction, and multiplication. This means that sums, differences, and products of integers are integers.
- Of course, most quotients of integers are not integers. For example, $3 \div 2$, which equals 3/2, is not an integer, and $3 \div 0$ is not even a number.

The mathematical content of this section primarily concerns even and odd integers and prime and composite numbers.

Definitions

In order to evaluate the truth or falsity of a statement, you must understand what the statement is about. In other words, you must know the meanings of all terms that occur in the statement. Mathematicians define terms very carefully and precisely and consider it important to learn definitions virtually word for word.

• Definitions

An integer n is **even** if, and only if, n equals twice some integer. An integer n is **odd** if, and only if, n equals twice some integer plus 1.

Symbolically, if n is an integer, then

$$n \text{ is even} \quad \Leftrightarrow \quad \exists \text{ an integer } k \text{ such that } n = 2k.$$
$$n \text{ is odd} \quad \Leftrightarrow \quad \exists \text{ an integer } k \text{ such that } n = 2k + 1.$$

It follows from the definition that if you are doing a problem in which you happen to know that a certain integer is even, you can deduce that it has the form $2 \cdot (\text{some integer})$. Conversely, if you know in some situation that an integer equals $2 \cdot (\text{some integer})$, then you can deduce that the integer is even.

Know a particular integer n is even. $\xrightarrow{\text{deduce}}$ n has the form $2 \cdot (\text{some integer})$.

Know n has the form $2 \cdot (\text{some integer})$. $\xrightarrow{\text{deduce}}$ n is even.

Example 4.1.1 Even and Odd Integers

Use the definitions of *even* and *odd* to justify your answers to the following questions.

a. Is 0 even?

b. Is -301 odd?

c. If a and b are integers, is $6a^2b$ even?

d. If a and b are integers, is $10a + 8b + 1$ odd?

e. Is every integer either even or odd?

Solution

a. Yes, $0 = 2 \cdot 0$.

b. Yes, $-301 = 2(-151) + 1$.

c. Yes, $6a^2b = 2(3a^2b)$, and since a and b are integers, so is $3a^2b$ (being a product of integers).

d. Yes, $10a + 8b + 1 = 2(5a + 4b) + 1$, and since a and b are integers, so is $5a + 4b$ (being a sum of products of integers).

e. The answer is yes, although the proof is not obvious. (Try giving a reason yourself.) We will show in Section 4.4 that this fact results from another fact known as the quotient-remainder theorem. ∎

The integer 6, which equals $2 \cdot 3$, is a product of two smaller positive integers. On the other hand, 7 cannot be written as a product of two smaller positive integers; its only positive factors are 1 and 7. A positive integer, such as 7, that cannot be written as a product of two smaller positive integers is called *prime*.

• Definition

An integer n is **prime** if, and only if, $n > 1$ and for all positive integers r and s, if $n = rs$, then either r or s equals n. An integer n is **composite** if, and only if, $n > 1$ and $n = rs$ for some integers r and s with $1 < r < n$ and $1 < s < n$.

In symbols: For all integers n with $n > 1$,

$$n \text{ is prime} \quad \Leftrightarrow \quad \forall \text{ positive integers } r \text{ and } s, \text{ if } n = rs$$
$$\text{then either } r = 1 \text{ and } s = n \text{ or } r = n \text{ and } s = 1.$$

$$n \text{ is composite} \quad \Leftrightarrow \quad \exists \text{ positive integers } r \text{ and } s \text{ such that } n = rs$$
$$\text{and } 1 < r < n \text{ and } 1 < s < n.$$

Example 4.1.2 Prime and Composite Numbers

a. Is 1 prime?

b. Is every integer greater than 1 either prime or composite?

c. Write the first six prime numbers.

d. Write the first six composite numbers.

Solution

Note The reason for not allowing 1 to be prime is discussed in Section 4.3.

a. No. A prime number is required to be greater than 1.

b. Yes. Let n be any integer that is greater than 1. Consider all pairs of positive integers r and s such that $n = rs$. There exist at least two such pairs, namely $r = n$ and $s = 1$ and $r = 1$ and $s = n$. Moreover, since $n = rs$, all such pairs satisfy the inequalities $1 \le r \le n$ and $1 \le s \le n$. If n is prime, then the two displayed pairs are the only ways to write n as rs. Otherwise, there exists a pair of positive integers r and s such that $n = rs$ and neither r nor s equals either 1 or n. Therefore, in this case $1 < r < n$ and $1 < s < n$, and hence n is composite.

c. 2, 3, 5, 7, 11, 13

d. 4, 6, 8, 9, 10, 12 ∎

Proving Existential Statements

According to the definition given in Section 3.1, a statement in the form

$$\exists x \in D \text{ such that } Q(x)$$

is true if, and only if,

$$Q(x) \text{ is true for at least one } x \text{ in } D.$$

One way to prove this is to find an x in D that makes $Q(x)$ true. Another way is to give a set of directions for finding such an x. Both of these methods are called **constructive proofs of existence.**

Example 4.1.3 Constructive Proofs of Existence

    a. Prove the following: ∃ an even integer n that can be written in two ways as a sum of two prime numbers.

    b. Suppose that r and s are integers. Prove the following: ∃ an integer k such that $22r + 18s = 2k$.

Solution

    a. Let $n = 10$. Then $10 = 5 + 5 = 3 + 7$ and 3, 5, and 7 are all prime numbers.

    b. Let $k = 11r + 9s$. Then k is an integer because it is a sum of products of integers; and by substitution, $2k = 2(11r + 9s)$, which equals $22r + 18s$ by the distributive law of algebra. ∎

    A **nonconstructive proof of existence** involves showing either (a) that the existence of a value of x that makes $Q(x)$ true is guaranteed by an axiom or a previously proved theorem or (b) that the assumption that there is no such x leads to a contradiction. The disadvantage of a nonconstructive proof is that it may give virtually no clue about where or how x may be found. The widespread use of digital computers in recent years has led to some dissatisfaction with this aspect of nonconstructive proofs and to increased efforts to produce constructive proofs containing directions for computer calculation of the quantity in question.

Disproving Universal Statements by Counterexample

To disprove a statement means to show that it is false. Consider the question of disproving a statement of the form

$$\forall x \text{ in } D, \text{ if } P(x) \text{ then } Q(x).$$

Showing that this statement is false is equivalent to showing that its negation is true. The negation of the statement is existential:

$$\exists x \text{ in } D \text{ such that } P(x) \text{ and not } Q(x).$$

But to show that an existential statement is true, we generally give an example, and because the example is used to show that the original statement is false, we call it a *counterexample*. Thus the method of disproof by *counterexample* can be written as follows:

Disproof by Counterexample

To disprove a statement of the form "$\forall x \in D$, if $P(x)$ then $Q(x)$," find a value of x in D for which the hypothesis $P(x)$ is true and the conclusion $Q(x)$ is false. Such an x is called a **counterexample.**

Example 4.1.4 Disproof by Counterexample

Disprove the following statement by finding a counterexample:

$$\forall \text{ real numbers } a \text{ and } b, \text{ if } a^2 = b^2 \text{ then } a = b.$$

Solution   To disprove this statement, you need to find real numbers a and b such that the hypothesis $a^2 = b^2$ is true and the conclusion $a = b$ is false. The fact that both positive and negative integers have positive squares helps in the search. If you flip through some possibilities in your mind, you will quickly see that 1 and −1 will work (or 2 and −2, or 0.5 and −0.5, and so forth).

Statement: $\forall$ real numbers a and b, if $a^2 = b^2$, then $a = b$.

Counterexample: Let $a = 1$ and $b = -1$. Then $a^2 = 1^2 = 1$ and $b^2 = (-1)^2 = 1$, and so $a^2 = b^2$. But $a \neq b$ since $1 \neq -1$.

It is a sign of intelligence to make generalizations. Frequently, after observing a property to hold in a large number of cases, you may guess that it holds in all cases. You may, however, run into difficulty when you try to prove your guess. Perhaps you just have not figured out the key to the proof. But perhaps your guess is false. Consequently, when you are having serious difficulty proving a general statement, you should interrupt your efforts to look for a counterexample. Analyzing the kinds of problems you are encountering in your proof efforts may help in the search. It may even happen that if you find a counterexample and therefore prove the statement false, your understanding may be sufficiently clarified that you can formulate a more limited but true version of the statement. For instance, Example 4.1.4 shows that it is not always true that if the squares of two numbers are equal, then the numbers are equal. However, it is true that if the squares of two *positive* numbers are equal, then the numbers are equal.

Proving Universal Statements

The vast majority of mathematical statements to be proved are universal. In discussing how to prove such statements, it is helpful to imagine them in a standard form:

$$\forall x \in D, \text{ if } P(x) \text{ then } Q(x).$$

Sections 1.1 and 3.1 give examples showing how to write any universal statement in this form. When D is finite or when only a finite number of elements satisfy $P(x)$, such a statement can be proved by the method of exhaustion.

Example 4.1.5 The Method of Exhaustion

Use the method of exhaustion to prove the following statement:

$\forall n \in \mathbf{Z}$, if n is even and $4 \leq n \leq 26$, then n can be written as a sum of two prime numbers.

Solution

$4 = 2 + 2$	$6 = 3 + 3$	$8 = 3 + 5$	$10 = 5 + 5$
$12 = 5 + 7$	$14 = 11 + 3$	$16 = 5 + 11$	$18 = 7 + 11$
$20 = 7 + 13$	$22 = 5 + 17$	$24 = 5 + 19$	$26 = 7 + 19$

In most cases in mathematics, however, the method of exhaustion cannot be used. For instance, can you prove by exhaustion that *every* even integer greater than 2 can be written as a sum of two prime numbers? No. To do that you would have to check every even integer, and because there are infinitely many such numbers, this is an impossible task.

Even when the domain is finite, it may be infeasible to use the method of exhaustion. Imagine, for example, trying to check by exhaustion that the multiplication circuitry of a particular computer gives the correct result for every pair of numbers in the computer's range. Since a typical computer would require thousands of years just to compute all possible products of all numbers in its range (not to mention the time it would take to check the accuracy of the answers), checking correctness by the method of exhaustion is obviously impractical.

The most powerful technique for proving a universal statement is one that works regardless of the size of the domain over which the statement is quantified. It is called the *method of generalizing from the generic particular*. Here is the idea underlying the method:

Method of Generalizing from the Generic Particular

To show that every element of a set satisfies a certain property, suppose x is a *particular* but *arbitrarily chosen* element of the set, and show that x satisfies the property.

Example 4.1.6 Generalizing from the Generic Particular

At some time you may have been shown a "mathematical trick" like the following. You ask a person to pick any number, add 5, multiply by 4, subtract 6, divide by 2, and subtract twice the original number. Then you astound the person by announcing that their final result was 7. How does this "trick" work? Let an empty box $\square$ or the symbol x stand for the number the person picks. Here is what happens when the person follows your directions:

Step	Visual Result	Algebraic Result
Pick a number.	$\square$	x
Add 5.	$\square\mid\mid\mid\mid\mid$	$x + 5$
Multiply by 4.	$\square\mid\mid\mid\mid\mid$ $\square\mid\mid\mid\mid\mid$ $\square\mid\mid\mid\mid\mid$ $\square\mid\mid\mid\mid\mid$	$(x + 5) \cdot 4 = 4x + 20$
Subtract 6.	$\square\mid\mid$ $\square\mid\mid$ $\square\mid\mid\mid\mid\mid$ $\square\mid\mid\mid\mid\mid$	$(4x + 20) - 6 = 4x + 14$
Divide by 2.	$\square\mid\mid$ $\square\mid\mid\mid\mid\mid$	$\dfrac{4x + 14}{2} = 2x + 7$
Subtract twice the original number.	$\mid\mid$ $\mid\mid\mid\mid\mid$	$(2x + 7) - 2x = 7$

Thus no matter what number the person starts with, the result will always be 7. Note that the x in the analysis above is *particular* (because it represents a single quantity), but it is also *arbitrarily chosen* or *generic* (because any number whatsoever can be put in its place). This illustrates the process of drawing a general conclusion from a particular but generic object. ∎

The point of having x be arbitrarily chosen (or generic) is to make a proof that can be generalized to all elements of the domain. By choosing x arbitrarily, you are making no special assumptions about x that are not also true of all other elements of the domain. The word *generic* means "sharing all the common characteristics of a group or class." Thus everything you deduce about a generic element x of the domain is equally true of any other element of the domain.

When the method of generalizing from the generic particular is applied to a property of the form "If $P(x)$ then $Q(x)$," the result is the method of *direct proof*. Recall that the only way an if-then statement can be false is for the hypothesis to be true and the conclusion to be false. Thus, given the statement "If $P(x)$ then $Q(x)$," if you can show that the truth of $P(x)$ compels the truth of $Q(x)$, then you will have proved the statement. It follows by the method of generalizing from the generic particular that to show that "$\forall x$, if $P(x)$ then $Q(x)$," is true for *all* elements x in a set D, you suppose x is a particular but arbitrarily chosen element of D that makes $P(x)$ true, and then you show that x makes $Q(x)$ true.

Method of Direct Proof

1. Express the statement to be proved in the form "$\forall x \in D$, if $P(x)$ then $Q(x)$." (This step is often done mentally.)

2. Start the proof by supposing x is a particular but arbitrarily chosen element of D for which the hypothesis $P(x)$ is true. (This step is often abbreviated "Suppose $x \in D$ and $P(x)$.")

3. Show that the conclusion $Q(x)$ is true by using definitions, previously established results, and the rules for logical inference.

Example 4.1.7 A Direct Proof of a Theorem

Prove that the sum of any two even integers is even.

Caution! The word *two* in this statement does not necessarily refer to two distinct integers. If a choice of integers is made arbitrarily, the integers are very likely to be distinct, but they might be the same.

Solution Whenever you are presented with a statement to be proved, it is a good idea to ask yourself whether you believe it to be true. In this case you might imagine some pairs of even integers, say $2 + 4, 6 + 10, 12 + 12, 28 + 54$, and mentally check that their sums are even. However, since you cannot possibly check all pairs of even numbers, you cannot know for sure that the statement is true in general by checking its truth in these particular instances. Many properties hold for a large number of examples and yet fail to be true in general.

To prove this statement in general, you need to show that no matter what even integers are given, their sum is even. But given any two even integers, it is possible to represent them as $2r$ and $2s$ for some integers r and s. And by the distributive law of algebra, $2r + 2s = 2(r + s)$, which is even. Thus the statement is true in general.

Suppose the statement to be proved were much more complicated than this. What is the method you could use to derive a proof?

Formal Restatement: $\forall$ integers m and n, if m and n are even then $m + n$ is even.

This statement is universally quantified over an infinite domain. Thus to prove it in general, you need to show that no matter what two integers you might be given, if both of them are even then their sum will also be even.

Next ask yourself, "Where am I starting from?" or "What am I supposing?" The answer to such a question gives you the starting point, or first sentence, of the proof.

Starting Point: Suppose m and n are particular but arbitrarily chosen integers that are even.

Or, in abbreviated form:

Suppose m and n are any even integers.

Then ask yourself, "What conclusion do I need to show in order to complete the proof?"

To Show: $m + n$ is even.

At this point you need to ask yourself, "How do I get from the starting point to the conclusion?" Since both involve the term *even integer,* you must use the definition of this term—and thus you must know what it means for an integer to be even. It follows from the definition that since m and n are even, each equals twice some integer. One of the basic laws of logic, called *existential instantiation*, says, in effect, that if you know something exists, you can give it a name. However, you cannot use the same name to refer to two different things, both of which are currently under discussion.

Existential Instantiation

If the existence of a certain kind of object is assumed or has been deduced then it can be given a name, as long as that name is not currently being used to denote something else.

Caution! Because m and n are arbitrarily chosen, they could be any pair of even integers whatsoever. Once r is introduced to satisfy $m = 2r$, then r is not available to represent something else. If you had set $m = 2r$, and $n = 2r$, then m would equal n, which need not be the case.

Thus since m equals twice some integer, you can give that integer a name, and since n equals twice some integer, you can also give that integer a name:

$$m = 2r, \text{ for some integer } r \quad \text{and} \quad n = 2s, \text{ for some integer } s.$$

Now what you want to show is that $m + n$ is even. In other words, you want to show that $m + n$ equals $2 \cdot$(some integer). Having just found alternative representations for m (as $2r$) and n (as $2s$), it seems reasonable to substitute these representations in place of m and n:

$$m + n = 2r + 2s.$$

Your goal is to show that $m + n$ is even. By definition of even, this means that $m + n$ can be written in the form

$$2 \cdot \text{(some integer)}.$$

This analysis narrows the gap between the starting point and what is to be shown to showing that

$$2r + 2s = 2 \cdot \text{(some integer)}.$$

Why is this true? First, because of the distributive law from algebra, which says that

$$2r + 2s = 2(r + s),$$

and, second, because the sum of any two integers is an integer, which implies that $r + s$ is an integer.

This discussion is summarized by rewriting the statement as a theorem and giving a formal proof of it. (In mathematics, the word *theorem* refers to a statement that is known to be true because it has been proved.) The formal proof, as well as many others in this text, includes explanatory notes to make its logical flow apparent. Such comments are purely a convenience for the reader and could be omitted entirely. For this reason they are italicized and enclosed in italic square brackets: *[ ]*.

Courtesy of Donald Knuth

Donald Knuth
(born 1938)

Donald Knuth, one of the pioneers of the science of computing, has compared constructing a computer program from a set of specifications to writing a mathematical proof based on a set of axioms.* In keeping with this analogy, the bracketed comments can be thought of as similar to the explanatory documentation provided by a good programmer. Documentation is not necessary for a program to run, but it helps a human reader understand what is going on.

Theorem 4.1.1

The sum of any two even integers is even.

Proof:

Suppose m and n are *[particular but arbitrarily chosen]* even integers. *[We must show that $m + n$ is even.]* By definition of even, $m = 2r$ and $n = 2s$ for some integers r and s. Then

$$m + n = 2r + 2s \qquad \text{by substitution}$$
$$= 2(r + s) \qquad \text{by factoring out a 2.}$$

Let $t = r + s$. Note that t is an integer because it is a sum of integers. Hence

$$m + n = 2t \quad \text{where } t \text{ is an integer.}$$

It follows by definition of even that $m + n$ is even. *[This is what we needed to show.]*[†]

Note Introducing t to equal $r + s$ is another use of existential instantiation.

Most theorems, like the one above, can be analyzed to a point where you realize that as soon as a certain thing is shown, the theorem will be proved. When that thing has been shown, it is natural to end the proof with the words "this is what we needed to show." The Latin words for this are *quod erat demonstrandum,* or Q.E.D. for short. Proofs in older mathematics books end with these initials.

Note that both the *if* and the *only if* parts of the definition of even were used in the proof of Theorem 4.1.1. Since m and n were known to be even, the *only if* ($\Rightarrow$) part of the definition was used to deduce that m and n had a certain general form. Then, after some algebraic substitution and manipulation, the *if* ($\Leftarrow$) part of the definition was used to deduce that $m + n$ was even.

Directions for Writing Proofs of Universal Statements

Think of a proof as a way to communicate a convincing argument for the truth of a mathematical statement. When you write a proof, imagine that you will be sending it to a capable classmate who has had to miss the last week or two of your course. Try to be clear and complete. Keep in mind that your classmate will see only what you actually write down, not any unexpressed thoughts behind it. Ideally, your proof will lead your classmate to understand *why* the given statement is true.

---

*Donald E. Knuth, *The Art of Computer Programming,* 2nd ed., Vol. I (Reading, MA: Addison-Wesley, 1973), p. ix.

[†]See page 97 for a discussion of the role of universal modus ponens in this proof.

Over the years, the following rules of style have become fairly standard for writing the final versions of proofs:

1. **Copy the statement of the theorem to be proved on your paper.**

2. **Clearly mark the beginning of your proof with the word <u>Proof</u>.**

3. **Make your proof self-contained.**

    This means that you should explain the meaning of each variable used in your proof in the body of the proof. Thus you will begin proofs by introducing the initial variables and stating what kind of objects they are. The first sentence of your proof would be something like "Suppose m and n are any even integers" or "Let x be a real number such that x is greater than 2." This is similar to declaring variables and their data types at the beginning of a computer program.

    At a later point in your proof, you may introduce a new variable to represent a quantity that is known at that point to exist. For example, if you have assumed that a particular integer n is even, then you know that n equals 2 times some integer, and you can give this integer a name so that you can work with it concretely later in the proof. Thus if you decide to call the integer, say, s, you would write, "Since n is even, $n = 2s$ for some integer s," or "since n is even, there exists an integer s such that $n = 2s$."

4. **Write your proof in complete, gramatically correct sentences.**

    This does not mean that you should avoid using symbols and shorthand abbreviations, just that you should incorporate them into sentences. For example, the proof of Theorem 4.1.1 contains the sentence

    $$\text{Then } m + n = 2r + 2s$$
    $$= 2(r + s).$$

    To read such text as a sentence, read the first equals sign as "equals" and each subsequent equals sign as "which equals."

5. **Keep your reader informed about the status of each statement in your proof.**

    Your reader should never be in doubt about whether something in your proof has been assumed or established or is still to be deduced. If something is assumed, preface it with a word like *Suppose* or *Assume*. If it is still to be shown, preface it with words like, *We must show that* or *In other words, we must show that*. This is especially important if you introduce a variable in rephrasing what you need to show. (See Common Mistakes on the next page.)

6. **Give a reason for each assertion in your proof.**

    Each assertion in a proof should come directly from the hypothesis of the theorem, or follow from the definition of one of the terms in the theorem, or be a result obtained earlier in the proof, or be a mathematical result that has previously been established or is agreed to be assumed. Indicate the reason for each step of your proof using phrases such as *by hypothesis, by definition of . . .* , and *by theorem . . .* .

7. **Include the "little words and phrases" that make the logic of your arguments clear.**

    When writing a mathematical argument, especially a proof, indicate how each sentence is related to the previous one. Does it follow from the previous sentence or from a combination of the previous sentence and earlier ones? If so, start the sentence by stating the reason why it follows or by writing *Then,* or *Thus,* or *So,* or *Hence,* or *Therefore,* or *Consequently*, or *It follows that,* and include the reason at the end of the sentence. For instance, in the proof of Theorem 4.1.1, once you know that m is even, you can write: "By definition of even, $m = 2r$ for some integer r," or you can write, "Then $m = 2r$ for some integer r by definition of even."

If a sentence expresses a new thought or fact that does not follow as an immediate consequence of the preceding statement but is needed for a later part of a proof, introduce it by writing *Observe that,* or *Note that,* or *But,* or *Now.*

Sometimes in a proof it is desirable to define a new variable in terms of previous variables. In such a case, introduce the new variable with the word *Let.* For instance, in the proof of Theorem 4.1.1, once it is known that $m + n = 2(r + s)$, where r and s are integers, a new variable t is introduced to represent $r + s$. The proof goes on to say, "Let $t = r + s$. Then t is an integer because it is a sum of two integers."

8. **Display equations and inequalities.**

The convention is to display equations and inequalities on separate lines to increase readability, both for other people and for ourselves so that we can more easily check our work for accuracy. We follow the convention in the text of this book, but in order to save space, we violate it in a few of the exercises and in many of the solutions contained in Appendix B. So you may need to copy out some parts of solutions on scratch paper to understand them fully. Please follow the convention in your own work. Leave plenty of empty space, and don't be stingy with paper!

Variations among Proofs

It is rare that two proofs of a given statement, written by two different people, are identical. Even when the basic mathematical steps are the same, the two people may use different notation or may give differing amounts of explanation for their steps, or may choose different words to link the steps together into paragraph form. An important question is how detailed to make the explanations for the steps of a proof. This must ultimately be worked out between the writer of a proof and the intended reader, whether they be student and teacher, teacher and student, student and fellow student, or mathematician and colleague. Your teacher may provide explicit guidelines for you to use in your course. Or you may follow the example of the proofs in this book (which are generally explained rather fully in order to be understood by students at various stages of mathematical development). Remember that the phrases written inside brackets [ ] are intended to elucidate the logical flow or underlying assumptions of the proof and need not be written down at all. It is entirely your decision whether to include such phrases in your own proofs.

Common Mistakes

The following are some of the most common mistakes people make when writing mathematical proofs.

1. **Arguing from examples.**

Looking at examples is one of the most helpful practices a problem solver can engage in and is encouraged by all good mathematics teachers. However, it is a mistake to think that a general statement can be proved by showing it to be true for some special cases. A property referred to in a universal statement may be true in many instances without being true in general.

Here is an example of this mistake. It is an incorrect "proof" of the fact that the sum of any two even integers is even. (Theorem 4.1.1).

This is true because if $m = 14$ and $n = 6$, which are both even,
then $m + n = 20$, which is also even.

Some people find this kind of argument convincing because it does, after all, consist of evidence in support of a true conclusion. But remember that when we discussed valid arguments, we pointed out that an argument may be invalid and yet have a true conclusion. In the same way, an argument from examples may be mistakenly used to "prove" a true statement. In the previous example, it is not sufficient to show that the conclusion "$m + n$ is even" is true for $m = 14$ and $n = 6$. You must give an argument to show that the conclusion is true for any even integers m and n.

2. **Using the same letter to mean two different things.**

 Some beginning theorem provers give a new variable quantity the same letter name as a previously introduced variable. Consider the following "proof" fragment:

> Suppose m and n are any odd integers. Then by definition of odd, $m = 2k + 1$ and $n = 2k + 1$ for some integer k.

 This is incorrect. Using the same symbol, k, in the expressions for both m and n implies that $m = 2k + 1 = n$. It follows that the rest of the proof applies only to integers m and n that equal each other. This is inconsistent with the supposition that m and n are arbitrarily chosen odd integers. For instance, the proof would not show that the sum of 3 and 5 is even.

3. **Jumping to a conclusion.**

 To jump to a conclusion means to allege the truth of something without giving an adequate reason. Consider the following "proof" that the sum of any two even integers is even.

> Suppose m and n are any even integers. By definition of even, $m = 2r$ and $n = 2s$ for some integers r and s. Then $m + n = 2r + 2s$. So $m + n$ is even.

 The problem with this "proof" is that the crucial calculation

$$2r + 2s = 2(r + s)$$

 is missing. The author of the "proof" has jumped prematurely to a conclusion.

4. **Assuming what is to be proved.**

 To assume what is to be proved is a variation of jumping to a conclusion. As an example, consider the following "proof" of the fact that the product of any two odd integers is odd:

> Suppose m and n are any odd integers. When any odd integers are multiplied, their product is odd. Hence mn is odd.

5. **Confusion between what is known and what is still to be shown**.

 A more subtle way to engage in circular reasoning occurs when the conclusion to be shown is restated using a variable. Here is an example in a "proof" that the product of any two odd integers is odd:

 Suppose m and n are any odd integers. We must show that mn is odd. This means that there exists an integer s such that

$$mn = 2s + 1.$$

 Also by definition of odd, there exist integers a and b such that

$$m = 2a + 1 \text{ and } n = 2b + 1.$$

 Then

$$mn = (2a + 1)(2b + 1) = 2s + 1.$$

 So, since s is an integer, mn is odd by definition of odd.

 In this example, when the author restated the conclusion to be shown (that mn is odd), the author wrote "there exists an integer s such that $mn = 2s + 1$." Later the author jumped to an unjustified conclusion by assuming the existence of this s when that had not, in fact, been established. This mistake might have been avoided if the author had written "This means that we must show that there exists an integer s such that

$$mn = 2s + 1.$$

An even better way to avoid this kind of error is not to introduce a variable into a proof unless it is either part of the hypothesis or deducible from it.

6. **Use of *any* rather than *some*.**

There are a few situations in which the words *any* and *some* can be used interchangeably. For instance, in starting a proof that the square of any odd integer is odd, one could correctly write "Suppose m is any odd integer" or "Suppose m is some odd integer." In most situations, however, the words *any* and *some* are not interchangeable. Here is the start of a "proof" that the square of any odd integer is odd, which uses *any* when the correct word is *some*:

Suppose m is a particular but arbitrarily chosen odd integer.
By definition of odd, $m = 2a + 1$ for any integer a.

In the second sentence it is incorrect to say that "$m = 2a + 1$ for any integer a" because a cannot be just "any" integer; in fact, solving $m = 2a + 1$ for a shows that the only possible value for a is $(m - 1)/2$. The correct way to finish the second sentence is, "$m = 2a + 1$ for some integer a" or "there exists an integer a such that $m = 2a + 1$."

7. **Misuse of the word *if*.**

Another common error is not serious in itself, but it reflects imprecise thinking that sometimes leads to problems later in a proof. This error involves using the word *if* when the word *because* is really meant. Consider the following proof fragment:

Suppose p is a prime number. If p is prime, then p cannot be
written as a product of two smaller positive integers.

The use of the word *if* in the second sentence is inappropriate. It suggests that the primeness of p is in doubt. But p is known to be prime by the first sentence. It cannot be written as a product of two smaller positive integers *because* it is prime. Here is a correct version of the fragment:

Suppose p is a prime number. Because p is prime, p cannot be
written as a product of two smaller positive integers.

Getting Proofs Started

Believe it or not, once you understand the idea of generalizing from the generic particular and the method of direct proof, you can write the beginnings of proofs even for theorems you do not understand. The reason is that the starting point and what is to be shown in a proof depend only on the linguistic form of the statement to be proved, not on the content of the statement.

Example 4.1.8 Identifying the "Starting Point" and the "Conclusion to Be Shown"

Note You are not expected to know anything about complete, bipartite graphs.

Write the first sentence of a proof (the "starting point") and the last sentence of a proof (the "conclusion to be shown") for the following statement:

Every complete, bipartite graph is connected.

Solution   It is helpful to rewrite the statement formally using a quantifier and a variable:

$$\overbrace{\phantom{\forall \text{ graphs } G,}}^{\text{domain}} \quad \overbrace{\phantom{\text{if } G \text{ is complete and bipartite,}}}^{\text{hypothesis}} \quad \overbrace{\phantom{\text{then } G \text{ is connected.}}}^{\text{conclusion}}$$

Formal Restatement: $\forall$ graphs G, if G is complete and bipartite, then G is connected.

The first sentence, or starting point, of a proof supposes the existence of an object (in this case G) in the domain (in this case the set of all graphs) that satisfies the hypothesis of the if-then part of the statement (in this case that G is complete and bipartite). The conclusion to be shown is just the conclusion of the if-then part of the statement (in this case that G is connected).

Starting Point: Suppose G is a *[particular but arbitrarily chosen]* graph such that G is complete and bipartite.

Conclusion to Be Shown: G is connected.

Thus the proof has the following shape:

Proof:

Suppose G is a *[particular but arbitrarily chosen]* graph such that G is complete and bipartite.

$\vdots$

Therefore, G is connected.   ∎

Showing That an Existential Statement Is False

Recall that the negation of an existential statement is universal. It follows that to prove an existential statement is false, you must prove a universal statement (its negation) is true.

Example 4.1.9 Disproving an Existential Statement

Show that the following statement is false:

There is a positive integer n such that $n^2 + 3n + 2$ is prime.

Solution   Proving that the given statement is false is equivalent to proving its negation is true. The negation is

For all positive integers n, $n^2 + 3n + 2$ is not prime.

Because the negation is universal, it is proved by generalizing from the generic particular.

Claim: The statement "There is a positive integer n such that $n^2 + 3n + 2$ is prime" is false.

Proof:

Suppose n is any *[particular but arbitrarily chosen]* positive integer. *[We will show that $n^2 + 3n + 2$ is not prime.]* We can factor $n^2 + 3n + 2$ to obtain $n^2 + 3n + 2 = (n + 1)(n + 2)$. We also note that $n + 1$ and $n + 2$ are integers (because they are sums of integers) and that both $n + 1 > 1$ and $n + 2 > 1$ (because $n \geq 1$). Thus $n^2 + 3n + 2$ is a product of two integers each greater than 1, and so $n^2 + 3n + 2$ is not prime.   ∎

Conjecture, Proof, and Disproof

More than 350 years ago, the French mathematician Pierre de Fermat claimed that it is impossible to find positive integers x, y, and z with $x^n + y^n = z^n$ if n is an integer that is at least 3. (For $n = 2$, the equation has many integer solutions, such as $3^2 + 4^2 = 5^2$ and $5^2 + 12^2 = 13^2$.) Fermat wrote his claim in the margin of a book, along with the comment "I have discovered a truly remarkable PROOF of this theorem which this margin

Pierre de Fermat
(1601–1665)

Andrew Wiles
(born 1953)

is too small to contain." No proof, however, was found among his papers, and over the years some of the greatest mathematical minds tried and failed to discover a proof or a counterexample, for what came to be known as Fermat's last theorem.

In 1986 Kenneth Ribet of the University of California at Berkeley showed that if a certain other statement, the Taniyama–Shimura conjecture, could be proved, then Fermat's theorem would follow. Andrew Wiles, an English mathematician and faculty member at Princeton University, had become intrigued by Fermat's claim while still a child and, as an adult, had come to work in the branch of mathematics to which the Taniyama–Shimura conjecture belonged. As soon as he heard of Ribet's result, Wiles immediately set to work to prove the conjecture. In June of 1993, after 7 years of concentrated effort, he presented a proof to worldwide acclaim.

During the summer of 1993, however, while every part of the proof was being carefully checked to prepare for formal publication, Wiles found that he could not justify one step and that that step might actually be wrong. He worked unceasingly for another year to resolve the problem, finally realizing that the gap in the proof was a genuine error but that an approach he had worked on years earlier and abandoned provided a way around the difficulty. By the end of 1994, the revised proof had been thoroughly checked and pronounced correct in every detail by experts in the field. It was published in the *Annals of Mathematics* in 1995. Several books and an excellent documentary television show have been produced that convey the drama and excitement of Wiles's discovery.*

One of the oldest problems in mathematics that remains unsolved is the Goldbach conjecture. In Example 4.1.5 it was shown that every even integer from 4 to 26 can be represented as a sum of two prime numbers. More than 250 years ago, Christian Goldbach (1690–1764) conjectured that every even integer greater than 2 can be so represented. Explicit computer-aided calculations have shown the conjecture to be true up to at least 10^{18}. But there is a huge chasm between 10^{18} and infinity. As pointed out by James Gleick of the *New York Times*, many other plausible conjectures in number theory have proved false. Leonhard Euler (1707–1783), for example, proposed in the eighteenth century that $a^4 + b^4 + c^4 = d^4$ had no nontrivial whole number solutions. In other words, no three perfect fourth powers add up to another perfect fourth power. For small numbers, Euler's conjecture looked good. But in 1987 a Harvard mathematician, Noam Elkies, proved it wrong. One counterexample, found by Roger Frye of Thinking Machines Corporation in a long computer search, is $95{,}800^4 + 217{,}519^4 + 414{,}560^4 = 422{,}481^4$.[†]

In May 2000, "to celebrate mathematics in the new millennium," the Clay Mathematics Institute of Cambridge, Massachusetts, announced that it would award prizes of $1 million each for the solutions to seven longstanding, classical mathematical questions. One of them, "P vs. NP," asks whether problems belonging to a certain class can be solved on a computer using more efficient methods than the very inefficient methods that are presently known to work for them.

Test Yourself

Answers to Test Yourself questions are located at the end of each section.

1. An integer is even if, and only if, _____.

2. An integer is odd if, and only if, _____.

3. An integer n is prime if, and only if, _____.

4. The most common way to disprove a universal statement is to find _____.

*"The Proof," produced in 1997, for the series *Nova* on the Public Broadcasting System; *Fermat's Enigma: The Epic Quest to Solve the World's Greatest Mathematical Problem,* by Simon Singh and John Lynch (New York: Bantam Books, 1998); *Fermat's Last Theorem: Unlocking the Secret of an Ancient Mathematical Problem* by Amir D. Aczel (New York: Delacorte Press, 1997).
[†]James Gleick, "Fermat's Last Theorem Still Has Zero Solutions," *New York Times,* 17 April 1988.

5. According to the method of generalizing from the generic particular, to show that every element of a set satisfies a certain property, suppose x is a _____, and show that _____.

6. To use the method of direct proof to prove a statement of the form, "For all x in a set D, if $P(x)$ then $Q(x)$," one supposes that _____ and one shows that _____.

Exercise Set 4.1*

In 1–3, use the definitions of even, odd, prime, and composite to justify each of your answers.

1. Assume that k is a particular integer.
   a. Is -17 an odd integer?   b. Is 0 an even integer?
   c. Is $2k - 1$ odd?

2. Assume that m and n are particular integers.
   a. Is $6m + 8n$ even?   b. Is $10mn + 7$ odd?
   c. If $m > n > 0$, is $m^2 - n^2$ composite?

3. Assume that r and s are particular integers.
   a. Is $4rs$ even?   b. Is $6r + 4s^2 + 3$ odd?
   c. If r and s are both positive, is $r^2 + 2rs + s^2$ composite?

Prove the statements in 4–10.

4. There are integers m and n such that $m > 1$ and $n > 1$ and $\frac{1}{m} + \frac{1}{n}$ is an integer.

5. There are distinct integers m and n such that $\frac{1}{m} + \frac{1}{n}$ is an integer.

6. There are real numbers a and b such that
$$\sqrt{a + b} = \sqrt{a} + \sqrt{b}.$$

7. There is an integer $n > 5$ such that $2^n - 1$ is prime.

8. There is a real number x such that $x > 1$ and $2^x > x^{10}$.

> **Definition:** An integer n is called a **perfect square** if, and only if, $n = k^2$ for some integer k.

9. There is a perfect square that can be written as a sum of two other perfect squares.

10. There is an integer n such that $2n^2 - 5n + 2$ is prime.

Disprove the statements in 11–13 by giving a counterexample.

11. For all real numbers a and b, if $a < b$ then $a^2 < b^2$.

12. For all integers n, if n is odd then $\frac{n-1}{2}$ is odd.

13. For all integers m and n, if $2m + n$ is odd then m and n are both odd.

In 14–16, determine whether the property is true for all integers, true for no integers, or true for some integers and false for other integers. Justify your answers.

14. $(a + b)^2 = a^2 + b^2$ **H** 15. $-a^n = (-a)^n$

16. The average of any two odd integers is odd.

Prove the statements in 17 and 18 by the method of exhaustion.

17. Every positive even integer less than 26 can be expressed as a sum of three or fewer perfect squares. (For instance, $10 = 1^2 + 3^2$ and $16 = 4^2$.)

18. For each integer n with $1 \le n \le 10$, $n^2 - n + 11$ is a prime number.

19. a. Rewrite the following theorem in three different ways: as $\forall$ _____, if _____ then _____, as $\forall$ _____ , _____ (without using the words *if* or *then*), and as If _____, then _____ (without using an explicit universal quantifier).

    b. Fill in the blanks in the proof of the theorem.

    Theorem: The sum of any even integer and any odd integer is odd.

    Proof: Suppose m is any even integer and n is (a) . By definition of even, $m = 2r$ for some (b) , and by definition of odd, $n = 2s + 1$ for some integer s. By substitution and algebra,
$$m + n = \underline{\text{(c)}} = 2(r + s) + 1.$$
    Since r and s are both integers, so is their sum $r + s$. Hence $m + n$ has the form twice some integer plus one, and so (d) by definition of odd.

Each of the statements in 20–23 is true. For each, (a) rewrite the statement with the quantification implicit as If _____, then _____, and (b) write the first sentence of a proof (the "starting point") and the last sentence of a proof (the "conclusion to be shown"). Note that you do not need to understand the statements in order to be able to do these exercises.

20. For all integers m, if $m > 1$ then $0 < \frac{1}{m} < 1$.

21. For all real numbers x, if $x > 1$ then $x^2 > x$.

22. For all integers m and n, if $mn = 1$ then $m = n = 1$ or $m = n = -1$.

23. For all real numbers x, if $0 < x < 1$ then $x^2 < x$.

*For exercises with blue numbers, solutions are given in Appendix B. The symbol **H** indicates that only a hint or partial solution is given. The symbol ✶ signals that an exercise is more challenging than usual.

Prove the statements in 24–34. In each case use only the definitions of the terms and the Assumptions listed on page 110, not any previously established properties of odd and even integers. Follow the directions given in this section for writing proofs of universal statements.

24. The negative of any even integer is even.

25. The difference of any even integer minus any odd integer is odd.

H **26.** The difference between any odd integer and any even integer is odd. (*Note:* The "proof" shown in exercise 39 contains an error. Can you spot it?)

27. The sum of any two odd integers is even.

28. For all integers n, if n is odd then n^2 is odd.

29. For all integers n, if n is odd then $3n + 5$ is even.

30. For all integers m, if m is even then $3m + 5$ is odd.

31. If k is any odd integer and m is any even integer, then, $k^2 + m^2$ is odd.

32. If a is any odd integer and b is any even integer, then, $2a + 3b$ is even.

33. If n is any even integer, then $(-1)^n = 1$.

34. If n is any odd integer, then $(-1)^n = -1$.

Prove that the statements in 35–37 are false.

35. There exists an integer $m \geq 3$ such that $m^2 - 1$ is prime.

36. There exists an integer n such that $6n^2 + 27$ is prime.

37. There exists an integer $k \geq 4$ such that $2k^2 - 5k + 2$ is prime.

Find the mistakes in the "proofs" shown in 38–42.

38. Theorem: For all integers k, if $k > 0$ then $k^2 + 2k + 1$ is composite.
"**Proof:** For $k = 2$, $k^2 + 2k + 1 = 2^2 + 2 \cdot 2 + 1 = 9$. But $9 = 3 \cdot 3$, and so 9 is composite. Hence the theorem is true."

39. Theorem: The difference between any odd integer and any even integer is odd.
"**Proof:** Suppose n is any odd integer, and m is any even integer. By definition of odd, $n = 2k + 1$ where k is an integer, and by definition of even, $m = 2k$ where k is an integer. Then

$$n - m = (2k + 1) - 2k = 1.$$

But 1 is odd. Therefore, the difference between any odd integer and any even integer is odd."

40. Theorem: For all integers k, if $k > 0$ then $k^2 + 2k + 1$ is composite.
"**Proof:** Suppose k is any integer such that $k > 0$. If $k^2 + 2k + 1$ is composite, then $k^2 + 2k + 1 = rs$ for some integers r and s such that

$$1 < r < (k^2 + 2k + 1)$$

and
$$1 < s < (k^2 + 2k + 1).$$

Since
$$k^2 + 2k + 1 = rs$$

and both r and s are strictly between 1 and $k^2 + 2k + 1$, then $k^2 + 2k + 1$ is not prime. Hence $k^2 + 2k + 1$ is composite as was to be shown."

41. Theorem: The product of an even integer and an odd integer is even.
"**Proof:** Suppose m is an even integer and n is an odd integer. If $m \cdot n$ is even, then by definition of even there exists an integer r such that $m \cdot n = 2r$. Also since m is even, there exists an integer p such that $m = 2p$, and since n is odd there exists an integer q such that $n = 2q + 1$. Thus

$$mn = (2p)(2q + 1) = 2r,$$

where r is an integer. By definition of even, then, $m \cdot n$ is even, as was to be shown."

42. Theorem: The sum of any two even integers equals $4k$ for some integer k.
"**Proof:** Suppose m and n are any two even integers. By definition of even, $m = 2k$ for some integer k and $n = 2k$ for some integer k. By substitution,

$$m + n = 2k + 2k = 4k.$$

This is what was to be shown."

In 43–60 determine whether the statement is true or false. Justify your answer with a proof or a counterexample, as appropriate. In each case use only the definitions of the terms and the Assumptions listed on page 110 not any previously established properties.

43. The product of any two odd integers is odd.

44. The negative of any odd integer is odd.

45. The difference of any two odd integers is odd.

46. The product of any even integer and any integer is even.

47. If a sum of two integers is even, then one of the summands is even. (In the expression $a + b$, a and b are called **summands.**)

48. The difference of any two even integers is even.

49. The difference of any two odd integers is even.

50. For all integers n and m, if $n - m$ is even then $n^3 - m^3$ is even.

51. For all integers n, if n is prime then $(-1)^n = -1$.

52. For all integers m, if $m > 2$ then $m^2 - 4$ is composite.

53. For all integers n, $n^2 - n + 11$ is a prime number.

54. For all integers n, $4(n^2 + n + 1) - 3n^2$ is a perfect square.

55. Every positive integer can be expressed as a sum of three or fewer perfect squares.

H ✱ 56. (Two integers are **consecutive** if, and only if, one is one more than the other.) Any product of four consecutive integers is one less than a perfect square.

57. If m and n are positive integers and mn is a perfect square, then m and n are perfect squares.

58. The difference of the squares of any two consecutive integers is odd.

59. For all nonnegative real numbers a and b, $\sqrt{ab} = \sqrt{a}\sqrt{b}$. (Note that if x is a nonnegative real number, then there is a unique nonnegative real number y, denoted $\sqrt{x}$, such that $y^2 = x$.)

60. For all nonnegative real numbers a and b,
$$\sqrt{a+b} = \sqrt{a} + \sqrt{b}.$$

61. Suppose that integers m and n are perfect squares. Then $m + n + 2\sqrt{mn}$ is also a perfect square. Why?

H ✱ 62. If p is a prime number, must $2^p - 1$ also be prime? Prove or give a counterexample.

✱ 63. If n is a nonnegative integer, must $2^{2^n} + 1$ be prime? Prove or give a counterexample.

Answers for Test Yourself

1. it equals twice some integer   2. it equals twice some integer plus 1   3. n is greater than 1 and if n equals the product of any two positive integers, then one of the integers equals 1 and the other equals n.   4. a counterexample   5. particular but arbitrarily chosen element of the set; x satisfies the given property   6. x is a particular but arbitrarily chosen element of the set D that makes the hypothesis $P(x)$ true; x makes the conclusion $Q(x)$ true.

4.2 Direct Proof and Counterexample II: Rational Numbers

Such, then, is the whole art of convincing. It is contained in two principles: to define all notations used, and to prove everything by replacing mentally the defined terms by their definitions. — Blaise Pascal, 1623–1662

Sums, differences, and products of integers are integers. But most quotients of integers are not integers. Quotients of integers are, however, important; they are known as *rational numbers.*

ALL DEF. ARE "IF & ONLY IF" !!

> • **Definition**
>
> A real number r is **rational** if, and only if, it can be expressed as a quotient of two integers with a nonzero denominator. A real number that is not rational is **irrational.** More formally, if r is a real number, then
>
> $$r \text{ is rational} \iff \exists \text{ integers } a \text{ and } b \text{ such that } r = \frac{a}{b} \text{ and } b \neq 0.$$

The word *rational* contains the word *ratio,* which is another word for quotient. A rational number can be written as a ratio of integers.

Example 4.2.1 Determining Whether Numbers Are Rational or Irrational

a. Is 10/3 a rational number?

b. Is $-\frac{5}{39}$ a rational number?

c. Is 0.281 a rational number?

d. Is 7 a rational number?

    e. Is 0 a rational number?

    f. Is 2/0 a rational number?

    g. Is 2/0 an irrational number?

    h. Is 0.12121212... a rational number (where the digits 12 are assumed to repeat forever)?

    i. If m and n are integers and neither m nor n is zero, is $(m + n)/mn$ a rational number?

Solution

    a. Yes, 10/3 is a quotient of the integers 10 and 3 and hence is rational.

    b. Yes, $-\frac{5}{39} = \frac{-5}{39}$, which is a quotient of the integers -5 and 39 and hence is rational.

    c. Yes, $0.281 = 281/1000$. Note that the real numbers represented on a typical calculator display are all finite decimals. An explanation similar to the one in this example shows that any such number is rational. It follows that a calculator with such a display can represent only rational numbers.

    d. Yes, $7 = 7/1$.

    e. Yes, $0 = 0/1$.

    f. No, 2/0 is not a number (division by 0 is not allowed).

    g. No, because every irrational number is a number, and 2/0 is not a number. We discuss additional techniques for determining whether numbers are irrational in Sections 4.6, 4.7, and 9.4.

    h. Yes. Let $x = 0.12121212\ldots$. Then $100x = 12.12121212\ldots$. Thus

$$100x - x = 12.12121212\ldots - 0.12121212\ldots = 12.$$

But also
$$100x - x = 99x \quad \text{by basic algebra}$$

Hence
$$99x = 12,$$

and so
$$x = \frac{12}{99}.$$

Therefore, $0.12121212\ldots = 12/99$, which is a ratio of two nonzero integers and thus is a rational number.

    Note that you can use an argument similar to this one to show that any repeating decimal is a rational number. In Section 9.4 we show that any rational number can be written as a repeating or terminating decimal.

    i. Yes, since m and n are integers, so are $m + n$ and mn (because sums and products of integers are integers). Also $mn \neq 0$ by the *zero product property*. One version of this property says the following:

Zero Product Property

If neither of two real numbers is zero, then their product is also not zero.

(See Theorem T11 in Appendix A and exercise 8 at the end of this section.) It follows that $(m + n)/mn$ is a quotient of two integers with a nonzero denominator and hence is a rational number. ∎

More on Generalizing from the Generic Particular

Some people like to think of the method of generalizing from the generic particular as a challenge process. If you claim a property holds for all elements in a domain, then someone can challenge your claim by picking any element in the domain whatsoever and asking you to prove that that element satisfies the property. To prove your claim, you must be able to meet all such challenges. That is, you must have a way to convince the challenger that the property is true for an *arbitrarily chosen* element in the domain.

For example, suppose "A" claims that every integer is a rational number. "B" challenges this claim by asking "A" to prove it for $n = 7$. "A" observes that

$$7 = \frac{7}{1} \qquad \text{which is a quotient of integers and hence rational.}$$

"B" accepts this explanation but challenges again with $n = -12$. "A" responds that

$$-12 = \frac{-12}{1} \qquad \text{which is a quotient of integers and hence rational.}$$

Next "B" tries to trip up "A" by challenging with $n = 0$, but "A" answers that

$$0 = \frac{0}{1} \qquad \text{which is a quotient of integers and hence rational.}$$

As you can see, "A" is able to respond effectively to all "B"s challenges because "A" has a general procedure for putting integers into the form of rational numbers: "A" just divides whatever integer "B" gives by 1. That is, no matter what integer n "B" gives "A", "A" writes

$$n = \frac{n}{1} \qquad \text{which is a quotient of integers and hence rational.}$$

This discussion proves the following theorem.

Theorem 4.2.1

Every integer is a rational number.

In exercise 11 at the end of this section you are asked to condense the above discussion into a formal proof.

Proving Properties of Rational Numbers

The next example shows how to use the method of generalizing from the generic particular to prove a property of rational numbers.

Example 4.2.2 A Sum of Rationals Is Rational

Prove that the sum of any two rational numbers is rational.

Solution   Begin by mentally or explicitly rewriting the statement to be proved in the form "∀ _____, if _____ then _____."

Formal Restatement: ∀ real numbers r and s, if r and s are rational then $r + s$ is rational.

Next ask yourself, "Where am I starting from?" or "What am I supposing?" The answer gives you the starting point, or first sentence, of the proof.

Starting Point: Suppose r and s are particular but arbitrarily chosen real numbers such that r and s are rational; or, more simply,

Suppose r and s are rational numbers.

Then ask yourself, "What must I show to complete the proof?"

To Show: $r + s$ is rational.

Finally ask, "How do I get from the starting point to the conclusion?" or "Why must $r + s$ be rational if both r and s are rational?" The answer depends in an essential way on the definition of rational.

Rational numbers are quotients of integers, so to say that r and s are rational means that

$$r = \frac{a}{b} \quad \text{and} \quad s = \frac{c}{d} \quad \text{for some integers } a, b, c, \text{ and } d$$
$$\text{where } b \neq 0 \text{ and } d \neq 0.$$

It follows by substitution that

$$r + s = \frac{a}{b} + \frac{c}{d}.$$

You need to show that $r + s$ is rational, which means that $r + s$ can be written as a single fraction or ratio of two integers with a nonzero denominator. But the right-hand side of equation (4.2.1) in

$$\frac{a}{b} + \frac{c}{d} = \frac{ad}{bd} + \frac{bc}{bd} \qquad \text{rewriting the fraction with a common denominator}$$

$$= \frac{ad + bc}{bd} \qquad \text{adding fractions with a common denominator.}$$

Is this fraction a ratio of integers? Yes. Because products and sums of integers are integers, $ad + bc$ and bd are both integers. Is the denominator $bd \neq 0$? Yes, by the zero product property (since $b \neq 0$ and $d \neq 0$). Thus $r + s$ is a rational number.

This discussion is summarized as follows:

Theorem 4.2.2

The sum of any two rational numbers is rational.

Proof:

Suppose r and s are rational numbers. *[We must show that $r + s$ is rational.]* Then, by definition of rational, $r = a/b$ and $s = c/d$ for some integers $a, b, c,$ and d with $b \neq 0$ and $d \neq 0$. Thus

$$r + s = \frac{a}{b} + \frac{c}{d} \qquad \text{by substitution}$$

$$= \frac{ad + bc}{bd} \qquad \text{by basic algebra.}$$

Let $p = ad + bc$ and $q = bd$. Then p and q are integers because products and sums of integers are integers and because a, b, c, and d are all integers. Also $q \neq 0$ by the zero product property. Thus

$$r + s = \frac{p}{q} \text{ where } p \text{ and } q \text{ are integers and } q \neq 0.$$

Therefore, $r + s$ is rational by definition of a rational number. *[This is what was to be shown.]*

■

Deriving New Mathematics from Old

Section 4.1 focused on establishing truth and falsity of mathematical theorems using only the basic algebra normally taught in secondary school; the fact that the integers are closed under addition, subtraction, and multiplication; and the definitions of the terms in the theorems themselves. In the future, when we ask you to **prove something directly from the definitions,** we will mean that you should restrict yourself to this approach. However, once a collection of statements has been proved directly from the definitions, another method of proof becomes possible. The statements in the collection can be used to derive additional results.

Example 4.2.3 Deriving Additional Results about Even and Odd Integers

Suppose that you have already proved the following properties of even and odd integers:

1. The sum, product, and difference of any two even integers are even.

2. The sum and difference of any two odd integers are even.

3. The product of any two odd integers is odd.

4. The product of any even integer and any odd integer is even.

5. The sum of any odd integer and any even integer is odd.

6. The difference of any odd integer minus any even integer is odd.

7. The difference of any even integer minus any odd integer is odd.

Use the properties listed above to prove that if a is any even integer and b is any odd integer, then $\frac{a^2+b^2+1}{2}$ is an integer.

Solution   Suppose a is any even integer and b is any odd integer. By property 3, b^2 is odd, and by property 1, a^2 is even. Then by property 5, $a^2 + b^2$ is odd, and because 1 is also odd, the sum $(a^2 + b^2) + 1 = a^2 + b^2 + 1$ is even by property 2. Hence, by definition of even, there exists an integer k such that $a^2 + b^2 + 1 = 2k$. Dividing both sides by 2 gives $\frac{a^2+b^2+1}{2} = k$, which is an integer. Thus $\frac{a^2+b^2+1}{2}$ is an integer *[as was to be shown].* ■

A **corollary** is a statement whose truth can be immediately deduced from a theorem that has already been proved.

Example 4.2.4 The Double of a Rational Number

Derive the following as a corollary of Theorem 4.2.2.

Corollary 4.2.3

The double of a rational number is rational.

Solution   The double of a number is just its sum with itself. But since the sum of any two rational numbers is rational (Theorem 4.2.2), the sum of a rational number with itself is rational. Hence the double of a rational number is rational. Here is a formal version of this argument:

Proof:

Suppose r is any rational number. Then $2r = r + r$ is a sum of two rational numbers. So, by Theorem 4.2.2, $2r$ is rational. ■

Test Yourself

1. To show that a real number is rational, we must show that we can write it as _____.

2. An irrational number is a _____ that is _____.

3. Zero is a rational number because _____.

Exercise Set 4.2

The numbers in 1–7 are all rational. Write each number as a ratio of two integers.

1. $-\dfrac{35}{6}$ 2. 4.6037   3. $\dfrac{4}{5} + \dfrac{2}{9}$

4. 0.37373737...

5. 0.56565656...

6. 320.5492492492...

7. 52.4672167216721...

8. The zero product property, says that if a product of two real numbers is 0, then one of the numbers must be 0.
   a. Write this property formally using quantifiers and variables.
   b. Write the contrapositive of your answer to part (a).
   c. Write an informal version (without quantifier symbols or variables) for your answer to part (b).

9. Assume that a and b are both integers and that $a \neq 0$ and $b \neq 0$. Explain why $(b-a)/(ab^2)$ must be a rational number.

10. Assume that m and n are both integers and that $n \neq 0$. Explain why $(5m + 12n)/(4n)$ must be a rational number.

11. Prove that every integer is a rational number.

12. Fill in the blanks in the following proof that the square of any rational number is rational:

    Proof: Suppose that r is __(a)__. By definition of rational, $r = a/b$ for some __(b)__ with $b \neq 0$. By substitution,
    $$r^2 = \underline{\text{(c)}} = a^2/b^2.$$
    Since a and b are both integers, so are the products a^2 and __(d)__. Also $b^2 \neq 0$ by the __(e)__. Hence r^2 is a ratio of two integers with a nonzero denominator, and so __(f)__ by definition of rational.

13. Consider the statement: The negative of any rational number is rational.
    a. Write the statement formally using a quantifier and a variable.
    b. Determine whether the statement is true or false and justify your answer.

14. Consider the statement: The cube of any rational number is a rational number.
    a. Write the statement formally using a quantifier and a variable.
    b. Determine whether the statement is true or false and justify your answer.

Determine which of the statements in 15–20 are true and which are false. Prove each true statement directly from the definitions, and give a counterexample for each false statement.

In case the statement is false, determine whether a small change would make it true. If so, make the change and prove the new statement. Follow the directions for writing proofs on page 118.

15. The product of any two rational numbers is a rational number.

H 16. The quotient of any two rational numbers is a rational number.

H 17. The difference of any two rational numbers is a rational number.

H 18. If r and s are any two rational numbers, then $\frac{r+s}{2}$ is rational.

H 19. For all real numbers a and b, if $a < b$ then $a < \frac{a+b}{2} < b$. (You may use the properties of inequalities in T17–T27 of Appendix A.)

20. Given any two rational numbers r and s with $r < s$, there is another rational number between r and s. (*Hint:* Use the results of exercises 18 and 19.)

Use the properties of even and odd integers that are listed in Example 4.2.3 to do exercises 21–23. Indicate which properties you use to justify your reasoning.

21. True or false? If m is any even integer and n is any odd integer, then $m^2 + 3n$ is odd. Explain.

22. True or false? If a is any odd integer, then $a^2 + a$ is even. Explain.

23. True or false? If k is any even integer and m is any odd integer, then $(k + 2)^2 - (m - 1)^2$ is even. Explain.

Derive the statements in 24–26 as corollaries of Theorems 4.2.1, 4.2.2, and the results of exercises 12, 13, 14, 15, and 17.

24. For any rational numbers r and s, $2r + 3s$ is rational.

25. If r is any rational number, then $3r^2 - 2r + 4$ is rational.

26. For any rational number s, $5s^3 + 8s^2 - 7$ is rational.

27. It is a fact that if n is any nonnegative integer, then

$$1 + \frac{1}{2} + \frac{1}{2^2} + \frac{1}{2^3} + \cdots + \frac{1}{2^n} = \frac{1 - (1/2^{n+1})}{1 - (1/2)}.$$

(A more general form of this statement is proved in Section 5.2). Is the right-hand side of this equation rational? If so, express it as a ratio of two integers.

28. Suppose a, b, c, and d are integers and $a \neq c$. Suppose also that x is a real number that satisfies the equation

$$\frac{ax + b}{cx + d} = 1.$$

Must x be rational? If so, express x as a ratio of two integers.

⭑29. Suppose a, b, and c are integers and x, y, and z are nonzero real numbers that satisfy the following equations:

$$\frac{xy}{x+y} = a \quad \text{and} \quad \frac{xz}{x+z} = b \quad \text{and} \quad \frac{yz}{y+z} = c.$$

Is x rational? If so, express it as a ratio of two integers.

30. Prove that if one solution for a quadratic equation of the form $x^2 + bx + c = 0$ is rational (where b and c are rational), then the other solution is also rational. (Use the fact that if the solutions of the equation are r and s, then $x^2 + bx + c = (x - r)(x - s)$.)

31. Prove that if a real number c satisfies a polynomial equation of the form

$$r_3 x^3 + r_2 x^2 + r_1 x + r_0 = 0,$$

where r_0, r_1, r_2, and r_3 are rational numbers, then c satisfies an equation of the form

$$n_3 x^3 + n_2 x^2 + n_1 x + n_0 = 0,$$

where n_0, n_1, n_2, and n_3 are integers.

Definition: A number c is called a **root** of a polynomial $p(x)$ if, and only if, $p(c) = 0$.

⭑32. Prove that for all real numbers c, if c is a root of a polynomial with rational coefficients, then c is a root of a polynomial with integer coefficients.

Use the properties of even and odd integers that are listed in Example 4.2.3 to do exercises 33 and 34.

33. When expressions of the form $(x - r)(x - s)$ are multiplied out, a quadratic polynomial is obtained. For instance, $(x - 2)(x - (-7)) = (x - 2)(x + 7) = x^2 + 5x - 14$.

H a. What can be said about the coefficients of the polynomial obtained by multiplying out $(x - r)(x - s)$ when both r and s are odd integers? when both r and s are even integers? when one of r and s is even and the other is odd?

b. It follows from part (a) that $x^2 - 1253x + 255$ cannot be written as a product of two polynomials with integer coefficients. Explain why this is so.

⭑34. Observe that $(x - r)(x - s)(x - t)$
$$= x^3 - (r + s + t)x^2 + (rs + rt + st)x - rst.$$

a. Derive a result for cubic polynomials similar to the result in part (a) of exercise 33 for quadratic polynomials.

b. Can $x^3 + 7x^2 - 8x - 27$ be written as a product of three polynomials with integer coefficients? Explain.

In 35–39 find the mistakes in the "proofs" that the sum of any two rational numbers is a rational number.

35. "**Proof:** Any two rational numbers produce a rational number when added together. So if r and s are particular but arbitrarily chosen rational numbers, then $r + s$ is rational."

36. "**Proof:** Let rational numbers $r = \frac{1}{4}$ and $s = \frac{1}{2}$ be given. Then $r + s = \frac{1}{4} + \frac{1}{2} = \frac{3}{4}$, which is a rational number. This is what was to be shown."

37. "**Proof:** Suppose r and s are rational numbers. By definition of rational, $r = a/b$ for some integers a and b with $b \neq 0$, and $s = a/b$ for some integers a and b with $b \neq 0$. Then

$$r + s = \frac{a}{b} + \frac{a}{b} = \frac{2a}{b}.$$

Let $p = 2a$. Then p is an integer since it is a product of integers. Hence $r + s = p/b$, where p and b are integers and $b \neq 0$. Thus $r + s$ is a rational number by definition of rational. This is what was to be shown."

38. "**Proof:** Suppose r and s are rational numbers. Then $r = a/b$ and $s = c/d$ for some integers $a, b, c,$ and d with $b \neq 0$ and $d \neq 0$ (by definition of rational). Then

$$r + s = \frac{a}{b} + \frac{c}{d}.$$

But this is a sum of two fractions, which is a fraction. So $r + s$ is a rational number since a rational number is a fraction."

39. "**Proof:** Suppose r and s are rational numbers. If $r + s$ is rational, then by definition of rational $r + s = a/b$ for some integers a and b with $b \neq 0$. Also since r and s are rational, $r = i/j$ and $s = m/n$ for some integers $i, j, m,$ and n with $j \neq 0$ and $n \neq 0$. It follows that

$$r + s = \frac{i}{j} + \frac{m}{n} = \frac{a}{b},$$

which is a quotient of two integers with a nonzero denominator. Hence it is a rational number. This is what was to be shown."

Answers for Test Yourself

1. a ratio of integers with a nonzero denominator   2. real number; not rational   3. $0 = \dfrac{0}{1}$

4.3 Direct Proof and Counterexample III: Divisibility

The essential quality of a proof is to compel belief. — Pierre de Fermat

When you were first introduced to the concept of division in elementary school, you were probably taught that 12 divided by 3 is 4 because if you separate 12 objects into groups of 3, you get 4 groups with nothing left over.

$$\boxed{\text{XXX}} \quad \boxed{\text{XXX}} \quad \boxed{\text{XXX}} \quad \boxed{\text{XXX}}$$

You may also have been taught to describe this fact by saying that "12 is evenly divisible by 3" or "3 divides 12 evenly."

The notion of divisibility is the central concept of one of the most beautiful subjects in advanced mathematics: **number theory,** the study of properties of integers.

• Definition

If n and d are integers and $d \neq 0$ then

> n is **divisible by** d if, and only if, n equals d times some integer.

Instead of "n is divisible by d," we can say that

> n **is a multiple of** d, or
> d **is a factor of** n, or
> d **is a divisor of** n, or
> d **divides** n.

The notation $\mathbf{d \mid n}$ is read "d divides n." Symbolically, if n and d are integers and $d \neq 0$:

 $d \mid n \quad \Leftrightarrow \quad \exists$ an integer k such that $n = dk$.

Example 4.3.1 Divisibility

a. Is 21 divisible by 3? b. Does 5 divide 40? c. Does 7 | 42?

d. Is 32 a multiple of -16? e. Is 6 a factor of 54? f. Is 7 a factor of -7?

Solution

a. Yes, $21 = 3 \cdot 7$. b. Yes, $40 = 5 \cdot 8$. c. Yes, $42 = 7 \cdot 6$.

d. Yes, $32 = (-16) \cdot (-2)$. e. Yes, $54 = 6 \cdot 9$. f. Yes, $-7 = 7 \cdot (-1)$. ■

Example 4.3.2 Divisors of Zero

If k is any nonzero integer, does k divide 0?

Solution Yes, because $0 = k \cdot 0$. ■

Two useful properties of divisibility are (1) that if one positive integer divides a second positive integer, then the first is less than or equal to the second, and (2) that the only divisors of 1 are 1 and -1.

Theorem 4.3.1 A Positive Divisor of a Positive Integer

For all integers a and b, if a and b are positive and a divides b, then $a \le b$.

Proof:

Suppose a and b are positive integers and a divides b. *[We must show that $a \le b$.]* Then there exists an integer k so that $b = ak$. By property T25 of Appendix A, k must be positive because both a and b are positive. It follows that

$$1 \le k$$

because every positive integer is greater than or equal to 1. Multiplying both sides by a gives

$$a \le ka = b$$

because multiplying both sides of an inequality by a positive number preserves the inequality by property T20 of Appendix A. Thus $a \le b$ *[as was to be shown]*.

■

Theorem 4.3.2 Divisors of 1

The only divisors of 1 are 1 and -1.

Proof:

Since $1 \cdot 1 = 1$ and $(-1)(-1) = 1$, both 1 and -1 are divisors of 1. Now suppose m is any integer that divides 1. Then there exists an integer n such that $1 = mn$. By Theorem T25 in Appendix A, either both m and n are positive or both m and n are negative. If both m and n are positive, then m is a positive integer divisor of 1. By Theorem 4.3.1, $m \le 1$, and, since the only positive integer that is less than or equal

continued on page 136

to 1 is 1 itself, it follows that $m = 1$. On the other hand, if both m and n are negative, then, by Theorem T12 in Appendix A, $(-m)(-n) = mn = 1$. In this case $-m$ is a positive integer divisor of 1, and so, by the same reasoning, $-m = 1$ and thus $m = -1$. Therefore there are only two possibilities: either $m = 1$ or $m = -1$. So the only divisors of 1 are 1 and -1.

Example 4.3.3 Divisibility of Algebraic Expressions

a. If a and b are integers, is $3a + 3b$ divisible by 3?

b. If k and m are integers, is $10km$ divisible by 5?

Solution

a. Yes. By the distributive law of algebra, $3a + 3b = 3(a + b)$ and $a + b$ is an integer because it is a sum of two integers.

b. Yes. By the associative law of algebra, $10km = 5 \cdot (2km)$ and $2km$ is an integer because it is a product of three integers. ∎

When the definition of divides is rewritten formally using the existential quantifier, the result is

$$d \mid n \quad \Leftrightarrow \quad \exists \text{ an integer } k \text{ such that } n = dk.$$

Since the negation of an existential statement is universal, it follows that d does not divide n (denoted $d \nmid n$) if, and only if, $\forall$ integers k, $n \neq dk$, or, in other words, the quotient n/d is not an integer.

For all integers n and d, $\quad d \nmid n \quad \Leftrightarrow \quad \dfrac{n}{d}$ is not an integer.

Example 4.3.4 Checking Nondivisibility

Does $4 \mid 15$?

Solution   No, $\frac{15}{4} = 3.75$, which is not an integer. ∎

Caution!
$a \mid b$ denotes the *sentence* "*a* divides *b*," whereas a/b denotes the *number* a divided by b.

Be careful to distinguish between the notation $a \mid b$ and the notation a/b. The notation $a \mid b$ stands for the sentence "a divides b," which means that there is an integer k such that $b = ak$. Dividing both sides by a gives $b/a = k$, an integer. Thus, when $a \neq 0$, $a \mid b$ if, and only if, b/a is an integer. On the other hand, the notation a/b stands for the number a/b which is the result of dividing a by b and which may or may not be an integer. In particular, be sure to avoid writing things like

$$4 \mid \cancel{(3 + 5)} = 4 \mid 8.$$

If read out loud, this becomes, "4 divides the quantity 3 plus 5 equals 4 divides 8," which is nonsense.

Example 4.3.5 Prime Numbers and Divisibility

An alternative way to define a prime number is to say that an integer $n > 1$ is prime if, and only if, its only positive integer divisors are 1 and itself. ∎

Proving Properties of Divisibility

One of the most useful properties of divisibility is that it is transitive. If one number divides a second and the second number divides a third, then the first number divides the third.

Example 4.3.6   Transitivity of Divisibility

Prove that for all integers a, b, and c, if $a \mid b$ and $b \mid c$, then $a \mid c$.

Solution   Since the statement to be proved is already written formally, you can immediately pick out the starting point, or first sentence of the proof, and the conclusion that must be shown.

Starting Point: Suppose a, b, and c are particular but arbitrarily chosen integers such that $a \mid b$ and $b \mid c$.

To Show: $a \mid c$.

You need to show that $a \mid c$, or, in other words, that

$$c = a \cdot (\text{some integer}).$$

But since $a \mid b$,

$$b = ar \quad \text{for some integer } r. \tag{4.3.1}$$

And since $b \mid c$,

$$c = bs \quad \text{for some integer } s. \tag{4.3.2}$$

Equation 4.3.2 expresses c in terms of b, and equation 4.3.1 expresses b in terms of a. Thus if you substitute 4.3.1 into 4.3.2, you will have an equation that expresses c in terms of a.

$$
\begin{aligned}
c &= bs & &\text{by equation 4.3.2} \\
  &= (ar)s & &\text{by equation 4.3.1.}
\end{aligned}
$$

But $(ar)s = a(rs)$ by the associative law for multiplication. Hence

$$c = a(rs).$$

Now you are almost finished. You have expressed c as $a \cdot (\text{something})$. It remains only to verify that that something is an integer. But of course it is, because it is a product of two integers.

This discussion is summarized as follows:

Theorem 4.3.3 Transitivity of Divisibility

For all integers a, b, and c, if a divides b and b divides c, then a divides c.

Proof:

Suppose a, b, and c are *[particular but arbitrarily chosen]* integers such that a divides b and b divides c. *[We must show that a divides c.]* By definition of divisibility,

$$b = ar \quad \text{and} \quad c = bs \quad \text{for some integers } r \text{ and } s.$$

continued on page 138

By substitution

$$c = bs$$
$$= (ar)s$$
$$= a(rs) \qquad \text{by basic algebra.}$$

Let $k = rs$. Then k is an integer since it is a product of integers, and therefore

$$c = ak \quad \text{where } k \text{ is an integer.}$$

Thus a divides c by definition of divisibility. *[This is what was to be shown.]*

∎

It would appear from the definition of prime that to show that an integer is prime you would need to show that it is not divisible by any integer greater than 1 and less than itself. In fact, you need only check whether it is divisible by a prime number less than or equal to itself. This follows from Theorems 4.3.1, 4.3.3, and the following theorem, which says that any integer greater than 1 is divisible by a prime number. The idea of the proof is quite simple. You start with a positive integer. If it is prime, you are done; if not, it is a product of two smaller positive factors. If one of these is prime, you are done; if not, you can pick one of the factors and write it as a product of still smaller positive factors. You can continue in this way, factoring the factors of the number you started with, until one of them turns out to be prime. This must happen eventually because all the factors can be chosen to be positive and each is smaller than the preceding one.

Theorem 4.3.4 Divisibility by a Prime

Any integer $n > 1$ is divisible by a prime number.

Proof:

Suppose n is a *[particular but arbitrarily chosen]* integer that is greater than 1. *[We must show that there is a prime number that divides n.]* If n is prime, then n is divisible by a prime number (namely itself), and we are done. If n is not prime, then, as discussed in Example 4.1.2b,

$$n = r_0 s_0 \qquad \text{where } r_0 \text{ and } s_0 \text{ are integers and}$$
$$1 < r_0 < n \text{ and } 1 < s_0 < n.$$

It follows by definition of divisibility that $r_0 \mid n$.

If r_0 is prime, then r_0 is a prime number that divides n, and we are done. If r_0 is not prime, then

$$r_0 = r_1 s_1 \qquad \text{where } r_1 \text{ and } s_1 \text{ are integers and}$$
$$1 < r_1 < r_0 \text{ and } 1 < s_1 < r_0.$$

It follows by the definition of divisibility that $r_1 \mid r_0$. But we already know that $r_0 \mid n$. Consequently, by transitivity of divisibility, $r_1 \mid n$.

If r_1 is prime, then r_1 is a prime number that divides n, and we are done. If r_1 is not prime, then

$$r_1 = r_2 s_2 \qquad \text{where } r_2 \text{ and } s_2 \text{ are integers and}$$
$$1 < r_2 < r_1 \text{ and } 1 < s_2 < r_1.$$

It follows by definition of divisibility that $r_2 \mid r_1$. But we already know that $r_1 \mid n$. Consequently, by transitivity of divisibility, $r_2 \mid n$.

If r_2 is prime, then r_2 is a prime number that divides n, and we are done. If r_2 is not prime, then we may repeat the previous process by factoring r_2 as $r_3 s_3$.

We may continue in this way, factoring successive factors of n until we find a prime factor. We must succeed in a finite number of steps because each new factor is both less than the previous one (which is less than n) and greater than 1, and there are fewer than n integers strictly between 1 and n.* Thus we obtain a sequence

$$r_0, r_1, r_2, \ldots, r_k,$$

where $k \geq 0$, $1 < r_k < r_{k-1} < \cdots < r_2 < r_1 < r_0 < n$, and $r_i \mid n$ for each $i = 0, 1, 2, \ldots, k$. The condition for termination is that r_k should be prime. Hence r_k is a prime number that divides n. *[This is what we were to show.]*

Counterexamples and Divisibility

To show that a proposed divisibility property is not universally true, you need only find one pair of integers for which it is false.

Example 4.3.7 Checking a Proposed Divisibility Property

Is the following statement true or false? For all integers a and b, if $a \mid b$ and $b \mid a$ then $a = b$.

Solution This statement is false. Can you think of a counterexample just by concentrating for a minute or so?

The following discussion describes a mental process that may take just a few seconds. It is helpful to be able to use it consciously, however, to solve more difficult problems.

To discover the truth or falsity of a statement such as the one given above, start off much as you would if you were trying to prove it.

Starting Point: Suppose a and b are integers such that $a \mid b$ and $b \mid a$.

Ask yourself, "*Must* it follow that $a = b$, or *could* it happen that $a \neq b$ for some a and b?" Focus on the supposition. What does it mean? By definition of divisibility, the conditions $a \mid b$ and $b \mid a$ mean that

$$b = ka \quad \text{and} \quad a = lb \quad \text{for some integers } k \text{ and } l.$$

Must it follow that $a = b$, or can you find integers a and b that satisfy these equations for which $a \neq b$? The equations imply that

$$b = ka = k(lb) = (kl)b.$$

Since $b \mid a$, $b \neq 0$, and so you can cancel b from the extreme left and right sides to obtain

$$1 = kl.$$

In other words, k and l are divisors of 1. But, by Theorem 4.3.2, the only divisors of 1 are 1 and -1. Thus k and l are both 1 or are both -1. If $k = l = 1$, then $b = a$. But

---

*Strictly speaking, this statement is justified by an axiom for the integers called the well-ordering principle, which is discussed in Section 5.4. Theorem 4.3.4 can also be proved using strong mathematical induction, as shown in Example 5.4.1.

if $k = l = -1$, then $b = -a$ and so $a \neq b$. This analysis suggests that you can find a counterexample by taking $b = -a$. Here is a formal answer:

Proposed Divisibility Property: For all integers a and b, if $a \mid b$ and $b \mid a$ then $a = b$.

Counterexample: Let $a = 2$ and $b = -2$. Then

$$a \mid b \text{ since } 2 \mid (-2) \text{ and } b \mid a \text{ since } (-2) \mid 2, \text{ but } a \neq b \text{ since } 2 \neq -2.$$

Therefore, the statement is false.

■

The search for a proof will frequently help you discover a counterexample (provided the statement you are trying to prove is, in fact, false). Conversely, in trying to find a counterexample for a statement, you may come to realize the reason why it is true (if it is, in fact, true). The important thing is to keep an open mind until you are convinced by the evidence of your own careful reasoning.

The Unique Factorization of Integers Theorem

The most comprehensive statement about divisibility of integers is contained in the *unique factorization of integers theorem*. Because of its importance, this theorem is also called the *fundamental theorem of arithmetic*. Although Euclid, who lived about 300 B.C., seems to have been acquainted with the theorem, it was first stated precisely by the great German mathematician Carl Friedrich Gauss (rhymes with *house*) in 1801.

The unique factorization of integers theorem says that any integer greater than 1 either is prime or can be written as a product of prime numbers in a way that is unique except, perhaps, for the order in which the primes are written. For example,

$$72 = 2 \cdot 2 \cdot 2 \cdot 3 \cdot 3 = 2 \cdot 3 \cdot 3 \cdot 2 \cdot 2 = 3 \cdot 2 \cdot 2 \cdot 3 \cdot 2$$

and so forth. The three 2's and two 3's may be written in any order, but any factorization of 72 as a product of primes must contain exactly three 2's and two 3's—no other collection of prime numbers besides three 2's and two 3's multiplies out to 72.

Note This theorem is the reason the number 1 is not allowed to be prime. If 1 were prime, then factorizations would not be unique. For example, $6 = 2 \cdot 3 = 1 \cdot 2 \cdot 3$, and so forth.

Theorem 4.3.5 Unique Factorization of Integers Theorem (Fundamental Theorem of Arithmetic)

Given any integer $n > 1$, there exist a positive integer k, distinct prime numbers $p_1, p_2, \ldots, p_k$, and positive integers $e_1, e_2, \ldots, e_k$ such that

$$n = p_1^{e_1} p_2^{e_2} p_3^{e_3} \cdots p_k^{e_k},$$

and any other expression for n as a product of prime numbers is identical to this except, perhaps, for the order in which the factors are written.

The proof of the unique factorization of integers theorem is outlined in the exercises for Sections 5.4 and 8.5.

Because of the unique factorization of integers theorem, any integer $n > 1$ can be put into a *standard factored form* in which the prime factors are written in ascending order from left to right.

> **• Definition**
>
> Given any integer $n > 1$, the **standard factored form** of n is an expression of the form
>
> $$n = p_1^{e_1} p_2^{e_2} p_3^{e_3} \cdots p_k^{e_k},$$
>
> where k is a positive integer; $p_1, p_2, \ldots, p_k$ are prime numbers; $e_1, e_2, \ldots, e_k$ are positive integers; and $p_1 < p_2 < \cdots < p_k$.

Example 4.3.8 Writing Integers in Standard Factored Form

Write 3,300 in standard factored form.

Solution First find all the factors of 3,300. Then write them in ascending order:

$$3,300 = 100 \cdot 33 = 4 \cdot 25 \cdot 3 \cdot 11$$
$$= 2 \cdot 2 \cdot 5 \cdot 5 \cdot 3 \cdot 11 = 2^2 \cdot 3^1 \cdot 5^2 \cdot 11^1. \blacksquare$$

Example 4.3.9 Using Unique Factorization to Solve a Problem

Suppose m is an integer such that

$$8 \cdot 7 \cdot 6 \cdot 5 \cdot 4 \cdot 3 \cdot 2 \cdot m = 17 \cdot 16 \cdot 15 \cdot 14 \cdot 13 \cdot 12 \cdot 11 \cdot 10.$$

Does $17 \mid m$?

Solution Since 17 is one of the prime factors of the right-hand side of the equation, it is also a prime factor of the left-hand side (by the unique factorization of integers theorem). But 17 does not equal any prime factor of 8, 7, 6, 5, 4, 3, or 2 (because it is too large). Hence 17 must occur as one of the prime factors of m, and so $17 \mid m$. $\blacksquare$

Test Yourself

1. To show that a nonzero integer d divides an integer n, we must show that _____.

2. To say that d divides n means the same as saying that _____ is divisible by _____.

3. If a and b are positive integers and $a \mid b$, then _____ is less than or equal to _____.

4. For all integers n and d, $d \nmid n$ if, and only if, _____.

5. If a and b are integers, the notation $a \mid b$ denotes _____ and the notation a/b denotes _____.

6. The transitivity of divisibility theorem says that for all integers a, b, and c, if _____ then _____.

7. The divisibility by a prime theorem says that every integer greater than 1 is _____.

8. The unique factorization of integers theorem says that any integer greater than 1 is either _____ or can be written as _____ in a way that is unique except possibly for the _____ in which the numbers are written.

Exercise Set 4.3

Give a reason for your answer in each of 1–13. Assume that all variables represent integers.

1. Is 52 divisible by 13?   2. Does $7 \mid 56$?

3. Does $5 \mid 0$?

4. Does 3 divide $(3k + 1)(3k + 2)(3k + 3)$?

5. Is $6m(2m + 10)$ divisible by 4?

6. Is 29 a multiple of 3?   7. Is -3 a factor of 66?

8. Is $6a(a + b)$ a multiple of $3a$?

9. Is 4 a factor of $2a \cdot 34b$?

10. Does $7 \mid 34$? 11. Does $13 \mid 73$?

12. If $n = 4k + 1$, does 8 divide $n^2 - 1$?

13. If $n = 4k + 3$, does 8 divide $n^2 - 1$?

14. Fill in the blanks in the following proof that for all integers a and b, if $a \mid b$ then $a \mid (-b)$.

   Proof: Suppose a and b are any integers such that ___(a)___. By definition of divisibility, there exists an integer r such that ___(b)___. By substitution.

   $$-b = -ar = a(-r).$$

   Let $t = $ ___(c)___. Then t is an integer because $t = (-1) \cdot r$, and both -1 and r are integers. Thus, by substitution, $-b = at$, where t is an integer, and so by definition of divisibility, ___(d)___, as was to be shown.

Prove statements 15 and 16 directly from the definition of divisibility.

15. For all integers a, b, and c, if $a \mid b$ and $a \mid c$ then $a \mid (b + c)$.

H 16. For all integers a, b, and c, if $a \mid b$ and $a \mid c$ then $a \mid (b - c)$.

17. Consider the following statement: The negative of any multiple of 3 is a multiple of 3.
   a. Write the statement formally using a quantifier and a variable.
   b. Determine whether the statement is true or false and justify your answer.

18. Show that the following statement is false: For all integers a and b, if $3 \mid (a + b)$ then $3 \mid (a - b)$.

For each statement in 19–31, determine whether the statement is true or false. Prove the statement directly from the definitions if it is true, and give a counterexample if it is false.

H 19. For all integers a, b, and c, if a divides b then a divides bc.

20. The sum of any three consecutive integers is divisible by 3. (Two integers are **consecutive** if, and only if, one is one more than the other.)

21. The product of any two even integers is a multiple of 4.

H 22. A necessary condition for an integer to be divisible by 6 is that it be divisible by 2.

23. A sufficient condition for an integer to be divisible by 8 is that it be divisible by 16.

24. For all integers a, b, and c, if $a \mid b$ and $a \mid c$ then $a \mid (2b - 3c)$.

25. For all integers a, b, and c, if a is a factor of c then ab is a factor of c.

H 26. For all integers a, b, and c, if $ab \mid c$ then $a \mid c$ and $b \mid c$.

H 27. For all integers a, b, and c, if $a \mid (b + c)$ then $a \mid b$ or $a \mid c$.

28. For all integers a, b, and c, if $a \mid bc$ then $a \mid b$ or $a \mid c$.

29. For all integers a and b, if $a \mid b$ then $a^2 \mid b^2$.

30. For all integers a and n, if $a \mid n^2$ and $a \leq n$ then $a \mid n$.

31. For all integers a and b, if $a \mid 10b$ then $a \mid 10$ or $a \mid b$.

32. A fast-food chain has a contest in which a card with numbers on it is given to each customer who makes a purchase. If some of the numbers on the card add up to 100, then the customer wins $100. A certain customer receives a card containing the numbers

   72, 21, 15, 36, 69, 81, 9, 27, 42, and 63.

   Will the customer win $100? Why or why not?

33. Is it possible to have a combination of nickels, dimes, and quarters that add up to $4.72? Explain.

34. Is it possible to have 50 coins, made up of pennies, dimes, and quarters, that add up to $3? Explain.

35. Two athletes run a circular track at a steady pace so that the first completes one round in 8 minutes and the second in 10 minutes. If they both start from the same spot at 4 P.M., when will be the first time they return to the start together?

36. It can be shown (see exercises 44–48) that an integer is divisible by 3 if, and only if, the sum of its digits is divisible by 3. An integer is divisible by 9 if, and only if, the sum of its digits is divisible by 9. An integer is divisible by 5 if, and only if, its right-most digit is a 5 or a 0. And an integer is divisible by 4 if, and only if, the number formed by its right-most two digits is divisible by 4. Check the following integers for divisibility by 3, 4, 5 and 9.
   a. 637,425,403,705,125 b. 12,858,306,120,312
   c. 517,924,440,926,512 d. 14,328,083,360,232

37. Use the unique factorization theorem to write the following integers in standard factored form.
   a. 1,176 b. 5,733 c. 3,675

38. Suppose that in standard factored form $a = p_1^{e_1} p_2^{e_2} \cdots p_k^{e_k}$, where k is a positive integer; $p_1, p_2, \ldots, p_k$ are prime numbers; and $e_1, e_2, \ldots, e_k$ are positive integers.
   a. What is the standard factored form for a^2?
   b. Find the least positive integer n such that $2^5 \cdot 3 \cdot 5^2 \cdot 7^3 \cdot n$ is a perfect square. Write the resulting product as a perfect square.
   c. Find the least positive integer m such that $2^2 \cdot 3^5 \cdot 7 \cdot 11 \cdot m$ is a perfect square. Write the resulting product as a perfect square.

39. Suppose that in standard factored form $a = p_1^{e_1} p_2^{e_2} \cdots p_k^{e_k}$, where k is a positive integer; $p_1, p_2, \ldots, p_k$ are prime numbers; and $e_1, e_2, \ldots, e_k$ are positive integers.
   a. What is the standard factored form for a^3?
   b. Find the least positive integer k such that $2^4 \cdot 3^5 \cdot 7 \cdot 11^2 \cdot k$ is a perfect cube (i.e., equals an integer to the third power). Write the resulting product as a perfect cube.

40. **a.** If a and b are integers and $12a = 25b$, does $12 \mid b$? does $25 \mid a$? Explain.
    b. If x and y are integers and $10x = 9y$, does $10 \mid y$? does $9 \mid x$? Explain.

H 41. How many zeros are at the end of $45^8 \cdot 88^5$? Explain how you can answer this question without actually computing the number. (*Hint:* $10 = 2 \cdot 5$.)

42. If n is an integer and $n > 1$, then $n!$ is the product of n and every other positive integer that is less than n. For example, $5! = 5 \cdot 4 \cdot 3 \cdot 2 \cdot 1$.
    a. Write 6! in standard factored form.
    b. Write 20! in standard factored form.
    c. Without computing the value of $(20!)^2$ determine how many zeros are at the end of this number when it is written in decimal form. Justify your answer.

$*$ 43. In a certain town 2/3 of the adult men are married to 3/5 of the adult women. Assume that all marriages are monogamous (no one is married to more than one other person). Also assume that there are at least 100 adult men in the town. What is the least possible number of adult men in the town? of adult women in the town?

> **Definition:** Given any nonnegative integer n, the **decimal representation** of n is an expression of the form
>
> $$d_k d_{k-1} \cdots d_2 d_1 d_0,$$
>
> where k is a nonnegative integer; $d_0, d_1, d_2, \ldots, d_k$ (called the **decimal digits** of n) are integers from 0 to 9 inclusive; $d_k \neq 0$ unless $n = 0$ and $k = 0$; and
>
> $$n = d_k \cdot 10^k + d_{k-1} \cdot 10^{k-1} + \cdots + d_2 \cdot 10^2 + d_1 \cdot 10 + d_0.$$
>
> (For example, $2{,}503 = 2 \cdot 10^3 + 5 \cdot 10^2 + 0 \cdot 10 + 3$.)

44. Prove that if n is any nonnegative integer whose decimal representation ends in 0, then $5 \mid n$. (*Hint:* If the decimal representation of a nonnegative integer n ends in d_0, then $n = 10m + d_0$ for some integer m.)

45. Prove that if n is any nonnegative integer whose decimal representation ends in 5, then $5 \mid n$.

46. Prove that if the decimal representation of a nonnegative integer n ends in $d_1 d_0$ and if $4 \mid (10d_1 + d_0)$, then $4 \mid n$. (*Hint:* If the decimal representation of a nonnegative integer n ends in $d_1 d_0$, then there is an integer s such that $n = 100s + 10d_1 + d_0$.)

H $*$ 47. Observe that

$$\begin{aligned}
7524 &= 7 \cdot 1000 + 5 \cdot 100 + 2 \cdot 10 + 4 \\
&= 7(999 + 1) + 5(99 + 1) + 2(9 + 1) + 4 \\
&= (7 \cdot 999 + 7) + (5 \cdot 99 + 5) + (2 \cdot 9 + 2) + 4 \\
&= (7 \cdot 999 + 5 \cdot 99 + 2 \cdot 9) + (7 + 5 + 2 + 4) \\
&= (7 \cdot 111 \cdot 9 + 5 \cdot 11 \cdot 9 + 2 \cdot 9) + (7 + 5 + 2 + 4) \\
&= (7 \cdot 111 + 5 \cdot 11 + 2) \cdot 9 + (7 + 5 + 2 + 4) \\
&= \text{(an integer divisible by 9)} \\
&\qquad + \text{(the sum of the digits of 7524)}.
\end{aligned}$$

Since the sum of the digits of 7524 is divisible by 9, 7524 can be written as a sum of two integers each of which is divisible by 9. It follows from exercise 15 that 7524 is divisible by 9.

Generalize the argument given in this example to any nonnegative integer n. In other words, prove that for any nonnegative integer n, if the sum of the digits of n is divisible by 9, then n is divisible by 9.

$*$ 48. Prove that for any nonnegative integer n, if the sum of the digits of n is divisible by 3, then n is divisible by 3.

$*$ 49. Given a positive integer n written in decimal form, the alternating sum of the digits of n is obtained by starting with the right-most digit, subtracting the digit immediately to its left, adding the next digit to the left, subtracting the next digit, and so forth. For example, the alternating sum of the digits of 180,928 is $8 - 2 + 9 - 0 + 8 - 1 = 22$. Justify the fact that for any nonnegative integer n, if the alternating sum of the digits of n is divisible by 11, then n is divisible by 11.

Answers for Test Yourself

1. n equals d times some integer (Or: there is an integer r such that $n = dr$)   2. n; d 3. a; b 4. $\frac{n}{d}$ is not an integer   5. the sentence "a divides b"; the number obtained when a is divided by b 6. a divides b and b divides c; a divides c 7. divisible by some prime number   8. prime; a product of prime numbers; order

4.4 Direct Proof and Counterexample IV: Division into Cases and the Quotient-Remainder Theorem

Be especially critical of any statement following the word "obviously."
— Anna Pell Wheeler 1883–1966

When you divide 11 by 4, you get a quotient of 2 and a remainder of 3.

$$
\begin{array}{r}
2 \leftarrow \text{quotient} \\
4\overline{)11} \\
\underline{8} \\
3 \leftarrow \text{remainder}
\end{array}
$$

Another way to say this is that 11 equals 2 groups of 4 with 3 left over:

$$
\underbrace{\boxed{\text{XXXX}} \quad \boxed{\text{XXXX}}}_{\text{2 groups of 4}} \quad \underbrace{\text{XXX}}_{\text{3 left over}}
$$

Or,

$$
11 = 2\cdot 4 + 3.
$$
$$
\underbrace{}_{\text{2 groups of 4}} \quad \underbrace{}_{\text{3 left over}}
$$

Of course, the number left over (3) is less than the size of the groups (4) because if 4 or more were left over, another group of 4 could be separated off.

The quotient-remainder theorem says that when any integer n is divided by any positive integer d, the result is a quotient q and a nonnegative remainder r that is smaller than d.

Theorem 4.4.1 The Quotient-Remainder Theorem

Given any integer n and positive integer d, there exist unique integers q and r such that

$$
n = dq + r \quad \text{and} \quad 0 \le r < d.
$$

The proof that there exist integers q and r with the given properties is in Section 5.4; the proof that q and r are unique is outlined in exercise 18 in Section 4.6.

If n is positive, the quotient-remainder theorem can be illustrated on the number line as follows:

If n is negative, the picture changes. Since $n = dq + r$, where r is nonnegative, d must be multiplied by a negative integer q to go below n. Then the nonnegative integer r is added to come back up to n. This is illustrated as follows:

Example 4.4.1 The Quotient-Remainder Theorem

For each of the following values of n and d, find integers q and r such that $n = dq + r$ and $0 \leq r < d$.

a. $n = 54$, $d = 4$ b. $n = -54$, $d = 4$ c. $n = 54$, $d = 70$

Solution

a. $54 = 4 \cdot 13 + 2$; hence $q = 13$ and $r = 2$.

b. $-54 = 4 \cdot (-14) + 2$; hence $q = -14$ and $r = 2$.

c. $54 = 70 \cdot 0 + 54$; hence $q = 0$ and $r = 54$.   ■

div and mod

A number of computer languages have built-in functions that enable you to compute many values of q and r for the quotient-remainder theorem. These functions are called **div** and **mod** in Pascal, are called / and % in C and C++, are called / and % in Java, and are called / (or \\) and **mod** in .NET. The functions give the values that satisfy the quotient-remainder theorem when a *nonnegative* integer n is divided by a positive integer d and the result is assigned to an integer variable. However, they do not give the values that satisfy the quotient-remainder theorem when a negative integer n is divided by a positive integer d.

• Definition

Given an integer n and a positive integer d,

> n *div* d = the integer quotient obtained
> when n is divided by d, and

> n *mod* d = the nonnegative integer remainder obtained
> when n is divided by d.

Symbolically, if n and d are integers and $d > 0$, then

$$n \text{ } div \text{ } d = q \quad \text{and} \quad n \text{ } mod \text{ } d = r \quad \Leftrightarrow \quad n = dq + r$$

where q and r are integers and $0 \leq r < d$.

Note that it follows from the quotient-remainder theorem that n *mod* d equals one of the integers from 0 through $d - 1$ (since the remainder of the division of n by d must be one of these integers). Note also that a necessary and sufficient condition for an integer n to be divisible by an integer d is that n *mod* $d = 0$. You are asked to prove this in the exercises at the end of this section.

You can also use a calculator to compute values of *div* and *mod*. For instance, to compute n *div* d for a nonnegative integer n and a positive integer d, you just divide n by d and ignore the part of the answer to the right of the decimal point. To find n *mod* d, you can use the fact that if $n = dq + r$, then $r = n - dq$. Thus $n = d \cdot (n \text{ } div \text{ } d) + n \text{ } mod \text{ } d$, and so

$$n \text{ } mod \text{ } d = n - d \cdot (n \text{ } div \text{ } d).$$

Hence, to find n *mod* d compute n *div* d, multiply by d, and subtract the result from n.

Example 4.4.2 Computing *div* and *mod*

Compute 32 *div* 9 and 32 *mod* 9 by hand and with a calculator.

Solution   Performing the division by hand gives the following results:

$$
\begin{array}{r}
3 \quad \leftarrow 32\ div\ 9 \\
9\overline{)32} \\
27 \\
\overline{5} \quad \leftarrow 32\ mod\ 9
\end{array}
$$

If you use a four-function calculator to divide 32 by 9, you obtain an expression like 3.555555556. Discarding the fractional part gives 32 *div* 9 = 3, and so

$$32\ mod\ 9 = 32 - 9\cdot(32\ div\ 9) = 32 - 27 = 5.$$

A calculator with a built-in integer-part function iPart allows you to input a single expression for each computation:

$$32\ div\ 9 = \text{iPart}(32/9)$$
$$\text{and}\quad 32\ mod\ 9 = 32 - 9\cdot\text{iPart}\,(32/9) = 5. \qquad\blacksquare$$

Example 4.4.3 Computing the Day of the Week

Suppose today is Tuesday, and neither this year nor next year is a leap year. What day of the week will it be 1 year from today?

Solution   There are 365 days in a year that is not a leap year, and each week has 7 days. Now

$$365\ div\ 7 = 52 \quad\text{and}\quad 365\ mod\ 7 = 1$$

because $365 = 52\cdot 7 + 1$. Thus 52 weeks, or 364 days, from today will be a Tuesday, and so 365 days from today will be 1 day later, namely Wednesday.

More generally, if *DayT* is the day of the week today and *DayN* is the day of the week in N days, then

$$DayN = (DayT + N)\ mod\ 7, \qquad\qquad 4.4.1$$

where Sunday = 0, Monday = 1, ... , Saturday = 6. $\qquad\blacksquare$

Example 4.4.4 Solving a Problem about *mod*

Suppose m is an integer. If $m\ mod\ 11 = 6$, what is $4m\ mod\ 11$?

Solution   Because $m\ mod\ 11 = 6$, the remainder obtained when m is divided by 11 is 6. This means that there is some integer q so that

$$m = 11q + 6.$$

Thus $\qquad\qquad 4m = 44q + 24 = 44q + 22 + 2 = 11(4q + 2) + 2.$

Since $4q + 2$ is an integer (because products and sums of integers are integers) and since $2 < 11$, the remainder obtained when $4m$ is divided by 11 is 2. Therefore,

$$4m\ mod\ 11 = 2. \qquad\blacksquare$$

Representations of Integers

In Section 4.1 we defined an even integer to have the form twice some integer. At that time we could have defined an odd integer to be one that was not even. Instead, because it was more useful for proving theorems, we specified that an odd integer has the form twice some integer plus one. The quotient-remainder theorem brings these two ways of describing odd integers together by guaranteeing that any integer is either even or odd. To see why, let n be any integer, and consider what happens when n is divided by 2. By the quotient-remainder theorem (with $d = 2$), there exist unique integers q and r such that

$$n = 2q + r \quad \text{and} \quad 0 \leq r < 2.$$

But the only integers that satisfy $0 \leq r < 2$ are $r = 0$ and $r = 1$. It follows that given any integer n, there exists an integer q with

$$n = 2q + 0 \quad \text{or} \quad n = 2q + 1.$$

In the case that $n = 2q + 0 = 2q$, n is even. In the case that $n = 2q + 1$, n is odd. Hence n is either even or odd, and, because of the uniqueness of q and r, n cannot be both even and odd.

The *parity* of an integer refers to whether the integer is even or odd. For instance, 5 has odd parity and 28 has even parity. We call the fact that any integer is either even or odd the **parity property.**

Example 4.4.5 Consecutive Integers Have Opposite Parity

Prove that given any two consecutive integers, one is even and the other is odd.

Solution   Two integers are called *consecutive* if, and only if, one is one more than the other. So if one integer is m, the next consecutive integer is $m + 1$.

To prove the given statement, start by supposing that you have two particular but arbitrarily chosen consecutive integers. If the smaller is m, then the larger will be $m + 1$. How do you know for sure that one of these is even and the other is odd? You might imagine some examples: 4, 5; 12, 13; 1,073, 1,074. In the first two examples, the smaller of the two integers is even and the larger is odd; in the last example, it is the reverse. These observations suggest dividing the analysis into two cases.

Case 1: The smaller of the two integers is even.

Case 2: The smaller of the two integers is odd.

In the first case, when m is even, it appears that the next consecutive integer is odd. Is this always true? If an integer m is even, must $m + 1$ necessarily be odd? Of course the answer is yes. Because if m is even, then $m = 2k$ for some integer k, and so $m + 1 = 2k + 1$, which is odd.

In the second case, when m is odd, it appears that the next consecutive integer is even. Is this always true? If an integer m is odd, must $m + 1$ necessarily be even? Again, the answer is yes. For if m is odd, then $m = 2k + 1$ for some integer k, and so $m + 1 = (2k + 1) + 1 = 2k + 2 = 2(k + 1)$, which is even.

This discussion is summarized on the following page.

Theorem 4.4.2 The Parity of Consecutive Integers

Any two consecutive integers have opposite parity.

Proof:

Suppose that two *[particular but arbitrarily chosen]* consecutive integers are given; call them m and $m + 1$. *[We must show that one of m and $m + 1$ is even and that the other is odd.]* By the parity property, either m is even or m is odd. *[We break the proof into two cases depending on whether m is even or odd.]*

Case 1 (m is even): In this case, $m = 2k$ for some integer k, and so $m + 1 = 2k + 1$, which is odd *[by definition of odd]*. Hence in this case, one of m and $m + 1$ is even and the other is odd.

Case 2 (m is odd): In this case, $m = 2k + 1$ for some integer k, and so $m + 1 = (2k + 1) + 1 = 2k + 2 = 2(k + 1)$. But $k + 1$ is an integer because it is a sum of two integers. Therefore, $m + 1$ equals twice some integer, and thus $m + 1$ is even. Hence in this case also, one of m and $m + 1$ is even and the other is odd.

It follows that regardless of which case actually occurs for the particular m and $m + 1$ that are chosen, one of m and $m + 1$ is even and the other is odd. *[This is what was to be shown.]*

■

The division into cases in a proof is like the transfer of control for an **if-then-else** statement in a computer program. If m is even, control transfers to case 1; if not, control transfers to case 2. For any given integer, only one of the cases will apply. You must consider both cases, however, to obtain a proof that is valid for an arbitrarily given integer whether even or not.

There are times when division into more than two cases is called for. Suppose that at some stage of developing a proof, you know that a statement of the form

$$A_1 \text{ or } A_2 \text{ or } A_3 \text{ or } \ldots \text{ or } A_n$$

is true, and suppose you want to deduce a conclusion C. By definition of *or*, you know that at least one of the statements A_i is true (although you may not know which). In this situation, you should use the method of division into cases. First assume A_1 is true and deduce C; next assume A_2 is true and deduce C; and so forth until you have assumed A_n is true and deduced C. At that point, you can conclude that regardless of which statement A_i happens to be true, the truth of C follows.

Method of Proof by Division into Cases

To prove a statement of the form "If A_1 or A_2 or $\ldots$ or A_n, then C," prove all of the following:

$$\text{If } A_1, \text{ then } C,$$
$$\text{If } A_2, \text{ then } C,$$
$$\vdots$$
$$\text{If } A_n, \text{ then } C.$$

This process shows that C is true regardless of which of $A_1, A_2, \ldots, A_n$ happens to be the case.

Proof by division into cases is a generalization of the argument form shown in Example 2.3.7, whose validity you were asked to establish in exercise 21 of Section 2.3. This method of proof was combined with the quotient-remainder theorem for $d = 2$ to prove Theorem 4.4.2. Allowing d to take on additional values makes it possible to obtain a variety of other results. We begin by showing what happens when $a = 4$.

Example 4.4.6 Representations of Integers Modulo 4

Show that any integer can be written in one of the four forms

$$n = 4q \quad \text{or} \quad n = 4q + 1 \quad \text{or} \quad n = 4q + 2 \quad \text{or} \quad n = 4q + 3$$

for some integer q.

Solution   Given any integer n, apply the quotient-remainder theorem to n with $d = 4$. This implies that there exist an integer quotient q and a remainder r such that

$$n = 4q + r \quad \text{and} \quad 0 \le r < 4.$$

But the only nonnegative remainders r that are less than 4 are 0, 1, 2, and 3. Hence

$$n = 4q \quad \text{or} \quad n = 4q + 1 \quad \text{or} \quad n = 4q + 2 \quad \text{or} \quad n = 4q + 3$$

for some integer q. ∎

The next example illustrates how the alternative representations for integers modulo 4 can help establish a result in number theory. The solution is broken into two parts: a discussion and a formal proof. These correspond to the stages of actual proof development. Very few people, when asked to prove an unfamiliar theorem, immediately write down the kind of formal proof you find in a mathematics text. Most need to experiment with several possible approaches before they find one that works. A formal proof is much like the ending of a mystery story—the part in which the action of the story is systematically reviewed and all the loose ends are carefully tied together.

Example 4.4.7 The Square of an Odd Integer

Note Another way to state this fact is that if you square an odd integer and divide by 8, you will always get a remainder of 1. Try a few examples!

Prove: The square of any odd integer has the form $8m + 1$ for some integer m.

Solution   Begin by asking yourself, "Where am I starting from?" and "What do I need to show?" To help answer these questions, introduce variables to represent the quantities in the statement to be proved.

Formal Restatement: ∀ odd integers n, ∃ an integer m such that $n^2 = 8m + 1$.

From this, you can immediately identify the starting point and what is to be shown.

Starting Point: Suppose n is a particular but arbitrarily chosen odd integer.

To Show: ∃ an integer m such that $n^2 = 8m + 1$.

This looks tough. Why should there be an integer m with the property that $n^2 = 8m + 1$? That would say that $(n^2 - 1)/8$ is an integer, or that 8 divides $n^2 - 1$. Perhaps you could make use of the fact that $n^2 - 1 = (n - 1)(n + 1)$. Does 8 divide $(n - 1)(n + 1)$? Since n is odd, both $(n - 1)$ and $(n + 1)$ are even. That means that their product is divisible by 4. But that's not enough. You need to show that the product is divisible by 8. This seems to be a blind alley.

You could try another tack. Since n is odd, you could represent n as $2q + 1$ for some integer q. Then $n^2 = (2q + 1)^2 = 4q^2 + 4q + 1 = 4(q^2 + q) + 1$. It is clear from this analysis that n^2 can be written in the form $4m + 1$, but it may not be clear that it can be written as $8m + 1$. This also seems to be a blind alley.*

Yet another possibility is to use the result of Example 4.4.6. That example showed that any integer can be written in one of the four forms $4q$, $4q + 1$, $4q + 2$, or $4q + 3$. Two of these, $4q + 1$ and $4q + 3$, are odd. Thus any odd integer can be written in the form $4q + 1$ or $4q + 3$ for some integer q. You could try breaking into cases based on these two different forms.

It turns out that this last possibility works! In each of the two cases, the conclusion follows readily by direct calculation. The details are shown in the following formal proof:

Note Desperation can spur creativity. When you have tried all the obvious approaches without success and you really care about solving a problem, you reach into the odd corners of your memory for *anything* that may help.

Theorem 4.4.3

The square of any odd integer has the form $8m + 1$ for some integer m.

Proof:

Suppose n is a *[particular but arbitrarily chosen]* odd integer. By the quotient-remainder theorem, n can be written in one of the forms

$$4q \quad \text{or} \quad 4q + 1 \quad \text{or} \quad 4q + 2 \quad \text{or} \quad 4q + 3$$

for some integer q. In fact, since n is odd and $4q$ and $4q + 2$ are even, n must have one of the forms

$$4q + 1 \quad \text{or} \quad 4q + 3.$$

Case 1 ($n = 4q + 1$ for some integer q): *[We must find an integer m such that $n^2 = 8m + 1$.]* Since $n = 4q + 1$,

$$
\begin{aligned}
n^2 &= (4q + 1)^2 && \text{by substitution} \\
&= (4q + 1)(4q + 1) && \text{by definition of square} \\
&= 16q^2 + 8q + 1 \\
&= 8(2q^2 + q) + 1 && \text{by the laws of algebra.}
\end{aligned}
$$

Let $m = 2q^2 + q$. Then m is an integer since 2 and q are integers and sums and products of integers are integers. Thus, substituting,

$$n^2 = 8m + 1 \quad \text{where } m \text{ is an integer.}$$

Case 2 ($n = 4q + 3$ for some integer q): *[We must find an integer m such that $n^2 = 8m + 1$.]* Since $n = 4q + 3$,

$$
\begin{aligned}
n^2 &= (4q + 3)^2 && \text{by substitution} \\
&= (4q + 3)(4q + 3) && \text{by definition of square} \\
&= 16q^2 + 24q + 9 \\
&= 16q^2 + 24q + (8 + 1) \\
&= 8(2q^2 + 3q + 1) + 1 && \text{by the laws of algebra.}
\end{aligned}
$$

[The motivation for the choice of algebra steps was the desire to write the expression in the form $8 \cdot (some\ integer) + 1$.]

---

*See exercise 18 for a different perspective.

Let $m = 2q^2 + 3q + 1$. Then m is an integer since 1, 2, 3, and q are integers and sums and products of integers are integers. Thus, substituting,

$$n^2 = 8m + 1 \quad \text{where } m \text{ is an integer.}$$

Cases 1 and 2 show that given any odd integer, whether of the form $4q + 1$ or $4q + 3$, $n^2 = 8m + 1$ for some integer m. *[This is what we needed to show.]*

■

Note that the result of Theorem 4.4.3 can also be written, "For any odd integer n, $n^2 \bmod 8 = 1$."

In general, according to the quotient-remainder theorem, if an integer n is divided by an integer d, the possible remainders are 0, 1, 2, ..., $(d - 1)$. This implies that n can be written in one of the forms

$$dq, \ dq + 1, \ dq + 2,, \ \ldots, \ dq + (d - 1) \qquad \text{for some integer } q.$$

Many properties of integers can be obtained by giving d a variety of different values and analyzing the cases that result.

Absolute Value and the Triangle Inequality

The triangle inequality is one of the most important results involving absolute value. It has applications in many areas of mathematics.

• Definition

For any real number x, the **absolute value of x**, denoted $|x|$, is defined as follows:

$$|x| = \begin{cases} x & \text{if } x \geq 0 \\ -x & \text{if } x < 0 \end{cases}.$$

The triangle inequality says that the absolute value of the sum of two numbers is less than or equal to the sum of their absolute values. We give a proof based on the following two facts, both of which are derived using division into cases. We state both as lemmas. A **lemma** is a statement that does not have much intrinsic interest but is helpful in deriving other results.

Lemma 4.4.4

For all real numbers r, $-|r| \leq r \leq |r|$.

Proof:

Suppose r is any real number. We divide into cases according to whether $r \geq 0$ or $r < 0$.
Case 1 ($r \geq 0$): In this case, by definition of absolute value, $|r| = r$. Also, since r is positive and $-|r|$ is negative, $-|r| < r$. Thus it is true that

$$-|r| \leq r \leq |r|.$$

continued on page 152

Case 2 (r < 0): In this case, by definition of absolute value, $|r| = -r$. Multiplying both sides by -1 gives that $-|r| = r$. Also, since r is negative and $|r|$ is positive, $r < |r|$. Thus it is also true in this case that

$$-|r| \leq r \leq |r|.$$

Hence, in either case,

$$-|r| \leq r \leq |r|$$

[as was to be shown].

Lemma 4.4.5

For all real numbers r, $|-r| = |r|$.

Proof:

Suppose r is any real number. By Theorem T23 in Appendix A, if $r > 0$, then $-r < 0$, and if $r < 0$, then $-r > 0$. Thus

$$|-r| = \begin{cases} -r & \text{if } -r > 0 \\ 0 & \text{if } -r = 0 \\ -(-r) & \text{if } -r < 0 \end{cases} \quad \text{by definition of absolute value}$$

$$= \begin{cases} -r & \text{if } -r > 0 \\ 0 & \text{if } -r = 0 \\ r & \text{if } -r < 0 \end{cases} \quad \begin{array}{l} \text{because } -(-r) = r \text{ by Theorem T4} \\ \text{in Appendix A} \end{array}$$

$$= \begin{cases} -r & \text{if } r < 0 \\ 0 & \text{if } -r = 0 \\ r & \text{if } r > 0 \end{cases} \quad \begin{array}{l} \text{because, by Theorem T24 in Appendix A, when} \\ -r > 0, \text{ then } r < 0, \text{ when } -r < 0, \text{ then } r > 0, \\ \text{and when } -r = 0, \text{ then } r = 0 \end{array}$$

$$= \begin{cases} r & \text{if } r \geq 0 \\ -r & \text{if } r < 0 \end{cases} \quad \text{by reformatting the previous result}$$

$$= |r| \quad \text{by definition of absolute value.}$$

Lemmas 4.4.4 and 4.4.5 now provide a basis for proving the triangle inequlity.

Theorem 4.4.6 The Triangle Inequality

For all real numbers x and y, $|x + y| \leq |x| + |y|$.

Proof:

Suppose x and y, are any real numbers.

Case 1 (x + y ≥ 0): In this case, $|x + y| = x + y$, and so, by Lemma 4.4.4,

$$x \leq |x| \quad \text{and} \quad y \leq |y|.$$

Hence, by Theorem T26 of Appendix A,

$$|x + y| = x + y \leq |x| + |y|.$$

Case 2 (x + y < 0): In this case, $|x + y| = -(x + y) = (-x) + (-y)$, and so, by Lemmas 4.4.4 and 4.4.5,

$$-x \leq |-x| = |x| \quad \text{and} \quad -y \leq |-y| = |y|.$$

It follows, by Theorem T26 of Appendix A, that

$$|x + y| = (-x) + (-y) \leq |x| + |y|.$$

Hence in both cases $|x + y| \leq |x| + |y|$ *[as was to be shown]*.

Test Yourself

1. The quotient-remainder theorem says that for all integers n and d with $d > 0$, there exist _____ q and r such that _____ and _____.

2. If n and d are integers with $d > 0$, n div d is _____ and n mod d is _____.

3. The parity of an integer indicates whether the integer is _____.

4. According to the quotient-remainder theorem, if an integer n is divided by a positive integer d, the possible remainders are _____. This implies that n can be written in one of the forms _____ for some integer q.

5. To prove a statement of the form "If A_1 or A_2 or A_3, then C," prove _____ and _____ and _____.

6. The triangle inequality says that for all real numbers x and y, _____.

Exercise Set 4.4

For each of the values of n and d given in 1–6, find integers q and r such that $n = dq + r$ and $0 \leq r < d$.

1. $n = 70, d = 9$
2. $n = 62, d = 7$
3. $n = 36, d = 40$
4. $n = 3, d = 11$
5. $n = -45, d = 11$
6. $n = -27, d = 8$

Evaluate the expressions in 7–10.

7. a. 43 div 9   b. 43 mod 9

8. a. 50 div 7   b. 50 mod 7

9. a. 28 div 5   b. 28 mod 5

10. a. 30 div 2   b. 30 mod 2

11. Check the correctness of formula (4.4.1) given in Example 4.4.3 for the following values of *DayT* and N.
    a. *DayT* = 6 (Saturday) and $N = 15$
    b. *DayT* = 0 (Sunday) and $N = 7$
    c. *DayT* = 4 (Thursday) and $N = 12$

✱ 12. Justify formula (4.4.1) for general values of *DayT* and N.

13. On a Monday a friend says he will meet you again in 30 days. What day of the week will that be?

H 14. If today is Tuesday, what day of the week will it be 1,000 days from today?

15. January 1, 2000, was a Saturday, and 2000 was a leap year. What day of the week will January 1, 2050, be?

16. Suppose d is a positive integer and n is any integer. If $d \mid n$, what is the remainder obtained when the quotient-remainder theorem is applied to n with divisor d?

17. Prove that the product of any two consecutive integers is even.

18. The result of exercise 17 suggests that the second apparent blind alley in the discussion of Example 4.4.7 might not be a blind alley after all. Write a new proof of Theorem 4.4.3 based on this observation.

19. Prove that for all integers n, $n^2 - n + 3$ is odd.

20. Suppose a is an integer. If a mod 7 = 4, what is $5a$ mod 7? In other words, if division of a by 7 gives a remainder of 4, what is the remainder when $5a$ is divided by 7?

21. Suppose b is an integer. If b mod 12 = 5, what is $8b$ mod 12? In other words, if division of b by 12 gives a remainder of 5, what is the remainder when $8b$ is divided by 12?

22. Suppose c is an integer. If c mod 15 = 3, what is $10c$ mod 15? In other words, if division of c by 15 gives a remainder of 3, what is the remainder when $10c$ is divided by 15?

23. Prove that for all integers n, if n mod 5 = 3 then n^2 mod 5 = 4.

24. Prove that for all integers m and n, if m mod 5 = 2 and n mod 5 = 1 then mn mod 5 = 2.

25. Prove that for all integers a and b, if a mod 7 = 5 and b mod 7 = 6 then ab mod 7 = 2.

H 26. Prove that a necessary and sufficient condition for a non-negative integer n to be divisible by a positive integer d is that n mod d = 0.

27. Show that any integer n can be written in one of the three forms

$$n = 3q \quad \text{or} \quad n = 3q + 1 \quad \text{or} \quad n = 3q + 2$$

for some integer q.

28. a. Use the quotient-remainder theorem with $d = 3$ to prove that the product of any three consecutive integers is divisible by 3.
    b. Use the *mod* notation to rewrite the result of part (a).

H 29. a. Use the quotient-remainder theorem with $d = 3$ to prove that the square of any integer has the form $3k$ or $3k + 1$ for some integer k.

    b. Use the *mod* notation to rewrite the result of part (a).

30. a. Use the quotient-remainder theorem with $d = 3$ to prove that the product of any two consecutive integers has the form $3k$ or $3k + 2$ for some integer k.

    b. Use the *mod* notation to rewrite the result of part (a).

In 31–33, you may use the properties listed in Example 4.2.3.

31. a. Prove that for all integers m and n, $m + n$ and $m - n$ are either both odd or both even.
    b. Find all solutions to the equation $m^2 - n^2 = 56$ for which both m and n are positive integers.
    c. Find all solutions to the equation $m^2 - n^2 = 88$ for which both m and n are positive integers.

32. Given any integers a, b, and c, if $a - b$ is even and $b - c$ is even, what can you say about the parity of $2a - (b + c)$? Prove your answer.

33. Given any integers a, b, and c, if $a - b$ is odd and $b - c$ is even, what can you say about the parity of $a - c$? Prove your answer.

H 34. Given any integer n, if $n > 3$, could $n, n + 2$, and $n + 4$ all be prime? Prove or give a counterexample.

Prove each of the statements in 35–46.

35. The fourth power of any integer has the form $8m$ or $8m + 1$ for some integer m.

H 36. The product of any four consecutive integers is divisible by 8.

37. The square of any integer has the form $4k$ or $4k + 1$ for some integer k.

H 38. For any integer n, $n^2 + 5$ is not divisible by 4.

H 39. The sum of any four consecutive integers has the form $4k + 2$ for some integer k.

40. For any integer n, $n(n^2 - 1)(n + 2)$ is divisible by 4.

41. For all integers m, $m^2 = 5k$, or $m^2 = 5k + 1$, or $m^2 = 5k + 4$ for some integer k.

H 42. Every prime number except 2 and 3 has the form $6q + 1$ or $6q + 5$ for some integer q.

43. If n is an odd integer, then $n^4 \bmod 16 = 1$.

H 44. For all real numbers x and y, $|x| \cdot |y| = |xy|$.

45. For all real numbers r and c with $c \geq 0$, if $-c \leq r \leq c$, then $|r| \leq c$.

46. For all real numbers r and c with $c \geq 0$, if $|r| \leq c$, then $-c \leq r \leq c$.

Answers for Test Yourself

1. integers; $n = dq + r$; $0 \leq r < d$ 2. the quotient obtained when n is divided by d; the nonnegative remainder obtained when n is divided by d 3. odd or even   4. $0, 1, 2, \ldots, (d - 1)$; $dq, dq + 1, dq + 2, \ldots, dq + (d - 1)$ 5. If A_1, then C; If A_2, then C; If A_3, then C 6. $|x + y| \leq |x| + |y|$

4.5 Indirect Argument: Contradiction and Contraposition

Reductio ad absurdum is one of a mathematician's finest weapons. It is a far finer gambit than any chess gambit: a chess player may offer the sacrifice of a pawn or even a piece, but the mathematician offers the game. — G. H. Hardy, 1877–1947

In a direct proof you start with the hypothesis of a statement and make one deduction after another until you reach the conclusion. Indirect proofs are more roundabout. One kind of indirect proof, *argument by contradiction,* is based on the fact that either a statement is true or it is false but not both. So if you can show that the assumption that a given statement is not true leads logically to a contradiction, impossibility, or absurdity, then that assumption must be false: and, hence, the given statement must be true. This method of proof is also known as *reductio ad impossible* or *reductio ad absurdum* because it relies on reducing a given assumption to an impossibility or absurdity.

Argument by contradiction occurs in many different settings. For example, if a man accused of holding up a bank can prove that he was some place else at the time the crime was committed, he will certainly be acquitted. The logic of his defense is as follows:

Suppose I did commit the crime. Then at the time of the crime, I would have had to be at the scene of the crime. In fact, at the time of the crime I was in a meeting with 20 people far from the crime scene, as they will testify. This contradicts the assumption that I committed the crime since it is impossible to be in two places at one time. Hence that assumption is false.

Another example occurs in debate. One technique of debate is to say, "Suppose for a moment that what my opponent says is correct." Starting from this supposition, the debater then deduces one statement after another until finally arriving at a statement that is completely ridiculous and unacceptable to the audience. By this means the debater shows the opponent's statement to be false.

The point of departure for a proof by contradiction is the supposition that the statement to be proved is false. The goal is to reason to a contradiction. Thus proof by contradiction has the following outline:

Method of Proof by Contradiction

1. Suppose the statement to be proved is false. That is, suppose that the negation of the statement is true.

2. Show that this supposition leads logically to a contradiction.

3. Conclude that the statement to be proved is true.

Note Be very careful when writing the negation!

There are no clear-cut rules for when to try a direct proof and when to try a proof by contradiction, but there are some general guidelines. Proof by contradiction is indicated if you want to show that there is no object with a certain property, or if you want to show that a certain object does not have a certain property. The next two examples illustrate these situations.

Example 4.5.1 There Is No Greatest Integer

Use proof by contradiction to show that there is no greatest integer.

Solution   Most small children believe there is a greatest integer—they often call it a "zillion." But with age and experience, they change their belief. At some point they realize that if there were a greatest integer, they could add 1 to it to obtain an integer that was greater still. Since that is a contradiction, no greatest integer can exist. This line of reasoning is the heart of the formal proof.

For the proof, the "certain property" is the property of being the greatest integer. To prove that there is no object with this property, begin by supposing the negation: that there is an object with the property.

Starting Point: Suppose not. Suppose there is a greatest integer; call it N.

This means that $N \geq n$ for all integers n.

To Show: This supposition leads logically to a contradiction.

Theorem 4.5.1

There is no greatest integer.

Proof:

[We take the negation of the theorem and suppose it to be true.] Suppose not. That is, suppose there is a greatest integer N. *[We must deduce a contradiction.]* Then

---

continued on page 156

$N \geq n$ for every integer n. Let $M = N + 1$. Now M is an integer since it is a sum of integers. Also $M > N$ since $M = N + 1$. Thus M is an integer that is greater than N. So N is the greatest integer and N is not the greatest integer, which is a contradiction. *[This contradiction shows that the supposition is false and, hence, that the theorem is true.]*

∎

After a contradiction has been reached, the logic of the argument is always the same: "This is a contradiction. Hence the supposition is false and the theorem is true." Because of this, most mathematics texts end proofs by contradiction at the point at which the contradiction has been obtained.

The contradiction in the next example is based on the fact that $1/2$ is not an integer.

Example 4.5.2 No Integer Can Be Both Even and Odd

The fact that no integer can be both even and odd follows from the uniqueness part of the quotient-remainder theorem. A full proof of this part of the theorem is outlined in exercise 18 of Section 4.6. This example shows how to use proof by contradiction to prove one specific case.

Theorem 4.5.2

There is no integer that is both even and odd.

Proof:

[We take the negation of the theorem and suppose it to be true.] Suppose not. That is, suppose there is at least one integer n that is both even and odd. *[We must deduce a contradiction.]* By definition of even, $n = 2a$ for some integer a, and by definition of odd, $n = 2b + 1$ for some integer b. Consequently,

$$2a = 2b + 1 \qquad \text{by equating the two expressions for } n$$

and so

$$2a - 2b = 1$$
$$2(a - b) = 1$$
$$a - b = 1/2 \qquad \text{by algebra.}$$

Now since a and b are integers, the difference $a - b$ must also be an integer. But $a - b = 1/2$, and $1/2$ is not an integer. Thus $a - b$ is an integer and $a - b$ is not an integer, which is a contradiction. *[This contradiction shows that the supposition is false and, hence, that the theorem is true.]*

∎

The next example asks you to show that the sum of any rational number and any irrational number is irrational. One way to think of this is in terms of a certain object (the sum of a rational and an irrational) not having a certain property (the property of being rational). This suggests trying a proof by contradiction: suppose the object has the property and deduce a contradiction.

Example 4.5.3 The Sum of a Rational Number and an Irrational Number

Use proof by contradiction to show that the sum of any rational number and any irrational number is irrational.

Caution! The negation of "The sum of any irrational number and any rational number is irrational" is NOT "The sum of any irrational number and any rational number is rational."

Solution Begin by supposing the negation of what you are to prove. Be very careful when writing down what this means. If you take the negation incorrectly, the entire rest of the proof will be flawed. In this example, the statement to be proved can be written formally as

$$\forall \text{ real numbers } r \text{ and } s, \text{ if } r \text{ is rational and}$$
$$s \text{ is irrational, then } r + s \text{ is irrational.}$$

From this you can see that the negation is

$$\exists \text{ a rational number } r \text{ and an irrational}$$
$$\text{number } s \text{ such that } r + s \text{ is rational.}$$

It follows that the starting point and what is to be shown are as follows:

Starting Point: Suppose not. That is, suppose there is a rational number r and an irrational number s such that $r + s$ is rational.

To Show: This supposition leads to a contradiction.

To derive a contradiction, you need to understand what you are supposing: that there are numbers r and s such that r is rational, s is irrational, and $r + s$ is rational. By definition of rational and irrational, this means that s cannot be written as a quotient of any two integers but that r and $r + s$ can:

$$r = \frac{a}{b} \quad \text{for some integers } a \text{ and } b \text{ with } b \neq 0, \text{ and} \qquad 4.5.1$$

$$r + s = \frac{c}{d} \quad \text{for some integers } c \text{ and } d \text{ with } d \neq 0. \qquad 4.5.2$$

If you substitute (4.5.1) into (4.5.2), you obtain

$$\frac{a}{b} + s = \frac{c}{d}.$$

Subtracting a/b from both sides gives

$$s = \frac{c}{d} - \frac{a}{b}$$

$$= \frac{bc}{bd} - \frac{ad}{bd} \qquad \text{by rewriting } c/d \text{ and } a/b \text{ as equivalent fractions}$$

$$= \frac{bc - ad}{bd} \qquad \text{by the rule for subtracting fractions with the same denominator.}$$

But both $bc - ad$ and bd are integers because products and differences of integers are integers, and $bd \neq 0$ by the zero product property. Hence s can be expressed as a quotient of two integers with a nonzero denominator, and so s is rational, which contradicts the supposition that it is irrational.

This discussion is summarized in a formal proof.

Theorem 4.5.3

The sum of any rational number and any irrational number is irrational.

Proof:

[We take the negation of the theorem and suppose it to be true.] Suppose not. That is, suppose there is a rational number r and an irrational number s such that $r + s$ is

continued on page 158

rational. *[We must deduce a contradiction.]* By definition of rational, $r = a/b$ and $r + s = c/d$ for some integers $a, b, c,$ and d with $b \neq 0$ and $d \neq 0$. By substitution,

$$\frac{a}{b} + s = \frac{c}{d},$$

and so

$$s = \frac{c}{d} - \frac{a}{b} \qquad \text{by subtracting } a/b \text{ from both sides}$$

$$= \frac{bc - ad}{bd} \qquad \text{by the laws of algebra.}$$

Now $bc - ad$ and bd are both integers *[since $a, b, c,$ and d are integers and since products and differences of integers are integers]*, and $bd \neq 0$ *[by the zero product property]*. Hence s is a quotient of the two integers $bc - ad$ and bd with $bd \neq 0$. Thus, by definition of rational, s is rational, which contradicts the supposition that s is irrational. *[Hence the supposition is false and the theorem is true.]*

■

Argument by Contraposition

A second form of indirect argument, *argument by contraposition,* is based on the logical equivalence between a statement and its contrapositive. To prove a statement by contraposition, you take the contrapositive of the statement, prove the contrapositive by a direct proof, and conclude that the original statement is true. The underlying reasoning is that since a conditional statement is logically equivalent to its contrapositive, if the contrapositive is true then the statement must also be true.

Method of Proof by Contraposition

1. Express the statement to be proved in the form

   $$\forall x \text{ in } D, \text{ if } P(x) \text{ then } Q(x).$$

   (This step may be done mentally.)

2. Rewrite this statement in the contrapositive form

   $$\forall x \text{ in } D, \text{ if } Q(x) \text{ is false then } P(x) \text{ is false.}$$

   (This step may also be done mentally.)

3. Prove the contrapositive by a direct proof.

   a. Suppose x is a (particular but arbitrarily chosen) element of D such that $Q(x)$ is false.

   b. Show that $P(x)$ is false.

Example 4.5.4 If the Square of an Integer Is Even, Then the Integer Is Even

Prove that for all integers n, if n^2 is even then n is even.

Solution   First form the contrapositive of the statement to be proved.

Contrapositive: For all integers n, if n is not even then n^2 is not even.

By the quotient-remainder theorem with $d = 2$, any integer is even or odd, so any integer that is not even is odd. Also by Theorem 4.5.2, no integer can be both even and odd. So if an integer is odd, then it is not even. Thus the contrapositive can be restated as follows:

Contrapositive: For all integers n, if n is odd then n^2 is odd.

A straightforward computation is the heart of a direct proof for this statement, as shown below.

Proposition 4.5.4

For all integers n, if n^2 is even then n is even.

Proof (by contraposition):

Suppose n is any odd integer. *[We must show that n^2 is odd.]* By definition of odd, $n = 2k + 1$ for some integer k. By substitution and algebra,

$$n^2 = (2k + 1)^2 = 4k^2 + 4k + 1 = 2(2k^2 + 2k) + 1.$$

But $2k^2 + 2k$ is an integer because products and sums of integers are integers. So $n^2 = 2 \cdot (\text{an integer}) + 1$, and thus, by definition of odd, n^2 is odd *[as was to be shown]*.

We used the word *proposition* here rather than *theorem* because although the word *theorem* can refer to any statement that has been proved, mathematicians often restrict it to especially important statements that have many and varied consequences. Then they use the word **proposition** to refer to a statement that is somewhat less consequential but nonetheless worth writing down. We will use Proposition 4.5.4 in Section 4.6 to prove that $\sqrt{2}$ is irrational. ■

Relation between Proof by Contradiction and Proof by Contraposition

Observe that any proof by contraposition can be recast in the language of proof by contradiction. In a proof by contraposition, the statement

$$\forall x \text{ in } D, \text{ if } P(x) \text{ then } Q(x)$$

is proved by giving a direct proof of the equivalent statement

$$\forall x \text{ in } D, \text{ if } \sim Q(x) \text{ then } \sim P(x).$$

To do this, you suppose you are given an arbitrary element x of D such that $\sim Q(x)$. You then show that $\sim P(x)$. This is illustrated in Figure 4.5.1.

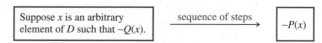

Figure 4.5.1 Proof by Contraposition

Exactly the same sequence of steps can be used as the heart of a proof by contradiction for the given statement. The only thing that changes is the context in which the steps are written down.

To rewrite the proof as a proof by contradiction, you suppose there is an x in D such that $P(x)$ and $\sim Q(x)$. You then follow the steps of the proof by contraposition to deduce the statement $\sim P(x)$. But $\sim P(x)$ is a contradiction to the supposition that $P(x)$ and $\sim Q(x)$. (Because to contradict a conjunction of two statements, it is only necessary to contradict one of them.) This process is illustrated in Figure 4.5.2.

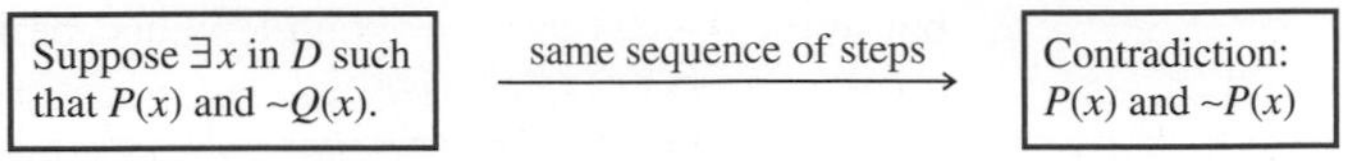

Suppose $\exists x$ in D such that $P(x)$ and $\sim Q(x)$.	$\xrightarrow{\text{same sequence of steps}}$	Contradiction: $P(x)$ and $\sim P(x)$

Figure 4.5.2 Proof by Contradiction

As an example, here is a proof by contradiction of Proposition 4.5.4, namely that for any integer n, if n^2 is even then n is even.

Proposition 4.5.4

For all integers n, if n^2 is even then n is even.

Proof (by contradiction):

[We take the negation of the theorem and suppose it to be true.] Suppose not. That is, suppose there is an integer n such that n^2 is even and n is not even. *[We must deduce a contradiction.]* By the quotient-remainder theorem with $d = 2$, any integer is even or odd. Hence, since n is not even it is odd, and thus, by definition of odd, $n = 2k + 1$ for some integer k. By substitution and algebra:

$$n^2 = (2k + 1)^2 = 4k^2 + 4k + 1 = 2(2k^2 + 2k) + 1.$$

But $2k^2 + 2k$ is an integer because products and sums of integers are integers. So $n^2 = 2 \cdot (\text{an integer}) + 1$, and thus, by definition of odd, n^2 is odd. Therefore, n^2 is both even and odd. This contradicts Theorem 4.5.2, which states that no integer can be both even and odd. *[This contradiction shows that the supposition is false and, hence, that the proposition is true.]*

Note that when you use proof by contraposition, you know exactly what conclusion you need to show, namely the negation of the hypothesis; whereas in proof by contradiction, it may be difficult to know what contradiction to head for. On the other hand, when you use proof by contradiction, once you have deduced any contradiction whatsoever, you are done. The main advantage of contraposition over contradiction is that you avoid having to take (possibly incorrectly) the negation of a complicated statement. The disadvantage of contraposition as compared with contradiction is that you can use contraposition only for a specific class of statements—those that are universal and conditional. The previous discussion shows that any statement that can be proved by contraposition can be proved by contradiction. But the converse is not true. Statements such as "$\sqrt{2}$ is irrational" (discussed in the next section) can be proved by contradiction but not by contraposition.

Proof as a Problem-Solving Tool

Direct proof, disproof by counterexample, proof by contradiction, and proof by contraposition are all tools that may be used to help determine whether statements are true or false. Given a statement of the form

For all elements in a domain, if (hypothesis) then (conclusion), imagine elements in the domain that satisfy the hypothesis. Ask yourself: Must they satisfy the conclusion? If you can see that the answer is "yes" in all cases, then the statement is true and your insight will form the basis for a direct proof. If after some thought it is not clear that the answer is "yes," ask yourself whether there are elements of the domain that satisfy the hypothesis and *not* the conclusion. If you are successful in finding some, then the statement is false and you have a counterexample. On the other hand, if you are not successful in finding such elements, perhaps none exist. Perhaps you can show that assuming the existence of elements in the domain that satisfy the hypothesis and not the conclusion leads logically to a contradiction. If so, then the given statement is true and you have the basis for a proof by contradiction. Alternatively, you could imagine elements of the domain for which the conclusion is false and ask whether such elements also fail to satisfy the hypothesis. If the answer in all cases is "yes," then you have a basis for a proof by contraposition.

Solving problems, especially difficult problems, is rarely a straightforward process. At any stage of following the guidelines above, you might want to try the method of a previous stage again. If, for example, you fail to find a counterexample for a certain statement, your experience in trying to find it might help you decide to reattempt a direct argument rather than trying an indirect one. Psychologists who have studied problem solving have found that the most successful problem solvers are those who are flexible and willing to use a variety of approaches without getting stuck in any one of them for very long. Mathematicians sometimes work for months (or longer) on difficult problems. Don't be discouraged if some problems in this book take you quite a while to solve.

Learning the skills of proof and disproof is much like learning other skills, such as those used in swimming, tennis, or playing a musical instrument. When you first start out, you may feel bewildered by all the rules, and you may not feel confident as you attempt new things. But with practice the rules become internalized and you can use them in conjunction with all your other powers—of balance, coordination, judgment, aesthetic sense—to concentrate on winning a meet, winning a match, or playing a concert successfully.

Now that you have worked through the first five sections of this chapter, return to the idea that, above all, a proof or disproof should be a convincing argument. You need to know how direct and indirect proofs and counterexamples are structured. But to use this knowledge effectively, you must use it in conjunction with your imaginative powers, your intuition, and especially your common sense.

Test Yourself

1. To prove a statement by contradiction, you suppose that _____ and you show that _____.

2. A proof by contraposition of a statement of the form "$\forall x \in D$, if $P(x)$ then $Q(x)$" is a direct proof of _____.

3. To prove a statement of the form "$\forall x \in D$, if $P(x)$ then $Q(x)$" by contraposition, you suppose that _____ and you show that _____.

Exercise Set 4.5

1. Fill in the blanks in the following proof by contradiction that there is no least positive real number.

   Proof: Suppose not. That is, suppose that there is a least positive real number x. [*We must deduce* (a)] Consider the number $x/2$. Since x is a positive real number, $x/2$ is also (b) . In addition, we can deduce that $x/2 < x$ by multiplying both sides of the inequality $1 < 2$ by (c) and dividing (d) . Hence $x/2$ is a positive real number that is less than the least positive real number. This is a (e) [*Thus the supposition is false, and so there is no least positive real number.*]

2. Is $\dfrac{1}{0}$ an irrational number? Explain.

3. Use proof by contradiction to show that for all integers n, $3n + 2$ is not divisible by 3.

4. Use proof by contradiction to show that for all integers m, $7m + 4$ is not divisible by 7.

Carefully formulate the negations of each of the statements in 5–7. Then prove each statement by contradiction.

5. There is no greatest even integer.

6. There is no greatest negative real number.

7. There is no least positive rational number.

8. Fill in the blanks for the following proof that the difference of any rational number and any irrational number is irrational.
   Proof: Suppose not. That is, suppose that there exist __(a)__ x and __(b)__ y such that $x - y$ is rational. By definition of rational, there exist integers a, b, c, and d with $b \neq 0$ and $d \neq 0$ so that $x = $ __(c)__ and $x - y = $ __(d)__. By substitution,

$$\frac{a}{b} - y = \frac{c}{d}$$

Adding y and subtracting $\dfrac{c}{d}$ on both sides gives

$$y = \text{(e)}$$
$$= \frac{ad}{bd} - \frac{bc}{bd}$$
$$= \frac{ad - bc}{bd} \qquad \text{by algebra.}$$

Now both $ad - bc$ and bd are integers because products and differences of __(f)__ are __(g)__. And $bd \neq 0$ by the __(h)__. Hence y is a ratio of integers with a nonzero denominator, and thus y is __(i)__ by definition of rational. We therefore have both that y is irrational and that y is rational, which is a contradiction. *[Thus the supposition is false and the statement to be proved is true.]*

9. **a.** When asked to prove that the difference of any irrational number and any rational number is irrational, a student began, "Suppose not. That is, suppose the difference of any irrational number and any rational number is rational." What is wrong with beginning the proof in this way? (*Hint*: Review the answer to exercise 11 in Section 3.2.)
   b. Prove that the difference of any irrational number and any rational number is irrational.

Prove each statement in 10–17 by contradiction.

10. The square root of any irrational number is irrational.

11. The product of any nonzero rational number and any irrational number is irrational.

12. If a and b are rational numbers, $b \neq 0$, and r is an irrational number, then $a + br$ is irrational.

H 13. For any integer n, $n^2 - 2$ is not divisible by 4.

H 14. For all prime numbers a, b, and c, $a^2 + b^2 \neq c^2$.

H 15. If a, b, and c are integers and $a^2 + b^2 = c^2$, then at least one of a and b is even.

H ✱ 16. For all odd integers a, b, and c, if z is a solution of $ax^2 + bx + c = 0$ then z is irrational. (In the proof, use the properties of even and odd integers that are listed in Example 4.2.3.)

17. For all integers a, if $a \bmod 6 = 3$, then $a \bmod 3 \neq 2$.

18. Fill in the blanks in the following proof by contraposition that for all integers n, if $5 \nmid n^2$ then $5 \nmid n$.
    Proof (by contraposition): *[The contrapositive is: For all integers n, if $5 \mid n$ then $5 \mid n^2$.]* Suppose n is any integer such that __(a)__. *[We must show that __(b)__.]* By definition of divisibility, $n = $ __(c)__ for some integer k. By substitution, $n^2 = $ __(d)__ $= 5(5k^2)$. But $5k^2$ is an integer because it is a product of integers. Hence $n^2 = 5 \cdot$ (an integer), and so __(e)__ *[as was to be shown]*.

Prove the statements in 19 and 20 by contraposition.

19. If a product of two positive real numbers is greater than 100, then at least one of the numbers is greater than 10.

20. If a sum of two real numbers is less than 50, then at least one of the numbers is less than 25.

21. Consider the statement "For all integers n, if n^2 is odd then n is odd."
    a. Write what you would suppose and what you would need to show to prove this statement by contradiction.
    b. Write what you would suppose and what you would need to show to prove this statement by contraposition.

22. Consider the statement "For all real numbers r, if r^2 is irrational then r is irrational."
    a. Write what you would suppose and what you would need to show to prove this statement by contradiction.
    b. Write what you would suppose and what you would need to show to prove this statement by contraposition.

Prove each of the statements in 23–29 in two ways: (a) by contraposition and (b) by contradiction.

23. The negative of any irrational number is irrational.

24. The reciprocal of any irrational number is irrational. (The **reciprocal** of a nonzero real number x is $1/x$.)

H 25. For all integers n, if n^2 is odd then n is odd.

26. For all integers a, b, and c, if $a \nmid bc$ then $a \nmid b$. (Recall that the symbol $\nmid$ means "does not divide.")

H 27. For all integers m and n, if $m + n$ is even then m and n are both even or m and n are both odd.

28. For all integers m and n, if mn is even then m is even or n is even.

29. For all integers a, b, and c, if $a \mid b$ and $a \nmid c$, then $a \nmid (b + c)$. (*Hint:* To prove $p \to q \vee r$, it suffices to prove either $p \wedge {\sim}q \to r$ or $p \wedge {\sim}r \to q$. See exercise 14 in Section 2.2.)

30. The following "proof" that every integer is rational is incorrect. Find the mistake.

"**Proof (by contradiction):** Suppose not. Suppose every integer is irrational. Then the integer 1 is irrational. But $1 = 1/1$, which is rational. This is a contradiction. *[Hence the supposition is false and the theorem is true.]*"

31. a. Prove by contraposition: For all positive integers n, r, and s, if $rs \leq n$, then $r \leq \sqrt{n}$ or $s \leq \sqrt{n}$.
 b. Prove: For all integers $n > 1$, if n is not prime, then there exists a prime number p such that $p \leq \sqrt{n}$ and n is divisible by p. (*Hints:* Use the result of part (a) and Theorems 4.3.1, 4.3.3, and 4.3.4.)
 c. State the contrapositive of the result of part (b).
 The results of exercise 31 provide a way to test whether an integer is prime.

Test for Primality
Given an integer $n > 1$, to test whether n is prime check to see if it is divisible by a prime number less than or equal to its square root. If it is not divisible by any of these numbers, then it is prime.

---

32. Use the test for primality to determine whether the following numbers are prime or not.
 a. 667 **b.** 557 **c.** 527 **d.** 613

33. The sieve of Eratosthenes, named after its inventor, the Greek scholar Eratosthenes (276–194 B.C.E.), provides a way to find all prime numbers less than or equal to some fixed number n. To construct it, write out all the integers from 2 to n. Cross out all multiples of 2 except 2 itself, then all multiples of 3 except 3 itself, then all multiples of 5 except 5 itself, and so forth. Continue crossing out the multiples of each successive prime number up to $\sqrt{n}$. The numbers that are not crossed out are all the prime numbers from 2 to n. Here is a sieve of Eratosthenes that includes the numbers from 2 to 27. The multiples of 2 are crossed out with a /, the multiples of 3 with a \, and the multiples of 5 with a —.

2 3 4̸ 5 6̸ 7 8̸ 9̸ 1̸0̸ 11 1̸2̸ 13 1̸4̸
1̸5̸ 1̸6̸ 17 1̸8̸ 19 2̸0̸ 2̸1̸ 2̸2̸ 23 2̸4̸ 2̸5̸ 2̸6̸ 2̸7̸

Use the sieve of Eratosthenes to find all prime numbers less than 100.

34. Use the test for primality and the result of exercise 33 to determine whether the following numbers are prime.
 a. 9,269 **b.** 9,103 **c.** 8,623 **d.** 7,917

H ✱ **35.** Use proof by contradiction to show that every integer greater than 11 is a sum of two composite numbers.

Answers for Test Yourself

1. the statement is false; this supposition leads to a contradiction 2. the contrapositive of the statement, namely, $\forall x \in D$, if $\sim Q(x)$ then $\sim P(x)$ 3. x is any *[particular but arbitrarily chosen]* element of D for which $Q(x)$ is false; $P(x)$ is false

4.6 Indirect Argument: Two Classical Theorems

He is unworthy of the name of man who does not know that the diagonal of a square is incommensurable with its side.—Plato (ca. 428–347 B.C.E.)

This section contains proofs of two of the most famous theorems in mathematics: that $\sqrt{2}$ is irrational and that there are infinitely many prime numbers. Both proofs are examples of indirect arguments and were well known more than 2,000 years ago, but they remain exemplary models of mathematical argument to this day.

The Irrationality of $\sqrt{2}$

When mathematics flourished at the time of the ancient Greeks, mathematicians believed that given any two line segments, say $A:$ —— and $B:$ ————, a certain unit of length could be found so that segment A was exactly a units long and segment B was exactly b units long. (The segments were said to be *commensurable* with respect to this special unit of length.) Then the ratio of the lengths of A and B would be in the same proportion as the ratio of the integers a and b. Symbolically:

$$\frac{\text{length } A}{\text{length } B} = \frac{a}{b}.$$

Now it is easy to find a line segment of length $\sqrt{2}$; just take the diagonal of the unit square:

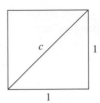

By the Pythagorean theorem, $c^2 = 1^2 + 1^2 = 2$, and so $c = \sqrt{2}$. If the belief of the ancient Greeks were correct, there would be integers a and b such that

$$\frac{\text{length (diagonal)}}{\text{length (side)}} = \frac{a}{b}.$$

And this would imply that

$$\frac{c}{1} = \frac{\sqrt{2}}{1} = \sqrt{2} = \frac{a}{b}.$$

But then $\sqrt{2}$ would be a ratio of two integers, or, in other words, $\sqrt{2}$ would be rational.

In the fourth or fifth century B.C.E., the followers of the Greek mathematician and philosopher Pythagoras discovered that $\sqrt{2}$ was not rational. This discovery was very upsetting to them, for it undermined their deep, quasi-religious belief in the power of whole numbers to describe phenomena.

The following proof of the irrationality of $\sqrt{2}$ was known to Aristotle and is similar to that in the tenth book of Euclid's *Elements of Geometry*. The Greek mathematician Euclid is best known as a geometer. In fact, knowledge of the geometry in the first six books of his *Elements* has been considered an essential part of a liberal education for more than 2,000 years. Books 7–10 of his *Elements*, however, contain much that we would now call number theory.

The proof begins by supposing the negation: $\sqrt{2}$ is rational. This means that there exist integers m and n such that $\sqrt{2} = m/n$. Now if m and n have any common factors, these may be factored out to obtain a new fraction, equal to m/n, in which the numerator and denominator have no common factors. (For example, $18/12 = (6 \cdot 3)/(6 \cdot 2) = 3/2$, which is a fraction whose numerator and denominator have no common factors.) Thus, without loss of generality, we may assume that m and n had no common factors in the first place. We will then derive the contradiction that m and n *do* have a common factor of 2. The argument makes use of Proposition 4.5.4. If the square of an integer is even, then that integer is even.

Euclid
(fl. 300 B.C.E.)

Note Strictly speaking, being able to assume that m and n have no common factors is a consequence of the "well-ordering principle for the integers," which is discussed in Section 5.4.

Theorem 4.6.1 Irrationality of $\sqrt{2}$

$\sqrt{2}$ is irrational.

Proof:

[We take the negation and suppose it to be true.] Suppose not. That is, suppose $\sqrt{2}$ is rational. Then there are integers m and n with no common factors such that

$$\sqrt{2} = \frac{m}{n} \qquad\qquad 4.6.1$$

Bettmann/CORBIS

[by dividing m and n by any common factors if necessary]. [We must derive a contradiction.] Squaring both sides of equation (4.6.1) gives

$$2 = \frac{m^2}{n^2}.$$

Or, equivalently,

$$m^2 = 2n^2. \tag{4.6.2}$$

Note that equation (4.6.2) implies that m^2 is even (by definition of even). It follows that m is even (by Proposition 4.5.4). We file this fact away for future reference and also deduce (by definition of even) that

$$m = 2k \quad \text{for some integer } k. \tag{4.6.3}$$

Substituting equation (4.6.3) into equation (4.6.2), we see that

$$m^2 = (2k)^2 = 4k^2 = 2n^2.$$

Dividing both sides of the right-most equation by 2 gives

$$n^2 = 2k^2.$$

Consequently, n^2 is even, and so n is even (by Proposition 4.5.4). But we also know that m is even. *[This is the fact we filed away.]* Hence both m and n have a common factor of 2. But this contradicts the supposition that m and n have no common factors. *[Hence the supposition is false and so the theorem is true.]*

Now that you have seen the proof that $\sqrt{2}$ is irrational, you can use the irrationality of $\sqrt{2}$ to derive the irrationality of certain other real numbers.

Example 4.6.1 Irrationality of $1 + 3\sqrt{2}$

Prove by contradiction that $1 + 3\sqrt{2}$ is irrational.

Solution   The essence of the argument is the observation that if $1 + 3\sqrt{2}$ could be written as a ratio of integers, then so could $\sqrt{2}$. But by Theorem 4.6.1, we know that to be impossible.

Proposition 4.6.2

$1 + 3\sqrt{2}$ is irrational.

Proof:

Suppose not. Suppose $1 + 3\sqrt{2}$ is rational. *[We must derive a contradiction.]* Then by definition of rational,

$$1 + 3\sqrt{2} = \frac{a}{b} \quad \text{for some integers } a \text{ and } b \text{ with } b \neq 0.$$

continued on page 166

It follows that

$$3\sqrt{2} = \frac{a}{b} - 1 \qquad \text{by subtracting 1 from both sides}$$

$$= \frac{a}{b} - \frac{b}{b} \qquad \text{by substitution}$$

$$= \frac{a - b}{b} \qquad \text{by the rule for subtracting fractions with a common denominator.}$$

Hence

$$\sqrt{2} = \frac{a - b}{3b} \qquad \text{by dividing both sides by 3.}$$

But $a - b$ and $3b$ are integers (since a and b are integers and differences and products of integers are integers), and $3b \neq 0$ by the zero product property. Hence $\sqrt{2}$ is a quotient of the two integers $a - b$ and $3b$ with $3b \neq 0$, and so $\sqrt{2}$ is rational (by definition of rational.) This contradicts the fact that $\sqrt{2}$ is irrational. *[This contradiction shows that the supposition is false.]* Hence $1 + 3\sqrt{2}$ is irrational.

■

Are There Infinitely Many Prime Numbers?

You know that a prime number is a positive integer that cannot be factored as a product of two smaller positive integers. Is the set of all such numbers infinite, or is there a largest prime number? The answer was known to Euclid, and a proof that the set of all prime numbers is infinite appears in Book 9 of his *Elements of Geometry*.

Euclid's proof requires one additional fact we have not yet established: If a prime number divides an integer, then it does not divide the next successive integer.

Proposition 4.6.3

For any integer a and any prime number p, if $p \mid a$ then $p \nmid (a + 1)$.

Proof:

Suppose not. That is, suppose there exists an integer a and a prime number p such that $p \mid a$ and $p \mid (a + 1)$. Then, by definition of divisibility, there exist integers r and s such that $a = pr$ and $a + 1 = ps$. It follows that

$$1 = (a + 1) - a = ps - pr = p(s - r),$$

and so (since $s - r$ is an integer) $p \mid 1$. But, by Theorem 4.3.2, the only integer divisors of 1 are 1 and -1, and $p > 1$ because p is prime. Thus $p \leq 1$ and $p > 1$, which is a contradiction. *[Hence the supposition is false, and the proposition is true.]*

The idea of Euclid's proof is this: Suppose the set of prime numbers were finite. Then you could take the product of all the prime numbers and add one. By Theorem 4.3.4 this number must be divisible by some prime number. But by Proposition 4.6.3, this number is not divisible by any of the prime numbers in the set. Hence there must be a prime number that is not in the set of all prime numbers, which is impossible.

The following formal proof fills in the details of this outline.

Theorem 4.6.4 Infinitude of the Primes

The set of prime numbers is infinite.

Proof (by contradiction):

Suppose not. That is, suppose the set of prime numbers is finite. *[We must deduce a contradiction.]* Then some prime number p is the largest of all the prime numbers, and hence we can list the prime numbers in ascending order:

$$2, \; 3, \; 5, \; 7, 11,\ldots,p.$$

Let N be the product of all the prime numbers plus 1:

$$N = (2\cdot3\cdot5\cdot7\cdot11\cdots p) + 1$$

Then $N > 1$, and so, by Theorem 4.3.4, N is divisible by some prime number q. Because q is prime, q must equal one of the prime numbers 2, 3, 5, 7, 11, $\ldots$, p. Thus, by definition of divisibility, q divides $2\cdot3\cdot5\cdot7\cdot11\cdots p$, and so, by Proposition 4.6.3, q does not divide $(2\cdot3\cdot5\cdot7\cdot11\cdots p) + 1$, which equals N. Hence N is divisible by q and N is not divisible by q, and we have reached a contradiction. *[Therefore, the supposition is false and the theorem is true.]*

∎

The proof of Theorem 4.6.4 shows that if you form the product of all prime numbers up to a certain point and add one, the result, N, is divisible by a prime number not on the list. The proof does not show that N is, itself, prime. In the exercises at the end of this section you are asked to find an example of an integer N constructed in this way that is not prime.

When to Use Indirect Proof

The examples in this section and Section 4.5 have not provided a definitive answer to the question of when to prove a statement directly and when to prove it indirectly. Many theorems can be proved either way. Usually, however, when both types of proof are possible, indirect proof is clumsier than direct proof. In the absence of obvious clues suggesting indirect argument, try first to prove a statement directly. Then, if that does not succeed, look for a counterexample. If the search for a counterexample is unsuccessful, look for a proof by contradiction or contraposition.

Open Questions in Number Theory

In this section we proved that there are infinitely many prime numbers. There is no known formula for obtaining primes, but a few formulas have been found to be more successful at producing them than other formulas. One such is due to Marin Mersenne, a French monk who lived from 1588–1648. *Mersenne primes* have the form $2^p - 1$, where p is prime. Not all numbers of this form are prime, but because Mersenne primes are easier to test for primality than are other numbers, most of the largest known prime numbers are Mersenne primes.

An interesting question is whether there are infinitely many Mersenne primes. As of the date of publication of this book, the answer is not known, but new mathematical discoveries are being made every day and by the time you read this someone may have discovered the answer. Another formula that seems to produce a relatively large number

*Ben Joseph Green
(born 1977)*

*Terence Chi-Shen Tao
(born 1975)*

*Marie-Sophie Germain
(1776–1831)*

of prime numbers is due to Fermat. *Fermat primes* are prime numbers of the form $2^{2^n} + 1$, where n is a positive integer. Are there infinitely many Fermat primes? Again, as of now, no one knows. Similarly unknown are whether there are infinitely many primes of the form $n^2 + 1$, where n is a positive integer, and whether there is always a prime number between integers n^2 and $(n + 1)^2$.

Another famous open question involving primes is the *twin primes conjecture,* which states that there are infinitely many pairs of prime numbers of the form p and $p + 2$. As with other well-known problems in number theory, this conjecture has withstood computer testing up to extremely large numbers, and some progress has been made toward a proof. In 2004, Ben Green and Terence Tao showed that for any integer $m > 1$, there is a sequence of m equally spaced integers all of which are prime. In other words, there are are positive integers n and k so that the following numbers are all prime:

$$n, n + k, \ n + 2k, \ n + 3k, \ \ldots, \ n + (m - 1)k.$$

Related to the twin primes conjecture is a conjecture made by Sophie Germain, a French mathematician born in 1776, who made significant progress toward a proof of Fermat's Last Theorem. Germain conjectured that there are infinitely many prime number pairs of the form p and $2p + 1$. Initial values of p with this property are 2, 3, 5, 11, 23, 29, 41, and 53, and computer testing has verified the conjecture for many additional values. In fact, as of the writing of this book, the largest prime p for which $2p + 1$ is also known to be prime is $183027 \cdot 2^{265440} - 1$. This is a number with 79911 decimal digits! But compared with infinity, any number, no matter how large, is less than a drop in the bucket.

In 1844, the Belgian mathematician Eugène Catalan conjectured that the only solutions to the equation $x^n - y^m = 1$, where x, y, n, and m are all integers greater than 1, is $3^2 - 2^3 = 1$. This conjecture was finally proved by Preda Mihăilescu in 2002.

In 1993, while trying to prove Fermat's last theorem, an amateur number theorist, Andrew Beal, became intrigued by the equation $x^m + y^n = z^k$, where no two of x, y, or z have any common factor other than ± 1. When diligent effort, first by hand and then by computer, failed to reveal any solutions, Beal conjectured that no solutions exist. His conjecture has become known as *Beal's conjecture,* and he has offered a prize of $100,000 to anyone who can either prove or disprove it.

These are just a few of a large number of open questions in number theory. Many people believe that mathematics is a fixed subject that changes very little from one century to the next. In fact, more mathematical questions are being raised and more results are being discovered now than ever before in history.

Test Yourself

1. The ancient Greeks discovered that in a right triangle where both legs have length 1, the ratio of the length of the hypotenuse to the length of one of the legs is not equal to a ratio of _____.

2. One way to prove that $\sqrt{2}$ is an irrational number is to assume that $\sqrt{2} = a/b$ for some integers a and b that have no common factor greater than 1, use the lemma that says

that if the square of an integer is even then _____, and eventually show that a and b _____.

3. One way to prove that there are infinitely many prime numbers is to assume that there is a largest prime number p, construct the number _____, and then show that this number has to be divisible by a prime number that is greater than _____.

Exercise Set 4.6

1. A calculator display shows that $\sqrt{2} = 1.414213562$, and $1.414213562 = \dfrac{1414213562}{1000000000}$. This suggests that $\sqrt{2}$ is a rational number, which contradicts Theorem 4.6.1. Explain the discrepancy.

2. Example 4.2.1(h) illustrates a technique for showing that any repeating decimal number is rational. A calculator display shows the result of a certain calculation as 40.72727272727. Can you be sure that the result of the calculation is a rational number? Explain.

Determine which statements in 3–13 are true and which are false. Prove those that are true and disprove those that are false.

3. $6 - 7\sqrt{2}$ is irrational.

4. $3\sqrt{2} - 7$ is irrational.

5. $\sqrt{4}$ is irrational.

6. $\sqrt{2}/6$ is rational.

7. The sum of any two irrational numbers is irrational.

8. The difference of any two irrational numbers is irrational.

9. The positive square root of a positive irrational number is irrational.

10. If r is any rational number and s is any irrational number, then r/s is irrational.

11. The sum of any two positive irrational numbers is irrational.

12. The product of any two irrational numbers is irrational.

H **13.** If an integer greater than 1 is a perfect square, then its cube root is irrational.

14. Consider the following sentence: If x is rational then $\sqrt{x}$ is irrational. Is this sentence always true, sometimes true and sometimes false, or always false? Justify your answer.

15. a. Prove that for all integers a, if a^3 is even then a is even.
   b. Prove that $\sqrt[3]{2}$ is irrational.

16. a. Use proof by contradiction to show that for any integer n, it is impossible for n to equal both $3q_1 + r_1$ and $3q_2 + r_2$, where q_1, q_2, r_1, and r_2, are integers, $0 \leq r_1 < 3, 0 \leq r_2 < 3$, and $r_1 \neq r_2$.
   b. Use proof by contradiction, the quotient-remainder theorem, division into cases, and the result of part (a) to prove that for all integers n, if n^2 is divisible by 3 then n is divisible by 3.
   c. Prove that $\sqrt{3}$ is irrational.

17. Give an example to show that if d is not prime and n^2 is divisible by d, then n need not be divisible by d.

H **18.** The quotient-remainder theorem says not only that there exist quotients and remainders but also that the quotient and remainder of a division are unique. Prove the uniqueness. That is, prove that if a and d are integers with $d > 0$ and if q_1, r_1, q_2, and r_2 are integers such that

$$a = dq_1 + r_1 \quad \text{where } 0 \leq r_1 < d$$

and

$$a = dq_2 + r_2 \quad \text{where } 0 \leq r_2 < d,$$

then

$$q_1 = q_2 \quad \text{and} \quad r_1 = r_2.$$

H **19.** Prove that $\sqrt{5}$ is irrational.

H **20.** Prove that for any integer a, $9 \nmid (a^2 - 3)$.

21. An alternative proof of the irrationality of $\sqrt{2}$ counts the number of 2's on the two sides of the equation $2n^2 = m^2$ and uses the unique factorization of integers theorem

to deduce a contradiction. Write a proof that uses this approach.

22. Use the proof technique illustrated in exercise 21 to prove that if n is any positive integer that is not a perfect square, then $\sqrt{n}$ is irrational.

H **23.** Prove that $\sqrt{2} + \sqrt{3}$ is irrational.

✱ **24.** Prove that $\log_5(2)$ is irrational. (*Hint:* Use the unique factorisation of integers theorem.)

H **25.** Let $N = 2 \cdot 3 \cdot 5 \cdot 7 + 1$. What remainder is obtained when N is divided by 2? 3? 5? 7? Is N prime? Justify your answer.

H **26.** Suppose a is an integer and p is a prime number such that $p \mid a$ and $p \mid (a + 3)$. What can you deduce about p? Why?

27. Let $p_1, p_2, p_3, \ldots$ be a list of all prime numbers in ascending order. Here is a table of the first six:

p_1	p_2	p_3	p_4	p_5	p_6
2	3	5	7	11	13

H **a.** For each $i = 1, 2, 3, 4, 5, 6$, let $N_i = p_1 p_2 \cdots p_i + 1$. Calculate N_1, N_2, N_3, N_4, N_5, and N_6.
   b. For each $i = 1, 2, 3, 4, 5, 6$, find the smallest prime number q_i such that q_i divides N_i. (*Hint:* Use the test for primality from exercise 31 in Section 4.5 to determine your answers.)

For exercises 28 and 29, use the fact that for all integers n,

$$n! = n(n-1) \ldots 3 \cdot 2 \cdot 1.$$

28. An alternative proof of the infinitude of the prime numbers begins as follows:

Proof: Suppose there are only finitely many prime numbers. Then one is the largest. Call it p. Let $M = p! + 1$. We will show that there is a prime number q such that $q > p$. Complete this proof.

H ✱ **29.** Prove that for all integers n, if $n > 2$ then there is a prime number p such that $n < p < n!$.

H ✱ **30.** Prove that if $p_1, p_2, \ldots$, and p_n are distinct prime numbers with $p_1 = 2$ and $n > 1$, then $p_1 p_2 \cdots p_n + 1$ can be written in the form $4k + 3$ for some integer k.

H **31. a.** Fermat's last theorem says that for all integers $n > 2$, the equation $x^n + y^n = z^n$ has no positive integer solution (solution for which x, y, and z are positive integers). Prove the following: If for all prime numbers $p > 2$, $x^p + y^p = z^p$ has no positive integer solution, then for any integer $n > 2$ that is not a power of 2, $x^n + y^n = z^n$ has no positive integer solution.
   b. Fermat proved that there are no integers x, y, and z such that $x^4 + y^4 = z^4$. Use this result to remove the restriction in part (a) that n not be a power of 2. That is, prove that if n is a power of 2 and $n > 4$, then $x^n + y^n = z^n$ has no positive integer solution.

For exercises 32–35 note that to show there is a unique object with a certain property, show that (1) there is an object with the property and (2) if objects A and B have the property, then $A = B$.

32. Prove that there exists a unique prime number of the form $n^2 - 1$, where n is an integer that is greater than or equal to 2.

33. Prove that there exists a unique prime number of the form $n^2 + 2n - 3$, where n is a positive integer.

34. Prove that there is at most one real number a with the property that $a + r = r$ for all real numbers r. (Such a number is called an *additive identity*.)

35. Prove that there is at most one real number b with the property that $br = r$ for all real numbers r. (Such a number is called a *multiplicative identity*.)

Answers for Test Yourself

1. two integers   2. the integer is even; have a common factor greater than 1   3. $2 \cdot 3 \cdot 5 \cdot 7 \cdot 11 \cdots p + 1$; p

CHAPTER 5

SEQUENCES, MATHEMATICAL INDUCTION, AND RECURSION

One of the most important tasks of mathematics is to discover and characterize regular patterns, such as those associated with processes that are repeated. The main mathematical structure used in the study of repeated processes is the *sequence,* and the main mathematical tool used to verify conjectures about sequences is *mathematical induction*. In this chapter we introduce the notation and terminology of sequences, show how to use both ordinary and strong mathematical induction to prove properties about them, illustrate the various ways recursively defined sequences arise, describe a method for obtaining an explicit formula for a recursively defined sequence, and explain how to verify the correctness of such a formula. We also discuss a principle—the well-ordering principle for the integers—that is logically equivalent to the two forms of mathematical induction.

5.1 Sequences

A mathematician, like a painter or poet, is a maker of patterns.
— G. H. Hardy, *A Mathematician's Apology,* 1940

Imagine that a person decides to count his ancestors. He has two parents, four grandparents, eight great-grandparents, and so forth, These numbers can be written in a row as

$$2, 4, 8, 16, 32, 64, 128, \ldots$$

The symbol ".." is called an *ellipsis*. It is shorthand for "and so forth."

To express the pattern of the numbers, suppose that each is labeled by an integer giving its position in the row.

Position in the row	1	2	3	4	5	6	7...
Number of ancestors	2	4	8	16	32	64	128...

The number corresponding to position 1 is 2, which equals 2^1. The number corresponding to position 2 is 4, which equals 2^2. For positions 3, 4, 5, 6, and 7, the corresponding numbers are 8, 16, 32, 64, and 128, which equal 2^3, 2^4, 2^5, 2^6, and 2^7, respectively. For

Note Strictly speaking, the true value of A_k is less than 2^k when k is large, because ancestors from one branch of the family tree may also appear on other branches of the tree.

a general value of k, let A_k be the number of ancestors in the kth generation back. The pattern of computed values strongly suggests the following for each k:

$$A_k = 2^k.$$

• **Definition**

A **sequence** is a function whose domain is either all the integers between two given integers or all the integers greater than or equal to a given integer.

We typically represent a sequence as a set of elements written in a row. In the sequence denoted

$$a_m, a_{m+1}, a_{m+2}, \ldots, a_n,$$

each individual element a_k (read "a sub k") is called a **term.** The k in a_k is called a **subscript** or **index,** m (which may be any integer) is the subscript of the **initial term,** and n (which must be greater than or equal to m) is the subscript of the **final term.** The notation

$$a_m, a_{m+1}, a_{m+2}, \ldots$$

denotes an **infinite sequence.** An **explicit formula** or **general formula** for a sequence is a rule that shows how the values of a_k depend on k.

The following example shows that it is possible for two different formulas to give sequences with the same terms.

Example 5.1.1 Finding Terms of Sequences Given by Explicit Formulas

Define sequences $a_1, a_2, a_3, \ldots$ and $b_2, b_3, b_4, \ldots$ by the following explicit formulas:

$$a_k = \frac{k}{k+1} \quad \text{for all integers } k \geq 1,$$

$$b_i = \frac{i-1}{i} \quad \text{for all integers } i \geq 2.$$

Compute the first five terms of both sequences.

Solution

$$a_1 = \frac{1}{1+1} = \frac{1}{2} \qquad\qquad b_2 = \frac{2-1}{2} = \frac{1}{2}$$

$$a_2 = \frac{2}{2+1} = \frac{2}{3} \qquad\qquad b_3 = \frac{3-1}{3} = \frac{2}{3}$$

$$a_3 = \frac{3}{3+1} = \frac{3}{4} \qquad\qquad b_4 = \frac{4-1}{4} = \frac{3}{4}$$

$$a_4 = \frac{4}{4+1} = \frac{4}{5} \qquad\qquad b_5 = \frac{5-1}{5} = \frac{4}{5}$$

$$a_5 = \frac{5}{5+1} = \frac{5}{6} \qquad\qquad b_6 = \frac{6-1}{6} = \frac{5}{6}$$

As you can see, the first terms of both sequences are $\frac{1}{2}, \frac{2}{3}, \frac{3}{4}, \frac{4}{5}, \frac{5}{6}$; in fact, it can be shown that all terms of both sequences are identical. ∎

The next example shows that an infinite sequence may have a finite number of values.

Example 5.1.2   An Alternating Sequence

Compute the first six terms of the sequence $c_0, c_1, c_2, \ldots$ defined as follows:

$$c_j = (-1)^j \quad \text{for all integers } j \geq 0.$$

Solution

$$
\begin{aligned}
c_0 &= (-1)^0 = 1 \\
c_1 &= (-1)^1 = -1 \\
c_2 &= (-1)^2 = 1 \\
c_3 &= (-1)^3 = -1 \\
c_4 &= (-1)^4 = 1 \\
c_5 &= (-1)^5 = -1
\end{aligned}
$$

Thus the first six terms are $1, -1, 1, -1, 1, -1$. By exercises 33 and 34 of Section 4.1, even powers of -1 equal 1 and odd powers of -1 equal -1. It follows that the sequence oscillates endlessly between 1 and -1. ∎

In Examples 5.1.1 and 5.1.2 the task was to compute the first few values of a sequence given by an explicit formula. The next example treats the question of how to find an explicit formula for a sequence with given initial terms. Any such formula is a guess, but it is very useful to be able to make such guesses.

Example 5.1.3   Finding an Explicit Formula to Fit Given Initial Terms

Find an explicit formula for a sequence that has the following initial terms:

$$1, \quad -\frac{1}{4}, \quad \frac{1}{9}, \quad -\frac{1}{16}, \quad \frac{1}{25}, \quad -\frac{1}{36}, \ldots$$

Solution   Denote the general term of the sequence by a_k and suppose the first term is a_1. Then observe that the denominator of each term is a perfect square. Thus the terms can be rewritten as

$$
\frac{1}{1^2}, \quad \frac{(-1)}{2^2}, \quad \frac{1}{3^2}, \quad \frac{(-1)}{4^2}, \quad \frac{1}{5^2}, \quad \frac{(-1)}{6^2}.
$$
$$
\updownarrow \qquad \updownarrow \qquad \updownarrow \qquad \updownarrow \qquad \updownarrow \qquad \updownarrow
$$
$$
a_1 \qquad a_2 \qquad a_3 \qquad a_4 \qquad a_5 \qquad a_6
$$

Note that the denominator of each term equals the square of the subscript of that term, and that the numerator equals ± 1. Hence

$$a_k = \frac{\pm 1}{k^2}.$$

Also the numerator oscillates back and forth between $+1$ and -1; it is $+1$ when k is odd and -1 when k is even. To achieve this oscillation, insert a factor of $(-1)^{k+1}$ (or $(-1)^{k-1}$) into the formula for a_k. *[For when k is odd, $k+1$ is even and thus $(-1)^{k+1} = +1$; and when k is even, $k+1$ is odd and thus $(-1)^{k+1} = -1$.]* Consequently, an explicit formula that gives the correct first six terms is

$$a_k = \frac{(-1)^{k+1}}{k^2} \quad \text{for all integers } k \geq 1.$$

Caution! It is also possible for two sequences to start off with the same initial values but diverge later on. See exercise 5 at the end of this section.

Note that making the first term a_0 would have led to the alternative formula

$$a_k = \frac{(-1)^k}{(k+1)^2} \quad \text{for all integers } k \geq 0.$$

You should check that this formula also gives the correct first six terms. ■

Summation Notation

Consider again the example in which $A_k = 2^k$ represents the number of ancestors a person has in the kth generation back. What is the total number of ancestors for the past six generations? The answer is

$$A_1 + A_2 + A_3 + A_4 + A_5 + A_6 = 2^1 + 2^2 + 2^3 + 2^4 + 2^5 + 2^6 = 126.$$

It is convenient to use a shorthand notation to write such sums. In 1772 the French mathematician Joseph Louis Lagrange introduced the capital Greek letter sigma, Σ, to denote the word *sum* (or *summation*), and defined the summation notation as follows:

Joseph Louis Lagrange (1736–1813)

• Definition

If m and n are integers and $m \leq n$, the symbol $\sum\limits_{k=m}^{n} a_k$, read the **summation from k equals m to n of a-sub-k**, is the sum of all the terms $a_m,\ a_{m+1},\ a_{m+2},\ \ldots,\ a_n$. We say that $a_m + a_{m+1} + a_{m+2} + \ldots + a_n$ is the **expanded form** of the sum, and we write

$$\sum_{k=m}^{n} a_k = a_m + a_{m+1} + a_{m+2} + \cdots + a_n.$$

We call k the **index** of the summation, m the **lower limit** of the summation, and n the **upper limit** of the summation.

Example 5.1.4 Computing Summations

Let $a_1 = -2$, $a_2 = -1$, $a_3 = 0$, $a_4 = 1$, and $a_5 = 2$. Compute the following:

a. $\sum\limits_{k=1}^{5} a_k$ b. $\sum\limits_{k=2}^{2} a_k$ c. $\sum\limits_{k=1}^{2} a_{2k}$

Solution

a. $\sum\limits_{k=1}^{5} a_k = a_1 + a_2 + a_3 + a_4 + a_5 = (-2) + (-1) + 0 + 1 + 2 = 0$

b. $\sum\limits_{k=2}^{2} a_k = a_2 = -1$

c. $\sum\limits_{k=1}^{2} a_{2k} = a_{2 \cdot 1} + a_{2 \cdot 2} = a_2 + a_4 = -1 + 1 = 0$ ■

Oftentimes, the terms of a summation are expressed using an explicit formula. For instance, it is common to see summations such as

$$\sum_{k=1}^{5} k^2 \quad \text{or} \quad \sum_{i=0}^{8} \frac{(-1)^i}{i+1}.$$

Example 5.1.5 When the Terms of a Summation Are Given by a Formula

Compute the following summation:

$$\sum_{k=1}^{5} k^2.$$

Solution
$$\sum_{k=1}^{5} k^2 = 1^2 + 2^2 + 3^2 + 4^2 + 5^2 = 55.$$ ∎

When the upper limit of a summation is a variable, an ellipsis is used to write the summation in expanded form.

Example 5.1.6 Changing from Summation Notation to Expanded Form

Write the following summation in expanded form:

$$\sum_{i=0}^{n} \frac{(-1)^i}{i+1}.$$

Solution
$$\sum_{i=0}^{n} \frac{(-1)^i}{i+1} = \frac{(-1)^0}{0+1} + \frac{(-1)^1}{1+1} + \frac{(-1)^2}{2+1} + \frac{(-1)^3}{3+1} + \cdots + \frac{(-1)^n}{n+1}$$
$$= \frac{1}{1} + \frac{(-1)}{2} + \frac{1}{3} + \frac{(-1)}{4} + \cdots + \frac{(-1)^n}{n+1}$$
$$= 1 - \frac{1}{2} + \frac{1}{3} - \frac{1}{4} + \cdots + \frac{(-1)^n}{n+1}$$ ∎

Example 5.1.7 Changing from Expanded Form to Summation Notation

Express the following using summation notation:

$$\frac{1}{n} + \frac{2}{n+1} + \frac{3}{n+2} + \cdots + \frac{n+1}{2n}.$$

Solution   The general term of this summation can be expressed as $\dfrac{k+1}{n+k}$ for integers k from 0 to n. Hence

$$\frac{1}{n} + \frac{2}{n+1} + \frac{3}{n+2} + \cdots + \frac{n+1}{2n} = \sum_{k=0}^{n} \frac{k+1}{n+k}.$$ ∎

For small values of n, the expanded form of a sum may appear ambiguous. For instance, consider

$$1^2 + 2^2 + 3^2 + \cdots + n^2.$$

This expression is intended to represent the sum of squares of consecutive integers starting with 1^2 and ending with n^2. Thus, if $n = 1$ the sum is just 1^2, if $n = 2$ the sum is $1^2 + 2^2$, and if $n = 3$ the sum is $1^2 + 2^2 + 3^2$.

Example 5.1.8   Evaluating $a_1, a_2, a_3, \ldots, a_n$ for Small n

What is the value of the expression $\dfrac{1}{1 \cdot 2} + \dfrac{1}{2 \cdot 3} + \dfrac{1}{3 \cdot 4} + \cdots + \dfrac{1}{n \cdot (n + 1)}$ when $n = 1$? $n = 2$? $n = 3$?

Caution!   Do not write that for $n = 1$, the sum is

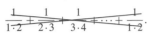

This is crossed out because it is incorrect.

Solution

When $n = 1$, the expression equals $\dfrac{1}{1 \cdot 2} = \dfrac{1}{2}$.

When $n = 2$, it equals $\dfrac{1}{1 \cdot 2} + \dfrac{1}{2 \cdot 3} = \dfrac{1}{2} + \dfrac{1}{6} = \dfrac{2}{3}$.

When $n = 3$, it is $\dfrac{1}{1 \cdot 2} + \dfrac{1}{2 \cdot 3} + \dfrac{1}{3 \cdot 4} = \dfrac{1}{2} + \dfrac{1}{6} + \dfrac{1}{12} = \dfrac{3}{4}$.   ∎

A more mathematically precise definition of summation, called a *recursive definition*, is the following:* If m is any integer, then

$$\sum_{k=m}^{m} a_k = a_m \qquad \text{and} \qquad \sum_{k=m}^{n} a_k = \sum_{k=m}^{n-1} a_k + a_n \quad \text{for all integers } n > m.$$

When solving problems, it is often useful to rewrite a summation using the recursive form of the definition, either by separating off the final term of a summation or by adding a final term to a summation.

Example 5.1.9   Separating Off a Final Term and Adding On a Final Term

a. Rewrite $\displaystyle\sum_{i=1}^{n+1} \dfrac{1}{i^2}$ by separating off the final term.

b. Write $\displaystyle\sum_{k=0}^{n} 2^k + 2^{n+1}$ as a single summation.

Solution

a. $\displaystyle\sum_{i=1}^{n+1} \dfrac{1}{i^2} = \sum_{i=1}^{n} \dfrac{1}{i^2} + \dfrac{1}{(n+1)^2}$ 　　　　 b. $\displaystyle\sum_{k=0}^{n} 2^k + 2^{n+1} = \sum_{k=0}^{n+1} 2^k$ ∎

In certain sums each term is a difference of two quantities. When you write such sums in expanded form, you sometimes see that all the terms cancel except the first and the last. Successive cancellation of terms collapses the sum like a telescope.

Example 5.1.10   A Telescoping Sum

Some sums can be transformed into telescoping sums, which then can be rewritten as a simple expression. For instance, observe that

$$\frac{1}{k} - \frac{1}{k+1} = \frac{(k+1) - k}{k(k+1)} = \frac{1}{k(k+1)}.$$

Use this identity to find a simple expression for $\displaystyle\sum_{k=1}^{n} \dfrac{1}{k(k+1)}$.

---

*Other recursively defined sequences are discussed later in this section and, in greater detail, in Section 5.5.

Solution

$$\sum_{k=1}^{n} \frac{1}{k(k+1)} = \sum_{k=1}^{n} \left(\frac{1}{k} - \frac{1}{k+1} \right)$$

$$= \left(\frac{1}{1} - \frac{1}{2} \right) + \left(\frac{1}{2} - \frac{1}{3} \right) + \left(\frac{1}{3} - \frac{1}{4} \right) + \cdots + \left(\frac{1}{n-1} - \frac{1}{n} \right) + \left(\frac{1}{n} - \frac{1}{n+1} \right)$$

$$= 1 - \frac{1}{n+1}.$$ ∎

Product Notation

The notation for the product of a sequence of numbers is analogous to the notation for their sum. The Greek capital letter pi, Π, denotes a product. For example,

$$\prod_{k=1}^{5} a_k = a_1 a_2 a_3 a_4 a_5.$$

> **• Definition**
>
> If m and n are integers and $m \le n$, the symbol $\prod_{k=m}^{n} a_k$, read the **product from k equals m to n of a-sub-k**, is the product of all the terms a_m, a_{m+1}, a_{m+2}, ..., a_n. We write
>
> $$\prod_{k=m}^{n} a_k = a_m \cdot a_{m+1} \cdot a_{m+2} \cdots a_n.$$

A recursive definition for the product notation is the following: If m is any integer, then

$$\prod_{k=m}^{m} a_k = a_m \quad \text{and} \quad \prod_{k=m}^{n} a_k = \left(\prod_{k=m}^{n-1} a_k \right) \cdot a_n \quad \text{for all integers } n > m.$$

Example 5.1.11 Computing Products

Compute the following products:

a. $\prod_{k=1}^{5} k$

b. $\prod_{k=1}^{1} \frac{k}{k+1}$

Solution

a. $\prod_{k=1}^{5} k = 1 \cdot 2 \cdot 3 \cdot 4 \cdot 5 = 120$

b. $\prod_{k=1}^{1} \frac{k}{k+1} = \frac{1}{1+1} = \frac{1}{2}$ ∎

Properties of Summations and Products

The following theorem states general properties of summations and products. The proof of the theorem is discussed in Section 5.5.

Theorem 5.1.1

If $a_m, a_{m+1}, a_{m+2}, \ldots$ and $b_m, b_{m+1}, b_{m+2}, \ldots$ are sequences of real numbers and c is any real number, then the following equations hold for any integer $n \geq m$:

1. $\displaystyle\sum_{k=m}^{n} a_k + \sum_{k=m}^{n} b_k = \sum_{k=m}^{n}(a_k + b_k)$

2. $\displaystyle c \cdot \sum_{k=m}^{n} a_k = \sum_{k=m}^{n} c \cdot a_k$ generalized distributive law

3. $\displaystyle \left(\prod_{k=m}^{n} a_k\right) \cdot \left(\prod_{k=m}^{n} b_k\right) = \prod_{k=m}^{n}(a_k \cdot b_k).$

Example 5.1.12 Using Properties of Summation and Product

Let $a_k = k + 1$ and $b_k = k - 1$ for all integers k. Write each of the following expressions as a single summation or product:

a. $\displaystyle\sum_{k=m}^{n} a_k + 2 \cdot \sum_{k=m}^{n} b_k$ b. $\displaystyle\left(\prod_{k=m}^{n} a_k\right) \cdot \left(\prod_{k=m}^{n} b_k\right)$

Solution

a. $\displaystyle\sum_{k=m}^{n} a_k + 2 \cdot \sum_{k=m}^{n} b_k = \sum_{k=m}^{n}(k+1) + 2 \cdot \sum_{k=m}^{n}(k-1)$ by substitution

$\displaystyle = \sum_{k=m}^{n}(k+1) + \sum_{k=m}^{n} 2 \cdot (k-1)$ by Theorem 5.1.1 (2)

$\displaystyle = \sum_{k=m}^{n}((k+1) + 2 \cdot (k-1))$ by Theorem 5.1.1 (1)

$\displaystyle = \sum_{k=m}^{n}(3k - 1)$ by algebraic simplification

b. $\displaystyle\left(\prod_{k=m}^{n} a_k\right) \cdot \left(\prod_{k=m}^{n} b_k\right) = \left(\prod_{k=m}^{n}(k+1)\right) \cdot \left(\prod_{k=m}^{n}(k-1)\right)$ by substitution

$\displaystyle = \prod_{k=m}^{n}(k+1) \cdot (k-1)$ by Theorem 5.1.1 (3)

$\displaystyle = \prod_{k=m}^{n}(k^2 - 1)$ by algebraic simplification ∎

Change of Variable

Observe that

$$\sum_{k=1}^{3} k^2 = 1^2 + 2^2 + 3^2$$

and also that

$$\sum_{i=1}^{3} i^2 = 1^2 + 2^2 + 3^2.$$

Hence
$$\sum_{k=1}^{3} k^2 = \sum_{i=1}^{3} i^2.$$

This equation illustrates the fact that the symbol used to represent the index of a summation can be replaced by any other symbol as long as the replacement is made in each location where the symbol occurs. As a consequence, the index of a summation is called a dummy variable. A **dummy variable** is a symbol that derives its entire meaning from its local context. Outside of that context (both before and after), the symbol may have another meaning entirely.

The appearance of a summation can be altered by more complicated changes of variable as well. For example, observe that

$$\sum_{j=2}^{4} (j-1)^2 = (2-1)^2 + (3-1)^2 + (4-1)^2$$
$$= 1^2 + 2^2 + 3^2$$
$$= \sum_{k=1}^{3} k^2.$$

A general procedure to transform the first summation into the second is illustrated in Example 5.1.13.

Example 5.1.13 Transforming a Sum by a Change of Variable

Transform the following summation by making the specified change of variable.

$$\text{summation: } \sum_{k=0}^{6} \frac{1}{k+1} \qquad \text{change of variable: } j = k+1$$

Solution First calculate the lower and upper limits of the new summation:

$$\text{When } k = 0, \quad j = k+1 = 0+1 = 1.$$
$$\text{When } k = 6, \quad j = k+1 = 6+1 = 7.$$

Thus the new sum goes from $j = 1$ to $j = 7$.

Next calculate the general term of the new summation. You will need to replace each occurrence of k by an expression in j:

$$\text{Since } j = k+1, \text{ then } k = j-1.$$
$$\text{Hence } \frac{1}{k+1} = \frac{1}{(j-1)+1} = \frac{1}{j}.$$

Finally, put the steps together to obtain

$$\sum_{k=0}^{6} \frac{1}{k+1} = \sum_{j=1}^{7} \frac{1}{j}. \qquad\qquad 5.1.1$$

∎

Equation (5.1.1) can be given an additional twist by noting that because the j in the right-hand summation is a dummy variable, it may be replaced by any other variable name, as long as the substitution is made in every location where j occurs. In particular, it is legal to substitute k in place of j to obtain

$$\sum_{j=1}^{7} \frac{1}{j} = \sum_{k=1}^{7} \frac{1}{k}.$$

5.1.2

Putting equations (5.1.1) and (5.1.2) together gives

$$\sum_{k=0}^{6} \frac{1}{k+1} = \sum_{k=1}^{7} \frac{1}{k}.$$

Sometimes it is necessary to shift the limits of one summation in order to add it to another. An example is the algebraic proof of the binomial theorem, given in Section 9.7. A general procedure for making such a shift when the upper limit is part of the summand is illustrated in the next example.

Example 5.1.14 When the Upper Limit Appears in the Expression to Be Summed

a. Transform the following summation by making the specified change of variable.

$$\textit{summation: } \sum_{k=1}^{n+1} \left(\frac{k}{n+k} \right) \qquad \textit{change of variable: } j = k - 1$$

b. Transform the summation obtained in part (a) by changing all j's to k's.

Solution

a. When $k = 1$, then $j = k - 1 = 1 - 1 = 0$. (So the new lower limit is 0.) When $k = n + 1$, then $j = k - 1 = (n + 1) - 1 = n$. (So the new upper limit is n.)

Since $j = k - 1$, then $k = j + 1$. Also note that n is a constant as far as the terms of the sum are concerned. It follows that

$$\frac{k}{n+k} = \frac{j+1}{n+(j+1)}$$

and so the general term of the new summation is

$$\frac{j+1}{n+(j+1)}.$$

Therefore,

$$\sum_{k=1}^{n+1} \frac{k}{n+k} = \sum_{j=0}^{n} \frac{j+1}{n+(j+1)}.$$

5.1.3

b. Changing all the j's to k's in the right-hand side of equation (5.1.3) gives

$$\sum_{j=0}^{n} \frac{j+1}{n+(j+1)} = \sum_{k=0}^{n} \frac{k+1}{n+(k+1)}$$

5.1.4

Combining equations (5.1.3) and (5.1.4) results in

$$\sum_{k=1}^{n+1} \frac{k}{n+k} = \sum_{k=0}^{n} \frac{k+1}{n+(k+1)}.$$

∎

Factorial and "n Choose r" Notation

The product of all consecutive integers up to a given integer occurs so often in mathematics that it is given a special notation—*factorial* notation.

• Definition

For each positive integer n, the quantity n **factorial** denoted $n!$, is defined to be the product of all the integers from 1 to n:

$$n! = n \cdot (n - 1) \cdots 3 \cdot 2 \cdot 1.$$

Zero factorial, denoted 0!, is defined to be 1:

$$0! = 1.$$

The definition of zero factorial as 1 may seem odd, but, as you will see when you read Chapter 9, it is convenient for many mathematical formulas.

Example 5.1.15 The First Ten Factorials

$$
\begin{aligned}
0! &= 1 & 1! &= 1 \\
2! &= 2 \cdot 1 = 2 & 3! &= 3 \cdot 2 \cdot 1 = 6 \\
4! &= 4 \cdot 3 \cdot 2 \cdot 1 = 24 & 5! &= 5 \cdot 4 \cdot 3 \cdot 2 \cdot 1 = 120 \\
6! &= 6 \cdot 5 \cdot 4 \cdot 3 \cdot 2 \cdot 1 = 720 & 7! &= 7 \cdot 6 \cdot 5 \cdot 4 \cdot 3 \cdot 2 \cdot 1 = 5{,}040 \\
8! &= 8 \cdot 7 \cdot 6 \cdot 5 \cdot 4 \cdot 3 \cdot 2 \cdot 1 & 9! &= 9 \cdot 8 \cdot 7 \cdot 6 \cdot 5 \cdot 4 \cdot 3 \cdot 2 \cdot 1 \\
&= 40{,}320 & &= 362{,}880
\end{aligned}
$$

∎

As you can see from the example above, the values of $n!$ grow very rapidly. For instance, $40! \cong 8.16 \times 10^{47}$, which is a number that is too large to be computed exactly using the standard integer arithmetic of the machine-specific implementations of many computer languages. (The symbol $\cong$ means "is approximately equal to.")

A recursive definition for factorial is the following: Given any nonnegative integer n,

$$n! = \begin{cases} 1 & \text{if } n = 0 \\ n \cdot (n - 1)! & \text{if } n \geq 1. \end{cases}$$

Caution! Note that $n \cdot (n - 1)!$ is to be interpreted as $n \cdot [(n - 1)!]$.

The next example illustrates the usefulness of the recursive definition for making computations.

Example 5.1.16 Computing with Factorials

Simplify the following expressions:

a. $\dfrac{8!}{7!}$ b. $\dfrac{5!}{2! \cdot 3!}$ c. $\dfrac{1}{2! \cdot 4!} + \dfrac{1}{3! \cdot 3!}$ d. $\dfrac{(n + 1)!}{n!}$ e. $\dfrac{n!}{(n - 3)!}$

Solution

a. $\dfrac{8!}{7!} = \dfrac{8 \cdot 7!}{7!} = 8$

b. $\dfrac{5!}{2! \cdot 3!} = \dfrac{5 \cdot 4 \cdot 3!}{2! \cdot 3!} = \dfrac{5 \cdot 4}{2 \cdot 1} = 10$

c. $\dfrac{1}{2!\cdot 4!} + \dfrac{1}{3!\cdot 3!} = \dfrac{1}{2!\cdot 4!}\cdot\dfrac{3}{3} + \dfrac{1}{3!\cdot 3!}\cdot\dfrac{4}{4}$ by multiplying each numerator and denominator by just what is necessary to obtain a common denominator

$\qquad\qquad\qquad = \dfrac{3}{3\cdot 2!\cdot 4!} + \dfrac{4}{3!\cdot 4\cdot 3!}$ by rearranging factors

$\qquad\qquad\qquad = \dfrac{3}{3!\cdot 4!} + \dfrac{4}{3!\cdot 4!}$ because $3\cdot 2! = 3!$ and $4\cdot 3! = 4!$

$\qquad\qquad\qquad = \dfrac{7}{3!\cdot 4!}$ by the rule for adding fractions with a common denominator

$\qquad\qquad\qquad = \dfrac{7}{144}$

d. $\dfrac{(n+1)!}{n!} = \dfrac{(n+1)\cdot\cancel{n!}}{\cancel{n!}} = n+1$

e. $\dfrac{n!}{(n-3)!} = \dfrac{n\cdot(n-1)\cdot(n-2)\cdot\cancel{(n-3)!}}{\cancel{(n-3)!}} = n\cdot(n-1)\cdot(n-2)$

$\qquad\qquad\quad = n^3 - 3n^2 + 2n$ ∎

An important use for the factorial notation is in calculating values of quantities, called *n choose r*, that occur in many branches of mathematics, especially those connected with the study of counting techniques and probability.

• Definition

Let n and r be integers with $0 \le r \le n$. The symbol

$$\binom{n}{r}$$

is read "***n* choose *r***" and represents the number of subsets of size r that can be chosen from a set with n elements.

Observe that the definition implies that $\binom{n}{r}$ will always be an integer because it is a number of subsets. In Section 9.5 we will explore many uses of *n* choose *r* for solving problems involving counting, and we will prove the following computational formula:

• Formula for Computing $\binom{n}{r}$

For all integers n and r with $0 \le r \le n$,

$$\binom{n}{r} = \dfrac{n!}{r!(n-r)!}.$$

In the meantime, we will provide a few experiences with using it. Because *n* choose *r* is always an integer, you can be sure that all the factors in the denominator of the formula will be canceled out by factors in the numerator. Many electronic calculators have keys for computing values of $\binom{n}{r}$. These are denoted in various ways such as nCr, $C(n,r)$, ^nC_r, and $C_{n,r}$. The letter C is used because the quantities $\binom{n}{r}$ are also called *combinations*. Sometimes they are referred to as *binomial coefficients* because of the connection with the binomial theorem discussed in Section 9.6.

Example 5.1.17 Computing $\binom{n}{r}$ by Hand

Use the formula for computing $\binom{n}{r}$ to evaluate the following expressions:

a. $\binom{8}{5}$ b. $\binom{4}{0}$ c. $\binom{n+1}{n}$

Solution

a. $\binom{8}{5} = \dfrac{8!}{5!(8-5)!}$

$= \dfrac{8 \cdot 7 \cdot \cancel{6} \cdot \cancel{5} \cdot \cancel{4} \cdot \cancel{3} \cdot \cancel{2} \cdot \cancel{1}}{(\cancel{5} \cdot \cancel{4} \cdot \cancel{3} \cdot \cancel{2} \cdot 1) \cdot (\cancel{3} \cdot \cancel{2} \cdot \cancel{1})}$ always cancel common factors before multiplying

$= 56.$

b. $\binom{4}{4} = \dfrac{4!}{4!(4-4)!} = \dfrac{4!}{4!0!} = \dfrac{\cancel{4 \cdot 3 \cdot 2 \cdot 1}}{\cancel{(4 \cdot 3 \cdot 2 \cdot 1)}(1)} = 1$

The fact that $0! = 1$ makes this formula computable. It gives the correct value because a set of size 4 has exactly one subset of size 4, namely itself.

c. $\binom{n+1}{n} = \dfrac{(n+1)!}{n!((n+1)-n)!} = \dfrac{(n+1)!}{n!1!} = \dfrac{(n+1) \cdot \cancel{n!}}{\cancel{n!}} = n+1$ ∎

Test Yourself

Answers to Test Yourself questions are located at the end of each section.

1. The notation $\sum_{k=m}^{n} a_k$ is read "_____."

2. The expanded form of $\sum_{k=m}^{n} a_k$ is _____.

3. The value of $a_1 + a_2 + a_3 + \cdots + a_n$ when $n = 2$ is "_____."

4. The notation $\prod_{k=m}^{n} a_k$ is read "_____."

5. If n is a positive integer, then $n! = $ _____.

6. $\sum_{k=m}^{n} a_k + c \sum_{k=m}^{n} b_k = $ _____.

7. $\left(\prod_{k=m}^{n} a_k \right) \left(\prod_{k=m}^{n} b_k \right) = $ _____.

Exercise Set 5.1*

Write the first four terms of the sequences defined by the formulas in 1–6.

1. $a_k = \dfrac{k}{10+k}$, for all integers $k \geq 1$.

2. $b_j = \dfrac{5-j}{5+j}$, for all integers $j \geq 1$.

3. $c_i = \dfrac{(-1)^i}{3^i}$, for all integers $i \geq 0$.

4. $d_m = 1 + \left(\dfrac{1}{2} \right)^m$ for all integers $m \geq 0$.

5. Let $a_k = 2k + 1$ and $b_k = (k-1)^3 + k + 2$ for all integers $k \geq 0$. Show that the first three terms of these sequences are identical but that their fourth terms differ.

*For exercises with blue numbers or letters, solutions are given in Appendix B. The symbol **H** indicates that only a hint or a partial solution is given. The symbol ✶ signals that an exercise is more challenging than usual.

Find explicit formulas for sequences of the form $a_1, a_2, a_3, \ldots$ with the initial terms given in 6–12.

6. $-1, 1, -1, 1, -1, 1$ **7.** $0, 1, -2, 3, -4, 5$

8. $\dfrac{1}{4}, \dfrac{2}{9}, \dfrac{3}{16}, \dfrac{4}{25}, \dfrac{5}{36}, \dfrac{6}{49}$

9. $1 - \dfrac{1}{2}, \dfrac{1}{2} - \dfrac{1}{3}, \dfrac{1}{3} - \dfrac{1}{4}, \dfrac{1}{4} - \dfrac{1}{5}, \dfrac{1}{5} - \dfrac{1}{6}, \dfrac{1}{6} - \dfrac{1}{7}$

10. $\dfrac{1}{3}, \dfrac{4}{9}, \dfrac{9}{27}, \dfrac{16}{81}, \dfrac{25}{243}, \dfrac{36}{729}$

11. $0, -\dfrac{1}{2}, \dfrac{2}{3}, -\dfrac{3}{4}, \dfrac{4}{5}, -\dfrac{5}{6}, \dfrac{6}{7}$

12. $3, 6, 12, 24, 48, 96$

13. Let $a_0 = 2$, $a_1 = 3$, $a_2 = -2$, $a_3 = 1$, $a_4 = 0$, $a_5 = -1$, and $a_6 = -2$. Compute each of the summations and products below.

a. $\displaystyle\sum_{i=0}^{6} a_i$ **b.** $\displaystyle\sum_{i=0}^{0} a_i$ **c.** $\displaystyle\sum_{j=1}^{3} a_{2j}$ **d.** $\displaystyle\prod_{k=0}^{6} a_k$ **e.** $\displaystyle\prod_{k=2}^{2} a_k$

Compute the summations and products in 14–23.

14. $\displaystyle\sum_{k=1}^{5}(k+1)$ **15.** $\displaystyle\prod_{k=2}^{4} k^2$ **16.** $\displaystyle\sum_{m=0}^{3} \dfrac{1}{2^m}$

17. $\displaystyle\prod_{j=0}^{4}(-1)^j$ **18.** $\displaystyle\sum_{i=1}^{1} i(i+1)$ **19.** $\displaystyle\sum_{j=0}^{0}(j+1)\cdot 2^j$

20. $\displaystyle\prod_{k=2}^{2}\left(1 - \dfrac{1}{k}\right)$ **21.** $\displaystyle\sum_{k=-1}^{1}(k^2 + 3)$

22. $\displaystyle\sum_{n=1}^{10}\left(\dfrac{1}{n} - \dfrac{1}{n+1}\right)$ **23.** $\displaystyle\prod_{i=2}^{5} \dfrac{i(i+2)}{(i-1)\cdot(i+1)}$

Write the summations in 24–27 in expanded form.

24. $\displaystyle\sum_{i=1}^{n}(-2)^i$ **25.** $\displaystyle\sum_{j=1}^{n} j(j+1)$ **26.** $\displaystyle\sum_{k=0}^{n+1}\dfrac{1}{k!}$ **27.** $\displaystyle\sum_{i=1}^{k+1} i(i!)$

Evaluate the summations and products in 28–31 for the indicated values of the variable.

28. $\dfrac{1}{1^2} + \dfrac{1}{2^2} + \dfrac{1}{3^2} + \ldots + \dfrac{1}{n^2}$; $n = 1$

29. $1(1!) + 2(2!) + 3(3!) + \ldots + m(m!)$; $m = 2$

30. $\left(\dfrac{1}{1+1}\right)\left(\dfrac{2}{2+1}\right)\left(\dfrac{3}{3+1}\right)\cdots\left(\dfrac{k}{k+1}\right)$; $k = 3$

31. $\left(\dfrac{1\cdot 2}{3\cdot 4}\right)\left(\dfrac{4\cdot 5}{6\cdot 7}\right)\left(\dfrac{6\cdot 7}{8\cdot 9}\right)\cdots\left(\dfrac{m\cdot(m+1)}{(m+2)\cdot(m+3)}\right)$; $m = 1$

Rewrite 32–34 by separating off the final term.

32. $\displaystyle\sum_{i=1}^{k+1} i(i!)$ **33.** $\displaystyle\sum_{k=1}^{m+1} k^2$ **34.** $\displaystyle\sum_{m=1}^{n+1} m(m+1)$

Write each of 35–37 as a single summation.

35. $\displaystyle\sum_{i=1}^{k} i^3 + (k+1)^3$ **36.** $\displaystyle\sum_{k=1}^{m}\dfrac{k}{k+1} + \dfrac{m+1}{m+2}$

37. $\displaystyle\sum_{m=0}^{n}(m+1)2^m + (n+2)2^{n+1}$

Write each of 38–47 using summation or product notation.

38. $1^2 - 2^2 + 3^2 - 4^2 + 5^2 - 6^2 + 7^2$

39. $(1^3 - 1) - (2^3 - 1) + (3^3 - 1) - (4^3 - 1) + (5^3 - 1)$

40. $(2^2 - 1)\cdot(3^2 - 1)\cdot(4^2 - 1)$

41. $\dfrac{2}{3\cdot 4} - \dfrac{3}{4\cdot 5} + \dfrac{4}{5\cdot 6} - \dfrac{5}{6\cdot 7} + \dfrac{6}{7\cdot 8}$

42. $1 - r + r^2 - r^3 + r^4 - r^5$

43. $(1 - t)\cdot(1 - t^2)\cdot(1 - t^3)\cdot(1 - t^4)$

44. $1^3 + 2^3 + 3^3 + \cdots + n^3$

45. $\dfrac{1}{2!} + \dfrac{2}{3!} + \dfrac{3}{4!} + \cdots + \dfrac{n}{(n+1)!}$

46. $n + (n-1) + (n-2) + \cdots + 1$

47. $n + \dfrac{n-1}{2!} + \dfrac{n-2}{3!} + \dfrac{n-3}{4!} + \cdots + \dfrac{1}{n!}$

Transform each of 48 and 49 by making the change of variable $i = k + 1$.

48. $\displaystyle\sum_{k=0}^{5} k(k-1)$ **49.** $\displaystyle\prod_{k=1}^{n}\dfrac{k}{k^2 + 4}$

Transform each of 50–53 by making the change of variable $j = i - 1$.

50. $\displaystyle\sum_{i=1}^{n+1}\dfrac{(i-1)^2}{i\cdot n}$ **51.** $\displaystyle\sum_{i=3}^{n}\dfrac{i}{i+n-1}$

52. $\displaystyle\sum_{i=1}^{n-1}\dfrac{i}{(n-i)^2}$ **53.** $\displaystyle\prod_{i=n}^{2n}\dfrac{n-i+1}{n+i}$

Write each of 54–56 as a single summation or product.

54. $3\cdot\displaystyle\sum_{k=1}^{n}(2k-3) + \sum_{k=1}^{n}(4-5k)$

55. $2\cdot\displaystyle\sum_{k=1}^{n}(3k^2 + 4) + 5\cdot\sum_{k=1}^{n}(2k^2 - 1)$

56. $\left(\displaystyle\prod_{k=1}^{n}\dfrac{k}{k+1}\right)\cdot\left(\prod_{k=1}^{n}\dfrac{k+1}{k+2}\right)$

Compute each of 57–72. Assume the values of the variables are restricted so that the expressions are defined.

57. $\dfrac{4!}{3!}$ **58.** $\dfrac{6!}{8!}$ **59.** $\dfrac{4!}{0!}$

60. $\dfrac{n!}{(n-1)!}$ **61.** $\dfrac{(n-1)!}{(n+1)!}$ **62.** $\dfrac{n!}{(n-2)!}$

63. $\dfrac{((n+1)!)^2}{(n!)^2}$ **64.** $\dfrac{n!}{(n-k)!}$ **65.** $\dfrac{n!}{(n-k+1)!}$

66. a. Prove that $n! + 2$ is divisible by 2, for all integers $n \geq 2$.
   b. Prove that $n! + k$ is divisible by k, for all integers $n \geq 2$ and $k = 2, 3, \ldots, n$.

H **c.** Given any integer $m \geq 2$, is it possible to find a sequence of $m - 1$ consecutive positive integers none of which is prime? Explain your answer.

67. $\dbinom{5}{3}$ **68.** $\dbinom{7}{4}$ **69.** $\dbinom{3}{0}$

70. $\dbinom{5}{5}$ **71.** $\dbinom{n}{n-1}$ **72.** $\dbinom{n+1}{n-1}$

73. Prove that for all nonnegative integers n and r with $r + 1 \leq n$, $\dbinom{n}{r+1} = \dfrac{n-r}{r+1} \dbinom{n}{r}$.

74. Prove that if p is a prime number and r is an integer with $0 < r < p$, then $\dbinom{p}{r}$ is divisible by p.

Answers for Test Yourself

1. the summation from k equals m to n of a-sub-k 2. $a_m + a_{m+1} + a_{m+2} + \cdots + a_n$ 3. $a_1 + a_2$ 4. the product from k equals m to n of a-sub-k 5. $n \cdot (n-1) \cdots 3 \cdot 2 \cdot 1$ (*Or:* $n \cdot (n-1)!$)   6. $\displaystyle\sum_{k=m}^{n} (a_k + cb_k)$ 7. $\displaystyle\prod_{k=m}^{n} a_k b_k$

5.2 *Mathematical Induction I*

> [*Mathematical induction is*] *the standard proof technique in computer science.*
> — Anthony Ralston, 1984

Mathematical induction is one of the more recently developed techniques of proof in the history of mathematics. It is used to check conjectures about the outcomes of processes that occur repeatedly and according to definite patterns. We introduce the technique with an example.

Some people claim that the United States penny is such a small coin that it should be abolished. They point out that frequently a person who drops a penny on the ground does not even bother to pick it up. Other people argue that abolishing the penny would not give enough flexibility for pricing merchandise. What prices could still be paid with exact change if the penny were abolished and another coin worth 3¢ were introduced? The answer is that the only prices that could not be paid with exact change would be 1¢, 2¢, 4¢, and 7¢. In other words,

> Any whole number of cents of at least 8¢ can be obtained using 3¢ and 5¢ coins.

More formally:

> For all integers $n \geq 8$, n cents can be obtained using 3¢ and 5¢ coins.

Even more formally:

> For all integers $n \geq 8$, $P(n)$ is true, where $P(n)$ is the sentence "n cents can be obtained using 3¢ and 5¢ coins."

You could check that $P(n)$ is true for a few particular values of n, as is done in the table below.

Number of Cents	How to Obtain It
8¢	3¢ + 5¢
9¢	3¢ + 3¢ + 3¢
10¢	5¢ + 5¢
11¢	3¢ + 3¢ + 5¢
12¢	3¢ + 3¢ + 3¢ + 3¢
13¢	3¢ + 5¢ + 5¢
14¢	3¢ + 3¢ + 3¢ + 5¢
15¢	5¢ + 5¢ + 5¢
16¢	3¢ + 3¢ + 5¢ + 5¢
17¢	3¢ + 3¢ + 3¢ + 3¢ + 5¢

The cases shown in the table provide inductive evidence to support the claim that $P(n)$ is true for general n. Indeed, $P(n)$ *is true for all $n \geq 8$ if, and only if, it is possible to continue filling in the table for arbitrarily large values of n.*

The kth line of the table gives information about how to obtain k¢ using 3¢ and 5¢ coins. To continue the table to the next row, directions must be given for how to obtain $(k + 1)$¢ using 3¢ and 5¢ coins. The secret is to observe first that if k¢ can be obtained using at least one 5¢ coin, then $(k + 1)$¢ can be obtained by replacing the 5¢ coin by two 3¢ coins, as shown in Figure 5.2.1.

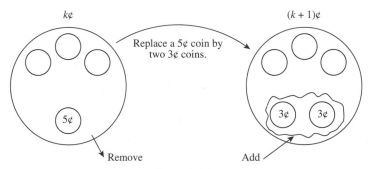

Figure 5.2.1

If, on the other hand, k¢ is obtained without using a 5¢ coin, then 3¢ coins are used exclusively. And since the total is at least 8¢, three or more 3¢ coins must be included. Three of the 3¢ coins can be replaced by two 5¢ coins to obtain a total of $(k + 1)$¢, as shown in Figure 5.2.2.

The structure of the argument above can be summarized as follows: To show that $P(n)$ is true for all integers $n \geq 8$, (1) show that $P(8)$ is true, and (2) show that the truth of $P(k + 1)$ follows necessarily from the truth of $P(k)$ for each $k \geq 8$.

Any argument of this form is an argument by *mathematical induction*. In general, mathematical induction is a method for proving that a property defined for integers n is true for all values of n that are greater than or equal to some initial integer.

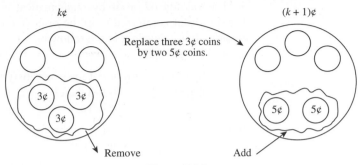

Figure 5.2.2

Principle of Mathematical Induction

Let $P(n)$ be a property that is defined for integers n, and let a be a fixed integer. Suppose the following two statements are true:

1. $P(a)$ is true.

2. For all integers $k \geq a$, if $P(k)$ is true then $P(k+1)$ is true.

Then the statement

$$\text{for all integers } n \geq a, \, P(n)$$

is true.

The first known use of mathematical induction occurs in the work of the Italian scientist Francesco Maurolico in 1575. In the seventeenth century both Pierre de Fermat and Blaise Pascal used the technique, Fermat calling it the "method of infinite descent." In 1883 Augustus De Morgan (best known for De Morgan's laws) described the process carefully and gave it the name *mathematical induction*.

To visualize the idea of mathematical induction, imagine an infinite collection of dominoes positioned one behind the other in such a way that if any given domino falls backward, it makes the one behind it fall backward also. (See Figure 5.2.3) Then imagine that the first domino falls backward. What happens? . . . They all fall down!

Figure 5.2.3 If the kth domino falls backward, it pushes the $(k + 1)$st domino backward also.

To see the connection between this image and the principle of mathematical induction, let $P(n)$ be the sentence "The nth domino falls backward." It is given that for each $k \geq 1$, if $P(k)$ is true (the kth domino falls backward), then $P(k + 1)$ is also true (the $(k + 1)$st domino falls backward). It is also given that $P(1)$ is true (the first domino falls backward). Thus by the principle of mathematical induction, $P(n)$ (the nth domino falls backward) is true for every integer $n \geq 1$.

The validity of proof by mathematical induction is generally taken as an axiom. That is why it is referred to as the *principle* of mathematical induction rather than as a theorem. It is equivalent to the following property of the integers, which is easy to accept on intuitive grounds:

> Suppose S is any set of integers satisfying (1) $a \in S$, and (2) for all integers $k \geq a$, if $k \in S$ then $k + 1 \in S$. Then S must contain every integer greater than or equal to a.

To understand the equivalence of this formulation and the one given earlier, just let S be the set of all integers for which $P(n)$ is true.

Proving a statement by mathematical induction is a two-step process. The first step is called the *basis step,* and the second step is called the *inductive step*.

Method of Proof by Mathematical Induction

Consider a statement of the form, "For all integers $n \geq a$, a property $P(n)$ is true." To prove such a statement, perform the following two steps:

Step 1 (basis step): Show that $P(a)$ is true.

Step 2 (inductive step): Show that for all integers $k \geq a$, if $P(k)$ is true then $P(k + 1)$ is true. To perform this step,

> **suppose** that $P(k)$ is true, where k is any particular but arbitrarily chosen integer with $k \geq a$.
> *[This supposition is called the **inductive hypothesis**.]*

Then

> **show** that $P(k + 1)$ is true.

Here is a formal version of the proof about coins previously developed informally.

Proposition 5.2.1

For all integers $n \geq 8$, n¢ can be obtained using 3¢ and 5¢ coins.

Proof (by mathematical induction):

Let the property $P(n)$ be the sentence

$\qquad n$¢ can be obtained using 3¢ and 5¢ coins. $\quad \leftarrow P(n)$

Show that P(8) is true:
$P(8)$ is true because 8¢ can be obtained using one 3¢ coin and one 5¢ coin.

Show that for all integers k ≥ 8, if P(k) is true then P(k+1) is also true:

[Suppose that P(k) is true for a particular but arbitrarily chosen integer $k \geq 8$. That is:]
Suppose that k is any integer with $k \geq 8$ such that

$\qquad k$¢ can be obtained using 3¢ and 5¢ coins. $\quad \leftarrow P(k)$
$\qquad\qquad\qquad\qquad\qquad\qquad\qquad\qquad\qquad\qquad$ inductive hypothesis

[We must show that P(k + 1) is true. That is:] We must show that

$\qquad (k + 1)$¢ can be obtained using 3¢ and 5¢ coins. $\leftarrow P(k + 1)$

Case 1 (There is a 5¢ coin among those used to make up the k¢.): In this case replace the 5¢ coin by two 3¢ coins; the result will be $(k + 1)$¢.

Case 2 (There is not a 5¢ coin among those used to make up the k¢.): In this case, because $k \geq 8$, at least three 3¢ coins must have been used. So remove three 3¢ coins and replace them by two 5¢ coins; the result will be $(k + 1)¢$.

Thus in either case $(k + 1)¢$ can be obtained using 3¢ and 5¢ coins *[as was to be shown].*

[Since we have proved the basis step and the inductive step, we conclude that the proposition is true.]

The following example shows how to use mathematical induction to prove a formula for the sum of the first n integers.

Example 5.2.1 Sum of the First n Integers

Use mathematical induction to prove that

$$1 + 2 + \cdots + n = \frac{n(n + 1)}{2} \quad \text{for all integers } n \geq 1.$$

Solution To construct a proof by induction, you must first identify the property $P(n)$. In this case, $P(n)$ is the equation

$$1 + 2 + \cdots + n = \frac{n(n + 1)}{2}. \qquad \leftarrow \text{the property } (P(n))$$

[To see that $P(n)$ is a sentence, note that its subject is "the sum of the integers from 1 to n" and its verb is "equals."]

In the basis step of the proof, you must show that the property is true for $n = 1$, or, in other words that $P(1)$ is true. Now $P(1)$ is obtained by substituting 1 in place of n in $P(n)$. The left-hand side of $P(1)$ is the sum of all the successive integers starting at 1 and ending at 1. This is just 1. Thus $P(1)$ is

Note To write $P(1)$, just copy $P(n)$ and replace each n by 1.

$$1 = \frac{1(1 + 1)}{2}. \qquad \leftarrow \text{basis } (P(1))$$

Of course, this equation is true because the right-hand side is

$$\frac{1(1 + 1)}{2} = \frac{1 \cdot 2}{2} = 1,$$

which equals the left-hand side.

In the inductive step, you assume that $P(k)$ is true, for a particular but arbitrarily chosen integer k with $k \geq 1$. *[This assumption is the inductive hypothesis.]* You must then show that $P(k + 1)$ is true. What are $P(k)$ and $P(k + 1)$? $P(k)$ is obtained by substituting k for every n in $P(n)$. Thus $P(k)$ is

Note To write $P(k)$, just copy $P(n)$ and replace each n by k.

$$1 + 2 + \cdots + k = \frac{k(k + 1)}{2}. \qquad \leftarrow \text{inductive hypothesis } (P(k))$$

Similarly, $P(k + 1)$ is obtained by substituting the quantity $(k + 1)$ for every n that appears in $P(n)$. Thus $P(k + 1)$ is

Note To write $P(k + 1)$, just copy $P(n)$ and replace each n by $(k + 1)$.

$$1 + 2 + \cdots + (k + 1) = \frac{(k + 1)((k + 1) + 1)}{2},$$

or, equivalently,

$$1 + 2 + \cdots + (k + 1) = \frac{(k + 1)(k + 2)}{2}.$$ $\leftarrow$ to show $(P(k + 1))$

Now the inductive hypothesis is the supposition that $P(k)$ is true. How can this supposition be used to show that $P(k + 1)$ is true? $P(k + 1)$ is an equation, and the truth of an equation can be shown in a variety of ways. One of the most straightforward is to use the inductive hypothesis along with algebra and other known facts to transform separately the left-hand and right-hand sides until you see that they are the same. In this case, the left-hand side of $P(k + 1)$ is

$$1 + 2 + \cdots + (k + 1),$$

which equals

$$(1 + 2 + \cdots + k) + (k + 1)$$

The next-to-last term is k because the terms are successive integers and the last term is $k + 1$.

But by substitution from the inductive hypothesis,

$$(1 + 2 + \cdots + k) + (k + 1)$$

$$= \frac{k(k + 1)}{2} + (k + 1)$$ since the inductive hypothesis says that $1 + 2 + \cdots + k = \frac{k(k + 1)}{2}$

$$= \frac{k(k + 1)}{2} + \frac{2(k + 1)}{2}$$ by multiplying the numerator and denominator of the second term by 2 to obtain a common denominator

$$= \frac{k^2 + k}{2} + \frac{2k + 2}{2}$$ by multiplying out the two numerators

$$= \frac{k^2 + 3k + 2}{2}$$ by adding fractions with the same denominator and combining like terms.

So the left-hand side of $P(k + 1)$ is $\dfrac{k^2 + 3k + 2}{2}$. Now the right-hand side of $P(k + 1)$ is

$$\frac{(k + 1)(k + 2)}{2} = \frac{k^2 + 3k + 2}{2}$$ by multiplying out the numerator.

Thus the two sides of $P(k + 1)$ are equal to each other, and so the equation $P(k + 1)$ is true.

This discussion is summarized as follows:

Theorem 5.2.2 Sum of the First n Integers

For all integers $n \geq 1$,

$$1 + 2 + \cdots + n = \frac{n(n + 1)}{2}.$$

Proof (by mathematical induction):

Let the property $P(n)$ be the equation

$$1 + 2 + 3 + \cdots + n = \frac{n(n + 1)}{2}.$$ $\leftarrow P(n)$

Show that P(1) is true:

To establish $P(1)$, we must show that

$$1 = \frac{1(1+1)}{2} \qquad \leftarrow \quad P(1)$$

But the left-hand side of this equation is 1 and the right-hand side is

$$\frac{1(1+1)}{2} = \frac{2}{2} = 1$$

also. Hence $P(1)$ is true.

Show that for all integers k ≥ 1, if P(k) is true then P(k + 1) is also true:
[Suppose that $P(k)$ is true for a particular but arbitrarily chosen integer $k \geq 1$. That is:] Suppose that k is any integer with $k \geq 1$ such that

$$1 + 2 + 3 + \cdots + k = \frac{k(k+1)}{2} \qquad \begin{array}{l} \leftarrow P(k) \\ \text{inductive hypothesis} \end{array}$$

[We must show that $P(k + 1)$ is true. That is:] We must show that

$$1 + 2 + 3 + \cdots + (k+1) = \frac{(k+1)[(k+1)+1]}{2},$$

or, equivalently, that

$$1 + 2 + 3 + \cdots + (k+1) = \frac{(k+1)(k+2)}{2}. \quad \leftarrow P(k+1)$$

[We will show that the left-hand side and the right-hand side of $P(k + 1)$ are equal to the same quantity and thus are equal to each other.]

The left-hand side of $P(k + 1)$ is

$$1 + 2 + 3 + \cdots + (k+1)$$

$$= 1 + 2 + 3 + \cdots + k + (k+1) \qquad \begin{array}{l} \text{by making the next-to-last} \\ \text{term explicit} \end{array}$$

$$= \frac{k(k+1)}{2} + (k+1) \qquad \begin{array}{l} \text{by substitution from the} \\ \text{inductive hypothesis} \end{array}$$

$$= \frac{k(k+1)}{2} + \frac{2(k+1)}{2}$$

$$= \frac{k^2 + k}{2} + \frac{2k+2}{2}$$

$$= \frac{k^2 + 3k + 2}{2} \qquad \text{by algebra.}$$

And the right-hand side of $P(k + 1)$ is

$$\frac{(k+1)(k+2)}{2} = \frac{k^2 + 3k + 1}{2}.$$

Thus the two sides of $P(k + 1)$ are equal to the same quantity and so they are equal to each other. Therefore the equation $P(k + 1)$ is true *[as was to be shown]*.
[Since we have proved both the basis step and the inductive step, we conclude that the theorem is true.]

The story is told that one of the greatest mathematicians of all time, Carl Friedrich Gauss (1777–1855), was given the problem of adding the numbers from 1 to 100 by his teacher when he was a young child. The teacher had asked his students to compute the sum, supposedly to gain himself some time to grade papers. But after just a few moments, Gauss produced the correct answer. Needless to say, the teacher was dumbfounded. How could young Gauss have calculated the quantity so rapidly? In his later years, Gauss explained that he had imagined the numbers paired according to the following schema.

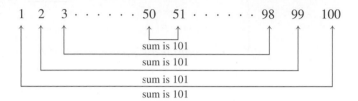

The sum of the numbers in each pair is 101, and there are 50 pairs in all; hence the total sum is $50 \cdot 101 = 5{,}050$.

• Definition Closed Form

If a sum with a variable number of terms is shown to be equal to a formula that does not contain either an ellipsis or a summation symbol, we say that it is written **in closed form**.

For example, writing $1 + 2 + 3 + \cdots + n = \dfrac{n(n+1)}{2}$ expresses the sum $1 + 2 + 3 + \cdots + n$ in closed form.

Example 5.2.2 Applying the Formula for the Sum of the First n Integers

a. Evaluate $2 + 4 + 6 + \cdots + 500$.

b. Evaluate $5 + 6 + 7 + 8 + \cdots + 50$.

c. For an integer $h \geq 2$, write $1 + 2 + 3 + \cdots + (h-1)$ in closed form.

Solution

a. $2 + 4 + 6 + \cdots + 500 = 2 \cdot (1 + 2 + 3 + \cdots + 250)$

$$= 2 \cdot \left(\frac{250 \cdot 251}{2} \right) \qquad \text{by applying the formula for the sum of the first } n \text{ integers with } n = 250$$

$$= 62{,}750.$$

b. $5 + 6 + 7 + 8 + \cdots + 50 = (1 + 2 + 3 + \cdots + 50) - (1 + 2 + 3 + 4)$

$$= \frac{50 \cdot 51}{2} - 10 \qquad \text{by applying the formula for the sum of the first } n \text{ integers with } n = 50$$

$$= 1{,}265$$

c. $1 + 2 + 3 + \cdots + (h-1) = \dfrac{(h-1) \cdot [(h-1)+1]}{2} \qquad \text{by applying the formula for the sum of the first } n \text{ integers with } n = h - 1$

$$= \frac{(h-1) \cdot h}{2} \qquad \text{since } (h-1)+1 = h. \qquad \blacksquare$$

The next example asks for a proof of another famous and important formula in mathematics—the formula for the sum of a geometric sequence. In a **geometric sequence,** each term is obtained from the preceding one by multiplying by a constant factor. If the first term is 1 and the constant factor is r, then the sequence is $1, r, r^2, r^3, \ldots, r^n, \ldots$. The sum of the first n terms of this sequence is given by the formula

$$\sum_{i=0}^{n} r^i = \frac{r^{n+1} - 1}{r - 1}$$

for all integers $n \geq 0$ and real numbers r not equal to 1. The expanded form of the formula is

$$r^0 + r^1 + r^2 + \cdots + r^n = \frac{r^{n+1} - 1}{r - 1},$$

and because $r^0 = 1$ and $r^1 = r$, the formula for $n \geq 1$ can be rewritten as

$$1 + r + r^2 + \cdots + r^n = \frac{r^{n+1} - 1}{r - 1}.$$

Example 5.2.3 Sum of a Geometric Sequence

Prove that $\sum_{i=0}^{n} r^i = \frac{r^{n+1} - 1}{r - 1}$, for all integers $n \geq 0$ and all real numbers r except 1.

Solution   In this example the property $P(n)$ is again an equation, although in this case it contains a real variable r:.

$$\boxed{\sum_{i=0}^{n} r^i = \frac{r^{n+1} - 1}{r - 1}.}$$ $\leftarrow$ the property ($P(n)$)

Because r can be any real number other than 1, the proof begins by supposing that r is a particular but arbitrarily chosen real number not equal to 1. Then the proof continues by mathematical induction on n, starting with $n = 0$. In the basis step, you must show that $P(0)$ is true; that is, you show the property is true for $n = 0$. So you substitute 0 for each n in $P(n)$:

$$\boxed{\sum_{i=0}^{0} r^i = \frac{r^{0+1} - 1}{r - 1}.}$$ $\leftarrow$ basis ($P(0)$)

In the inductive step, you suppose k is any integer with $k \geq 0$ for which $P(k)$ is true; that is, you suppose the property is true for $n = k$. So you substitute k for each n in $P(n)$:

$$\boxed{\sum_{i=0}^{k} r^i = \frac{r^{k+1} - 1}{r - 1}.}$$ $\leftarrow$ inductive hypothesis ($P(k)$)

Then you show that $P(k + 1)$ is true; that is, you show the property is true for $n = k + 1$. So you substitute $k + 1$ for each n in $P(n)$:

$$\sum_{i=0}^{k+1} r^i = \frac{r^{(k+1)+1} - 1}{r - 1},$$

or, equivalently,

$$\boxed{\sum_{i=0}^{k+1} r^i = \frac{r^{k+2} - 1}{r - 1}.} \qquad \leftarrow \text{to show } (P(k+1))$$

In the inductive step for this proof we use another common technique for showing that an equation is true: We start with the left-hand side and transform it step-by-step into the right-hand side using the inductive hypothesis together with algebra and other known facts.

Theorem 5.2.3 Sum of a Geometric Sequence

For any real number r except 1, and any integer $n \geq 0$,

$$\sum_{i=0}^{n} r^i = \frac{r^{n+1} - 1}{r - 1}.$$

Proof (by mathematical induction):

Suppose r is a particular but arbitrarily chosen real number that is not equal to 1, and let the property $P(n)$ be the equation

$$\sum_{i=0}^{n} r^i = \frac{r^{n+1} - 1}{r - 1} \qquad \leftarrow P(n)$$

We must show that $P(n)$ is true for all integers $n \geq 0$. We do this by mathematical induction on n.

Show that P(0) is true:

To establish $P(0)$, we must show that

$$\sum_{i=0}^{0} r^i = \frac{r^{0+1} - 1}{r - 1} \qquad \leftarrow P(0)$$

The left-hand side of this equation is $r^0 = 1$ and the right-hand side is

$$\frac{r^{0+1} - 1}{r - 1} = \frac{r - 1}{r - 1} = 1$$

also because $r^1 = r$ and $r \neq 1$. Hence $P(0)$ is true.

Show that for all integers $k \geq 0$, if $P(k)$ is true then $P(k+1)$ is also true:
[Suppose that $P(k)$ is true for a particular but arbitrarily chosen integer $k \geq 0$. That is:]
Let k be any integer with $k \geq 0$, and suppose that

$$\sum_{i=0}^{k} r^i = \frac{r^{k+1} - 1}{r - 1} \qquad \begin{array}{l} \leftarrow P(k) \\ \text{inductive hypothesis} \end{array}$$

[We must show that $P(k+1)$ is true. That is:] We must show that

$$\sum_{i=0}^{k+1} r^i = \frac{r^{(k+1)+1} - 1}{r - 1},$$

or, equivalently, that

$$\sum_{i=0}^{k+1} r^i = \frac{r^{k+2} - 1}{r - 1}. \quad \leftarrow P(k+1)$$

[We will show that the left-hand side of $P(k+1)$ equals the right-hand side.]
The left-hand side of $P(k+1)$ is

$$\sum_{i=0}^{k+1} r^i = \sum_{i=0}^{k} r^i + r^{k+1} \qquad \text{by writing the } (k+1)\text{st term separately from the first } k \text{ terms}$$

$$= \frac{r^{k+1} - 1}{r - 1} + r^{k+1} \qquad \text{by substitution from the inductive hypothesis}$$

$$= \frac{r^{k+1} - 1}{r - 1} + \frac{r^{k+1}(r - 1)}{r - 1} \qquad \text{by multiplying the numerator and denominator of the second term by } (r-1) \text{ to obtain a common denominator}$$

$$= \frac{(r^{k+1} - 1) + r^{k+1}(r - 1)}{r - 1} \qquad \text{by adding fractions}$$

$$= \frac{r^{k+1} - 1 + r^{k+2} - r^{k+1}}{r - 1} \qquad \text{by multiplying out and using the fact that } r^{k+1} \cdot r = r^{k+1} \cdot r^1 = r^{k+2}$$

$$= \frac{r^{k+2} - 1}{r - 1} \qquad \text{by canceling the } r^{k+1}\text{'s.}$$

which is the right-hand side of $P(k+1)$ *[as was to be shown.]*
[Since we have proved the basis step and the inductive step, we conclude that the theorem is true.]

Proving an Equality

The proofs of the basis and inductive steps in Examples 5.2.1 and 5.2.3 illustrate two different ways to show that an equation is true: (1) transforming the left-hand side and the right-hand side independently until they are seen to be equal, and (2) transforming one side of the equation until it is seen to be the same as the other side of the equation.

Sometimes people use a method that they believe proves equality but that is actually invalid. For example, to prove the basis step for Theorem 5.2.3, they perform the following steps:

Caution!   Don't do this!

$$\sum_{i=0}^{0} r^i = \frac{r^{0+1} - 1}{r - 1}$$

$$r^0 = \frac{r^1 - 1}{r - 1}$$

$$1 = \frac{r - 1}{r - 1}$$

$$1 = 1$$

The problem with this method is that starting from a statement and deducing a true conclusion does not prove that the statement is true. A true conclusion can also be deduced from a false statement. For instance, the steps below show how to deduce the true conclusion that $1 = 1$ from the false statement that $1 = 0$:

$$1 = 0 \qquad \leftarrow \text{false}$$
$$0 = 1$$
$$1 + 0 = 0 + 1$$
$$1 = 1 \qquad \leftarrow \text{true}$$

When using mathematical induction to prove formulas, be sure to use a method that avoids invalid reasoning, both for the basis step and for the inductive step.

Deducing Additional Formulas

The formula for the sum of a geometric sequence can be thought of as a family of different formulas in r, one for each real number r except 1.

Example 5.2.4 Applying the Formula for the Sum of a Geometric Sequence

In each of (a) and (b) below, assume that m is an integer that is greater than or equal to 3. Write each of the sums in closed form.

a. $1 + 3 + 3^2 + \cdots + 3^{m-2}$

b. $3^2 + 3^3 + 3^4 + \cdots + 3^m$

Solution

a. $1 + 3 + 3^2 + \cdots + 3^{m-2} = \dfrac{3^{(m-2)+1} - 1}{3 - 1}$ by applying the formula for the sum of a geometric sequence with $r = 3$ and $n = m - 2$

$$= \frac{3^{m-1} - 1}{2}.$$

b. $3^2 + 3^3 + 3^4 + \cdots + 3^m = 3^2 \cdot (1 + 3 + 3^2 + \cdots + 3^{m-2})$ by factoring out 3^2

$$= 9 \cdot \left(\frac{3^{m-1} - 1}{2} \right) \qquad \text{by part (a).} \qquad \blacksquare$$

As with the formula for the sum of the first n integers, there is a way to think of the formula for the sum of the terms of a geometric sequence that makes it seem simple and intuitive. Let

$$S_n = 1 + r + r^2 + \cdots + r^n.$$

Then

$$r S_n = r + r^2 + r^3 + \cdots + r^{n+1},$$

and so

$$r S_n - S_n = (r + r^2 + r^3 + \cdots + r^{n+1}) - (1 + r + r^2 + \cdots + r^n)$$
$$= r^{n+1} - 1. \qquad\qquad 5.2.1$$

But

$$r S_n - S_n = (r - 1) S_n. \qquad\qquad 5.2.2$$

Equating the right-hand sides of equations (5.2.1) and (5.2.2) and dividing by $r - 1$ gives

$$S_n = \frac{r^{n+1} - 1}{r - 1}.$$

This derivation of the formula is attractive and is quite convincing. However, it is not as logically airtight as the proof by mathematical induction. To go from one step to another in the previous calculations, the argument is made that each term among those indicated by the ellipsis (...) has such-and-such an appearance and when these are canceled such-and-such occurs. But it is impossible actually to see each such term and each such calculation, and so the accuracy of these claims cannot be fully checked. With mathematical induction it is possible to focus exactly on what happens in the middle of the ellipsis and verify without doubt that the calculations are correct.

Test Yourself

1. Mathematical induction is a method for proving that a property defined for integers n is true for all values of n that are _____.

2. Let $P(n)$ be a property defined for integers n and consider constructing a proof by mathematical induction for the statement "$P(n)$ is true for all $n \geq a$."

(a) In the basis step one must show that _____.

(b) In the inductive step one supposes that _____ for some particular but arbitrarily chosen value of an integer $k \geq a$. This supposition is called the _____. One then has to show that _____.

Exercise Set 5.2

1. Use mathematical induction (and the proof of Proposition 5.2.1 as a model) to show that any amount of money of at least 14¢ can be made up using 3¢ and 8¢ coins.

2. Use mathematical induction to show that any postage of at least 12¢ can be obtained using 3¢ and 7¢ stamps.

3. For each positive integer n, let $P(n)$ be the formula

$$1^2 + 2^2 + \cdots + n^2 = \frac{n(n + 1)(2n + 1)}{6}.$$

a. Write $P(1)$. Is $P(1)$ true?
b. Write $P(k)$.
c. Write $P(k + 1)$.
d. In a proof by mathematical induction that the formula holds for all integers $n \geq 1$, what must be shown in the inductive step?

4. For each integer n with $n \geq 2$, let $P(n)$ be the formula

$$\sum_{i=1}^{n-1} i(i + 1) = \frac{n(n - 1)(n + 1)}{3}.$$

a. Write $P(2)$. Is $P(2)$ true?
b. Write $P(k)$.
c. Write $P(k + 1)$.
d. In a proof by mathematical induction that the formula holds for all integers $n \geq 2$, what must be shown in the inductive step?

5. Fill in the missing pieces in the following proof that

$$1 + 3 + 5 + \cdots + (2n - 1) = n^2$$

for all integers $n \geq 1$.

Proof: Let the property $P(n)$ be the equation

$$1 + 3 + 5 + \cdots + (2n - 1) = n^2. \quad \leftarrow P(n)$$

Show that $P(1)$ is true: To establish $P(1)$, we must show that when 1 is substituted in place of n, the left-hand side equals the right-hand side. But when $n = 1$, the left-hand side is the sum of all the odd integers from 1 to $2 \cdot 1 - 1$, which is the sum of the odd integers from 1 to 1, which is just 1. The right-hand side is __(a)__, which also equals 1. So $P(1)$ is true.

Show that for all integers $k \geq 1$, if $P(k)$ is true then $P(k + 1)$ is true: Let k be any integer with $k \geq 1$.

  [Suppose $P(k)$ is true. That is:]

Suppose $1 + 3 + 5 + \cdots + (2k - 1) = $ __(b)__. $\leftarrow P(k)$
  [This is the inductive hypothesis.]

  [We must show that $P(k + 1)$ is true. That is:]

We must show that

$$\underline{\text{(c)}} = \underline{\text{(d)}}. \qquad \leftarrow P(k + 1)$$

continued on page 198

But the left-hand side of $P(k + 1)$ is

$1 + 3 + 5 + \cdots + (2(k + 1) - 1)$

$\begin{aligned} &= 1 + 3 + 5 + \cdots + (2k + 1) \quad \text{by algebra} \\ &= [1 + 3 + 5 + \cdots + (2k - 1)] + (2k + 1) \\ &\quad \text{the next-to-last term is } 2k - 1 \text{ because } \underline{\text{(e)}} \\ &= k^2 + (2k + 1) \quad \text{by } \underline{\text{(f)}} \\ &= (k + 1)^2 \quad\quad \text{by algebra} \end{aligned}$

which is the right-hand side of $P(k + 1)$ *[as was to be shown.]*

[Since we have proved the basis step and the inductive step, we conclude that the given statement is true.]

The previous proof was annotated to help make its logical flow more obvious. In standard mathematical writing, such annotation is omitted.

Prove each statement in 6–9 using mathematical induction. Do not derive them from Theorem 5.2.2 or Theorem 5.2.3.

6. For all integers $n \geq 1$, $2 + 4 + 6 + \cdots + 2n = n^2 + n$.

7. For all integers $n \geq 1$,

$$1 + 6 + 11 + 16 + \cdots + (5n - 4) = \frac{n(5n - 3)}{2}.$$

8. For all integers $n \geq 0$, $1 + 2 + 2^2 + \cdots + 2^n = 2^{n+1} - 1$.

9. For all integers $n \geq 3$,

$$4^3 + 4^4 + 4^5 + \cdots + 4^n = \frac{4(4^n - 16)}{3}.$$

Prove each of the statements in 10–17 by mathematical induction.

10. $1^2 + 2^2 + \cdots + n^2 = \dfrac{n(n + 1)(2n + 1)}{6}$, for all integers $n \geq 1$.

11. $1^3 + 2^3 + \cdots + n^3 = \left[\dfrac{n(n + 1)}{2}\right]^2$, for all integers $n \geq 1$.

12. $\dfrac{1}{1 \cdot 2} + \dfrac{1}{2 \cdot 3} + \cdots + \dfrac{1}{n(n + 1)} = \dfrac{n}{n + 1}$, for all integers $n \geq 1$.

13. $\displaystyle\sum_{i=1}^{n-1} i(i + 1) = \dfrac{n(n - 1)(n + 1)}{3}$, for all integers $n \geq 2$.

14. $\displaystyle\sum_{i=1}^{n+1} i \cdot 2^i = n \cdot 2^{n+2} + 2$, for all integers $n \geq 0$.

H 15. $\displaystyle\sum_{i=1}^{n} i(i!) = (n + 1)! - 1$, for all integers $n \geq 1$.

16. $\left(1 - \dfrac{1}{2^2}\right)\left(1 - \dfrac{1}{3^2}\right) \cdots \left(1 - \dfrac{1}{n^2}\right) = \dfrac{n + 1}{2n}$, for all integers $n \geq 2$.

17. $\displaystyle\prod_{i=0}^{n}\left(\dfrac{1}{2i + 1} \cdot \dfrac{1}{2i + 2}\right) = \dfrac{1}{(2n + 2)!}$, for all integers $n \geq 0$.

H ✶ 18. If x is a real number not divisible by π, then for all integers $n \geq 1$,

$$\sin x + \sin 3x + \sin 5x + \cdots + \sin (2n - 1)x = \frac{1 - \cos 2nx}{2 \sin x}.$$

19. (For students who have studied calculus) Use mathematical induction, the product rule from calculus, and the facts that $\dfrac{d(x)}{dx} = 1$ and that $x^{k+1} = x \cdot x^k$ to prove that for all integers $n \geq 1$, $\dfrac{d(x^n)}{dx} = nx^{n-1}$.

Use the formula for the sum of the first n integers and/or the formula for the sum of a geometric sequence to evaluate the sums in 20–29 or to write them in closed form.

20. $4 + 8 + 12 + 16 + \cdots + 200$

21. $5 + 10 + 15 + 20 + \cdots + 300$

22. $3 + 4 + 5 + 6 + \cdots + 1000$

23. $7 + 8 + 9 + 10 + \cdots + 600$

24. $1 + 2 + 3 + \cdots + (k - 1)$, where k is an integer and $k \geq 2$.

25. a. $1 + 2 + 2^2 + \cdots + 2^{25}$
  b. $2 + 2^2 + 2^3 + \cdots + 2^{26}$

26. $3 + 3^2 + 3^3 + \cdots + 3^n$, where n is an integer with $n \geq 1$

27. $5^3 + 5^4 + 5^5 + \cdots + 5^k$, where k is any integer with $k \geq 3$.

28. $1 + \dfrac{1}{2} + \dfrac{1}{2^2} + \cdots + \dfrac{1}{2^n}$, where n is a positive integer

29. $1 - 2 + 2^2 - 2^3 + \cdots + (-1)^n 2^n$, where n is a positive integer

H 30. Find a formula in n, a, m, and d for the sum $(a + md) + (a + (m + 1)d) + (a + (m + 2)d) + \cdots + (a + (m + n)d)$, where m and n are integers, $n \geq 0$, and a and d are real numbers. Justify your answer.

31. Find a formula in a, r, m, and n for the sum

$$ar^m + ar^{m+1} + ar^{m+2} + \cdots + ar^{m+n}$$

where m and n are integers, $n \geq 0$, and a and r are real numbers. Justify your answer.

32. You have two parents, four grandparents, eight great-grandparents, and so forth.
  a. If all your ancestors were distinct, what would be the total number of your ancestors for the past 40 generations (counting your parents' generation as number one)? (*Hint:* Use the formula for the sum of a geometric sequence.)
  b. Assuming that each generation represents 25 years, how long is 40 generations?
  c. The total number of people who have ever lived is approximately 10 billion, which equals 10^{10} people. Compare this fact with the answer to part (a). What do you deduce?

Find the mistakes in the proof fragments in 33–35.

33. Theorem: For any integer $n \geq 1$,

$$1^2 + 2^2 + \cdots + n^2 = \frac{n(n+1)(2n+1)}{6}.$$

"**Proof (by mathematical induction):** Certainly the theorem is true for $n = 1$ because $1^2 = 1$ and

$$\frac{1(1+1)(2 \cdot 1 + 1)}{6} = 1.$$ So the basis step is true.

For the inductive step, suppose that for some integer $k \geq 1$,

$k^2 = \dfrac{k(k+1)(2k+1)}{6}$. We must show that

$$(k+1)^2 = \frac{(k+1)((k+1)+1)(2(k+1)+1)}{6}."$$

H 34. Theorem: For any integer $n \geq 0$,

$$1 + 2 + 2^2 + \cdots + 2^n = 2^{n+1} - 1.$$

"**Proof (by mathematical induction): Let the property** $P(n)$ be $1 + 2 + 2^2 + \cdots + 2^n = 2^{n+1} - 1$.

Show that P(0) is true:
The left-hand side of $P(0)$ is $1 + 2 + 2^2 + \cdots + 2^0 = 1$ and the right-hand side is $2^{0+1} - 1 = 2 - 1 = 1$ also. So $P(0)$ is true."

H 35. Theorem: For any integer $n \geq 1$,

$$\sum_{i=1}^{n} i(i!) = (n+1)! - 1.$$

"**Proof (by mathematical induction): Let the property** $P(n)$ be $\displaystyle\sum_{i=1}^{n} i(i!) = (n+1)! - 1$.

Show that P(1) is true: When $n = 1$

$$\sum_{i=1}^{1} i(i!) = (1+1)! - 1$$

So $$1(1!) = 2! - 1$$

and $$1 = 1$$

Thus $P(1)$ is true."

✱ 36. Use Theorem 5.2.2 to prove that if m and n are any positive integers and m is odd, then $\sum_{k=0}^{m-1}(n+k)$ is divisible by m. Does the conclusion hold if m is even? Justify your answer.

H ✱ 37. Use Theorem 5.2.2 and the result of exercise 10 to prove that if p is any prime number with $p \geq 5$, then the sum of squares of any p consecutive integers is divisible by p.

Answers for Test Yourself

1. greater than or equal to some initial value   2. (a) $P(a)$ is true   (b) $P(k)$ is true; inductive hypothesis; $P(k+1)$ is true

5.3 *Mathematical Induction II*

A good proof is one which makes us wiser. — I. Manin, *A Course in Mathematical Logic*, 1977

In natural science courses, deduction and induction are presented as alternative modes of thought—deduction being to infer a conclusion from general principles using the laws of logical reasoning, and induction being to enunciate a general principle after observing it to hold in a large number of specific instances. In this sense, then, *mathematical* induction is not inductive but deductive. Once proved by mathematical induction, a theorem is known just as certainly as if it were proved by any other mathematical method. Inductive reasoning, in the natural sciences sense, *is* used in mathematics, but only to make conjectures, not to prove them. For example, observe that

$$1 - \frac{1}{2} = \frac{1}{2}$$

$$\left(1 - \frac{1}{2}\right)\left(1 - \frac{1}{3}\right) = \frac{1}{3}$$

$$\left(1 - \frac{1}{2}\right)\left(1 - \frac{1}{3}\right)\left(1 - \frac{1}{4}\right) = \frac{1}{4}$$

This pattern seems so unlikely to occur by pure chance that it is reasonable to conjecture (though it is by no means certain) that the pattern holds true in general. In a case like this, a proof by mathematical induction (which you are asked to write in exercise 1 at the end of this section) gets to the essence of why the pattern holds in general. It reveals the mathematical mechanism that necessitates the truth of each successive case from the previous one. For instance, in this example observe that if

$$\left(1 - \frac{1}{2}\right)\left(1 - \frac{1}{3}\right)\cdots\left(1 - \frac{1}{k}\right) = \frac{1}{k},$$

then by substitution

$$\left(1 - \frac{1}{2}\right)\left(1 - \frac{1}{3}\right)\cdots\left(1 - \frac{1}{k}\right)\left(1 - \frac{1}{k+1}\right)$$

$$= \frac{1}{k}\left(1 - \frac{1}{k+1}\right) = \frac{1}{k}\left(\frac{k+1-1}{k+1}\right) = \frac{1}{k}\left(\frac{k}{k+1}\right) = \frac{1}{k+1}.$$

Thus mathematical induction makes knowledge of the general pattern a matter of mathematical certainty rather than vague conjecture.

In the remainder of this section we show how to use mathematical induction to prove additional kinds of statements such as divisibility properties of the integers and inequalities. The basic outlines of the proofs are the same in all cases, but the details of the basis and inductive steps differ from one to another.

Example 5.3.1 Proving a Divisibility Property

Use mathematical induction to prove that for all integers $n \geq 0$, $2^{2n} - 1$ is divisible by 3.

Solution   As in the previous proofs by mathematical induction, you need to identify the property $P(n)$. In this example, $P(n)$ is the sentence

$2^{2n} - 1$ is divisible by 3.   ← the property $(P(n))$

By substitution, the statement for the basis step, $P(0)$, is

$2^{2 \cdot 0} - 1$ is divisible by 3.   ← basis $(P(0))$

The supposition for the inductive step, $P(k)$, is

$2^{2k} - 1$ is divisible by 3,   ← inductive hypothesis $(P(k))$

and the conclusion to be shown, $P(k + 1)$, is

$2^{2(k+1)} - 1$ is divisible by 3.   ← to show $(P(k + 1))$

Recall that an integer m is divisible by 3 if, and only if, $m = 3r$ for some integer r. Now the statement $P(0)$ is true because $2^{2 \cdot 0} - 1 = 2^0 - 1 = 1 - 1 = 0$, which is divisible by 3 because $0 = 3 \cdot 0$.

To prove the inductive step, you suppose that k is any integer greater than or equal to 0 such that $P(k)$ is true. This means that $2^{2k} - 1$ is divisible by 3. You must then prove the truth of $P(k + 1)$. Or, in other words, you must show that $2^{2(k+1)} - 1$ is divisible by 3. But

$$\begin{aligned}
2^{2(k+1)} - 1 &= 2^{2k+2} - 1 \\
&= 2^{2k} \cdot 2^2 - 1 \qquad \text{by the laws of exponents} \\
&= 2^{2k} \cdot 4 - 1.
\end{aligned}$$

The aim is to show that this quantity, $2^{2k} \cdot 4 - 1$, is divisible by 3. Why should that be so? By the inductive hypothesis, $2^{2k} - 1$ is divisible by 3, and $2^{2k} \cdot 4 - 1$ resembles $2^{2k} - 1$. Observe what happens, if you subtract $2^{2k} - 1$ from $2^{2k} \cdot 4 - 1$:

$$\underbrace{2^{2k} \cdot 4 - 1}_{\substack{\uparrow \\ \text{divisible by 3?}}} - \underbrace{(2^{2k} - 1)}_{\substack{\uparrow \\ \text{divisible by 3}}} = \underbrace{2^{2k} \cdot 3}_{\substack{\uparrow \\ \text{divisible by 3}}}.$$

Adding $2^{2k} - 1$ to both sides gives

$$\underbrace{2^{2k} \cdot 4 - 1}_{\substack{\uparrow \\ \text{divisible by 3?}}} = \underbrace{2^{2k} \cdot 3}_{\substack{\uparrow \\ \text{divisible by 3}}} + \underbrace{2^{2k} - 1}_{\substack{\uparrow \\ \text{divisible by 3}}}.$$

Both terms of the sum on the right-hand side of this equation are divisible by 3; hence the sum is divisible by 3. (See exercise 15 of Section 4.3.) Therefore, the left-hand side of the equation is also divisible by 3, which is what was to be shown.

This discussion is summarized as follows:

Proposition 5.3.1

For all integers $n \geq 0$, $2^{2n} - 1$ is divisible by 3.

Proof (by mathematical induction):

Let the property $P(n)$ be the sentence "$2^{2n} - 1$ is divisible by 3."

$$2^{2n} - 1 \text{ is divisible by 3.} \qquad \leftarrow P(n)$$

Show that $P(0)$ is true:
To establish $P(0)$, we must show that

$$2^{2 \cdot 0} - 1 \text{ is divisible by 3.} \qquad \leftarrow P(0)$$

But

$$2^{2 \cdot 0} - 1 = 2^0 - 1 = 1 - 1 = 0$$

and 0 is divisible by 3 because $0 = 3 \cdot 0$. Hence $P(0)$ is true.

Show that for all integers $k \geq 0$, if $P(k)$ is true then $P(k + 1)$ is also true:
[Suppose that $P(k)$ is true for a particular but arbitrarily chosen integer $k \geq 0$. That is:]
Let k be any integer with $k \geq 0$, and suppose that

$$2^{2k} - 1 \text{ is divisible by 3.} \qquad \substack{\leftarrow P(k) \\ \text{inductive hypothesis}}$$

By definition of divisibility, this means that

$$2^{2k} - 1 = 3r \quad \text{for some integer } r.$$

[We must show that $P(k + 1)$ is true. That is:] We must show that

$$2^{2(k+1)} - 1 \text{ is divisible by 3.} \quad \leftarrow P(k + 1)$$

continued on page 202

But

$$2^{2(k+1)} - 1 = 2^{2k+2} - 1$$

$$= 2^{2k} \cdot 2^2 - 1 \qquad \text{by the laws of exponents}$$

$$= 2^{2k} \cdot 4 - 1$$

$$= 2^{2k}(3+1) - 1$$

$$= 2^{2k} \cdot 3 + (2^{2k} - 1) \qquad \text{by the laws of algebra}$$

$$= 2^{2k} \cdot 3 + 3r \qquad \text{by inductive hypothesis}$$

$$= 3(2^{2k} + r) \qquad \text{by factoring out the 3.}$$

But $2^{2k} + r$ is an integer because it is a sum of products of integers, and so, by definition of divisibility, $2^{2(k+1)} - 1$ is divisible by 3 *[as was to be shown].*
[Since we have proved the basis step and the inductive step, we conclude that the proposition is true.]

■

The next example illustrates the use of mathematical induction to prove an inequality.

Example 5.3.2 Proving an Inequality

Use mathematical induction to prove that for all integers $n \geq 3$,

$$2n + 1 < 2^n.$$

Solution   In this example the property $P(n)$ is the inequality

$$\boxed{2n + 1 < 2^n.} \qquad \leftarrow \text{the property } (P(n))$$

By substitution, the statement for the basis step, $P(3)$, is

$$\boxed{2 \cdot 3 + 1 < 2^3.} \qquad \leftarrow \text{basis } (P(3))$$

The supposition for the inductive step, $P(k)$, is

$$\boxed{2k + 1 < 2^k,} \qquad \leftarrow \text{inductive hypothesis } (P(k))$$

and the conclusion to be shown is

$$\boxed{2(k+1) + 1 < 2^{k+1}.} \qquad \leftarrow \text{to show } (P(k+1))$$

To prove the basis step, observe that the statement $P(3)$ is true because $2 \cdot 3 + 1 = 7$, $2^3 = 8$, and $7 < 8$.

To prove the inductive step, suppose the inductive hypothesis, that $P(k)$ is true for an integer $k \geq 3$. This means that $2k + 1 < 2^k$ is assumed to be true for a particular but arbitrarily chosen integer $k \geq 3$. Then derive the truth of $P(k + 1)$. Or, in other words, show that the inequality $2(k + 1) + 1 < 2^{k+1}$ is true. But by multiplying out and regrouping,

$$2(k + 1) + 1 = 2k + 3 = (2k + 1) + 2, \qquad \text{5.3.1}$$

and by substitution from the inductive hypothesis,

$$(2k + 1) + 2 < 2^k + 2. \qquad \text{5.3.2}$$

Hence

$$2(k + 1) + 1 < 2^k + 2 \qquad \text{The left-most part of equation (5.3.1) is less than the right-most part of inequality (5.3.2).}$$

Note Properties of order are listed in Appendix A.

If it can be shown that $2^k + 2$ is less than 2^{k+1}, then the desired inequality will have been proved. But since the quantity 2^k can be added to or subtracted from an inequality without changing its direction,

$$2^k + 2 < 2^{k+1} \quad \Leftrightarrow \quad 2 < 2^{k+1} - 2^k = 2^k(2 - 1) = 2^k.$$

And since multiplying or dividing an inequality by 2 does not change its direction,

$$2 < 2^k \quad \Leftrightarrow \quad 1 = \frac{2}{2} < \frac{2^k}{2} = 2^{k-1} \qquad \text{by the laws of exponents.}$$

This last inequality is clearly true for all $k \geq 2$. Hence it is true that $2(k + 1) + 1 < 2^{k+1}$.

This discussion is made more flowing (but less intuitive) in the following formal proof:

Proposition 5.3.2

For all integers $n \geq 3$, $2n + 1 < 2^n$.

Proof (by mathematical induction):

Let the property $P(n)$ be the inequality

$$2n + 1 < 2^n. \qquad \leftarrow P(n)$$

Show that P(3) is true:
To establish $P(3)$, we must show that

$$2 \cdot 3 + 1 < 2^3. \qquad \leftarrow P(3)$$

But

$$2 \cdot 3 + 1 = 7 \quad \text{and} \quad 2^3 = 8 \quad \text{and} \quad 7 < 8.$$

Hence $P(3)$ is true.

Show that for all integers $k \geq 3$, if $P(k)$ is true then $P(k + 1)$ is also true:
[Suppose that $P(k)$ is true for a particular but arbitrarily chosen integer $k \geq 3$. That is:]
Suppose that k is any integer with $k \geq 3$ such that

$$2k + 1 < 2^k. \qquad \leftarrow P(k)$$
$$\text{inductive hypothesis}$$

[We must show that $P(k + 1)$ is true. That is:] We must show that

$$2(k + 1) + 1 < 2^{(k+1)},$$

or, equivalently,

$$2k + 3 < 2^{(k+1)}. \qquad \leftarrow P(k + 1)$$

continued on page 204

But

$$2k + 3 = (2k + 1) + 2 \qquad \text{by algebra}$$

$$< 2^k + 2^k \qquad \text{because } 2k + 1 < 2^k \text{ by the inductive hypothesis} \\ \text{and because } 2 < 2^k \text{ for all integers } k \geq 2$$

$$\therefore 2k + 3 < 2 \cdot 2^k = 2^{k+1} \qquad \text{by the laws of exponents.}$$

[This is what we needed to show.]
[Since we have proved the basis step and the inductive step, we conclude that the proposition is true.]

■

The next example demonstrates how to use mathematical induction to show that the terms of a sequence satisfy a certain explicit formula.

Example 5.3.3 Proving a Property of a Sequence

Define a sequence $a_1, a_2, a_3, \ldots$ as follows.*

$$a_1 = 2$$
$$a_k = 5a_{k-1} \quad \text{for all integers } k \geq 2.$$

a. Write the first four terms of the sequence.

b. It is claimed that for each integer $n \geq 1$, the nth term of the sequence has the same value as that given by the formula $2 \cdot 5^{n-1}$. In other words, the claim is that the terms of the sequence satisfy the equation $a_n = 2 \cdot 5^{n-1}$. Prove that this is true.

Solution

a. $a_1 = 2$.
$a_2 = 5a_{2-1} = 5a_1 = 5 \cdot 2 = 10$
$a_3 = 5a_{3-1} = 5a_2 = 5 \cdot 10 = 50$
$a_4 = 5a_{4-1} = 5a_3 = 5 \cdot 50 = 250.$

b. To use mathematical induction to show that every term of the sequence satisfies the equation, begin by showing that the first term of the sequence satisfies the equation. Then suppose that an arbitrarily chosen term a_k satisfies the equation and prove that the next term a_{k+1} also satisfies the equation.

Proof:

Let $a_1, a_2, a_3, \ldots$ be the sequence defined by specifying that $a_1 = 2$ and $a_k = 5a_{k-1}$ for all integers $k \geq 2$, and let the property $P(n)$ be the equation

$$a_n = 2 \cdot 5^{n-1}. \qquad \leftarrow P(n)$$

We will use mathematical induction to prove that for all integers $n \geq 1$, $P(n)$ is true.

Show that P(1) is true:
To establish $P(1)$, we must show that

$$a_1 = 2 \cdot 5^{1-1}. \qquad \leftarrow P(1)$$

---

*This is another example of a recursive definition. The general subject of recursion is discussed in Section 5.5.

But the left-hand side of $P(1)$ is

$$a_1 = 2 \qquad \text{by definition of } a_1, a_2, a_3, \ldots,$$

and the right-hand side of $P(1)$ is

$$2 \cdot 5^{1-1} = 2 \cdot 5^0 = 2 \cdot 1 = 2.$$

Thus the two sides of $P(1)$ are equal to the same quantity, and hence $P(1)$ is true.

Show that for all integers $k \geq 1$, if $P(k)$ is true then $P(k + 1)$ is also true:
[Suppose that $P(k)$ is true for a particular but arbitrarily chosen integer $k \geq 1$. That is:] Let k be any integer with $k \geq 1$, and suppose that

$$a_k = 2 \cdot 5^{k-1}. \qquad \leftarrow P(k) \\ \text{inductive hypothesis}$$

[We must show that $P(k + 1)$ is true. That is:] We must show that

$$a_{k+1} = 2 \cdot 5^{(k+1)-1},$$

or, equivalently,

$$a_{k+1} = 2 \cdot 5^k. \qquad \leftarrow P(k+1)$$

But the left-hand side of $P(k + 1)$ is

$$\begin{aligned}
a_{k+1} &= 5a_{(k+1)-1} &&\text{by definition of } a_1, a_2, a_3, \ldots \\
&= 5a_k &&\text{since } (k+1) - 1 = k \\
&= 5 \cdot (2 \cdot 5^{k-1}) &&\text{by inductive hypothesis} \\
&= 2 \cdot (5 \cdot 5^{k-1}) &&\text{by regrouping} \\
&= 2 \cdot 5^k &&\text{by the laws of exponents}
\end{aligned}$$

which is the right-hand side of the equation *[as was to be shown.]*

[Since we have proved the basis step and the inductive step, we conclude that the formula holds for all terms of the sequence.] ■

A Problem with Trominoes

The word *polyomino*, a generalization of *domino*, was introduced by Solomon Golomb in 1954 when he was a 22-year-old student at Harvard. Subsequently, he and others proved many interesting properties about them, and they became the basis for the popular computer game Tetris. A particular type of polyomino, called a *tromino*, is made up of three attached squares, which can be of two types:

straight 　　　　　and L-shaped

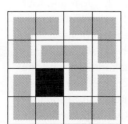

Call a checkerboard that is formed using m squares on a side an $m \times m$ ("m by m") checkerboard. Observe that if one square is removed from a 4×4 checkerboard, the remaining squares can be completely covered by L-shaped trominoes. For instance, a covering for one such board is illustrated in the figure to the left.

In his first article about polyominoes, Golomb included a proof of the following theorem. It is a beautiful example of an argument by mathematical induction.

Theorem Covering a Board with Trominoes

For any integer $n \geq 1$, if one square is removed from a $2^n \times 2^n$ checkerboard, the remaining squares can be completely covered by L-shaped trominoes.

The main insight leading to a proof of this theorem is the observation that because $2^{k+1} = 2 \cdot 2^k$, when a $2^{k+1} \times 2^{k+1}$ board is split in half both vertically and horizontally, each half side will have length 2^k and so each resulting quadrant will be a $2^k \times 2^k$ checkerboard.

Proof (by mathematical induction):

Let the property $P(n)$ be the sentence

> If any square is removed from a $2^n \times 2^n$ checkerboard,
> then the remaining squares can be completely covered.   ← $P(n)$
> by L-shaped trominoes

Show that P(1) is true:

A $2^1 \times 2^1$ checkerboard just consists of four squares. If one square is removed, the remaining squares form an L, which can be covered by a single L-shaped tromino, as illustrated in the figure to the left. Hence $P(1)$ is true.

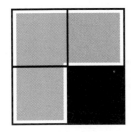

Show that for all integers k ≥ 1, if P(k) is true then P(k + 1) is also true:

[Suppose that $P(k)$ is true for a particular but arbitrarily chosen integer $k \geq 1$. That is:] Let k be any integer such that $k \geq 1$, and suppose that

> If any square is removed from a $2^k \times 2^k$ checkerboard,
> then the remaining squares can be completely covered   ← $P(k)$
> by L-shaped trominoes.

$P(k)$ is the inductive hypothesis.
   [We must show that $P(k + 1)$ is true. That is:] We must show that

> If any square is removed from a $2^{k+1} \times 2^{k+1}$ checkerboard,
> then the remaining squares can be completely covered   ← $P(k + 1)$
> by L-shaped trominoes.

$$2^k + 2^k = 2^{k+1}$$

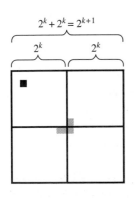

Consider a $2^{k+1} \times 2^{k+1}$ checkerboard with one square removed. Divide it into four equal quadrants: Each will consist of a $2^k \times 2^k$ checkerboard. In one of the quadrants, one square will have been removed, and so, by inductive hypothesis, all the remaining squares in this quadrant can be completely covered by L-shaped trominoes. The other three quadrants meet at the center of the checkerboard, and the center of the checkerboard serves as a corner of a square from each of those quadrants. An L-shaped tromino can, therefore, be placed on those three central squares. This situation is illustrated in the figure to the left. By inductive hypothesis, the remaining squares in each of the three quadrants can be completely covered by L-shaped trominoes. Thus every square in the $2^{k+1} \times 2^{k+1}$ checkerboard except the one that was removed can be completely covered by L-shaped trominoes *[as was to be shown]*.

Test Yourself

1. Mathematical induction differs from the kind of induction used in the natural sciences because it is actually a form of _____ reasoning.

2. Mathematical induction can be used to _____ conjectures that have been made using inductive reasoning.

Exercise Set 5.3

1. Based on the discussion of the product $\left(1 - \frac{1}{2}\right)\left(1 - \frac{1}{3}\right)$ $\left(1 - \frac{1}{4}\right)\cdots\left(1 - \frac{1}{n}\right)$ at the beginning of this section, conjecture a formula for general n. Prove your conjecture by mathematical induction.

2. Experiment with computing values of the product $\left(1 + \frac{1}{1}\right)\left(1 + \frac{1}{2}\right)\left(1 + \frac{1}{3}\right)\cdots\left(1 + \frac{1}{n}\right)$ for small values of n to conjecture a formula for this product for general n. Prove your conjecture by mathematical induction.

3. Observe that

$$\frac{1}{1\cdot 3} = \frac{1}{3}$$

$$\frac{1}{1\cdot 3} + \frac{1}{3\cdot 5} = \frac{2}{5}$$

$$\frac{1}{1\cdot 3} + \frac{1}{3\cdot 5} + \frac{1}{5\cdot 7} = \frac{3}{7}$$

$$\frac{1}{1\cdot 3} + \frac{1}{3\cdot 5} + \frac{1}{5\cdot 7} + \frac{1}{7\cdot 9} = \frac{4}{9}$$

Guess a general formula and prove it by mathematical induction.

H 4. Observe that

$$1 = 1,$$
$$1 - 4 = -(1 + 2),$$
$$1 - 4 + 9 = 1 + 2 + 3,$$
$$1 - 4 + 9 - 16 = -(1 + 2 + 3 + 4),$$
$$1 - 4 + 9 - 16 + 25 = 1 + 2 + 3 + 4 + 5.$$

Guess a general formula and prove it by mathematical induction.

5. Evaluate the sum $\sum_{k=1}^{n} \frac{k}{(k+1)!}$ for $n = 1, 2, 3, 4,$ and 5. Make a conjecture about a formula for this sum for general n, and prove your conjecture by mathematical induction.

6. For each positive integer n, let $P(n)$ be the property

$$5^n - 1 \text{ is divisible by } 4.$$

a. Write $P(0)$. Is $P(0)$ true?
b. Write $P(k)$.
c. Write $P(k + 1)$.
d. In a proof by mathematical induction that this divisibility property holds for all integers $n \geq 0$, what must be shown in the inductive step?

7. For each positive integer n, let $P(n)$ be the property

$$2^n < (n + 1)!.$$

a. Write $P(2)$. Is $P(2)$ true?
b. Write $P(k)$.
c. Write $P(k + 1)$.
d. In a proof by mathematical induction that this inequality holds for all integers $n \geq 2$, what must be shown in the inductive step?

Prove each statement in 8–23 by mathematical induction.

8. $5^n - 1$ is divisible by 4, for each integer $n \geq 0$.

9. $7^n - 1$ is divisible by 6, for each integer $n \geq 0$.

10. $n^3 - 7n + 3$ is divisible by 3, for each integer $n \geq 0$.

11. $3^{2n} - 1$ is divisible by 8, for each integer $n \geq 0$.

12. For any integer $n \geq 0$, $7^n - 2^n$ is divisible by 5.

H 13. For any integer $n \geq 0$, $x^n - y^n$ is divisible by $x - y$, where x and y are any integers with $x \neq y$.

H 14. $n^3 - n$ is divisible by 6, for each integer $n \geq 0$.

15. $n(n^2 + 5)$ is divisible by 6, for each integer $n \geq 0$.

16. $2^n < (n + 1)!$, for all integers $n \geq 2$.

17. $1 + 3n \leq 4^n$, for every integer $n \geq 0$.

18. $5^n + 9 < 6^n$, for all integers $n \geq 2$.

19. $n^2 < 2^n$, for all integers $n \geq 5$.

20. $2^n < (n + 2)!$, for all integers $n \geq 0$.

21. $\sqrt{n} < \frac{1}{\sqrt{1}} + \frac{1}{\sqrt{2}} + \cdots + \frac{1}{\sqrt{n}}$, for all integers $n \geq 2$.

22. $1 + nx \leq (1 + x)^n$, for all real numbers $x > -1$ and integers $n \geq 2$.

23. a. $n^3 > 2n + 1$, for all integers $n \geq 2$.
    b. $n! > n^2$, for all integers $n \geq 4$.

24. A sequence $a_1, a_2, a_3, \ldots$ is defined by letting $a_1 = 3$ and $a_k = 7a_{k-1}$ for all integers $k \geq 2$. Show that $a_n = 3 \cdot 7^{n-1}$ for all integers $n \geq 1$.

25. A sequence $b_0, b_1, b_2, \ldots$ is defined by letting $b_0 = 5$ and $b_k = 4 + b_{k-1}$ for all integers $k \geq 1$. Show that $b_n > 4n$ for all integers $n \geq 0$.

26. A sequence $c_0, c_1, c_2, \ldots$ is defined by letting $c_0 = 3$ and $c_k = (c_{k-1})^2$ for all integers $k \geq 1$. Show that $c_n = 3^{2^n}$ for all integers $n \geq 0$.

27. A sequence $d_1, d_2, d_3, \ldots$ is defined by letting $d_1 = 2$ and $d_k = \dfrac{d_{k-1}}{k}$ for all integers $k \geq 2$. Show that for all integers $n \geq 1, d_n = \dfrac{2}{n!}$.

28. Prove that for all integers $n \geq 1$,

$$\frac{1}{3} = \frac{1+3}{5+7} = \frac{1+3+5}{7+9+11} = \cdots$$
$$= \frac{1+3+\cdots+(2n-1)}{(2n+1)+\cdots+(4n-1)}.$$

29. As each of a group of businesspeople arrives at a meeting, each shakes hands with all the other people present. Use mathematical induction to show that if n people come to the meeting then $[n(n-1)]/2$ handshakes occur.

In order for a proof by mathematical induction to be valid, the basis statement must be true for $n = a$ and the argument of the inductive step must be correct for every integer $k \geq a$. In 30 and 31 find the mistakes in the "proofs" by mathematical induction.

30. **"Theorem:"** For any integer $n \geq 1$, all the numbers in a set of n numbers are equal to each other.

    "Proof (by mathematical induction): It is obviously true that all the numbers in a set consisting of just one number are equal to each other, so the basis step is true. For the inductive step, let $A = \{a_1, a_2, \ldots, a_k, a_{k+1}\}$ be any set of $k + 1$ numbers. Form two subsets each of size k:

$$B = \{a_1, a_2, a_3, \ldots, a_k\} \quad \text{and}$$
$$C = \{a_1, a_3, a_4, \ldots, a_{k+1}\}.$$

    (B consists of all the numbers in A except a_{k+1}, and C consists of all the numbers in A except a_2.) By inductive hypothesis, all the numbers in B equal a_1 and all the numbers in C equal a_1 (since both sets have only k numbers). But every number in A is in B or C, so all the numbers in A equal a_1; hence all are equal to each other."

H 31. **"Theorem:"** For all integers $n \geq 1, 3^n - 2$ is even.

    "Proof (by mathematical induction): Suppose the theorem is true for an integer k, where $k \geq 1$. That is, suppose that $3^k - 2$ is even. We must show that $3^{k+1} - 2$ is even. But

$$3^{k+1} - 2 = 3^k \cdot 3 - 2 = 3^k(1+2) - 2$$
$$= (3^k - 2) + 3^k \cdot 2.$$

    Now $3^k - 2$ is even by inductive hypothesis and $3^k \cdot 2$ is even by inspection. Hence the sum of the two quantities is even (by Theorem 4.1.1). It follows that $3^{k+1} - 2$ is even, which is what we needed to show."

H 32. Some 5×5 checkerboards with one square removed can be completely covered by L-shaped trominoes, whereas other 5×5 checkerboards cannot. Find examples of both kinds of checkerboards. Justify your answers.

33. Consider a 4×6 checkerboard. Draw a covering of the board by L-shaped trominoes.

H 34. **a.** Use mathematical induction to prove that any checkerboard with dimensions $2 \times 3n$ can be completely covered with L-shaped trominoes for any integer $n \geq 1$.
    b. Let n be any integer greater than or equal to 1. Use the result of part (a) to prove by mathematical induction that for all integers m, any checkerboard with dimensions $2m \times 3n$ can be completely covered with L-shaped trominoes.

35. Let m and n be any integers that are greater than or equal to 1.
    a. Prove that a necessary condition for an $m \times n$ checkerboard to be completely coverable by L-shaped trominoes is that mn be divisible by 3.
    H b. Prove that having mn be divisible by 3 is not a sufficient condition for an $m \times n$ checkerboard to be completely coverable by L-shaped trominoes.

36. In a round-robin tournament each team plays every other team exactly once. If the teams are labeled $T_1, T_2, \ldots, T_n$, then the outcome of such a tournament can be represented by a drawing, called a *directed graph*, in which the teams are represented as dots and an arrow is drawn from one dot to another if, and only if, the team represented by the first dot beats the team represented by the second dot. For example, the directed graph below shows one outcome of a round-robin tournament involving five teams, A, B, C, D, and E.

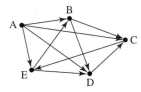

Use mathematical induction to show that in any round-robin tournament involving n teams, where $n \geq 2$, it is possible to label the teams $T_1, T_2, \ldots, T_n$ so that T_i beats T_{i+1} for all $i = 1, 2, \ldots, n-1$. (For instance, one such labeling in the example above is $T_1 = A, T_2 = B, T_3 = C, T_4 = E, T_5 = D$.) (*Hint:* Given $k + 1$ teams, pick one— say T'—and apply the inductive hypothesis to the remaining teams to obtain an ordering $T_1, T_2, \ldots, T_k$. Consider three cases: T' beats T_1, T' loses to the first m teams (where $1 \leq m \leq k - 1$) and beats the $(m+1)$st team, and T' loses to all the other teams.)

H ✶ 37. On the outside rim of a circular disk the integers from 1 through 30 are painted in random order. Show that no matter what this order is, there must be three successive integers whose sum is at least 45.

H **38.** Suppose that *n* *a*'s and *n* *b*'s are distributed around the outside of a circle. Use mathematical induction to prove that for all integers $n \geq 1$, given any such arrangement, it is possible to find a starting point so that if one travels around the circle in a clockwise direction, the number of *a*'s one has passed is never less than the number of *b*'s one has passed. For example, in the diagram shown below, one could start at the *a* with an asterisk.

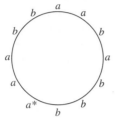

39. For a polygon to be **convex** means that given any two points on or inside the polygon, the line joining the points lies entirely inside the polygon. Use mathematical induction to prove that for all integers $n \geq 3$, the angles of any *n*-sided convex polygon add up to $180(n-2)$ degrees.

40. a. Prove that in an 8×8 checkerboard with alternating black and white squares, if the squares in the top right and bottom left corners are removed the remaining board cannot be covered with dominoes. (*Hint*: Mathematical induction is not needed for this proof.)

    H b. Use mathematical induction to prove that for all integers *n*, if a $2n \times 2n$ checkerboard with alternating black and white squares has one white square and one black square removed anywhere on the board, the remaining squares can be covered with dominoes.

Answers for Test Yourself

1. deductive   2. prove

5.4 Strong Mathematical Induction and the Well-Ordering Principle for the Integers

Mathematics takes us still further from what is human into the region of absolute necessity, to which not only the actual world, but every possible world, must conform.
— Bertrand Russell, 1902

Strong mathematical induction is similar to ordinary mathematical induction in that it is a technique for establishing the truth of a sequence of statements about integers. Also, a proof by strong mathematical induction consists of a basis step and an inductive step. However, the basis step may contain proofs for several initial values, and in the inductive step the truth of the predicate $P(n)$ is assumed not just for one value of *n* but for *all* values through *k*, and then the truth of $P(k+1)$ is proved.

Principle of Strong Mathematical Induction

Let $P(n)$ be a property that is defined for integers *n*, and let *a* and *b* be fixed integers with $a \leq b$. Suppose the following two statements are true:

1. $P(a)$, $P(a+1)$, ..., and $P(b)$ are all true. (**basis step**)

2. For any integer $k \geq b$, if $P(i)$ is true for all integers *i* from *a* through *k*, then $P(k+1)$ is true. (**inductive step**)

Then the statement

$$\text{for all integers } n \geq a, \; P(n)$$

is true. (The supposition that $P(i)$ is true for all integers *i* from *a* through *k* is called the **inductive hypothesis.** Another way to state the inductive hypothesis is to say that $P(a)$, $P(a+1)$, ..., $P(k)$ are all true.)

Any statement that can be proved with ordinary mathematical induction can be proved with strong mathematical induction. The reason is that given any integer $k \geq b$, if the truth of $P(k)$ alone implies the truth of $P(k+1)$, then certainly the truth of $P(a)$, $P(a+1), \ldots$, and $P(k)$ implies the truth of $P(k+1)$. It is also the case that any statement that can be proved with strong mathematical induction can be proved with ordinary mathematical induction. A proof is sketched in exercise 27 at the end of this section.

Strictly speaking, the principle of strong mathematical induction can be written without a basis step if the inductive step is changed to "$\forall k \geq a - 1$, if $P(i)$ is true for all integers i from a through k, then $P(k+1)$ is true." The reason for this is that the statement "$P(i)$ is true for all integers i from a through k" is vacuously true for $k = a-1$. Hence, if the implication in the inductive step is true, then the conclusion $P(a)$ must also be true,* which proves the basis step. However, in many cases the proof of the implication for $k > b$ does not work for $a \leq k \leq b$. So it is a good idea to get into the habit of thinking separately about the cases where $a \leq k \leq b$ by explicitly including a basis step.

The principle of strong mathematical induction is known under a variety of different names including the *second principle of induction*, the *second principle of finite induction*, and the *principle of complete induction*.

Applying Strong Mathematical Induction

The divisibility-by-a-prime theorem states that any integer greater than 1 is divisible by a prime number. We prove this theorem using strong mathematical induction.

Example 5.4.1 Divisibility by a Prime

Prove Theorem 4.3.4: Any integer greater than 1 is divisible by a prime number.

Solution   The idea for the inductive step is this: If a given integer greater than 1 is not itself prime, then it is a product of two smaller positive integers, each of which is greater than 1. Since you are assuming that each of these smaller integers is divisible by a prime number, by transitivity of divisibility, those prime numbers also divide the integer you started with.

Proof (by strong mathematical induction):

Let the property $P(n)$ be the sentence

$\qquad\qquad$ n is divisible by a prime number. $\qquad$ ← $P(n)$

Show that P(2) is true:
To establish $P(2)$, we must show that

$\qquad\qquad$ 2 is divisible by a prime number. $\qquad$ ← $P(2)$

But this is true because 2 is divisible by 2 and 2 is a prime number.

Show that for all integers $k \geq 2$, if $P(i)$ is true for all integers i from 2 through k, then $P(k+1)$ is also true:

---

*If you have proved that a certain if-then statement is true and if you also know that the hypothesis is true, then the conclusion must be true.

Let k be any integer with $k \geq 2$ and suppose that

i is divisible by a prime number for all integers
i from 2 through k.   ← inductive hypothesis

We must show that

$k + 1$ is divisible by a prime number.   ← $P(k + 1)$

Case 1 ($k + 1$ is prime): In this case $k + 1$ is divisible by a prime number, namely itself.

Case 2 ($k + 1$ is not prime): In this case $k + 1 = ab$ where a and b are integers with $1 < a < k + 1$ and $1 < b < k + 1$. Thus, in particular, $2 \leq a \leq k$, and so by inductive hypothesis, a is divisible by a prime number p. In addition because $k + 1 = ab$, we have that $k + 1$ is divisible by a. Hence, since $k + 1$ is divisible by a and a is divisible by p, by transitivity of divisibility, $k + 1$ is divisible by the prime number p.

Therefore, regardless of whether $k + 1$ is prime or not, it is divisible by a prime number *[as was to be shown].*
[Since we have proved both the basis and the inductive step of the strong mathematical induction, we conclude that the given statement is true.]

∎

Both ordinary and strong mathematical induction can be used to show that the terms of certain sequences satisfy certain properties. The next example shows how this is done using strong induction.

Example 5.4.2 Proving a Property of a Sequence with Strong Induction

Define a sequence $s_0, s_1, s_2, \ldots$ as follows:

$$s_0 = 0, \quad s_1 = 4, \quad s_k = 6a_{k-1} - 5a_{k-2} \quad \text{for all integers } k \geq 2.$$

a. Find the first four terms of this sequence.

b. It is claimed that for each integer $n \geq 0$, the nth term of the sequence has the same value as that given by the formula $5^n - 1$. In other words, the claim is that all the terms of the sequence satisfy the equation $s_n = 5^n - 1$. Prove that this is true.

Solution

a. $s_0 = 0,$ $s_1 = 4,$ $s_2 = 6s_1 - 5s_0 = 6 \cdot 4 - 5 \cdot 0 = 24,$
$s_3 = 6s_2 - 5s_1 = 6 \cdot 24 - 5 \cdot 4 = 144 - 20 = 124$

b. To use strong mathematical induction to show that every term of the sequence satisfies the equation, the basis step must show that the first two terms satisfy it. This is necessary because, according to the definition of the sequence, computing values of later terms requires knowing the values of the *two* previous terms. So if the basis step only shows that the first term satisfies the equation, it would not be possible to use the inductive step to deduce that the second term satisfies the equation. In the inductive step you suppose that for an arbitrarily chosen integer $k \geq 1$, all the terms of the sequence from s_0 through s_k satisfy the given equation and you then deduce that s_{k+1} must also satisfy the equation.

Proof:

Let $s_0, s_1, s_2, \ldots$ be the sequence defined by specifying that $s_0 = 0$, $s_1 = 4$, and $s_k = 6a_{k-1} - 5a_{k-2}$ for all integers $k \geq 2$, and let the property $P(n)$ be the formula

$$s_n = 5^n - 1 \qquad \leftarrow P(n)$$

We will use strong mathematical induction to prove that for all integers $n \geq 0$, $P(n)$ is true.

Show that $P(0)$ and $P(1)$ are true:
To establish $P(0)$ and $P(1)$, we must show that

$$s_0 = 5^0 - 1 \quad \text{and} \quad s_1 = 5^1 - 1. \qquad \leftarrow P(0) \quad \text{and} \quad P(1)$$

But, by definition of $s_0, s_1, s_2, \ldots$, we have that $s_0 = 0$ and $s_1 = 4$. Since $5^0 - 1 = 1 - 1 = 0$ and $5^1 - 1 = 5 - 1 = 4$, the values of s_0 and s_1 agree with the values given by the formula.

Show that for all integers $k \geq 1$, if $P(i)$ is true for all integers i from 0 through k, then $P(k + 1)$ is also true:
Let k be any integer with $k \geq 1$ and suppose that

$$s_i = 5^i - 1 \text{ for all integers } i \text{ with } 0 \leq i \leq k. \quad \leftarrow \text{inductive hypothesis}$$

We must show that

$$s_{k+1} = 5^{k+1} - 1. \qquad \leftarrow P(k + 1)$$

But since $k \geq 1$, we have that $k + 1 \geq 2$, and so

$$
\begin{aligned}
s_{k+1} &= 6s_k - 5s_{k-1} && \text{by definition of } s_0, s_1, s_2, \ldots \\
&= 6(5^k - 1) - 5(5^{k-1} - 1) && \text{by definition hypothesis} \\
&= 6 \cdot 5^k - 6 - 5^k + 5 && \text{by multiplying out and applying} \\
& && \text{a law of exponents} \\
&= (6 - 1)5^k - 1 && \text{by factoring out 6 and arithmetic} \\
&= 5 \cdot 5^k - 1 && \text{by arithmetic} \\
&= 5^{k+1} - 1 && \text{by applying a law of exponents,}
\end{aligned}
$$

[as was to be shown].

[Since we have proved both the basis and the inductive step of the strong mathematical induction, we conclude that the given statement is true.]

∎

Another use of strong induction concerns the computation of products. A product of four numbers may be computed in a variety of different ways as indicated by the placement of parentheses. For instance,

$((x_1 x_2) x_3) x_4$ means multiply x_1 and x_2, multiply the result by x_3, and then multiply that number by x_4.

And

$(x_1 x_2)(x_3 x_4)$ means multiply x_1 and x_2, multiply x_3 and x_4, and then take the product of the two.

Note that in both examples above, although the factors are multiplied in a different order, the number of multiplications—three—is the same. Strong mathematical induction is used to prove a generalization of this fact.

Note Like many definitions, for extreme cases this may look strange but it makes things work out nicely.

Convention

Let us agree to say that a single number x_1 is a product with one factor and can be computed with zero multiplications.

Example 5.4.3 The Number of Multiplications Needed to Multiply n Numbers

Prove that for any integer $n \geq 1$, if $x_1, x_2, \ldots, x_n$ are n numbers, then no matter how the parentheses are inserted into their product, the number of multiplications used to compute the product is $n - 1$.

Solution   The truth of the basis step follows immediately from the convention about a product with one factor. The inductive step is based on the fact that when several numbers are multiplied together, each step of the process involves multiplying two individual quantities. For instance, the final step for computing $((x_1 x_2) x_3)(x_4 x_5)$ is to multiply $(x_1 x_2) x_3$ and $x_4 x_5$. In general, if $k + 1$ numbers are multiplied, the two quantities in the final step each consist of fewer than $k + 1$ factors. This is what makes it possible to use the inductive hypothesis.

Proof (by strong mathematical induction):

Let the property $P(n)$ be the sentence

> If $x_1, x_2, \ldots, x_n$ are n numbers, then no matter how parentheses are inserted into their product, the number of multiplications used to compute the product is $n - 1$. $\leftarrow P(n)$

Show that $P(1)$ is true:
To establish $P(1)$, we must show that

> The number of multiplications needed to compute the product of x_1 is $1 - 1$. $\leftarrow P(1)$

This is true because, by convention, x_1 is a product that can be computed with 0 multiplications, and $0 = 1 - 1$.

Show that for all integers $k \geq 1$, if $P(i)$ is true for all integers i from 1 through k, then $P(k + 1)$ is also true:
Let k by any integer with $k \geq 1$ and suppose that

> For all integers i from 1 through k, if $x_1, x_2, \ldots, x_i$ are numbers, then no matter how parentheses are inserted into their product, the number of multiplications used to compute the product is $i - 1$. $\leftarrow$ inductive hypothesis

We must show that

> If $x_1, x_2, \ldots, x_{k+1}$ are $k + 1$ numbers, then no matter how parentheses are inserted into their product, the number of multiplications used to compute the product is $(k + 1) - 1 = k$. $\leftarrow P(k + 1)$

continued on page 214

Consider a product of $k + 1$ factors: $x_1, x_2 \ldots, x_{k+1}$. When parentheses are inserted in order to compute the product, some multiplication is the final one and each of the two factors making up the final multiplication is a product of fewer than $k + 1$ factors. Let L be the product of the left-hand factors and R be the product of the right-hand factors, and suppose that L is composed of l factors and R is composed of r factors. Then $l + r = k + 1$, the total number of factors in the product, and

$$1 \leq l \leq k \quad \text{and} \quad 1 \leq r \leq k.$$

By inductive hypothesis, evaluating L takes $l - 1$ multiplications and evaluating R takes $r - 1$ multiplications. Because one final multiplication is needed to evaluate $L \cdot R$, the number of multiplications needed to evaluate the product of all $k + 1$ factors is

$$(l - 1) + (r - 1) + 1 = (l + r) - 1 = (k + 1) - 1 = k.$$

[This is what was to be shown.]
[Since we have proved the basis step and the inductive step of the strong mathematical induction, we conclude that the given statement is true.]

Binary Representation of Integers

In elementary school, you learned the meaning of decimal notation: that to interpret a string of decimal digits as a number, you mentally multiply each digit by its place value. For instance, 5,049 has a 5 in the thousands place, a 0 in the hundreds place, a 4 in the tens place, and a 9 in the ones place. Thus

$$5{,}049 = 5 \cdot (1{,}000) + 0 \cdot (100) + 4 \cdot (10) + 9 \cdot (1).$$

Using exponential notation, this equation can be rewritten as

$$5{,}049 = 5 \cdot 10^3 + 0 \cdot 10^2 + 4 \cdot 10^1 + 9 \cdot 10^0.$$

More generally, decimal notation is based on the fact that any positive integer can be written uniquely as a sum of products of the form

$$d \cdot 10^n,$$

where each n is a nonnegative integer and each d is one of the decimal digits 0, 1, 2, 3, 4, 5, 6, 7, 8, or 9. The word *decimal* comes from the Latin root *deci*, meaning "ten." Decimal (or base 10) notation expresses a number as a string of digits in which each digit's position indicates the power of 10 by which it is multiplied. The right-most position is the ones place (or 10^0 place), to the left of that is the tens place (or 10^1 place), to the left of that is the hundreds place (or 10^2 place), and so forth, as illustrated below.

Place	10^3 thousands	10^2 hundreds	10^1 tens	10^0 ones
Decimal Digit	5	0	4	9

There is nothing sacred about the number 10; we use 10 as a base for our usual number system because we happen to have ten fingers. In fact, any integer greater than 1 can serve as a base for a number system. In computer science, **base 2 notation,** or **binary notation,** is of special importance because the signals used in modern electronics are always in one of only two states. (The Latin root *bi* means "two.")

In binary notation, the fundamental term has the form

$$d \cdot 2^n,$$

where n is an integer and d is one of the binary digits (or bits) 0 or 1. For example,

$$27 = 16 + 8 + 2 + 1$$
$$= 1 \cdot 2^4 + 1 \cdot 2^3 + 0 \cdot 2^2 + 1 \cdot 2^1 + 1 \cdot 2^0.$$

In binary notation, as in decimal notation, we write just the binary digits, and not the powers of the base. Thus, in binary notation:

$$1 \cdot 2^4 + 1 \cdot 2^3 + 0 \cdot 2^2 + 1 \cdot 2^1 + 1 \cdot 2^0$$

$$27_{10} \qquad = \qquad 1\,1\,0\,1\,1_2$$

where the subscripts indicate the base, whether 10 or 2, in which the number is written. The places in binary notation correspond to the various powers of 2. The right-most position is the one's place (or 2^0 place), to the left of that is the twos place (or 2^1 place), to the left of that is the fours place (or 2^2 place), and so forth, as illustrated below.

Place	2^4 sixteens	2^3 eights	2^2 fours	2^1 twos	2^0 ones
Binary Digit	1	1	0	1	1

As in the decimal notation, leading zeros may be added or dropped as desired. For example,

$$003_{10} = 3_{10} = 1 \cdot 2^1 + 1 \cdot 2^0 = 11_2 = 011_2.$$

Example 5.4.4 Converting from Binary to Decimal Notation

Represent 110101_2 in decimal notation.

Solution $110101_2 = 1 \cdot 2^5 + 1 \cdot 2^4 + 0 \cdot 2^3 + 1 \cdot 2^2 + 0 \cdot 2^1 + 1 \cdot 2^0$
$$= 32 + 16 + 4 + 1$$
$$= 53_{10}$$

Alternatively, the schema below may be used.

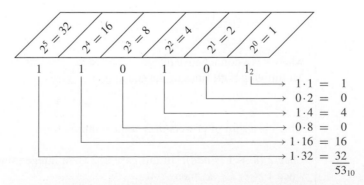

Strong mathematical induction makes possible a proof of the fact used frequently in computer science that every positive integer n has a unique binary integer representation. The proof looks complicated because of all the notation needed to write down the various steps. But the idea of the proof is simple. It is that if smaller integers than n have unique representations as sums of powers of 2, then the unique representation for n as a sum of powers of 2 can be found by taking the representation for $n/2$ (or for $(n-1)/2$ if n is odd) and multiplying it by 2.

Theorem 5.4.1 Existence and Uniqueness of Binary Integer Representations

Given any positive integer n, n has a unique representation in the form

$$n = c_r \cdot 2^r + c_{r-1} \cdot 2^{r-1} + \cdots + c_2 \cdot 2^2 + c_1 \cdot 2 + c_0,$$

where r is a nonnegative integer, $c_r = 1$, and $c_j = 1$ or 0 for all $j = 0, 1, 2, \ldots, r-1$.

Proof:

We give separate proofs by strong mathematical induction to show first the existence and second the uniqueness of the binary representation.

Existence (proof by strong mathematical induction): Let the property $P(n)$ be the equation

$$n = c_r \cdot 2^r + c_{r-1} \cdot 2^{r-1} + \cdots + c_2 \cdot 2^2 + c_1 \cdot 2 + c_0, \qquad \leftarrow P(n)$$

where r is a nonnegative integer, $c_r = 1$, and $c_j = 1$ or 0 for all $j = 0, 1, 2, \ldots, r-1$.

Show that $P(1)$ is true:
Let $r = 0$ and $c_0 = 1$. Then $1 = c_r \cdot 2^r$, and so $n = 1$ can be written in the required form.

Show that for all integers $k \geq 1$, if $P(i)$ is true for all integers i from 1 through k, then $P(k+1)$ is also true:
Let k be an integer with $k \geq 1$. Suppose that for all integers i from 1 through k,

$$i = c_r \cdot 2^r + c_{r-1} \cdot 2^{r-1} + \cdots + c_2 \cdot 2^2 + c_1 \cdot 2 + c_0, \quad \leftarrow \text{inductive hypothesis}$$

where r is a nonnegative integer, $c_r = 1$, and $c_j = 1$ or 0 for all $j = 0, 1, 2, \ldots, r-1$. We must show that $k+1$ can be written as a sum of powers of 2 in the required form.

Case 1 ($k+1$ is even): In this case $(k+1)/2$ is an integer, and by inductive hypothesis, since $1 \leq (k+1)/2 \leq k$, then,

$$\frac{k+1}{2} = c_r \cdot 2^r + c_{r-1} \cdot 2^{r-1} + \cdots + c_2 \cdot 2^2 + c_1 \cdot 2 + c_0,$$

where r is a nonnegative integer, $c_r = 1$, and $c_j = 1$ or 0 for all $j = 0, 1, 2, \ldots, r-1$. Multiplying both sides of the equation by 2 gives

$$k+1 = c_r \cdot 2^{r+1} + c_{r-1} \cdot 2^r + \cdots + c_2 \cdot 2^3 + c_1 \cdot 2^2 + c_0 \cdot 2,$$

which is a sum of powers of 2 of the required form.

Case 2 ($k+1$ is odd): In this case $k/2$ is an integer, and by inductive hypothesis, since $1 \leq k/2 \leq k$, then

$$\frac{k}{2} = c_r \cdot 2^r + c_{r-1} \cdot 2^{r-1} + \cdots + c_2 \cdot 2^2 + c_1 \cdot 2 + c_0,$$

where r is a nonnegative integer, $c_r = 1$, and $c_j = 1$ or 0 for all $j = 0, 1, 2, \ldots, r - 1$. Multiplying both sides of the equation by 2 and adding 1 gives

$$k + 1 = c_r \cdot 2^{r+1} + c_{r-1} \cdot 2^r + \cdots + c_2 \cdot 2^3 + c_1 \cdot 2^2 + c_0 \cdot 2 + 1,$$

which is also a sum of powers of 2 of the required form.

The preceding arguments show that regardless of whether $k + 1$ is even or odd, $k + 1$ has a representation of the required form. *[Or, in other words, $P(k + 1)$ is true as was to be shown.]*

[Since we have proved the basis step and the inductive step of the strong mathematical induction, the existence half of the theorem is true.]

***Uniqueness*:** To prove uniqueness, suppose that there is an integer n with two different representations as a sum of nonnegative integer powers of 2. Equating the two representations and canceling all identical terms gives

$$2^r + c_{r-1} \cdot 2^{r-1} + \cdots + c_1 \cdot 2 + c_0 = 2^s + d_{s-1} \cdot 2^{s-1} + \cdots + d_1 \cdot 2 + d_0 \qquad 5.4.1$$

where r and s are nonnegative integers, and each c_i and each d_i equal 0 or 1. Without loss of generality, we may assume that $r < s$. But by the formula for the sum of a geometric sequence (Theorem 5.2.3) and because $r < s$,

$$2^r + c_{r-1} \cdot 2^{r-1} + \cdots + c_1 \cdot 2 + c_0 \leq 2^r + 2^{r-1} + \cdots + 2 + 1 = 2^{r+1} - 1$$
$$< 2^s.$$

Thus

$$2^r + c_{r-1} \cdot 2^{r-1} + \cdots + c_1 \cdot 2 + c_0 < 2^s + d_{s-1} \cdot 2^{s-1} + \cdots + d_1 \cdot 2 + d_0,$$

which contradicts equation (5.4.1). Hence the supposition is false, so any integer n has only one representation as a sum of nonnegative integer powers of 2.

The Well-Ordering Principle for the Integers

The well-ordering principle for the integers looks very different from both the ordinary and the strong principles of mathematical induction, but it can be shown that all three principles are equivalent. That is, if any one of the three is true, then so are both of the others.

Well-Ordering Principle for the Integers

Let S be a set of integers containing one or more integers all of which are greater than some fixed integer. Then S has a least element.

Note that when the context makes the reference clear, we will write simply "the well-ordering principle" rather than "the well-ordering principle for the integers."

Example 5.4.5 Finding Least Elements

In each case, if the set has a least element, state what it is. If not, explain why the well-ordering principle is not violated.

a. The set of all positive real numbers.

b. The set of all nonnegative integers n such that $n^2 < n$.

c. The set of all nonnegative integers of the form $46 - 7k$, where k is an integer.

Solution

a. There is no least positive real number. For if x is any positive real number, then $x/2$ is a positive real number that is less than x. No violation of the well-ordering principle occurs because the well-ordering principle refers only to sets of integers, and this set is not a set of integers.

b. There is no *least* nonnegative integer n such that $n^2 < n$ because there is *no* nonnegative integer that satisfies this inequality. The well-ordering principle is not violated because the well-ordering principle refers only to sets that contain at least one element.

c. The following table shows values of $46 - 7k$ for various values of k.

k	0	1	2	3	4	5	6	7	$\cdots$	-1	-2	-3	$\cdots$
$46 - 7k$	46	39	32	25	18	11	4	-3	$\cdots$	53	60	67	$\cdots$

The table suggests, and you can easily confirm, that $46 - 7k < 0$ for $k \geq 7$ and that $46 - 7k \geq 46$ for $k \leq 0$. Therefore, from the other values in the table it is clear that 4 is the least nonnegative integer of the form $46 - 7k$. This corresponds to $k = 6$. ■

Another way to look at the analysis of Example 5.4.5(c) is to observe that subtracting six 7's from 46 leaves 4 left over and this is the least nonnegative integer obtained by repeated subtraction of 7's from 46. In other words, 6 is the quotient and 4 is the remainder for the division of 46 by 7. More generally, in the division of any integer n by any positive integer d, the remainder r is the least nonnegative integer of the form $n - dk$. This is the heart of the following proof of the existence part of the quotient-remainder theorem (the part that guarantees the existence of a quotient and a remainder of the division of an integer by a positive integer). For a proof of the uniqueness of the quotient and remainder, see exercise 18 of Section 4.6.

Quotient-Remainder Theorem (Existence Part)

Given any integer n and any positive integer d, there exist integers q and r such that

$$n = dq + r \quad \text{and} \quad 0 \leq r < d.$$

Proof:

Let S be the set of all nonnegative integers of the form

$$n - dk,$$

where k is an integer. This set has at least one element. *[For if n is nonnegative, then*

$$n - 0 \cdot d = n \geq 0,$$

and so $n - 0 \cdot d$ is in S. And if n is negative, then

$$n - nd = n\underbrace{(1 - d)}_{} \geq 0,$$

$\uparrow$
$<0 \qquad \leq 0$ *since d is a positive integer*

and so $n - nd$ is in S.] It follows by the well-ordering principle for the integers that S contains a least element r. Then, for some specific integer $k = q$,

$$n - dq = r$$

[because every integer in S can be written in this form]. Adding dq to both sides gives

$$n = dq + r.$$

Furthermore, $r < d$. *[For suppose $r \geq d$. Then*

$$n - d(q + 1) = n - dq - d = r - d \geq 0,$$

and so $n - d(q + 1)$ would be a nonnegative integer in S that would be smaller than r. But r is the smallest integer in S. This contradiction shows that the supposition $r \geq d$ must be false.] The preceding arguments prove that there exist integers r and q for which

$$n = dq + r \quad \text{and} \quad 0 \leq r < d.$$

[This is what was to be shown.]

Another consequence of the well-ordering principle is the fact that any strictly decreasing sequence of nonnegative integers is finite. That is, if $r_1, r_2, r_3, \ldots$ is a sequence of nonnegative integers satisfying

$$r_i > r_{i+1}$$

for all $i \geq 1$, then $r_1, r_2, r_3, \ldots$ is a finite sequence. *[For by the well-ordering principle such a sequence would have to have a least element r_k. It follows that r_k must be the final term of the sequence because if there were a term r_{k+1}, then since the sequence is strictly decreasing, $r_{k+1} < r_k$, which would be a contradiction.]* This fact is frequently used in computer science to prove that algorithms terminate after a finite number of steps.

Test Yourself

1. In a proof by strong mathematical induction the basis step may require checking a property $P(n)$ for more _____ value of n.

2. Suppose that in the basis step for a proof by strong mathematical induction the property $P(n)$ was checked for all integers n from a through b. Then in the inductive step one assumes that for any integer $k \geq b$, the property $P(n)$ is true for all values of i from _____ through _____ and one shows that _____ is true.

3. According to the well-ordering principle for the integers, if a set S of integers contains at least _____ and if there is some integer that is less than or equal to every _____, then _____.

Exercise Set 5.4

1. Suppose $a_1, a_2, a_3, \ldots$ is a sequence defined as follows:

$$a_1 = 1, \ a_2 = 3,$$
$$a_k = a_{k-2} + 2a_{k-1} \quad \text{for all integers } k \geq 3.$$

Prove that a_n is odd for all integers $n \geq 1$.

2. Suppose $b_1, b_2, b_3, \ldots$ is a sequence defined as follows:

$$b_1 = 4, \ b_2 = 12$$
$$b_k = b_{k-2} + b_{k-1} \quad \text{for all integers } k \geq 3.$$

Prove that b_n is divisible by 4 for all integers $n \geq 1$.

3. Suppose that $c_0, c_1, c_2, \ldots$ is a sequence defined as follows:

$$c_0 = 2, \ c_1 = 2, \ c_2 = 6,$$
$$c_k = 3c_{k-3} \quad \text{for all integers } k \geq 3.$$

Prove that c_n is even for all integers $n \geq 0$.

4. Suppose that $d_1, d_2, d_3, \ldots$ is a sequence defined as follows:

$$d_1 = \frac{9}{10}, \ d_2 = \frac{10}{11},$$
$$d_k = d_{k-1} \cdot d_{k-2} \quad \text{for all integers } k \geq 3.$$

Prove that $0 < d_n \leq 1$ for all integers $n \geq 1$.

5. Suppose that $e_0, e_1, e_2, \ldots$ is a sequence defined as follows:

$$e_0 = 12, \ e_1 = 29$$
$$e_k = 5e_{k-1} - 6e_{k-2} \quad \text{for all integers } k \geq 2.$$

Prove that $e_n = 5 \cdot 3^n + 7 \cdot 2^n$ for all integers $n \geq 0$.

6. Suppose that $f_0, f_1, f_2, \ldots$ is a sequence defined as follows:

$$f_0 = 5, \ f_1 = 16$$
$$f_k = 7f_{k-1} - 10f_{k-2} \quad \text{for all integers } k \geq 2.$$

Prove that $f_n = 3 \cdot 2^n + 2 \cdot 5^n$ for all integers $n \geq 0$.

7. Suppose that $g_1, g_2, g_3, \ldots$ is a sequence defined as follows:

$$g_1 = 3, \quad g_2 = 5$$
$$g_k = 3g_{k-1} - 2g_{k-2} \quad \text{for all integers } k \geq 3.$$

Prove that $g_n = 2^n + 1$ for all integers $n \geq 1$.

8. Suppose that $h_0, h_1, h_2, \ldots$ is a sequence defined as follows:

$$h_0 = 1, \quad h_1 = 2, \quad h_2 = 3,$$
$$h_k = h_{k-1} + h_{k-2} + h_{k-3} \quad \text{for all integers } k \geq 3.$$

a. Prove that $h_n \leq 3^n$ for all integers $n \geq 0$.
b. Suppose that s is any real number such that $s^3 \geq s^2 + s + 1$. (This implies that $s > 1.83$.) Prove that $h_n \leq s^n$ for all $n \geq 2$.

9. Define a sequence $a_1, a_2, a_3, \ldots$ as follows: $a_1 = 1, a_2 = 3$, and $a_k = a_{k-1} + a_{k-2}$ for all integers $k \geq 3$. (This sequence is known as the Lucas sequence.) Use strong mathematical induction to prove that $a_n \leq \left(\frac{7}{4}\right)^n$ for all integers $n \geq 1$.

H 10. The problem that was used to introduce ordinary mathematical induction in Section 5.2 can also be solved using strong mathematical induction. Let $P(n)$ be "$n\cent$ can be obtained using a combination of $3\cent$ and $5\cent$ coins." Use strong mathematical induction to prove that $P(n)$ is true for all integers $n \geq 8$.

11. You begin solving a jigsaw puzzle by finding two pieces that match and fitting them together. Each subsequent step of the solution consists of fitting together two blocks made up of one or more pieces that have previously been assembled. Use strong mathematical induction to prove that the number of steps required to put together all n pieces of a jigsaw puzzle is $n - 1$.

H 12. The sides of a circular track contain a sequence of cans of gasoline. The total amount in the cans is sufficient to enable a certain car to make one complete circuit of the track, and it could all fit into the car's gas tank at one time. Use mathematical induction to prove that it is possible to find an initial location for placing the car so that it will be able to traverse the entire track by using the various amounts of gasoline in the cans that it encounters along the way.

H 13. Use strong mathematical induction to prove the existence part of the unique factorization of integers (Theorem 4.3.5): Every integer greater than 1 is either a prime number or a product of prime numbers.

14. Any product of two or more integers is a result of successive multiplications of two integers at a time. For instance, here are a few of the ways in which $a_1 a_2 a_3 a_4$ might be computed: $(a_1 a_2)(a_3 a_4)$ or $((a_1 a_2)a_3)a_4)$ or $a_1((a_2 a_3)a_4)$. Use strong mathematical induction to prove that any product of two or more odd integers is odd.

15. Define the "sum" of one integer to be that integer, and use strong mathematical induction to prove that for all integers $n \geq 1$, any sum of n even integers is even.

H 16. Use strong mathematical induction to prove that for any integer $n \geq 2$, if n is even, then any sum of n odd integers is even, and if n is odd, then any sum of n odd integers is odd.

17. Compute $4^1, 4^2, 4^3, 4^4, 4^5, 4^6, 4^7$, and 4^8. Make a conjecture about the units digit of 4^n where n is a positive integer. Use strong mathematical induction to prove your conjecture.

18. Compute $9^0, 9^1, 9^2, 9^3, 9^4$, and 9^5. Make a conjecture about the units digit of 9^n where n is a positive integer. Use strong mathematical induction to prove your conjecture.

19. Find the mistake in the following "proof" that purports to show that every nonnegative integer power of every nonzero real number is 1.

"**Proof:** Let r be any nonzero real number and let the property $P(n)$ be the equation $r^n = 1$.

Show that $P(0)$ is true: $P(0)$ is true because $r^0 = 1$ by definition of zeroth power.

Show that for all integers $k \geq 0$, if $P(i)$ is true for all integers i from 0 through k, then $P(k + 1)$ is also true: Let k be any integer with $k \geq 0$ and suppose that $r^i = 1$ for all integers i from 0 through k. This is the inductive hypothesis. We must show that $r^{k+1} = 1$. Now

$$
\begin{aligned}
r^{k+1} &= r^{k+k-(k-1)} && \text{because } k+k-(k-1) \\
&&& = k+k-k+1 = k+1 \\
&= \frac{r^k \cdot r^k}{r^{k-1}} && \text{by the laws of exponents} \\
&= \frac{1 \cdot 1}{1} && \text{by inductive hypothesis} \\
&= 1.
\end{aligned}
$$

Thus $r^{k+1} = 1$ *[as was to be shown]*.

[Since we have proved the basis step and the inductive step of the strong mathematical induction, we conclude that the given statement is true.]"

20. Use the well-ordering principle for the integers to prove Theorem 4.3.4: Every integer greater than 1 is divisible by a prime number.

21. Use the well-ordering principle for the integers to prove the existence part of the unique factorization of integers theorem: Every integer greater than 1 is either prime or a product of prime numbers.

22. **a.** The Archimedean property for the rational numbers states that for all rational numbers r, there is an integer n such that $n > r$. Prove this property.
   b. Prove that given any rational number r, the number $-r$ is also rational.
   c. Use the results of parts (a) and (b) to prove that given any rational number r, there is an integer m such that $m < r$.

H 23. Use the results of exercise 22 and the well-ordering principle for the integers to show that given any rational number r, there is an integer m such that $m \leq r < m + 1$.

24. Use the well-ordering principle to prove that given any integer $n \geq 1$, there exists an odd integer m and a nonnegative integer k such that $n = 2^k \cdot m$.

25. Imagine a situation in which eight people, numbered consecutively 1–8, are arranged in a circle. Starting from person #1, every second person in the circle is eliminated. The elimination process continues until only one person remains. In the first round the people numbered 2, 4, 6, and 8 are eliminated, in the second round the people numbered 3 and 7 are eliminated, and in the third round person #5 is eliminated. So after the third round only person #1 remains, as shown below.

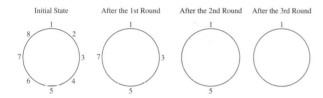

   a. Given a set of sixteen people arranged in a circle and numbered, consecutively 1–16, list the numbers of the people who are eliminated in each round if every second person is eliminated and the elimination process continues until only one person remains. Assume that the starting point is person #1.
   b. Use mathematical induction to prove that for all integers $n \geq 1$, given any set of 2^n people arranged in a circle and numbered consecutively 1 through 2^n, if one starts from person #1 and goes repeatedly around the circle successively eliminating every second person, eventually only person #1 will remain.
   c. Use the result of part (b) to prove that for any nonnegative integers n and m with $2^n \leq 2^n + m < 2^{n+1}$, if $r = 2^n + m$, then given any set of r people arranged in a circle and numbered consecutively 1 through r, if

one starts from person #1 and goes repeatedly around the circle successively eliminating every second person, eventually only person #$(2m + 1)$ will remain.

26. Suppose $P(n)$ is a property such that
   1. $P(0)$, $P(1)$, $P(2)$ are all true,
   2. for all integers $k \geq 0$, if $P(k)$ is true, then $P(3k)$ is true. Must it follow that $P(n)$ is true for all integers $n \geq 0$? If yes, explain why; if no, give a counterexample.

27. Prove that if a statement can be proved by strong mathematical induction, then it can be proved by ordinary mathematical induction. To do this, let $P(n)$ be a property that is defined for integers n, and suppose the following two statements are true:
   1. $P(a)$, $P(a + 1)$, $P(a + 2)$, ..., $P(b)$.
   2. For any integer $k \geq b$, if $P(i)$ is true for all integers i from a through k, then $P(k + 1)$ is true.
   The principle of strong mathematical induction would allow us to conclude immediately that $P(n)$ is true for all integers $n \geq a$. Can we reach the same conclusion using the principle of ordinary mathematical induction? Yes! To see this, let $Q(n)$ be the property

   $P(j)$ is true for all integers j with $a \leq j \leq n$.

   Then use ordinary mathematical induction to show that $Q(n)$ is true for all integers $n \geq b$. That is, prove
   1. $Q(b)$ is true.
   2. For any integer $k \geq b$, if $Q(k)$ is true then $Q(k + 1)$ is true.

28. Give examples to illustrate the proof of Theorem 5.4.1.

29. Write the following numbers in decimal notation:
   a. 1110_2 **b.** 10111_2 c. 110110_2
   d. 1100101_2 e. 1000111_2 f. 1011011_2

H 30. It is a fact that every integer $n \geq 1$ can be written in the form

$$c_r \cdot 3^r + c_{r-1} \cdot 3^{r-1} + \cdots + c_2 \cdot 3^2 + c_1 \cdot 3 + c_0,$$

where $c_r = 1$ or 2 and $c_i = 0, 1,$ or 2 for all integers $i = 0, 1, 2, \ldots, r - 1$. Sketch a proof of this fact.

H✶ 31. Use mathematical induction to prove the existence part of the quotient-remainder theorem for integers $n \geq 0$.

H✶ 32. Prove that if a statement can be proved by ordinary mathematical induction, then it can be proved by the well-ordering principle.

H 33. Use the principle of ordinary mathematical induction to prove the well-ordering principle for the integers.

Answers for Test Yourself

1. than one   2. a; k; $P(k + 1)$ 3. one integer; integer in S; S contains a least element

5.5 Defining Sequences Recursively

So, Nat'ralists observe, a Flea/Hath smaller Fleas that on him prey,/And these have smaller Fleas to bite 'em,/And so proceed ad infinitum. — Jonathan Swift, 1733

A sequence can be defined in a variety of different ways. One informal way is to write the first few terms with the expectation that the general pattern will be obvious. We might say, for instance, "consider the sequence 3, 5, 7, ..." Unfortunately, misunderstandings can occur when this approach is used. The next term of the sequence could be 9 if we mean a sequence of odd integers, or it could be 11 if we mean the sequence of odd prime numbers.

The second way to define a sequence is to give an explicit formula for its nth term. For example, a sequence $a_0, a_1, a_2 \ldots$ can be specified by writing

$$a_n = \frac{(-1)^n}{n+1} \quad \text{for all integers } n \geq 0.$$

The advantage of defining a sequence by such an explicit formula is that each term of the sequence is uniquely determined and can be computed in a fixed, finite number of steps, by substitution.

The third way to define a sequence is to use recursion, as was done in Examples 5.3.3 and 5.4.2. This requires giving both an equation, called a *recurrence relation*, that defines each later term in the sequence by reference to earlier terms and also one or more initial values for the sequence.

Sometimes it is very difficult or impossible to find an explicit formula for a sequence, but it *is* possible to define the sequence using recursion. Note that defining sequences recursively is similar to proving theorems by mathematical induction. The recurrence relation is like the inductive step and the initial conditions are like the basis step. Indeed, the fact that sequences can be defined recursively is equivalent to the fact that mathematical induction works as a method of proof.

> **• Definition**
>
> A **recurrence relation** for a sequence $a_0, a_1, a_2, \ldots$ is a formula that relates each term a_k to certain of its predecessors $a_{k-1}, a_{k-2}, \ldots, a_{k-i}$, where i is an integer with $k - i \geq 0$. The **initial conditions** for such a recurrence relation specify the values of $a_0, a_1, a_2, \ldots, a_{i-1}$, if i is a fixed integer, or $a_0, a_1, \ldots, a_m$, where m is an integer with $m \geq 0$, if i depends on k.

Example 5.5.1 Computing Terms of a Recursively Defined Sequence

Define a sequence $c_0, c_1, c_2, \ldots$ recursively as follows: For all integers $k \geq 2$,

$$(1) \quad c_k = c_{k-1} + kc_{k-2} + 1 \quad \text{recurrence relation}$$
$$(2) \quad c_0 = 1 \quad \text{and} \quad c_1 = 2 \quad \text{initial conditions.}$$

Find c_2, c_3, and c_4.

Solution
$$\begin{aligned} c_2 &= c_1 + 2c_0 + 1 &&\text{by substituting } k = 2 \text{ into (1)}\\ &= 2 + 2\cdot 1 + 1 &&\text{since } c_1 = 2 \text{ and } c_0 = 1 \text{ by (2)} \end{aligned}$$

$$(3) \therefore c_2 \quad = 5$$
$$c_3 \quad = c_2 + 3c_1 + 1 \qquad \text{by substituting } k = 3 \text{ into (1)}$$
$$= 5 + 3 \cdot 2 + 1 \qquad \text{since } c_2 = 5 \text{ by (3) and } c_1 = 2 \text{ by (2)}$$
$$(4) \therefore c_3 \quad = 12$$
$$c_4 \quad = c_3 + 4c_2 + 1 \qquad \text{by substituting } k = 4 \text{ into (1)}$$
$$= 12 + 4 \cdot 5 + 1 \qquad \text{since } c_3 = 12 \text{ by (4) and } c_2 = 5 \text{ by (3)}$$
$$(5) \therefore c_4 \quad = 33$$

∎

A given recurrence relation may be expressed in several different ways.

Example 5.5.2 Writing a Recurrence Relation in More Than One Way

Note   Think of the recurrence relation as $s_\square = 3s_{\square-1} - 1$, where any positive integer expression may be placed in the box.

Let $s_0, s_1, s_2, \ldots$ be a sequence that satisfies the following recurrence relation:

$$\text{for all integers } k \geq 1, \quad s_k = 3s_{k-1} - 1.$$

Explain why the following statement is true:

$$\text{for all integers } k \geq 0, \quad s_{k+1} = 3s_k - 1.$$

Solution   In informal language, the recurrence relation says that any term of the sequence equals 3 times the previous term minus 1. Now for any integer $k \geq 0$, the term previous to s_{k+1} is s_k. Thus for any integer $k \geq 0$, $s_{k+1} = 3s_k - 1$. ∎

A sequence defined recursively need not start with a subscript of zero. Also, a given recurrence relation may be satisfied by many different sequences; the actual values of the sequence are determined by the initial conditions.

Example 5.5.3 Sequences That Satisfy the Same Recurrence Relation

Let $a_1, a_2, a_3, \ldots$ and $b_1, b_2, b_3, \ldots$ satisfy the recurrence relation that the kth term equals 3 times the $(k-1)$st term for all integers $k \geq 2$:

$$(1) \quad a_k = 3a_{k-1} \quad \text{and} \quad b_k = 3b_{k-1}.$$

But suppose that the initial conditions for the sequences are different:

$$(2) \quad a_1 = 2 \quad \text{and} \quad b_1 = 1.$$

Find (a) a_2, a_3, a_4 and (b) b_2, b_3, b_4.

Solution

$$\begin{aligned} \text{a.} \quad & a_2 = 3a_1 = 3 \cdot 2 = 6 & \text{b.} \quad & b_2 = 3b_1 = 3 \cdot 1 = 3 \\ & a_3 = 3a_2 = 3 \cdot 6 = 18 & & b_3 = 3b_2 = 3 \cdot 3 = 9 \\ & a_4 = 3a_3 = 3 \cdot 18 = 54 & & b_4 = 3b_3 = 3 \cdot 9 = 27 \end{aligned}$$

Thus

$$a_1, a_2, a_3, \ldots \text{ begins } 2, 6, 18, 54, \ldots \text{ and}$$
$$b_1, b_2, b_3, \ldots \text{ begins } 1, 3, 9, 27, \ldots.$$

∎

Example 5.5.4 Showing That a Sequence Given by an Explicit Formula Satisfies a Certain Recurrence Relation

The sequence of **Catalan numbers**, named after the Belgian mathematician Eugène Catalan (1814–1894), arises in a remarkable variety of different contexts in discrete mathematics. It can be defined as follows: For each integer $n \geq 1$,

$$C_n = \frac{1}{n+1}\binom{2n}{n}.$$

a. Find $C_1, C_2,$ and C_3.

b. Show that this sequence satisfies the recurrence relation $C_k = \dfrac{4k-2}{k+1}C_{k-1}$ for all integers $k \geq 2$

Eugène Catalan
(1814–1894)

Academie Royale de Belgique

Solution

a. $C_1 = \dfrac{1}{2}\binom{2}{1} = \dfrac{1}{2}\cdot 2 = 1,\quad C_2 = \dfrac{1}{3}\binom{4}{2} = \dfrac{1}{3}\cdot 6 = 2,\quad C_3 = \dfrac{1}{4}\binom{6}{3} = \dfrac{1}{4}\cdot 20 = 5$

b. To obtain the kth and $(k-1)$st terms of the sequence, just substitute k and $k-1$ in place of n in the explicit formula for $C_1, C_2, C_3, \ldots$.

$$C_k = \frac{1}{k+1}\binom{2k}{k}$$

$$C_{k-1} = \frac{1}{(k-1)+1}\binom{2(k-1)}{k-1} = \frac{1}{k}\binom{2k-2}{k-1}.$$

Then start with the right-hand side of the recurrence relation and transform it into the left-hand side: For each integer $k \geq 2$,

$$\frac{4k-2}{k+1}C_{k-1} = \frac{4k-2}{k+1}\left[\frac{1}{k}\binom{2k-2}{k-1}\right] \qquad \text{by substituting}$$

$$= \frac{2(2k-1)}{k+1}\cdot\frac{1}{k}\cdot\frac{(2k-2)!}{(k-1)!(2k-2-(k-1))!} \qquad \text{by the formula for } n \text{ choose } r$$

$$= \frac{1}{k+1}\cdot(2(2k-1))\cdot\frac{(2k-2)!}{(k(k-1)!)(k-1)!} \qquad \text{by rearranging the factors}$$

$$= \frac{1}{k+1}\cdot(2(2k-1))\cdot\frac{1}{k!(k-1)!}\cdot(2k-2)!\cdot\frac{1}{2}\cdot\frac{1}{k}\cdot 2k. \qquad \text{because } \frac{1}{2}\cdot\frac{1}{k}\cdot 2k = 1$$

$$= \frac{1}{k+1}\cdot\frac{2}{2}\cdot\frac{1}{k!}\cdot\frac{1}{(k-1)!}\cdot\frac{1}{k}\cdot(2k)\cdot(2k-1)\cdot(2k-2)! \qquad \text{by rearranging the factors}$$

$$= \frac{1}{k+1}\cdot\frac{(2k)!}{k!k!} \qquad \begin{array}{l}\text{because } k(k-1)! = k!, \\ \frac{2}{2} = 1, \text{ and} \\ 2k\cdot(2k-1)\cdot(2k-2)! = (2k)!\end{array}$$

$$= \frac{1}{k+1}\binom{2k}{k} \qquad \text{by the formula for } n \text{ choose } r$$

$$= C_k \qquad \text{by definition of } C_1, C_2, C_3, \ldots.$$

■

Examples of Recursively Defined Sequences

Recursion is one of the central ideas of computer science. To solve a problem recursively means to find a way to break it down into smaller subproblems each having the same form as the original problem—and to do this in such a way that when the process is repeated many times, the last of the subproblems are small and easy to solve and the solutions of the subproblems can be woven together to form a solution to the original problem.

Probably the most difficult part of solving problems recursively is to figure out how knowing the solution to smaller subproblems of the same type as the original problem will give you a solution to the problem as a whole. You *suppose* you know the solutions to smaller subproblems and ask yourself how you would best make use of that knowledge to solve the larger problem. The supposition that the smaller subproblems have already been solved has been called the *recursive paradigm* or the *recursive leap of faith*. Once you take this leap, you are right in the middle of the most difficult part of the problem, but generally, the path to a solution from this point, though difficult, is short. The recursive leap of faith is similar to the inductive hypothesis in a proof by mathematical induction.

Example 5.5.5   The Tower of Hanoi

Édouard Lucas
(1842–1891)

In 1883 a French mathematician, Édouard Lucas, invented a puzzle that he called The Tower of Hanoi (La Tour D'Hanoï). The puzzle consisted of eight disks of wood with holes in their centers, which were piled in order of decreasing size on one pole in a row of three. A facsimile of the cover of the box is shown in Figure 5.5.1. Those who played the game were supposed to move all the disks one by one from one pole to another, never placing a larger disk on top of a smaller one. The directions to the puzzle claimed it was based on an old Indian legend:

On the steps of the altar in the temple of Benares, for many, many years Brahmins have been moving a tower of 64 golden disks from one pole to another; one by one, never placing a larger on top of a smaller. When all the disks have been transferred the Tower and the Brahmins will fall, and it will be the end of the world.

Figure 5.5.1

The puzzle offered a prize of ten thousand francs (about $34,000 US today) to anyone who could move a tower of 64 disks by hand while following the rules of the game. (See Figure 5.5.2 on the following page.) Assuming that you transferred the disks as efficiently as possible, how many moves would be required to win the prize?

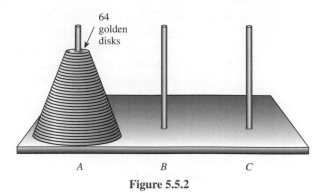

Figure 5.5.2

Solution An elegant and efficient way to solve this problem is to think recursively. Suppose that you, somehow or other, have found the most efficient way possible to transfer a tower of $k-1$ disks one by one from one pole to another, obeying the restriction that you never place a larger disk on top of a smaller one. What is the most efficient way to transfer a tower of k disks from one pole to another? The answer is sketched in Figure 5.5.3, where pole A is the initial pole and pole C is the target pole, and is described as follows:

Step 1 : Transfer the top $k-1$ disks from pole A to pole B. If $k > 2$, execution of this step will require a number of moves of individual disks among the three poles. But the point of thinking recursively is not to get caught up in imagining the details of how those moves will occur.

Step 2 : Move the bottom disk from pole A to pole C.

Step 3 : Transfer the top $k-1$ disks from pole B to pole C. (Again, if $k > 2$, execution of this step will require more than one move.)

To see that this sequence of moves is most efficient, observe that to move the bottom disk of a stack of k disks from one pole to another, you must first transfer the top $k-1$ disks to a third pole to get them out of the way. Thus transferring the stack of k disks from pole A to pole C requires at least two transfers of the top $k-1$ disks: one to transfer them off the bottom disk to free the bottom disk so that it can be moved and another to transfer them back on top of the bottom disk after the bottom disk has been moved to pole C. If the bottom disk were not moved directly from pole A to pole C but were moved to pole B first, at least two additional transfers of the top $k-1$ disks would be necessary: one to move them from pole A to pole C so that the bottom disk could be moved from pole A to pole B and another to move them off pole C so that the bottom disk could be moved onto pole C. This would increase the total number of moves and result in a less efficient transfer.

Thus the minimum sequence of moves must include going from the initial position (a) to position (b) to position (c) to position (d). It follows that

Note Defining the sequence is a crucial step in solving the problem. The recurrence relation and initial conditions are specified in terms of the sequence.

$$\begin{bmatrix} \text{the minimum} \\ \text{number of moves} \\ \text{needed to transfer} \\ \text{a tower of } k \text{ disks} \\ \text{from pole } A \text{ to} \\ \text{pole } C \end{bmatrix} = \begin{bmatrix} \text{the minimum} \\ \text{number of} \\ \text{moves needed} \\ \text{to go from} \\ \text{position (a)} \\ \text{to position (b)} \end{bmatrix} + \begin{bmatrix} \text{The minimum} \\ \text{number of} \\ \text{moves needed} \\ \text{to go from} \\ \text{position (b)} \\ \text{to position (c)} \end{bmatrix} + \begin{bmatrix} \text{the minimum} \\ \text{number of} \\ \text{moves needed} \\ \text{to go from} \\ \text{position (c)} \\ \text{to position (d)} \end{bmatrix} \quad 5.5.1$$

For each integer $n \geq 1$, let

$$m_n = \begin{bmatrix} \text{the minimum number of moves needed to transfer} \\ \text{a tower of } n \text{ disks from one pole to another} \end{bmatrix}$$

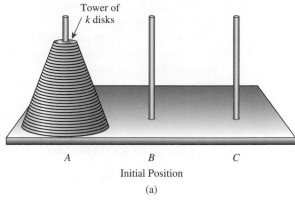

Initial Position

(a)

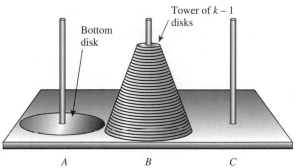

Position after Transferring $k - 1$ Disks from A to B

(b)

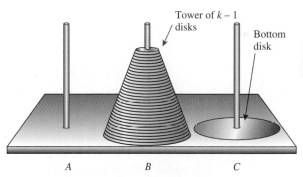

Position after Moving the Bottom Disk from A to C

(c)

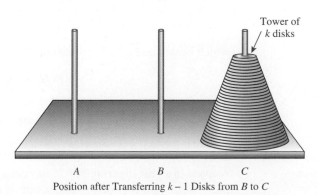

Position after Transferring $k - 1$ Disks from B to C

(d)

Figure 5.5.3 Moves for the Tower of Hanoi

Note that the numbers m_n are independent of the labeling of the poles; it takes the same minimum number of moves to transfer n disks from pole A to pole C as to transfer n disks from pole A to pole B, for example. Also the values of m_n are independent of the number of larger disks that may lie below the top n, provided these remain stationary while the top n are moved. Because the disks on the bottom are all larger than the ones on the top, the top disks can be moved from pole to pole as though the bottom disks were not present.

Going from position (a) to position (b) requires m_{k-1} moves, going from position (b) to position (c) requires just one move, and going from position (c) to position (d) requires m_{k-1} moves. By substitution into equation (5.5.1), therefore,

$$m_k = m_{k-1} + 1 + m_{k-1}$$
$$= 2m_{k-1} + 1 \qquad \text{for all integers } k \geq 2.$$

The initial condition, or base, of this recursion is found by using the definition of the sequence. Because just one move is needed to move one disk from one pole to another,

$$m_1 = \left[\begin{array}{c} \text{the minimum number of moves needed to move} \\ \text{a tower of one disk from one pole to another} \end{array} \right] = 1.$$

Hence the complete recursive specification of the sequence $m_1, m_2, m_3, \ldots$ is as follows: For all integers $k \geq 2$,

$$(1) \quad m_k = 2m_{k-1} + 1 \quad \text{recurrence relation}$$

$$(2) \quad m_1 = 1 \quad \text{initial conditions}$$

Here is a computation of the next five terms of the sequence:

$$(3) \quad m_2 = 2m_1 + 1 = 2 \cdot 1 + 1 = 3 \qquad \text{by (1) and (2)}$$

$$(4) \quad m_3 = 2m_2 + 1 = 2 \cdot 3 + 1 = 7 \qquad \text{by (1) and (3)}$$

$$(5) \quad m_4 = 2m_3 + 1 = 2 \cdot 7 + 1 = 15 \qquad \text{by (1) and (4)}$$

$$(6) \quad m_5 = 2m_4 + 1 = 2 \cdot 15 + 1 = 31 \qquad \text{by (1) and (5)}$$

$$(7) \quad m_6 = 2m_5 + 1 = 2 \cdot 31 + 1 = 63 \qquad \text{by (1) and (6)}$$

Going back to the legend, suppose the priests work rapidly and move one disk every second. Then the time from the beginning of creation to the end of the world would be m_{64} seconds. In the next section we derive an explicit formula for m_n. Meanwhile, we can compute m_{64} on a calculator or a computer by continuing the process started above (Try it!). The approximate result is

$$1.844674 \times 10^{19} \text{ seconds} \cong 5.84542 \times 10^{11} \text{ years}$$

$$\cong 584.5 \text{ billion years},$$

which is obtained by the estimate of

$$60 \cdot 60 \cdot 24 \cdot (365.25) = 31{,}557{,}600$$

$$\uparrow \quad \uparrow \quad \nwarrow \quad \nwarrow \quad \uparrow$$

seconds per minute	minutes per hour	hours per day	days per year	seconds per year

seconds in a year (figuring 365.25 days in a year to take leap years into account). Surprisingly, this figure is close to some scientific estimates of the life of the universe! ∎

Example 5.5.6 The Fibonacci Numbers

Fibonacci (Leonardo of Pisa)
(ca. 1175–1250)

One of the earliest examples of a recursively defined sequence arises in the writings of Leonardo of Pisa, commonly known as Fibonacci, who was the greatest European mathematician of the Middle Ages. In 1202 Fibonacci posed the following problem:

A single pair of rabbits (male and female) is born at the beginning of a year. Assume the following conditions:

1. Rabbit pairs are not fertile during their first month of life but thereafter give birth to one new male/female pair at the end of every month.

2. No rabbits die.

How many rabbits will there be at the end of the year?

Solution One way to solve this problem is to plunge right into the middle of it using recursion. Suppose you know how many rabbit pairs there were at the ends of previous months. How many will there be at the end of the current month?

The crucial observation is that the number of rabbit pairs born at the end of month k is the same as the number of pairs alive at the end of month $k - 2$. Why? Because it is exactly the rabbit pairs that were alive at the end of month $k - 2$ that were fertile during month k. The rabbits born at the end of month $k - 1$ were not.

$$
\begin{array}{ccccc}
\text{month} & k - 2 & & k - 1 & k \\
\hline
\end{array}
$$

Each pair alive here ↑   gives birth to a pair here ↑

Now the number of rabbit pairs alive at the end of month k equals the ones alive at the end of month $k - 1$ plus the pairs newly born at the end of the month. Thus

Note It is essential to rephrase this observation in terms of a sequence.

$$
\left[\begin{array}{l}\text{the number}\\\text{of rabbit}\\\text{pairs alive}\\\text{at the end}\\\text{of month } k\end{array}\right]
=
\left[\begin{array}{l}\text{the number}\\\text{of rabbit}\\\text{pairs alive}\\\text{at the end}\\\text{of month } k - 1\end{array}\right]
+
\left[\begin{array}{l}\text{the number}\\\text{of rabbit}\\\text{pairs born}\\\text{at the end}\\\text{of month } k\end{array}\right]
$$

$$
=
\left[\begin{array}{l}\text{the number}\\\text{of rabbit}\\\text{pairs alive}\\\text{at the end}\\\text{of month } k - 1\end{array}\right]
+
\left[\begin{array}{l}\text{the number}\\\text{of rabbit}\\\text{pairs alive}\\\text{at the end}\\\text{of month } k - 2\end{array}\right]
\qquad 5.5.2
$$

For each integer $n \geq 1$, let

$$
F_n = \left[\begin{array}{l}\text{the number of rabbit pairs}\\\text{alive at the end of month } n\end{array}\right]
$$

and let

$$
F_0 = \text{the initial number of rabbit pairs}
$$
$$
= 1.
$$

Then by substitution into equation (5.5.2), for all integers $k \geq 2$,

$$
F_k = F_{k-1} + F_{k-2}.
$$

Now $F_0 = 1$, as already noted, and $F_1 = 1$ also, because the first pair of rabbits is not fertile until the second month. Hence the complete specification of the Fibonacci sequence is as follows: For all integers $k \geq 2$,

$$(1) \quad F_k = F_{k-1} + F_{k-2} \qquad \text{recurrence relation}$$
$$(2) \quad F_0 = 1, \quad F_1 = 1 \qquad \text{initial conditions.}$$

To answer Fibonacci's question, compute F_2, F_3, and so forth through F_{12}:

$$(3) \quad F_2 = F_1 + F_0 = 1 + 1 \quad = 2 \qquad \text{by (1) and (2)}$$
$$(4) \quad F_3 = F_2 + F_1 = 2 + 1 \quad = 3 \qquad \text{by (1), (2) and (3)}$$
$$(5) \quad F_4 = F_3 + F_2 = 3 + 2 \quad = 5 \qquad \text{by (1), (3) and (4)}$$
$$(6) \quad F_5 = F_4 + F_3 = 5 + 3 \quad = 8 \qquad \text{by (1), (4) and (5)}$$
$$(7) \quad F_6 = F_5 + F_4 = 8 + 5 \quad = 13 \qquad \text{by (1), (5) and (6)}$$
$$(8) \quad F_7 = F_6 + F_5 = 13 + 8 \quad = 21 \qquad \text{by (1), (6) and (7)}$$
$$(9) \quad F_8 = F_7 + F_6 = 21 + 13 = 34 \qquad \text{by (1), (7) and (8)}$$
$$(10) \quad F_9 = F_8 + F_7 = 34 + 21 = 55 \qquad \text{by (1), (8) and (9)}$$
$$(11) \quad F_{10} = F_9 + F_8 = 55 + 34 = 89 \qquad \text{by (1), (9) and (10)}$$
$$(12) \quad F_{11} = F_{10} + F_9 = 89 + 55 = 144 \qquad \text{by (1), (10) and (11)}$$
$$(13) \quad F_{12} = F_{11} + F_{10} = 144 + 89 = 233 \qquad \text{by (1), (11) and (12)}$$

At the end of the twelfth month there are 233 rabbit pairs, or 466 rabbits in all. ∎

Example 5.5.7 Compound Interest

On your twenty-first birthday you get a letter informing you that on the day you were born an eccentric rich aunt deposited $100,000 in a bank account earning 4% interest compounded annually and she now intends to turn the account over to you, provided you can figure out how much it is worth. What is the amount currently in the account?

Solution To approach this problem recursively, observe that

$$\begin{bmatrix} \text{the amount in} \\ \text{the account at} \\ \text{the end of any} \\ \text{particular year} \end{bmatrix} = \begin{bmatrix} \text{the amount in} \\ \text{the account at} \\ \text{the end of the} \\ \text{previous year} \end{bmatrix} + \begin{bmatrix} \text{the interest} \\ \text{earned on the} \\ \text{account during} \\ \text{the year} \end{bmatrix}.$$

Now the interest earned during the year equals the interest rate, $4\% = 0.04$ times the amount in the account at the end of the previous year. Thus

$$\begin{bmatrix} \text{the amount in} \\ \text{the account at} \\ \text{the end of any} \\ \text{particular year} \end{bmatrix} = \begin{bmatrix} \text{the amount in} \\ \text{the account at} \\ \text{the end of the} \\ \text{previous year} \end{bmatrix} + (0.04) \cdot \begin{bmatrix} \text{the amount in} \\ \text{the account at} \\ \text{the end of the} \\ \text{previous year} \end{bmatrix}. \qquad 5.5.3$$

For each positive integer n, let

Note Again, a crucial step is to define the sequence explicitly.

$$A_n = \begin{bmatrix} \text{the amount in the account} \\ \text{at the end of year } n \end{bmatrix}$$

and let

$$A_0 = \begin{bmatrix} \text{the initial amount} \\ \text{in the account} \end{bmatrix} = \$100,000.$$

Then for any particular year k, substitution into equation (5.5.3) gives

$$A_k = A_{k-1} + (0.04) \cdot A_{k-1}$$
$$= (1 + 0.04) \cdot A_{k-1} = (1.04) \cdot A_{k-1} \qquad \text{by factoring out } A_{k-1}.$$

Consequently, the values of the sequence $A_0, A_1, A_2, \ldots$ are completely specified as follows: for all integers $k \geq 1$,

$$(1) \quad A_k = (1.04) \cdot A_{k-1} \qquad \text{recurrence relation}$$
$$(2) \quad A_0 = \$100{,}000 \qquad \text{initial condition.}$$

The number 1.04 is called the *growth factor* of the sequence.

In the next section we derive an explicit formula for the value of the account in any year n. The value on your twenty-first birthday can also be computed by repeated substitution as follows:

$$(3) \quad A_1 \; = 1.04 \cdot A_0 \; = (1.04) \cdot \$100{,}000 \quad = \$104{,}000 \qquad \text{by (1) and (2)}$$
$$(4) \quad A_2 \; = 1.04 \cdot A_1 \; = (1.04) \cdot \$104{,}000 \quad = \$108{,}160 \qquad \text{by (1) and (3)}$$
$$(5) \quad A_3 \; = 1.04 \cdot A_2 \; = (1.04) \cdot \$108{,}160 \quad = \$112{,}486.40 \;\; \text{by (1) and (4)}$$

$$\vdots \qquad\qquad\qquad\qquad \vdots$$

$$(22) \quad A_{20} \; = 1.04 \cdot A_{19} \cong (1.04) \cdot \$210{,}684.92 \cong \$219{,}112.31 \;\; \text{by (1) and (21)}$$
$$(23) \quad A_{21} \; = 1.04 \cdot A_{20} \cong (1.04) \cdot \$219{,}112.31 \cong \$227{,}876.81 \;\; \text{by (1) and (22)}$$

The amount in the account is \$227,876.81 (to the nearest cent). Fill in the dots (to check the arithmetic) and collect your money! ∎

Example 5.5.8 Compound Interest with Compounding Several Times a Year

When an annual interest rate of i is compounded m times per year, the interest rate paid per period is i/m. For instance, if $3\% = 0.03$ annual interest is compounded quarterly, then the interest rate paid per quarter is $0.03/4 = 0.0075$.

For each integer $k \geq 1$, let $P_k =$ the amount on deposit at the end of the kth period, assuming no additional deposits or withdrawals. Then the interest earned during the kth period equals the amount on deposit at the end of the $(k-1)$st period times the interest rate for the period:

$$\text{interest earned during } k\text{th period} = P_{k-1}\left(\frac{i}{m}\right).$$

The amount on deposit at the end of the kth period, P_k, equals the amount at the end of the $(k-1)$st period, P_{k-1}, plus the interest earned during the kth period:

$$P_k = P_{k-1} + P_{k-1}\left(\frac{i}{m}\right) = P_{k-1}\left(1 + \frac{i}{m}\right). \qquad\qquad 5.5.4$$

Suppose \$10,000 is left on deposit at 3% compounded quarterly.

a. How much will the account be worth at the end of one year, assuming no additional deposits or withdrawals?

b. The **annual percentage rate (APR)** is the percentage increase in the value of the account over a one-year period. What is the APR for this account?

Solution

a. For each integer $n \geq 1$, let $P_n =$ the amount on deposit after n consecutive quarters, assuming no additional deposits or withdrawals, and let P_0 be the initial \$10,000. Then by equation (5.5.4) with $i = 0.03$ and $m = 4$, a recurrence relation for the sequence $P_0, P_1, P_2, \ldots$ is

(1) $P_k = P_{k-1}(1 + 0.0075) = (1.0075) \cdot P_{k-1}$ for all integers k≥ 1.

The amount on deposit at the end of one year (four quarters), P_4, can be found by successive substitution:

(2) $P_0 = \$10,000$

(3) $P_1 = 1.0075 \cdot P_0 = (1.0075) \cdot \$10,000.00 = \$10,075.00$ by (1) and (2)

(4) $P_2 = 1.0075 \cdot P_1 = (1.0075) \cdot \$10,075.00 = \$10,150.56$ by (1) and (3)

(5) $P_3 = 1.0075 \cdot P_2 \cong (1.0075) \cdot \$10,150.56 = \$10,226.69$ by (1) and (4)

(6) $P_4 = 1.0075 \cdot P_3 \cong (1.0075) \cdot \$10,226.69 = \$10,303.39$ by (1) and (5)

Hence after one year there is \$10,303.39 (to the nearest cent) in the account.

b. The percentage increase in the value of the account, or APR, is

$$\frac{10303.39 - 10000}{10000} = 0.03034 = 3.034\%. \qquad \blacksquare$$

Recursive Definitions of Sum and Product

Addition and multiplication are called *binary* operations because only two numbers can be added or multiplied at a time. Careful definitions of sums and products of more than two numbers use recursion.

• Definition

Given numbers $a_1, a_2, \ldots, a_n$, where n is a positive integer, the **summation from $i = 1$ to n of the a_i**, denoted $\sum_{i=1}^{n} a_i$, is defined as follows:

$$\sum_{i=1}^{1} a_i = a_1 \quad \text{and} \quad \sum_{i=1}^{n} a_i = \left(\sum_{i=1}^{n-1} a_i \right) + a_n, \quad \text{if } n > 1.$$

The **product from $i = 1$ to n of the a_i**, denoted $\prod_{i=1}^{n} a_i$, is defined by

$$\prod_{i=1}^{1} a_i = a_1 \quad \text{and} \quad \prod_{i=1}^{n} a_i = \left(\prod_{i=1}^{n-1} a_i \right) \cdot a_n, \quad \text{if } n > 1.$$

The effect of these definitions is to specify an *order* in which sums and products of more than two numbers are computed. For example,

$$\sum_{i=1}^{4} a_i = \left(\sum_{i=1}^{3} a_i \right) + a_4 = \left(\left(\sum_{i=1}^{2} a_i \right) + a_3 \right) + a_4 = ((a_1 + a_2) + a_3) + a_4.$$

The recursive definitions are used with mathematical induction to establish various properties of general finite sums and products.

Example 5.5.9 A Sum of Sums

Prove that for any positive integer n, if $a_1, a_2, \ldots, a_n$ and $b_1, b_2, \ldots, b_n$ are real numbers, then

$$\sum_{i=1}^{n}(a_i + b_i) = \sum_{i=1}^{n}a_i + \sum_{i=1}^{n}b_i.$$

Solution The proof is by mathematical induction. Let the property $P(n)$ be the equation

$$\sum_{i=1}^{n}(a_i + b_i) = \sum_{i=1}^{n}a_i + \sum_{i=1}^{n}b_i. \qquad \leftarrow P(n)$$

We must show that $P(n)$ is true for all integers $n \geq 0$. We do this by mathematical induction on n.

Show that P(1) is true: To establish $P(1)$, we must show that

$$\sum_{i=1}^{1}(a_i + b_i) = \sum_{i=1}^{1}a_i + \sum_{i=1}^{1}b_i. \qquad \leftarrow P(1)$$

But

$$\sum_{i=1}^{1}(a_i + b_i) = a_1 + b_1 \qquad \text{by definition of } \Sigma$$

$$= \sum_{i=1}^{1}a_i + \sum_{i=1}^{1}b_i \qquad \text{also by definition of } \Sigma.$$

Hence $P(1)$ is true.

Show that for all integers $k \geq 1$, if $P(k)$ is true then $P(k+1)$ is also true:
Suppose $a_1, a_2, \ldots, a_k, a_{k+1}$ and $b_1, b_2, \ldots, b_k, b_{k+1}$ are real numbers and that for some $k \geq 1$

$$\sum_{i=1}^{k}(a_i + b_i) = \sum_{i=1}^{k}a_i + \sum_{i=1}^{k}b_i. \qquad \begin{array}{l} \leftarrow P(k) \\ \text{inductive hypothesis} \end{array}$$

We must show that

$$\sum_{i=1}^{k+1}(a_i + b_i) = \sum_{i=1}^{k+1}a_i + \sum_{i=1}^{k+1}b_i. \qquad \leftarrow P(k+1)$$

[We will show that the left-hand side of this equation equals the right-hand side.]

But the left-hand side of the equation is

$$\sum_{i=1}^{k+1}(a_i + b_i) = \sum_{i=1}^{k}(a_i + b_i) + (a_{k+1} + b_{k+1}) \qquad \text{by definition of } \Sigma$$

$$= \left(\sum_{i=1}^{k}a_i + \sum_{i=1}^{k}b_i\right) + (a_{k+1} + b_{k+1}) \qquad \text{by inductive hypothesis}$$

$$= \left(\sum_{i=1}^{k}a_i + a_{k+1}\right) + \left(\sum_{i=1}^{k}b_i + b_{k+1}\right) \qquad \begin{array}{l} \text{by the associative and cummutative} \\ \text{laws of algebra} \end{array}$$

$$= \sum_{i=1}^{k+1}a_i + \sum_{i=1}^{k+1}b_i \qquad \text{by definition of } \Sigma$$

which equals the right-hand side of the equation. *[This is what was to be shown.]* ∎

Test Yourself

1. A recursive definition for a sequence consists of a _____ and _____.

2. A recurrence relation is an equation that defines each later term of a sequence by reference to _____ in the sequence.

3. Initial conditions for a recursive definition of a sequence consist of one or more of the _____ of the sequence.

4. To solve a problem recursively means to divide the problem into smaller subproblems of the same type as the initial problem, to suppose _____, and to figure out how to use the supposition to _____.

5. A crucial step for solving a problem recursively is to define a _____ in terms of which the recurrence relation and initial conditions can be specified.

Exercise Set 5.5

Find the first four terms of each of the recursively defined sequences in 1–8.

1. $a_k = 2a_{k-1} + k$, for all integers $k \geq 2$
   $a_1 = 1$

2. $b_k = b_{k-1} + 3k$, for all integers $k \geq 2$
   $b_1 = 1$

3. $c_k = k(c_{k-1})^2$, for all integers $k \geq 1$
   $c_0 = 1$

4. $d_k = k(d_{k-1})^2$, for all integers $k \geq 1$
   $d_0 = 3$

5. $s_k = s_{k-1} + 2s_{k-2}$, for all integers $k \geq 2$
   $s_0 = 1$, $s_1 = 1$

6. $t_k = t_{k-1} + 2t_{k-2}$, for all integers $k \geq 2$
   $t_0 = -1$, $t_1 = 2$

7. $u_k = ku_{k-1} - u_{k-2}$, for all integers $k \geq 3$
   $u_1 = 1$, $u_2 = 1$

8. $v_k = v_{k-1} + v_{k-2} + 1$, for all integers $k \geq 3$
   $v_1 = 1$, $v_2 = 3$

9. Let $a_0, a_1, a_2, \ldots$ be defined by the formula $a_n = 3n + 1$, for all integers $n \geq 0$. Show that this sequence satisfies the recurrence relation $a_k = a_{k-1} + 3$, for all integers $k \geq 1$.

10. Let $b_0, b_1, b_2, \ldots$ be defined by the formula $b_n = 4^n$, for all integers $n \geq 0$. Show that this sequence satisfies the recurrence relation $b_k = 4b_{k-1}$, for all integers $k \geq 1$.

11. Let $c_0, c_1, c_2, \ldots$ be defined by the formula $c_n = 2^n - 1$ for all integers $n \geq 0$. Show that this sequence satisfies the recurrence relation
$$c_k = 2c_{k-1} + 1.$$

12. Let $s_0, s_1, s_2, \ldots$ be defined by the formula $s_n = \dfrac{(-1)^n}{n!}$ for all integers $n \geq 0$. Show that this sequence satisfies the recurrence relation
$$s_k = \frac{-s_{k-1}}{k}.$$

13. Let $t_0, t_1, t_2, \ldots$ be defined by the formula $t_n = 2 + n$ for all integers $n \geq 0$. Show that this sequence satisfies the recurrence relation
$$t_k = 2t_{k-1} - t_{k-2}.$$

14. Let $d_0, d_1, d_2, \ldots$ be defined by the formula $d_n = 3^n - 2^n$ for all integers $n \geq 0$. Show that this sequence satisfies the recurrence relation
$$d_k = 5d_{k-1} - 6d_{k-2}.$$

H 15. For the sequence of Catalan numbers defined in Example 5.5.4, prove that for all integers $n \geq 1$,
$$C_n = \frac{1}{4n+2} \binom{2n+2}{n+1}.$$

16. Use the recurrence relation and values for the Tower of Hanoi sequence $m_1, m_2, m_3, \ldots$ discussed in Example 5.5.5 to compute m_7 and m_8.

17. *Tower of Hanoi with Adjacency Requirement*: Suppose that in addition to the requirement that they never move a larger disk on top of a smaller one, the priests who move the disks of the Tower of Hanoi are also allowed only to move disks one by one from one pole to an *adjacent* pole. Assume poles A and C are at the two ends of the row and pole B is in the middle. Let

$$a_n = \begin{bmatrix} \text{the minimum number of moves} \\ \text{needed to transfer a tower of } n \\ \text{disks from pole } A \text{ to pole } C \end{bmatrix}.$$

   a. Find a_1, a_2, and a_3.   **b.** Find a_4.
   c. Find a recurrence relation for $a_1, a_2, a_3, \ldots$.

18. *Tower of Hanoi with Adjacency Requirement*: Suppose the same situation as in exercise 17. Let

$$b_n = \begin{bmatrix} \text{the minimum number of moves} \\ \text{needed to transfer a tower of } n \\ \text{disks from pole } A \text{ to pole } B \end{bmatrix}.$$

   a. Find b_1, b_2, and b_3.   **b.** Find b_4.

c. Show that $b_k = a_{k-1} + 1 + b_{k-1}$ for all integers $k \geq 2$, where $a_1, a_2, a_3, \ldots$ is the sequence defined in exercise 17.

d. Show that $b_k \leq 3b_{k-1} + 1$ for all integers $k \geq 2$.

H ✱ e. Show that $b_k = 3b_{k-1} + 1$ for all integers $k \geq 2$.

19. *Four-Pole Tower of Hanoi*: Suppose that the Tower of Hanoi problem has four poles in a row instead of three. Disks can be transferred one by one from one pole to any other pole, but at no time may a larger disk be placed on top of a smaller disk. Let s_n be the minimum number of moves needed to transfer the entire tower of n disks from the left-most to the right-most pole.

a. Find $s_1, s_2,$ and s_3.   b. Find s_4.

c. Show that $s_k \leq 2s_{k-2} + 3$ for all integers $k \geq 3$.

20. *Tower of Hanoi Poles in a Circle:* Suppose that instead of being lined up in a row, the three poles for the original Tower of Hanoi are placed in a circle. The monks move the disks one by one from one pole to another, but they may only move disks one over in a clockwise direction and they may never move a larger disk on top of a smaller one. Let c_n be the minimum number of moves needed to transfer a pile of n disks from one pole to the next adjacent pole in the clockwise direction.

a. Justify the inequality $c_k \leq 4c_{k-1} + 1$ for all integers $k \geq 2$.

b. The expression $4c_{k-1} + 1$ is not the minimum number of moves needed to transfer a pile of k disks from one pole to another. Explain, for example, why $c_3 \neq 4c_2 + 1$.

21. *Double Tower of Hanoi*: In this variation of the Tower of Hanoi there are three poles in a row and $2n$ disks, two of each of n different sizes, where n is any positive integer. Initially one of the poles contains all the disks placed on top of each other in pairs of decreasing size. Disks are transferred one by one from one pole to another, but at no time may a larger disk be placed on top of a smaller disk. However, a disk may be placed on top of one of the same size. Let t_n be the minimum number of moves needed to transfer a tower of $2n$ disks from one pole to another.

a. Find t_1 and t_2.   b. Find t_3.

c. Find a recurrence relation for $t_1, t_2, t_3, \ldots$.

22. *Fibonacci Variation*: A single pair of rabbits (male and female) is born at the beginning of a year. Assume the following conditions (which are more realistic than Fibonacci's):

(1) Rabbit pairs are not fertile during their first month of life but thereafter give birth to four new male/female pairs at the end of every month.

(2) No rabbits die.

a. Let $r_n =$ the number of pairs of rabbits alive at the end of month n, for each integer $n \geq 1$, and let $r_0 = 1$. Find a recurrence relation for $r_0, r_1, r_2, \ldots$.

b. Compute $r_0, r_1, r_2, r_3, r_4, r_5,$ and r_6.

c. How many rabbits will there be at the end of the year?

23. *Fibonacci Variation*: A single pair of rabbits (male and female) is born at the beginning of a year. Assume the following conditions:

(1) Rabbit pairs are not fertile during their first *two* months of life, but thereafter give birth to three new male/female pairs at the end of every month.

(2) No rabbits die.

a. Let $s_n =$ the number of pairs of rabbits alive at the end of month n, for each integer $n \geq 1$, and let $s_0 = 1$. Find a recurrence relation for $s_0, s_1, s_2, \ldots$.

b. Compute $s_0, s_1, s_2, s_3, s_4,$ and s_5.

c. How many rabbits will there be at the end of the year?

In 24–34, $F_0, F_1, F_2, \ldots$ is the Fibonacci sequence.

24. Use the recurrence relation and values for $F_0, F_1, F_2, \ldots$ given in Example 5.5.6 to compute F_{13} and F_{14}.

25. The Fibonacci sequence satisfies the recurrence relation $F_k = F_{k-1} + F_{k-2}$, for all integers $k \geq 2$.

a. Explain why the following is true:

$$F_{k+1} = F_k + F_{k-1} \quad \text{for all integers } k \geq 1.$$

b. Write an equation expressing F_{k+2} in terms of F_{k+1} and F_k.

c. Write an equation expressing F_{k+3} in terms of F_{k+2} and F_{k+1}

26. Prove that $F_k = 3F_{k-3} + 2F_{k-4}$ for all integers $k \geq 4$.

27. Prove that $F_k^2 - F_{k-1}^2 = F_k F_{k+1} - F_{k-1} F_{k+1}$, for all integers $k \geq 1$.

28. Prove that $F_{k+1}^2 - F_k^2 - F_{k-1}^2 = 2F_k F_{k-1}$, for all integers $k \geq 1$.

29. Prove that $F_{k+1}^2 - F_k^2 = F_{k-1} F_{k+2}$, for all integers $k \geq 1$.

30. Use mathematical induction to prove that for all integers $n \geq 0$, $F_{n+2} F_n - F_{n+1}^2 = (-1)^n$.

✱31. Use strong mathematical induction to prove that $F_n < 2^n$ for all integers $n \geq 1$.

32. It turns out that the Fibonacci sequence satisfies the following explicit formula: For all integers $F_n \geq 0$,

$$F_n = \frac{1}{\sqrt{5}} \left[\left(\frac{1 + \sqrt{5}}{2} \right)^{n+1} - \left(\frac{1 - \sqrt{5}}{2} \right)^{n+1} \right]$$

Verify that the sequence defined by this formula satisfies the recurrence relation $F_k = F_{k-1} + F_{k-2}$ for all integers $k \geq 2$.

H 33. (For students who have studied calculus) Find $\lim\limits_{n \to \infty} \left(\dfrac{F_{n+1}}{F_n} \right)$, assuming that the limit exists.

H ✱ **34.** (For students who have studied calculus) Prove that $\lim\limits_{n \to \infty} \left(\dfrac{F_{n+1}}{F_n} \right)$ exists.

35. (For students who have studied calculus) Define $x_0, x_1, x_2, \ldots$ as follows:

$$x_k = \sqrt{2 + x_{k-1}} \qquad \text{for all integers } k \geq 1$$
$$x_0 = 0$$

Find $\lim_{n \to \infty} x_n$. (Assume that the limit exists.)

36. *Compound Interest*: Suppose a certain amount of money is deposited in an account paying 4% annual interest compounded quarterly. For each positive integer n, let R_n = the amount on deposit at the end of the nth quarter, assuming no additional deposits or withdrawals, and let R_0 be the initial amount deposited.
  a. Find a recurrence relation for $R_0, R_1, R_2, \ldots$.
  b. If $R_0 = \$5000$, find the amount of money on deposit at the end of one year.
  c. Find the APR for the account.

37. *Compound Interest*: Suppose a certain amount of money is deposited in an account paying 3% annual interest compounded monthly. For each positive integer n, let S_n = the amount on deposit at the end of the nth month, and let S_0 be the initial amount deposited.
  a. Find a recurrence relation for $S_0, S_1, S_2, \ldots$, assuming no additional deposits or withdrawals during the year.
  b. If $S_0 = \$10,000$, find the amount of money on deposit at the end of one year.
  c. Find the APR for the account.

38. With each step you take when climbing a staircase, you can move up either one stair or two stairs. As a result, you can climb the entire staircase taking one stair at a time, taking two at a time, or taking a combination of one- and two-stair increments. For each integer $n \geq 1$, if the staircase consists of n stairs, let c_n be the number of different ways to climb the staircase. Find a recurrence relation for $c_1, c_2, c_3, \ldots$.

39. A set of blocks contains blocks of heights 1, 2, and 4 centimeters. Imagine constructing towers by piling blocks of different heights directly on top of one another. (A tower of height 6 cm could be obtained using six 1-cm blocks, three 2-cm blocks one 2-cm block with one 4-cm block on top, one 4-cm block with one 2-cm block on top, and so forth.) Let t_n be the number of ways to construct a tower of height n cm using blocks from the set. (Assume an unlimited supply of blocks of each size.) Find a recurrence relation for $t_1, t_2, t_3, \ldots$.

40. Use the recursive definition of summation, together with mathematical induction, to prove the generalized distributive law that for all positive integers n, if $a_1, a_2, \ldots, a_n$ and c are real numbers, then

$$\sum_{i=1}^{n} ca_i = c \left(\sum_{i=1}^{n} a_i \right).$$

41. Use the recursive definition of product, together with mathematical induction, to prove that for all positive integers n, if $a_1, a_2, \ldots, a_n$ and $b_1, b_2, \ldots, b_n$ are real numbers, then

$$\prod_{i=1}^{n}(a_i b_i) = \left(\prod_{i=1}^{n} a_i \right)\left(\prod_{i=1}^{n} b_i \right).$$

42. Use the recursive definition of product, together with mathematical induction, to prove that for all positive integers n, if $a_1, a_2, \ldots, a_n$ and c are real numbers, then

$$\prod_{i=1}^{n}(ca_i) = c^n \left(\prod_{i=1}^{n} a_i \right).$$

H **43.** The triangle inequality for absolute value states that for all real numbers a and b, $|a + b| \leq |a| + |b|$. Use the recursive definition of summation, the triangle inequality, the definition of absolute value, and mathematical induction to prove that for all positive integers n, if $a_1, a_2, \ldots, a_n$ are real numbers, then

$$\left| \sum_{i=1}^{n} a_i \right| \leq \sum_{i=1}^{n} |a_i|.$$

Answers for Test Yourself

1. recurrence relation; initial conditions   2. earlier terms   3. values of the first few terms   4. that the smaller subproblems have already been solved; solve the initial problem   5. sequence

5.6 Solving Recurrence Relations by Iteration

The keener one's sense of logical deduction, the less often one makes hard and fast inferences. — Bertrand Russell, 1872–1970

Suppose you have a sequence that satisfies a certain recurrence relation and initial conditions. It is often helpful to know an explicit formula for the sequence, especially if you need to compute terms with very large subscripts or if you need to examine general properties of the sequence. Such an explicit formula is called a **solution** to the recurrence relation. In this section, we discuss methods for solving recurrence relations. For example, in the text and exercises of this section, we will show that the Tower of Hanoi sequence of Example 5.5.5 satisfies the formula

$$m_n = 2^n - 1,$$

and that the compound interest sequence of Example 5.5.7 satisfies

$$A_n = (1.04)^n \cdot \$100{,}000.$$

The Method of Iteration

The most basic method for finding an explicit formula for a recursively defined sequence is **iteration**. Iteration works as follows: Given a sequence $a_0, a_1, a_2, \ldots$ defined by a recurrence relation and initial conditions, you start from the initial conditions and calculate successive terms of the sequence until you see a pattern developing. At that point you guess an explicit formula.

Example 5.6.1 Finding an Explicit Formula

Let $a_0, a_1, a_2, \ldots$ be the sequence defined recursively as follows: For all integers $k \geq 1$,

(1) $a_k = a_{k-1} + 2$ recurrence relation

(2) $a_0 = 1$ initial condition.

Use iteration to guess an explicit formula for the sequence.

Solution   Recall that to say

$$a_k = a_{k-1} + 2 \quad \text{for all integers } k \geq 1$$

means

$$a_\square = a_{\square-1} + 2 \qquad \begin{array}{l}\text{no matter what positive integer is}\\\text{placed into the box } \square.\end{array}$$

In particular,

$$a_1 = a_0 + 2,$$
$$a_2 = a_1 + 2,$$
$$a_3 = a_2 + 2,$$

and so forth. Now use the initial condition to begin a process of successive substitutions into these equations, not just of numbers (as was done in Section 5.5) but of *numerical expressions*.

The reason for using numerical expressions rather than numbers is that in these problems you are seeking a numerical pattern that underlies a general formula. The secret of success is to leave most of the arithmetic undone. However, you do need to eliminate parentheses as you go from one step to the next. Otherwise, you will soon end up with a bewilderingly large nest of parentheses. Also, it is nearly always helpful to use shorthand notations for regrouping additions, subtractions, and multiplications of numbers that repeat. Thus, for instance, you would write

$$5 \cdot 2 \qquad \text{instead of } 2 + 2 + 2 + 2 + 2$$

and 2^5 instead of $2 \cdot 2 \cdot 2 \cdot 2 \cdot 2$.

Notice that you don't lose any information about the number patterns when you use these shorthand notations.

Here's how the process works for the given sequence:

$a_0 = 1$ — the initial condition

$a_1 = a_0 + 2 = 1 + 2$ — by substitution

$a_2 = a_1 + 2 = (1 + 2) + 2 \qquad = 1 + 2 + 2$ — eliminate parentheses

$a_3 = a_2 + 2 = (1 + 2 + 2) + 2 \qquad = 1 + 2 + 2 + 2$ — eliminate parentheses again; write $3 \cdot 2$ instead of $2 + 2 + 2$?

$a_4 = a_3 + 2 = (1 + 2 + 2 + 2) + 2 = 1 + 2 + 2 + 2 + 2$ — eliminate parentheses again; definitely write $4 \cdot 2$ instead of $2 + 2 + 2 + 2$—the length of the string of 2's is getting out of hand.

Tip Do no arithmetic *except*

- replace $n \cdot 1$ and $1 \cdot n$ by n
- reformat repeated numbers
- get rid of parentheses

Since it appears helpful to use the shorthand $k \cdot 2$ in place of $2 + 2 + \cdots + 2$ (k times), we do so, starting again from a_0.

$a_0 = 1 \qquad\qquad\qquad = 1 + 0 \cdot 2$ — the initial condition

$a_1 = a_0 + 2 = 1 + 2 \qquad = 1 + 1 \cdot 2$ — by substitution

$a_2 = a_1 + 2 = (1 + 2) + 2 \quad = 1 + 2 \cdot 2$

$a_3 = a_2 + 2 = (1 + 2 \cdot 2) + 2 = 1 + 3 \cdot 2$

$a_4 = a_3 + 2 = (1 + 3 \cdot 2) + 2 = 1 + 4 \cdot 2$ — At this point it certainly seems likely that the general pattern is $1 + n \cdot 2$; check whether the next calculation supports this.

$a_5 = a_4 + 2 = (1 + 4 \cdot 2) + 2 = 1 + 5 \cdot 2$ — It does! So go ahead and write an answer. It's only a guess, after all.

$\vdots$

Guess: $\quad a_n = 1 + n \cdot 2 = 1 + 2n$

The answer obtained for this problem is just a guess. To be sure of the correctness of this guess, you will need to check it by mathematical induction. Later in this section, we will show how to do this. ■

A sequence like the one in Example 5.6.1, in which each term equals the previous term plus a fixed constant, is called an *arithmetic sequence*. In the exercises at the end of this section you are asked to show that the nth term of an arithmetic sequence always equals the initial value of the sequence plus n times the fixed constant.

> **• Definition**
>
> A sequence a_0, a_1, a_2, ... is called an **arithmetic sequence** if, and only if, there is a constant d such that
>
> $$a_k = a_{k-1} + d \quad \text{for all integers } k \geq 1.$$
>
> It follows that,
>
> $$a_n = a_0 + dn \quad \text{for all integers } n \geq 0.$$

Example 5.6.2 An Arithmetic Sequence

Under the force of gravity, an object falling in a vacuum falls about 9.8 meters per second (m/sec) faster each second than it fell the second before. Thus, neglecting air resistance, a skydiver's speed upon leaving an airplane is approximately 9.8 m/sec one second after departure, $9.8 + 9.8 = 19.6$ m/sec two seconds after departure, and so forth. If air resistance is neglected, how fast would the skydiver be falling 60 seconds after leaving the airplane?

Solution   Let s_n be the skydiver's speed in m/sec n seconds after exiting the airplane if there were no air resistance. Thus s_0 is the initial speed, and since the diver would travel 9.8 m/sec faster each second than the second before,

$$s_k = s_{k-1} + 9.8 \text{ m/sec} \quad \text{for all integers } k \geq 1.$$

It follows that $s_0, s_1, s_2, \ldots$ is an arithmetic sequence with a fixed constant of 9.8, and thus

$$s_n = s_0 + (9.8)n \quad \text{for each integer } n \geq 0.$$

Hence sixty seconds after exiting and neglecting air resistance, the skydiver would travel at a speed of

$$s_{60} = 0 + (9.8)(60) = 588 \text{ m/sec}.$$

Note that 588 m/sec is over half a kilometer per second or over a third of a mile per second, which is very fast for a human being to travel. Happily for the skydiver, taking air resistance into account cuts the speed considerably. ■

In an arithmetic sequence, each term equals the previous term plus a fixed constant. In a geometric sequence, each term equals the previous term *times* a fixed constant. Geometric sequences arise in a large variety of applications, such as compound interest certain models of population growth, radioactive decay, and the number of operations needed to execute certain computer algorithms.

Example 5.6.3 The Explicit Formula for a Geometric Sequence

Let r be a fixed nonzero constant, and suppose a sequence $a_0, a_1, a_2, \ldots$ is defined recursively as follows:

$$a_k = r a_{k-1} \quad \text{for all integers } k \geq 1,$$
$$a_0 = a.$$

Use iteration to guess an explicit formula for this sequence.

Solution

$$a_0 = a$$

$$a_1 = ra_0 = ra$$

$$a_2 = ra_1 = r\,(ra) = r^2a$$

$$a_3 = ra_2 = r\,(r^2a) = r^3a$$

$$a_4 = ra_3 = r\,(r^3a) = r^4a$$

$$\vdots$$

Guess: $\quad a_n = r^na = ar^n \quad$ for any arbitrary integer $n \geq 0$

In the exercises at the end of this section, you are asked to prove that this formula is correct. ∎

• Definition

A sequence $a_0, a_1, a_2, \ldots$ is called a **geometric sequence** if, and only if, there is a constant r such that

$$a_k = ra_{k-1} \quad \text{for all integers } k \geq 1.$$

It follows that,

$$a_n = a_0r^n \quad \text{for all integers } n \geq 0.$$

Example 5.6.4 A Geometric Sequence

As shown in Example 5.5.7, if a bank pays interest at a rate of 4% per year compounded annually and A_n denotes the amount in the account at the end of year n, then $A_k = (1.04)A_{k-1}$, for all integers $k \geq 1$, assuming no deposits or withdrawals during the year. Suppose the initial amount deposited is $100,000, and assume that no additional deposits or withdrawals are made.

a. How much will the account be worth at the end of 21 years?

b. In how many years will the account be worth $1,000,000?

Solution

a. $A_0, A_1, A_2, \ldots$ is a geometric sequence with initial value 100,000 and constant multiplier 1.04. Hence,

$$A_n = \$100{,}000 \cdot (1.04)^n \quad \text{for all integers } n \geq 0.$$

After 21 years, the amount in the account will be

$$A_{21} = \$100{,}000 \cdot (1.04)^{21} \cong \$227{,}876.81.$$

This is the same answer as that obtained in Example 5.5.7 but is computed much more easily (at least if a calculator with a powering key, such as $\boxed{\wedge}$ or $\boxed{x^y}$, is used).

b. Let t be the number of years needed for the account to grow to \$1,000,000. Then

$$\$1,000,000 = \$100,000 \cdot (1.04)^t.$$

Dividing both sides by 100,000 gives

$$10 = (1.04)^t,$$

and taking natural logarithms of both sides results in

$$\ln(10) = \ln(1.04)^t.$$

Note Properties of logarithms are reviewed in Section 7.2.

Then

$$\ln(10) \cong t \, \ln(1.04) \qquad \text{because } \log_b(x^a) = a \log_b(x) \\ \text{(see exercise 29 of Section 7.2)}$$

and so

$$t = \frac{\ln(10)}{\ln(1.04)} \cong 58.7$$

Hence the account will grow to \$1,000,000 in approximately 58.7 years. ∎

An important property of a geometric sequence with constant multiplier greater than 1 is that its terms increase very rapidly in size as the subscripts get larger and larger. For instance, the first ten terms of a geometric sequence with a constant multiplier of 10 are

$$1, \ 10, \ 10^2, \ 10^3, \ 10^4, \ 10^5, \ 10^6, \ 10^7, \ 10^8, \ 10^9.$$

Thus, by its tenth term, the sequence already has the value $10^9 = 1,000,000,000 = 1$ billion. The following box indicates some quantities that are approximately equal to certain powers of 10.

$10^7 \cong$ number of seconds in a year

$10^9 \cong$ number of bytes of memory in a personal computer

$10^{11} \cong$ number of neurons in a human brain

$10^{17} \cong$ age of the universe in seconds (according to one theory)

$10^{31} \cong$ number of seconds to process all possible positions of a checkers game if moves are processed at a rate of 1 per billionth of a second

$10^{81} \cong$ number of atoms in the universe

$10^{111} \cong$ number of seconds to process all possible positions of a chess game if moves are processed at a rate of 1 per billionth of a second

---

Using Formulas to Simplify Solutions Obtained by Iteration

Explicit formulas obtained by iteration can often be simplified by using formulas such as those developed in Section 5.2. For instance, according to the formula for the sum of a geometric sequence with initial term 1 (Theorem 5.2.3), for each real number r except $r = 1$,

$$1 + r + r^2 + \cdots + r^n = \frac{r^{n+1} - 1}{r - 1} \qquad \text{for all integers } n \geq 0.$$

And according to the formula for the sum of the first n integers (Theorem 5.2.2),

$$1 + 2 + 3 + \cdots + n = \frac{n(n + 1)}{2} \qquad \text{for all integers } n \geq 1.$$

Example 5.6.5 An Explicit Formula for the Tower of Hanoi Sequence

Recall that the Tower of Hanoi sequence $m_1, \ m_2, \ m_3, \ldots$ of Example 5.5.5 satisfies the recurrence relation

$$m_k = 2m_{k-1} + 1 \quad \text{for all integers } k \geq 2$$

and has the initial condition

$$m_1 = 1.$$

Use iteration to guess an explicit formula for this sequence, and make use of a formula from Section 5.2 to simplify the answer.

Solution   By iteration

$$m_1 = \underbrace{1}$$

$$m_{②} = 2m_1 + 1 = 2 \cdot 1 + 1 \qquad\qquad = \underbrace{2^{①} + 1},$$

$$m_{③} = 2m_2 + 1 = 2 \overbrace{(2+1)} + 1 \qquad = \underbrace{2^{②} + 2 + 1},$$

$$m_{④} = 2m_3 + 1 = 2 \overbrace{(2^2 + 2 + 1)} + 1 \qquad = \underbrace{2^{③} + 2^2 + 2 + 1},$$

$$m_{⑤} = 2m_4 + 1 = 2 \overbrace{(2^3 + 2^2 + 2 + 1)} + 1 = 2^{④} + 2^3 + 2^2 + 2 + 1.$$

These calculations show that each term up to m_5 is a sum of successive powers of 2, starting with $2^0 = 1$ and going up to 2^k, where k is 1 less than the subscript of the term. The pattern would seem to continue to higher terms because each term is obtained from the preceding one by multiplying by 2 and adding 1; multiplying by 2 raises the exponent of each component of the sum by 1, and adding 1 adds back the 1 that was lost when the previous 1 was multiplied by 2. For instance, for $n = 6$,

$$m_6 = 2m_5 + 1 = 2(2^4 + 2^3 + 2^2 + 2 + 1) + 1 = 2^5 + 2^4 + 2^3 + 2^2 + 2 + 1.$$

Thus it seems that, in general,

$$m_n = 2^{n-1} + 2^{n-2} + \cdots + 2^2 + 2 + 1.$$

By the formula for the sum of a geometric sequence (Theorem 5.2.3),

$$2^{n-1} + 2^{n-2} + \cdots + 2^2 + 2 + 1 = \frac{2^n - 1}{2 - 1} = 2^n - 1.$$

Hence the explicit formula seems to be

$$m_n = 2^n - 1 \quad \text{for all integers } n \geq 1. \qquad\blacksquare$$

A common mistake people make when doing problems such as this is to misuse the laws of algebra. For instance, by the distributive law,

$$a \cdot (b + c) = a \cdot b + a \cdot c \quad \text{for all real numbers } a, \ b, \text{ and } c.$$

Caution!   It is not true that

This is crossed out because it is false.

Thus, in particular, for $a = 2$, $b = 2$, and $c = 1$,

$$2 \cdot (2 + 1) = 2 \cdot 2 + 2 \cdot 1 = 2^2 + 2.$$

It follows that

$$2 \cdot (2 + 1) + 1 = (2^2 + 2) + 1 = 2^2 + 2 + 1.$$

Example 5.6.6 Using the Formula for the Sum of the First *n* Positive Integers

Let K_n be the picture obtained by drawing n dots (which we call *vertices*) and joining each pair of vertices by a line segment (which we call an *edge*). (In Chapter 10 we discuss these objects in a more general context.) Then K_1, K_2, K_3, and K_4 are as follows:

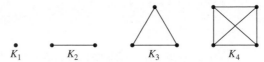

Observe that K_5 may be obtained from K_4 by adding one vertex and drawing edges between this new vertex and all the vertices of K_4 (the old vertices). The reason this procedure gives the correct result is that each pair of old vertices is already joined by an edge, and adding the new edges joins each pair of vertices consisting of an old one and the new one.

New vertex

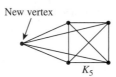

K_5

Thus   the number of edges of $K_5 = 4 +$ the number of edges of K_4.

By the same reasoning, for all integers $k \geq 2$, the number of edges of K_k is $k - 1$ more than the number of edges of K_{k-1}. That is, if for each integer $n \geq 1$

$$s_n = \text{the number of edges of } K_n,$$

then $s_k = s_{k-1} + (k - 1)$ for all integers $k \geq 2$.

Note that s_1, is the number of edges in K_1, which is 0, and use iteration to find an explicit formula for $s_1,\ s_2,\ s_3, \ldots$.

Solution   Because

$$s_k = s_{k-1} + (k - 1) \quad \text{for } \textit{all} \text{ integers } k \geq 2$$

and

$$s_① = ⓪ \overset{1-1}{}$$

then, in particular,

$$s_② = s_1 + 1 = \overbrace{0} + ① \overset{2-1}{};$$

$$s_③ = s_2 + 2 = \overline{(0 + 1)} + 2 = \underbrace{0 + 1} + ② \overset{3-1}{};$$

$$s_④ = s_3 + 3 = \overline{(0 + 1 + 2)} + 3 = \underbrace{0 + 1 + 2} + ③ \overset{4-1}{};$$

$$s_⑤ = s_4 + 4 = \overline{(0 + 1 + 2 + 3)} + 4 = 0 + 1 + 2 + 3 + ④ \overset{5-1}{},$$

$$\vdots$$

Guess: $s_ⓝ = 0 + 1 + 2 + \cdots + (n - 1)$.

But by Theorem 5.2.2,

$$0 + 1 + 2 + 3 + \cdots + (n-1) = \frac{(n-1)n}{2} = \frac{n(n-1)}{2}.$$

Hence it appears that

$$s_n = \frac{n(n-1)}{2}.$$

■

Checking the Correctness of a Formula by Mathematical Induction

As you can see from some of the previous examples, the process of solving a recurrence relation by iteration can involve complicated calculations. It is all too easy to make a mistake and come up with the wrong formula. That is why it is important to confirm your calculations by checking the correctness of your formula. The most common way to do this is to use mathematical induction.

Example 5.6.7 Using Mathematical Induction to Verify the Correctness of a Solution to a Recurrence Relation

In Example 5.5.5 we obtained a formula for the Tower of Hanoi sequence. Use mathematical induction to show that this formula is correct.

Solution   What does it mean to show the correctness of a formula for a recursively defined sequence? Given a sequence of numbers that satisfies a certain recurrence relation and initial condition, your job is to show that each term of the sequence satisfies the proposed explicit formula. In this case, you need to prove the following statement:

If $m_1, m_2, m_3, \ldots$ is the sequence defined by

$\quad m_k = 2m_{k-1} + 1 \quad$ for all integers $k \geq 2$, and

$\quad m_1 = 1,$

then $m_n = 2^n - 1$ for all integers $n \geq 1$.

Proof of Correctness:

Let $m_1, m_2, m_3, \ldots$ be the sequence defined by specifying that $m_1 = 1$ and $m_k = 2m_{k+1} + 1$ for all integers $k \geq 2$, and let the property $P(n)$ be the equation

$$m_n = 2^n - 1 \qquad \leftarrow P(n)$$

We will use mathematical induction to prove that for all integers $n \geq 1$, $P(n)$ is true.

Show that P(1) is true:
To establish $P(1)$, we must show that

$$m_1 = 2^1 - 1. \qquad \leftarrow P(1)$$

But the left-hand side of $P(1)$ is

$$m_1 = 1 \qquad \qquad \text{by definition of } m_1, m_2, m_3, \ldots,$$

and the right-hand side of $P(1)$ is

$$2^1 - 1 = 2 - 1 = 1.$$

Thus the two sides of $P(1)$ equal the same quantity, and hence $P(1)$ is true.

Show that for all integers $k \geq 1$, if $P(k)$ is true then $P(k+1)$ is also true:
[Suppose that $P(k)$ is true for a particular but arbitrarily chosen integer $k \geq 1$. That is:]
Suppose that k is any integer with $k \geq 1$ such that

$$m_k = 2^k - 1. \qquad \leftarrow P(k)$$
$$\text{inductive hypothesis}$$

[We must show that $P(k+1)$ is true. That is:] We must show that

$$m_{k+1} = 2^{k+1} - 1. \qquad \leftarrow P(k+1)$$

But the left-hand side of $P(k+1)$ is

$$
\begin{aligned}
m_{k+1} &= 2m_{(k+1)-1} + 1 && \text{by definition of } m_1, \ m_2, \ m_3, \ldots \\
&= 2m_k + 1 \\
&= 2(2^k - 1) + 1 && \text{by substitution from the inductive hypothesis} \\
&= 2^{k+1} - 2 + 1 && \text{by the distributive law and the fact that } 2 \cdot 2^k = 2^{k-1} \\
&= 2^{k+1} - 1 && \text{by basic algebra}
\end{aligned}
$$

which equals the right-hand side of $P(k+1)$. *[Since the basis and inductive steps have been proved, it follows by mathematical induction that the given formula holds for all integers $n \geq 1$.]* ∎

Discovering That an Explicit Formula Is Incorrect

The following example shows how the process of trying to verify a formula by mathematical induction may reveal a mistake.

Example 5.6.8 Using Verification by Mathematical Induction to Find a Mistake

Let $c_0, \ c_1, \ c_2, \ldots$ be the sequence defined as follows:

$$c_k = 2c_{k-1} + k \qquad \text{for all integers } k \geq 1,$$
$$c_0 = 1.$$

Suppose your calculations suggest that $c_0, \ c_1, \ c_2, \ldots$ satisfies the following explicit formula:

$$c_n = 2^n + n \qquad \text{for all integers } n \geq 0.$$

Is this formula correct?

Solution   Start to prove the statement by mathematical induction and see what develops. The proposed formula passes the basis step of the inductive proof with no trouble, for on the one hand, $c_0 = 1$ by definition and on the other hand, $2^0 + 0 = 1 + 0 = 1$ also.
    In the inductive step, you suppose

$$c_k = 2^k + k \qquad \text{for some integer } k \geq 0 \qquad \text{This is the inductive hypothesis.}$$

and then you must show that

$$c_{k+1} = 2^{k+1} + (k+1).$$

To do this, you start with c_{k+1}, substitute from the recurrence relation, and then use the inductive hypothesis as follows:

$$c_{k+1} = 2c_k + (k+1) \qquad \text{by the recurrence relation}$$
$$= 2(2^k + k) + (k+1) \qquad \text{by substitution from the inductive hypothesis}$$
$$= 2^{(k+1)} + 3k + 1 \qquad \text{by basic algebra}$$

To finish the verification, therefore, you need to show that

$$2^{k+1} + 3k + 1 = 2^{k+1} + (k+1).$$

Now this equation is equivalent to

$$2k = 0 \qquad \text{by subtracting } 2^{k+1} + k + 1 \text{ from both sides.}$$

which is equivalent to

$$k = 0 \qquad \text{by dividing both sides by 2.}$$

But this is false since k may be *any* nonnegative integer.
Observe that when $k = 0$, then $k + 1 = 1$, and

$$c_1 = 2 \cdot 1 + 1 = 3 \quad \text{and} \quad 2^1 + 1 = 3.$$

Thus the formula gives the correct value for c_1. However, when $k = 1$, then $k + 1 = 2$, and

$$c_2 = 2 \cdot 3 + 2 = 8 \quad \text{whereas} \quad 2^2 + 2 = 4 + 2 = 6.$$

So the formula does not give the correct value for c_2. Hence the sequence c_0, c_1, c_2, ... does not satisfy the proposed formula. ■

Once you have found a proposed formula to be false, you should look back at your calculations to see where you made a mistake, correct it, and try again.

Test Yourself

1. To use iteration to find an explicit formula for a recursively defined sequence, start with the _____ and use successive substitution into the _____ to look for a numerical pattern.

2. At every step of the iteration process, it is important to eliminate _____.

3. If a single number, say a, is added to itself k times in one of the steps of the iteration, replace the sum by the expression _____.

4. If a single number, say a, is multiplied by itself k times in one of the steps of the iteration, replace the product by the expression _____.

5. A general arithmetic sequence a_0, a_1, a_2, ... with initial value a_0 and fixed constant d satisfies the recurrence relation _____ and has the explicit formula _____.

6. A general geometric sequence a_0, a_1, a_2, ... with initial value a_0 and fixed constant r satisfies the recurrence relation _____ and has the explicit formula _____.

7. When an explicit formula for a recursively defined sequence has been obtained by iteration, its correctness can be checked by _____.

Exercise Set 5.6

1. The formula

$$1 + 2 + 3 + \cdots + n = \frac{n(n+1)}{2}$$

is true for all integers $n \geq 1$. Use this fact to solve each of the following problems:

a. If k is an integer and $k \geq 2$, find a formula for the expression $1 + 2 + 3 + \cdots + (k - 1)$.

b. If n is an integer and $n \geq 1$, find a formula for the expression $3 + 2 + 4 + 6 + 8 + \cdots + 2n$.

c. If n is an integer and $n \geq 1$, find a formula for the expression $3 + 3 \cdot 2 + 3 \cdot 3 + \cdots + 3 \cdot n + n$.

2. The formula

$$1 + r + r^2 + \cdots + r^n = \frac{r^{n+1} - 1}{r - 1}$$

is true for all real numbers r except $r = 1$ and for all integers $n \geq 0$. Use this fact to solve each of the following problems:

a. If i is an integer and $i \geq 1$, find a formula for the expression $1 + 2 + 2^2 + \cdots + 2^{i-1}$.

b. If n is an integer and $n \geq 1$, find a formula for the expression $3^{n-1} + 3^{n-2} + \cdots + 3^2 + 3 + 1$.

c. If n is an integer and $n \geq 2$, find a formula for the expression $2^n + 2^{n-2} \cdot 3 + 2^{n-3} \cdot 3 + \cdots + 2^2 \cdot 3 + 2 \cdot 3 + 3$

d. If n is an integer and $n \geq 1$, find a formula for the expression

$$2^n - 2^{n-1} + 2^{n-2} - 2^{n-3} + \cdots + (-1)^{n-1} \cdot 2 + (-1)^n.$$

In each of 3–15 a sequence is defined recursively. Use iteration to guess an explicit formula for the sequence. Use the formulas from Section 5.2 to simplify your answers whenever possible.

3. $a_k = ka_{k-1}$, for all integers $k \geq 1$
   $a_0 = 1$

4. $b_k = \dfrac{b_{k-1}}{1 + b_{k-1}}$, for all integers $k \geq 1$
   $b_0 = 1$

5. $c_k = 3c_{k-1} + 1$, for all integers $k \geq 2$
   $c_1 = 1$

H **6.** $d_k = 2d_{k-1} + 3$, for all integers $k \geq 2$
   $d_1 = 2$

7. $e_k = 4e_{k-1} + 5$, for all integers $k \geq 1$
   $e_0 = 2$

8. $f_k = f_{k-1} + 2^k$, for all integers $k \geq 2$
   $f_1 = 1$

H **9.** $g_k = \dfrac{g_{k-1}}{g_{k-1} + 2}$, for all integers $k \geq 2$
   $g_1 = 1$

10. $h_k = 2^k - h_{k-1}$, for all integers $k \geq 1$
    $h_0 = 1$

11. $p_k = p_{k-1} + 2 \cdot 3^k$
    $p_1 = 2$

12. $s_k = s_{k-1} + 2k$, for all integers $k \geq 1$
    $s_0 = 3$

13. $t_k = t_{k-1} + 3k + 1$, for all integers $k \geq 1$
    $t_0 = 0$

★ 14. $x_k = 3x_{k-1} + k$, for all integers $k \geq 2$
    $x_1 = 1$

15. $y_k = y_{k-1} + k^2$, for all integers $k \geq 2$
    $y_1 = 1$

16. Solve the recurrence relation obtained as the answer to exercise 17(c) of Section 5.5.

17. Solve the recurrence relation obtained as the answer to exercise 21(c) of Section 5.5.

18. Suppose d is a fixed constant and $a_0, a_1, a_2, \ldots$ is a sequence that satisfies the recurrence relation $a_k = a_{k-1} + d$, for all integers $k \geq 1$. Use mathematical induction to prove that $a_n = a_0 + nd$, for all integers $n \geq 0$.

19. A worker is promised a bonus if he can increase his productivity by 2 units a day every day for a period of 30 days. If on day 0 he produces 170 units, how many units must he produce on day 30 to qualify for the bonus?

20. A runner targets herself to improve her time on a certain course by 3 seconds a day. If on day 0 she runs the course in 3 minutes, how fast must she run it on day 14 to stay on target?

21. Suppose r is a fixed constant and $a_0, a_1, a_2 \ldots$ is a sequence that satisfies the recurrence relation $a_k = ra_{k-1}$, for all integers $k \geq 1$ and $a_0 = a$. Use mathematical induction to prove that $a_n = ar^n$, for all integers $n \geq 0$.

22. As shown in Example 5.5.8, if a bank pays interest at a rate of i compounded m times a year, then the amount of money P_k at the end of k time periods (where one time period $= 1/m$th of a year) satisfies the recurrence relation $P_k = [1 + (i/m)]P_{k-1}$ with initial condition $P_0 =$ the initial amount deposited. Find an explicit formula for P_n.

23. Suppose the population of a country increases at a steady rate of 3% per year. If the population is 50 million at a certain time, what will it be 25 years later?

24. A chain letter works as follows: One person sends a copy of the letter to five friends, each of whom sends a copy to five friends, each of whom sends a copy to five friends, and so forth. How many people will have received copies of the letter after the twentieth repetition of this process, assuming no person receives more than one copy?

25. A certain computer algorithm executes twice as many operations when it is run with an input of size k as when it is run with an input of size $k - 1$ (where k is an integer that is greater than 1). When the algorithm is run with an input of size 1, it executes seven operations. How many operations does it execute when it is run with an input of size 25?

26. A person saving for retirement makes an initial deposit of $1,000 to a bank account earning interest at a rate of 3% per year compounded monthly, and each month she adds an additional $200 to the account.

   a. For each nonnegative integer n, let A_n be the amount in the account at the end of n months. Find a recurrence relation relating A_k to A_{k-1}.

   H **b.** Use iteration to find an explicit formula for A_n.

   c. Use mathematical induction to prove the correctness of the formula you obtained in part (b).

   d. How much will the account be worth at the end of 20 years? At the end of 40 years?

   H **e.** In how many years will the account be worth $10,000?

27. A person borrows $3,000 on a bank credit card at a nominal rate of 18% per year, which is actually charged at a rate of 1.5% per month.

 H **a.** What is the annual percentage rate (APR) for the card? (See Example 5.5.8 for a definition of APR.)

 b. Assume that the person does not place any additional charges on the card and pays the bank $150 each month to pay off the loan. Let B_n be the balance owed on the card after n months. Find an explicit formula for B_n.

 H **c.** How long will be required to pay off the debt?

 d. What is the total amount of money the person will have paid for the loan?

In 28–42 use mathematical induction to verify the correctness of the formula you obtained in the referenced exercise.

28. Exercise 3   29. Exercise 4   **30.** Exercise 5

31. Exercise 6   32. Exercise 7   **33.** Exercise 8

34. Exercise 9   **H 35.** Exercise 10   36. Exercise 11

H 37. Exercise 12   38. Exercise 13   **39.** Exercise 14

40. Exercise 15   41. Exercise 16   42. Exercise 17

In each of 43–49 a sequence is defined recursively. (a) Use iteration to guess an explicit formula for the sequence. (b) Use strong mathematical induction to verify that the formula of part (a) is correct.

43. $a_k = \dfrac{a_{k-1}}{2a_{k-1} - 1}$, for all integers $k \geq 1$
  $a_0 = 2$

44. $b_k = \dfrac{2}{b_{k-1}}$, for all integers $k \geq 2$
  $b_1 = 1$

45. $v_k = v_{\lfloor k/2 \rfloor} + v_{\lfloor (k+1)/2 \rfloor} + 2$, for all integers $k \geq 2$,
  $v_1 = 1$.

H 46. $s_k = 2s_{k-2}$, for all integers $k \geq 2$,
  $s_0 = 1, s_1 = 2$.

47. $t_k = k - t_{k-1}$, for all integers $k \geq 1$,
  $t_0 = 0$.

H 48. $w_k = w_{k-2} + k$, for all integers $k \geq 3$,
  $w_1 = 1, w_2 = 2$.

H 49. $u_k = u_{k-2} \cdot u_{k-1}$, for all integers $k \geq 2$,
  $u_0 = u_1 = 2$.

In 50 and 51 determine whether the given recursively defined sequence satisfies the explicit formula $a_n = (n-1)^2$, for all integers $n \geq 1$.

50. $a_k = 2a_{k-1} + k - 1$, for all integers $k \geq 2$
  $a_1 = 0$

51. $a_k = (a_{k-1} + 1)^2$, for all integers $k \geq 2$
  $a_1 = 0$

52. A single line divides a plane into two regions. Two lines (by crossing) can divide a plane into four regions; three lines can divide it into seven regions (see the figure). Let P_n be the maximum number of regions into which n lines divide a plane, where n is a positive integer.

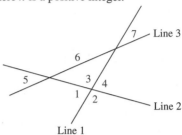

 a. Derive a recurrence relation for P_k in terms of P_{k-1}, for all integers $k \geq 2$.

 b. Use iteration to guess an explicit formula for P_n.

H 53. Compute $\begin{bmatrix} 1 & 1 \\ 1 & 0 \end{bmatrix}^n$ for small values of n (up to about 5 or 6). Conjecture explicit formulas for the entries in this matrix, and prove your conjecture using mathematical induction.

54. In economics the behavior of an economy from one period to another is often modeled by recurrence relations. Let Y_k be the income in period k and C_k be the consumption in period k. In one economic model, income in any period is assumed to be the sum of consumption in that period plus investment and government expenditures (which are assumed to be constant from period to period), and consumption in each period is assumed to be a linear function of the income of the preceding period. That is,

 $Y_k = C_k + E$ where E is the sum of investment plus government expenditures

 $C_k = c + mY_{k-1}$ where c and m are constants.

 Substituting the second equation into the first gives $Y_k = E + c + mY_{k-1}$.

 a. Use iteration on the above recurrence relation to obtain

 $$Y_n = (E + c)\left(\frac{m^n - 1}{m - 1}\right) + m^n Y_0$$

 for all integers $n \geq 1$.

 b. (For students who have studied calculus) Show that if $0 < m < 1$, then $\displaystyle\lim_{n \to \infty} Y_n = \frac{E + c}{1 - m}$.

Answers for Test Yourself

1. initial conditions; recurrence relation   2. parentheses   3. $k \cdot a$ 4. a^k 5. $a_k = a_{k-1} + d$; $a_n = a_0 + dn$ 6. $a_k = ra_{k-1}$; $a_n = a_0 r^n$ 7. mathematical induction

CHAPTER 6

SET THEORY

*Georg Cantor
(1845–1918)*

In the late nineteenth century, Georg Cantor was the first to realize the potential usefulness of investigating properties of sets in general as distinct from properties of the elements that comprise them. Many mathematicians of his time resisted accepting the validity of Cantor's work. Now, however, abstract set theory is regarded as the foundation of mathematical thought. All mathematical objects (even numbers!) can be defined in terms of sets, and the language of set theory is used in every mathematical subject.

In this chapter we add to the basic definitions and notation of set theory introduced in Chapter 1 and show how to establish properties of sets through the use of proofs and counterexamples. We also introduce the notion of a Boolean algebra, explain how to derive its properties, and discuss their relationships to logical equivalencies and set identities. The chapter ends with a discussion of a famous "paradox" of set theory and its relation to computer science.

6.1 Set Theory: Definitions and the Element Method of Proof

The introduction of suitable abstractions is our only mental aid to organize and master complexity. —E. W. Dijkstra, 1930–2002

The words *set* and *element* are undefined terms of set theory just as *sentence, true,* and *false* are undefined terms of logic. The founder of set theory, Georg Cantor, suggested imagining a set as a "collection into a whole M of definite and separate objects of our intuition or our thought. These objects are called the elements of M." Cantor used the letter M because it is the first letter of the German word for set: *Menge.*

Following the spirit of Cantor's notation (though not the letter), let S denote a set and a an element of S. Then, as indicated in Section 1.2, $a \in S$ means that a is an element of S, $a \notin S$ means that a is not an element of S, $\{1, 2, 3\}$ refers to the set whose elements are 1, 2, and 3, and $\{1, 2, 3, \ldots\}$ refers to the set of all positive integers. If S is a set and $P(x)$ is a property that elements of S may or may not satisfy, then a set A may be defined by writing

Caution! Don't forget to include the words "the set of all."

$$A = \{x \in S \mid P(x)\},$$

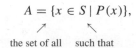

the set of all   such that

which is read "the set of all x in S such that P of x."

Subsets: Proof and Disproof

We begin by rewriting what it means for a set A to be a subset of a set B as a formal universal conditional statement:

$$A \subseteq B \iff \forall x, \text{ if } x \in A \text{ then } x \in B.$$

The negation is, therefore, existential:

$$A \nsubseteq B \iff \exists x \text{ such that } x \in A \text{ and } x \notin B.$$

A *proper subset* of a set is a subset that is not equal to its containing set. Thus

A is a **proper subset** of B $\iff$

(1) $A \subseteq B$, and
(2) there is at least one element in B that is not in A.

Example 6.1.1 Testing Whether One Set Is a Subset of Another

Let $A = \{1\}$ and $B = \{\mathbf{1}, \{1\}\}$.

a. Is $A \subseteq B$?

b. If so, is A a proper subset of B?

Solution

Note A set like $\{1\}$, with just one element, is called a **singleton set**.

a. Because $A = \{1\}$, A has only one element, namely the symbol 1. This element is also one of the elements in set B. Hence every element in A is in B, and so $A \subseteq B$.

b. B has two distinct elements, the symbol 1 and the set $\{1\}$ whose only element is 1. Since $1 \neq \{1\}$, the set $\{1\}$ is not an element of A, and so there is an element of B that is not an element of A. Hence A is a proper subset of B. ∎

Because we define what it means for one set to be a subset of another by means of a universal conditional statement, we can use the method of direct proof to establish a subset relationship. Such a proof is called an *element argument* and is the fundamental proof technique of set theory.

Element Argument: The Basic Method for Proving That One Set Is a Subset of Another

Let sets X and Y be given. To prove that $X \subseteq Y$,

1. **suppose** that x is a particular but arbitrarily chosen element of X,

2. **show** that x is an element of Y.

Example 6.1.2 **Proving and Disproving Subset Relations**

Define sets A and B as follows:

$$A = \{m \in \mathbf{Z} \mid m = 6r + 12 \text{ for some } r \in \mathbf{Z}\}$$
$$B = \{n \in \mathbf{Z} \mid n = 3s \text{ for some } s \in \mathbf{Z}\}.$$

a. Outline a proof that $A \subseteq B$.　　b. Prove that $A \subseteq B$.　　c. Disprove that $B \subseteq A$.

Solution

a. **Proof Outline**:

Suppose x is a particular but arbitrarily chosen element of A.

.

.

.

Therefore, x is an element of B.

b. **Proof**:

Suppose x is a particular but arbitrarily chosen element of A.
　　　　[We must show that $x \in B$. By definition of B, this means
　　　　we must show that $x = 3 \cdot (\text{some integer})$.]

By definition of A, there is an integer r such that $x = 6r + 12$.
　　　　[Given that $x = 6r + 12$, can we express x as $3 \cdot (\text{some integer})$?
　　　　I.e., does $6r + 12 = 3 \cdot (\text{some integer})$? Yes, $6r + 12 = 3 \cdot (2r + 4)$.]

Let $s = 2r + 4$.
　　　　[We must check that s is an integer.]

Then s is an integer because products and sums of integers are integers.
　　　　[Now we must check that $x = 3s$.]

Also $3s = 3(2r + 4) = 6r + 12 = x$,
Thus, by definition of B, x is an element of B,
　　　　[which is what was to be shown].

c. To disprove a statement means to show that it is false, and to show it is false that $B \subseteq A$, you must find an element of B that is not an element of A. By the definitions of A and B, this means that you must find an integer x of the form $3 \cdot (\text{some integer})$ that cannot be written in the form $6 \cdot (\text{some integer}) + 12$. A little experimentation reveals that various numbers do the job. For instance, you could let $x = 3$. Then $x \in B$ because $3 = 3 \cdot 1$, but $x \notin A$ because there is no integer r such that $3 = 6r + 12$. For if there were such an integer, then

Note Recall that the notation $P(x) \Rightarrow Q(x)$ means that every element that makes $P(x)$ true also makes $Q(x)$ true.

$$
\begin{aligned}
& 6r + 12 = 3 && \text{by assumption} \\
\Rightarrow\quad & 2r + 4 = 1 && \text{by dividing both sides by 3} \\
\Rightarrow\quad & 2r = -3 && \text{by subtracting 4 from both sides} \\
\Rightarrow\quad & r = -3/2 && \text{by dividing both sides by 2,}
\end{aligned}
$$

but $-3/2$ is not an integer. Thus $3 \in B$ but $3 \notin A$, and so $B \nsubseteq A$. ■

Set Equality

Recall that by the axiom of extension, sets A and B are equal if, and only if, they have exactly the same elements. We restate this as a definition that uses the language of subsets.

• Definition

Given sets A and B, A **equals** B, written $A = B$, if, and only if, every element of A is in B and every element of B is in A.

Symbolically:

$$A = B \quad \Leftrightarrow \quad A \subseteq B \text{ and } B \subseteq A.$$

This version of the definition of equality implies the following:

> To know that a set A equals a set B, you must know that $A \subseteq B$ and you must also know that $B \subseteq A$.

Example 6.1.3 Set Equality

Define sets A and B as follows:

$$A = \{m \in \mathbf{Z} \mid m = 2a \text{ for some integer } a\}$$
$$B = \{n \in \mathbf{Z} \mid n = 2b - 2 \text{ for some integer } b\}$$

Is $A = B$?

Solution   Yes. To prove this, both subset relations $A \subseteq B$ and $B \subseteq A$ must be proved.

Part 1, Proof That $A \subseteq B$:

Suppose x is a particular but arbitrarily chosen element of A.
> *[We must show that $x \in B$. By definition of B, this means we must show that $x = 2 \cdot (\text{some integer}) - 2$.]*

By definition of A, there is an integer a such that $x = 2a$.
> *[Given that $x = 2a$, can x also be expressed as $2 \cdot (\text{some integer}) - 2$? I.e., is there an integer, say b, such that $2a = 2b - 2$? Solve for b to obtain $b = (2a + 2)/2 = a + 1$. Check to see if this works.]*

Let $b = a + 1$.
> *[First check that b is an integer.]*

Then b is an integer because it is a sum of integers.
> *[Then check that $x = 2b - 2$.]*

Also $2b - 2 = 2(a + 1) - 2 = 2a + 2 - 2 = 2a = x$,
Thus, by definition of B, x is an element of B
> *[which is what was to be shown].*

Part 2, Proof That $B \subseteq A$: This part of the proof is left as exercise 2 at the end of this section. ∎

Venn Diagrams

John Venn
(1834–1923)

If sets *A* and *B* are represented as regions in the plane, relationships between *A* and *B* can be represented by pictures, called **Venn diagrams,** that were introduced by the British mathematician John Venn in 1881. For instance, the relationship $A \subseteq B$ can be pictured in one of two ways, as shown in Figure 6.1.1.

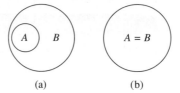

(a)   (b)

Figure 6.1.1 $A \subseteq B$

The relationship $A \nsubseteq B$ can be represented in three different ways with Venn diagrams, as shown in Figure 6.1.2.

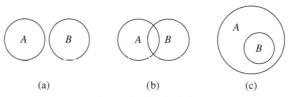

(a)   (b)   (c)

Figure 6.1.2 $A \nsubseteq B$

If we allow the possibility that some subregions of Venn diagrams do not contain any points, then in Figure 6.1.1 diagram (b) can be viewed as a special case of diagram (a) by imagining that the part of *B* outside *A* does not contain any points. Similarly, diagrams (a) and (c) of Figure 6.1.2 can be viewed as special cases of diagram (b). To obtain (a) from (b), imagine that the region of overlap between *A* and *B* does not contain any points. To obtain (c), imagine that the part of *B* that lies outside *A* does not contain any points. However, in all three diagrams it would be necessary to specify that there is a point in *A* that is not in *B*.

Example 6.1.4 Relations among Sets of Numbers

Since **Z**, **Q**, and **R** denote the sets of integers, rational numbers, and real numbers, respectively, **Z** is a subset of **Q** because every integer is rational (any integer *n* can be written in the form $\frac{n}{1}$), and **Q** is a subset of **R** because every rational number is real (any rational number can be represented as a length on the number line). **Z** is a proper subset of **Q** because there are rational numbers that are not integers (for example, $\frac{1}{2}$), and **Q** is a proper subset of **R** because there are real numbers that are not rational (for example, $\sqrt{2}$). This is shown diagrammatically in Figure 6.1.3. ■

Figure 6.1.3

Operations on Sets

Most mathematical discussions are carried on within some context. For example, in a certain situation all sets being considered might be sets of real numbers. In such a situation, the set of real numbers would be called a **universal set** or a **universe of discourse** for the discussion.

• Definition

Let A and B be subsets of a universal set U.

1. The **union** of A and B, denoted $A \cup B$, is the set of all elements that are in at least one of A or B.

2. The **intersection** of A and B, denoted $A \cap B$, is the set of all elements that are common to both A and B.

3. The **difference** of B minus A (or **relative complement** of A in B), denoted $B - A$, is the set of all elements that are in B and not A.

4. The **complement** of A, denoted A^c, is the set of all elements in U that are not in A.

Symbolically:
$$A \cup B = \{x \in U \mid x \in A \text{ or } x \in B\},$$
$$A \cap B = \{x \in U \mid x \in A \text{ and } x \in B\},$$
$$B - A = \{x \in U \mid x \in B \text{ and } x \notin A\},$$
$$A^c = \{x \in U \mid x \notin A\}.$$

Giuseppe Peano
(1858–1932)

Stock Montage

The symbols $\in$, $\cup$, and $\cap$ were introduced in 1889 by the Italian mathematician Giuseppe Peano.

Venn diagram representations for union, intersection, difference, and complement are shown in Figure 6.1.4.

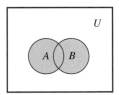
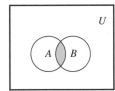
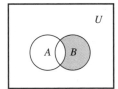
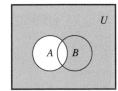

Shaded region
represents $A \cup B$.

Shaded region
represents $A \cap B$.

Shaded region
represents $B - A$.

Shaded region
represents A^c.

Figure 6.1.4

Example 6.1.5 Unions, Intersections, Differences, and Complements

Let the universal set be the set $U = \{a, b, c, d, e, f, g\}$ and let $A = \{a, c, e, g\}$ and $B = \{d, e, f, g\}$. Find $A \cup B$, $A \cap B$, $B - A$, and A^c.

Solution
$$A \cup B = \{a, c, d, e, f, g\} \qquad A \cap B = \{e, g\}$$
$$B - A = \{d, f\} \qquad\qquad A^c = \{b, d, f\}$$

∎

There is a convenient notation for subsets of real numbers that are intervals.

• **Notation**

Given real numbers a and b with $a \leq b$:

$$(a, b) = \{x \in \mathbf{R} \mid a < x < b\} \qquad [a, b] = \{x \in \mathbf{R} \mid a \leq x \leq b\}$$
$$(a, b] = \{x \in \mathbf{R} \mid a < x \leq b\} \qquad [a, b) = \{x \in \mathbf{R} \mid a \leq x < b\}.$$

The symbols ∞ and $-\infty$ are used to indicate intervals that are unbounded either on the right or on the left:

$$(a, \infty) = \{x \in \mathbf{R} \mid x > a\} \qquad [a, \infty) = \{x \in \mathbf{R} \mid x \geq a\}$$
$$(-\infty, b) = \{x \in \mathbf{R} \mid x < b\} \qquad (-\infty, b) = \{x \in \mathbf{R} \mid x \leq b\}.$$

Note The symbol ∞ does not represent a number. It just indicates the unboundedness of the interval.

Observe that the notation for the interval (a, b) is identical to the notation for the ordered pair (a, b). However, context makes it unlikely that the two will be confused.

Example 6.1.6 An Example with Intervals

Let the universal set be the set $\mathbf{R}$ of all real numbers and let

$$A = (-1, 0] = \{x \in \mathbf{R} \mid -1 < x \leq 0\} \text{ and } B = [0, 1) = \{x \in \mathbf{R} \mid 0 \leq x < 1\}.$$

These sets are shown on the number lines below.

Find $A \cup B$, $A \cap B$, $B - A$, and A^c.

Solution

$$A \cup B = \{x \in \mathbf{R} \mid x \in (-1, 0] \text{ or } x \in [0, 1)\} = \{x \in \mathbf{R} \mid x \in (-1, 1)\} = (-1, 1).$$

$$A \cap B = \{x \in \mathbf{R} \mid x \in (-1, 0] \text{ and } x \in [0, 1)\} = \{0\}.$$

$$B - A = \{x \in \mathbf{R} \mid x \in [0, 1) \text{ and } x \notin (-1, 0]\} = \{x \in \mathbf{R} \mid 0 < x < 1\} = (0, 1)$$

$$A^c = \{x \in \mathbf{R} \mid \text{it is not the case that } x \in (-1, 0]\}$$
$$= \{x \in \mathbf{R} \mid \text{it is not the case that } (-1 < x \text{ and } x \leq 0)\} \quad \text{by definition of the double inequality}$$
$$= \{x \in \mathbf{R} \mid x \leq -1 \text{ or } x > 0\} = (-\infty, -1] \cup (0, \infty) \quad \text{by De Morgan's law}$$

■

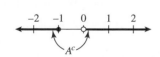

The definitions of unions and intersections for more than two sets are very similar to the definitions for two sets.

• **Definition**

Unions and Intersections of an Indexed Collection of Sets
Given sets $A_0, A_1, A_2, \ldots$ that are subsets of a universal set U and given a nonnegative integer n,

$$\bigcup_{i=0}^{n} A_i = \{x \in U \mid x \in A_i \text{ for at least one } i = 0, 1, 2, \ldots, n\}$$

$$\bigcup_{i=0}^{\infty} A_i = \{x \in U \mid x \in A_i \text{ for at least one nonnegative integer } i\}$$

$$\bigcap_{i=0}^{n} A_i = \{x \in U \mid x \in A_i \text{ for all } i = 0, 1, 2, \ldots, n\}$$

$$\bigcap_{i=0}^{\infty} A_i = \{x \in U \mid x \in A_i \text{ for all nonnegative integers } i\}.$$

An alternative notation for $\bigcup\limits_{i=0}^{n} A_i$ is $A_0 \cup A_1 \cup \ldots \cup A_n$, and an alternative notation for $\bigcap\limits_{i=0}^{n} A_i$ is $A_0 \cap A_1 \cap \ldots \cap A_n$.

Example 6.1.7 Finding Unions and Intersections of More than Two Sets

For each positive integer i, let $A_i = \left\{x \in \mathbf{R} \mid -\frac{1}{i} < x < \frac{1}{i}\right\} = A_i = \left(-\frac{1}{i}, \frac{1}{i}\right)$.

a. Find $A_1 \cup A_2 \cup A_3$ and $A_1 \cap A_2 \cap A_3$.   b. Find $\bigcup\limits_{i=1}^{\infty} A_i$ and $\bigcap\limits_{i=1}^{\infty} A_i$.

Solution

a. $A_1 \cup A_2 \cup A_3 = \{x \in \mathbf{R} \mid x$ is in at least one of the intervals $(-1, 1)$,

$$\text{or } \left(-\frac{1}{2}, \frac{1}{2}\right), \text{ or } \left(-\frac{1}{3}, \frac{1}{3}\right)\}$$

$= \{x \in \mathbf{R} \mid -1 < x < 1\}$ because all the elements in $\left(-\frac{1}{2}, \frac{1}{2}\right)$

$= (-1, 1)$ and $\left(-\frac{1}{3}, \frac{1}{3}\right)$ are in $(-1, 1)$

$A_1 \cap A_2 \cap A_3 = \{x \in \mathbf{R} \mid x$ is in all of the intervals $(-1, 1)$,

$$\text{and } \left(-\frac{1}{2}, \frac{1}{2}\right), \text{ and } \left(-\frac{1}{3}, \frac{1}{3}\right)\}$$

$= \left\{x \in \mathbf{R} \mid -\frac{1}{3} < x < \frac{1}{3}\right\}$ because $\left(-\frac{1}{3}, \frac{1}{3}\right) \subseteq \left(-\frac{1}{2}, \frac{1}{2}\right) \subseteq (-1, 1)$

$= \left(-\frac{1}{3}, \frac{1}{3}\right)$

b. $\bigcup\limits_{i=1}^{\infty} A_i = \{x \in \mathbf{R} \mid x$ is in at least one of the intervals $\left(-\frac{1}{i}, \frac{1}{i}\right)$,

where i is a positive integer}

$= \{x \in \mathbf{R} \mid -1 < x < 1\}$ because all the elements in every interval

$= (-1, 1)$ $\left(-\frac{1}{i}, \frac{1}{i}\right)$ are in $(-1, 1)$

$\bigcap\limits_{i=1}^{\infty} A_i = \{x \in \mathbf{R} \mid x$ is in all of the intervals $\left(-\frac{1}{i}, \frac{1}{i}\right)$, where i is a positive integer}

$= \{0\}$ because the only element in every interval is 0   ∎

The Empty Set

We have stated that a set is defined by the elements that compose it. This being so, can there be a set that has no elements? It turns out that it is convenient to allow such a set. Otherwise, every time we wanted to take the intersection of two sets or to define a set by specifying a property, we would have to check that the result had elements and hence qualified for "sethood." For example, if $A = \{1, 3\}$ and $B = \{2, 4\}$, then $A \cap B$ has no elements. Neither does $\{x \in \mathbf{R} \mid x^2 = -1\}$ because no real numbers have negative squares.

It is somewhat unsettling to talk about a set with no elements, but it often happens in mathematics that the definitions formulated to fit one set of circumstances are satisfied by some extreme cases not originally anticipated. Yet changing the definitions to exclude those cases would seriously undermine the simplicity and elegance of the theory taken as a whole.

In Section 6.2 we will show that there is only one set with no elements. Because it is unique, we can give it a special name. We call it the **empty set** (or **null set**) and denote it by the symbol $\emptyset$. Thus $\{1, 3\} \cap \{2, 4\} = \emptyset$ and $\{x \in \mathbf{R} \mid x^2 = -1\} = \emptyset$.

Example 6.1.8 A Set with No Elements

Describe the set $D = \{x \in \mathbf{R} \mid 3 < x < 2\}$.

Solution Recall that $a < x < b$ means that $a < x$ and $x < b$. So D consists of all real numbers that are both greater than 3 and less than 2. Since there are no such numbers, D has no elements and so $D = \emptyset$. ∎

Partitions of Sets

In many applications of set theory, sets are divided up into nonoverlapping (or *disjoint*) pieces. Such a division is called a *partition*.

• Definition

Two sets are called **disjoint** if, and only if, they have no elements in common. Symbolically:

$$A \text{ and } B \text{ are disjoint} \quad \Leftrightarrow \quad A \cap B = \emptyset.$$

Example 6.1.9 Disjoint Sets

Let $A = \{1, 3, 5\}$ and $B = \{2, 4, 6\}$. Are A and B disjoint?

Solution Yes. By inspection A and B have no elements in common, or, in other words, $\{1, 3, 5\} \cap \{2, 4, 6\} = \emptyset$. ∎

> **• Definition**
>
> Sets $A_1, A_2, A_3 \ldots$ are **mutually disjoint** (or **pairwise disjoint** or **nonoverlapping**) if, and only if, no two sets A_i and A_j with distinct subscripts have any elements in common. More precisely, for all $i, j = 1, 2, 3, \ldots$
>
> $$A_i \cap A_j = \emptyset \quad \text{whenever } i \neq j.$$

Example 6.1.10 Mutually Disjoint Sets

a. Let $A_1 = \{3, 5\}$, $A_2 = \{1, 4, 6\}$, and $A_3 = \{2\}$. Are A_1, A_2, and A_3 mutually disjoint?

b. Let $B_1 = \{2, 4, 6\}$, $B_2 = \{3, 7\}$, and $B_3 = \{4, 5\}$. Are B_1, B_2, and B_3 mutually disjoint?

Solution

a. Yes. A_1 and A_2 have no elements in common, A_1 and A_3 have no elements in common, and A_2 and A_3 have no elements in common.

b. No. B_1 and B_3 both contain 4. ∎

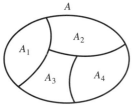

Figure 6.1.5 A Partition of a Set

Suppose A, A_1, A_2, A_3, and A_4 are the sets of points represented by the regions shown in Figure 6.1.5. Then A_1, A_2, A_3, and A_4 are subsets of A, and $A = A_1 \cup A_2 \cup A_3 \cup A_4$. Suppose further that boundaries are assigned to the regions representing A_1, A_2, A_3, and A_4 in such a way that these sets are mutually disjoint. Then A is called a *union of mutually disjoint subsets*, and the collection of sets $\{A_1, A_2, A_3, A_4\}$ is said to be a *partition* of A.

> **• Definition**
>
> A finite or infinite collection of nonempty sets $\{A_1, A_2, A_3 \ldots\}$ is a **partition** of a set A if, and only if,
>
> 1. A is the union of all the A_i
>
> 2. The sets $A_1, A_2, A_3, \ldots$ are mutually disjoint.

Example 6.1.11 Partitions of Sets

a. Let $A = \{1, 2, 3, 4, 5, 6\}$, $A_1 = \{1, 2\}$, $A_2 = \{3, 4\}$, and $A_3 = \{5, 6\}$. Is $\{A_1, A_2, A_3\}$ a partition of A?

b. Let $\mathbf{Z}$ be the set of all integers and let

$$T_0 = \{n \in \mathbf{Z} \mid n = 3k, \text{ for some integer } k\},$$
$$T_1 = \{n \in \mathbf{Z} \mid n = 3k + 1, \text{ for some integer } k\}, \text{ and}$$
$$T_2 = \{n \in \mathbf{Z} \mid n = 3k + 2, \text{ for some integer } k\}.$$

Is $\{T_0, T_1, T_2\}$ a partition of $\mathbf{Z}$?

Solution

    a. Yes. By inspection, $A = A_1 \cup A_2 \cup A_3$ and the sets A_1, A_2, and A_3 are mutually disjoint.

    b. Yes. By the quotient-remainder theorem, every integer n can be represented in exactly one of the three forms

$$n = 3k \quad \text{or} \quad n = 3k + 1 \quad \text{or} \quad n = 3k + 2,$$

for some integer k. This implies that no integer can be in any two of the sets T_0, T_1, or T_2. So T_0, T_1, and T_2 are mutually disjoint. It also implies that every integer is in one of the sets T_0, T_1, or T_2. So $\mathbf{Z} = T_0 \cup T_1 \cup T_2$. ∎

Power Sets

There are various situations in which it is useful to consider the set of all subsets of a particular set. The **power set axiom** guarantees that this is a set.

> **• Definition**
>
> Given a set A, the **power set** of A, denoted $\mathscr{P}(A)$, is the set of all subsets of A.

Example 6.1.12 Power Set of a Set

Find the power set of the set $\{x, y\}$. That is, find $\mathscr{P}(\{x, y\})$.

Solution $\mathscr{P}(\{x, y\})$ is the set of all subsets of $\{x, y\}$. In Section 6.2 we will show that $\emptyset$ is a subset of every set, and so $\emptyset \in \mathscr{P}(\{x, y\})$. Also any set is a subset of itself, so $\{x, y\} \in \mathscr{P}(\{x, y\})$. The only other subsets of $\{x, y\}$ are $\{x\}$ and $\{y\}$, so

$$\mathscr{P}(\{x, y\}) = \{\emptyset, \{x\}, \{y\}, \{x, y\}\}.$$ ∎

Cartesian Products

Recall that the definition of a set is unaffected by the order in which its elements are listed or the fact that some elements may be listed more than once. Thus $\{a, b\}$, $\{b, a\}$, and $\{a, a, b\}$ all represent the same set. The notation for an *ordered n-tuple* is a generalization of the notation for an ordered pair. (See Section 1.2.) It takes both order and multiplicity into account.

> **• Definition**
>
> Let n be a positive integer and let $x_1, x_2, \ldots, x_n$ be (not necessarily distinct) elements. The **ordered n-tuple, $(x_1, x_2, \ldots, x_n)$,** consists of $x_1, x_2, \ldots, x_n$ together with the ordering: first x_1, then x_2, and so forth up to x_n. An ordered 2-tuple is called an **ordered pair,** and an ordered 3-tuple is called an **ordered triple.**
>
>   Two ordered n-tuples $(x_1, x_2, \ldots, x_n)$ and $(y_1, y_2, \ldots, y_n)$ are **equal** if, and only if, $x_1 = y_1, x_2 = y_2, \ldots, x_n = y_n$.
>   Symbolically:
>
> $$(x_1, x_2, \ldots, x_n) = (y_1, y_2, \ldots, y_n) \quad \Leftrightarrow \quad x_1 = y_1, x_2 = y_2, \ldots, x_n = y_n.$$
>
> In particular,
>
> $$(a, b) = (c, d) \quad \Leftrightarrow \quad a = c \text{ and } b = d.$$

Example 6.1.13 Ordered *n*-tuples

    a. Is $(1, 2, 3, 4) = (1, 2, 4, 3)$?

    b. Is $\left(3, (-2)^2, \frac{1}{2}\right) = \left(\sqrt{9}, 4, \frac{3}{6}\right)$?

Solution

    a. No. By definition of equality of ordered 4-tuples,

$$(1,\ 2,\ 3,\ 4) = (1,\ 2,\ 4,\ 3) \Leftrightarrow 1 = 1,\ 2 = 2,\ 3 = 4,\ \text{and } 4 = 3$$

    But $3 \neq 4$, and so the ordered 4-tuples are not equal.

    b. Yes. By definition of equality of ordered triples,

$$\left(3, (-2)^2, \tfrac{1}{2}\right) = \left(\sqrt{9}, 4, \tfrac{3}{6}\right) \quad \Leftrightarrow \quad 3 = \sqrt{9} \text{ and } (-2)^2 = 4 \text{ and } \tfrac{1}{2} = \tfrac{3}{6}.$$

    Because these equations are all true, the two ordered triples are equal. ∎

• Definition

Given sets $A_1, A_2, \ldots, A_n$, the **Cartesian product** of $A_1, A_2, \ldots, A_n$ denoted $A_1 \times A_2 \times \ldots \times A_n$, is the set of all ordered *n*-tuples $(a_1, a_2, \ldots, a_n)$ where $a_1 \in A_1, a_2 \in A_2, \ldots, a_n \in A_n$.
    Symbolically:

$$A_1 \times A_2 \times \cdots \times A_n = \{(a_1, a_2, \ldots, a_n) \mid a_1 \in A_1, a_2 \in A_2, \ldots, a_n \in A_n\}.$$

In particular,

$$A_1 \times A_2 = \{(a_1, a_2) \mid a_1 \in A_1 \text{ and } a_2 \in A_2\}$$

is the Cartesian product of A_1 and A_2.

Example 6.1.14 Cartesian Products

    Let $A_1 = \{x, y\}$, $A_2 = \{1, 2, 3\}$, and $A_3 = \{a, b\}$.

    a. Find $A_1 \times A_2$.   b. Find $(A_1 \times A_2) \times A_3$.   c. Find $A_1 \times A_2 \times A_3$.

Solution

    a. $A_1 \times A_2 = \{(x, 1), (x, 2), (x, 3), (y, 1), (y, 2), (y, 3)\}$

    b. The Cartesian product of A_1 and A_2 is a set, so it may be used as one of the sets making up another Cartesian product. This is the case for $(A_1 \times A_2) \times A_3$.

$$
\begin{aligned}
(A_1 \times A_2) \times A_3 = \{(u, v) \mid u \in A_1 \times A_2 \text{ and } v \in A_3\} \quad &\text{by definition of Cartesian product}\\
= \{((x, 1), a), ((x, 2), a), ((x, 3), a), ((y, 1), a),&\\
((y, 2), a), ((y, 3), a), ((x, 1), b), ((x, 2), b), ((x, 3), b),&\\
((y, 1), b), ((y, 2), b), ((y, 3), b)\}&
\end{aligned}
$$

c. The Cartesian product $A_1 \times A_2 \times A_3$ is superficially similar to, but is not quite the same mathematical object as, $(A_1 \times A_2) \times A_3$. $(A_1 \times A_2) \times A_3$ is a set of ordered pairs of which one element is itself an ordered pair, whereas $A_1 \times A_2 \times A_3$ is a set of ordered triples. By definition of Cartesian product,

$$A_1 \times A_2 \times A_3 = \{(u, v, w) \mid u \in A_1, v \in A_2, \text{and } w \in A_3\}$$
$$= \{(x, 1, a), (x, 2, a), (x, 3, a), (y, 1, a), (y, 2, a),$$
$$(y, 3, a), (x, 1, b), (x, 2, b), (x, 3, b), (y, 1, b),$$
$$(y, 2, b), (y, 3, b)\}. \qquad \blacksquare$$

Test Yourself

Answers to Test Yourself questions are located at the end of each section.

1. The notation $A \subseteq B$ is read "_____" and means that _____.

2. To use an element argument for proving that a set X is a subset of a set Y, you suppose that _____ and show that _____.

3. To disprove that a set X is a subset of a set Y, you show that there is _____.

4. An element x is in $A \cup B$ if, and only if, _____.

5. An element x is in $A \cap B$ if, and only if, _____.

6. An element x is in $B - A$ if, and only if, _____.

7. An element x is in A^c if, and only if, _____.

8. The empty set is a set with _____.

9. The power set of a set A is _____.

10. Sets A and B are disjoint if, and only if, _____.

11. A collection of nonempty sets $A_1, A_2, A_3, \ldots$ is a partition of a set A if, and only if, _____.

12. Given sets $A_1, A_2, \ldots, A_n$, the Cartesian product $A_1 \times A_2 \times \ldots \times A_n$ is _____.

Exercise Set 6.1*

1. In each of (a)–(f), answer the following questions: Is $A \subseteq B$? Is $B \subseteq A$? Is either A or B a proper subset of the other?

   a. $A = \{2, \{2\}, (\sqrt{2})^2\}$, $B = \{2, \{2\}, \{\{2\}\}\}$
   b. $A = \{3, \sqrt{5^2 - 4^2}, 24 \bmod 7\}$, $B = \{8 \bmod 5\}$
   c. $A = \{\{1, 2\}, \{2, 3\}\}$, $B = \{1, 2, 3\}$
   d. $A = \{a, b, c\}$, $B = \{\{a\}, \{b\}, \{c\}\}$
   e. $A = \{\sqrt{16}, \{4\}\}$, $B = \{4\}$
   f. $A = \{x \in \mathbf{R} \mid \cos x \in \mathbf{Z}\}$, $B = \{x \in \mathbf{R} \mid \sin x \in \mathbf{Z}\}$

2. Complete the proof from Example 6.1.3: Prove that $B \subseteq A$ where
   $$A = \{m \in \mathbf{Z} \mid m = 2a \text{ for some integer } a\}$$
   and
   $$B = \{n \in \mathbf{Z} \mid n = 2b - 2 \text{ for some integer } b\}$$

3. Let sets R, S, and T be defined as follows:
   $$R = \{x \in \mathbf{Z} \mid x \text{ is divisible by 2}\}$$
   $$S = \{y \in \mathbf{Z} \mid y \text{ is divisible by 3}\}$$
   $$T = \{z \in \mathbf{Z} \mid z \text{ is divisible by 6}\}$$
   a. Is $R \subseteq T$? Explain.
   b. Is $T \subseteq R$? Explain.
   c. Is $T \subseteq S$? Explain.

4. Let $A = \{n \in \mathbf{Z} \mid n = 5r \text{ for some integer } r\}$ and $B = \{m \in \mathbf{Z} \mid m = 20s \text{ for some integer } s\}$.
   a. Is $A \subseteq B$? Explain. b. Is $B \subseteq A$? Explain.

5. Let $C = \{n \in \mathbf{Z} \mid n = 6r - 5 \text{ for some integer } r\}$ and $D = \{m \in \mathbf{Z} \mid m = 3s + 1 \text{ for some integer } s\}$.
   Prove or disprove each of the following statements.
   a. $C \subseteq D$ b. $D \subseteq C$

* For exercises with blue numbers or letters, solutions are given in Appendix B. The symbol **H** indicates that only a hint or a partial solution is given. The symbol ✶ signals that an exercise is more challenging than usual.

6. Let $A = \{x \in \mathbf{Z} \mid x = 5a + 2 \text{ for some integer } a\}$,
   $B = \{y \in \mathbf{Z} \mid y = 10b - 3 \text{ for some integer } b\}$, and
   $C = \{z \in \mathbf{Z} \mid z = 10c + 7 \text{ for some integer } c\}$.
   Prove or disprove each of the following statements.
   a. $A \subseteq B$ b. $B \subseteq A$ **H** c. $B = C$

7. Let $A = \{x \in \mathbf{Z} \mid x = 6a + 4 \text{ for some integer } a\}$,
   $B = \{y \in \mathbf{Z} \mid y = 18b - 2 \text{ for some integer } b\}$, and
   $C = \{z \in \mathbf{Z} \mid z = 18c + 16 \text{ for some integer } c\}$.
   Prove or disprove each of the following statements.
   a. $A \subseteq B$ b. $B \subseteq A$ c. $B = C$

8. Write in words how to read each of the following out loud.
   Then write the shorthand notation for each set.
   a. $\{x \in U \mid x \in A \text{ and } x \in B\}$
   b. $\{x \in U \mid x \in A \text{ or } x \in B\}$
   c. $\{x \in U \mid x \in A \text{ and } x \notin B\}$
   d. $\{x \in U \mid x \notin A\}$

9. Complete the following sentences without using the symbols $\cup$, $\cap$, or $-$.
   a. $x \notin A \cup B$ if, and only if, _____.
   b. $x \notin A \cap B$ if, and only if, _____.
   c. $x \notin A - B$ if, and only if, _____.

10. Let $A = \{1, 3, 5, 7, 9\}$, $B = \{3, 6, 9\}$, and $C = \{2, 4, 6, 8\}$. Find each of the following:
    a. $A \cup B$ b. $A \cap B$ c. $A \cup C$ d. $A \cap C$
    e. $A - B$ f. $B - A$ g. $B \cup C$ h. $B \cap C$

11. Let the universal set be the set $\mathbf{R}$ of all real numbers and let
    $A = \{x \in \mathbf{R} \mid 0 < x \leq 2\}$, $B = \{x \in \mathbf{R} \mid 1 \leq x < 4\}$, and
    $C = \{x \in \mathbf{R} \mid 3 \leq x < 9\}$. Find each of the following:
    a. $A \cup B$ b. $A \cap B$ c. A^c d. $A \cup C$
    e. $A \cap C$ f. B^c g. $A^c \cap B^c$
    h. $A^c \cup B^c$ i. $(A \cap B)^c$ j. $(A \cup B)^c$

12. Let the universal set be the set $\mathbf{R}$ of all real numbers and
    let $A = \{x \in \mathbf{R} \mid -3 \leq x \leq 0\}$, $B = \{x \in \mathbf{R} \mid -1 < x < 2\}$,
    and $C = \{x \in \mathbf{R} \mid 6 < x \leq 8\}$. Find each of the following:
    a. $A \cup B$ b. $A \cap B$ c. A^c d. $A \cup C$
    e. $A \cap C$ f. B^c g. $A^c \cap B^c$
    h. $A^c \cup B^c$ i. $(A \cap B)^c$ j. $(A \cup B)^c$

13. Indicate which of the following relationships are true and which are false:
    a. $\mathbf{Z}^+ \subseteq \mathbf{Q}$ b. $\mathbf{R}^- \subseteq \mathbf{Q}$
    c. $\mathbf{Q} \subseteq \mathbf{Z}$ d. $\mathbf{Z}^- \cup \mathbf{Z}^+ = \mathbf{Z}$
    e. $\mathbf{Z}^- \cap \mathbf{Z}^+ = \emptyset$ f. $\mathbf{Q} \cap \mathbf{R} = \mathbf{Q}$
    g. $\mathbf{Q} \cup \mathbf{Z} = \mathbf{Q}$ h. $\mathbf{Z}^+ \cap \mathbf{R} = \mathbf{Z}^+$
    i. $\mathbf{Z} \cup \mathbf{Q} = \mathbf{Z}$

14. In each of the following, draw a Venn diagram for sets A, B, and C that satisfy the given conditions:
    a. $A \subseteq B$; $C \subseteq B$; $A \cap C = \emptyset$
    b. $C \subseteq A$; $B \cap C = \emptyset$

15. Draw Venn diagrams to describe sets A, B, and C that satisfy the given conditions.
    a. $A \cap B = \emptyset$, $A \subseteq C$, $C \cap B \neq \emptyset$
    b. $A \subseteq B$, $C \subseteq B$, $A \cap C \neq \emptyset$
    c. $A \cap B \neq \emptyset$, $B \cap C \neq \emptyset$, $A \cap C = \emptyset$, $A \nsubseteq B$, $C \nsubseteq B$

16. Let $A = \{a, b, c\}$, $B = \{b, c, d\}$, and $C = \{b, c, e\}$.
    a. Find $A \cup (B \cap C)$, $(A \cup B) \cap C$, and $(A \cup B) \cap (A \cup C)$. Which of these sets are equal?
    b. Find $A \cap (B \cup C)$, $(A \cap B) \cup C$, and $(A \cap B) \cup (A \cap C)$. Which of these sets are equal?
    c. Find $(A - B) - C$ and $A - (B - C)$. Are these sets equal?

17. Consider the Venn diagram shown below. For each of (a)–(f), copy the diagram and shade the region corresponding to the indicated set.
    a. $A \cap B$ b. $B \cup C$ c. A^c
    d. $A - (B \cup C)$ e. $(A \cup B)^c$ f. $A^c \cap B^c$

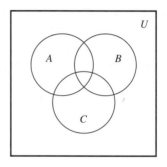

18. a. Is the number 0 in $\emptyset$? Why?   b. Is $\emptyset = \{\emptyset\}$? Why?
    c. Is $\emptyset \in \{\emptyset\}$? Why?   d. Is $\emptyset \in \emptyset$? Why?

19. Let $A_i = \{i, i^2\}$ for all integers $i = 1, 2, 3, 4$.
    a. $A_1 \cup A_2 \cup A_3 \cup A_4 = ?$
    b. $A_1 \cap A_2 \cap A_3 \cap A_4 = ?$
    c. Are A_1, A_2, A_3, and A_4 mutually disjoint? Explain.

20. Let $B_i = \{x \in \mathbf{R} \mid 0 \leq x \leq i\}$ for all integers $i = 1, 2, 3, 4$.
    a. $B_1 \cup B_2 \cup B_3 \cup B_4 = ?$
    b. $B_1 \cap B_2 \cap B_3 \cap B_4 = ?$
    c. Are B_1, B_2, B_3, and B_4 mutually disjoint? Explain.

21. Let $C_i = \{i, -i\}$ for all nonnegative integers i.
    a. $\bigcup_{i=0}^{4} C_i = ?$ b. $\bigcap_{i=0}^{4} C_i = ?$
    c. Are $C_0, C_1, C_2, \ldots$ mutually disjoint? Explain.
    d. $\bigcup_{i=0}^{n} C_i = ?$ e. $\bigcap_{i=0}^{n} C_i = ?$
    f. $\bigcup_{i=0}^{\infty} C_i = ?$ g. $\bigcap_{i=0}^{\infty} C_i = ?$

22. Let $D_i = \{x \in \mathbf{R} \mid -i \leq x \leq i\} = [-i, i]$ for all nonnegative integers i.
    a. $\bigcup_{i=0}^{4} D_i = ?$ b. $\bigcap_{i=0}^{4} D_i = ?$
    c. Are $D_0, D_1, D_2, \ldots$ mutually disjoint? Explain.

d. $\bigcup\limits_{i=0}^{n} D_i = ?$ e. $\bigcap\limits_{i=0}^{n} D_i = ?$

f. $\bigcup\limits_{i=0}^{\infty} D_i = ?$ g. $\bigcap\limits_{i=0}^{\infty} D_i = ?$

23. Let $V_i = \left\{ x \in \mathbf{R} \mid -\frac{1}{i} \le x \le \frac{1}{i} \right\} = \left[-\frac{1}{i}, \frac{1}{i} \right]$ for all positive integers i.

a. $\bigcup\limits_{i=1}^{4} V_i = ?$ b. $\bigcap\limits_{i=1}^{4} V_i = ?$

c. Are $V_1, V_2, V_3, \ldots$ mutually disjoint? Explain.

d. $\bigcup\limits_{i=1}^{n} V_i = ?$ e. $\bigcap\limits_{i=1}^{n} V_i = ?$

f. $\bigcup\limits_{i=1}^{\infty} V_i = ?$ g. $\bigcap\limits_{i=1}^{\infty} V_i = ?$

24. Let $W_i = \{x \in \mathbf{R} \mid x > i\} = (i, \infty)$ for all nonnegative integers i.

a. $\bigcup\limits_{i=0}^{4} W_i = ?$ b. $\bigcap\limits_{i=0}^{4} W_i = ?$

c. Are $W_0, W_1, W_2, \ldots$ mutually disjoint? Explain.

d. $\bigcup\limits_{i=0}^{n} W_i = ?$ e. $\bigcap\limits_{i=0}^{n} W_i = ?$

f. $\bigcup\limits_{i=0}^{\infty} W_i = ?$ g. $\bigcap\limits_{i=0}^{\infty} W_i = ?$

25. Let $R_i = \left\{ x \in \mathbf{R} \mid 1 \le x \le 1 + \frac{1}{i} \right\} = \left[1, 1 + \frac{1}{i} \right]$ for all positive integers i.

a. $\bigcup\limits_{i=1}^{4} R_i = ?$ b. $\bigcap\limits_{i=1}^{4} R_i = ?$

c. Are $R_1, R_2, R_3, \ldots$ mutually disjoint? Explain.

d. $\bigcup\limits_{i=1}^{n} R_i = ?$ e. $\bigcap\limits_{i=1}^{n} R_i = ?$

f. $\bigcup\limits_{i=1}^{\infty} R_i = ?$ g. $\bigcap\limits_{i=1}^{\infty} R_i = ?$

26. Let $S_i = \left\{ x \in \mathbf{R} \mid 1 < x < 1 + \frac{1}{i} \right\} = \left(1, 1 + \frac{1}{i} \right)$ for all positive integers i.

a. $\bigcup\limits_{i=1}^{4} S_i = ?$ b. $\bigcap\limits_{i=1}^{4} S_i = ?$

c. Are $S_1, S_2, S_3, \ldots$ mutually disjoint? Explain.

d. $\bigcup\limits_{i=1}^{n} S_i = ?$ e. $\bigcap\limits_{i=1}^{n} S_i = ?$

f. $\bigcup\limits_{i=1}^{\infty} S_i = ?$ g. $\bigcap\limits_{i=1}^{\infty} S_i = ?$

27. **a.** Is $\{\{a, d, e\}, \{b, c\}, \{d, f\}\}$ a partition of $\{a, b, c, d, e, f\}$?
   b. Is $\{\{w, x, v\}, \{u, y, q\}, \{p, z\}\}$ a partition of $\{p, q, u, v, w, x, y, z\}$?
   c. Is $\{\{5, 4\}, \{7, 2\}, \{1, 3, 4\}, \{6, 8\}\}$ a partition of $\{1, 2, 3, 4, 5, 6, 7, 8\}$?
   d. Is $\{\{3, 7, 8\}, \{2, 9\}, \{1, 4, 5\}\}$ a partition of $\{1, 2, 3, 4, 5, 6, 7, 8, 9\}$?
   e. Is $\{\{1, 5\}, \{4, 7\}, \{2, 8, 6, 3\}\}$ a partition of $\{1, 2, 3, 4, 5, 6, 7, 8\}$?

28. Let E be the set of all even integers and O the set of all odd integers. Is $\{E, O\}$ a partition of $\mathbf{Z}$, the set of all integers? Explain your answer.

29. Let $\mathbf{R}$ be the set of all real numbers. Is $\{\mathbf{R}^+, \mathbf{R}^-, \{0\}\}$ a partition of $\mathbf{R}$? Explain your answer.

30. Let $\mathbf{Z}$ be the set of all integers and let

$A_0 = \{n \in \mathbf{Z} \mid n = 4k, \text{ for some integer } k\},$

$A_1 = \{n \in \mathbf{Z} \mid n = 4k + 1, \text{ for some integer } k\},$

$A_2 = \{n \in \mathbf{Z} \mid n = 4k + 2, \text{ for some integer } k\}, \text{ and}$

$A_3 = \{n \in \mathbf{Z} \mid n = 4k + 3, \text{ for some integer } k\}.$

Is $\{A_0, A_1, A_2, A_3\}$ a partition of $\mathbf{Z}$? Explain your answer.

31. Suppose $A = \{1, 2\}$ and $B = \{2, 3\}$. Find each of the following:
   a. $\mathscr{P}(A \cap B)$ b. $\mathscr{P}(A)$
   c. $\mathscr{P}(A \cup B)$ d. $\mathscr{P}(A \times B)$

32. **a.** Suppose $A = \{1\}$ and $B = \{u, v\}$. Find $\mathscr{P}(A \times B)$.
   b. Suppose $X = \{a, b\}$ and $Y = \{x, y\}$. Find $\mathscr{P}(X \times Y)$.

33. a. Find $\mathscr{P}(\emptyset)$. **b.** Find $\mathscr{P}(\mathscr{P}(\emptyset))$.
   c. Find $\mathscr{P}(\mathscr{P}(\mathscr{P}(\emptyset)))$.

34. Let $A_1 = \{1, 2, 3\}$, $A_2 = \{u, v\}$, and $A_3 = \{m, n\}$. Find each of the following sets:
   a. $A_1 \times (A_2 \times A_3)$ b. $(A_1 \times A_2) \times A_3$
   c. $A_1 \times A_2 \times A_3$

35. Let $A = \{a, b\}$, $B = \{1, 2\}$, and $C = \{2, 3\}$. Find each of the following sets.
   a. $A \times (B \cup C)$ **b.** $(A \times B) \cup (A \times C)$
   c. $A \times (B \cap C)$ d. $(A \times B) \cap (A \times C)$

Answers for Test Yourself

1. the set A is a subset of the set B; for all x, if $x \in A$ then $x \in B$ (*Or*: every element of A is also an element of B) 2. x is any *[particular but arbitrarily chosen]* element of X; x is an element of Y 3. an element in X that is not in Y 4. x is in A or x is in B (*Or*: x is in at least one of the sets A and B) 5. x is in A and x is in B (*Or*: x is in both A and B) 6. x is in B and x is not in A 7. x is in the universal set and is not in A 8. no elements 9. the set of all subsets of A 10. $A \cap B = \emptyset$ (*Or*: A and B have no elements in common) 11. A is the union of all the sets $A_1, A_2, A_3, \ldots$ and $A_i \cap A_j = \emptyset$ whenever $i \ne j$. 12. the set of all ordered n-tuples $(a_1, a_2, \ldots, a_n)$, where a_i is in A_i for all $i = 1, 2, \ldots, n$

6.2 Properties of Sets

... only the last line is a genuine theorem here—everything else is in the fantasy.
—Douglas Hofstadter, *Gödel, Escher, Bach,* 1979

It is possible to list many relations involving unions, intersections, complements, and differences of sets. Some of these are true for all sets, whereas others fail to hold in some cases. In this section we show how to establish basic set properties using *element arguments*, and we discuss a variation used to prove that a set is empty. In the next section we will show how to disprove a proposed set property by constructing a counterexample and how to use an algebraic technique to derive new set properties from set properties already known to be true.

We begin by listing some set properties that involve subset relations. As you read them, keep in mind that the operations of union, intersection, and difference take precedence over set inclusion. Thus, for example, $A \cap B \subseteq C$ means $(A \cap B) \subseteq C$.

Theorem 6.2.1 Some Subset Relations

1. *Inclusion of Intersection:* For all sets A and B,

   (a) $A \cap B \subseteq A$ and   (b) $A \cap B \subseteq B$.

2. *Inclusion in Union:* For all sets A and B,

   (a) $A \subseteq A \cup B$ and   (b) $B \subseteq A \cup B$.

3. *Transitive Property of Subsets:* For all sets A, B, and C,

   if $A \subseteq B$ and $B \subseteq C$, then $A \subseteq C$.

The conclusion of each part of Theorem 6.2.1 states that one set x is a subset of another set Y and so to prove them, you suppose that x is any *[particular but arbitrarily chosen]* element of X and you show that x is an element of Y.

In most proofs of set properties, the secret of getting from the assumption that x is in X to the conclusion that x is in Y is to think of the definitions of basic set operations in procedural terms. For example, the union of sets X and Y, $X \cup Y$, is defined as

$$X \cup Y = \{x \mid x \in X \text{ or } x \in Y\}.$$

This means that any time you know an element x is in $X \cup Y$, you can conclude that x must be in X *or* x must be in Y. Conversely, any time you know that a particular x is in some set X or is in some set Y, you can conclude that x is in $X \cup Y$. Thus, for any sets X and Y and any element x,

$$x \in X \cup Y \quad \text{if, and only if,} \quad x \in X \text{ or } x \in Y.$$

Procedural versions of the definitions of the other set operations are derived similarly and are summarized on the next page.

> ### Procedural Versions of Set Definitions
>
> Let X and Y be subsets of a universal set U and suppose x and y are elements of U.
>
> 1. $x \in X \cup Y \quad \Leftrightarrow \quad x \in X$ or $x \in Y$
>
> 2. $x \in X \cap Y \quad \Leftrightarrow \quad x \in X$ and $x \in Y$
>
> 3. $x \in X - Y \quad \Leftrightarrow \quad x \in X$ and $x \notin Y$
>
> 4. $x \in X^c \quad \Leftrightarrow \quad x \notin X$
>
> 5. $(x, y) \in X \times Y \quad \Leftrightarrow \quad x \in X$ and $y \in Y$

Example 6.2.1 Proof of a Subset Relation

Prove Theorem 6.2.1(1)(a): For all sets A and B, $A \cap B \subseteq A$.

Solution   We start by giving a proof of the statement and then explain how you can obtain such a proof yourself.

Proof:

> Suppose A and B are any sets and suppose x is any element of $A \cap B$.
> Then $x \in A$ and $x \in B$ by definition of intersection. In particular, $x \in A$.
> Thus $A \cap B \subseteq A$.

The underlying structure of this proof is not difficult, but it is more complicated than the brevity of the proof suggests. The first important thing to realize is that the statement to be proved is universal (it says that for *all* sets A and B, $A \cap B \subseteq A$). The proof, therefore, has the following outline:

Starting Point: Suppose A and B are any (particular but arbitrarily chosen) sets.

To Show: $A \cap B \subseteq A$

Now to prove that $A \cap B \subseteq A$, you must show that

$$\forall x, \text{ if } x \in A \cap B \text{ then } x \in A.$$

But this statement also is universal. So to prove it, you

suppose x is an element in $A \cap B$

and then you

show that x is in A.

Filling in the gap between the "suppose" and the "show" is easy if you use the procedural version of the definition of intersection: To say that x is in $A \cap B$ means that

x is in A and x is in B.

This allows you to complete the proof by deducing that, in particular,

x is in A,

as was to be shown. Note that this deduction is just a special case of the valid argument form

$$p \wedge q$$
$$\therefore p.$$ ∎

In his book *Gödel, Escher, Bach,*[*] Douglas Hofstadter introduces the fantasy rule for mathematical proof. Hofstadter points out that when you start a mathematical argument with *if, let,* or *suppose,* you are stepping into a fantasy world where not only are all the facts of the real world true but whatever you are supposing is also true. Once you are in that world, you can suppose something else. That sends you into a subfantasy world where not only is everything in the fantasy world true but also the new thing you are supposing. Of course you can continue stepping into new subfantasy worlds in this way indefinitely. You return one level closer to the real world each time you derive a conclusion that makes a whole if-then or universal statement true. Your aim in a proof is to continue deriving such conclusions until you return to the world from which you made your first supposition.

Occasionally, mathematical problems are stated in the following form:

Suppose (*statement* 1). Prove that (*statement* 2).

When this phrasing is used, the author intends the reader to add statement 1 to his or her general mathematical knowledge and not to make explicit reference to it in the proof. In Hofstadter's terms, the author invites the reader to enter a fantasy world where statement 1 is known to be true and to prove statement 2 in this fantasy world. Thus the solver of such a problem would begin a proof with the starting point for a proof of statement 2. Consider, for instance, the following restatement of Example 6.2.1:

Suppose A and B are arbitrarily chosen sets.
Prove that $A \cap B \subseteq A$.

The proof would begin "Suppose $x \in A \cap B$," it being *understood* that sets A and B have already been chosen arbitrarily.

The proof of Example 6.2.1 is called an element argument because it shows one set to be a subset of another by demonstrating that every element in the one set is also an element in the other. In higher mathematics, element arguments are the standard method of establishing relations among sets. High school students are often allowed to justify set properties by using Venn diagrams. This method is appealing, but for it to be mathematically rigorous may be more complicated than you might expect. Appropriate Venn diagrams can be drawn for two or three sets, but the verbal explanations needed to justify conclusions inferred from them are normally as long as a straightforward element proof.

In general, Venn diagrams are not very helpful when the number of sets is four or more. For instance, if the requirement is made that a Venn diagram must show every possible intersection of the sets, it is impossible to draw a symmetric Venn diagram for four sets, or, in fact, for any nonprime number of sets. In 2002, computer scientists/mathematicians Carla Savage and Jerrold Griggs and undergraduate student Charles Killian solved a longstanding open problem by proving that it *is* possible to draw such a symmetric Venn diagram for any prime number of sets. For $n > 5$, however, the resulting pictures are very complicated! The existence of such symmetric diagrams has applications in the area of computer science called coding theory.

---

[*]*Gödel, Escher, Bach: An Eternal Golden Braid* (New York: Basic Books, 1979).

Set Identities

An **identity** is an equation that is universally true for all elements in some set. For example, the equation $a + b = b + a$ is an identity for real numbers because it is true for all real numbers a and b. The collection of set properties in the next theorem consists entirely of set identities. That is, they are equations that are true for all sets in some universal set.

Theorem 6.2.2 Set Identities

Let all sets referred to below be subsets of a universal set U.

1. *Commutative Laws:* For all sets A and B,
$$\text{(a) } A \cup B = B \cup A \quad \text{and} \quad \text{(b) } A \cap B = B \cap A.$$

2. *Associative Laws:* For all sets A, B, and C,
$$\text{(a) } (A \cup B) \cup C = A \cup (B \cup C) \quad \text{and}$$
$$\text{(b) } (A \cap B) \cap C = A \cap (B \cap C).$$

3. *Distributive Laws:* For all sets, A, B, and C,
$$\text{(a) } A \cup (B \cap C) = (A \cup B) \cap (A \cup C) \quad \text{and}$$
$$\text{(b) } A \cap (B \cup C) = (A \cap B) \cup (A \cap C).$$

4. *Identity Laws:* For all sets A,
$$\text{(a) } A \cup \emptyset = A \quad \text{and} \quad \text{(b) } A \cap U = A.$$

5. *Complement Laws:*
$$\text{(a) } A \cup A^c = U \quad \text{and} \quad \text{(b) } A \cap A^c = \emptyset.$$

6. *Double Complement Law:* For all sets A,
$$(A^c)^c = A.$$

7. *Idempotent Laws:* For all sets A,
$$\text{(a) } A \cup A = A \quad \text{and} \quad \text{(b) } A \cap A = A.$$

8. *Universal Bound Laws:* For all sets A,
$$\text{(a) } A \cup U = U \quad \text{and} \quad \text{(b) } A \cap \emptyset = \emptyset.$$

9. *De Morgan's Laws:* For all sets A and B,
$$\text{(a) } (A \cup B)^c = A^c \cap B^c \quad \text{and} \quad \text{(b) } (A \cap B)^c = A^c \cup B^c.$$

10. *Absorption Laws:* For all sets A and B,
$$\text{(a) } A \cup (A \cap B) = A \quad \text{and} \quad \text{(b) } A \cap (A \cup B) = A.$$

11. *Complements of U and $\emptyset$:*
$$\text{(a) } U^c = \emptyset \quad \text{and} \quad \text{(b) } \emptyset^c = U.$$

12. *Set Difference Law:* For all sets A and B,
$$A - B = A \cap B^c.$$

Proving Set Identities

The conclusion of each part of Theorem 6.2.2 is that one set equals another set. As we noted in Section 6.1,

Two sets are equal ⇔ each is a subset of the other.

The method derived from this fact is the most basic way to prove equality of sets.

Basic Method for Proving That Sets Are Equal

Let sets X and Y be given. To prove that $X = Y$:

1. Prove that $X \subseteq Y$.

2. Prove that $Y \subseteq X$.

Example 6.2.2 Proof of a Distributive Law

Prove that for all sets A, B, and C,

$$A \cup (B \cap C) = (A \cup B) \cap (A \cup C).$$

Solution The proof of this fact is somewhat more complicated than the proof in Example 6.2.1, so we first derive its logical structure, then find the core arguments, and end with a formal proof as a summary. As in Example 6.2.1, the statement to be proved is universal, and so, by the method of generalizing from the generic particular, the proof has the following outline:

Starting Point: Suppose A, B, and C are arbitrarily chosen sets.

To Show: $A \cup (B \cap C) = (A \cup B) \cap (A \cup C)$.

Now two sets are equal if, and only if, each is a subset of the other. Hence, the following two statements must be proved:

$$A \cup (B \cap C) \subseteq (A \cup B) \cap (A \cup C)$$

and $\quad\quad (A \cup B) \cap (A \cup C) \subseteq A \cup (B \cap C)$.

Showing the first containment requires showing that

$$\forall x, \text{ if } x \in A \cup (B \cap C) \text{ then } x \in (A \cup B) \cap (A \cup C).$$

Showing the second containment requires showing that

$$\forall x, \text{ if } x \in (A \cup B) \cap (A \cup C) \text{ then } x \in A \cup (B \cap C).$$

Note that both of these statements are universal. So to prove the first containment, you

suppose you have any element x in $A \cup (B \cap C)$,

and then you **show** that $x \in (A \cup B) \cap (A \cup C)$.

And to prove the second containment, you

suppose you have any element x in $(A \cup B) \cap (A \cup C)$,

and then you **show** that $x \in A \cup (B \cap C)$.

In Figure 6.2.1, the structure of the proof is illustrated by the kind of diagram that is often used in connection with structured programs. The analysis in the diagram reduces the proof to two concrete tasks: filling in the steps indicated by dots in the two center boxes of Figure 6.2.1.

Suppose A, B, and C are sets. *[Show $A \cup (B \cap C) = (A \cup B) \cap (A \cup C)$. That is, show $A \cup (B \cap C) \subseteq (A \cup B) \cap (A \cup C)$ and $(A \cup B) \cap (A \cup C) \subseteq A \cup (B \cap C)$.]*

Show $A \cup (B \cap C) \subseteq (A \cup B) \cap (A \cup C)$. *[That is, show $\forall x$, if $x \in A \cup (B \cap C)$ then $x \in (A \cup B) \cap (A \cup C)$.]*

Suppose $x \in A \cup (B \cap C)$. *[Show $x \in (A \cup B) \cap (A \cup C)$.]*
$\vdots$
Thus $x \in (A \cup B) \cap (A \cup C)$.

Hence $A \cup (B \cap C) \subseteq (A \cup B) \cap (A \cup C)$.

Show $(A \cup B) \cap (A \cup C) \subseteq A \cup (B \cap C)$. *[That is, show $\forall x$, if $x \in (A \cup B) \cap (A \cup C)$ then $x \in A \cup (B \cap C)$.]*

Suppose $x \in (A \cup B) \cap (A \cup C)$. *[Show $x \in A \cup (B \cap C)$.]*
$\vdots$
Thus $x \in A \cup (B \cap C)$.

Hence $(A \cup B) \cap (A \cup C) \subseteq A \cup (B \cap C)$.

Thus $(A \cup B) \cap (A \cup C) = A \cup (B \cap C)$.

Figure 6.2.1

Filling in the missing steps in the top box:
To fill in these steps, you go from the supposition that $x \in A \cup (B \cap C)$ to the conclusion that $x \in (A \cup B) \cap (A \cup C)$.

Now when $x \in A \cup (B \cap C)$, then by definition of union, $x \in A$ or $x \in B \cap C$. But either of these possibilities might be the case because x is assumed to be chosen arbitrarily from the set $A \cup (B \cap C)$. So you have to show you can reach the conclusion that $x \in (A \cup B) \cap (A \cup C)$ regardless of whether x happens to be in A or x happens to be in $B \cap C$. This leads you to break your analysis into two cases: $x \in A$ and $x \in B \cap C$.

In case $x \in A$, your goal is to show that $x \in (A \cup B) \cap (A \cup C)$, which means that $x \in A \cup B$ and $x \in A \cup C$ (by definition of intersection). But when $x \in A$, both statements $x \in A \cup B$ and $x \in A \cup C$ are true by virtue of x's being in A.

Similarly, in case $x \in B \cap C$, your goal is also to show that $x \in (A \cup B) \cap (A \cup C)$, which means that $x \in A \cup B$ and $x \in A \cup C$. But when $x \in B \cap C$, then $x \in B$ and $x \in C$ (by definition of intersection), and so $x \in A \cup B$ (by virtue of being in B) and $x \in A \cup C$ (by virtue of being in C).

This analysis shows that regardless of whether $x \in A$ or $x \in B \cap C$, the conclusion $x \in (A \cup B) \cap (A \cup C)$ follows. So you can fill in the steps in the top inner box.

Filling in the missing steps in the bottom box:

To fill in these steps, you need to go from the supposition that $x \in (A \cup B) \cap (A \cup C)$ to the conclusion that $x \in A \cup (B \cap C)$.

When $x \in (A \cup B) \cap (A \cup C)$ it is natural to consider the two cases $x \in A$ and $x \notin A$ because when x happens to be in A, then the statement "$x \in A$ or $x \in B \cap C$" is certainly true, and so x is in $A \cup (B \cap C)$ by definition of union. Thus it remains only to show that even in the case when x is not in A, and $x \in (A \cup B) \cap (A \cup C)$, then $x \in A \cup (B \cap C)$.

So suppose x is not in A. Now to say that $x \in (A \cup B) \cap (A \cup C)$ means that $x \in A \cup B$ and $x \in A \cup C$ (by definition of intersection). But when $x \in A \cup B$, then x is in at least one of A or B, so since x is not in A, then x must be in B. Similarly, when $x \in A \cup C$, then x is in at least one of A or C, so since x is not in A, then x must be in C. Thus, when x is not in A and $x \in (A \cup B) \cap (A \cup C)$, then x is in both B and C, which means that $x \in B \cap C$. It follows that the statement "$x \in A$ or $x \in B \cap C$" is true, and so $x \in A \cup (B \cap C)$ by definition of union.

This analysis shows that if $x \in (A \cup B) \cap (A \cup C)$, then regardless of whether $x \in A$ or $x \notin A$, you can conclude that $x \in A \cup (B \cap C)$. Hence you can fill in the steps of the bottom inner box.

A formal proof is shown below.

Theorem 6.2.2(3)(a) A Distributive Law for Sets

For all sets A, B, and C,

$$A \cup (B \cap C) = (A \cup B) \cap (A \cup C).$$

Proof:

Suppose A and B are sets.

Proof that $A \cup (B \cap C) \subseteq (A \cup B) \cap (A \cup C)$:

Suppose $x \in A \cup (B \cap C)$. By definition of union, $x \in A$ or $x \in B \cap C$.

Case 1 ($x \in A$): Since $x \in A$, then $x \in A \cup B$ by definition of union and also $x \in A \cup C$ by definition of union. Hence $x \in (A \cup B) \cap (A \cup C)$ by definition of intersection.

Case 2 ($x \in B \cap C$): Since $x \in B \cap C$, then $x \in B$ and $x \in C$ by definition of intersection. Since $x \in B$, $x \in A \cup B$ and since $x \in C$, $x \in A \cup C$, both by definition of union. Hence $x \in (A \cup B) \cap (A \cup C)$ by definition of intersection.

In both cases, $x \in (A \cup B) \cap (A \cup C)$. Hence $A \cup (B \cap C) \subseteq (A \cup B) \cap (A \cup C)$ by definition of subset.

Proof that $(A \cup B) \cap (A \cup C) \subseteq A \cup (B \cap C)$:

Suppose $x \in (A \cup B) \cap (A \cup C)$. By definition of intersection, $x \in A \cup B$ and $x \in A \cup C$. Consider the two cases $x \in A$ and $x \notin A$.

Case 1 ($x \in A$): Since $x \in A$, we can immediately conclude that $x \in A \cup (B \cap C)$ by definition of union.

Case 2 (x ∉ A): Since $x \in A \cup B$, x is in at least one of A or B. But x is not in A; hence x is in B. Similarly, since $x \in A \cup C$, x is in at least one of A or C. But x is not in A; hence x is in C. We have shown that both $x \in B$ and $x \in C$, and so by definition of intersection, $x \in B \cap C$. It follows by definition of union that $x \in A \cup (B \cap C)$.

In both cases $x \in A \cup (B \cap C)$. Hence, by definition of subset, $(A \cup B) \cap (A \cup C) \subseteq A \cup (B \cap C)$.

Conclusion: Since both subset relations have been proved, it follows by definition of set equality that $A \cup (B \cap C) = (A \cup B) \cap (A \cup C)$.

■

In the study of artificial intelligence, the types of reasoning used previously to derive the proof of the distributive law are called *forward chaining* and *backward chaining*. First what is to be shown is viewed as a goal to be reached starting from a certain initial position: the starting point. Analysis of this goal leads to the realization that if a certain job is accomplished, then the goal will be reached. Call this job subgoal 1: SG_1. (For instance, if the goal is to show that $A \cup (B \cap C) = (A \cup B) \cap (A \cup C)$, then SG_1 would be to show that each set is a subset of the other.) Analysis of SG_1 shows that when yet another job is completed, SG_1 will be reached. Call this job subgoal 2: SG_2. Continuing in this way, a chain of argument leading backward from the goal is constructed.

$$\boxed{\text{starting point}} \qquad \rightarrow SG_3 \rightarrow SG_2 \rightarrow SG_1 \rightarrow \boxed{\text{goal}}$$

At a certain point, backward chaining becomes difficult, but analysis of the current subgoal suggests it may be reachable by a direct line of argument, called forward chaining, beginning at the starting point. Using the information contained in the starting point, another piece of information, I_1, is deduced; from that another piece of information, I_2, is deduced; and so forth until finally one of the subgoals is reached. This completes the chain and proves the theorem. A completed chain is illustrated below.

$$\boxed{\text{starting point}} \rightarrow I_1 \rightarrow I_2 \rightarrow I_3 \rightarrow I_4 \rightarrow SG_3 \rightarrow SG_2 \rightarrow SG_1 \rightarrow \boxed{\text{goal}}$$

Since set complement is defined in terms of *not,* and since unions and intersections are defined in terms of *or* and *and,* it is not surprising that there are analogues of De Morgan's laws of logic for sets.

Example 6.2.3 Proof of a De Morgan's Law for Sets

Prove that for all sets A and B, $(A \cup B)^c = A^c \cap B^c$.

Solution   As in previous examples, the statement to be proved is universal, and so the starting point of the proof and the conclusion to be shown are as follows:

Starting Point: Suppose A and B are arbitrarily chosen sets.

To Show: $(A \cup B)^c = A^c \cap B^c$

To do this, you must show that $(A \cup B)^c \subseteq A^c \cap B^c$ and that $A^c \cap B^c \subseteq (A \cup B)^c$. To show the first containment means to show that

$$\forall x, \text{ if } x \in (A \cup B)^c \text{ then } x \in A^c \cap B^c.$$

And to show the second containment means to show that

$$\forall x, \text{ if } x \in A^c \cap B^c \text{ then } x \in (A \cup B)^c.$$

Since each of these statements is universal and conditional, for the first containment, you

suppose $x \in (A \cup B)^c$,

and then you   **show** that $x \in A^c \cap B^c$.

And for the second containment, you

suppose $x \in A^c \cap B^c$,

and then you   **show** that $x \in (A \cup B)^c$.

To fill in the steps of these arguments, you use the procedural versions of the definitions of complement, union, and intersection, and at crucial points you use De Morgan's laws of logic.

Theorem 6.2.2(9)(a) A De Morgan's Law for Sets

For all sets A and B, $(A \cup B)^c = A^c \cap B^c$.

Proof:
Suppose A and B are sets.
Proof that $(A \cup B)^c \subseteq A^c \cap B^c$:
[We must show that $\forall x$, if $x \in (A \cup B)^c$ then $x \in A^c \cap B^c$.]
    Suppose $x \in (A \cup B)^c$. *[We must show that $x \in A^c \cap B^c$.]* By definition of complement,

$$x \notin A \cup B.$$

But to say that $x \notin A \cup B$ means that

it is false that (x is in A or x is in B).

By De Morgan's laws of logic, this implies that

x is not in A and x is not in B,

which can be written $x \notin A$ and $x \notin B$.

Hence $x \in A^c$ and $x \in B^c$ by definition of complement. It follows, by definition of intersection, that $x \in A^c \cap B^c$ *[as was to be shown]*. So $(A \cup B)^c \subseteq A^c \cap B^c$ by definition of subset.
Proof that $A^c \cap B^c \subseteq (A \cup B)^c$:
[We must show that $\forall x$, if $x \in A^c \cap B^c$ then $x \in (A \cup B)^c$.]
    Suppose $x \in A^c \cap B^c$. *[We must show that $x \in (A \cup B)^c$.]* By definition of intersection, $x \in A^c$ and $x \in B^c$, and by definition of complement,

$$x \notin A \quad \text{and} \quad x \notin B.$$

In other words, x is not in A and x is not in B.

By De Morgan's laws of logic this implies that

it is false that (x is in A or x is in B),

which can be written $x \notin A \cup B$

by definition of union. Hence, by definition of complement, $x \in (A \cup B)^c$ *[as was to be shown]*. It follows that $A^c \cap B^c \subseteq (A \cup B)^c$ by definition of subset.

Conclusion: Since both set containments have been proved, $(A \cup B)^c = A^c \cap B^c$ by definition of set equality.

The set property given in the next theorem says that if one set is a subset of another, then their intersection is the smaller of the two sets and their union is the larger of the two sets.

Theorem 6.2.3 Intersection and Union with a Subset

For any sets A and B, if $A \subseteq B$, then

(a) $A \cap B = A$ and   (b) $A \cup B = B$.

Proof:

Part (a): Suppose A and B are sets with $A \subseteq B$. To show part (a) we must show both that $A \cap B \subseteq A$ and that $A \subseteq A \cap B$. We already know that $A \cap B \subseteq A$ by the inclusion of intersection property. To show that $A \subseteq A \cap B$, let $x \in A$. *[We must show that $x \in A \cap B$.]* Since $A \subseteq B$, then $x \in B$ also. Hence

$x \in A$ and $x \in B$,

and thus $x \in A \cap B$

by definition of intersection *[as was to be shown]*.

Proof:

Part (b): The proof of part (b) is left as an exercise.

The Empty Set

In Section 6.1 we introduced the concept of a set with no elements and promised that in this section we would show that there is only one such set. To do so, we start with the most basic—and strangest—property of a set with no elements: It is a subset of every set. To see why this is true, just ask yourself, "Could it possibly be false? Could there be a set without elements that is *not* a subset of some given set?" The crucial fact is that the negation of a universal statement is existential: If a set B is not a subset of a set A, then there exists an element in B that is not in A. But if B has no elements, then no such element can exist.

Theorem 6.2.4 A Set with No Elements Is a Subset of Every Set

If E is a set with no elements and A is any set, then $E \subseteq A$.

Proof (by contradiction):

Suppose not. *[We take the negation of the theorem and suppose it to be true.]* Suppose there exists a set E with no elements and a set A such that $E \nsubseteq A$. *[We must deduce a contradiction.]* Then there would be an element of E that is not an element of A *[by definition of subset]*. But there can be no such element since E has no elements. This is a contradiction. *[Hence the supposition that there are sets E and A, where E has no elements and $E \nsubseteq A$, is false, and so the theorem is true.]*

The truth of Theorem 6.2.4 can also be understood by appeal to the notion of vacuous truth. If E is a set with no elements and A is any set, then to say that $E \subseteq A$ is the same as saying that

$$\forall x, \text{ if } x \in E, \text{ then } x \in A.$$

But since E has no elements, this conditional statement is vacuously true.

How many sets with no elements are there? Only one.

Corollary 6.2.5 Uniqueness of the Empty Set

There is only one set with no elements.

Proof:

Suppose E_1 and E_2 are both sets with no elements. By Theorem 6.2.4, $E_1 \subseteq E_2$ since E_1 has no elements. Also $E_2 \subseteq E_1$ since E_2 has no elements. Thus $E_1 = E_2$ by definition of set equality.

It follows from Corollary 6.2.5 that the set of pink elephants is equal to the set of all real numbers whose square is -1 because each set has no elements! Since there is only one set with no elements, we are justified in calling it by a special name, the empty set (or null set) and in denoting it by the special symbol $\emptyset$.

Note that whereas $\emptyset$ is the set with no elements, the set $\{\emptyset\}$ has one element, the empty set. This is similar to the convention in the computer programming languages LISP and Scheme, in which ( ) denotes the empty list and (( )) denotes the list whose one element is the empty list.

Suppose you need to show that a certain set equals the empty set. By Corollary 6.2.5 it suffices to show that the set has no elements. For since there is only one set with no elements (namely $\emptyset$), if the given set has no elements, then it must equal $\emptyset$.

Element Method for Proving a Set Equals the Empty Set

To prove that a set X is equal to the empty set $\emptyset$, prove that X has no elements. To do this, suppose X has an element and derive a contradiction.

Example 6.2.4 Proving That a Set Is Empty

Prove Theorem 6.2.2(8)(b). That is, prove that for any set A, $A \cap \emptyset = \emptyset$.

Solution Let A be a *[particular, but arbitrarily chosen]* set. To show that $A \cap \emptyset = \emptyset$, it suffices to show that $A \cap \emptyset$ has no elements *[by the element method for proving a set equals the empty set]*. Suppose not. That is, suppose there is an element x such that $x \in A \cap \emptyset$. Then, by definition of intersection, $x \in A$ and $x \in \emptyset$. In particular, $x \in \emptyset$. But this is impossible since $\emptyset$ has no elements. *[This contradiction shows that the supposition that there is an element x in $A \cap \emptyset$ is false. So $A \cap \emptyset$ has no elements, as was to be shown.]* Thus $A \cap \emptyset = \emptyset$. ∎

Example 6.2.5 A Proof for a Conditional Statement

Prove that for all sets A, B, and C, if $A \subseteq B$ and $B \subseteq C^c$, then $A \cap C = \emptyset$.

Solution Since the statement to be proved is both universal and conditional, you start with the method of direct proof:

> **Suppose** A, B, and C are arbitrarily chosen sets
> that satisfy the condition: $A \subseteq B$ and $B \subseteq C^c$.
>
> **Show** that $A \cap C = \emptyset$.

Since the conclusion to be shown is that a certain set is empty, you can use the principle for proving that a set equals the empty set. A complete proof is shown below.

Proposition 6.2.6

For all sets A, B, and C, if $A \subseteq B$ and $B \subseteq C^c$, then $A \cap C = \emptyset$.

Proof:

Suppose A, B, and C are any sets such that $A \subseteq B$ and $B \subseteq C^c$. We must show that $A \cap C = \emptyset$. Suppose not. That is, suppose there is an element x in $A \cap C$. By definition of intersection, $x \in A$ and $x \in C$. Then, since $A \subseteq B$, $x \in B$ by definition of subset. Also, since $B \subseteq C^c$, then $x \in C^c$ by definition of subset again. It follows by definition of complement that $x \notin C$. Thus $x \in C$ and $x \notin C$, which is a contradiction. So the supposition that there is an element x in $A \cap C$ is false, and thus $A \cap C = \emptyset$ *[as was to be shown]*.

∎

Example 6.2.6 A Generalized Distributive Law

Prove that for all sets A and B_1, B_2, B_3, ..., B_n,

$$A \cup \left(\bigcap_{i=1}^{n} B_i \right) = \bigcap_{i=1}^{n} (A \cup B_i).$$

Solution Compare this proof to the one given in Example 6.2.2. Although the notation is more complex, the basic ideas are the same.

Proof:

Suppose A and B_1, B_2, B_3, ..., B_n are any sets.

Part 1, Proof that $A \cup \left(\bigcap_{i=1}^{n} B_i \right) \subseteq \bigcap_{i=1}^{n} (A \cup B_i)$:

Suppose x is any element in $A \cup \left(\bigcap_{i=1}^{n} B_i \right)$. *[We must show that x is in $\bigcap_{i=1}^{n} (A \cup B_i)$.]*

By definition of union, $x \in A$ or $x \in \bigcap_{i=1}^{n} B_i$.

Case 1, $x \in A$: In this case, it is true by definition of union that for all $i = 1, 2, \ldots, n, x \in A \cup B_i$. Hence $x \in \bigcap_{i=1}^{n} (A \cup B_i)$.

Case 2, $x \in \bigcap_{i=1}^{n} B_i$: In this case, by definition of the general intersection, we have that for all integers $i = 1, 2, \ldots, n, x \in B_i$. Hence, by definition of union, for all integers $i = 1, 2, \ldots, n, x \in A \cup B_i$, and so, by definition of general intersection, $x \in \bigcap_{i=1}^{n} (A \cup B_i)$.

Thus, in either case, $x \in \bigcap_{i=1}^{n} (A \cup B_i)$ *[as was to be shown]*.

Part 2, Proof that $\bigcap_{i=1}^{n} (A \cup B_i) \subseteq A \cup \left(\bigcap_{i=1}^{n} B_i \right)$:

Suppose x is any element in $\bigcap_{i=1}^{n} (A \cup B_i)$. *[We must show that x is in $A \cup \left(\bigcap_{i=1}^{n} B_i \right)$.]*

By definition of intersection, $x \in A \cup B_i$ for all integers $i = 1, 2, \ldots, n$. Either $x \in A$ or $x \notin A$.

Case 1, $x \in A$: In this case, $x \in A \cup \left(\bigcap_{i=1}^{n} B_i \right)$ by definition of union.

Case 2, $x \notin A$: By definition of intersection, $x \in A \cup B_i$ for all integers $i = 1, 2, \ldots, n$. Since $x \notin A$, x must be in each B_i for every integer $i = 1, 2, \ldots, n$. Hence, by definition of intersection, $x \in \bigcap_{i=1}^{n} B_i$, and so, by definition of union, $x \in A \cup \left(\bigcap_{i=1}^{n} B_i \right)$.

Conclusion: Since both set containments have been proved, it follows by definition of set equality that $A \cup \left(\bigcap_{i=1}^{n} B_i \right) = \bigcap_{i=1}^{n} (A \cup B_i)$. ∎

Test Yourself

1. To prove that a set X is a subset of a set $A \cap B$, you suppose that x is any element of X and you show that $x \in A$ _____ $x \in B$.

2. To prove that a set X is a subset of a set $A \cup B$, you suppose that x is any element of X and you show that $x \in A$ _____ $x \in B$.

3. To prove that a set $A \cup B$ is a subset of a set X, you start with any element x in $A \cup B$ and consider the two cases _____ and _____. You then show that in either case _____.

4. To prove that a set $A \cap B$ is a subset of a set X, you suppose that _____ and you show that _____.

5. To prove that a set X equals a set Y, you prove that _____ and that _____.

6. To prove that a set X does not equal a set Y, you need to find an element that is in _____ and not _____ or that is in _____ and not _____.

Exercise Set 6.2

1. **a.** To say that an element is in $A \cap (B \cup C)$ means that it is in _(1)_ and in _(2)_.
   b. To say that an element is in $(A \cap B) \cup C$ means that it is in _(1)_ or in _(2)_.
   c. To say that an element is in $A - (B \cap C)$ means that it is in _(1)_ and not in _(2)_.

2. The following are two proofs that for all sets A and B, $A - B \subseteq A$. The first is less formal, and the second is more formal. Fill in the blanks.
   a. Proof: Suppose A and B are any sets. To show that $A - B \subseteq A$, we must show that every element in _(1)_ is in _(2)_. But any element in $A - B$ is in _(3)_ and not

in __(4)__ (by definition of $A - B$). In particular, such an element is in A.

b. **Proof:** Suppose A and B are any sets and $x \in A - B$. *[We must show that __(1)__ .]* By definition of set difference, $x \in$ __(2)__ and $x \notin$ __(3)__ . In particular, $x \in$ __(4)__ *[which is what was to be shown].*

3. The following is a proof that for all sets A, B, and C, if $A \subseteq B$ and $B \subseteq C$, then $A \subseteq C$. Fill in the blanks.

   Proof: Suppose A, B, and C are sets and $A \subseteq B$ and $B \subseteq C$. To show that $A \subseteq C$, we must show that every element in __(a)__ is in __(b)__ . But given any element in A, that element is in __(c)__ (because $A \subseteq B$), and so that element is also in __(d)__ (because __(e)__). Hence $A \subseteq C$.

4. The following is a proof that for all sets A and B, if $A \subseteq B$, then $A \cup B \subseteq B$. Fill in the blanks.

   Proof: Suppose A and B are any sets and $A \subseteq B$. *[We must show that __(a)__ .]* Let $x \in$ __(b)__ . *[We must show that __(c)__ .]* By definition of union, $x \in$ __(d)__ __(e)__ $x \in$ __(f)__ . In case $x \in$ __(g)__ , then since $A \subseteq B$, $x \in$ __(h)__ . In case $x \in B$, then clearly $x \in B$. So in either case, $x \in$ __(i)__ *[as was to be shown].*

5. Prove that for all sets A and B, $(B - A) = B \cap A^c$.

H 6. The following is a proof that for any sets A, B, and C, $A \cap (B \cup C) = (A \cap B) \cup (A \cap C)$. Fill in the blanks.

   Proof: Suppose A, B, and C are any sets.

   (1) *Proof that $A \cap (B \cup C) \subseteq (A \cap B) \cup (A \cap C)$:*
   Let $x \in A \cap (B \cup C)$. *[We must show that $x \in$ __(a)__ .]* By definition of intersection, $x \in$ __(b)__ and $x \in$ __(c)__ . Thus $x \in A$ and, by definition of union, $x \in B$ or __(d)__ .

   Case 1 ($x \in A$ and $x \in B$): In this case, by definition of intersection, $x \in$ __(e)__ , and so, by definition of union, $x \in (A \cap B) \cup (A \cap C)$.

   Case 2 ($x \in A$ and $x \in C$): In this case, __(f)__ . Hence in either case, $x \in (A \cap B) \cup (A \cap C)$ *[as was to be shown].*
   [So $A \cap (B \cup C) \subseteq (A \cap B) \cup (A \cap C)$ by definition of subset.]

   (2) $(A \cap B) \cup (A \cap C) \subseteq A \cap (B \cup C)$:
   Let $x \in (A \cap B) \cup (A \cap C)$. *[We must show that __(a)__ .]* By definition of union, $x \in A \cap B$ __(b)__ $x \in A \cap C$.

   Case 1 ($x \in A \cap B$): In this case, by definition of intersection, $x \in A$ __(c)__ $x \in B$. Since $x \in B$, then by definition of union, $x \in B \cup C$. Hence $x \in A$ and $x \in B \cup C$, and so, by definition of intersection, $x \in$ __(d)__ .

   Case 2 ($x \in A \cap C$): In this case, __(e)__ .
   In either case, $x \in A \cap (B \cup C)$ *[as was to be shown]. [Thus $(A \cap B) \cup (A \cap C) \subseteq A \cap (B \cup C)$ by definition of subset.]*

   (3) *Conclusion:* *[Since both subset relations have been proved, it follows, by definition of set equality, that __(a)__ .]*

Use an element argument to prove each statement in 7–19. Assume that all sets are subsets of a universal set U.

H 7. For all sets A and B, $(A \cap B)^c = A^c \cup B^c$.

8. For all sets A and B, $(A \cap B) \cup (A \cap B^c) = A$.

H 9. For all sets A, B, and C,
$$(A - B) \cup (C - B) = (A \cup C) - B.$$

10. For all sets A, B, and C,
$$(A - B) \cap (C - B) = (A \cap C) - B.$$

H 11. For all sets A and B, $A \cup (A \cap B) = A$.

12. For all sets A, $A \cup \emptyset = A$.

13. For all sets A, B, and C, if $A \subseteq B$ then $A \cap C \subseteq B \cap C$.

14. For all sets A, B, and C, if $A \subseteq B$ then $A \cup C \subseteq B \cup C$.

15. For all sets A and B, if $A \subseteq B$ then $B^c \subseteq A^c$.

H 16. For all sets A, B, and C, if $A \subseteq B$ and $A \subseteq C$ then $A \subseteq B \cap C$.

17. For all sets A, B, and C, if $A \subseteq C$ and $B \subseteq C$ then $A \cup B \subseteq C$.

18. For all sets A, B, and C,
$$A \times (B \cup C) = (A \times B) \cup (A \times C).$$

19. For all sets A, B, and C,
$$A \times (B \cap C) = (A \times B) \cap (A \times C).$$

20. Find the mistake in the following "proof" that for all sets A, B, and C, if $A \subseteq B$ and $B \subseteq C$ then $A \subseteq C$.

   "**Proof:** Suppose A, B, and C are sets such that $A \subseteq B$ and $B \subseteq C$. Since $A \subseteq B$, there is an element x such that $x \in A$ and $x \in B$. Since $B \subseteq C$, there is an element x such that $x \in B$ and $x \in C$. Hence there is an element x such that $x \in A$ and $x \in C$ and so $A \subseteq C$."

H 21. Find the mistake in the following "proof."

   "**Theorem:**" For all sets A and B, $A^c \cup B^c \subseteq (A \cup B)^c$.

   "**Proof:** Suppose A and B are sets, and $x \in A^c \cup B^c$. Then $x \in A^c$ or $x \in B^c$ by definition of union. It follows that $x \notin A$ or $x \notin B$ by definition of complement, and so $x \notin A \cup B$ by definition of union. Thus $x \in (A \cup B)^c$ by definition of complement, and hence $A^c \cup B^c \subseteq (A \cup B)^c$."

22. Find the mistake in the following "proof" that for all sets A and B, $(A - B) \cup (A \cap B) \subseteq A$.

   "**Proof:** Suppose A and B are sets, and suppose $x \in (A - B) \cup (A \cap B)$. If $x \in A$ then $x \in A - B$. Then, by definition of difference, $x \in A$ and $x \notin B$. Hence $x \in A$, and so $(A - B) \cup (A \cap B) \subseteq A$ by definition of subset."

23. Consider the Venn diagram below.

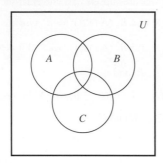

a. Illustrate one of the distributive laws by shading in the region corresponding to $A \cup (B \cap C)$ on one copy of the diagram and $(A \cup B) \cap (A \cup C)$ on another.

b. Illustrate the other distributive law by shading in the region corresponding to $A \cap (B \cup C)$ on one copy of the diagram and $(A \cap B) \cup (A \cap C)$ on another.

c. Illustrate one of De Morgan's laws by shading in the region corresponding to $(A \cup B)^c$ on one copy of the diagram and $A^c \cap B^c$ on the other. (Leave the set C out of your diagrams.)

d. Illustrate the other De Morgan's law by shading in the region corresponding to $(A \cap B)^c$ on one copy of the diagram and $A^c \cup B^c$ on the other. (Leave the set C out of your diagrams.)

24. Fill in the blanks in the following proof that for all sets A and B, $(A - B) \cap (B - A) = \emptyset$.

Proof: Let A and B be any sets and supppose $(A - B) \cap (B - A) \neq \emptyset$. That is, suppose there were an element x in _(a)_. By definition of _(b)_, $x \in A - B$ and $x \in$ _(c)_. Then by definition of set difference, $x \in A$ and $x \notin B$ and $x \in$ _(d)_ and $x \notin$ _(e)_. In particular $x \in A$ and $x \notin$ _(f)_, which is a contradiction. Hence *[the supposition that $(A - B) \cap (B - A) \neq \emptyset$ is false, and so]* _(g)_.

Use the element method for proving a set equals the empty set to prove each statement in 25–35. Assume that all sets are subsets of a universal set U.

25. For all sets A and B, $(A \cap B) \cap (A \cap B^c) = \emptyset$.

26. For all sets A, B, and C,

$$(A - C) \cap (B - C) \cap (A - B) = \emptyset.$$

27. For all subsets A of a universal set U, $A \cap A^c = \emptyset$.

28. If U denotes a universal set, then $U^c = \emptyset$.

29. For all sets A, $A \times \emptyset = \emptyset$.

30. For all sets A and B, if $A \subseteq B$ then $A \cap B^c = \emptyset$.

31. For all sets A and B, if $B \subseteq A^c$ then $A \cap B = \emptyset$.

32. For all sets A, B, and C, if $A \subseteq B$ and $B \cap C = \emptyset$ then $A \cap C = \emptyset$.

33. For all sets A, B, and C, if $C \subseteq B - A$, then $A \cap C = \emptyset$.

34. For all sets A, B, and C,

$$\text{if } B \cap C \subseteq A, \text{ then } (C - A) \cap (B - A) = \emptyset.$$

35. For all sets A, B, C, and D,

$$\text{if } A \cap C = \emptyset \text{ then } (A \times B) \cap (C \times D) = \emptyset.$$

Prove each statement in 36–41.

H 36. For all sets A and B,
a. $(A - B) \cup (B - A) \cup (A \cap B) = A \cup B$
b. The sets $(A - B)$, $(B - A)$, and $(A \cap B)$ are mutually disjoint.

37. For all integers $n \geq 1$, if A and $B_1, B_2, B_3, \ldots$ are any sets, then

$$A \cap \left(\bigcup_{i=1}^{n} B_i \right) = \bigcup_{i=1}^{n} (A \cap B_i).$$

H 38. For all integers $n \geq 1$, if $A_1, A_2, A_3, \ldots$ and B are any sets, then

$$\bigcup_{i=1}^{n} (A_i - B) = \left(\bigcup_{i=1}^{n} A_i \right) - B.$$

39. For all integers $n \geq 1$, if $A_1, A_2, A_3, \ldots$ and B are any sets, then

$$\bigcap_{i=1}^{n} (A_i - B) = \left(\bigcap_{i=1}^{n} A_i \right) - B.$$

40. For all integers $n \geq 1$, if A and $B_1, B_2, B_3, \ldots$ are any sets, then

$$\bigcup_{i=1}^{n} (A \times B_i) = A \times \left(\bigcup_{i=1}^{n} B_i \right).$$

41. For all integers $n \geq 1$, if A and $B_1, B_2, B_3, \ldots$ are any sets, then

$$\bigcap_{i=1}^{n} (A \times B_i) = A \times \left(\bigcap_{i=1}^{n} B_i \right).$$

Answers for Test Yourself

1. and   2. or   3. $x \in A$; $x \in B$; $x \in X$ 4. $x \in A \cap B$ *(Or: x is an element of both A and B)*; $x \in X$ 5. $X \subseteq Y$; $Y \subseteq X$ 6. X; in Y; Y; in X

6.3 Disproofs and Algebraic Proofs

If a fact goes against common sense, and we are nevertheless compelled to accept and deal with this fact, we learn to alter our notion of common sense.
—Phillip J. Davis and Reuben Hersh, *The Mathematical Experience*, 1981

In Section 6.2 we gave examples only of set properties that were true. Occasionally, however, a proposed set property is false. We begin this section by discussing how to disprove such a proposed property. Then we prove an important theorem about the power set of a set and go on to discuss an "algebraic" method for deriving new set properties from set properties already known to be true. We finish the section with an introduction to Boolean algebras.

Disproving an Alleged Set Property

Recall that to show a universal statement is false, it suffices to find one example (called a counterexample) for which it is false.

Example 6.3.1 Finding a Counterexample for a Set Identity

Is the following set property true?

$$\text{For all sets } A, B, \text{ and } C, (A - B) \cup (B - C) = A - C.$$

Solution   Observe that the property is true if, and only if,

the given equality holds for *all* sets A, B, and C.

So it is false if, and only if,

there are sets A, B, and C for which the equality does *not* hold.

One way to solve this problem is to picture sets A, B, and C by drawing a Venn diagram such as that shown in Figure 6.3.1. If you assume that any of the eight regions of the diagram may be empty of points, then the diagram is quite general.

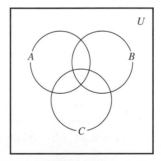

Figure 6.3.1

Find and shade the region corresponding to $(A - B) \cup (B - C)$. Then shade the region corresponding to $A - C$. These are shown in Figure 6.3.2 on the next page.

Comparing the shaded regions seems to indicate that the property is false. For instance, if there is an element in B that is not in either A or C then this element would be in $(A - B) \cup (B - C)$ (because of being in B and not C) but it would not be in $A - C$ since $A - C$ contains nothing outside A. Similarly, an element that is in both A and C but not B would be in $(A - B) \cup (B - C)$ (because of being in A and not B), but it would not be in $A - C$ (because of being in both A and C).

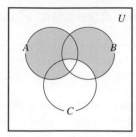 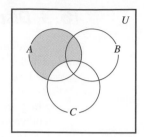

Figure 6.3.2

Construct a concrete counterexample in order to confirm your answer and make sure that you did not make a mistake either in drawing or analyzing your diagrams. One way is to put one of the integers from 1–7 into each of the seven subregions enclosed by the circles representing A, B, and C. If the proposed set property had involved set complements, it would also be helpful to label the region outside the circles, and so we place the number 8 there. (See Figure 6.3.3.) Then define discrete sets A, B, and C to consist of all the numbers in their respective subregions.

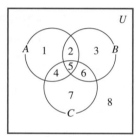

Figure 6.3.3

> **Counterexample 1:** Let $A = \{1, 2, 4, 5\}$, $B = \{2, 3, 5, 6\}$, and $C = \{4, 5, 6, 7\}$. Then
>
> $$A - B = \{1, 4\}, \quad B - C = \{2, 3\}, \quad \text{and} \quad A - C = \{1, 2\}.$$
>
> Hence
>
> $$(A - B) \cup (B - C) = \{1, 4\} \cup \{2, 3\} = \{1, 2, 3, 4\}, \quad \text{whereas} \quad A - C = \{1, 2\}.$$
>
> Since $\{1, 2, 3, 4\} \neq \{1, 2\}$, we have that $(A - B) \cup (B - C) \neq A - C$.

A more economical counterexample can be obtained by observing that as long as the set B contains an element, such as 3, that is not in either A or C, then regardless of whether B contains any other elements and regardless of whether A and C contain any elements at all, $(A - B) \cup (B - C) \neq A - C$.

> **Counterexample 2:** Let $A = \emptyset$, $B = \{3\}$, and $C = \emptyset$. Then
>
> $$A - B = \emptyset, \quad B - C = \{3\}, \quad \text{and} \quad A - C = \emptyset.$$
>
> Hence $\quad (A - B) \cup (B - C) = \emptyset \cup \{3\} = \{3\}, \quad$ whereas $\quad A - C = \emptyset$.
>
> Since $\{3\} \neq \emptyset$, we have that $(A - B) \cup (B - C) \neq A - C$.

Note Check that when $A = C = \{4\}$ and $B = \emptyset$, $(A - B) \cup (B - C) \neq A - C$.

Another economical counterexample requires only that $A = C = $ a singleton set, such as $\{4\}$, while B is the empty set.

Problem-Solving Strategy

How can you discover whether a given universal statement about sets is true or false? There are two basic approaches: the optimistic and the pessimistic. In the optimistic approach, you simply plunge in and start trying to prove the statement, asking yourself, "What do I need to show?" and "How do I show it?" In the pessimistic approach, you start by searching your mind for a set of conditions that must be fulfilled to construct a counterexample. With either approach you may have clear sailing and be immediately successful or you may run into difficulty. The trick is to be ready to switch to the other approach if the one you are trying does not look promising. For more difficult questions, you may alternate several times between the two approaches before arriving at the correct answer.

The Number of Subsets of a Set

The following theorem states the important fact that if a set has n elements, then its power set has 2^n elements. The proof uses mathematical induction and is based on the following observations. Suppose X is a set and z is an element of X.

1. The subsets of X can be split into two groups: those that do not contain z and those that do contain z.

2. The subsets of X that do not contain z are the same as the subsets of $X - \{z\}$.

3. The subsets of X that do not contain z can be matched up one for one with the subsets of X that do contain z by matching each subset A that does not contain z to the subset $A \cup \{z\}$ that contains z. Thus there are as many subsets of X that contain z as there are subsets of X that do not contain z. For instance, if $X = \{x, y, z\}$, the following table shows the correspondence between subsets of X that do not contain z and subsets of X that contain z.

Subsets of X That Do Not Contain z		Subsets of X That Contain z
$\emptyset$	$\longleftrightarrow$	$\emptyset \cup \{z\} = \{z\}$
$\{x\}$	$\longleftrightarrow$	$\{x\} \cup \{z\} = \{x, z\}$
$\{y\}$	$\longleftrightarrow$	$\{y\} \cup \{z\} = \{y, z\}$
$\{x, y\}$	$\longleftrightarrow$	$\{x, y\} \cup \{z\} = \{x, y, z\}$

Theorem 6.3.1

For all integers $n \geq 0$, if a set X has n elements, then $\mathscr{P}(X)$ has 2^n elements.

Proof (by mathematical induction):

Let the property $P(n)$ be the sentence

$$\text{Any set with } n \text{ elements has } 2^n \text{ subsets.} \qquad \leftarrow P(n)$$

Show that P(0) is true:
To establish $P(0)$, we must show that

$$\text{Any set with 0 elements has } 2^0 \text{ subsets.} \qquad \leftarrow P(0)$$

continued on page 282

But the only set with zero elements is the empty set, and the only subset of the empty set is itself. Thus a set with zero elements has one subset. Since $1 = 2^0$, we have that $P(0)$ is true.

Show that for all integers $k \geq 0$, if $P(k)$ is true then $P(k + 1)$ is also true:
[Suppose that $P(k)$ is true for a particular but arbitrarily chosen integer $k \geq 0$. That is:] Suppose that k is any integer with $k \geq 0$ such that

Any set with k elements has 2^k subsets.

$\leftarrow P(k)$
inductive hypothesis

[We must show that $P(k + 1)$ is true. That is:] We must show that

Any set with $k + 1$ elements has 2^{k+1} subsets. $\leftarrow P(k + 1)$

Let X be a set with $k + 1$ elements. Since $k + 1 \geq 1$, we may pick an element z in X. Observe that any subset of X either contains z or not. Furthermore, any subset of X that does not contain z is a subset of $X - \{z\}$. And any subset A of $X - \{z\}$ can be matched up with a subset B, equal to $A \cup \{z\}$, of X that contains z. Consequently, there are as many subsets of X that contain z as do not, and thus there are twice as many subsets of X as there are subsets of $X - \{z\}$. But $X - \{z\}$ has k elements, and so

the number of subsets of $X - \{z\} = 2^k$ by inductive hypothesis.

Therefore,

the number of subsets of $X = 2 \cdot$ (the number of subsets of $X - \{z\}$)
$= 2 \cdot (2^k)$ by substitution
$= 2^{k+1}$ by basic algebra.

[This is what was to be shown.]
[Since we have proved both the basis step and the inductive step, we conclude that the theorem is true.]

"Algebraic" Proofs of Set Identities

Let U be a universal set and consider the power set of U, $\mathscr{P}(U)$. The set identities given in Theorem 6.2.2 hold for all elements of $\mathscr{P}(U)$. Once a certain number of identities and other properties have been established, new properties can be derived from them algebraically without having to use element method arguments. It turns out that only identities (1–5) of Theorem 6.2.2 are needed to prove any other identity involving only unions, intersections, and complements. With the addition of identity (12), the set difference law, any set identity involving unions, intersections, complements, and set differences can be established.

To use known properties to derive new ones, you need to use the fact that such properties are universal statements. Like the laws of algebra for real numbers, they apply to a wide variety of different situations. Assume that all sets are subsets of $\mathscr{P}(U)$, then, for instance, one of the distributive laws states that

for all sets A, B, and C, $A \cap (B \cup C) = (A \cap B) \cup (A \cap C)$.

This law can be viewed as a general template into which *any* three particular sets can be placed. Thus, for example, if A_1, A_2, and A_3 represent particular sets, then

$$\underbrace{A_1}_{A} \cap (\underbrace{A_2}_{(B} \cup \underbrace{A_3}_{\cup~~~C)})=(\underbrace{A_1}_{(A} \cap \underbrace{A_2}_{\cap~~~B)}) \cup (\underbrace{A_1}_{(A} \cap \underbrace{A_3}_{\cap~~~C)}),$$

where A_1 plays the role of A, A_2 plays the role of B, and A_3 plays the role of C. Similarly, if W, X, Y, and Z are any particular sets, then, by the distributive law,

$$(\underbrace{W \cap X}_{A}) \cap (\underbrace{Y}_{\cap~(B} \cup \underbrace{Z}_{\cup~~Z)}) = ((\underbrace{W \cap X}_{(A}) \cap \underbrace{Y}_{\cap~B)}) \cup ((\underbrace{W \cap X}_{(A}) \cap \underbrace{Z}_{\cap~C)}),$$

where $W \cap X$ plays the role of A, Y plays the role of B, and Z plays the role of C.

Example 6.3.2 Deriving a Set Difference Property

Construct an algebraic proof that for all sets A, B, and C,

$$(A \cup B) - C = (A - C) \cup (B - C).$$

Cite a property from Theorem 6.2.2 for every step of the proof.

Solution Let A, B, and C be any sets. Then

$$
\begin{aligned}
(A \cup B) - C &= (A \cup B) \cap C^c && \text{by the set difference law} \\
&= C^c \cap (A \cup B) && \text{by the commutative law for } \cap \\
&= (C^c \cap A) \cup (C^c \cap B) && \text{by the distributive law} \\
&= (A \cap C^c) \cup (B \cap C^c) && \text{by the commutative law for } \cap \\
&= (A - C) \cup (B - C) && \text{by the set difference law.}
\end{aligned}
$$

∎

Example 6.3.3 Deriving a Set Identity Using Properties of ∅

Construct an algebraic proof that for all sets A and B,

$$A - (A \cap B) = A - B.$$

Cite a property from Theorem 6.2.2 for every step of the proof.

Solution Suppose A and B are any sets. Then

$$
\begin{aligned}
A - (A \cap B) &= A \cap (A \cap B)^c && \text{by the set difference law} \\
&= A \cap (A^c \cup B^c) && \text{by De Morgan's laws} \\
&= (A \cap A^c) \cup (A \cap B^c) && \text{by the distributive law} \\
&= \emptyset \cup (A \cap B^c) && \text{by the complement law} \\
&= (A \cap B^c) \cup \emptyset && \text{by the commutative law for } \cup \\
&= A \cap B^c && \text{by the identity law for } \cup \\
&= A - B && \text{by the set difference law.}
\end{aligned}
$$

∎

To many people an algebraic proof seems more attractive than an element proof, but often an element proof is actually simpler. For instance, in Example 6.3.3 above, you could see immediately that $A - (A \cap B) = A - B$ because for an element to be in $A - (A \cap B)$ means that it is in A and not in both A and B, and this is equivalent to saying that it is in A and not in B.

Example 6.3.4 Deriving a Generalized Associative Law

Prove that for any sets A_1, A_2, A_3, and A_4,

$$((A_1 \cup A_2) \cup A_3) \cup A_4 = A_1 \cup ((A_2 \cup A_3) \cup A_4).$$

Cite a property from Theorem 6.2.2 for every step of the proof.

Solution Let A_1, A_2, A_3, and A_4 be any sets. Then

$$((A_1 \cup A_2) \cup A_3) \cup A_4 = (A_1 \cup (A_2 \cup A_3)) \cup A_4$$

by the associative law for $\cup$ with A_1 playing the role of A, A_2 playing the role of B, and A_3 playing the role of C

$$= A_1 \cup ((A_2 \cup A_3) \cup A_4)$$

by the associative law for $\cup$ with A_1 playing the role of A, $A_2 \cup A_3$ playing the role of B, and A_4 playing the role of C. ■

Test Yourself

1. Given a proposed set identity involving set variables A, B, and C, the most common way to show that the equation does not hold in general is to find concrete sets A, B, and C that, when substituted for the set variables in the equation, _____.

2. When using the algebraic method for proving a set identity, it is important to _____ for every step.

3. When applying a property from Theorem 6.2.2, it must be used _____ as it is stated.

Exercise Set 6.3

For each of 1–4 find a counterexample to show that the statement is false. Assume all sets are subsets of a universal set U.

1. For all sets A, B, and C, $(A \cap B) \cup C = A \cap (B \cup C)$.

2. For all sets A and B, $(A \cup B)^c = A^c \cup B^c$.

3. For all sets A, B, and C, if $A \nsubseteq B$ and $B \nsubseteq C$ then $A \nsubseteq C$.

4. For all sets A, B, and C, if $B \cap C \subseteq A$ then
$(A - B) \cap (A - C) = \emptyset$.

For each of 5–21 prove each statement that is true and find a counterexample for each statement that is false. Assume all sets are subsets of a universal set U.

5. For all sets A, B, and C, $A - (B - C) = (A - B) - C$.

6. For all sets A and B, $A \cap (A \cup B) = A$.

7. For all sets A, B, and C,

$$(A - B) \cap (C - B) = A - (B \cup C).$$

8. For all sets A and B, if $A^c \subseteq B$ then $A \cup B = U$.

9. For all sets A, B, and C, if $A \subseteq C$ and $B \subseteq C$ then $A \cup B \subseteq C$.

10. For all sets A and B, if $A \subseteq B$ then $A \cap B^c = \emptyset$.

H 11. For all sets A, B, and C, if $A \subseteq B$ then $A \cap (B \cap C)^c = \emptyset$.

H 12. For all sets A, B, and C,

$$A \cap (B - C) = (A \cap B) - (A \cap C).$$

13. For all sets A, B, and C,

$$A \cup (B - C) = (A \cup B) - (A \cup C).$$

H 14. For all sets A, B, and C, if $A \cap C \subseteq B \cap C$ and $A \cup C \subseteq B \cup C$, then $A \subseteq B$.

H 15. For all sets A, B, and C, if $A \cap C = B \cap C$ and $A \cup C = B \cup C$, then $A = B$.

16. For all sets A and B, if $A \cap B = \emptyset$ then $A \times B = \emptyset$.

17. For all sets A and B, if $A \subseteq B$ then $\mathscr{P}(A) \subseteq \mathscr{P}(B)$.

18. For all sets A and B, $\mathscr{P}(A \cup B) \subseteq \mathscr{P}(A) \cup \mathscr{P}(B)$.

H 19. For all sets A and B, $\mathscr{P}(A) \cup \mathscr{P}(B) \subseteq \mathscr{P}(A \cup B)$.

20. For all sets A and B, $\mathscr{P}(A \cap B) = \mathscr{P}(A) \cap \mathscr{P}(B)$.

21. For all sets A and B, $\mathscr{P}(A \times B) = \mathscr{P}(A) \times \mathscr{P}(B)$.

22. Write a negation for each of the following statements. Indicate which is true, the statement or its negation. Justify your answers.
   a. $\forall$ sets S, $\exists$ a set T such that $S \cap T = \emptyset$.
   b. $\exists$ a set S such that $\forall$ sets T, $S \cup T = \emptyset$.

H 23. Let $S = \{a, b, c\}$ and for each integer $i = 0, 1, 2, 3$, let S_i be the set of all subsets of S that have i elements. List the elements in S_0, S_1, S_2, and S_3. Is $\{S_0, S_1, S_2, S_3\}$ a partition of $\mathscr{P}(S)$?

24. Let $S = \{a, b, c\}$ and let S_a be the set of all subsets of S that contain a, let S_b be the set of all subsets of S that contain b, let S_c be the set of all subsets of S that contain c, and let $S_\emptyset$ be the set whose only element is $\emptyset$. Is $\{S_a, S_b, S_c, S_\emptyset\}$ a partition of $\mathscr{P}(S)$?

25. Let $A = \{t, u, v, w\}$ and let S_1 be the set of all subsets of A that do not contain w and S_2 the set of all subsets of A that contain w.
    a. Find S_1.   b. Find S_2.   c. Are S_1 and S_2 disjoint?
    d. Compare the sizes of S_1 and S_2.
    e. How many elements are in $S_1 \cup S_2$?
    f. What is the relation between $S_1 \cup S_2$ and $\mathscr{P}(A)$?

H✶ 26. The following problem, devised by Ginger Bolton, appeared in the January 1989 issue of the *College Mathematics Journal* (Vol. 20, No. 1, p. 68): Given a positive integer $n \geq 2$, let S be the set of all nonempty subsets of $\{2, 3, \ldots, n\}$. For each $S_i \in S$, let P_i be the product of the elements of S_i. Prove or disprove that

$$\sum_{i=1}^{2^{n-1}-1} P_i = \frac{(n+1)!}{2} - 1.$$

In 27 and 28 supply a reason for each step in the derivation.

27. For all sets A, B, and C,

$$(A \cup B) \cap C = (A \cap C) \cup (B \cap C).$$

Proof: Suppose A, B, and C are any sets. Then

$$\begin{aligned}
(A \cup B) \cap C &= C \cap (A \cup B) & \text{by } \underline{\quad(a)\quad} \\
&= (C \cap A) \cup (C \cap B) & \text{by } \underline{\quad(b)\quad} \\
&= (A \cap C) \cup (B \cap C) & \text{by } \underline{\quad(c)\quad}.
\end{aligned}$$

H 28. For all sets A, B, and C,
$$(A \cup B) - (C - A) = A \cup (B - C).$$

Proof: Suppose A, B, and C are any sets. Then

$$\begin{aligned}
(A \cup B) - (C - A) &= (A \cup B) \cap (C - A)^c & \text{by } \underline{\quad(a)\quad} \\
&= (A \cup B) \cap (C \cap A^c)^c & \text{by } \underline{\quad(b)\quad} \\
&= (A \cup B) \cap (A^c \cap C)^c & \text{by } \underline{\quad(c)\quad} \\
&= (A \cup B) \cap ((A^c)^c \cup C^c) & \text{by } \underline{\quad(d)\quad} \\
&= (A \cup B) \cap (A \cup C^c) & \text{by } \underline{\quad(e)\quad} \\
&= A \cup (B \cap C^c) & \text{by } \underline{\quad(f)\quad} \\
&= A \cup (B - C) & \text{by } \underline{\quad(g)\quad}.
\end{aligned}$$

H 29. Some steps are missing from the following proof that for all sets $(A \cup B) - C = (A - C) \cup (B - C)$. Indicate what they are, and then write the proof correctly.

Proof: Let A, B, and C be any sets. Then

$$\begin{aligned}
(A \cup B) - C &= (A \cup B) \cap C^c & \text{by the set difference law} \\
&= (A \cap C^c) \cup (B \cap C^c) & \text{by the distributive law} \\
&= (A - C) \cup (B - C) & \text{by the set difference law}
\end{aligned}$$

In 30–40, construct an algebraic proof for the given statement. Cite a property from Theorem 6.2.2 for every step.

30. For all sets A, B, and C,

$$(A \cap B) \cup C = (A \cup C) \cap (B \cup C).$$

31. For all sets A and B, $A \cup (B - A) = A \cup B$.

32. For all sets A and B, $(A - B) \cup (A \cap B) = A$.

33. For all sets A and B, $(A - B) \cap (A \cap B) = \emptyset$.

34. For all sets A, B, and C,

$$(A - B) - C = A - (B \cup C).$$

35. For all sets A and B, $A - (A - B) = A \cap B$.

36. For all sets A and B, $((A^c \cup B^c) - A)^c = A$.

37. For all sets A and B, $(B^c \cup (B^c - A))^c = B$.

38. For all sets A and B, $A - (A \cap B) = A - B$.

H 39. For all sets A and B,

$$(A - B) \cup (B - A) = (A \cup B) - (A \cap B).$$

40. For all sets A, B, and C,

$$(A - B) - (B - C) = A - B.$$

In 41–43 simplify the given expression. Cite a property from Theorem 6.2.2 for every step.

H **41.** $A \cap ((B \cup A^c) \cap B^c)$

42. $(A - (A \cap B)) \cap (B - (A \cap B))$

43. $((A \cap (B \cup C)) \cap (A - B)) \cap (B \cup C^c)$

44. Consider the following set property: For all sets A and B, $A - B$ and B are disjoint.
    a. Use an element argument to derive the property.
    b. Use an algebraic argument to derive the property (by applying properties from Theorem 6.2.2).
    c. Comment on which method you found easier.

45. Consider the following set property: For all sets A, B, and C, $(A - B) \cup (B - C) = (A \cup B) - (B \cap C)$.
    a. Use an element argument to derive the property.
    b. Use an algebraic argument to derive the property (by applying properties from Theorem 6.2.2).
    c. Comment on which method you found easier.

> **Definition:** Given sets A and B, the **symmetric difference of A and B,** denoted $A \triangle B$, is
>
> $$A \triangle B = (A - B) \cup (B - A).$$

46. Let $A = \{1, 2, 3, 4\}$, $B = \{3, 4, 5, 6\}$, and $C = \{5, 6, 7, 8\}$. Find each of the following sets:
    a. $A \triangle B$ b. $B \triangle C$
    c. $A \triangle C$ d. $(A \triangle B) \triangle C$

Refer to the definition of symmetric difference given above. Prove each of 47–52, assuming that A, B, and C are all subsets of a universal set U.

47. $A \triangle B = B \triangle A$ 48. $A \triangle \emptyset = A$

49. $A \triangle A^c = U$ 50. $A \triangle A = \emptyset$

H **51.** If $A \triangle C = B \triangle C$, then $A = B$.

H **52.** $(A \triangle B) \triangle C = A \triangle (B \triangle C)$.

H **53.** Derive the set identity $A \cup (A \cap B) = A$ from the properties listed in Theorem 6.2.2(1)–(9). Start by showing that for all subsets B of a universal set U, $U \cup B = U$. Then intersect both sides with A and deduce the identity.

54. Derive the set identity $A \cap (A \cup B) = A$ from the properties listed in Theorem 6.2.2(1)–(9). Start by showing that for all subsets B of a universal set U, $\emptyset = \emptyset \cap B$. Then take the union of both sides with A and deduce the identity.

Answers for Test Yourself

1. make the left-hand side unequal to the right-hand side (*Or: result in different values on the two sides of the equation*)   2. *cite one of the properties from Theorem 6.2.2 (Or: give a reason)*   3. *exactly*

6.4 Boolean Algebras and Russell's Paradox

From the paradise created for us by Cantor, no one will drive us out.
— David Hilbert (1862–1943)

Table 6.4.1 summarizes the main features of the logical equivalences from Theorem 2.1.1 and the set properties from Theorem 6.2.2. Notice how similar the entries in the two columns are.

Logical Equivalences	Set Properties
For all statement variables p, q, and r:	For all sets A, B, and C:
a. $p \vee q \equiv q \vee p$ b. $p \wedge q \equiv q \wedge p$	a. $A \cup B = B \cup A$ b. $A \cap B = B \cap A$
a. $p \wedge (q \wedge r) \equiv p \wedge (q \wedge r)$ b. $p \vee (q \vee r) \equiv p \vee (q \vee r)$	a. $A \cup (B \cup C) = A \cup (B \cup C)$ b. $A \cap (B \cap C) = A \cap (B \cap C)$
a. $p \wedge (q \vee r) \equiv (p \wedge q) \vee (p \wedge r)$ b. $p \vee (q \wedge r) \equiv (p \vee q) \wedge (p \vee r)$	a. $A \cap (B \cup C) = (A \cap B) \cup (A \cap C)$ b. $A \cup (B \cap C) = (A \cup B) \cap (A \cup C)$
a. $p \vee \mathbf{c} \equiv p$ b. $p \wedge \mathbf{t} \equiv p$	a. $A \cup \emptyset = A$ b. $A \cap U = A$
a. $p \vee \sim p \equiv \mathbf{t}$ b. $p \wedge \sim p \equiv \mathbf{c}$	a. $A \cup A^c = U$ b. $A \cap A^c = \emptyset$
$\sim (\sim p) \equiv p$	$(A^c)^c = A$
a. $p \vee p \equiv p$ b. $p \wedge p \equiv p$	a. $A \cup A = A$ b. $A \cap A = A$
a. $p \vee \mathbf{t} \equiv \mathbf{t}$ b. $p \wedge \mathbf{c} \equiv \mathbf{c}$	a. $A \cup U = U$ b. $A \cap \emptyset = \emptyset$
a. $\sim (p \vee q) \equiv \sim p \wedge \sim q$ b. $\sim (p \wedge q) \equiv \sim p \vee \sim q$	a. $(A \cup B)^c = A^c \cap B^c$ b. $(A \cap B)^c = A^c \cup B^c$
a. $p \vee (p \wedge q) \equiv p$ b. $p \wedge (p \vee q) \equiv p$	a. $A \cup (A \cap B) = A$ b. $A \cap (A \cup B) = A$
a. $\sim \mathbf{t} \equiv \mathbf{c}$ b. $\sim \mathbf{c} \equiv \mathbf{t}$	a. $U^c = \emptyset$ b. $\emptyset^c = U$

Table 6.4.1

George Boole
(1815–1864)

CORBIS

If you let ∨ (*or*) correspond to ∪ (union), ∧ (*and*) correspond to ∩ (intersection), **t** (a tautology) correspond to U (a universal set), **c** (a contradiction) correspond to ∅ (the empty set), and ∼ (negation) correspond to ^{c} (complementation), then you can see that the structure of the set of statement forms with operations ∨ and ∧ is essentially identical to the structure of the set of subsets of a universal set with operations ∪ and ∩. In fact, both are special cases of the same general structure, known as a *Boolean algebra*. The essential idea of a Boolean algebra was introduced by the self-taught English mathematician/logician George Boole in 1847 in a book entitled *The Mathematical Analysis of Logic*. During the remainder of the nineteenth century, Boole and others amplified and clarified the concept until it reached the form in which we use it today.

In this section we show how to derive the various properties associated with a Boolean algebra from a set of just five axioms.

• Definition: Boolean Algebra

A **Boolean algebra** is a set B together with two operations, generally denoted $+$ and $\cdot$, such that for all a and b in B both $a + b$ and $a \cdot b$ are in B and the following properties hold:

1. *Commutative Laws:* For all a and b in B,

$$\text{(a) } a + b = b + a \quad \text{and} \quad \text{(b) } a \cdot b = b \cdot a.$$

2. *Associative Laws:* For all a, b, and c in B,

$$\text{(a) } (a + b) + c = a + (b + c) \quad \text{and} \quad \text{(b) } (a \cdot b) \cdot c = a \cdot (b \cdot c).$$

3. *Distributive Laws:* For all a, b, and c in B,

$$\text{(a) } a + (b \cdot c) = (a + b) \cdot (a + c) \quad \text{and} \quad \text{(b) } a \cdot (b + c) = (a \cdot b) + (a \cdot c).$$

4. *Identity Laws:* There exist distinct elements 0 and 1 in B such that for all a in B,

$$\text{(a) } a + 0 = a \quad \text{and} \quad \text{(b) } a \cdot 1 = a.$$

5. *Complement Laws:* For each a in B, there exists an element in B, denoted $\overline{a}$ and called the **complement** or **negation** of a, such that

$$\text{(a) } a + \overline{a} = 1 \quad \text{and} \quad \text{(b) } a \cdot \overline{a} = 0.$$

In any Boolean algebra, the complement of each element is unique, the quantities 0 and 1 are unique, and identities analogous to those in Theorem 2.1.1 and Theorem 6.2.2 can be deduced.

Theorem 6.4.1 Properties of a Boolean Algebra

Let B be any Boolean algebra.

1. *Uniqueness of the Complement Law:* For all a and x in B, if $a + x = 1$ and $a \cdot x = 0$ then $x = \overline{a}$.

2. *Uniqueness of 0 and 1:* If there exists x in B such that $a + x = a$ for all a in B, then $x = 0$, and if there exists y in B such that $a \cdot y = a$ for all a in B, then $y = 1$.

3. *Double Complement Law:* For all $a \in B$, $\overline{(\overline{a})} = a$.

continued on page 288

4. *Idempotent Law:* For all $a \in B$,

$$\text{(a) } a + a = a \quad \text{and} \quad \text{(b) } a \cdot a = a.$$

5. *Universal Bound Law:* For all $a \in B$,

$$\text{(a) } a + 1 = 1 \quad \text{and} \quad \text{(b) } a \cdot 0 = 0.$$

6. *De Morgan's Laws:* For all a and $b \in B$,

$$\text{(a) } \overline{a + b} = \overline{a} \cdot \overline{b} \quad \text{and} \quad \text{(b) } \overline{a \cdot b} = \overline{a} + \overline{b}.$$

7. *Absorption Laws:* For all a and $b \in B$,

$$\text{(a) } (a + b) \cdot a = a \quad \text{and} \quad \text{(b) } (a \cdot b) + a = a.$$

8. *Complements of* 0 *and* 1:

$$\text{(a) } \overline{0} = 1 \quad \text{and} \quad \text{(b) } \overline{1} = 0.$$

Proof:

Part 1: Uniqueness of the Complement Law

Suppose a and x are particular, but arbitrarily chosen, elements of B that satisfy the following hypothesis: $a + x = 1$ and $a \cdot x = 0$. Then

$x = x \cdot 1$	because 1 is an identity for $\cdot$
$= x \cdot (a + \overline{a})$	by the complement law for $+$
$= x \cdot a + x \cdot \overline{a}$	by the distributive law for $\cdot$ over $+$
$= a \cdot x + x \cdot \overline{a}$	by the commutative law for $\cdot$
$= 0 + x \cdot \overline{a}$	by hypothesis
$= a \cdot \overline{a} + x \cdot \overline{a}$	by the complement law for $\cdot$
$= (\overline{a} \cdot a) + (\overline{a} \cdot x)$	by the commutative law for $\cdot$
$= \overline{a} \cdot (a + x)$	by the distributive law for $\cdot$ over $+$
$= \overline{a} \cdot 1$	by hypothesis
$= \overline{a}$	because 1 is an identity for $\cdot$.

Proofs of the other parts of the theorem are discussed in the examples that follow and in the exercises.

You may notice that all parts of the definition of a Boolean algebra and most parts of Theorem 6.4.1 contain paired statements. For instance, the distributive laws state that for all a, b, and c in B,

$$\text{(a) } a + (b \cdot c) = (a + b) \cdot (a + c) \quad \text{and} \quad \text{(b) } a \cdot (b + c) = (a \cdot b) + (a \cdot c),$$

and the identity laws state that for all a in B,

$$\text{(a) } a + 0 = a \quad \text{and} \quad \text{(b) } a \cdot 1 = a.$$

Note that each of the paired statements can be obtained from the other by interchanging all the $+$ and $\cdot$ signs and interchanging 1 and 0. Such interchanges transform any Boolean identity into its **dual** identity. It can be proved that the dual of any Boolean identity is also an identity. This fact is often called the **duality principle** for a Boolean algebra.

Example 6.4.1 Proof of the Double Complement Law

Prove that for all elements a in a Boolean algebra B, $\overline{(\overline{a})} = a$.

Solution   Start by supposing that B is a Boolean algebra and a is any element of B. The basis for the proof is the uniqueness of the complement law: that each element in B has a unique complement that satisfies certain equations with respect to it. So if a can be shown to satisfy those equations with respect to $\overline{a}$, then a must be the complement of $\overline{a}$.

Theorem 6.4.1(3) Double Complement Law

For all elements a in a Boolean algebra B, $\overline{(\overline{a})} = a$.

Proof:

Suppose B is a Boolean algebra and a is any element of B. Then

$$\overline{a} + a = a + \overline{a} \qquad \text{by the commutative law}$$
$$= 1 \qquad \text{by the complement law for 1}$$

and

$$\overline{a} \cdot a = a \cdot \overline{a} \qquad \text{by the commutative law}$$
$$= 0 \qquad \text{by the complement law for 0.}$$

Thus a satisfies the two equations with respect to $\overline{a}$ that are satisfied by the complement of $\overline{a}$. From the fact that the complement of a is unique, we conclude that $\overline{(\overline{a})} = a$.

Example 6.4.2 Proof of an Idempotent Law

Fill in the blanks in the following proof that for all elements a in a Boolean algebra B, $a + a = a$.

Proof:

Suppose B is a Boolean algebra and a is any element of B. Then

$$
\begin{aligned}
a &= a + 0 & \underline{\text{(a)}} \\
&= a + (a \cdot \overline{a}) & \underline{\text{(b)}} \\
&= (a + a) \cdot (a + \overline{a}) & \underline{\text{(c)}} \\
&= (a + a) \cdot 1 & \underline{\text{(d)}} \\
&= a + a & \underline{\text{(e)}}.
\end{aligned}
$$

Solution

a. because 0 is an identity for $+$

b. by the complement law for $\cdot$

c. by the distributive law for $+$ over $\cdot$

d. by the complement law for $+$

e. because 1 is an identity for $\cdot$

Russell's Paradox

*Bertrand Russell
(1872–1970)*

By the beginning of the twentieth century, abstract set theory had gained such wide acceptance that a number of mathematicians were working hard to show that all of mathematics could be built upon a foundation of set theory. In the midst of this activity, the English mathematician and philosopher Bertrand Russell discovered a "paradox" (really a genuine contradiction) that seemed to shake the very core of the foundation. The paradox assumes Cantor's definition of set as "any collection into a whole of definite and separate objects of our intuition or our thought."

Russell's Paradox: Most sets are not elements of themselves. For instance, the set of all integers is not an integer and the set of all horses is not a horse. However, we can imagine the possibility of a set's being an element of itself. For instance, the set of all abstract ideas might be considered an abstract idea. If we are allowed to use any description of a property as the defining property of a set, we can let S be the set of all sets that are not elements of themselves:

$$S = \{A \mid A \text{ is a set and } A \notin A\}.$$

Is S an element of itself?

The answer is neither yes nor no. For if $S \in S$, then S satisfies the defining property for S, and hence $S \notin S$. But if $S \notin S$, then S is a set such that $S \notin S$ and so S satisfies the defining property for S, which implies that $S \in S$. Thus neither is $S \in S$ nor is $S \notin S$, which is a contradiction.

To help explain his discovery to laypeople, Russell devised a puzzle, the barber puzzle, whose solution exhibits the same logic as his paradox.

Example 6.4.3 The Barber Puzzle

In a certain town there is a male barber who shaves all those men, and only those men, who do not shave themselves. *Question:* Does the barber shave himself?

Solution   Neither yes nor no. If the barber shaves himself, he is a member of the class of men who shave themselves. But no member of this class is shaved by the barber, and so the barber does *not* shave himself. On the other hand, if the barber does not shave himself, he belongs to the class of men who do not shave themselves. But the barber shaves every man in this class, so the barber *does* shave himself. ■

But how can the answer be neither yes nor no? Surely any barber either does or does not shave himself. You might try to think of circumstances that would make the paradox disappear. For instance, maybe the barber happens to have no beard and never shaves. But a condition of the puzzle is that the barber is a man who shaves *all* those men who do not shave themselves. If he does not shave, then he does not shave himself, in which case he is shaved by the barber and the contradiction is as present as ever. Other attempts at resolving the paradox by considering details of the barber's situation are similarly doomed to failure.

So let's accept the fact that the paradox has no easy resolution and see where that thought leads. Since the barber neither shaves himself nor doesn't shave himself, the sentence "The barber shaves himself" is neither true nor false. But the sentence arose in a natural way from a description of a situation. If the situation actually existed, then the sentence would have to be true or false. Thus we are forced to conclude that the situation described in the puzzle simply cannot exist in the world as we know it.

In a similar way, the conclusion to be drawn from Russell's paradox itself is that the object S is not a set. Because if it actually were a set, in the sense of satisfying the general properties of sets that we have been assuming, then it either would be an element of itself or not.

In the years following Russell's discovery, several ways were found to define the basic concepts of set theory so as to avoid his contradiction. The way used in this text requires that, except for the power set whose existence is guaranteed by an axiom, whenever a set is defined using a predicate as a defining property, the stipulation must also be made that the set is a subset of a known set. This method does not allow us to talk about "the set of all sets that are not elements of themselves." We can speak only of "the set of all sets that are subsets of some known set and that are not elements of themselves." When this restriction is made, Russell's paradox ceases to be contradictory. Here is what happens:

Let U be a universal set and suppose that all sets under discussion are subsets of U. Let

$$S = \{A \mid A \subseteq U \text{ and } A \notin A\}.$$

In Russell's paradox, both implications

$$S \in S \rightarrow S \notin S \quad \text{and} \quad S \notin S \rightarrow S \in S$$

are proved, and the contradictory conclusion

$$\text{neither } S \in S \quad \text{nor} \quad S \notin S$$

is therefore deduced. In the situation in which all sets under discussion are subsets of U, the implication $S \in S \rightarrow S \notin S$ is proved in almost the same way as it is for Russell's paradox: (Suppose $S \in S$. Then by definition of S, $S \subseteq U$ and $S \notin S$. In particular, $S \notin S$.) On the other hand, from the supposition that $S \notin S$ we can only deduce that the statement "$S \subseteq U$ and $S \notin S$" is false. By one of De Morgan's laws, this means that "$S \not\subseteq U$ or $S \in S$." Since $S \in S$ would contradict the supposition that $S \notin S$, we eliminate it and conclude that $S \not\subseteq U$. In other words, the only conclusion we can draw is that the seeming "definition" of S is faulty—that is, that S is not a set in U.

Kurt Gödel
(1906–1978)

Russell's discovery had a profound impact on mathematics because even though his contradiction could be made to disappear by more careful definitions, its existence caused people to wonder whether other contradictions remained. In 1931 Kurt Gödel showed that it is not possible to prove, in a mathematically rigorous way, that mathematics is free of contradictions. You might think that Gödel's result would have caused mathematicians to give up their work in despair, but that has not happened. On the contrary, there has been more mathematical activity since 1931 than in any other period in history.

Test Yourself

1. In the comparison between the structure of the set of statement forms and the set of subsets of a universal set, the *or* operation $\vee$ corresponds to _____, the *and* operation $\wedge$ corresponds to _____, a tautology **t** corresponds to _____, a contradiction **c** corresponds to _____, and the negation operation, denoted $\sim$, corresponds to _____.

2. The operations of $+$ and $\cdot$ in a Boolean algebra are general-izations of the operations of _____ and _____ in the set of all statement forms in a given finite number of variables and the operations of _____ and _____ in the set of all subsets of a given set.

3. Russell showed that the following proposed "set definition" could not actually define a set: _____.

Exercise Set 6.4

In 1–3 assume that B is a Boolean algebra with operations $+$ and $\cdot$. Give the reasons needed to fill in the blanks in the proofs, but do not use any parts of Theorem 6.4.1 unless they have already been proved. You may use any part of the definition of a Boolean algebra and the results of previous exercises, however.

1. For all a in B, $a \cdot a = a$.

    Proof: Let a be any element of B. Then

$$
\begin{aligned}
a &= a \cdot 1 & \text{(a)} \\
  &= a \cdot (a + \overline{a}) & \text{(b)} \\
  &= (a \cdot a) + (a \cdot \overline{a}) & \text{(c)} \\
  &= (a \cdot a) + 0 & \text{(d)} \\
  &= a \cdot a & \text{(e)}.
\end{aligned}
$$

2. For all a in B, $a + 1 = 1$.

    Proof: Let a be any element of B. Then

$$
\begin{aligned}
a + 1 &= a + (a + \overline{a}) & \text{(a)} \\
      &= (a + a) + \overline{a} & \text{(b)} \\
      &= a + \overline{a} & \text{by Example 6.4.2} \\
      &= 1 & \text{(c)}.
\end{aligned}
$$

3. For all a and b in B, $(a + b) \cdot a = a$.

    Proof: Let a and b be any elements of B. Then

$$
\begin{aligned}
(a + b) \cdot a &= a \cdot (a + b) & \text{(a)} \\
                &= a \cdot a + a \cdot b & \text{(b)} \\
                &= a + a \cdot b & \text{(c)} \\
                &= a \cdot 1 + a \cdot b & \text{(d)} \\
                &= a \cdot (1 + b) & \text{(e)} \\
                &= a \cdot (b + 1) & \text{(f)} \\
                &= a \cdot 1 & \text{by exercise 2} \\
                &= a & \text{(g)}.
\end{aligned}
$$

In 4–10 assume that B is a Boolean algebra with operations $+$ and $\cdot$. Prove each statement without using any parts of Theorem 6.4.1 unless they have already been proved. You may use any part of the definition of a Boolean algebra and the results of previous exercises, however.

4. For all a in B, $a \cdot 0 = 0$.

5. For all a and b in B, $(a \cdot b) + a = a$.

6. a. $\overline{0} = 1$.
    b. $\overline{1} = 0$

7. a. There is only one element of B that is an identity for $+$.
  H b. There is only one element of B that is an identity for $\cdot$.

8. For all a and b in B, $\overline{a \cdot b} = \overline{a} + \overline{b}$. (*Hint:* Prove that $(a \cdot b) + (\overline{a} + \overline{b}) = 1$ and that $(a \cdot b) \cdot (\overline{a} + \overline{b}) = 0$, and use the fact that $a \cdot b$ has a unique complement.)

9. For all a and b in B, $\overline{a + b} = \overline{a} \cdot \overline{b}$.

H 10. For all x, y, and z in B, if $x + y = x + z$ and $x \cdot y = x \cdot z$, then $y = z$.

11. Let $S = \{0, 1\}$, and define operations $+$ and $\cdot$ on S by the following tables:

$+$	0	1
0	0	1
1	1	1

$\cdot$	0	1
0	0	0
1	0	1

  a. Show that the elements of S satisfy the following properties:
    (i) the commutative law for $+$
    (ii) the commutative law for $\cdot$
    (iii) the associative law for $+$
    (iv) the associative law for $\cdot$
  H **(v)** the distributive law for $+$ over $\cdot$
    (vi) the distributive law for $\cdot$ over $+$
 H b. Show that 0 is an identity element for $+$ and that 1 is an identity element for $\cdot$.
  c. Define $\overline{0} = 1$ and $\overline{1} = 0$. Show that for all a in S, $a + \overline{a} = 1$ and $a \cdot \overline{a} = 0$. It follows from parts (a)–(c) that S is a Boolean algebra with the operations $+$ and $\cdot$.

H ✱ 12. Prove that the associative laws for a Boolean algebra can be omitted from the definition. That is, prove that the associative laws can be derived from the other laws in the definition.

In 13–18 determine whether each sentence is a statement. Explain your answers.

13. This sentence is false.

14. If $1 + 1 = 3$, then $1 = 0$.

15. The sentence in this box is a lie.

16. All positive integers with negative squares are prime.

17. This sentence is false or $1 + 1 = 3$.

18. This sentence is false and $1 + 1 = 2$.

19. a. Assuming that the following sentence is a statement, prove that $1 + 1 = 3$:

      If this sentence is true, then $1 + 1 = 3$.

  b. What can you deduce from part (a) about the status of "This sentence is true"? Why? (This example is known as **Löb's paradox**.)

H 20. The following two sentences were devised by the logician Saul Kripke. While not intrinsically paradoxical, they could be paradoxical under certain circumstances. Describe such circumstances.

    (i) Most of Nixon's assertions about Watergate are false.
    (ii) Everything Jones says about Watergate is true.

    (*Hint:* Suppose Nixon says (ii) and the only utterance Jones makes about Watergate is (i).)

21. Can there exist a computer program that has as output a list of all the computer programs that do not list themselves in their output? Explain your answer.

22. Can there exist a book that refers to all those books and only those books that do not refer to themselves? Explain your answer.

23. Some English adjectives are descriptive of themselves (for instance, the word *polysyllabic* is polysyllabic) whereas others are not (for instance, the word *monosyllabic* is not monosyllabic). The word *heterological* refers to an adjective that does not describe itself. Is *heterological* heterological? Explain your answer.

24. As strange as it may seem, it is possible to give a precise-looking verbal definition of an integer that, in fact, is not a definition at all. The following was devised by an English librarian, G. G. Berry, and reported by Bertrand Russell. Explain how it leads to a contradiction. Let n be "the smallest integer not describable in fewer than 12 English words." (Note that the total number of strings consisting of 11 or fewer English words is finite.)

25. Use a technique similar to that used to derive Russell's paradox to prove that for any set A, $\mathscr{P}(A) \nsubseteq A$.

Answers for Test Yourself

1. the operation of union $\cup$; the operation of intersection $\cap$; a universal set U; the empty set $\emptyset$; the operation of complementation, denoted ^{c} 2. $\vee$; $\wedge$; $\cup$; $\cap$ 3. the set of all sets that are not elements of themselves

CHAPTER 7

FUNCTIONS

In this chapter we consider a wide variety of functions, focusing on those defined on discrete sets (such as finite sets or sets of integers). We explore properties of functions such as one-to-one and onto, define inverse functions and composition of functions, and end the chapter with the surprising result that there are different sizes of infinity.

7.1 Functions Defined on General Sets

The theory that has had the greatest development in recent times is without any doubt the theory of functions. — Vito Volterra, 1888

As used in ordinary language, the word *function* indicates dependence of one varying quantity on another. If your teacher tells you that your grade in a course will be a function of your performance on the exams, you interpret this to mean that the teacher has some rule for translating exam scores into grades. To each collection of exam scores there corresponds a certain grade.

In Section 1.3 we defined a function as a certain type of relation. In this chapter we focus on the more dynamic way functions are used in mathematics. The following is a restatement of the definition of function that includes additional terminology associated with the concept.

• Definition

A **function f from a set X to a set Y**, denoted $f\colon X \to Y$, is a relation from X, the **domain**, to Y, the **co-domain**, that satisfies two properties: (1) every element in X is related to some element in Y, and (2) no element in X is related to more than one element in Y. Thus, given any element x in X, there is a unique element in Y that is related to x by f. If we call this element y, then we say that "f sends x to y" or "f maps x to y" and write $x \xrightarrow{f} y$ or $f\colon x \to y$. The unique element to which f sends x is denoted

$f(x)$ and is called   **f of x**, or
the output of f for the input x, or
the value of f at x, or
the image of x under f.

> The set of all values of f taken together is called the *range of f* or the *image of X under f.* Symbolically,
>
> **range of f = image of X under f** $= \{y \in Y \mid y = f(x), \text{for some } x \text{ in } X\}$.
>
> Given an element y in Y, there may exist elements in X with y as their image. If $f(x) = y$, then x is called **a preimage of y** or **an inverse image of y.** The set of all inverse images of y is called *the inverse image of y.* Symbolically,
>
> **the inverse image of y** $= \{x \in X \mid f(x) = y\}$.

Caution! Use $f(x)$ to refer to the value of the function f at x. Generally avoid using $f(x)$ to refer to the function f itself.

In some mathematical contexts, the notation $f(x)$ is used to refer both to the value of f at x and to the function f itself. Because using the notation this way can lead to confusion, we avoid it whenever possible. In this book, unless explicitly stated otherwise, the symbol $f(x)$ always refers to the value of the function f at x and not to the function f itself.

The concept of function was developed over a period of centuries. A definition similar to that given previously was first formulated for sets of numbers by the German mathematician Lejeune Dirichlet (DEER-ish-lay) in 1837.

Johann Peter Gustav Lejeune Dirichlet (1805–1859)

Arrow Diagrams

Recall from Section 1.3 that if X and Y are finite sets, you can define a function f from X to Y by drawing an arrow diagram. You make a list of elements in X and a list of elements in Y, and draw an arrow from each element in X to the corresponding element in Y, as shown in Figure 7.1.1.

This arrow diagram does define a function because

1. Every element of X has an arrow coming out of it.

2. No element of X has two arrows coming out of it that point to two different elements of Y.

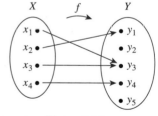

Figure 7.1.1

Example 7.1.1 Functions and Nonfunctions

Which of the arrow diagrams in Figure 7.1.2 define functions from $X = \{a, b, c\}$ to $Y = \{1, 2, 3, 4\}$?

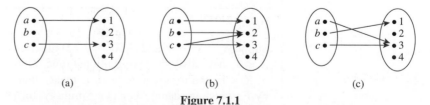

(a)   (b)   (c)

Figure 7.1.1

Solution Only (c) defines a function. In (a) there is an element of X, namely b, that is not sent to any element of Y; that is, there is no arrow coming out of b. And in (b) the element c is not sent to a *unique* element of Y; that is, there are two arrows coming out of c, one pointing to 2 and the other to 3. ∎

Example 7.1.2 A Function Defined by an Arrow Diagram

Let $X = \{a, b, c\}$ and $Y = \{1, 2, 3, 4\}$. Define a function f from X to Y by the arrow diagram in Figure 7.1.3.

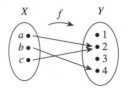

Figure 7.1.1

a. Write the domain and co-domain of f.

b. Find $f(a)$, $f(b)$, and $f(c)$.

c. What is the range of f?

d. Is c an inverse image of 2? Is b an inverse image of 3?

e. Find the inverse images of 2, 4, and 1.

f. Represent f as a set of ordered pairs.

Solution

a. domain of $f = \{a, b, c\}$, co-domain of $f = \{1, 2, 3, 4\}$

b. $f(a) = 2$, $f(b) = 4$, $f(c) = 2$

c. range of $f = \{2, 4\}$

d. Yes, No

e. inverse image of $2 = \{a, c\}$
   inverse image of $4 = \{b\}$
   inverse image of $1 = \emptyset$ (*since no arrows point to* 1)

f. $\{(a, 2), (b, 4), (c, 2)\}$ ∎

In Example 7.1.2 there are no arrows pointing to the 1 or the 3. This illustrates the fact that although each element of the domain of a function must have an arrow pointing out from it, there can be elements of the co-domain to which no arrows point. Note also that there are two arrows pointing to the 2—one coming from a and the other from c.

In Section 1.3 we gave a test for determining whether two functions with the same domain and co-domain are equal, saying that the test results from the definition of a function as a binary relation. We formalize this justification in Theorem 7.1.1.

Theorem 7.1.1 A Test for Function Equality

If $F: X \to Y$ and $G: X \to Y$ are functions, then $F = G$ if, and only if, $F(x) = G(x)$ for all $x \in X$.

Proof:

Note So $(x, y) \in F$ $\Leftrightarrow y = F(x)$ and $(x, y) \in G \Leftrightarrow y = G(x)$.

Suppose $F: X \to Y$ and $G: X \to Y$ are functions, that is, F and G are relations from X to Y that satisfy the two additional function properties. Then F and G are subsets of $X \times Y$, and for (x, y) to be in F means that y is the unique element related to x by F, which we denote as $F(x)$. Similarly, for (x, y) to be in G means that y is the unique element related to x by G, which we denote as $G(x)$.

Now suppose that $F(x) = G(x)$ for all $x \in X$. Then if x is any element of X,

$$(x, y) \in F \Leftrightarrow y = F(x) \Leftrightarrow y = G(x) \Leftrightarrow (x, y) \in G \qquad \text{because } F(x) = G(x)$$

So F and G consist of exactly the same elements and hence $F = G$.

Conversely, if $F = G$, then for all $x \in X$,

$$y = F(x) \Leftrightarrow (x, y) \in F \Leftrightarrow (x, y) \in G \Leftrightarrow y = G(x)$$

because F and G consist of exactly the same elements

Thus, since both $F(x)$ and $G(x)$ equal y, we have that

$$F(x) = G(x).$$

Example 7.1.3 Equality of Functions

a. Let $J_3 = \{0, 1, 2\}$, and define functions f and g from J_3 to J_3 as follows: For all x in J_3,

$$f(x) = (x^2 + x + 1) \bmod 3 \quad \text{and} \quad g(x) = (x + 2)^2 \bmod 3.$$

Does $f = g$?

b. Let $F: \mathbf{R} \to \mathbf{R}$ and $G: \mathbf{R} \to \mathbf{R}$ be functions. Define new functions $F + G: \mathbf{R} \to \mathbf{R}$ and $G + F: \mathbf{R} \to \mathbf{R}$ as follows: For all $x \in \mathbf{R}$,

$$(F + G)(x) = F(x) + G(x) \quad \text{and} \quad (G + F)(x) = G(x) + F(x).$$

Does $F + G = G + F$?

Solution

a. Yes, the table of values shows that $f(x) = g(x)$ for all x in J_3.

x	$x^2 + x + 1$	$f(x) = (x^2 + x + 1) \bmod 3$	$(x + 2)^2$	$g(x) = (x + 2)^2 \bmod 3$
0	1	$1 \bmod 3 = 1$	4	$4 \bmod 3 = 1$
1	3	$3 \bmod 3 = 0$	9	$9 \bmod 3 = 0$
2	7	$7 \bmod 3 = 1$	16	$16 \bmod 3 = 1$

b. Again the answer is yes. For all real numbers x,

$$\begin{aligned}
(F + G)(x) &= F(x) + G(x) && \text{by definition of } F + G \\
&= G(x) + F(x) && \text{by the commutative law for addition of real numbers} \\
&= (G + F)(x) && \text{by definition of } G + F
\end{aligned}$$

Hence $F + G = G + F$. ∎

Examples of Functions

The following examples illustrate some of the wide variety of different types of functions.

Example 7.1.4 The Identity Function on a Set

Given a set X, define a function I_X from X to X by

$$I_X(x) = x \quad \text{for all } x \text{ in } X.$$

The function I_X is called the **identity function on X** because it sends each element of X to the element that is identical to it. Thus the identity function can be pictured as a machine that sends each piece of input directly to the output chute without changing it in any way.

Let X be any set and suppose that a_{ij}^k and $\phi(z)$ are elements of X. Find $I_X\left(a_{ij}^k\right)$ and $I_X(\phi(z))$.

Solution Whatever is input to the identity function comes out unchanged, so $I_X\left(a_{ij}^k\right) = a_{ij}^k$ and $I_X(\phi(z)) = \phi(z)$. ∎

Example 7.1.5 Sequences

The formal definition of sequences specifies that an infinite sequence is a function defined on the set of integers that are greater than or equal to a particular integer. For example, the sequence denoted

$$1, -\frac{1}{2}, \frac{1}{3}, -\frac{1}{4}, \frac{1}{5}, \ldots, \frac{(-1)^n}{n+1}, \ldots$$

can be thought of as the function f from the nonnegative integers to the real numbers that associates $0 \to 1$, $1 \to -\frac{1}{2}$, $2 \to \frac{1}{3}$, $3 \to -\frac{1}{4}$, $4 \to \frac{1}{5}$, and, in general, $n \to \frac{(-1)^n}{n+1}$. In other words, $f\colon \mathbf{Z}^{nonneg} \to \mathbf{R}$ is the function defined as follows:

$$\text{Send each integer } n \geq 0 \text{ to } f(n) = \frac{(-1)^n}{n+1}.$$

In fact, there are many functions that can be used to define a given sequence. For instance, express the sequence above as a function from the set of *positive* integers to the set of real numbers.

Solution Define $g\colon \mathbf{Z}^+ \to \mathbf{R}$ by $g(n) = \frac{(-1)^{n+1}}{n}$, for each $n \in \mathbf{Z}^+$. Then $g(1) = 1$, $g(2) = -\frac{1}{2}$, $g(3) = \frac{1}{3}$, and in general

$$g(n+1) = \frac{(-1)^{n+2}}{n+1} = \frac{(-1)^n}{n+1} = f(n). \quad \blacksquare$$

Example 7.1.6 A Function Defined on a Power Set

Recall from Section 6.1 that $\mathscr{P}(A)$ denotes the set of all subsets of the set A. Define a function $F\colon \mathscr{P}(\{a, b, c\}) \to \mathbf{Z}^{nonneg}$ as follows: For each $X \in \mathscr{P}(\{a, b, c\})$,

$$F(X) = \text{the number of elements in } X.$$

Draw an arrow diagram for F.

Solution

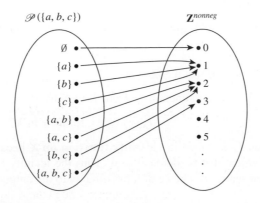

∎

Example 7.1.7 Functions Defined on a Cartesian Product

Define functions $M: \mathbf{R} \times \mathbf{R} \to \mathbf{R}$ and $R: \mathbf{R} \times \mathbf{R} \to \mathbf{R} \times \mathbf{R}$ as follows: For all ordered pairs (a, b) of integers,

$$M(a, b) = ab \quad \text{and} \quad R(a, b) = (-a, b).$$

Note It is customary to omit one set of parentheses when referring to functions defined on Cartesian products. For example, we write $M(a, b)$ rather than $M((a, b))$.

Then M is the multiplication function that sends each pair of real numbers to the product of the two, and R is the reflection function that sends each point in the plane that corresponds to a pair of real numbers to the mirror image of the point across the vertical axis. Find the following:

a. $M(-1, -1)$ b. $M\left(\frac{1}{2}, \frac{1}{2}\right)$ c. $M(\sqrt{2}, \sqrt{2})$

d. $R(2, 5)$ e. $R(-2, 5)$ f. $R(3, -4)$

Solution

a. $(-1)(-1) = 1$ b. $(1/2)(1/2) = 1/4$ c. $\sqrt{2} \cdot \sqrt{2} = 2$

d. $(-2, 5)$ e. $(-(-2), 5) = (2, 5)$ f. $(-3, -4)$ ■

• Definition Logarithms and Logarithmic Functions

Note It is not obvious, but it is true, that for any positive real number x there is a unique real number y such that $b^y = x$. Most calculus books contain a discussion of this result.

Let b be a positive real number with $b \neq 1$. For each positive real number x, the **logarithm with base b of x,** written $\log_b x$, is the exponent to which b must be raised to obtain x. Symbolically,

$$\log_b x = y \quad \Leftrightarrow \quad b^y = x.$$

The **logarithmic function with base b** is the function from $\mathbf{R}^+$ to $\mathbf{R}$ that takes each positive real number x to $\log_b x$.

Example 7.1.8 The Logarithmic Function with Base b

Find the following:

a. $\log_3 9$ b. $\log_2\left(\frac{1}{2}\right)$ c. $\log_{10}(1)$ d. $\log_2(2^m)$ (m is any real number)

e. $2^{\log_2 m}$ ($m > 0$)

Solution

a. $\log_3 9 = 2$ because $3^2 = 9$.   b. $\log_2\left(\frac{1}{2}\right) = -1$ because $2^{-1} = \frac{1}{2}$.

c. $\log_{10}(1) = 0$ because $10^0 = 1$.

d. $\log_2(2^m) = m$ because the exponent to which 2 must be raised to obtain 2^m is m.

e. $2^{\log_2 m} = m$ because $\log_2 m$ is the exponent to which 2 must be raised to obtain m. ■

Checking Whether a Function Is Well Defined

It can sometimes happen that what appears to be a function defined by a rule is not really a function at all. To give an example, suppose we wrote, "Define a function $f: \mathbf{R} \to \mathbf{R}$ by specifying that for all real numbers x,

$$f(x) \text{ is the real number } y \text{ such that } x^2 + y^2 = 1.$$

There are two distinct reasons why this description does not define a function. For almost all values of x, either (1) there is no y that satisfies the given equation or (2) there are two different values of y that satisfy the equation. For instance, when $x = 2$, there is no real number y such that $2^2 + y^2 = 1$, and when $x = 0$, both $y = -1$ and $y = 1$ satisfy the equation $0^2 + y^2 = 1$. In general, we say that a "function" is **not well defined** if it fails to satisfy at least one of the requirements for being a function.

Example 7.1.9 A Function That Is Not Well Defined

Recall that $\mathbf{Q}$ represents the set of all rational numbers. Suppose you read that a function $f : \mathbf{Q} \to \mathbf{Z}$ is to be defined by the formula

$$f\left(\frac{m}{n}\right) = m \quad \text{for all integers } m \text{ and } n \text{ with } n \neq 0.$$

That is, the integer associated by f to the number $\frac{m}{n}$ is m. Is f well defined? Why?

Solution   The function f is not well defined. The reason is that fractions have more than one representation as quotients of integers. For instance, $\frac{1}{2} = \frac{3}{6}$. Now if f were a function, then the definition of a function would imply that $f\left(\frac{1}{2}\right) = f\left(\frac{3}{6}\right)$ since $\frac{1}{2} = \frac{3}{6}$. But applying the formula for f, you find that

$$f\left(\frac{1}{2}\right) = 1 \quad \text{and} \quad f\left(\frac{3}{6}\right) = 3,$$

and so

$$f\left(\frac{1}{2}\right) \neq f\left(\frac{3}{6}\right).$$

This contradiction shows that f is not well defined and, therefore, is not a function. ■

Note that the phrase *well-defined function* is actually redundant; for a function to be well defined really means that it is worthy of being called a function.

Functions Acting on Sets

Given a function from a set X to a set Y, you can consider the set of images in Y of all the elements in a subset of X and the set of inverse images in X of all the elements in a subset of Y.

Note For $y \in Y$,
$f^{-1}(y) = f^{-1}(\{y\})$.

• Definition

If $f : X \to Y$ is a function and $A \subseteq X$ and $C \subseteq Y$, then

$$f(A) = \{y \in Y \mid y = f(x) \text{ for some } x \text{ in } A\}$$

and
$$f^{-1}(C) = \{x \in X \mid f(x) \in C\}.$$

$f(A)$ is called the **image of** A, and $f^{-1}(C)$ is called the **inverse image of** C.

Example 7.1.10 The Action of a Function on Subsets of a Set

Let $X = \{1, 2, 3, 4\}$ and $Y = \{a, b, c, d, e\}$, and define $F: X \rightarrow Y$ by the following arrow diagram:

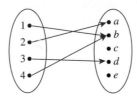

Let $A = \{1, 4\}$, $C = \{a, b\}$, and $D = \{c, e\}$. Find $F(A)$, $F(X)$, $F^{-1}(C)$, and $F^{-1}(D)$.

Solution

$$F(A) = \{b\} \quad F(X) = \{a, b, d\} \quad F^{-1}(C) = \{1, 2, 4\} \quad F^{-1}(D) = \emptyset$$ ■

Example 7.1.11 Interaction of a Function with Union

Let X and Y be sets, let F be a function from X to Y, and let A and B be any subsets of X. Prove that $F(A \cup B) \subseteq F(A) \cup F(B)$.

Solution

The fact that X, Y, F, A, and B were formally introduced prior to the word "Prove" allows you to regard their existence and relationships as part of your background knowledge. Thus to prove that $F(A \cup B) \subseteq F(A) \cup F(B)$, you only need show that if y is any element in $F(A \cup B)$, then y is an element of $F(A) \cup F(B)$.

Proof:

Suppose $y \in F(A \cup B)$. *[We must show that $y \in F(A) \cup F(B)$.]* By definition of function, $y = F(x)$ for some $x \in A \cup B$. By definition of union, $x \in A$ or $x \in B$.

Case 1, $x \in A$: In this case, $y = F(x)$ for some x in A. Hence $y \in F(A)$, and so by definition of union, $y \in F(A) \cup F(B)$.

Case 2, $x \in B$: In this case, $y = F(x)$ for some x in B. Hence $y \in F(B)$, and so by definition of union, $y \in F(A) \cup F(B)$.

Thus in either case $y \in F(A) \cup F(B)$ *[as was to be shown].* ■

Exercise 33 asks you to prove the opposite containment from the one in Example 7.1.11. Taken together, the example and the solution to the exercise establish the full equality that $F(A \cup B) = F(A) \cup F(B)$.

Test Yourself

Answers to Test Yourself questions are located at the end of each section.

1. Given a function f from a set X to a set Y, $f(x)$ is _____.

2. Given a function f from a set X to a set Y, if $f(x) = y$, then y is called _____ or _____ or ____.

3. Given a function f from a set X to a set Y, the range of f (or the image of X under f) is _____.

4. Given a function f from a set X to a set Y, if $f(x) = y$, then x is called _____ or _____.

5. Given a function f from a set X to a set Y, if $y \in Y$, then $f^{-1}(y) = $ _____ and is called _____.

6. Given functions f and g from a set X to a set Y, $f = g$ if, and only if, _____.

7. Given positive real numbers x and b with $b \neq 1$, $\log_b x = \underline{\qquad}$.

8. Given a function f from a set X to a set Y and a subset A of X, $f(A) = \underline{\qquad}$.

9. Given a function f from a set X to a set Y and a subset C of Y, $f^{-1}(C) = \underline{\qquad}$.

Exercise Set 7.1*

1. Let $X = \{1, 3, 5\}$ and $Y = \{s, t, u, v\}$. Define $f: X \to Y$ by the following arrow diagram.

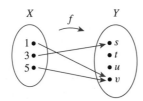

   a. Write the domain of f and the co-domain of f.
   b. Find $f(1)$, $f(3)$, and $f(5)$.
   c. What is the range of f?
   d. Is 3 an inverse image of s? Is 1 an inverse image of u?
   e. What is the inverse image of s? of u? of v?
   f. Represent f as a set of ordered pairs.

2. Let $X = \{1, 3, 5\}$ and $Y = \{a, b, c, d\}$. Define $g: X \to Y$ by the following arrow diagram.

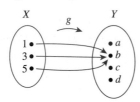

   a. Write the domain of g and the co-domain of g.
   b. Find $g(1)$, $g(3)$, and $g(5)$.
   c. What is the range of g?
   d. Is 3 an inverse image of a? Is 1 an inverse image of b?
   e. What is the inverse image of b? of c?
   f. Represent g as a set of ordered pairs.

3. Indicate whether the statements in parts (a)–(d) are true or false. Justify your answers.
   a. If two elements in the domain of a function are equal, then their images in the co-domain are equal.
   b. If two elements in the co-domain of a function are equal, then their preimages in the domain are also equal.
   c. A function can have the same output for more than one input.
   d. A function can have the same input for more than one output.

4. a. Find all functions from $X = \{a, b\}$ to $Y = \{u, v\}$.
   b. Find all functions from $X = \{a, b, c\}$ to $Y = \{u\}$.
   c. Find all functions from $X = \{a, b, c\}$ to $Y = \{u, v\}$.

5. Let $I_{\mathbf{Z}}$ be the identity function defined on the set of all integers, and suppose that $e, b_i^{jk}, K(t)$, and u_{kj} all represent integers. Find
   a. $I_{\mathbf{Z}}(e)$ b. $I_{\mathbf{Z}}\left(b_i^{jk}\right)$ c. $I_{\mathbf{Z}}(K(t))$ d. $I_{\mathbf{Z}}(u_{kj})$

6. Find functions defined on the set of nonnegative integers that define the sequences whose first six terms are given below.
   a. $1, -\dfrac{1}{3}, \dfrac{1}{5}, -\dfrac{1}{7}, \dfrac{1}{9}, -\dfrac{1}{11}$ b. $0, -2, 4, -6, 8, -10$

7. Let $A = \{1, 2, 3, 4, 5\}$ and define a function $F: \mathscr{P}(A) \to \mathbf{Z}$ as follows: For all sets X in $\mathscr{P}(A)$,

$$F(X) = \begin{cases} 0 & \text{if } X \text{ has an even number of elements} \\ 1 & \text{if } X \text{ has an odd number of elements.} \end{cases}$$

   Find the following:
   a. $F(\{1, 3, 4\})$ b. $F(\varnothing)$
   c. $F(\{2, 3\})$ d. $F(\{2, 3, 4, 5\})$

8. Let $J_5 = \{0, 1, 2, 3, 4\}$, and define a function $F: J_5 \to J_5$ as follows: For each $x \in J_5$, $F(x) = (x^3 + 2x + 4) \bmod 5$.
   Find the following:
   a. $F(0)$ b. $F(1)$ c. $F(2)$ d. $F(3)$ e. $F(4)$

9. Define a function $S: \mathbf{Z}^+ \to \mathbf{Z}^+$ as follows: For each positive integer n,

$$S(n) = \text{the sum of the positive divisors of } n.$$

   Find the following:
   a. $S(1)$ b. $S(15)$ c. $S(17)$
   d. $S(5)$ e. $S(18)$ f. $S(21)$

10. Let D be the set of all finite subsets of positive integers. Define a function $T: \mathbf{Z}^+ \to D$ as follows: For each positive integer n, $T(n) = $ the set of positive divisors of n.
    Find the following:
    a. $T(1)$ b. $T(15)$ c. $T(17)$
    d. $T(5)$ e. $T(18)$ f. $T(21)$

---

*For exercises with blue numbers or letters, solutions are given in Appendix B. The symbol **H** indicates that only a hint or a partial solution is given. The symbol ✶ signals that an exercise is more challenging than usual.

11. Define $F: \mathbf{Z} \times \mathbf{Z} \to \mathbf{Z} \times \mathbf{Z}$ as follows: For all ordered pairs (a, b) of integers, $F(a,\ b) = (2a + 1,\ 3b - 2)$.
    Find the following:
    a. $F(4, 4)$ **b.** $F(2, 1)$ **c.** $F(3, 2)$ **d.** $F(1, 5)$

12. Define $G: J_5 \times J_5 \to J_5 \times J_5$ as follows: For all $(a, b) \in J_5 \times J_5$,

    $$G(a, b) = ((2a + 1) \bmod 5, (3b - 2) \bmod 5).$$

    Find the following:
    a. $G(4, 4)$ **b.** $G(2, 1)$ **c.** $G(3, 2)$ **d.** $G(1, 5)$

13. Let $J_5 = \{0, 1, 2, 3, 4\}$, and define functions $f: J_5 \to J_5$ and $g: J_5 \to J_5$ as follows: For each $x \in J_5$,

    $$f(x) = (x + 4)^2 \bmod 5 \quad \text{and} \quad g(x) = (x^2 + 3x + 1) \bmod 5.$$

    Is $f = g$? Explain.

14. Let $J_5 = \{0, 1, 2, 3, 4\}$, and define functions $h: J_5 \to J_5$ and $k: J_5 \to J_5$ as follows: For each $x \in J_5$,

    $$h(x) = (x + 3)^3 \bmod 5 \quad \text{and} \quad k(x) = (x^3 + 4x^2 + 2x + 2) \bmod 5.$$

    Is $h = k$? Explain.

15. Let F and G be functions from the set of all real numbers to itself. Define the product functions $F \cdot G: \mathbf{R} \to \mathbf{R}$ and $G \cdot F: \mathbf{R} \to \mathbf{R}$ as follows: For all $x \in \mathbf{R}$,

    $$(F \cdot G)(x) = F(x) \cdot G(x)$$
    $$(G \cdot F)(x) = G(x) \cdot F(x)$$

    Does $F \cdot G = G \cdot F$? Explain.

16. Let F and G be functions from the set of all real numbers to itself. Define new functions $F - G: \mathbf{R} \to \mathbf{R}$ and $G - F: \mathbf{R} \to \mathbf{R}$ as follows: For all $x \in \mathbf{R}$,

    $$(F - G)(x) = F(x) - G(x)$$
    $$(G - F)(x) = G(x) - F(x)$$

    Does $F - G = G - F$? Explain.

17. Use the definition of logarithm to fill in the blanks below.
    a. $\log_2 8 = 3$ because _____.
    b. $\log_5 \left(\frac{1}{25}\right) = -2$ because _____.
    c. $\log_4 4 = 1$ because _____.
    d. $\log_3(3^n) = n$ because _____.
    e. $\log_4 1 = 0$ because _____.

18. Find exact values for each of the following quantities. Do not use a calculator.
    a. $\log_3 81$ **b.** $\log_2 1024$ **c.** $\log_3 \left(\frac{1}{27}\right)$ **d.** $\log_2 1$
    e. $\log_{10} \left(\frac{1}{10}\right)$ **f.** $\log_3 3$ **g.** $\log_2(2^k)$

19. Use the definition of logarithm to prove that for any positive real number b with $b \neq 1$, $\log_b b = 1$.

20. Use the definition of logarithm to prove that for any positive real number b with $b \neq 1$, $\log_b 1 = 0$.

21. If b is any positive real number with $b \neq 1$ and x is any real number, b^{-x} is defined as follows: $b^{-x} = \frac{1}{b^x}$. Use this definition and the definition of logarithm to prove that $\log_b \left(\frac{1}{u}\right) = -\log_b(u)$ for all positive real numbers u and b, with $b \neq 1$.

H 22. Use the unique factorization for the integers theorem (Section 4.3) and the definition of logarithm to prove that $\log_3(7)$ is irrational.

23. If b and y are positive real numbers such that $\log_b y = 3$, what is $\log_{1/b}(y)$? Why?

24. If b and y are positive real numbers such that $\log_b y = 2$, what is $\log_{b^2}(y)$? Why?

25. Let $A = \{2, 3, 5\}$ and $B = \{x, y\}$. Let p_1 and p_2 be the **projections of $A \times B$ onto the first and second coordinates.** That is, for each pair $(a, b) \in A \times B$, $p_1(a, b) = a$ and $p_2(a, b) = b$.

    a. Find $p_1(2, y)$ and $p_1(5, x)$. What is the range of p_1?
    b. Find $p_2(2, y)$ and $p_2(5, x)$. What is the range of p_2?

26. Observe that *mod* and *div* can be defined as functions from $\mathbf{Z}^{nonneg} \times \mathbf{Z}^+$ to $\mathbf{Z}$. For each ordered pair (n, d) consisting of a nonnegative integer n and a positive integer d, let

    $mod(n, d) = n \bmod d$ (the nonnegative remainder obtained when n is divided by d).

    $div(n, d) = n \operatorname{div} d$ (the integer quotient obtained when n is divided by d).

    Find each of the following:
    a. *mod* $(67, 10)$ and *div* $(67, 10)$
    b. *mod* $(59, 8)$ and *div* $(59, 8)$
    c. *mod* $(30, 5)$ and *div* $(30, 5)$

27. Student A tries to define a function $g: \mathbf{Q} \to \mathbf{Z}$ by the rule

    $$g\left(\frac{m}{n}\right) = m - n, \text{ for all integers } m \text{ and } n \text{ with } n \neq 0.$$

    Student B claims that g is not well defined. Justify student B's claim.

28. Student C tries to define a function $h: \mathbf{Q} \to \mathbf{Q}$ by the rule

    $$h\left(\frac{m}{n}\right) = \frac{m^2}{n}, \text{ for all integers } m \text{ and } n \text{ with } n \neq 0.$$

    Student D claims that h is not well defined. Justify student D's claim.

29. Let $J_5 = \{0, 1, 2, 3, 4\}$. Then $J_5 - \{0\} = \{1, 2, 3, 4\}$. Student A tries to define a function $R: J_5 - \{0\} \to \mathbf{Z}$ as follows: For each $x \in J_5 - \{0\}$,

    $$R(x) \text{ is the number } y \text{ so that } (xy) \bmod 5 = 1.$$

    Student B claims that R is not well defined. Who is right: student A or student B? Justify your answer.

30. Let $J_4 = \{0, 1, 2, 3\}$. Then $J_4 - \{0\} = \{1, 2, 3\}$. Student C tries to define a function $S\colon J_4 - \{0\} \to \mathbf{Z}$ as follows: For each $x \in J_4 - \{0\}$,

$$S(x) \text{ is the number } y \text{ so that } (xy) \bmod 4 = 1.$$

Student D claims that S is not well defined. Who is right: student C or student D? Justify your answer.

31. Let $X = \{a, b, c\}$ and $Y = \{r, s, t, u, v, w\}$. Define $f\colon X \to Y$ as follows: $f(a) = v$, $f(b) = v$, and $f(c) = t$.
    a. Draw an arrow diagram for f.
    b. Let $A = \{a, b\}$, $C = \{t\}$, $D = \{u, v\}$, and $E = \{r, s\}$. Find $f(A)$, $f(X)$, $f^{-1}(C)$, $f^{-1}(D)$, $f^{-1}(E)$, and $f^{-1}(Y)$.

32. Let $X = \{1, 2, 3, 4\}$ and $Y = \{a, b, c, d, e\}$. Define $g\colon X \to Y$ as follows: $g(1) = a$, $g(2) = a$, $g(3) = a$, and $g(4) = d$.
    a. Draw an arrow diagram for g.
    b. Let $A = \{2, 3\}$, $C = \{a\}$, and $D = \{b, c\}$. Find $g(A)$, $g(X)$, $g^{-1}(C)$, $g^{-1}(D)$, and $g^{-1}(Y)$.

H 33. Let X and Y be sets, let A and B be any subsets of X, and let F be a function from X to Y. Fill in the blanks in the following proof that $F(A) \cup F(B) \subseteq F(A \cup B)$.

Proof: Let y be any element in $F(A) \cup F(B)$. *[We must show that y is in $F(A \cup B)$.]* By definition of union, $\underline{(a)}$.

Case 1, $y \in F(A)$: In this case, by definition of $F(A)$, $y = F(x)$ for $\underline{(b)}$ $x \in A$. Since $A \subseteq A \cup B$, it follows from the definition of union that $x \in \underline{(c)}$. Hence, $y = F(x)$ for some $x \in A \cup B$, and thus, by definition of $F(A \cup B)$, $y \in \underline{(d)}$.

Case 2, $y \in F(B)$: In this case, by definition of $F(B)$, $\underline{(e)}$ $x \in B$. Since $B \subseteq A \cup B$ it follows from the definition of union that $\underline{(f)}$.

Therefore, regardless of whether $y \in F(A)$ or $y \in F(B)$, we have that $y \in F(A \cup B)$ *[as was to be shown]*.

In 34–41 let X and Y be sets, let A and B be any subsets of X, and let C and D be any subsets of Y. Determine which of the properties are true for all functions F from X to Y and which are false for at least one function F from X to Y. Justify your answers.

34. If $A \subseteq B$ then $F(A) \subseteq F(B)$.

35. $F(A \cap B) \subseteq F(A) \cap F(B)$

36. $F(A) \cap F(B) \subseteq F(A \cap B)$

37. For all subsets A and B of X, $F(A - B) = F(A) - F(B)$.

38. For all subsets C and D of Y, if $C \subseteq D$, then
    $$F^{-1}(C) \subseteq F^{-1}(D).$$

H 39. For all subsets C and D of Y,
    $$F^{-1}(C \cup D) = F^{-1}(C) \cup F^{-1}(D).$$

40. For all subsets C and D of Y,
    $$F^{-1}(C \cap D) = F^{-1}(C) \cap F^{-1}(D).$$

41. For all subsets C and D of Y,
    $$F^{-1}(C - D) = F^{-1}(C) - F^{-1}(D).$$

42. $F(F^{-1}(C)) \subseteq C$

43. Given a set S and a subset A, the **characteristic function of A**, denoted χ_A, is the function defined from S to $\mathbf{Z}$ with the property that for all $u \in S$,

    $$\chi_A(u) = \begin{cases} 1 & \text{if } u \in A \\ 0 & \text{if } u \notin A. \end{cases}$$

    Show that each of the following holds for all subsets A and B of S and all $u \in S$.
    a. $\chi_{A \cap B}(u) = \chi_A(u) \cdot \chi_B(u)$
    b. $\chi_{A \cup B}(u) = \chi_A(u) + \chi_B(u) - \chi_A(u) \cdot \chi_B(u)$

Each of exercises 44–46 refers to the Euler phi function, denoted ϕ, which is defined as follows: For each integer $n \geq 1$, $\phi(n)$ is the number of positive integers less than or equal to n that have no common factors with n except ± 1. For example, $\phi(10) = 4$ because there are four positive integers less than or equal to 10 that have no common factors with 10 except ± 1; namely, 1, 3, 7, and 9.

44. Find each of the following:
    a. $\phi(15)$ **b.** $\phi(2)$ **c.** $\phi(5)$
    d. $\phi(12)$ e. $\phi(11)$ f. $\phi(1)$

✶ 45. Prove that if p is a prime number and n is an integer with $n \geq 1$, then $\phi(p^n) = p^n - p^{n-1}$.

H 46. Prove that there are infinitely many integers n for which $\phi(n)$ is a perfect square.

Answers for Test Yourself

1. the unique output element in Y that is related to x by f 2. the value of f at x; the image of x under f; the output of f for the input x 3. the set of all y in Y such that $f(x) = y$ 4. an inverse image of y under f; a preimage of y 5. $\{x \in X \mid f(x) = y\}$; the inverse image of y 6. $f(x) = g(x)$ for all $x \in X$ 7. the exponent to which b must be raised to obtain x (*Or*: the real number y such that $x = b^y$)   8. $\{y \in Y \mid y = f(x) \text{ for some } x \in A\}$ (*Or*: $\{f(x) \mid x \in A\}$)   9. $\{x \in X \mid f(x) \in C\}$

7.2 One-to-One and Onto, Inverse Functions

Don't accept a statement just because it is printed. — Anna Pell Wheeler, 1883–1966

In this section we discuss two important properties that functions may satisfy: the property of being *one-to-one* and the property of being *onto*. Functions that satisfy both properties arc called *one-to-one correspondences* or *one-to-one onto functions*. When a function is a one-to-one correspondence, the elements of its domain and co-domain match up perfectly, and we can define an *inverse function* from the co-domain to the domain that "undoes" the action of the function.

One-to-One Functions

In Section 7.1 we noted that a function may send several elements of its domain to the same element of its co-domain. In terms of arrow diagrams, this means that two or more arrows that start in the domain can point to the same element in the co-domain. On the other hand, if no two arrows that start in the domain point to the same element of the co-domain then the function is called *one-to-one* or *injective*. For a one-to-one function, each element of the co-domain is the image of at most one element of the domain.

• Definition

Let F be a function from a set X to a set Y. F is **one-to-one** (or **injective**) if, and only if, for all elements x_1 and x_2 in X,

$$\text{if } F(x_1) = F(x_2), \text{ then } x_1 = x_2,$$

or, equivalently, $\text{if } x_1 \neq x_2, \text{ then } F(x_1) \neq F(x_2).$

Symbolically,

$$F: X \to Y \text{ is one-to-one} \quad \Leftrightarrow \quad \forall x_1, x_2 \in X, \text{ if } F(x_1) = F(x_2) \text{ then } x_1 = x_2.$$

To obtain a precise statement of what it means for a function *not* to be one-to-one, take the negation of one of the equivalent versions of the definition above. Thus:

A function $F: X \to Y$ is *not* one-to-one $\Leftrightarrow$ $\exists$ elements x_1 and x_2 in X with $F(x_1) = F(x_2)$ and $x_1 \neq x_2$.

That is, if elements x_1 and x_2 can be found that have the same function value but are not equal, then F is not one-to-one.

In terms of arrow diagrams, a one-to-one function can be thought of as a function that separates points. That is, it takes distinct points of the domain to distinct points of the co-domain. A function that is not one-to-one fails to separate points. That is, at least two points of the domain are taken to the same point of the co-domain. This is illustrated in Figure 7.2.1 on the next page.

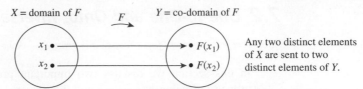

Figure 7.2.1(a) A One-to-One Function Separates Points

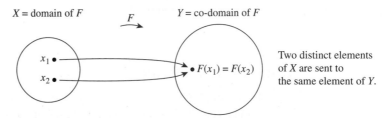

Figure 7.2.1(b) A Function That Is Not One-to-One Collapses Points Together

Example 7.2.1 Identifying One-to-One Functions Defined on Finite Sets

a. Do either of the arrow diagrams in Figure 7.2.2 define one-to-one functions?

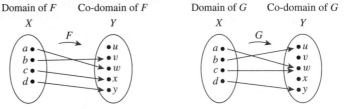

Figure 7.2.2

b. Let $X = \{1, 2, 3\}$ and $Y = \{a, b, c, d\}$. Define $H: X \rightarrow Y$ as follows: $H(1) = c$, $H(2) = a$, and $H(3) = d$. Define $K: X \rightarrow Y$ as follows: $K(1) = d$, $K(2) = b$, and $K(3) = d$. Is either H or K one-to-one?

Solution

a. F is one-to-one but G is not. F is one-to-one because no two different elements of X are sent by F to the same element of Y. G is not one-to-one because the elements a and c are both sent by G to the same element of Y: $G(a) = G(c) = w$ but $a \neq c$.

b. H is one-to-one but K is not. H is one-to-one because each of the three elements of the domain of H is sent by H to a different element of the co-domain: $H(1) \neq H(2)$, $H(1) \neq H(3)$, and $H(2) \neq H(3)$. K, however, is not one-to-one because $K(1) = K(3) = d$ but $1 \neq 3$. ∎

One-to-One Functions on Infinite Sets

Now suppose f is a function defined on an infinite set X. By definition, f is one-to-one if, and only if, the following universal statement is true:

$$\forall x_1, x_2 \in X, \text{ if } f(x_1) = f(x_2) \text{ then } x_1 = x_2.$$

Thus, to prove f is one-to-one, you will generally use the method of direct proof:

suppose x_1 and x_2 are elements of X such that $f(x_1) = f(x_2)$

and   **show** that $x_1 = x_2$.

To show that f is *not* one-to-one, you will ordinarily

find elements x_1 and x_2 in X so that $f(x_1) = f(x_2)$ but $x_1 \neq x_2$.

Example 7.2.2 Proving or Disproving That Functions Are One-to-One

Define $f: \mathbf{R} \to \mathbf{R}$ and $g: \mathbf{Z} \to \mathbf{Z}$ by the rules

$$f(x) = 4x - 1 \quad \text{for all} \quad x \in \mathbf{R}$$
$$g(n) = n^2 \quad \text{for all} \quad n \in \mathbf{Z}.$$

and

a. Is f one-to-one? Prove or give a counterexample.

b. Is g one-to-one? Prove or give a counterexample.

Solution   It is usually best to start by taking a positive approach to answering questions like these. Try to prove the given functions are one-to-one and see whether you run into difficulty. If you finish without running into any problems, then you have a proof. If you do encounter a problem, then analyzing the problem may lead you to discover a counterexample.

a. The function $f: \mathbf{R} \to \mathbf{R}$ is defined by the rule

$$\boxed{f(x) = 4x - 1 \quad \text{for all real numbers } x.}$$

To prove that f is one-to-one, you need to prove that

$\forall$ real numbers x_1 and x_2, if $f(x_1) = f(x_2)$ then $x_1 = x_2$.

Substituting the definition of f into the outline of a direct proof, you

suppose x_1 and x_2 are any real numbers such that $4x_1 - 1 = 4x_2 - 1$,

and   **show** that $x_1 = x_2$.

Can you reach what is to be shown from the supposition? Of course. Just add 1 to both sides of the equation in the supposition and then divide both sides by 4.

This discussion is summarized in the following formal answer.

Answer to (a):

If the function $f: \mathbf{R} \to \mathbf{R}$ is defined by the rule $f(x) = 4x - 1$, for all real numbers x, then f is one-to-one.

Proof:

Suppose x_1 and x_2 are real numbers such that $f(x_1) = f(x_2)$. *[We must show that $x_1 = x_2$.]* By definition of f,

$$4x_1 - 1 = 4x_2 - 1.$$

Adding 1 to both sides gives

$$4x_1 = 4x_2,$$

and dividing both sides by 4 gives

$$x_1 = x_2,$$

which is what was to be shown.

b. The function $g: \mathbf{Z} \to \mathbf{Z}$ is defined by the rule

$$g(n) = n^2 \quad \text{for all integers } n.$$

As done previously, you start as though you were going to prove that g is one-to-one. Substituting the definition of g into the outline of a direct proof, you

suppose n_1 and n_2 are integers such that $n_1^2 = n_2^2$,

and　　　　**try to show** that $n_1 = n_2$.

Can you reach what is to be shown from the supposition? No! It is quite possible for two numbers to have the same squares and yet be different. For example, $2^2 = (-2)^2$ but $2 \neq -2$.

Thus, in trying to prove that g is one-to-one, you run into difficulty. But analyzing this difficulty leads to the discovery of a counterexample, which shows that g is not one-to-one.

This discussion is summarized as follows:

Answer to (b):

If the function $g: \mathbf{Z} \to \mathbf{Z}$ is defined by the rule $g(n) = n^2$, for all $n \in \mathbf{Z}$, then g is not one-to-one.

Counterexample:

Let $n_1 = 2$ and $n_2 = -2$. Then by definition of g,

$$g(n_1) = g(2) = 2^2 = 4 \quad \text{and also}$$
$$g(n_2) = g(-2) = (-2)^2 = 4.$$

Hence　　　　$g(n_1) = g(n_2) \quad \text{but} \quad n_1 \neq n_2,$

and so g is not one-to-one.

Onto Functions

It was noted in Section 7.1 that there may be an element of the co-domain of a function that is not the image of any element in the domain. On the other hand, *every* element of a function's co-domain may be the image of some element of its domain. Such a function is called *onto* or *surjective*. When a function is onto, its range is equal to its co-domain.

• Definition

Let F be a function from a set X to a set Y. F is **onto** (or **surjective**) if, and only if, given any element y in Y, it is possible to find an element x in X with the property that $y = F(x)$.

Symbolically:

$$F: X \to Y \text{ is onto} \quad \Leftrightarrow \quad \forall y \in Y, \exists x \in X \text{ such that } F(x) = y.$$

To obtain a precise statement of what it means for a function *not* to be onto, take the negation of the definition of onto:

$$F: X \to Y \text{ is } not \text{ onto} \iff \exists y \text{ in } Y \text{ such that } \forall x \in X, F(x) \neq y.$$

That is, there is some element in Y that is *not* the image of *any* element in X.

In terms of arrow diagrams, a function is onto if each element of the co-domain has an arrow pointing to it from some element of the domain. A function is not onto if at least one element in its co-domain does not have an arrow pointing to it. This is illustrated in Figure 7.2.3.

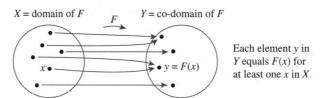

Figure 7.2.3(a) A Function That Is Onto

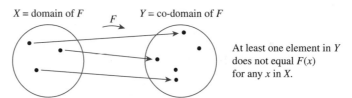

Figure 7.2.3(b) A Function That Is Not Onto

Example 7.2.3 Identifying Onto Functions Defined on Finite Sets

a. Do either of the arrow diagrams in Figure 7.2.4 define onto functions?

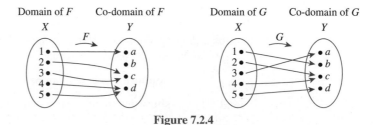

Figure 7.2.4

b. Let $X = \{1, 2, 3, 4\}$ and $Y = \{a, b, c\}$. Define $H: X \to Y$ as follows: $H(1) = c$, $H(2) = a$, $H(3) = c$, $H(4) = b$. Define $K: X \to Y$ as follows: $K(1) = c$, $K(2) = b$, $K(3) = b$, and $K(4) = c$. Is either H or K onto?

Solution

a. F is not onto because $b \neq F(x)$ for any x in X. G is onto because each element of Y equals $G(x)$ for some x in X: $a = G(3)$, $b = G(1)$, $c = G(2) = G(4)$, and $d = G(5)$.

b. H is onto but K is not. H is onto because each of the three elements of the co-domain of H is the image of some element of the domain of H: $a = H(2)$, $b = H(4)$, and $c = H(1) = H(3)$. K, however, is not onto because $a \neq K(x)$ for any x in $\{1, 2, 3, 4\}$.

∎

Onto Functions on Infinite Sets

Now suppose F is a function from a set X to a set Y, and suppose Y is infinite. By definition, F is onto if, and only if, the following universal statement is true:

$$\forall y \in Y, \exists x \in X \text{ such that } F(x) = y.$$

Thus to prove F is onto, you will ordinarily use the method of generalizing from the generic particular:

> **suppose** that y is any element of Y

and > **show** that there is an element X of X with $F(x) = y$.

To prove F is *not* onto, you will usually

> **find** an element y of Y such that $y \neq F(x)$ for *any* x in X.

Example 7.2.4 Proving or Disproving That Functions Are Onto

Define $f: \mathbf{R} \to \mathbf{R}$ and $h: \mathbf{Z} \to \mathbf{Z}$ by the rules

$$f(x) = 4x - 1 \quad \text{for all } x \in \mathbf{R}$$

and $$h(n) = 4n - 1 \quad \text{for all } n \in \mathbf{Z}.$$

a. Is f onto? Prove or give a counterexample.

b. Is h onto? Prove or give a counterexample.

Solution

a. The best approach is to start trying to prove that f is onto and be alert for difficulties that might indicate that it is not. Now $f: \mathbf{R} \to \mathbf{R}$ is the function defined by the rule

$$f(x) = 4x - 1 \quad \text{for all real numbers } x.$$

To prove that f is onto, you must prove

$$\forall y \in Y, \exists x \in X \text{ such that } f(x) = y.$$

Substituting the definition of f into the outline of a proof by the method of generalizing from the generic particular, you

> **suppose** y is a real number

and > **show** that there exists a real number x such that $y = 4x - 1$.

Scratch Work: **If** such a real number x exists, then

$$4x - 1 = y$$
$$4x = y + 1 \qquad \text{by adding 1 to both sides}$$
$$x = \frac{y + 1}{4} \qquad \text{by dividing both sides by 4.}$$

Caution!   This scratch work only proves what x has to be *if* it exists. The scratch work does not prove that x exists.

Thus *if* such a number x exists, it must equal $(y + 1)/4$. Does such a number exist? Yes. To show this, let $x = (y + 1)/4$, and then make sure that (1) x is a real number and that (2) f really does send x to y. The following formal answer summarizes this process.

Answer to (a):

If $f: \mathbf{R} \to \mathbf{R}$ is the function defined by the rule $f(x) = 4x - 1$ for all real numbers x, then f is onto.

Proof:

Let $y \in \mathbf{R}$. *[We must show that $\exists x$ in $\mathbf{R}$ such that $f(x) = y$.]* Let $x = (y+1)/4$. Then x is a real number since sums and quotients (other than by 0) of real numbers are real numbers. It follows that

$$f(x) = f\left(\frac{y+1}{4}\right) \qquad \text{by substitution}$$

$$= 4 \cdot \left(\frac{y+1}{4}\right) - 1 \quad \text{by definition of } f$$

$$= (y+1) - 1 = y \quad \text{by basic algebra.}$$

[This is what was to be shown.]

---

b. The function $h: \mathbf{Z} \to \mathbf{Z}$ is defined by the rule

$$h(n) = 4n - 1 \quad \text{for all integers } n.$$

To prove that h is onto, it would be necessary to prove that

$$\forall \text{ integers } m, \exists \text{ an integer } n \text{ such that } h(n) = m.$$

Substituting the definition of h into the outline of a proof by the method of generalizing from the generic particular, you

suppose m is any integer

and **try to show** that there is an integer n with $4n - 1 = m$.

Can you reach what is to be shown from the supposition? No! If $4n - 1 = m$, then

$$n = \frac{m+1}{4} \quad \text{by adding 1 and dividing by 4.}$$

But n must be an integer. And when, for example, $m = 0$, then

$$n = \frac{0+1}{4} = \frac{1}{4},$$

which is *not* an integer.

Thus, in trying to prove that h is onto, you run into difficulty, and this difficulty reveals a counterexample that shows h is not onto.

This discussion is summarized in the following formal answer.

Answer to (b):

If the function $h: \mathbf{Z} \to \mathbf{Z}$ is defined by the rule $h(n) = 4n - 1$ for all integers n, then h is not onto.

Counterexample:

The co-domain of h is $\mathbf{Z}$ and $0 \in \mathbf{Z}$. But $h(n) \neq 0$ for any integer n. For if $h(n) = 0$, then

$$4n - 1 = 0 \qquad \text{by definition of } h$$

which implies that

$$4n = 1 \qquad \text{by adding 1 to both sides}$$

and so

$$n = \frac{1}{4} \qquad \text{by dividing both sides by 4.}$$

But $1/4$ is not an integer. Hence there is no integer n for which $f(n) = 0$, and thus f is not onto.

∎

Relations between Exponential and Logarithmic Functions

Note That the quantity b^x is a real number for any real number x follows from the least-upper-bound property of the real number system. (See Appendix A.)

For positive numbers $b \neq 1$, the **exponential function with base b,** denoted $\exp_b$, is the function from $\mathbf{R}$ to $\mathbf{R}^+$ defined as follows: For all real numbers x,

$$\exp_b(x) = b^x$$

where $b^0 = 1$ and $b^{-x} = 1/b^x$.

When working with the exponential function, it is useful to recall the laws of exponents from elementary algebra.

Laws of Exponents

If b and c are any positive real numbers and u and v are any real numbers, the following laws of exponents hold true:

$$b^u b^v = b^{u+v} \qquad\qquad 7.2.1$$

$$(b^u)^v = b^{uv} \qquad\qquad 7.2.2$$

$$\frac{b^u}{b^v} = b^{u-v} \qquad\qquad 7.2.3$$

$$(bc)^u = b^u c^u \qquad\qquad 7.2.4$$

In Section 7.1 the logarithmic function with base b was defined for any positive number $b \neq 1$ to be the function from $\mathbf{R}^+$ to $\mathbf{R}$ with the property that for each positive real number x,

$$\log_b(x) = \text{ the exponent to which } b \text{ must be raised to obtain } x.$$

Or, equivalently, for each positive real number x and real number y,

$$\log_b x = y \quad \Leftrightarrow \quad b^y = x.$$

It can be shown using calculus that both the exponential and logarithmic functions are one-to-one and onto. Therefore, by definition of one-to-one, the following properties hold true:

For any positive real number b with $b \neq 1$,

$$\text{if } b^u = b^v \text{ then } u = v \quad \text{for all real numbers } u \text{ and } v, \qquad \text{7.2.5}$$

and

$$\text{if } \log_b u = \log_b v \text{ then } u = v \quad \text{for all positive real numbers } u \text{ and } v. \qquad \text{7.2.6}$$

These properties are used to derive many additional facts about exponents and logarithms. In particular we have the following properties of logarithms.

Theorem 7.2.1 Properties of Logarithms

For any positive real numbers $b, c, x,$ and y with $b \neq 1$ and $c \neq 1$ and for all real numbers a:

a. $\log_b(xy) = \log_b x + \log_b y$

b. $\log_b \left(\dfrac{x}{y} \right) = \log_b x - \log_b y$

c. $\log_b(x^a) = a \log_b x$

d. $\log_c x = \dfrac{\log_b x}{\log_b c}$

Theorem 7.2.1(d) is proved in the next example. You are asked to prove the remainder of the theorem in exercises 27–29 at the end of this section.

Example 7.2.5 Using the One-to-Oneness of the Exponential Function

Use the definition of logarithm, the laws of exponents, and the one-to-oneness of the exponential function (property 7.2.5) to prove part (d) of Theorem 7.2.1: For any positive real numbers $b, c,$ and x, with $b \neq 1$ and $c \neq 1$,

$$\log_c x = \frac{\log_b x}{\log_b c}.$$

Solution Suppose positive real numbers $b, c,$ and x are given with $b \neq 1$ and $c \neq 1$. Let

$$(1) \ \ u = \log_b c \qquad (2) \ \ v = \log_c x \qquad (3) \ \ w = \log_b x.$$

Then, by definition of logarithm,

$$(1') \ \ c = b^u \qquad (2') \ \ x = c^v \qquad (3') \ \ x = b^w.$$

Substituting $(1')$ into $(2')$ and using one of the laws of exponents gives

$$x = c^v = (b^u)^v = b^{uv} \qquad \text{by 7.2.2}$$

But by (3), $x = b^w$ also. Hence

$$b^{uv} = b^w,$$

and so by the one-to-oneness of the exponential function (property 7.2.5),

$$uv = w.$$

Substituting from (1), (2), and (3) gives that

$$(\log_b c)(\log_c x) = \log_b x.$$

And dividing both sides by $\log_b c$ (which is nonzero because $c \neq 1$) results in

$$\log_c x = \frac{\log_b x}{\log_b c}. \qquad \blacksquare$$

Example 7.2.6 Computing Logarithms with Base 2 on a Calculator

In computer science it is often necessary to compute logarithms with base 2. Most calculators do not have keys to compute logarithms with base 2 but do have keys to compute logarithms with base 10 (called **common logarithms** and often denoted simply log) and logarithms with base e (called **natural logarithms** and usually denoted ln). Suppose your calculator shows that $\ln 5 \cong 1.609437912$ and $\ln 2 \cong 0.6931471806$. Use Theorem 7.2.1(d) to find an approximate value for $\log_2 5$.

Solution   By Theorem 7.2.1(d),

$$\log_2 5 = \frac{\ln 5}{\ln 2} \cong \frac{1.609437912}{0.6931471806} \cong 2.321928095. \qquad \blacksquare$$

One-to-One Correspondences

Consider a function $F: X \to Y$ that is both one-to-one and onto. Given any element x in X, there is a unique corresponding element $y = F(x)$ in Y (since F is a function). Also given any element y in Y, there is an element x in X such that $F(x) = y$ (since F is onto) and there is only one such x (since F is one-to-one). Thus, a function that is one-to-one and onto sets up a pairing between the elements of X and the elements of Y that matches each element of X with exactly one element of Y and each element of Y with exactly one element of X. Such a pairing is called a *one-to-one correspondence* or *bijection* and is illustrated by the arrow diagram in Figure 7.2.5. One-to-one correspondences are often used as aids to counting. The pairing of Figure 7.2.5, for example, shows that there are five elements in the set X.

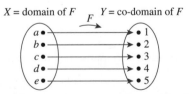

Figure 7.2.5 An Arrow Diagram for a One-to-One Correspondence

---

• **Definition**

A **one-to-one correspondence** (or **bijection**) from a set X to a set Y is a function $F: X \to Y$ that is both one-to-one and onto.

Example 7.2.7 A Function from a Power Set to a Set of Ordered Pairs

Let $\mathscr{P}(\{a, b\})$ be the set of all subsets of $\{a, b\}$ and let S be the set of all ordered pairs made up of 0's and 1's. Then $\mathscr{P}(\{a, b\}) = \{\emptyset, \{a\}, \{b\}, \{a, b\}\}$ and $S = \{(0, 0), (0, 1), (1, 0), (1, 1)\}$. Define a function h from $\mathscr{P}(\{a, b\})$ to S as follows: Given any subset A of $\{a, b\}$, a is either in A or not in A, and b is either in A or not in A. If a is in A, write a 1 in the first position of the ordered pair $h(A)$. If a is not in A, write a 0 in the first position of the ordered pair $h(A)$. Similarly, if b is in A, write a 1 in the second position of the ordered pair $h(A)$. If b is not in A, write a 0 in the second position of the ordered pair $h(A)$. This definition is summarized in the following table.

$$h$$

Subset of $\{a, b\}$	Status of a	Status of b	Ordered Pair
$\emptyset$	not in	not in	$(0, 0)$
$\{a\}$	in	not in	$(1, 0)$
$\{b\}$	not in	in	$(0, 1)$
$\{a, b\}$	in	in	$(1, 1)$

Note In generalizing this example to functions defined on the power set of a set with n elements, ordered n-tuples are used instead of ordered pairs.

Is h a one-to-one correspondence?

Solution   The arrow diagram shown in Figure 7.2.6 shows clearly that h is a one-to-one correspondence. It is onto because each element of S has an arrow pointing to it. It is one-to-one because each element of S has no more than one arrow pointing to it.

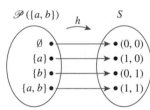

Figure 7.2.6

Example 7.2.8 A Function of Two Variables

Define a function $F: \mathbf{R} \times \mathbf{R} \to \mathbf{R} \times \mathbf{R}$ as follows: For all $(x, y) \in \mathbf{R} \times \mathbf{R}$,

$$F(x, y) = (x + y, x - y).$$

Is F a one-to-one correspondence from $\mathbf{R} \times \mathbf{R}$ to itself?

Solution   The answer is yes. To show that F is a one-to-one correspondence, you need to show both that F is one-to-one and that F is onto.

Proof that F is one-to-one: Suppose that (x_1, y_1) and (x_2, y_2) are any ordered pairs in $\mathbf{R} \times \mathbf{R}$ such that

$$F(x_1, y_1) = F(x_2, y_2).$$

[We must show that $(x_1, y_1) = (x_2, y_2)$.] By definition of F,

$$(x_1 + y_1, x_1 - y_1) = (x_2 + y_2, x_2 - y_2).$$

For two ordered pairs to be equal, both the first and second components must be equal. Thus x_1, y_1, x_2, and y_2 satisfy the following system of equations:

$$x_1 + y_1 = x_2 + y_2 \tag{1}$$
$$x_1 - y_1 = x_2 - y_2 \tag{2}$$

Adding equations (1) and (2) gives that

$$2x_1 = 2x_2, \quad \text{and so} \quad x_1 = x_2.$$

Substituting $x_1 = x_2$ into equation (1) yields

$$x_1 + y_1 = x_1 + y_2, \quad \text{and so} \quad y_1 = y_2.$$

Thus, by definition of equality of ordered pairs, $(x_1, y_1) = (x_2, y_2)$ *[as was to be shown].*

Caution! This scratch work only shows what (r, s) has to be *if* it exists. The scratch work does not prove that (r, s) exists.

Scratch Work for the Proof that F is onto: To prove that F is onto, you suppose you have any ordered pair in the co-domain $\mathbf{R} \times \mathbf{R}$, say (u, v), and then you show that there is an ordered pair in the domain that is sent to (u, v) by F. To do this, you suppose temporarily that you have found such an ordered pair, say (r, s). Then

$$F(r, s) = (u, v) \qquad \text{because you are supposing that } F \text{ sends}(r, s) \text{ to } (u, v),$$

and

$$F(r, s) = (r + s, r - s) \qquad \text{by definition of } F.$$

Equating the right-hand sides gives

$$(r + s, r - s) = (u, v).$$

By definition of equality of ordered pairs this means that

$$r + s = u \tag{1}$$
$$r - s = v \tag{2}$$

Adding equations (1) and (2) gives

$$2r = u + v, \quad \text{and so} \quad r = \frac{u+v}{2}.$$

Subtracting equation (2) from equation (1) yields

$$2s = u - v, \quad \text{and so} \quad s = \frac{u-v}{2}.$$

Thus, ***if*** F sends (r, s) to (u, v), then $r = (u + v)/2$ and $s = (u - v)/2$. To turn this scratch work into a proof, you need to make sure that (1) $\left(\frac{u+v}{2}, \frac{u-v}{2} \right)$ is in the domain of F, and (2) that F really does send $\left(\frac{u+v}{2}, \frac{u-v}{2} \right)$ to (u, v).

Proof that F is onto: Suppose (u, v) is any ordered pair in the co-domain of F. *[We will show that there is an ordered pair in the domain of F that is sent to (u, v) by F.]* Let

$$r = \frac{u+v}{2} \quad \text{and} \quad s = \frac{u-v}{2}.$$

Then (r, s) is an ordered pair of real numbers and so is in the domain of F. In addition:

$$
\begin{aligned}
F(r, s) &= F\left(\frac{u+v}{2}, \frac{u-v}{2} \right) & \text{by substitution} \\
&= \left(\frac{u+v}{2} + \frac{u-v}{2}, \frac{u+v}{2} - \frac{u-v}{2} \right) & \text{by definition of } F \\
&= \left(\frac{u+v+u-v}{2}, \frac{u+v-u+v}{2} \right) \\
&= \left(\frac{2u}{2}, \frac{2v}{2} \right) \\
&= (u, v) & \text{by algebra.}
\end{aligned}
$$

[This is what was to be shown.] ∎

Inverse Functions

If F is a one-to-one correspondence from a set X to a set Y, then there is a function from Y to X that "undoes" the action of F; that is, it sends each element of Y back to the element of X that it came from. This function is called the *inverse function* for F.

Theorem 7.2.2

Suppose $F: X \rightarrow Y$ is a one-to-one correspondence; that is, suppose F is one-to-one and onto. Then there is a function $\boldsymbol{F^{-1}: Y \rightarrow X}$ that is defined as follows:
  Given any element y in Y,

$$F^{-1}(y) = \text{that unique element } x \text{ in } X \text{ such that } F(x) \text{ equals } y.$$

In other words,

$$F^{-1}(y) = x \quad \Leftrightarrow \quad y = F(x).$$

The proof of Theorem 7.2.2 follows immediately from the definition of one-to-one and onto. Given an element y in Y, there is an element x in X with $F(x) = y$ because F is onto; x is unique because F is one-to-one.

• Definition

The function F^{-1} of Theorem 7.2.2 is called the **inverse function** for F.

Note that according to this definition, the logarithmic function with base $b > 0$ and $b \neq 1$ is the inverse of the exponential function with base b.

The diagram that follows illustrates the fact that an inverse function sends each element back to where it came from.

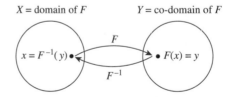

Example 7.2.9 Finding an Inverse Function for a Function Given by an Arrow Diagram

Define the inverse function for the one-to-one correspondence h given in Example 7.2.8.

Solution   The arrow diagram for h^{-1} is obtained by tracing the h-arrows back from S to $\mathscr{P}(\{a, b\})$ as shown below.

$$h^{-1}(0, 0) = \emptyset \quad h^{-1}(1, 0) = \{a\}$$
$$h^{-1}(0, 1) = \{b\} \quad h^{-1}(1, 1) = \{a, b\}$$

∎

Example 7.2.10 Finding an Inverse Function for a Function Given by a Formula

The function $f: \mathbf{R} \rightarrow \mathbf{R}$ defined by the formula

$$f(x) = 4x - 1 \quad \text{for all real numbers } x$$

was shown to be one-to-one in Example 7.2.2 and onto in Example 7.2.4. Find its inverse function.

Solution   For any *[particular but arbitrarily chosen]* y in **R**, by definition of f^{-1},

$$f^{-1}(y) = \text{that unique real number } x \text{ such that } f(x) = y.$$

But
$$f(x) = y$$
$$\Leftrightarrow \quad 4x - 1 = y \qquad \text{by definition of } f$$
$$\Leftrightarrow \qquad x = \frac{y+1}{4} \qquad \text{by algebra.}$$

Hence $f^{-1}(y) = \dfrac{y+1}{4}$.   ■

The following theorem follows easily from the definitions.

Theorem 7.2.3

If X and Y are sets and $F: X \to Y$ is one-to-one and onto, then $F^{-1}: Y \to X$ is also one-to-one and onto.

Proof:

F^{-1} **is one-to-one:** Suppose y_1 and y_2 are elements of Y such that $F^{-1}(y_1) = F^{-1}(y_2)$. *[We must show that $y_1 = y_2$.]* Let $x = F^{-1}(y_1) = F^{-1}(y_2)$. Then $x \in X$, and by definition of F^{-1},

$$F(x) = y_1 \quad \text{since } x = F^{-1}(y_1)$$
and
$$F(x) = y_2 \quad \text{since } x = F^{-1}(y_2).$$

Consequently, $y_1 = y_2$ since each is equal to $F(x)$. This is what was to be shown.

F^{-1} **is onto:** Suppose $x \in X$. *[We must show that there exists an element y in Y such that $F^{-1}(y) = x$.]* Let $y = F(x)$. Then $y \in Y$, and by definition of F^{-1}, $F^{-1}(y) = x$. This is what was to be shown.

Example 7.2.11 Finding an Inverse Function for a Function of Two Variables

Define the inverse function $F^{-1} : \mathbf{R} \times \mathbf{R} \to \mathbf{R} \times \mathbf{R}$ for the one-to-one correspondence given in Example 7.2.8.

Solution

The solution to Example 7.2.8 shows that $F\left(\frac{u+v}{2}, \frac{u-v}{2}\right) = (u, v)$. Because F is one-to-one, this means that

$\left(\frac{u+v}{2}, \frac{u-v}{2}\right)$ is the unique ordered pair in the domain of F that is sent to (u, v) by F.

Thus, F^{-1} is defined as follows: For all $(u, v) \in \mathbf{R} \times \mathbf{R}$,

$$F^{-1}(u, v) = \left(\frac{u+v}{2}, \frac{u-v}{2}\right).$$

■

Test Yourself

1. If F is a function from a set X to a set Y, then F is one-to-one if, and only if, _____.

2. If F is a function from a set X to a set Y, then F is not one-to-one if, and only if, _____.

3. If F is a function from a set X to a set Y, then F is onto if, and only if, _____.

4. If F is a function from a set X to a set Y, then F is not onto if, and only if, _____.

5. The following two statements are _____:

$$\forall u, v \in U, \text{ if } H(u) = H(v) \text{ then } u = v.$$
$$\forall u, v \in U, \text{ if } u \neq v \text{ then } H(u) \neq H(v).$$

6. Given a function $F: X \to Y$ and an infinite set X, to prove that F is one-to-one, you suppose that _____ and then you show that _____.

7. Given a function $F: X \to Y$ and an infinite set X, to prove that F is onto, you suppose that _____ and then you show that _____.

8. Given a function $F: X \to Y$, to prove that F is not one-to-one, you _____.

9. Given a function $F: X \to Y$, to prove that F is not onto, you _____.

10. A one-to-one correspondence from a set X to a set Y is a _____ that is _____.

11. If F is a one-to-one correspondence from a set X to a set Y and y is in Y, then $F^{-1}(y)$ is _____.

Exercise Set 7.2

1. The definition of one-to-one is stated in two ways:

$$\forall x_1, x_2 \in X, \text{ if } F(x_1) = F(x_2) \text{ then } x_1 = x_2$$

and $\quad \forall x_1, x_2 \in X, \text{ if } x_1 \neq x_2 \text{ then } F(x_1) \neq F(x_2).$

Why are these two statements logically equivalent?

2. Fill in each blank with the word *most* or *least*.
   a. A function F is one-to-one if, and only if, each element in the co-domain of F is the image of at _____ one element in the domain of F.
   b. A function F is onto if, and only if, each element in the co-domain of F is the image of at _____ one element in the domain of F.

H **3.** When asked to state the definition of one-to-one, a student replies, "A function f is one-to-one if, and only if, every element of X is sent by f to exactly one element of Y." Give a counterexample to show that the student's reply is incorrect.

H **4.** Let $f: X \to Y$ be a function. True or false? A sufficient condition for f to be one-to-one is that for all elements y in Y, there is at most one x in X with $f(x) = y$.

H **5.** All but two of the following statements are correct ways to express the fact that a function f is onto. Find the two that are incorrect.
   a. f is onto $\Leftrightarrow$ every element in its co-domain is the image of some element in its domain.
   b. f is onto $\Leftrightarrow$ every element in its domain has a corresponding image in its co-domain.
   c. f is onto $\Leftrightarrow \forall y \in Y, \exists x \in X$ such that $f(x) = y$.
   d. f is onto $\Leftrightarrow \forall x \in X, \exists y \in Y$ such that $f(x) = y$.
   e. f is onto $\Leftrightarrow$ the range of f is the same as the co-domain of f.

6. Let $X = \{1, 5, 9\}$ and $Y = \{3, 4, 7\}$.
   a. Define $f: X \to Y$ by specifying that
$$f(1) = 4, \quad f(5) = 7, \quad f(9) = 4.$$
   Is f one-to-one? Is f onto? Explain your answers.
   b. Define $g: X \to Y$ by specifying that
$$g(1) = 7, \quad g(5) = 3, \quad g(9) = 4.$$
   Is g one-to-one? Is g onto? Explain your answers.

7. Let $X = \{a, b, c, d\}$ and $Y = \{e, f, g\}$. Define functions F and G by the arrow diagrams below.

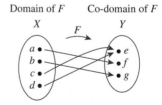

Domain of F Co-domain of F

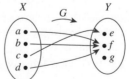

Domain of G Co-domain of G

   a. Is F one-to-one? Why or why not? Is it onto? Why or why not?
   b. Is G one-to-one? Why or why not? Is it onto? Why or why not?

8. Let $X = \{a, b, c\}$ and $Y = \{w, x, y, z\}$. Define functions H and K by the arrow diagrams below.

Domain of H Co-domain of H

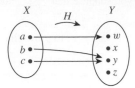

Domain of K Co-domain of K

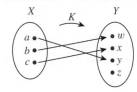

a. Is H one-to-one? Why or why not? Is it onto? Why or why not?

b. Is K one-to-one? Why or why not? Is it onto? Why or why not?

9. Let $X = \{1, 2, 3\}$, $Y = \{1, 2, 3, 4\}$, and $Z = \{1, 2\}$.

   a. Define a function $f: X \to Y$ that is one-to-one but not onto.

   b. Define a function $g: X \to Z$ that is onto but not one-to-one.

   c. Define a function $h: X \to X$ that is neither one-to-one nor onto.

   d. Define a function $k: X \to X$ that is one-to-one and onto but is not the identity function on X.

10. a. Define $f: \mathbf{Z} \to \mathbf{Z}$ by the rule $f(n) = 2n$, for all integers n.
    (i) Is f one-to-one? Prove or give a counterexample.
    (ii) Is f onto? Prove or give a counterexample.

    b. Let $2\mathbf{Z}$ denote the set of all even integers. That is, $2\mathbf{Z} = \{n \in \mathbf{Z} \mid n = 2k, \text{ for some integer } k\}$. Define $h: \mathbf{Z} \to 2\mathbf{Z}$ by the rule $h(n) = 2n$, for all integers n. Is h onto? Prove or give a counterexample.

H 11. a. Define $g: \mathbf{Z} \to \mathbf{Z}$ by the rule $g(n) = 4n - 5$, for all integers n.
    (i) Is g one-to-one? Prove or give a counterexample.
    (ii) Is g onto? Prove or give a counterexample.

    b. Define $G: \mathbf{R} \to \mathbf{R}$ by the rule $G(x) = 4x - 5$ for all real numbers x. Is G onto? Prove or give a counterexample.

12. a. Define $F: \mathbf{Z} \to \mathbf{Z}$ by the rule $F(n) = 2 - 3n$, for all integers n.
    (i) *Is F* one-to-one? Prove or give a counterexample.
    (ii) Is F onto? Prove or give a counterexample.

    b. Define $G: \mathbf{R} \to \mathbf{R}$ by the rule $G(x) = 2 - 3x$ for all real numbers x. Is G onto? Prove or give a counterexample.

13. **a.** Define $H: \mathbf{R} \to \mathbf{R}$ by the rule $H(x) = x^2$, for all real numbers x.
    (i) Is H one-to-one? Prove or give a counterexample.
    (ii) Is H onto? Prove or give a counterexample.

    b. Define $K: \mathbf{R}^{nonneg} \to \mathbf{R}^{nonneg}$ by the rule $K(x) = x^2$, for all nonnegative real numbers x. Is K onto? Prove or give a counterexample.

14. Explain the mistake in the following "proof."

    Theorem: The function $f: \mathbf{Z} \to \mathbf{Z}$ defined by the formula $f(n) = 4n + 3$, for all integers n, is one-to-one.

    "**Proof:** Suppose any integer n is given. Then by definition of f, there is only one possible value for $f(n)$, namely, $4n + 3$. Hence f is one-to-one."

In each of 15–18 a function f is defined on a set of real numbers. Determine whether or not f is one-to-one and justify your answer.

15. $f(x) = \dfrac{x + 1}{x}$, for all real numbers $x \neq 0$

16. $f(x) = \dfrac{x}{x^2 + 1}$, for all real numbers x

17. $f(x) = \dfrac{3x - 1}{x}$, for all real numbers $x \neq 0$

18. $f(x) = \dfrac{x + 1}{x - 1}$, for all real numbers $x \neq 1$

19. Define $F: \mathscr{P}(\{a, b, c\}) \to \mathbf{Z}$ as follows: For all A in $\mathscr{P}(\{a, b, c\})$,

$$F(A) = \text{the number of elements in } A.$$

    a. Is F one-to-one? Prove or give a counterexample.
    b. Is F onto? Prove or give a counterexample.

20. Define $S: \mathbf{Z}^+ \to \mathbf{Z}^+$ by the rule: For all integers n, $S(n) = $ the sum of the positive divisors of n.
    a. Is S one-to-one? Prove or give a counterexample.
    b. Is S onto? Prove or give a counterexample.

H 21. Let D be the set of all finite subsets of positive integers, and define $T: \mathbf{Z}^+ \to D$ by the rule: For all integers n, $T(n) = $ the set of all of the positive divisors of n.
    a. Is T one-to-one? Prove or give a counterexample.
    b. Is T onto? Prove or give a counterexample.

22. Define $G: \mathbf{R} \times \mathbf{R} \to \mathbf{R} \times \mathbf{R}$ as follows: $G(x, y) = (2y, -x)$ for all $(x, y) \in \mathbf{R} \times \mathbf{R}$.
    a. Is G one-to-one? Prove or give a counterexample.
    b. Is G onto? Prove or give a counterexample.

23. Define $H: \mathbf{R} \times \mathbf{R} \to \mathbf{R} \times \mathbf{R}$ as follows: $H(x, y) = (x + 1, 2 - y)$ for all $(x, y) \in \mathbf{R} \times \mathbf{R}$.
    a. Is H one-to-one? Prove or give a counterexample.
    b. Is H onto? Prove or give a counterexample.

24. Define $J: \mathbf{Q} \times \mathbf{Q} \to \mathbf{R}$ by the rule $J(r, s) = r + \sqrt{2}s$ for all $(r, s) \in \mathbf{Q} \times \mathbf{Q}$.
    a. Is J one-to-one? Prove or give a counterexample.
    b. Is J onto? Prove or give a counterexample.

✱ 25. Define $F: \mathbf{Z}^+ \times \mathbf{Z}^+ \to \mathbf{Z}^+$ and $G: \mathbf{Z}^+ \times \mathbf{Z}^+ \to \mathbf{Z}^+$ as follows: For all $(n, m) \in \mathbf{Z}^+ \times \mathbf{Z}^+$,

$$F(n, m) = 3^n 5^m \quad \text{and} \quad G(n, m) = 3^n 6^m.$$

H **a.** Is F one-to-one? Prove or give a counterexample.
   b. Is G one-to-one? Prove or give a counterexample.

26. **a.** Is $\log_8 27 = \log_2 3$? Why or why not?
    b. Is $\log_{16} 9 = \log_4 3$? Why or why not?

27. Prove that for all positive real numbers b, x, and y with $b \neq 1$,

$$\log_b\left(\frac{x}{y}\right) = \log_b x - \log_b y.$$

28. Prove that for all positive real numbers b, x, and y with $b \neq 1$,

$$\log_b(xy) = \log_b x + \log_b y.$$

H 29. Prove that for all real numbers a, b, and x with b and x positive and $b \neq 1$,

$$\log_b(x^a) = a \log_b x.$$

Exercises 30 and 31 use the following definition: If $f: \mathbf{R} \to \mathbf{R}$ and $g: \mathbf{R} \to \mathbf{R}$ are functions, then the function $(f + g): \mathbf{R} \to \mathbf{R}$ is defined by the formula $(f + g)(x) = f(x) + g(x)$ for all real numbers x.

30. If $f: \mathbf{R} \to \mathbf{R}$ and $g: \mathbf{R} \to \mathbf{R}$ are both one-to-one, is $f + g$ also one-to-one? Justify your answer.

31. If $f: \mathbf{R} \to \mathbf{R}$ and $g: \mathbf{R} \to \mathbf{R}$ are both onto, is $f + g$ also onto? Justify your answer.

Exercises 32 and 33 use the following definition: If $f: \mathbf{R} \to \mathbf{R}$ is a function and c is a nonzero real number, the function $(c \cdot f): \mathbf{R} \to \mathbf{R}$ is defined by the formula $(c \cdot f)(x) = c \cdot f(x)$ for all real numbers x.

32. Let $f: \mathbf{R} \to \mathbf{R}$ be a function and c a nonzero number. If f is one-to-one, is $c \cdot f$ also one-to-one? Justify your answer.

33. Let $f: \mathbf{R} \to \mathbf{R}$ be a function and c a nonzero real number. If f is onto, is $c \cdot f$ also onto? Justify your answer.

H 34. Suppose $F: X \to Y$ is one-to-one.
   a. Prove that for all subsets $A \subseteq X$, $F^{-1}(F(A)) = A$.
   b. Prove that for all subsets A_1 and A_2 in X, $F(A_1 \cap A_2) = F(A_1) \cap F(A_2)$.

35. Suppose $F: X \to Y$ is onto. Prove that for all subsets $B \subseteq Y$, $F(F^{-1}(B)) = B$.

Let $X = \{a, b, c, d, e\}$ and $Y = \{s, t, u, v, w\}$. In each of 36 and 37 a one-to-one correspondence $F: X \to Y$ is defined by an arrow diagram. In each case draw an arrow diagram for F^{-1}.

36.

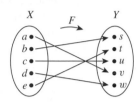

37.

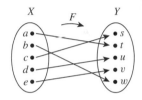

In 38–49 indicate which of the functions in the referenced exercise are one-to-one correspondences. For each function that is a one-to-one correspondence, find the inverse function.

38. Exercise 10a

39. Exercise 10b

40. Exercise 11a

41. Exercise 11b

42. Exercise 14b

43. Exercise 21

44. Exercise 22

45. Exercise 23

46. Exercise 15 with the co-domain taken to be the set of all real numbers not equal to 1.

H 47. Exercise 16 with the co-domain taken to be the set of all real numbers.

48. Exercise 17 with the co-domain taken to be the set of all real numbers not equal to 3.

49. Exercise 18 with the co-domain taken to be the set of all real numbers not equal to 1.

Answers for Test Yourself

1. for all x_1 and x_2 in X, if $F(x_1) = F(x_2)$ then $x_1 = x_2$ 2. there exist elements x_1 and x_2 in X such that $F(x_1) = F(x_2)$ and $x_1 \neq x_2$ 3. for all y in Y, there exists at least one element x in X such that $f(x) = y$ 4. there exists an element y in Y such that for all elements x in X, $f(x) \neq y$ 5. logically equivalent ways of expressing what it means for a function H to be one-to-one (The second is the contrapositive of the first.)   6. x_1 and x_2 are any *[particular but arbitrarily chosen]* elements in X with the property that $F(x_1) = F(x_2)$; $x_1 = x_2$ 7. y is any *[particular but arbitrarily chosen]* element in Y; there exists at least one element x in X such that $F(x) = y$ 8. show that there are concrete elements x_1 and x_2 in X with the property that $F(x_1) = F(x_2)$ and $x_1 \neq x_2$ 9. show that there is a concrete element y in Y with the property that $F(x) \neq y$ for any element x in X 10. function from X to Y; both one-to-one and onto   11. the unique element x in X such that $F(x) = y$ (in other words, $F^{-1}(y)$ is the unique preimage of y in X)

7.3 Composition of Functions

It is no paradox to say that in our most theoretical moods we may be nearest to our most practical applications. — Alfred North Whitehead

Consider two functions, the successor function and the squaring function, defined from **Z** (the set of integers) to **Z**, and imagine that each is represented by a machine. If the two machines are hooked up so that the output from the successor function is used as input to the squaring function, then they work together to operate as one larger machine. In this larger machine, an integer n is first increased by 1 to obtain $n + 1$; then the quantity $n + 1$ is squared to obtain $(n + 1)^2$. This is illustrated in the following drawing.

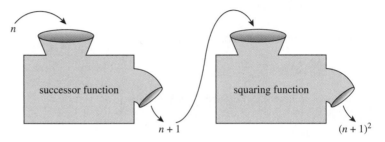

Combining functions in this way is called *composing* them; the resulting function is called the *composition* of the two functions. Note that the composition can be formed only if the output of the first function is acceptable input to the second function. That is, the range of the first function must be contained in the domain of the second function.

• Definition

Let $f: X \to Y$ and $g: Y \to Z$ be functions with the property that the range of f is a subset of the domain of g. Define a new function $g \circ f: X \to Z$ as follows:

$$(g \circ f)(x) = g(f(x)) \quad \text{for all } x \in X,$$

where $g \circ f$ is read "g circle f" and $g(f(x))$ is read "g of f of x." The function $g \circ f$ is called the **composition of f and g.**

Note We put the f first when we say "the composition of f and g" because an element x is acted upon first by f and then by g.

This definition is shown schematically below.

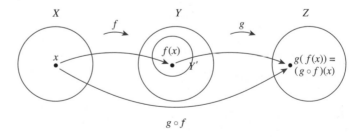

$$g \circ f$$

Example 7.3.1 Composition of Functions Defined by Formulas

Caution! Be careful not to confuse $g \circ f$ and $g(f(x))$: $g \circ f$ is the name of the function whereas $g(f(x))$ is the value of the function at x.

Let $f: \mathbf{Z} \to \mathbf{Z}$ be the successor function and let $g: \mathbf{Z} \to \mathbf{Z}$ be the squaring function. Then $f(n) = n + 1$ for all $n \in \mathbf{Z}$ and $g(n) = n^2$ for all $n \in \mathbf{Z}$.

a. Find the compositions $g \circ f$ and $f \circ g$.

b. Is $g \circ f = f \circ g$? Explain.

Solution

a. The functions $g \circ f$ and $f \circ g$ are defined as follows:

$$(g \circ f)(n) = g(f(n)) = g(n + 1) = (n + 1)^2 \quad \text{for all } n \in \mathbf{Z},$$

and

$$(f \circ g)(n) = f(g(n)) = f(n^2) = n^2 + 1 \quad \text{for all } n \in \mathbf{Z}.$$

b. Two functions from one set to another are equal if, and only if, they always take the same values. In this case,

$$(g \circ f)(1) = (1 + 1)^2 = 4, \text{ whereas } (f \circ g)(1) = 1^2 + 1 = 2.$$

Thus the two functions $g \circ f$ and $f \circ g$ are not equal:

$$g \circ f \neq f \circ g. \qquad\blacksquare$$

Example 7.3.1 illustrates the important fact that composition of functions is not a commutative operation: *For general functions F and G, F $\circ$ G need not necessarily equal G $\circ$ F* (although the two *may* be equal).

Example 7.3.2 Composition of Functions Defined on Finite Sets

Let $X = \{1, 2, 3\}$, $Y' = \{a, b, c, d\}$, $Y = \{a, b, c, d, e\}$, and $Z = \{x, y, z\}$. Define functions $f\colon X \to Y'$ and $g\colon Y \to Z$ by the arrow diagrams below.

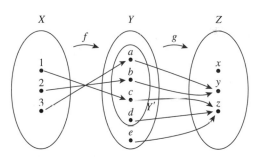

Draw the arrow diagram for $g \circ f$. What is the range of $g \circ f$?

Solution   To find the arrow diagram for $g \circ f$, just trace the arrows all the way across from X to Z through Y. The result is shown below.

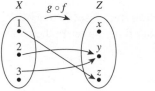

$$(g \circ f)(1) = g(f(1)) = g(c) = z$$
$$(g \circ f)(2) = g(f(2)) = g(b) = y$$
$$(g \circ f)(3) = g(f(3)) = g(a) = y$$

The range of $g \circ f$ is $\{y, z\}$. $\qquad\blacksquare$

Recall that the identity function on a set X, I_X, is the function from X to X defined by the formula

$$I_X(x) = x \quad \text{for all } x \in X.$$

That is, the identity function on X sends each element of X to itself. What happens when an identity function is composed with another function?

Example 7.3.3 Composition with the Identity Function

Let $X = \{a, b, c, d\}$ and $Y = \{u, v, w\}$, and suppose $f: X \to Y$ is given by the arrow diagram shown below.

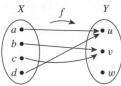

Find $f \circ I_X$ and $I_Y \circ f$.

Solution The values of $f \circ I_X$ are obtained by tracing through the arrow diagram shown below.

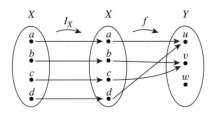

$$(f \circ I_X)(a) = f(I_X(a)) = f(a) = u$$
$$(f \circ I_X)(b) = f(I_X(b)) = f(b) = v$$
$$(f \circ I_X)(c) = f(I_X(c)) = f(c) = v$$
$$(f \circ I_X)(d) = f(I_X(d)) = f(d) = u$$

Note that for all elements x in X,

$$(f \circ I_X)(x) = f(x).$$

By definition of equality of functions, this means that $f \circ I_X = f$.

Similarly, the equality $I_Y \circ f = f$ can be verified by tracing through the arrow diagram below for each x in X and noting that in each case, $(I_Y \circ f)(x) = f(x)$.

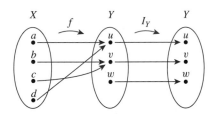

More generally, the composition of any function with an identity function equals the function.

Theorem 7.3.1 Composition with an Identity Function

If f is a function from a set X to a set Y, and I_X is the identity function on X, and I_Y is the identity function on Y, then

(a) $f \circ I_X = f$ and (b) $I_Y \circ f = f$.

Proof:

Part (a): Suppose f is a function from a set X to a set Y and I_X is the identity function on X. Then, for all x in X,

$$(f \circ I_X)(x) = f(I_X(x)) = f(x).$$

Hence, by definition of equality of functions, $f \circ I_X = f$, as was to be shown.

Part (b): This is exercise 13 at the end of this section.

Now let f be a function from a set X to a set Y, and suppose f has an inverse function f^{-1}. Recall that f^{-1} is the function from Y to X with the property that

$$f^{-1}(y) = x \quad \Leftrightarrow \quad f(x) = y.$$

What happens when f is composed with f^{-1}? Or when f^{-1} is composed with f?

Example 7.3.4 Composing a Function with Its Inverse

Let $X = \{a, b, c\}$ and $Y = \{x, y, z\}$. Define $f\colon X \to Y$ by the following arrow diagram.

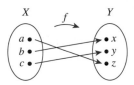

Then f is one-to-one and onto. Thus f^{-1} exists and is found by tracing the arrows backwards, as shown below.

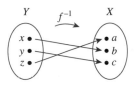

Now $f^{-1} \circ f$ is found by following the arrows from X to Y by f and back to X by f^{-1}. If you do this, you will see that

$$(f^{-1} \circ f)(a) = f^{-1}(f(a)) = f^{-1}(z) = a$$
$$(f^{-1} \circ f)(b) = f^{-1}(f(b)) = f^{-1}(x) = b$$

and $\qquad (f^{-1} \circ f)(c) = f^{-1}(f(c)) = f^{-1}(y) = c.$

Thus the composition of f and f^{-1} sends each element to itself. So by definition of the identity function,

$$f^{-1} \circ f = I_X.$$

In a similar way, you can see that

$$f \circ f^{-1} = I_Y. \qquad \blacksquare$$

More generally, the composition of any function with its inverse (if it has one) is an identity function. Intuitively, the function sends an element in its domain to an element in its co-domain and the inverse function sends it back again, so the composition of the two sends each element to itself. This reasoning is formalized in Theorem 7.3.2.

Theorem 7.3.2 Composition of a Function with Its Inverse

If $f: X \to Y$ is a one-to-one and onto function with inverse function $f^{-1}: Y \to X$, then

$$\text{(a) } f^{-1} \circ f = I_X \quad \text{and} \quad \text{(b) } f \circ f^{-1} = I_Y.$$

Proof:

Part (a): Suppose $f: X \to Y$ is a one-to-one and onto function with inverse function $f^{-1}: Y \to X$. *[To show that $f^{-1} \circ f = I_X$, we must show that for all $x \in X$, $(f^{-1} \circ f)(x) = x$.]* Let x be any element in X. Then

$$(f^{-1} \circ f)(x) = f^{-1}(f(x))$$

by definition of composition of functions. Now the inverse function f^{-1} satisfies the condition

$$f^{-1}(b) = a \quad \Leftrightarrow \quad f(a) = b \quad \text{for all } a \in X \text{ and } b \in Y. \qquad 7.3.1$$

Let

$$x' = f^{-1}(f(x)). \qquad 7.3.2$$

Apply property (7.3.1) with x' playing the role of a and $f(x)$ playing the role of b. Then

$$f(x') = f(x).$$

But since f is one-to-one, this implies that $x' = x$. Substituting x for x' in equation (7.3.2) gives

$$x = f^{-1}(f(x)).$$

Then by definition of composition of functions,

$$(f^{-1} \circ f)(x) = x,$$

as was to be shown.

Part (b): This is exercise 14 at the end of this section.

Composition of One-to-One Functions

The composition of functions interacts in interesting ways with the properties of being one-to-one and onto. What happens, for instance, when two one-to-one functions are composed? Must their composition be one-to-one? For example, let $X = \{a, b, c\}$, $Y = \{w, x, y, z\}$, and $Z = \{1, 2, 3, 4, 5\}$, and define one-to-one functions $f: X \to Y$ and $g: Y \to Z$ as shown in the arrow diagrams of Figure 7.3.1.

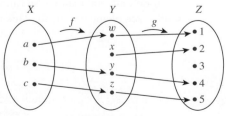

Figure 7.3.1

Then $g \circ f$ is the function with the arrow diagram shown in Figure 7.3.2.

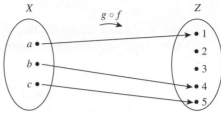

Figure 7.3.2

From the diagram it is clear that for these particular functions, the composition is one-to-one. This result is no accident. It turns out that the compositions of two one-to-one functions is always one-to-one.

Theorem 7.3.3

If $f: X \to Y$ and $g: Y \to Z$ are both one-to-one functions, then $g \circ f$ is one-to-one.

By the method of direct proof, the proof of Theorem 7.3.3 has the following starting point and conclusion to be shown.

Starting Point: Suppose f is a one-to-one function from X to Y and g is a one-to-one function from Y to Z.

To Show: $g \circ f$ is a one-to-one function from X to Z.

The conclusion to be shown says that a certain function is one-to-one. How do you show that? The crucial step is to realize that if you substitute $g \circ f$ into the definition of one-to-one, you see that

$$g \circ f \text{ is one-to-one} \quad \Leftrightarrow \quad \forall x_1, x_2 \in X, \text{ if } (g \circ f)(x_1) = (g \circ f)(x_2) \text{ then } x_1 = x_2.$$

By the method of direct proof, then, to show $g \circ f$ is one-to-one, you

suppose x_1 and x_2 are elements of X such that $(g \circ f)(x_1) = (g \circ f)(x_2)$,

and you

show that $x_1 = x_2$.

Now the heart of the proof begins. To show that $x_1 = x_2$, you work forward from the supposition that $(g \circ f)(x_1) = (g \circ f)(x_2)$, using the fact that f and g are both one-to-one. By definition of composition,

$$(g \circ f)(x_1) = g(f(x_1)) \quad \text{and} \quad (g \circ f)(x_2) = g(f(x_2)).$$

Since the left-hand sides of the equations are equal, so are the right-hand sides. Thus

$$g(f(x_1)) = g(f(x_2)).$$

Now just stare at the above equation for a moment. It says that

$$g(\text{something}) = g(\text{something else}).$$

Because g is a one-to-one function, any time g of one thing equals g of another thing, those two things are equal. Hence

$$f(x_1) = f(x_2).$$

But f is also a one-to-one function. Any time f of one thing equals f of another thing, those two things are equal. Therefore,

$$x_1 = x_2.$$

This is what was to be shown!

This discussion is summarized in the following formal proof.

Proof of Theorem 7.3.3:

Suppose $f: X \to Y$ and $g: Y \to Z$ are both one-to-one functions. *[We must show that $g \circ f$ is one-to-one.]* Suppose x_1 and x_2 are elements of X such that

$$(g \circ f)(x_1) = (g \circ f)(x_2).$$

[We must show that $x_1 = x_2$.] By definition of composition of functions,

$$g(f(x_1)) = g(f(x_2)).$$

Since g is one-to-one, $\qquad\qquad f(x_1) = f(x_2).$

And since f is one-to-one, $\qquad\qquad x_1 = x_2.$

[This is what was to be shown.] Hence $g \circ f$ is one-to-one.

Composition of Onto Functions

Now consider what happens when two onto functions are composed. For example, let $X = \{a, b, c, d, e\}$, $Y = \{w, x, y, z\}$, and $Z = \{1, 2, 3\}$. Define onto functions f and g by the following arrow diagrams.

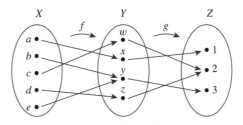

Then $g \circ f$ is the function with the arrow diagram shown below.

It is clear from the diagram that $g \circ f$ is onto.

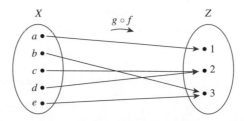

It turns out that the composition of any two onto functions (that can be composed) is onto.

> **Theorem 7.3.4**
>
> If $f: X \to Y$ and $g: Y \to Z$ are both onto functions, then $g \circ f$ is onto.

A direct proof of Theorem 7.3.4 has the following starting point and conclusion to be shown:

Starting Point: Suppose f is an onto function from X to Y, and g is an onto function from Y to Z.

To Show: $g \circ f$ is an onto function from X to Z.

The conclusion to be shown says that a certain function is onto. How do you show that? The crucial step is to realize that if you substitute $g \circ f$ into the definition of onto, you see that

> $g \circ f: X \to Z$ is onto $\Leftrightarrow$ given any element z of Z, it is possible to find an element x of X such that $(g \circ f)(x) = z$.

Since this statement is universal, to prove it you

suppose z is a [particular but arbitrarily chosen] element of Z

and   **show** that there is an element x in X such that $(g \circ f)(x) = z$.

Hence you must start the proof by supposing you are given a particular but arbitrarily chosen element in Z. Let us call it z. Your job is to find an element x in X such that $(g \circ f)(x) = z$.

    To find x, reason from the supposition that z is in Z, using the fact that both g and f are onto. Imagine arrow diagrams for the functions f and g.

Caution! To show that a function is onto, you *must* start with an arbitrary element of the co-domain and deduce that it is the image of some element in the domain.

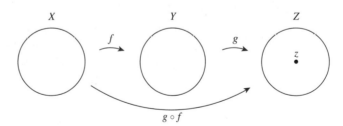

You have a particular element z in Z, and you need to find an element x in X such that when x is sent over to Z by $g \circ f$, its image will be z. Since g is onto, z is at the tip of some arrow coming from Y. That is, there is an element y in Y such that

$$g(y) = z. \qquad\qquad 7.3.3$$

This means that the arrow diagrams can be drawn as follows:

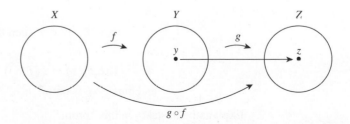

But f also is onto, so every element in Y is at the tip of an arrow coming from X. In particular, y is at the tip of some arrow. That is, there is an element x in X such that

$$f(x) = y. \qquad 7.3.4$$

The diagram, therefore, can be drawn as shown below.

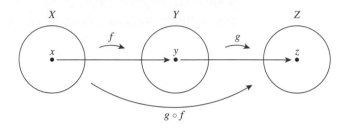

Now just substitute equation (7.3.4) into equation (7.3.3) to obtain

$$g(f(x)) = z.$$

But by definition of $g \circ f$,

$$g(f(x)) = (g \circ f)(x).$$

Hence

$$(g \circ f)(x) = z.$$

Thus x is an element of X that is sent by $g \circ f$ to z, and so x is the element you were supposed to find.

This discussion is summarized in the following formal proof.

Proof of Theorem 7.3.4:

Suppose $f: X \to Y$ and $g: Y \to Z$ are both onto functions. *[We must show that $g \circ f$ is onto.]* Let z be a *[particular but arbitrarily chosen]* element of Z. *[We must show the existence of an element x in X such that $(g \circ f)(x) = z$.]* Since g is onto, there is an element y in Y such that $g(y) = z$. And since f is onto, there is an element x in X such that $f(x) = y$. Hence there exists an element x in X such that

$$(g \circ f)(x) = g(f(x)) = g(y) = z$$

[as was to be shown]. It follows that $g \circ f$ is onto.

Example 7.3.5 An Incorrect "Proof" That a Function Is Onto

To prove that a composition of onto functions is onto, a student wrote,

"Suppose $f: X \to Y$ and $g: Y \to Z$ are both onto. Then

$$\forall y \in Y, \exists x \in X \text{ such that } f(x) = y \; (*)$$

and

$$\forall z \in Z, \exists y \in Y \text{ such that } f(y) = z.$$

So

$$(g \circ f)(x) = g(f(x)) = g(y) = z,$$

and thus $g \circ f$ is onto."

Explain the mistakes in this "proof."

Solution   To show that $g \circ f$ is onto, you must be able to meet the following challenge: If someone gives you an element z in Z (over which you have no control), you must be able to explain how to find an element x in X such that $(g \circ f)(x) = z$. Thus a proof that $g \circ f$ is onto must start with the assumption that you have been given a particular but arbitrarily chosen element of Z. This proof does not do that.

Moreover, note that statement (*) simply asserts that f is onto. An informal version of (*) is the following: Given any element in the co-domain of f, there is an element in the domain of f that is sent by f to the given element. Use of the symbols x and y to denote these elements is arbitrary. Any other two symbols could equally well have been used. Thus, if we replace the x and y in (*) by u and v, we obtain a logically equivalent statement, and the "proof" becomes the following:

"Suppose $f: X \to Y$ and $g: Y \to Z$ are both onto. Then
$$\forall v \in Y, \exists u \in X \text{ such that } f(u) = v$$
and
$$\forall z \in Z, \exists y \in Y \text{ such that } g(y) = z.$$
So (??!)
$$(g \circ f)(x) = g(f(x)) = g(y) = z,$$
and thus $g \circ f$ is onto."

From this logically equivalent version of the "proof," you can see that the statements leading up to the word *So* do not provide a rationale for the statement that follows it. The original reason for writing *So* was based on a misinterpretation of the meaning of the notation.   ■

Test Yourself

1. If f is a function from X to Y', g is a function from Y to Z, and $Y' \subseteq Y$, then $g \circ f$ is a function from _____ to _____, and $(g \circ f)(x) =$ _____ for all x in X.

2. If f is a function from X to Y and I_x and I_y are the identity functions from X to X and Y to Y, respectively, then $f \circ I_x =$ _____ and $I_y \circ f =$ _____.

3. If f is a one-to-one correspondence from X to Y, then $f^{-1} \circ f =$ _____ and $f \circ f^{-1} =$ _____.

4. If f is a one-to-one function from X to Y and g is a one-to-one function from Y to Z, you prove that $g \circ f$ is one-to-one by supposing that _____ and then showing that _____.

5. If f is an onto function from X to Y and g is an onto function from Y to Z, you prove that $g \circ f$ is onto by supposing that _____ and then showing that _____.

Exercise Set 7.3

In each of 1 and 2, functions f and g are defined by arrow diagrams. Find $g \circ f$ and $f \circ g$ and determine whether $g \circ f$ equals $f \circ g$.

1.

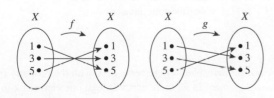

2.

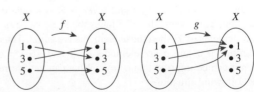

In 3 and 4, functions F and G are defined by formulas. Find $G \circ F$ and $F \circ G$ and determine whether $G \circ F$ equals $F \circ G$.

3. $F(x) = x^3$ and $G(x) = x - 1$, for all real numbers x.

4. $F(x) = x^5$ and $G(x) = x^{1/5}$ for all real numbers x.

5. Define $f: \mathbf{R} \to \mathbf{R}$ by the rule $f(x) = -x$ for all real numbers x. Find $(f \circ f)(x)$.

6. Define $F: \mathbf{Z} \to \mathbf{Z}$ and $G: \mathbf{Z} \to \mathbf{Z}$ by the rules $F(a) = 7a$ and $G(a) = a \bmod 5$ for all integers a. Find $(G \circ F)(0)$, $(G \circ F)(1)$, $(G \circ F)(2)$, $(G \circ F)(3)$, and $(G \circ F)(4)$.

7. Define $H: \mathbf{Z} \to \mathbf{Z}$ and $K: \mathbf{Z} \to \mathbf{Z}$ by the rules $H(a) = 6a$ and $K(a) = a \bmod 4$ for all integers a. Find $(K \circ H)(0)$, $(K \circ H)(1)$, $(K \circ H)(2)$, and $(K \circ H)(3)$.

8. Define $L: \mathbf{Z} \to \mathbf{Z}$ and $M: \mathbf{Z} \to \mathbf{Z}$ by the rules $L(a) = a^2$ and $M(a) = a \bmod 5$ for all integers a.
   a. Find $(L \circ M)(12)$, $(M \circ L)(12)$, $(L \circ M)(9)$, and $(M \circ L)(9)$.
   b. Is $L \circ M = M \circ L$?

The functions of each pair in 9–11 are inverse to each other. For each pair, check that both compositions give the identity function.

9. $F: \mathbf{R} \to \mathbf{R}$ and $F^{-1}: \mathbf{R} \to \mathbf{R}$ are defined by

$$F(x) = 3x + 2 \quad \text{and} \quad F^{-1}(y) = \frac{y - 2}{3},$$

for all $y \in \mathbf{R}$.

10. $G: \mathbf{R}^+ \to \mathbf{R}^+$ and $G^{-1}: \mathbf{R}^+ \to \mathbf{R}^+$ are defined by

$$G(x) = x^2 \quad \text{and} \quad G^{-1}(x) = \sqrt{x},$$

for all $x \in \mathbf{R}^+$.

11. H and H^{-1} are both defined from $\mathbf{R} - \{1\}$ to $\mathbf{R} - \{1\}$ by the formula

$$H(x) = H^{-1}(x) = \frac{x + 1}{x - 1}, \quad \text{for all } x \in \mathbf{R} - \{1\}.$$

12. Explain how it follows from the definition of logarithm that
   a. $\log_b(b^x) = x$, for all real numbers x.
   b. $b^{\log_b x} = x$, for all positive real numbers x.

H 13. Prove Theorem 7.3.1(b): If f is any function from a set X to a set Y, then $I_Y \circ f = f$, where I_Y is the identity function on Y.

14. Prove Theorem 7.3.2(b): If $f: X \to Y$ is a one-to-one and onto function with inverse function $f^{-1}: Y \to X$, then $f \circ f^{-1} = I_Y$, where I_Y is the identity function on Y.

15. Suppose Y and Z are sets and $g: Y \to Z$ is a one-to-one function. This means that if g takes the same value on any two elements of Y, then those elements are equal. Thus, for example, if a and b are elements of Y and $g(a) = g(b)$, then it can be inferred that $a = b$. What can be inferred in the following situations?

a. s_k and s_m are elements of Y and $g(s_k) = g(s_m)$.
b. $z/2$ and $t/2$ are elements of Y and $g(z/2) = g(t/2)$.
c. $f(x_1)$ and $f(x_2)$ are elements of Y and $g(f(x_1)) = g(f(x_2))$.

16. If $f: X \to Y$ and $g: Y \to Z$ are functions and $g \circ f$ is one-to-one, must g be one-to-one? Prove or give a counterexample.

17. If $f: X \to Y$ and $g: Y \to Z$ are functions and $g \circ f$ is onto, must f be onto? Prove or give a counterexample.

H 18. If $f: X \to Y$ and $g: Y \to Z$ are functions and $g \circ f$ is one-to-one, must f be one-to-one? Prove or give a counterexample.

H 19. If $f: X \to Y$ and $g: Y \to Z$ are functions and $g \circ f$ is onto, must g be onto? Prove or give a counterexample.

20. Let $f: W \to X$, $g: X \to Y$, and $h: Y \to Z$ be functions. Must $h \circ (g \circ f) = (h \circ g) \circ f$? Prove or give a counterexample.

21. True or False? Given any set X and given any functions $f: X \to X$, $g: X \to X$, and $h: X \to X$, if h is one-to-one and $h \circ f = h \circ g$, then $f = g$. Justify your answer.

22. True or False? Given any set X and given any functions $f: X \to X$, $g: X \to X$, and $h: X \to X$, if h is one-to-one and $f \circ h = g \circ h$, then $f = g$. Justify your answer.

In 23 and 24 find $g \circ f$, $(g \circ f)^{-1}$, g^{-1}, f^{-1}, and $f^{-1} \circ g^{-1}$, and state how $(g \circ f)^{-1}$ and $f^{-1} \circ g^{-1}$ are related.

23. Let $X = \{a, c, b\}$, $Y = \{x, y, z\}$, and $Z = \{u, v, w\}$. Define $f: X \to Y$ and $g: Y \to Z$ by the arrow diagrams below.

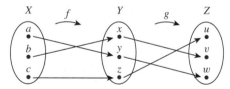

24. Define $f: \mathbf{R} \to \mathbf{R}$ and $g: \mathbf{R} \to \mathbf{R}$ by the formulas

$$f(x) = x + 3 \quad \text{and} \quad g(x) = -x \quad \text{for all } x \in \mathbf{R}.$$

25. Prove or give a counterexample: If $f: X \to Y$ and $g: Y \to X$ are functions such that $g \circ f = I_X$ and $f \circ g = I_Y$, then f and g are both one-to-one and onto and $g = f^{-1}$.

H 26. Suppose $f: X \to Y$ and $g: Y \to Z$ are both one-to-one and onto. Prove that $(g \circ f)^{-1}$ exists and that $(g \circ f)^{-1} = f^{-1} \circ g^{-1}$.

27. Let $f: X \to Y$ and $g: Y \to Z$. Is the following property true or false? For all subsets C in Z, $(g \circ f)^{-1}(C) = f^{-1}(g^{-1}(C))$. Justify your answer.

Answers for Test Yourself

1. X; Z; $g(f(x))$ 2. f; f 3. I_X; I_Y 4. x_1 and x_2 are any *[particular but arbitrarily chosen]* elements in X with the property that $(g \circ f)(x_1) = (g \circ f)(x_2)$; $x_1 = x_2$ 5. z is any *[particular but arbitrarily chosen]* element in Z; there exists at least one element x in X such that $(g \circ f)(x) = z$

7.4 Cardinality and Sizes of Infinity

There are as many squares as there are numbers because they are just as numerous as their roots. — Galileo Galilei, 1632

*Galileo Galilei
(1564–1642)*

Historically, the term *cardinal number* was introduced to describe the size of a set ("This set has *eight* elements") as distinguished from an *ordinal number* that refers to the order of an element in a sequence ("This is the *eighth* element in the row"). The definition of cardinal number derives from the primitive technique of representing numbers by fingers or tally marks. Small children, when asked how old they are, will often answer by holding up a certain number of fingers, each finger being paired with a year of their life. As was discussed in Section 7.2, a pairing of the elements of two sets is called a one-to-one correspondence. We say that two finite sets whose elements can be paired by a one-to-one correspondence have the *same size*. This is illustrated by the following diagram.

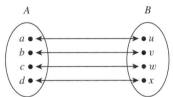

The elements of set *A* can be put into one-to-one correspondence with the elements of *B*.

Now a **finite set** is one that has no elements at all or that can be put into one-to-one correspondence with a set of the form $\{1, 2, \ldots, n\}$ for some positive integer n. By contrast, an **infinite set** is a nonempty set that cannot be put into one-to-one correspondence with $\{1, 2, \ldots, n\}$ for any positive integer n. Suppose that, as suggested by the quote from Galileo at the beginning of this section, we extend the concept of size to infinite sets by saying that one infinite set has the same size as another if, and only if, the first set can be put into one-to-one correspondence with the second. What consequences follow from such a definition? Do all infinite sets have the same size, or are some infinite sets larger than others? These are the questions we address in this section. The answers are sometimes surprising and have the interesting consequence that there are functions defined on the set of integers whose values cannot be computed on a computer.

• Definition

Let A and B be any sets. ***A* has the same cardinality as *B*** if, and only if, there is a one-to-one correspondence from A to B. In other words, A has the same cardinality as B if, and only if, there is a function f from A to B that is one-to-one and onto.

The following theorem gives some basic properties of cardinality, most of which follow from statements proved earlier about one-to-one and onto functions.

Theorem 7.4.1 Properties of Cardinality

For all sets A, B, and C:

a. **Reflexive property of cardinality:** A has the same cardinality as A.

b. **Symmetric property of cardinality:** If A has the same cardinality as B, then B has the same cardinality as A.

c. **Transitive property of cardinality:** If A has the same cardinality as B and B has the same cardinality as C, then A has the same cardinality as C.

---

continued on page 334

Proof:

Part (a), Reflexivity: Suppose A is any set. *[To show that A has the same cardinality as A, we must show there is a one-to-one correspondence from A to A.]* Consider the identity function I_A from A to A. This function is one-to-one because if x_1 and x_2 are any elements in A with $I_A(x_1) = I_A(x_2)$, then, by definition of I_A, $x_1 = x_2$. The identity function is also onto because if y is any element of A, then $y = I_A(y)$ by definition of I_A. Hence I_A is a one-to-one correspondence from A to A. *[So there exists a one-to-one correspondence from A to A, as was to be shown.]*

Part (b), Symmetry: Suppose A and B are any sets and A has the same cardinality as B. *[We must show that B has the same cardinality as A.]* Since A has the same cardinality as B, there is a function f from A to B that is one-to-one and onto. But then, by Theorems 7.2.2 and 7.2.3, there is a function f^{-1} from B to A that is also one-to-one and onto. Hence B has the same cardinality as A *[as was to be shown]*.

Part (c), Transitivity: Suppose A, B, and C are any sets and A has the same cardinality as B and B has the same cardinality as C. *[We must show that A has the same cardinality as C.]* Since A has the same cardinality as B, there is a function f from A to B that is one-to-one and onto, and since B has the same cardinality as C, there is a function g from B to C that is one-to-one and onto. But then, by Theorems 7.3.3 and 7.3.4, $g \circ f$ is a function from A to C that is one-to-one and onto. Hence A has the same cardinality as C *[as was to be shown]*.

Note that Theorem 7.4.1(b) makes it possible to say simply that two sets have the same cardinality instead of always having to say that one set has the same cardinality as another. That is, the following definition can be made.

• Definition

A and B **have the same cardinality** if, and only if, A has the same cardinality as B or B has the same cardinality as A.

The following example illustrates a very important property of infinite sets—namely, that an infinite set can have the same cardinality as a proper subset of itself. This property is sometimes taken as the definition of infinite set. The example shows that even though it may seem reasonable to say that there are twice as many integers as there are even integers, the elements of the two sets can be matched up exactly, and so, according to the definition, the two sets have the same cardinality.

Example 7.4.1 An Infinite Set and a Proper Subset Can Have the Same Cardinality

Let $2\mathbf{Z}$ be the set of all even integers. Prove that $2\mathbf{Z}$ and $\mathbf{Z}$ have the same cardinality.

Solution   Consider the function H from $\mathbf{Z}$ to $2\mathbf{Z}$ defined as follows:

$$H(n) = 2n \quad \text{for all } n \in \mathbf{Z}.$$

A (partial) arrow diagram for H is shown below.

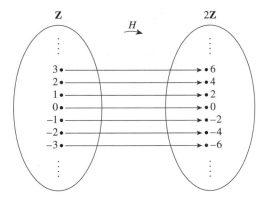

To show that H is one-to-one, suppose $H(n_1) = H(n_2)$ for some integers n_1 and n_2. Then $2n_1 = 2n_2$ by definition of H, and dividing both sides by 2 gives $n_1 = n_2$. Hence h is one-to-one.

To show that H is onto, suppose m is any element of $2\mathbf{Z}$. Then m is an even integer, and so $m = 2k$ for some integer k. It follows that $H(k) = 2k = m$. Thus there exists k in $\mathbf{Z}$ with $H(k) = m$, and hence H is onto.

Note So there are "as many" even integers as there are integers!

Therefore, by definition of cardinality, $\mathbf{Z}$ and $2\mathbf{Z}$ have the same cardinality. ∎

Countable Sets

The set $\mathbf{Z}^+$ of counting numbers $\{1, 2, 3, 4, \ldots\}$ is, in a sense, the most basic of all infinite sets. A set A having the same cardinality as this set is called *countably infinite*. The reason is that the one-to-one correspondence between the two sets can be used to "count" the elements of A: If F is a one-to-one and onto function from $\mathbf{Z}^+$ to A, then $F(1)$ can be designated as the first element of A, $F(2)$ as the second element of A, $F(3)$ as the third element of A, and so forth. This is illustrated graphically in Figure 7.4.1. Because F is one-to-one, no element is ever counted twice, and because it is onto, every element of A is counted eventually.

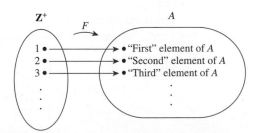

Figure 7.4.1 "Counting" a Countably Infinite Set

• Definition

A set is called **countably infinite** if, and only if, it has the same cardinality as the set of positive integers $\mathbf{Z}^+$. A set is called **countable** if, and only if, it is finite or countably infinite. A set that is not countable is called **uncountable.**

Example 7.4.2 Countability of Z, the Set of All Integers

Show that the set **Z** of all integers is countable.

Solution The set **Z** of all integers is certainly not finite, so if it is countable, it must be because it is countably infinite. To show that **Z** is countably infinite, find a function from the positive integers $\mathbf{Z}^+$ to **Z** that is one-to-one and onto. Looked at in one light, this contradicts common sense; judging from the diagram below, there appear to be more than twice as many integers as there are positive integers.

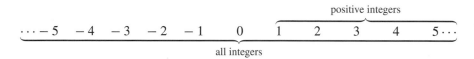

But you were alerted that results in this section might be surprising. Try to think of a way to "count" the set of all integers anyway.

The trick is to start in the middle and work outward systematically. Let the first integer be 0, the second 1, the third −1, the fourth 2, the fifth −2, and so forth as shown in Figure 7.4.2, starting at 0 and swinging outward in back-and-forth arcs from positive to negative integers and back again, picking up one additional integer at each swing.

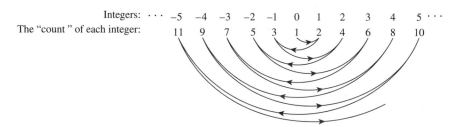

Figure 7.4.2 "Counting" the Set of All Integers

It is clear from the diagram that no integer is counted twice (so the function is one-to-one) and every integer is counted eventually (so the function is onto). Consequently, this diagram defines a function from $\mathbf{Z}^+$ to **Z** that is one-to-one and onto. Even though in one sense there seem to be more integers than positive integers, the elements of the two sets can be paired up one for one. It follows by definition of cardinality that $\mathbf{Z}^+$ has the same cardinality as **Z**. Thus **Z** is countably infinite and hence countable.

The diagrammatic description of the previous function is acceptable as given. You can check, however, that the function can also be described by the explicit formula

$$F(n) = \begin{cases} \dfrac{n}{2} & \text{if } n \text{ is an even positive integer} \\[2mm] -\dfrac{n-1}{2} & \text{if } n \text{ is an odd positive integer.} \end{cases} \qquad \blacksquare$$

In Section 9.4 we will show that a function from one finite set to another set of the same size is one-to-one if, and only if, it is onto. This result does not hold for infinite sets. Although it is true that for two infinite sets to have the same cardinality there must exist a function from one to the other that is both one-to-one and onto, it is also always the case that:

If A and B are infinite sets with the same cardinality, then there exist functions from A to B that are one-to-one but not onto and functions from A to B that are onto but not one-to-one.

For instance, since the function H in Example 7.4.2 is one-to-one and onto, $\mathbf{Z}$ and $\mathbf{Z}^+$ have the same cardinality. However, the "inclusion function" $i = \mathbf{Z}^+ \to \mathbf{Z}$, given by $i(n) = n$ for all positive integers n, is one-to-one but not onto. And the function $j = \mathbf{Z} \to \mathbf{Z}^+$ given by $j(n) = |n| + 1$, for all integers n, is onto but not one-to-one. (See exercise 6 at the end of this section.)

Example 7.4.3 Countability of 2**Z**, the Set of All Even Integers

Show that the set $2\mathbf{Z}$ of all even integers is countable.

Solution   Example 7.4.2 showed that $\mathbf{Z}^+$ has the same cardinality as $\mathbf{Z}$, and Example 7.4.1 showed that $\mathbf{Z}$ has the same cardinality as $2\mathbf{Z}$. Thus, by the transitive property of cardinality, $\mathbf{Z}^+$ has the same cardinality as $2\mathbf{Z}$. It follows by definition of countably infinite that $2\mathbf{Z}$ is countably infinite and hence countable. ∎

The Search for Larger Infinities: The Cantor Diagonalization Process

Every infinite set we have discussed so far has been countably infinite. Do any larger infinities exist? Are there uncountable sets? Here is one candidate.

Imagine the number line as shown below.

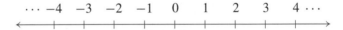

As noted in Section 1.2, the integers are spread along the number line at discrete intervals. The rational numbers, on the other hand, are *dense:* Between any two rational numbers, no matter how close, lies another rational number (the average of the two numbers, for instance; see exercise 17). This suggests the conjecture that the infinity of the set of rational numbers is larger than the infinity of the set of integers.

Amazingly, this conjecture is false. Despite the fact that the rational numbers are crowded onto the number line whereas the integers are quite separated, the set of all rational numbers can be put into one-to-one correspondence with the set of integers. The next example gives part of a proof of this fact. It shows that the set of all positive rational numbers can be put into one-to-one correspondence with the set of all positive integers. In exercise 16 at the end of this section you are asked to use this result, together with a technique similar to that of Example 7.4.2, to show that the set of *all* rational numbers is countable.

Example 7.4.4 The Set of All Positive Rational Numbers Is Countable

Show that the set $\mathbf{Q}^+$ of all positive rational numbers is countable.

Solution   Display the elements of the set $\mathbf{Q}^+$ of positive rational numbers in a grid as shown in Figure 7.4.3 on the next page.

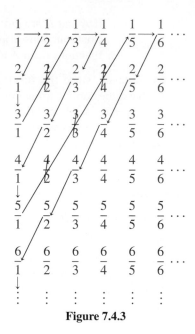

$$\frac{1}{1} \rightarrow \frac{1}{2} \quad \frac{1}{3} \rightarrow \frac{1}{4} \quad \frac{1}{5} \quad \frac{1}{6} \quad \cdots$$

$$\frac{2}{1} \quad \frac{2}{2} \quad \frac{2}{3} \quad \frac{2}{4} \quad \frac{2}{5} \quad \frac{2}{6} \quad \cdots$$

$$\frac{3}{1} \quad \frac{3}{2} \quad \frac{3}{3} \quad \frac{3}{4} \quad \frac{3}{5} \quad \frac{3}{6} \quad \cdots$$

$$\frac{4}{1} \quad \frac{4}{2} \quad \frac{4}{3} \quad \frac{4}{4} \quad \frac{4}{5} \quad \frac{4}{6} \quad \cdots$$

$$\frac{5}{1} \quad \frac{5}{2} \quad \frac{5}{3} \quad \frac{5}{4} \quad \frac{5}{5} \quad \frac{5}{6} \quad \cdots$$

$$\frac{6}{1} \quad \frac{6}{2} \quad \frac{6}{3} \quad \frac{6}{4} \quad \frac{6}{5} \quad \frac{6}{6} \quad \cdots$$

$$\vdots \qquad \vdots \qquad \vdots \qquad \vdots \qquad \vdots \qquad \vdots$$

Figure 7.4.3

Define a function F from $\mathbf{Z}^+$ to $\mathbf{Q}^+$ by starting to count at $\frac{1}{1}$ and following the arrows as indicated, skipping over any number that has already been counted.

To be specific: Set $F(1) = \frac{1}{1}$, $F(2) = \frac{1}{2}$, $F(3) = \frac{2}{1}$ and $F(4) = \frac{3}{1}$. Then skip $\frac{2}{2}$ since $\frac{2}{2} = \frac{1}{1}$, which was counted first. After that, set $F(5) = \frac{1}{3}$, $F(6) = \frac{1}{4}$, $F(7) = \frac{2}{3}$, $F(8) = \frac{3}{2}$, $F(9) = \frac{4}{1}$, and $F(10) = \frac{5}{1}$. Then skip $\frac{4}{2}$, $\frac{3}{3}$, and $\frac{2}{4}$ (since $\frac{4}{2} = \frac{2}{1}$, $\frac{3}{3} = \frac{1}{1}$, and $\frac{2}{4} = \frac{1}{2}$) and set $F(11) = \frac{1}{5}$. Continue in this way, defining $F(n)$ for each positive integer n.

Note that every positive rational number appears somewhere in the grid, and the counting procedure is set up so that every point in the grid is reached eventually. Thus the function F is onto. Also, skipping numbers that have already been counted ensures that no number is counted twice. Thus F is one-to-one. Consequently, F is a function from $\mathbf{Z}^+$ to $\mathbf{Q}^+$ that is one-to-one and onto, and so $\mathbf{Q}^+$ is countably infinite and hence countable. ∎

In 1874 the German mathematician Georg Cantor achieved success in the search for a larger infinity by showing that the set of all real numbers is uncountable. His method of proof was somewhat complicated, however. We give a proof of the uncountability of the set of all real numbers between 0 and 1 using a simpler technique introduced by Cantor in 1891 and now called the **Cantor diagonalization process.** Over the intervening years, this technique and variations on it have been used to establish a number of important results in logic and the theory of computation.

Before stating and proving Cantor's theorem, we note that every real number, which is a measure of location on a number line, can be represented by a decimal expansion of the form

$$a_0.a_1a_2a_3\ldots,$$

where a_0 is an integer (positive, negative, or zero) and for each $i \geq 1$, a_i is an integer from 0 through 9.

This way of thinking about numbers was developed over several centuries by mathematicians in the Chinese, Hindu, and Islamic worlds, culminating in the work of Ghiyāth al-Dīn Jamshīd al-Kashi in 1427. In Europe it was first clearly formulated and successfully promoted by the Flemish mathematician Simon Stevin in 1585. We illustrate the concept with an example.

al-Kashi
(1380–1429)

Simon Stevin
(1548–1620)

Bettmann/CORBIS

Consider the point P in Figure 7.4.4. Figure 7.4.4(a) shows P located between 1 and 2. When the interval from 1 to 2 is divided into ten equal subintervals (see Figure 7.4.4(b)) P is seen to lie between 1.6 and 1.7. If the interval from 1.6 to 1.7 is itself divided into ten equal subintervals (see Figure 7.4.4(c)), the P is seen to lie between 1.62 and 1.63 but closer to 1.62 than to 1.63. So the first three digits of the decimal expansion for P are 1.62.

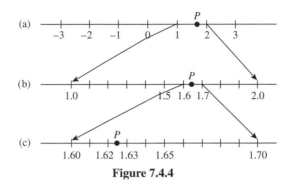

Figure 7.4.4

Assuming that any interval of real numbers, no matter how small, can be divided into ten equal subintervals, the process of obtaining additional digits in the decimal expansion for P can, in theory, be repeated indefinitely. If at any stage P is seen to be a subdivision point, then all further digits in the expansion may be taken to be 0. If not, then the process gives an expansion with an infinite number of digits.

The resulting decimal representation for P is unique except for numbers that end in infinitely repeating 9's or infinitely repeating 0's. For example (see exercise 25 at the end of this section),

$$0.199999\ldots = 0.200000\ldots.$$

Let us agree to express any such decimal in the form that ends in all 0's so that we will have a unique representation for every real number.

Theorem 7.4.2 (Cantor)

The set of all real numbers between 0 and 1 is uncountable.

Proof (by contradiction):

Suppose the set of all real numbers between 0 and 1 is countable. Then the decimal representations of these numbers can be written in a list as follows:

$$0.a_{11}a_{12}a_{13}\cdots a_{1n}\cdots$$
$$0.a_{21}a_{22}a_{23}\cdots a_{2n}\cdots$$
$$0.a_{31}a_{32}a_{33}\cdots a_{3n}\cdots$$
$$\vdots$$
$$0.a_{n1}a_{n2}a_{n3}\cdots a_{nn}\cdots$$
$$\vdots$$

[We will derive a contradiction by showing that there is a number between 0 and 1 that does not appear on this list.]

continued on page 340

For each pair of positive integers i and j, the jth decimal digit of the ith number on the list is a_{ij}. In particular, the first decimal digit of the first number on the list is a_{11}, the second decimal digit of the second number on the list is a_{22}, and so forth. As an example, suppose the list of real numbers between 0 and 1 starts out as follows:

$$
\begin{array}{llllllll}
0.\,②\,& 0 & 1 & 4 & 8 & 8 & 0 & 2\ldots \\
0.\,1 & ① & 6 & 6 & 6 & 0 & 2 & 1\ldots \\
0.\,0 & 3 & ③ & 5 & 3 & 3 & 2 & 0\ldots \\
0.\,9 & 6 & 7 & ⑦ & 6 & 8 & 0 & 9\ldots \\
0.\,0 & 0 & 0 & 3 & ① & 0 & 0 & 2\ldots \\
\end{array}
$$

$$\vdots$$

The diagonal elements are circled: a_{11} is 2, a_{22} is 1, a_{33} is 3, a_{44} is 7, a_{55} is 1, and so forth.

Construct a new decimal number $d = 0.d_1 d_2 d_3 \cdots d_n \cdots$ as follows:

$$
d_n = \begin{cases} 1 & \text{if } a_{nn} \neq 1 \\ 2 & \text{if } a_{nn} = 1 \end{cases}.
$$

In the previous example,

$$d_1 \text{ is } 1 \text{ because } a_{11} = 2 \neq 1,$$
$$d_2 \text{ is } 2 \text{ because } a_{22} = 1,$$
$$d_3 \text{ is } 1 \text{ because } a_{33} = 3 \neq 1,$$
$$d_4 \text{ is } 1 \text{ because } a_{44} = 7 \neq 1,$$
$$d_5 \text{ is } 2 \text{ because } a_{55} = 1,$$

and so forth. Hence d would equal $0.12112\ldots$.

The crucial observation is that for *each integer n, d differs in the nth decimal position from the nth number on the list*. But this implies that d is not on the list! In other words, d is a real number between 0 and 1 that is not on the list of *all* real numbers between 0 and 1. This contradiction shows the falseness of the supposition that the set of all numbers between 0 and 1 is countable. Hence the set of all real numbers between 0 and 1 is uncountable.

Along with demonstrating the existence of an uncountable set, Cantor developed a whole arithmetic theory of infinite sets of various sizes. One of the most basic theorems of the theory states that any subset of a countable set is countable.

Theorem 7.4.3

Any subset of any countable set is countable.

Proof:

Let A be a particular but arbitrarily chosen countable set and let B be any subset of A. *[We must show that B is countable.]* Either B is finite or it is infinite. If B is finite, then B is countable by definition of countable, and we are done. So suppose B is infinite. Since A is countable, the distinct elements of A can be represented as a sequence

$$a_1, a_2, a_3, \ldots.$$

Define a function $g: \mathbf{Z}^+ \to B$ inductively as follows:

1. Search sequentially through elements of $a_1, a_2, a_3, \ldots$ until an element of B is found. *[This must happen eventually since $B \subseteq A$ and $B \neq \emptyset$.]* Call that element $g(1)$.

2. For each integer $k \geq 2$, suppose $g(k-1)$ has been defined. Then $g(k-1) = a_i$ for some a_i in $\{a_1, a_2, a_3, \ldots\}$. Starting with a_{i+1}, search sequentially through $a_{i+1}, a_{i+2}, a_{i+3}, \ldots$ to find an element of B. One must be found eventually because B is infinite, and $\{g(1), g(2), \ldots, g(k-1)\}$ is a finite set. When an element of B is found, define it to be $g(k)$.

Note If $g(k-1) = a_i$, then $g(k)$ could also be defined by applying the well-ordering principle for the integers to the set $\{n \in \mathbf{Z} \mid n > i$ and $a_i \in B\}$.

By (1) and (2) above, the function g is defined for each positive integer.

Since the elements of $a_1, a_2, a_3, \ldots$ are all distinct, g is one-to-one. Furthermore, the searches for elements of B are sequential: Each picks up where the previous one left off. Thus every element of A is reached during some search. But all the elements of B are located somewhere in the sequence $a_1, a_2, a_3, \ldots$, and so every element of B is eventually found and made the image of some integer. Hence g is onto. These remarks show that g is a one-to-one correspondence from $\mathbf{Z}^+$ to B. So B is countably infinite and thus countable.

An immediate consequence of Theorem 7.4.3 is the following corollary.

Corollary 7.4.4

Any set with an uncountable subset is uncountable.

Proof:

Consider the following equivalent phrasing of Theorem 7.4.3: For all sets S and for all subsets A of S, if S is countable, then A is countable. The contrapositive of this statement is logically equivalent to it and states: For all sets S and for all subsets A of S, if A is uncountable then S is uncountable. But this is an equivalent phrasing for the corollary. So the corollary is proved.

Corollary 7.4.4 implies that the set of all real numbers is uncountable because the subset of numbers between 0 and 1 is uncountable. In fact, as Example 7.4.5 shows, the set of all real numbers has the same cardinality as the set of all real numbers between 0 and 1! This fact is further explored in exercises 13 and 14 at the end of this section.

Example 7.4.5 The Cardinality of the Set of All Real Numbers

Show that the set of all real numbers has the same cardinality as the set of real numbers between 0 and 1.

Solution   Let S be the open interval of real numbers between 0 and 1:

$$S = \{x \in \mathbf{R} \mid 0 < x < 1\}.$$

Imagine picking up S and bending it into a circle as shown on the next page. Since S does not include either endpoint 0 or 1, the top-most point of the circle is omitted from the drawing.

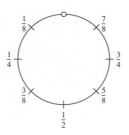

Define a function $F: S \to \mathbf{R}$ as follows:

Draw a number line and place the interval, S, somewhat enlarged and bent into a circle, tangent to the line above the point 0. This is shown below.

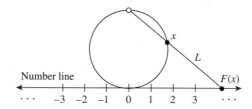

For each point x on the circle representing S, draw a straight line L through the topmost point of the circle and x. Let $F(x)$ be the point of intersection of L and the number line. ($F(x)$ is called the *projection* of x onto the number line.)

It is clear from the geometry of the situation that distinct points on the circle go to distinct points on the number line, so F is one-to-one. In addition, given any point y on the number line, a line can be drawn through y and the top-most point of the circle. This line must intersect the circle at some point x, and, by definition, $y = F(x)$. Thus F is onto. Hence F is a one-to-one correspondence from S to $\mathbf{R}$, and so S and $\mathbf{R}$ have the same cardinality. ∎

You know that every positive integer is a real number, so putting Example 7.4.5 together with Cantor's theorem (Theorem 7.4.2) shows that the infinity of the set of all real numbers is "greater" than the infinity of the set of all positive integers. In exercise 35, you are asked to show that any set and its power set have different cardinalities. Because there is a one-to-one function from any set to its power set (the function that takes each element a to the singleton set $\{a\}$), this implies that the cardinality of any set is "less than" the cardinality of its power set. As a result, you can create an infinite sequence of larger and larger infinities! For example, you could begin with $\mathbf{Z}$, the set of all integers, and take $\mathbf{Z}$, $\mathscr{P}(\mathbf{Z})$, $\mathscr{P}(\mathscr{P}(\mathbf{Z}))$, $\mathscr{P}(\mathscr{P}(\mathscr{P}(\mathbf{Z})))$, and so forth.

Test Yourself

1. A set is finite if, and only if, _____.

2. To prove that a set A has the same cardinality as a set B you must _____.

3. The reflexive property of cardinality says that given any set A, _____.

4. The symmetric property of cardinality says that given any sets A and B, _____.

5. The transitive property of cardinality says that given any sets A, B, and C, _____.

6. A set is called countably infinite if, and only if, _____.

7. A set is called countable if, and only if, _____.

8. In each of the following, fill in the blank with the word *countable* or the word *uncountable*.

   (a) The set of all integers is _____.

   (b) The set of all rational numbers is _____.

   (c) The set of all real numbers between 0 and 1 is _____.

   (d) The set of all real numbers is _____.

9. The Cantor diagonalization process is used to prove that _____.

Exercise Set 7.4

1. When asked what it means to say that set A has the same cardinality as set B, a student replies, "A and B are one-to-one and onto." What *should* the student have replied? Why?

2. Show that "there are as many squares as there are numbers" by exhibiting a one-to-one correspondence from the positive integers, $\mathbf{Z}^+$, to the set S of all squares of positive integers:
$$S = \{n \in \mathbf{Z}^+ \mid n = k^2, \text{ for some positive integer } k\}.$$

3. Let $3\mathbf{Z} = \{n \in \mathbf{Z} \mid n = 3k, \text{ for some integer } k\}$. Prove that $\mathbf{Z}$ and $3\mathbf{Z}$ have the same cardinality.

4. Let $\mathbf{O}$ be the set of all odd integers. Prove that $\mathbf{O}$ has the same cardinality as $2\mathbf{Z}$, the set of all even integers.

5. Let $25\mathbf{Z}$ be the set of all integers that are multiples of 25. Prove that $25\mathbf{Z}$ has the same cardinality as $2\mathbf{Z}$, the set of all even integers.

6. Check that the formula for F given at the end of Example 7.4.2 produces the correct values for $n = 1, 2, 3,$ and 4.

H **7.** Use the functions i and j defined in the paragraph following Example 7.4.2 to show that even though there is a one-to-one correspondence, F, from $\mathbf{Z}^+$ to $\mathbf{Z}$, there is also a function from $\mathbf{Z}^+$ to $\mathbf{Z}$ that is one-to-one but not onto and a function from $\mathbf{Z}$ to $\mathbf{Z}^+$ that is onto but not one-to-one. In other words, show that i is one-to-one but not onto, and show that j is onto but not one-to-one.

8. Use the result of exercise 3 to prove that $3\mathbf{Z}$ is countable.

9. Show that the set of all nonnegative integers is countable by exhibiting a one-to-one correspondence between $\mathbf{Z}^+$ and $\mathbf{Z}^{nonneg}$.

In 10–14, S denotes the set of real numbers strictly between 0 and 1. That is, $S = \{x \in \mathbf{R} \mid 0 < x < 1\}$.

10. Let $U = \{x \in \mathbf{R} \mid 0 < x < 2\}$. Prove that S and U have the same cardinality.

H **11.** Let $V = \{x \in \mathbf{R} \mid 2 < x < 5\}$. Prove that S and V have the same cardinality.

12. Let a and b be real numbers with $a < b$, and suppose that $W = \{x \in \mathbf{R} \mid a < x < b\}$. Prove that S and W have the same cardinality.

13. Draw the graph of the function f defined by the following formula:

For all real numbers x with $0 < x < 1$,
$$f(x) = \tan\left(\pi x - \frac{\pi}{2}\right).$$

Use the graph to explain why S and $\mathbf{R}$ have the same cardinality.

✱ **14.** Define a function g from the set of real numbers to S by the following formula:

For all real numbers x,
$$g(x) = \frac{1}{2} \cdot \left(\frac{x}{1 + |x|}\right) + \frac{1}{2}.$$

Prove that g is a one-to-one correspondence. (It is possible to prove this statement either with calculus or without it.) What conclusion can you draw from this fact?

15. Show that the set of all bit strings (strings of 0's and 1's) is countable.

16. Show that $\mathbf{Q}$, the set of all rational numbers, is countable.

17. Show that the set $\mathbf{Q}$ of all rational numbers is dense along the number line by showing that given any two rational numbers r_1 and r_2 with $r_1 < r_2$, there exists a rational number x such that $r_1 < x < r_2$.

H **18.** Must the average of two irrational numbers always be irrational? Prove or give a counterexample.

H✱ **19.** Show that the set of all irrational numbers is dense along the number line by showing that given any two real numbers, there is an irrational number in between.

20. Give two examples of functions from $\mathbf{Z}$ to $\mathbf{Z}$ that are one-to-one but not onto.

21. Give two examples of functions from $\mathbf{Z}$ to $\mathbf{Z}$ that are onto but not one-to-one.

H **22.** Define a function $g: \mathbf{Z}^+ \times \mathbf{Z}^+ \to \mathbf{Z}^+$ by the formula $g(m, n) = 2^m 3^n$ for all $(m, n) \in \mathbf{Z}^+ \times \mathbf{Z}^+$. Show that g is one-to-one and use this result to prove that $\mathbf{Z}^+ \times \mathbf{Z}^+$ is countable.

23. a. Explain how to use the following diagram to show that $\mathbf{Z}^{nonneg} \times \mathbf{Z}^{nonneg}$ and $\mathbf{Z}^{nonneg}$ have the same cardinality.

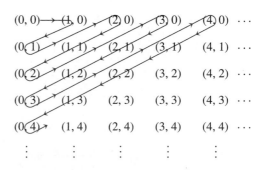

H✱ **b.** Define a function $H: \mathbf{Z}^{nonneg} \times \mathbf{Z}^{nonneg} \to \mathbf{Z}^{nonneg}$ by the formula
$$H(m, n) = n + \frac{(m + n)(m + n + 1)}{2}$$

for all nonnegative integers m and n. Interpret the action of H geometrically using the diagram of part (a).

✳ 24. Prove that the function H defined analytically in exercise 23b is a one-to-one correspondence.

H 25. Prove that $0.1999\ldots = 0.2$.

26. Prove that any infinite set contains a countably infinite subset.

27. If A is any countably infinite set, B is any set, and $g: A \rightarrow B$ is onto, then B is countable.

28. Prove that a disjoint union of any finite set and any countably infinite set is countably infinite.

H 29. Prove that a union of any two countably infinite sets is countably infinite.

H 30. Use the result of exercise 29 to prove that the set of all irrational numbers is uncountable.

H 31. Use the results of exercises 28 and 29 to prove that a union of any two countable sets is countable.

H 32. Prove that $\mathbf{Z} \times \mathbf{Z}$, the Cartesian product of the set of integers with itself, is countably infinite.

33. Use the results of exercises 27, 31, and 32 to prove the following: If R is the set of all solutions to all equations of the form $x^2 + bx + c = 0$, where b and c are integers, then R is countable.

H 34. Let $\mathscr{P}(S)$ be the set of all subsets of set S, and let T be the set of all functions from S to $\{0, 1\}$. Show that $\mathscr{P}(S)$ and T have the same cardinality.

H 35. Let S be a set and let $\mathscr{P}(S)$ be the set of all subsets of S. Show that S is "smaller than" $\mathscr{P}(S)$ in the sense that there is a one-to-one function from S to $\mathscr{P}(S)$ but there is no onto function from $\mathscr{P}(S)$ to S.

✳ 36. The Schroeder–Bernstein theorem states the following: If A and B are any sets with the property that there is a one-to-one function from A to B and a one-to-one function from B to A, then A and B have the same cardinality. Use this theorem to prove that there are as many functions from $\mathbf{Z}^+$ to $\{0, 1, 2, 3, 4, 5, 6, 7, 8, 9\}$ as there are functions from $\mathbf{Z}^+$ to $\{0, 1\}$.

H 37. Prove that if A and B are any countably infinite sets, then $A \times B$ is countably infinite.

✳ 38. Suppose $A_1, A_2, A_3, \ldots$ is an infinite sequence of countable sets. Recall that

$$\bigcup_{i=1}^{\infty} A_i = \{x \mid x \in A_i \text{ for some positive integer } i\}.$$

Prove that $\bigcup_{i=1}^{\infty} A_i$ is countable. (In other words, prove that a countably infinite union of countable sets is countable.)

Answers for Test Yourself

1. it is the empty set or there is a one-to-one correspondence from $\{1, 2, \ldots, n\}$ to it, where n is a positive integer   2. show that there exists a function from A to B that is one-to-one and onto (*Or*: show that there exists a one-to-one correspondence from A to B)   3. A has the same cardinality as A.   4. if A has the same cardinality as B, then B has the same cardinality as A 5. if A has the same cardinality as B and B has the same cardinality as C, then A has the same cardinality as C 6. it has the same cardinality as the set of all positive integers   7. it is finite or countably infinite   8. countable; countable; uncountable; uncountable   9. the set of all real numbers between 0 and 1 is uncountable

CHAPTER 8

RELATIONS

In this chapter we discuss the mathematics of relations defined on sets, focusing on ways to represent relations, exploring various properties they may have, and introducing the important concept of equivalence relation. Sections 8.4 and 8.5 extend the discussion to modular arithmetic, the Euclidean algorithm, and the solution of linear diophantine equations.

8.1 Relations on Sets

Strange as it may sound, the power of mathematics rests on its evasion of all unnecessary thought and on its wonderful saving of mental operations. — Ernst Mach, 1838–1916

Recall from Section 1.3 that if A and B are sets, then a (binary) relation from A to B is a subset of $A \times B$. Given an ordered pair (x, y) in $A \times B$, we say that ***x* is related to *y* by *R***, written ***x R y***, if, and only if, $(x, y) \in R$. Thus

$$(x, y) \in R \Leftrightarrow xRy$$

The term *binary* in the definition refers to the fact that the relation is a subset of the Cartesian product of two sets. Because we focus on binary relations in this text, when we use the term *relation* by itself, we will mean binary relation. A more general type of relation, called an *n-ary relation* forms the mathematical foundation for relational database theory.

Example 8.1.1 The Less-than Relation for Real Numbers

Define a relation L from **R** to **R** as follows: For all real numbers x and y,

$$x \, L \, y \Leftrightarrow x < y.$$

a. Is 57 L 53?   b. Is $(-17) \, L \, (-14)$?   c. Is 143 L 143?   d. Is $(-35) \, L \, 1$?

e. Draw the graph of L as a subset of the Cartesian plane **R** $\times$ **R**

Solution

a. No, $57 > 53$ b. Yes, $-17 < -14$ c. No, $143 = 143$ d. Yes, $-35 < 1$

e. For each value of x, all the points (x, y) with $y > x$ are on the graph. So the graph consists of all the points above the line $x = y$.

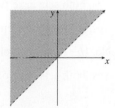

Example 8.1.2 The Congruence Modulo 2 Relation

Define a relation E from $\mathbf{Z}$ to $\mathbf{Z}$ as follows: For all $(m, n) \in \mathbf{Z} \times \mathbf{Z}$,

$$m \, E \, n \iff m - n \text{ is even.}$$

a. Is $4 \, E \, 0$? Is $2 \, E \, 6$? Is $3 \, E \, (-3)$? Is $5 \, E \, 2$?

b. List five integers that are related by E to 1.

c. Prove that if n is any odd integer, then $n \, E \, 1$.

Solution

a. Yes, $4 \, E \, 0$ because $4 - 0 = 4$ and 4 is even.
   Yes, $2 \, E \, 6$ because $2 - 6 = -4$ and -4 is even.
   Yes, $3 \, E \, (-3)$ because $3 - (-3) = 6$ and 6 is even.
   No, $5 \, \not{E} \, 2$ because $5 - 2 = 3$ and 3 is not even.

b. There are many such lists. One is

$$
\begin{array}{rl}
1 & \text{because } 1 - 1 = 0 \text{ is even,} \\
3 & \text{because } 3 - 1 = 2 \text{ is even,} \\
5 & \text{because } 5 - 1 = 4 \text{ is even,} \\
-1 & \text{because } -1 - 1 = -2 \text{ is even,} \\
-3 & \text{because } -3 - 1 = -4 \text{ is even.}
\end{array}
$$

c. **Proof:** Suppose n is any odd integer. Then $n = 2k + 1$ for some integer k. Now by definition of E, $n \, E \, 1$ if, and only if, $n - 1$ is even. But by substitution,

$$n - 1 = (2k + 1) - 1 = 2k,$$

and since k is an integer, $2k$ is even. Hence $n \, E \, 1$ *[as was to be shown].*

It can be shown (see exercise 2 at the end of this section) that integers m and n are related by E if, and only if, $m \bmod 2 = n \bmod 2$ (that is, both are even or both are odd). When this occurs m and n are said to be **congruent modulo 2.** ■

Example 8.1.3 A Relation on a Power Set

Let $X = \{a, b, c\}$. Then $\mathscr{P}(X) = \{\emptyset, \{a\}, \{b\}, \{c\}, \{a, b\}, \{a, c\}, \{b, c\}, \{a, b, c\}\}$. Define a relation $\mathbf{S}$ from $\mathscr{P}(X)$ to $\mathbf{Z}$ as follows: For all sets A and B in $\mathscr{P}(X)$ (i.e., for all subsets A and B of X),

$$A \, \mathbf{S} \, B \iff A \text{ has at least as many elements as } B.$$

a. Is $\{a, b\} \, \mathbf{S} \, \{b, c\}$?   b. Is $\{a\} \, \mathbf{S} \, \emptyset$?   c. Is $\{b, c\} \, \mathbf{S} \, \{a, b, c\}$?   d. Is $\{c\} \, \mathbf{S} \, \{a\}$?

Solution

a. Yes, both sets have two elements.

b. Yes, $\{a\}$ has one element and $\emptyset$ has zero elements, and $1 \geq 0$.

c. No, $\{b, c\}$ has two elements and $\{a, b, c\}$ has three elements and $2 < 3$.

d. Yes, both sets have one element. ■

The Inverse of a Relation

If R is a relation from A to B, then a relation R^{-1} from B to A can be defined by interchanging the elements of all the ordered pairs of R.

> **• Definition**
>
> Let R be a relation from A to B. Define the inverse relation R^{-1} from B to A as follows:
>
> $$R^{-1} = \{(y, x) \in B \times A \mid (x, y) \in R\}.$$

This definition can be written operationally as follows:

$$\text{For all } x \in A \text{ and } y \in B, \quad (y, x) \in R^{-1} \quad \Leftrightarrow \quad (x, y) \in R.$$

Example 8.1.4 The Inverse of a Finite Relation

Let $A = \{2, 3, 4\}$ and $B = \{2, 6, 8\}$ and let R be the "divides" relation from A to B: For all $(x, y) \in A \times B$,

$$x \, R \, y \quad \Leftrightarrow \quad x \mid y \qquad \qquad x \text{ divides } y.$$

a. State explicitly which ordered pairs are in R and R^{-1}, and draw arrow diagrams for R and R^{-1}.

b. Describe R^{-1} in words.

Solution

a. $R \quad = \{(2, 2), (2, 6), (2, 8), (3, 6), (4, 8)\}$
$R^{-1} = \{(2, 2), (6, 2), (8, 2), (6, 3), (8, 4)\}$

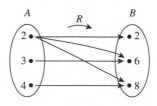

To draw the arrow diagram for R^{-1}, you can copy the arrow diagram for R but reverse the directions of the arrows.

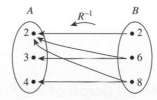

Or you can redraw the diagram so that B is on the left.

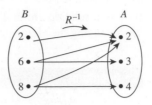

b. R^{-1} can be described in words as follows: For all $(y, x) \in B \times A$,

$$y \; R^{-1} \; x \quad \Leftrightarrow \quad y \text{ is a multiple of } x.$$

∎

Example 8.1.5 The Inverse of an Infinite Relation

Define a relation R from **R** to **R** as follows: For all $(x, y) \in \mathbf{R} \times \mathbf{R}$,

$$x \; R \; y \quad \Leftrightarrow \quad y = 2|x|.$$

Draw the graphs of R and R^{-1} in the Cartesian plane. Is R^{-1} a function?

Solution   A point (v, u) is on the graph of R^{-1} if, and only if, (u, v) is on the graph of R. Note that if $x \geq 0$, then the graph of $y = 2|x| = 2x$ is a straight line with slope 2. And if $x < 0$, then the graph of $y = 2|x| = 2(-x) = -2x$ is a straight line with slope -2. Some sample values are tabulated and the graphs are shown below.

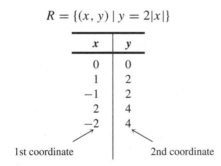

$R = \{(x, y) \mid y = 2|x|\}$

x	y
0	0
1	2
-1	2
2	4
-2	4

1st coordinate   2nd coordinate

$R^{-1} = \{(y, x) \mid y = 2|x|\}$

y	x
0	0
2	1
2	-1
4	2
4	-2

1st coordinate   2nd coordinate

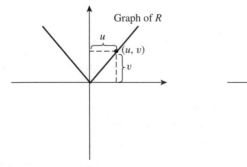

Graph of R

Graph of R^{-1}

R^{-1} is not a function because, for instance, both $(2, 1)$ and $(2, -1)$ are in R^{-1}. ∎

Directed Graph of a Relation

In the remaining sections of this chapter, we discuss important properties of relations that are defined from a set to itself.

Note   It is important to distinguish clearly between a relation and the set on which it is defined.

> ● **Definition**
>
> A **relation on a set** A is a relation from A to A.

When a relation R is defined *on* a set A, the arrow diagram of the relation can be modified so that it becomes a **directed graph.** Instead of representing A as two separate sets of points, represent A only once, and draw an arrow from each point of A to each related point. As with an ordinary arrow diagram,

> For all points x and y in A,
>
> there is an arrow from x to y $\Leftrightarrow$ $x\,R\,y$ $\Leftrightarrow$ $(x, y) \in R$.

If a point is related to itself, a loop is drawn that extends out from the point and goes back to it.

Example 8.1.6 Directed Graph of a Relation

Let $A = \{3, 4, 5, 6, 7, 8\}$ and define a relation R on A as follows: For all $x, y \in A$,

$$x\,R\,y \quad \Leftrightarrow \quad 2\,|\,(x - y).$$

Draw the directed graph of R.

Solution   Note that $3\,R\,3$ because $3 - 3 = 0$ and $2\,|\,0$ since $0 = 2 \cdot 0$. Thus there is a loop from 3 to itself. Similarly, there is a loop from 4 to itself, from 5 to itself, and so forth, since the difference of each integer with itself is 0, and $2\,|\,0$.

Note also that $3\,R\,5$ because $3 - 5 = -2 = 2 \cdot (-1)$. And $5\,R\,3$ because $5 - 3 = 2 = 2 \cdot 1$. Hence there is an arrow from 3 to 5 and also an arrow from 5 to 3. The other arrows in the directed graph, as shown below, are obtained by similar reasoning.

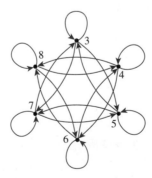

■

Test Yourself

Answers to Test Yourself questions are located at the end of each section.

1. If R is a relation from A to B, $x \in A$, and $y \in B$, the notation $x\,R\,y$ means that _____.

2. If R is a relation from A to B, $x \in A$, and $y \in B$, the notation $x\,\mathcal{R}\,y$ means that _____.

3. If R is a relation from A to B, $x \in A$, and $y \in B$, then $(y, x) \in R^{-1}$ if, and only if, _____.

4. A relation on a set A is a relation from _____ to _____.

5. If R is a relation on a set A, the directed graph of R has an arrow from x to y if, and only if, _____.

Exercise Set 8.1*

1. As in Example 8.1.2, the **congruence modulo 2** relation E is defined from $\mathbf{Z}$ to $\mathbf{Z}$ as follows: For all integers m and n,

$$m \, E \, n \quad \Leftrightarrow \quad m - n \text{ is even.}$$

   a. Is $0 \, E \, 0$? Is $5 \, E \, 2$? Is $(6, 6) \in E$? Is $(-1, 7) \in E$?
   b. Prove that for any even integer n, $n \, E \, 0$.

H **2.** Prove that for all integers m and n, $m - n$ is even if, and only if, both m and n are even or both m and n are odd.

3. The **congruence modulo 3** relation, T, is defined from $\mathbf{Z}$ to $\mathbf{Z}$ as follows: For all integers m and n,

$$m \, T \, n \quad \Leftrightarrow \quad 3 \mid (m - n).$$

   a. Is $10 \, T \, 1$? Is $1 \, T \, 10$? Is $(2, 2) \in T$? Is $(8, 1) \in T$?
   b. List five integers n such that $n \, T \, 0$.
   c. List five integers n such that $n \, T \, 1$.
   d. List five integers n such that $n \, T \, 2$.
   H **e.** Make and prove a conjecture about which integers are related by T to 0, which integers are related by T to 1, and which integers are related by T to 2.

4. Define a relation P on $\mathbf{Z}$ as follows: For all $m, n \in \mathbf{Z}$,

$$m \, P \, n \quad \Leftrightarrow \quad m \text{ and } n \text{ have a common prime factor.}$$

   a. Is $15 \, P \, 25$?　　**b.** $22 \, P \, 27$?
   c. Is $0 \, P \, 5$?　　　**d.** Is $8 \, P \, 8$?

5. Let $X = \{a, b, c\}$. Recall that $\mathscr{P}(X)$ is the power set of X. Define a relation **R** on $\mathscr{P}(X)$ as follows: For all $A, B \in \mathscr{P}(X)$,

$$A \, \mathbf{R} \, B \quad \Leftrightarrow \quad A \text{ has the same number of elements as } B.$$

   a. Is $\{a, b\} \, \mathbf{R} \, \{b, c\}$?　　**b.** Is $\{a\} \, \mathbf{R} \, \{a, b\}$?
   c. Is $\{c\} \, \mathbf{R} \, \{b\}$?

6. Let $X = \{a, b, c\}$. Define a relation **J** on $\mathscr{P}(X)$ as follows: For all $A, B \in \mathscr{P}(X)$,

$$A \, \mathbf{J} \, B \quad \Leftrightarrow \quad A \cap B \neq \varnothing.$$

   a. Is $\{a\} \, \mathbf{J} \, \{c\}$?　　**b.** Is $\{a, b\} \, \mathbf{J} \, \{b, c\}$?
   c. Is $\{a, b\} \, \mathbf{J} \, \{a, b, c\}$?

7. Define a relation R on $\mathbf{Z}$ as follows: For all integers m and n,

$$m \, R \, n \quad \Leftrightarrow \quad 5 \mid (m^2 - n^2).$$

   a. Is $1 \, R \, (-9)$?　　**b.** Is $2 \, R \, 13$?
   c. Is $2 \, R \, (-8)$?　　**d.** Is $(-8) \, R \, 2$?

8. Let A be the set of all strings of a's and b's of length 4. Define a relation R on A as follows: For all $s, t \in A$,

$$s \, R \, t \quad \Leftrightarrow \quad s \text{ has the same first two characters as } t.$$

   a. Is *abaa R abba*?　　**b.** Is *aabb R bbaa*?
   c. Is *aaaa R aaab*?　　**d.** Is *baaa R abaa*?

9. Let A be the set of all strings of 0's, 1's, and 2's of length 4. Define a relation R on A as follows: For all $s, t \in A$,

$$s \, R \, t \quad \Leftrightarrow \quad \begin{array}{l} \text{the sum of the characters in } s \text{ equals} \\ \text{the sum of the characters in } t. \end{array}$$

   a. Is $0121 \, R \, 2200$?　　**b.** Is $1011 \, R \, 2101$?
   c. Is $2212 \, R \, 2121$?　　**d.** Is $1220 \, R \, 2111$?

10. Let $A = \{3, 4, 5\}$ and $B = \{4, 5, 6\}$ and let R be the "less than" relation. That is, for all $(x, y) \in A \times B$,

$$x \, R \, y \quad \Leftrightarrow \quad x < y.$$

    State explicitly which ordered pairs are in R and R^{-1}.

11. Let $A = \{3, 4, 5\}$ and $B = \{4, 5, 6\}$ and let S be the "divides" relation. That is, for all $(x, y) \in A \times B$,

$$x \, S \, y \quad \Leftrightarrow \quad x \mid y.$$

    State explicitly which ordered pairs are in S and S^{-1}.

12. **a.** Suppose a function $F: X \to Y$ is one-to-one but not onto. Is F^{-1} (the inverse relation for F) a function? Explain your answer.
    b. Suppose a function $F: X \to Y$ is onto but not one-to-one. Is F^{-1} (the inverse relation for F) a function? Explain your answer.

Draw the directed graphs of the relations defined in 13–18.

13. Define a relation R on $A = \{0, 1, 2, 3\}$ by $R = \{(0, 0), (1, 2), (2, 2)\}$.

14. Define a relation S on $B = \{a, b, c, d\}$ by $S = \{(a, b), (a, c), (b, c), (d, d)\}$.

15. Let $A = \{2, 3, 4, 5, 6, 7, 8\}$ and define a relation R on A as follows: For all $x, y \in A$,

$$x \, R \, y \quad \Leftrightarrow \quad x \mid y.$$

H **16.** Let $A = \{5, 6, 7, 8, 9, 10\}$ and define a relation S on A as follows: For all $x, y \in A$,

$$x \, S \, y \quad \Leftrightarrow \quad 2 \mid (x - y).$$

17. Let $A = \{2, 3, 4, 5, 6, 7, 8\}$ and define a relation T on A as follows: For all $x, y \in A$,

$$x \, T \, y \quad \Leftrightarrow \quad 3 \mid (x - y).$$

---

*For exercises with blue numbers or letters, solutions are given in Appendix B. The symbol *H* indicates that only a hint or a partial solution is given. The symbol ∗ signals that an exercise is more challenging than usual.

18. Let $A = \{0, 1, 2, 3, 4, 5, 6, 7, 8\}$ and define a relation V on A as follows: For all $x, y \in A$,

$$x \, V \, y \Leftrightarrow 5 \mid (x^2 - y^2).$$

Exercises 19–20 refer to unions and intersections of relations. Since relations are subsets of Cartesian products, their unions and intersections can be calculated as for any subsets. Given two relations R and S from A to B,

$R \cup S = \{(x, y) \in A \times B \mid (x, y) \in R \text{ or } (x, y) \in S\}$
$R \cap S = \{(x, y) \in A \times B \mid (x, y) \in R \text{ and } (x, y) \in S\}.$

19. Let $A = \{2, 4\}$ and $B = \{6, 8, 10\}$ and define relations R and S from A to B as follows: For all $(x, y) \in A \times B$,

$$x \, R \, y \quad \Leftrightarrow \quad x \mid y \quad \text{and}$$
$$x \, S \, y \quad \Leftrightarrow \quad y - 4 = x.$$

State explicitly which ordered pairs are in $A \times B, R, S,$ $R \cup S$, and $R \cap S$.

20. Let $A = \{-1, 1, 2, 4\}$ and $B = \{1, 2\}$ and define relations R and S from A to B as follows: For all $(x, y) \in A \times B$,

$$x \, R \, y \quad \Leftrightarrow \quad |x| = |y| \quad \text{and}$$
$$x \, S \, y \quad \Leftrightarrow \quad x - y \text{ is even.}$$

State explicitly which ordered pairs are in $A \times B, R, S,$ $R \cup S$, and $R \cap S$.

21. Define relations R and S on **R** as follows:

$$R = \{(x, y) \in \mathbf{R} \times \mathbf{R} \mid x < y\} \quad \text{and}$$
$$S = \{(x, y) \in \mathbf{R} \times \mathbf{R} \mid x = y\}.$$

That is, R is the "less than" relation and S is the "equals" relation on **R**. Graph $R, S, R \cup S,$ and $R \cap S$ in the Cartesian plane.

Answers for Test Yourself

1. x is related to y by R 2. x is not related to y by R 3. $(x, y) \in R$ 4. $A; A$ 5. x is related to y by R

8.2 *Reflexivity, Symmetry, and Transitivity*

Mathematics is the tool specially suited for dealing with abstract concepts of any kind and there is no limit to its power in this field. — P. A. M. Dirac, 1902–1984

Let $A = \{2, 3, 4, 6, 7, 9\}$ and define a relation R on A as follows: For all $x, y \in A$,

$$x \, R \, y \quad \Leftrightarrow \quad 3 \mid (x - y).$$

Note For reference:
$x \, R \, y \quad \Leftrightarrow \quad 3 \mid (x - y).$

Then $2 \, R \, 2$ because $2 - 2 = 0$, and $3 \mid 0$. Similarly, $3 \, R \, 3, 4 \, R \, 4, 6 \, R \, 6, 7 \, R \, 7,$ and $9 \, R \, 9$. Also $6 \, R \, 3$ because $6 - 3 = 3$, and $3 \mid 3$. And $3 \, R \, 6$ because $3 - 6 = -(6 - 3) = -3$, and $3 \mid (-3)$. Similarly, $3 \, R \, 9, 9 \, R \, 3, 6 \, R \, 9, 9 \, R \, 6, 4 \, R \, 7,$ and $7 \, R \, 4$. Thus the directed graph for R has the appearance shown below.

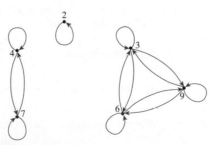

This graph has three important properties:

1. Each point of the graph has an arrow looping around from it back to itself.

2. In each case where there is an arrow going from one point to a second, there is an arrow going from the second point back to the first.

3. In each case where there is an arrow going from one point to a second and from the second point to a third, there is an arrow going from the first point to the third. That is, there are no "incomplete directed triangles" in the graph.

Properties (1), (2), and (3) correspond to properties of general relations called *reflexivity, symmetry,* and *transitivity.*

Caution! The definition of symmetric does not say that x is related to y by R; only that if it happens that x is related to y, then y must be related to x.

• Definition

Let R be a relation on a set A.

1. R is **reflexive** if, and only if, for all $x \in A$, $x\,R\,x$.

2. R is **symmetric** if, and only if, for all $x, y \in A$, *if* $x\,R\,y$ then $y\,R\,x$.

3. R is **transitive** if, and only if, for all $x, y, z \in A$, *if* $x\,R\,y$ and $y\,R\,z$ then $x\,R\,z$.

Because of the equivalence of the expressions $x\,R\,y$ and $(x, y) \in R$ for all x and y in A, the reflexive, symmetric, and transitive properties can also be written as follows:

1. R is reflexive $\quad\Leftrightarrow\quad$ for all x in A, $(x, x) \in R$.

2. R is symmetric $\quad\Leftrightarrow\quad$ for all x and y in A, *if* $(x, y) \in R$ then $(y, x) \in R$.

3. R is transitive $\quad\Leftrightarrow\quad$ for all x, y and z in A, *if* $(x, y) \in R$ and $(y, z) \in R$ then $(x, z) \in R$.

In informal terms, properties (1)–(3) say the following:

1. **Reflexive:** Each element is related to itself.

2. **Symmetric:** If any one element is related to any other element, then the second element is related to the first.

3. **Transitive:** If any one element is related to a second and that second element is related to a third, then the first element is related to the third.

Note that the definitions of reflexivity, symmetry, and transitivity are universal statements. This means that to prove a relation has one of the properties, you use either the method of exhaustion or the method of generalizing from the generic particular.

Now consider what it means for a relation *not* to have one of the properties defined previously. Recall that the negation of a universal statement is existential. Hence if R is a relation on a set A, then

Caution! The "first," "second," and "third" elements in the informal versions need not all be distinct. This is a disadvantage of informality: It may mask nuances that a formal definition makes clear.

1. R is **not reflexive** $\quad\Leftrightarrow\quad$ there is an element x in A such that $x\,\not{R}\,x$ *[that is, such that $(x, x) \notin R$].*

2. R is **not symmetric** $\quad\Leftrightarrow\quad$ there are elements x and y in A such that $x\,R\,y$ but $y\,\not{R}\,x$ *[that is, such that $(x, y) \in R$ but $(y, x) \notin R$].*

3. R is **not transitive** $\quad\Leftrightarrow\quad$ there are elements x, y and z in A such that $x\,R\,y$ and $y\,R\,z$ but $x\,\not{R}\,z$ *[that is, such that $(x, y) \in R$ and $(y, z) \in R$ but $(x, z) \notin R$].*

It follows that you can show that a relation does *not* have one of the properties by finding a counterexample.

Example 8.2.1 Properties of Relations on Finite Sets

Let $A = \{0, 1, 2, 3\}$ and define relations R, S, and T on A as follows:

$$R = \{(0, 0), (0, 1), (0, 3), (1, 0), (1, 1), (2, 2), (3, 0), (3, 3)\},$$
$$S = \{(0, 0), (0, 2), (0, 3), (2, 3)\},$$
$$T = \{(0, 1), (2, 3)\}.$$

a. Is R reflexive? symmetric? transitive?

b. Is S reflexive? symmetric? transitive?

c. Is T reflexive? symmetric? transitive?

Solution

a. The directed graph of R has the appearance shown below.

R is reflexive: There is a loop at each point of the directed graph. This means that each element of A is related to itself, so R is reflexive.

R is symmetric: In each case where there is an arrow going from one point of the graph to a second, there is an arrow going from the second point back to the first. This means that whenever one element of A is related by R to a second, then the second is related to the first. Hence R is symmetric.

R is not transitive: There is an arrow going from 1 to 0 and an arrow going from 0 to 3, but there is no arrow going from 1 to 3. This means that there are elements of A—0, 1, and 3—such that $1\ R\ 0$ and $0\ R\ 3$ but $1\ \not{R}\ 3$. Hence R is not transitive.

b. The directed graph of S has the appearance shown below.

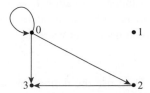

S is not reflexive: There is no loop at 1, for example. Thus $(1, 1) \notin S$, and so S is not reflexive.

S is not symmetric: There is an arrow from 0 to 2 but not from 2 to 0. Hence $(0, 2) \in S$ but $(2, 0) \notin S$, and so S is not symmetric.

S is transitive: There are three cases for which there is an arrow going from one point of the graph to a second and from the second point to a third: Namely, there are arrows going from 0 to 2 and from 2 to 3; there are arrows going from 0 to 0 and from 0 to 2; and there are arrows going from 0 to 0 and from 0 to 3. In each case there is an arrow going from the first point to the third. (Note again that the "first," "second," and "third" points need not be distinct.) This means that whenever $(x, y) \in S$ and $(y, z) \in S$, then $(x, z) \in S$, for all $x, y, z \in \{0, 1, 2, 3\}$, and so S is transitive.

c. The directed graph of T has the appearance shown below.

T is not reflexive: There is no loop at 0, for example. Thus $(0, 0) \notin T$, so T is not reflexive.

T is not symmetric: There is an arrow from 0 to 1 but not from 1 to 0. Thus $(0, 1) \in T$ but $(1, 0) \notin T$, and so T is not symmetric.

Note *T* is transitive by default because it is *not* not transitive!

T is transitive: The transitivity condition is vacuously true for T. To see this, observe that the transitivity condition says that

For all $x, y, z \in A$, if $(x, y) \in T$ and $(y, z) \in T$ then $(x, z) \in T$.

The only way for this to be false would be for there to exist elements of A that make the hypothesis true and the conclusion false. That is, there would have to be elements x, y, and z in A such that

$$(x, y) \in T \quad \text{and} \quad (y, z) \in T \quad \text{and} \quad (x, z) \notin T.$$

In other words, there would have to be two ordered pairs in T that have the potential to "link up" by having the *second* element of one pair be the *first* element of the other pair. But the only elements in T are $(0, 1)$ and $(2, 3)$, and these do not have the potential to link up. Hence the hypothesis is never true. It follows that it is impossible for T *not* to be transitive, and thus T is transitive. ■

Properties of Relations on Infinite Sets

Suppose a relation R is defined on an infinite set A. To prove the relation is reflexive, symmetric, or transitive, first write down what is to be proved. For instance, for symmetry you need to prove that

$$\forall x, y \in A, \text{ if } x \, R \, y \text{ then } y \, R \, x.$$

Then use the definitions of A and R to rewrite the statement for the particular case in question. For instance, for the "equality" relation on the set of real numbers, the rewritten statement is

$$\forall x, y \in \mathbf{R}, \text{ if } x = y \text{ then } y = x.$$

Sometimes the truth of the rewritten statement will be immediately obvious (as it is here). At other times you will need to prove it using the method of generalizing from the generic particular. We give examples of both cases in this section. We begin with the relation of equality, one of the simplest and yet most important relations.

Example 8.2.2 Properties of Equality

Define a relation R on **R** (the set of all real numbers) as follows: For all real numbers x and y.

$$x \, R \, y \quad \Leftrightarrow \quad x = y.$$

a. Is R reflexive?   b. Is R symmetric?   c. Is R transitive?

Solution

a. **R is reflexive**: R is reflexive if, and only if, the following statement is true:

For all $x \in \mathbf{R}$, $x \, R \, x$.

Since $x \, R \, x$ just means that $x = x$, this is the same as saying

For all $x \in \mathbf{R}$, $x = x$.

But this statement is certainly true; every real number is equal to itself.

b. **R is symmetric**: R is symmetric if, and only if, the following statement is true:

For all $x, y \in \mathbf{R}$,   **if** $x \, R \, y$ then $y \, R \, x$.

By definition of R, $x \, R \, y$ means that $x = y$ and $y \, R \, x$ means that $y = x$. Hence R is symmetric if, and only if,

For all $x, y \in \mathbf{R}$,   **if** $x = y$ then $y = x$.

But this statement is certainly true; if one number is equal to a second, then the second is equal to the first.

c. **R is transitive**: R is transitive if, and only if, the following statement is true:

For all $x, y, z \in \mathbf{R}$,   **if** $x \, R \, y$ and $y \, R \, z$ then $x \, R \, z$.

By definition of R, $x \, R \, y$ means that $x = y$, $y \, R \, z$ means that $y = z$, and $x \, R \, z$ means that $x = z$. Hence R is transitive if, and only if, the following statement is true:

For all $x, y, z \in \mathbf{R}$,   **if** $x = y$ and $y = z$ then $x = z$.

But this statement is certainly true: If one real number equals a second and the second equals a third, then the first equals the third. ∎

Example 8.2.3 Properties of "Less Than"

Define a relation R on **R** (the set of all real numbers) as follows: For all $x, y \in R$,

$$x \, R \, y \quad \Leftrightarrow \quad x < y.$$

a. Is R reflexive?   b. Is R symmetric?   c. Is R transitive?

Solution

a. **R is not reflexive**: R is reflexive if, and only if, $\forall x \in \mathbf{R}, x \, R \, x$. By definition of R, this means that $\forall x \in \mathbf{R}, x < x$. But this is false: $\exists x \in \mathbf{R}$ such that $x \not< x$. As a counterexample, let $x = 0$ and note that $0 \not< 0$. Hence R is not reflexive.

b. **R is not symmetric:** R is symmetric if, and only if, $\forall x, y \in \mathbf{R}$, if $x \, R \, y$ then $y \, R \, x$. By definition of R, this means that $\forall x, y \in \mathbf{R}$, if $x < y$ then $y < x$. But this is false: $\exists x, y \in \mathbf{R}$ such that $x < y$ and $y \not< x$. As a counterexample, let $x = 0$ and $y = 1$ and note that $0 < 1$ but $1 \not< 0$. Hence R is not symmetric.

c. **R is transitive:** R is transitive if, and only if, for all $x, y, z \in \mathbf{R}$, if $x \, R \, y$ and $y \, R \, z$ then $x \, R \, z$. By definition of R, this means that for all $x, y, z \in \mathbf{R}$, if $x < y$ and $y < z$, then $x < z$. But this statement is true by the transitive law of order for real numbers (Appendix A, T18). Hence R is transitive. ∎

Sometimes a property is "universally false" in the sense that it is false for *every* element of its domain. It follows immediately, of course, that the property is false for each particular element of the domain and hence counterexamples abound. In such a case, it may seem more natural to prove the universal falseness of the property rather than to give a single counterexample. In the previous example, for instance, you might find it natural to answer (a) and (b) as follows:

Alternative Answer to (a): R is not reflexive because $x \not< x$ for all real numbers x (by the trichotomy law—Appendix A, T17).

Alternative Answer to (b): R is not symmetric because for all x and y in A, if $x < y$, then $y \not< x$ (by the trichotomy law).

Example 8.2.4 Properties of Congruence Modulo 3

Define a relation T on **Z** (the set of all integers) as follows: For all integers m and n,

$$m \, T \, n \quad \Leftrightarrow \quad 3 \,|\, (m - n).$$

This relation is called **congruence modulo 3.**

a. Is T reflexive?   b. Is T symmetric?   c. Is T transitive?

Solution

a. **T is reflexive:** To show that T is reflexive, it is necessary to show that

$$\text{For all } m \in \mathbf{Z}, \quad m \, T \, m.$$

By definition of T, this means that

$$\text{For all } m \in \mathbf{Z}, \quad 3 \,|\, (m - m).$$

Or, since $m - m = 0$,   For all $m \in \mathbf{Z}$, $3 \,|\, 0$.

But this is true: $3 \,|\, 0$ since $0 = 3 \cdot 0$. Hence T is reflexive. This reasoning is formalized in the following proof.

> **Proof of Reflexivity:** Suppose m is a particular but arbitrarily chosen integer. *[We must show that $m \, T \, m$.]* Now $m - m = 0$. But $3 \,|\, 0$ since $0 = 3 \cdot 0$. Hence $3 \,|\, (m - m)$. Thus, by definition of T, $m \, T \, m$ *[as was to be shown].*

b. **T is symmetric:** To show that T is symmetric, it is necessary to show that

$$\text{For all } m, n \in \mathbf{Z}, \quad \text{if } m \, T \, n \text{ then } n \, T \, m.$$

By definition of T this means that

$$\text{For all } m, n \in \mathbf{Z}, \quad \text{if } 3 \mid (m - n) \text{ then } 3 \mid (n - m).$$

Is this true? Suppose m and n are particular but arbitrarily chosen integers such that $3 \mid (m - n)$. Must it follow that $3 \mid (n - m)$? *[In other words, can we find an integer so that $n - m = 3 \cdot$ (that integer)?]* By definition of "divides," since

$$3 \mid (m - n),$$

then $\qquad\qquad\qquad m - n = 3k \quad$ for some integer k.

The crucial observation is that $n - m = -(m - n)$. Hence, you can multiply both sides of this equation by -1 to obtain

$$-(m - n) = -3k,$$

which is equivalent to $\qquad\qquad n - m = 3(-k).$

[Thus we have found an integer, namely $-k$, so that $n - m = 3 \cdot$ (that integer).]
Since $-k$ is an integer, this equation shows that

$$3 \mid (n - m).$$

It follows that T is symmetric.

This reasoning is formalized in the following proof.

Proof of Symmetry: Suppose m and n are particular but arbitrarily chosen integers that satisfy the condition $m \ T \ n$. *[We must show that $n \ T \ m$.]* By definition of T, since $m \ T \ n$ then $3 \mid (m - n)$. By definition of "divides," this means that $m - n = 3k$, for some integer k. Multiplying both sides by -1 gives $n - m = 3(-k)$. Since $-k$ is an integer, this equation shows that $3 \mid (n - m)$. Hence, by definition of T, $n \ T \ m$ *[as was to be shown].*

---

c. **T is transitive:** To show that T is transitive, it is necessary to show that

$$\text{For all } m, n, p \in \mathbf{Z}, \quad \text{if } m \ T \ n \text{ and } n \ T \ p \text{ then } m \ T \ p.$$

By definition of T this means that

$$\text{For all } m, n \in \mathbf{Z}, \quad \text{if } 3 \mid (m - n) \text{ and } 3 \mid (n - p) \text{ then } 3 \mid (m - p).$$

Is this true? Suppose $m, n,$ and p are particular but arbitrarily chosen integers such that $3 \mid (m - n)$ and $3 \mid (n - p)$. Must it follow that $3 \mid (m - p)$? *[In other words, can we find an integer so that $m - p = 3 \cdot$ (that integer)?]* By definition of "divides," since

$$3 \mid (m - n) \quad \text{and} \quad 3 \mid (n - p),$$

then $\qquad\qquad\qquad m - n = 3r \quad$ for some integer $r,$

and $\qquad\qquad\qquad n - p = 3s \quad$ for some integer $s.$

The crucial observation is that $(m - n) + (n - p) = m - p$. Add these two equations together to obtain

$$(m - n) + (n - p) = 3r + 3s,$$

which is equivalent to $\qquad\qquad m - p = 3(r + s).$

[Thus we have found an integer so that $m - p = 3 \cdot$ (that integer).]
Since r and s are integers, $r + s$ is an integer. So this equation shows that

$$3 \mid (m - p).$$

It follows that T is transitive.

This reasoning is formalized in the following proof.

Proof of Transitivity: Suppose m, n, and p are particular but arbitrarily chosen integers that satisfy the condition $m\ T\ n$ and $n\ T\ p$. *[We must show that $m\ T\ p$.]* By definition of T, since $m\ T\ n$ and $n\ T\ p$, then $3 \mid (m - n)$ and $3 \mid (n - p)$. By definition of "divides," this means that $m - n = 3r$ and $n - p = 3s$, for some integers r and s. Adding the two equations gives $(m - n) + (n - p) = 3r + 3s$, and simplifying gives that $m - p = 3(r + s)$. Since $r + s$ is an integer, this equation shows that $3 \mid (m - p)$. Hence, by definition of T, $m\ T\ p$ *[as was to be shown]*.

■

Test Yourself

1. For a relation R on a set A to be reflexive means that _____.

2. For a relation R on a set A to be symmetric means that _____.

3. For a relation R on a set A to be transitive means that _____.

4. To show that a relation R on an infinite set A is reflexive, you suppose that _____ and you show that _____.

5. To show that a relation R on an infinite set A is symmetric, you suppose that _____ and you show that _____.

6. To show that a relation R on an infinite set A is transitive, you suppose that _____ and you show that _____.

7. To show that a relation R on a set A is not reflexive, you _____.

8. To show that a relation R on a set A is not symmetric, you _____.

9. To show that a relation R on a set A is not transitive, you _____.

Exercise Set 8.2

In 1–8 a number of relations are defined on the set $A = \{0, 1, 2, 3\}$. For each relation:
a. Draw the directed graph.
b. Determine whether the relation is reflexive.
c. Determine whether the relation is symmetric.
d. Determine whether the relation is transitive.
Give a counterexample in each case in which the relation does not satisfy one of the properties.

1. $R_1 = \{(0, 0), (0, 1), (0, 3), (1, 1), (1, 0), (2, 3), (3, 3)\}$

2. $R_2 = \{(0, 0), (0, 1), (1, 1), (1, 2), (2, 2), (2, 3)\}$

3. $R_3 = \{(2, 3), (3, 2)\}$

4. $R_4 = \{(1, 2), (2, 1), (1, 3), (3, 1)\}$

5. $R_5 = \{(0, 0), (0, 1), (0, 2), (1, 2)\}$

6. $R_6 = \{(0, 1), (0, 2)\}$

7. $R_7 = \{(0, 3), (2, 3)\}$

8. $R_8 = \{(0, 0), (1, 1)\}$

In 9–33 determine whether the given relation is reflexive, symmetric, transitive, or none of these. Justify your answers.

9. R is the "greater than or equal to" relation on the set of real numbers: For all $x, y \in \mathbf{R}$, $x\ R\ y \Leftrightarrow x \geq y$.

10. C is the circle relation on the set of real numbers: For all $x, y \in \mathbf{R}$, $x\ C\ y \Leftrightarrow x^2 + y^2 = 1$.

11. D is the relation defined on $\mathbf{R}$ as follows: For all $x, y \in \mathbf{R}$, $x\ D\ y \Leftrightarrow xy \geq 0$.

12. E is the congruence modulo 2 relation on $\mathbf{Z}$: For all $m, n \in \mathbf{Z}$, $m\ E\ n \Leftrightarrow 2 \mid (m - n)$.

13. F is the congruence modulo 5 relation on $\mathbf{Z}$: For all $m, n \in \mathbf{Z}$, $m\ F\ n \Leftrightarrow 5 \mid (m - n)$.

14. O is the relation defined on $\mathbf{Z}$ as follows: For all $m, n \in \mathbf{Z}$, $m\ O\ n \Leftrightarrow m - n$ is odd.

15. D is the "divides" relation on $\mathbf{Z}^+$: For all positive integers m and n, $m\ D\ n \Leftrightarrow m \mid n$.

16. A is the "absolute value" relation on $\mathbf{R}$: For all real numbers x and y, $x\ A\ y \Leftrightarrow |x| = |y|$.

17. Recall that a prime number is an integer that is greater than 1 and has no positive integer divisors other than 1 and itself. (In particular, 1 is not prime.) A relation P is defined on $\mathbf{Z}$ as follows: For all $m, n \in \mathbf{Z}$, $m\ P\ n \Leftrightarrow \exists$ a prime number p such that $p \mid m$ and $p \mid n$.

H 18. Define a relation Q on $\mathbf{R}$ as follows: For all real numbers x and y, $x\ Q\ y \Leftrightarrow x - y$ is rational.

19. Define a relation I on $\mathbf{R}$ as follows: For all real numbers x and y, $x \ I \ y \Leftrightarrow x - y$ is irrational.

20. Let $X = \{a, b, c\}$ and $\mathscr{P}(X)$ be the power set of X (the set of all subsets of X). A relation **E** is defined on $\mathscr{P}(X)$ as follows: For all A, B $\in \mathscr{P}(X)$, A **E** B $\Leftrightarrow$ the number of elements in A equals the number of elements in B.

21. Let $X = \{a, b, c\}$ and $\mathscr{P}(X)$ be the power set of X. A relation **L** is defined on $\mathscr{P}(X)$ as follows: For all A, B $\in \mathscr{P}(X)$, A **L** B $\Leftrightarrow$ the number of elements in A is less than the number of elements in B.

22. Let $X = \{a, b, c\}$ and $\mathscr{P}(X)$ be the power set of X. A relation **N** is defined on $\mathscr{P}(X)$ as follows: For all A, B $\in \mathscr{P}(X)$, A **N** B $\Leftrightarrow$ the number of elements in A is not equal to the number of elements in B.

23. Let X be a nonempty set and $\mathscr{P}(X)$ the power set of X. Define the "subset" relation **S** on $\mathscr{P}(X)$ as follows: For all A, B $\in \mathscr{P}(X)$, A **S** B $\Leftrightarrow$ A $\subseteq$ B.

24. Let X be a nonempty set and $\mathscr{P}(X)$ the power set of X. Define the "not equal to" relation **U** on $\mathscr{P}(X)$ as follows: For all A, B $\in \mathscr{P}(X)$, A **U** B $\Leftrightarrow$ A $\neq$ B.

25. Let A be the set of all strings of a's and b's of length 4. Define a relation R on A as follows: For all $s, t \in A$, $s \ R \ t \Leftrightarrow s$ has the same first two characters as t.

26. Let A be the set of all strings of 0's, 1's and 2's of length 4. Define a relation R on A as follows: For all $s, t \in A$, $s \ R \ t \Leftrightarrow$ the sum of the characters in s equals the sum of the characters in t.

27. Let A be the set of all English statements. A relation **I** is defined on A as follows: For all $p, q \in A$,
$$p \ \mathbf{I} \ q \Leftrightarrow p \rightarrow q \text{ is true.}$$

28. Let $A = \mathbf{R} \times \mathbf{R}$. A relation **F** is defined on A as follows: For all (x_1, y_1) and (x_2, y_2) in A,
$$(x_1, y_1) \ \mathbf{F} \ (x_2, y_2) \Leftrightarrow x_1 = x_2.$$

29. Let $A = \mathbf{R} \times \mathbf{R}$. A relation **S** is defined on A as follows: For all (x_1, y_1) and (x_2, y_2) in A,
$$(x_1, y_1) \ \mathbf{S} \ (x_2, y_2) \Leftrightarrow y_1 = y_2.$$

30. Let A be the "punctured plane"; that is, A is the set of all points in the Cartesian plane except the origin $(0, 0)$. A relation R is defined on A as follows: For all p_1 and p_2 in A, $p_1 \ R \ p_2 \Leftrightarrow p_1$ and p_2 lie on the same half line emanating from the origin.

31. Let A be the set of people living in the world today. A relation R is defined on A as follows: For all $p, q \in A$,
$$p \ R \ q \Leftrightarrow p \text{ lives within 100 miles of } q.$$

32. Let A be the set of all lines in the plane. A relation R is defined on A as follows: For all l_1 and l_2 in A, $l_1 \ R \ l_2 \Leftrightarrow l_1$ is parallel to l_2. (Assume that a line is parallel to itself.)

33. Let A be the set of all lines in the plane. A relation R is defined on A as follows: For all l_1 and l_2 in A,
$$l_1 \ R \ l_2 \Leftrightarrow l_1 \text{ is perpendicular to } l_2.$$

In 34–36, assume that R is a relation on a set A. Prove or disprove each statement.

34. If R is reflexive, then R^{-1} is reflexive.

35. If R is symmetric, then R^{-1} is symmetric.

36. If R is transitive, then R^{-1} is transitive.

In 37–42, assume that R and S are relations on a set A. Prove or disprove each statement.

37. If R and S are reflexive, is $R \cap S$ reflexive? Why?

H38. If R and S are symmetric, is $R \cap S$ symmetric? Why?

39. If R and S are transitive, is $R \cap S$ transitive? Why?

40. If R and S are reflexive, is $R \cup S$ reflexive? Why?

41. If R and S are symmetric, is $R \cup S$ symmetric? Why?

42. If R and S are transitive, is $R \cup S$ transitive? Why?

In 43–50 the following definitions are used: A relation on a set A is defined to be

irreflexive if, and only if, for all $x \in A$, $x \ \mathcal{R} \ x$;

asymmetric if, and only if, for all $x, y \in A$, if $x \ R \ y$ then $y \ \mathcal{R} \ x$;

intransitive if, and only if, for all $x, y, z \in A$, if $x \ R \ y$ and $y \ R \ z$ then $x \ \mathcal{R} \ z$.

For each of the relations in the referenced exercise, determine whether the relation is irreflexive, asymmetric, intransitive, or none of these.

43. Exercise 1   44. Exercise 2

45. Exercise 3   46. Exercise 4

47. Exercise 5   **48.** Exercise 6

49. Exercise 7   50. Exercise 8

Answers for Test Yourself

1. for all x in A, $x \ R \ x$ 2. for all x and y in A, if $x \ R \ y$ then $y \ R \ x$ 3. for all x, y, and z in A, if $x \ R \ y$ and $y \ R \ z$ then $x \ R \ z$ 4. x is any element of A; $x \ R \ x$ 5. x and y are any elements of A such that $x \ R \ y$; $y \ R \ x$ 6. x, y, and z are any elements of A such that $x \ R \ y$ and $y \ R \ z$; $x \ R \ z$ 7. show that there is an element x in A such that $x \ \mathcal{R} \ x$ 8. show that there are elements x and y in A such that $x \ R \ y$ but $y \ \mathcal{R} \ x$ 9. show that there are elements x, y, and z in A such that $x \ R \ y$ and $y \ R \ z$ but $x \ \mathcal{R} \ z$

8.3 Equivalence Relations

"You are sad" the Knight said in an anxious tone: "let me sing you a song to comfort you."

"Is it very long?" Alice asked, for she had heard a good deal of poetry that day.

"It's long," said the Knight, "but it's very, very beautiful. Everybody that hears me sing it—either it brings the tears into the eyes, or else—"

"Or else what?" said Alice, for the Knight had made a sudden pause.

"Or else it doesn't, you know. The name of the song is called 'Haddocks' Eyes.' "

"Oh, that's the name of the song, is it?" Alice said, trying to feel interested.

"No, you don't understand," the Knight said, looking a little vexed. "That's what the name is called. *The name really is 'The Aged Aged Man.' "*

"Then I ought to have said 'That's what the song is called'?" Alice corrected herself.

"No, you oughtn't: that's quite another thing! The song is called 'Ways and Means': but that's only what it's called, *you know!"*

"Well, what is the song, then?" said Alice, who was by this time completely bewildered.

"I was coming to that," the Knight said. "The song really is 'A-sitting on a Gate': and the tune's my own invention."

So saying, he stopped his horse and let the reins fall on its neck: then, slowly beating time with one hand, and with a faint smile lighting up his gentle foolish face, as if he enjoyed the music of his song, he began.

— Lewis Carroll, *Through the Looking Glass,* 1872

You know from your early study of fractions that each fraction has many equivalent forms. For example,

$$\frac{1}{2}, \frac{2}{4}, \frac{3}{6}, \frac{-1}{-2}, \frac{-3}{-6}, \frac{15}{30}, \ldots, \text{ and so on}$$

are all different ways to represent the same number. They may look different; they may be called different names; but they are all equal. The idea of grouping together things that "look different but are really the same" is the central idea of equivalence relations.

The Relation Induced by a Partition

A **partition** of a set A is a finite or infinite collection of nonempty, mutually disjoint subsets whose union is A. The diagram of Figure 8.3.1 illustrates a partition of a set A by subsets $A_1, A_2, \ldots, A_6$.

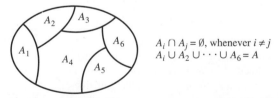

$A_i \cap A_j = \emptyset$, whenever $i \neq j$
$A_i \cup A_2 \cup \cdots \cup A_6 = A$

Figure 8.3.1 A Partition of a Set

• Definition

Given a partition of a set A, the **relation induced by the partition,** R, is defined on A as follows: For all $x, y \in A$,

$$x \, R \, y \quad \Leftrightarrow \quad \text{there is a subset } A_i \text{ of the partition} \\ \text{such that both } x \text{ and } y \text{ are in } A_i.$$

Example 8.3.1 Relation Induced by a Partition

Let $A = \{0, 1, 2, 3, 4\}$ and consider the following partition of A:

$$\{0, 3, 4\}, \{1\}, \{2\}.$$

Find the relation R induced by this partition.

Solution   Since $\{0, 3, 4\}$ is a subset of the partition,

$0 \, R \, 3$	because both 0 and 3 are in $\{0, 3, 4\}$,
$3 \, R \, 0$	because both 3 and 0 are in $\{0, 3, 4\}$,
$0 \, R \, 4$	because both 0 and 4 are in $\{0, 3, 4\}$,
$4 \, R \, 0$	because both 4 and 0 are in $\{0, 3, 4\}$,
$3 \, R \, 4$	because both 3 and 4 are in $\{0, 3, 4\}$, and
$4 \, R \, 3$	because both 4 and 3 are in $\{0, 3, 4\}$.

Note   These statements may seem strange, but, after all, they are not false!

Also,

$0 \, R \, 0$	because both 0 and 0 are in $\{0, 3, 4\}$
$3 \, R \, 3$	because both 3 and 3 are in $\{0, 3, 4\}$, and
$4 \, R \, 4$	because both 4 and 4 are in $\{0, 3, 4\}$.

Since $\{1\}$ is a subset of the partition,

$1 \, R \, 1$ because both 1 and 1 are in $\{1\}$,

and since $\{2\}$ is a subset of the partition,

$2 \, R \, 2$ because both 2 and 2 are in $\{2\}$.

Hence

$$R = \{(0, 0), (0, 3), (0, 4), (1, 1), (2, 2), (3, 0), (3, 3), (3, 4), (4, 0), (4, 3), (4, 4)\}. \quad \blacksquare$$

The fact is that a relation induced by a partition of a set satisfies all three properties studied in Section 8.2: reflexivity, symmetry, and transitivity.

Theorem 8.3.1

Let A be a set with a partition and let R be the relation induced by the partition. Then R is reflexive, symmetric, and transitive.

Proof:

Suppose A is a set with a partition. In order to simplify notation, we assume that the partition consists of only a finite number of sets. The proof for an infinite partition is identical except for notation. Denote the partition subsets by

$$A_1, A_2, \ldots, A_n.$$

Then $A_i \cap A_j = \emptyset$ whenever $i \neq j$, and $A_1 \cup A_2 \cup \cdots \cup A_n = A$. The relation R induced by the partition is defined as follows: For all $x, y \in A$,

$$x \, R \, y \quad \Leftrightarrow \quad \text{there is a set } A_i \text{ of the partition such that } x \in A_i \text{ and } y \in A_i.$$

continued on page 362

[Idea for the proof of reflexivity: For R to be reflexive means that each element of A is related by R to itself. But by definition of R, for an element x to be related to itself means that x is in the same subset of the partition as itself. Well, if x is in some subset of the partition, then it is certainly in the same subset as itself. But x is in some subset of the partition because the union of the subsets of the partition is all of A. This reasoning is formalized as follows.]

Proof that R is reflexive: Suppose $x \in A$. Since $A_1, A_2, \ldots, A_n$ is a partition of A, it follows that $x \in A_i$ for some i. But then the statement

> there is a set A_i of the partition such that $x \in A_i$ and $x \in A_i$

is true. Thus, by definition of R, $\ x \ R \ x$.

[Idea for the proof of symmetry: For R to be symmetric means that any time one element is related to a second, then the second is related to the first. Now for one element x to be related to a second element y means that x and y are in the same subset of the partition. But if this is the case, then y is in the same subset of the partition as x, so y is related to x by definition of R. This reasoning is formalized as follows.]

Proof that R is symmetric: Suppose x and y are elements of A such that $x \ R \ y$. Then

> there is a subset A_i of the partition such that $x \in A_i$ and $y \in A_i$

by definition of R. It follows that the statement

> there is a subset A_i of the partition such that $y \in A_i$ and $x \in A_i$

is also true. Hence, by definition of R, $y \ R \ x$.

[Idea for the proof of transitivity: For R to be transitive means that any time one element of A is related by R to a second and that second is related to a third, then the first element is related to the third. But for one element to be related to another means that there is a subset of the partition that contains both. So suppose x, y, and z are elements such that x is in the same subset as y and y is in the same subset as z. Must x be in the same subset as z? Yes, because the subsets of the partition are mutually disjoint. Since the subset that contains x and y has an element in common with the subset that contains y and z (namely y), the two subsets are equal. But this means that x, y, and z are all in the same subset, and so in particular, x and z are in the same subset. Hence x is related by R to z. This reasoning is formalized as follows.]

Proof that R is transitive: Suppose x, y, and z are in A and $x \ R \ y$ and $y \ R \ z$. By definition of R, there are subsets A_i and A_j of the partition such that

> x and y are in A_i $\quad$ and $\quad$ y and z are in A_j.

Suppose $A_i \neq A_j$. *[We will deduce a contradiction.]* Then $A_i \cap A_j = \emptyset$ since $\{A_1, A_2, A_3, \ldots, A_n\}$ is a partition of A. But y is in A_i and y is in A_j also. Hence $A_i \cap A_j \neq \emptyset$. *[This contradicts the fact that $A_i \cap A_j = \emptyset$.]* Thus $A_i = A_j$. It follows that x, y, and z are all in A_i, and so in particular,

> x and z are in A_i.

Thus, by definition of R, $x \ R \ z$.

Note The fact that $x \in A_i$ and $x \in A_i$ follows from the logical equivalence of the statement forms p and $p \wedge p$.

Note The fact that $y \in A_i$ and $x \in A_i$ follows from the logical equivalence of the statement forms $p \wedge q$ and $q \wedge p$.

Definition of an Equivalence Relation

A relation on a set that satisfies the three properties of reflexivity, symmetry, and transitivity is called an *equivalence relation*.

> **• Definition**
>
> Let A be a set and R a relation on A. R is an **equivalence relation** if, and only if, R is reflexive, symmetric, and transitive.

Thus, according to Theorem 8.3.1, the relation induced by a partition is an equivalence relation. Another example is congruence modulo 3. In Example 8.2.4 it was shown that this relation is reflexive, symmetric, and transitive. Hence, it also is an equivalence relation.

The following notation is used frequently when referring to congruence relations. It was introduced by Carl Friedrich Gauss in the first chapter of his book *Disquisitiones Arithmeticae*. This work, which was published when Gauss was only 24, laid the foundation for modern number theory.

*Carl Friedrich Gauss
(1777–1855)*

> **• Definition**
>
> Let m and n be integers and let d be a positive integer. We say that **m is congruent to n modulo d** and write
>
> $$m \equiv n \ (\text{mod } d)$$
>
> if, and only if, $\qquad\qquad d \mid (m - n)$.
>
> Symbolically: $\qquad m \equiv n \ (\text{mod } d) \quad \Leftrightarrow \quad d \mid (m - n)$

Exercise 15(b) at the end of this section asks you to show that $m \equiv n \ (\text{mod } d)$ if, and only if, $m \ mod \ d = n \ mod \ d$, where m, n, and d are integers and d is positive.

Example 8.3.2 Evaluating Congruences

Determine which of the following congruences are true and which are false.

a. $12 \equiv 7 \ (\text{mod } 5)$ b. $6 \equiv -8 \ (\text{mod } 4)$ c. $3 \equiv 3 \ (\text{mod } 7)$

Solution

a. True. $12 - 7 = 5 = 5 \cdot 1$. Hence $5 \mid (12 - 7)$, and so $12 \equiv 7 \ (\text{mod } 5)$.

b. False. $6 - (-8) = 14$, and $4 \nmid 14$ because $14 \neq 4 \cdot k$ for any integer k. Consequently, $6 \not\equiv -8 \ (\text{mod } 4)$.

c. True. $3 - 3 = 0 = 7 \cdot 0$. Hence $7 \mid (3 - 3)$, and so $3 \equiv 3 \ (\text{mod } 7)$. ∎

Example 8.3.3 An Equivalence Relation on a Set of Subsets

Let X be the set of all nonempty subsets of $\{1, 2, 3\}$. Then

$$X = \{\{1\}, \{2\}, \{3\}, \{1, \ 2\}, \{1, \ 3\}, \{2, \ 3\}, \{1, \ 2, \ 3\}\}$$

Define a relation **R** on X as follows: For all A and B in X,

$$A \ \mathbf{R} \ B \Leftrightarrow \text{the least element of } A \text{ equals the least element of } B.$$

Prove that **R** is an equivalence relation on X.

Solution

R *is reflexive*: Suppose *A* is a nonempty subset of {1, 2, 3}. *[We must show that A **R** A.]* It is true to say that the least element of *A* equals the least element of *A*. Thus, by definition of *R*, *A* **R** *A*.

R *is symmetric*: Suppose *A* and *B* are nonempty subsets of {1, 2, 3} and *A* **R** *B*. *[We must show that B **R** A.]* Since *A* **R** *B*, the least element of *A* equals the least element of *B*. But this implies that the least element of *B* equals the least element of *A*, and so, by definition of **R**, *B* **R** *A*.

R *is transitive*: Suppose *A*, *B*, and *C* are nonempty subsets of {1, 2, 3}, *A* **R** *B*, and *B R C*. *[We must show that A **R** C.]* Since *A* **R** *B*, the least element of *A* equals the least element of *B* and since *B* **R** *C*, the least element of *B* equals the least element of *C*. Thus the least element of *A* equals the least element of *C*, and so, by definition of **R**, *A* **R** *C*. ■

Equivalence Classes of an Equivalence Relation

Suppose there is an equivalence relation on a certain set. If *a* is any particular element of the set, then one can ask, "What is the subset of all elements that are related to *a*?" This subset is called the *equivalence class* of *a*.

Note Be careful to distinguish among the following: a relation on a set, the (underlying) set itself, and the equivalence class for an element of the (underlying) set.

> ● **Definition**
>
> Suppose *A* is a set and *R* is an equivalence relation on *A*. For each element *a* in *A*, the **equivalence class of *a*,** denoted [*a*] and called the **class of *a*** for short, is the set of all elements *x* in *A* such that *x* is related to *a* by *R*.
>
> In symbols:
>
> $$[a] = \{x \in A \mid x \, R \, a\}$$

When several equivalence relations on a set are under discussion, the notation $[a]_R$ is often used to denote the equivalence class of *a* under *R*.

The procedural version of this definition is

$$\text{for all } x \in A, \quad x \in [a] \quad \Leftrightarrow \quad x \, R \, a.$$

Example 8.3.4 Equivalence Classes of a Relation Given as a Set of Ordered Pairs

Let *A* = {0, 1, 2, 3, 4} and define a relation *R* on *A* as follows:

$$R = \{(0, 0), (0, 4), (1, 1), (1, 3), (2, 2), (3, 1), (3, 3), (4, 0), (4, 4)\}.$$

The directed graph for *R* is as shown below. As can be seen by inspection, *R* is an equivalence relation on *A*. Find the distinct equivalence classes of *R*.

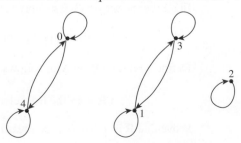

Solution   First find the equivalence class of every element of A.

$$[0] = \{x \in A \mid x\,R\,0\} = \{0, 4\}$$
$$[1] = \{x \in A \mid x\,R\,1\} = \{1, 3\}$$
$$[2] = \{x \in A \mid x\,R\,2\} = \{2\}$$
$$[3] = \{x \in A \mid x\,R\,3\} = \{1, 3\}$$
$$[4] = \{x \in A \mid x\,R\,4\} = \{0, 4\}$$

Note that $[0] = [4]$ and $[1] = [3]$. Thus the *distinct* equivalence classes of the relation are

$$\{0, 4\}, \{1, 3\}, \text{ and } \{2\}. \qquad \blacksquare$$

When a problem asks you to find the *distinct* equivalence classes of an equivalence relation, you will generally solve the problem in two steps. In the first step you either explicitly construct (as in Example 8.3.4) or imagine constructing (as in infinite cases) the equivalence class for every element of the domain A of the relation. Usually several of the classes will contain exactly the same elements, so in the second step you must take a careful look at the classes to determine which are the same. You then indicate the distinct equivalence classes by describing them without duplication.

Example 8.3.5 Equivalence Classes of a Relation on a Set of Subsets

In Example 8.3.3 it was shown that the relation **R** was an equivalence relation, where for nonempty subsets A and B of $\{1, 2, 3\}$ to be related by **R** means that they have the same least element. Describe the distinct equivalence classes of **R**.

Solution   The equivalence class of $\{1\}$ is the set of all the nonempty subsets of $\{1, 2, 3\}$ whose least element is 1. Thus

$$[\{1\}] = \{\{1\}, \{1, 2\}, \{1, 3\}, \{1, 2, 3\}\}.$$

The equivalence class of $\{2\}$ is the set of all the nonempty subsets of $\{1, 2, 3\}$ whose least element is 2. Thus

$$[\{2\}] = \{\{2\}, \{2, 3\}\}.$$

The equivalence class of $\{3\}$ is the set of all the nonempty subsets of $\{1, 2, 3\}$ whose least element is 3. There is only one such set, namely $\{3\}$ itself. Thus

$$[\{3\}] = \{\{3\}\}.$$

Since all the nonempty subsets of $\{1, 2, 3\}$ are in one of the equivalence classes, this is a complete listing. Moreover, these classes are all distinct. $\qquad \blacksquare$

Example 8.3.6 Equivalence Classes of the Identity Relation

Let A be any set and define a relation R on A as follows: For all x and y in A,

$$\boxed{x\,R\,y \quad \Leftrightarrow \quad x = y.}$$

Then R is an equivalence relation. *[To prove this, just generalize the argument used in Example 8.2.2.]* Describe the distinct equivalence classes of R.

Solution   Given any a in A, the class of a is

$$[a] = \{x \in A \mid x \, R \, a\}.$$

But by definition of R, $a \, R \, x$ if, and only if, $a = x$. So

$$[a] = \{x \in A \mid x = a\}$$
$$= \{a\} \qquad \text{since the only element of } A \text{ that equals } a \text{ is } a.$$

Hence, given any a in A,

$$[a] = \{a\},$$

and if $x \neq a$, then $\{x\} \neq \{a\}$. Consequently, all the classes of all the elements of A are distinct, and the distinct equivalence classes of R are all the single-element subsets of A.   ∎

In each of Examples 8.3.4, 8.3.5 and 8.3.6, the set of distinct equivalence classes of the relation consists of mutually disjoint subsets whose union is the entire domain A of the relation. This means that the set of equivalence classes of the relation forms a partition of the domain A. In fact, it is always the case that the equivalence classes of an equivalence relation partition the domain of the relation into a union of mutually disjoint subsets. We establish the truth of this statement in stages, first proving two lemmas and then proving the main theorem.

The first lemma says that if two elements of A are related by an equivalence relation R, then their equivalence classes are the same.

Lemma 8.3.2

Suppose A is a set, R is an equivalence relation on A, and a and b are elements of A. If $a \, R \, b$, then $[a] = [b]$.

This lemma says that if a certain condition is satisfied, then $[a] = [b]$. Now $[a]$ and $[b]$ are *sets,* and two sets are equal if, and only if, each is a subset of the other. Hence the proof of the lemma consists of two parts: first, a proof that $[a] \subseteq [b]$ and second, a proof that $[b] \subseteq [a]$. To show each subset relation, it is necessary to show that every element in the left-hand set is an element of the right-hand set.

Proof of Lemma 8.3.2:

Let A be a set, let R be an equivalence relation on A, and suppose

$$a \text{ and } b \text{ are elements of } A \text{ such that } a \, R \, b.$$

[We must show that $[a] = [b]$.]

Proof that $[a] \subseteq [b]$: Let $x \in [a]$. *[We must show that $x \in [b]$.]* Since

$$x \in [a]$$

then $$x \, R \, a$$

by definition of class. But $$a \, R \, b$$

by hypothesis. Thus, by transitivity of R,

$$x \ R \ b.$$

Hence $\quad\quad\quad\quad\quad\quad\quad\quad\quad x \in [b]$

by definition of class. *[This is what was to be shown.]*

Proof that $[b] \subseteq [a]$: Let $x \in [b]$. *[We must show that $x \in [a]$.]* Since

$$x \in [b]$$

then $\quad\quad\quad\quad\quad\quad\quad\quad\quad x \ R \ b$

by definition of class. Now $\quad\quad\quad a \ R \ b$

by hypothesis. Thus, since R is symmetric,

$$b \ R \ a$$

also. Then, since R is transitive and $x \ R \ b$ and $b \ R \ a$,

$$x \ R \ a.$$

Hence, $\quad\quad\quad\quad\quad\quad\quad\quad\quad x \in [a]$

by definition of class. *[This is what was to be shown.]*

Since $[a] \subseteq [b]$ and $[b] \subseteq [a]$, it follows that $[a] = [b]$ by definition of set equality.

The second lemma says that any two equivalence classes of an equivalence relation are either mutually disjoint or identical.

Lemma 8.3.3

If A is a set, R is an equivalence relation on A, and a and b are elements of A, then

$$\text{either} \quad [a] \cap [b] = \emptyset \quad \text{or} \quad [a] = [b].$$

The statement of Lemma 8.3.3 has the form

$$\text{if } p \text{ then } (q \text{ or } r),$$

Note You can always prove a statement of the form "if p then (q or r)" by proving one of the logically equivalent statements: "if (p and not q) then r" or "if (p and not r) then q."*

where p is the statement "A is a set, R is an equivalence relation on A, and a and b are elements of A," q is the statement "$[a] \cap [b] = \emptyset$," and r is the statement "$[a] = [b]$." To prove the lemma, we will prove the logically equivalent statement

$$\text{if } (p \text{ and not } q) \text{ then } r.$$

That is, we will prove the following:

If A is a set, R is an equivalence relation on A, a and b are
elements of A, and $[a] \cap [b] \neq \emptyset$, then $[a] = [b]$.

---

* See exercise 14 in Section 2.2.

Proof of Lemma 8.3.3:

Suppose A is a set, R is an equivalence relation on A, a and b are elements of A, and

$$[a] \cap [b] \neq \emptyset.$$

[We must show that $[a] = [b]$.] Since $[a] \cap [b] \neq \emptyset$, there exists an element x in A such that $x \in [a] \cap [b]$. By definition of intersection,

$$x \in [a] \quad \text{and} \quad x \in [b]$$

and so

$$x \, R \, a \quad \text{and} \quad x \, R \, b$$

by definition of class. Since R is symmetric *[being an equivalence relation]* and $x \, R \, a$, then $a \, R \, x$. But R is also transitive *[since it is an equivalence relation]*, and so, since $a \, R \, x$ and $x \, R \, b$,

$$a \, R \, b.$$

Now a and b satisfy the hypothesis of Lemma 8.3.2. Hence, by that lemma,

$$[a] = [b].$$

[This is what was to be shown.]

Theorem 8.3.4 The Partition Induced by an Equivalence Relation

If A is a set and R is an equivalence relation on A, then the distinct equivalence classes of R form a partition of A; that is, the union of the equivalence classes is all of A, and the intersection of any two distinct classes is empty.

The proof of Theorem 8.3.4 is divided into two parts: first, a proof that A is the union of the equivalence classes of R and second, a proof that the intersection of any two distinct equivalence classes is empty. The proof of the first part follows from the fact that the relation is reflexive. The proof of the second part follows from Lemma 8.3.3.

Proof of Theorem 8.3.4:

Suppose A is a set and R is an equivalence relation on A. For notational simplicity, we assume that R has only a finite number of distinct equivalence classes, which we denote

$$A_1, A_2, \ldots, A_n,$$

where n is a positive integer. (When the number of classes is infinite, the proof is identical except for notation.)

Proof that $A = A_1 \cup A_2 \cup \cdots \cup A_n$**:** *[We must show that $A \subseteq A_1 \cup A_2 \cup \cdots \cup A_n$ and that $A_1 \cup A_2 \cup \cdots \cup A_n \subseteq A$.]*

To show that $A \subseteq A_1 \cup A_2 \cup \cdots \cup A_n$, suppose x is any element of A. *[We must show that $x \in A_1 \cup A_2 \cup \cdots \cup A_n$.]* By reflexivity of R, $x \, R \, x$. But this implies that $x \in [x]$ by definition of class. Since x is in *some* equivalence class, it must be in one of the distinct equivalence classes $A_1, A_2, \ldots,$ or A_n. Thus $x \in A_i$ for some index i, and hence $x \in A_1 \cup A_2 \cup \cdots \cup A_n$ by definition of union *[as was to be shown]*.

To show that $A_1 \cup A_2 \cup \cdots \cup A_n \subseteq A$, suppose $x \in A_1 \cup A_2 \cup \cdots \cup A_n$. *[We must show that $x \in A$.]* Then $x \in A_i$ for some $i = 1, 2, \ldots, n$, by definition of union.

But each A_i is an equivalence class of R. And equivalence classes are subsets of A. Hence $A_i \subseteq A$ and so $x \in A$ *[as was to be shown].*

Since $A \subseteq A_1 \cup A_2 \cup \cdots \cup A_n$ and $A_1 \cup A_2 \cup \cdots \cup A_n \subseteq A$, then by definition of set equality, $A = A_1 \cup A_2 \cup \cdots \cup A_n$.

Proof that the distinct classes of R are mutually disjoint: Suppose that A_i and A_j are any two distinct equivalence classes of R. *[We must show that A_i and A_j are disjoint.]* Since A_i and A_j are distinct, then $A_i \neq A_j$. And since A_i and A_j are equivalence classes of R, there must exist elements a and b in A such that $A_i = [a]$ and $A_j = [b]$. By Lemma 8.3.3,

$$\text{either} \quad [a] \cap [b] = \emptyset \quad \text{or} \quad [a] = [b].$$

But $[a] \neq [b]$ because $A_i \neq A_j$. Hence $[a] \cap [b] = \emptyset$. Thus $A_i \cap A_j = \emptyset$, and so A_i and A_j are disjoint *[as was to be shown].*

Example 8.3.7 Equivalence Classes of Congruence Modulo 3

Let R be the relation of congruence modulo 3 on the set $\mathbf{Z}$ of all integers. That is, for all integers m and n,

$$m \, R \, n \quad \Leftrightarrow \quad 3 \mid (m - n) \quad \Leftrightarrow \quad m \equiv n \, (\text{mod } 3).$$

Describe the distinct equivalence classes of R.

Solution   For each integer a,

$$[a] = \{x \in \mathbf{Z} \mid x \, R \, a\}$$
$$= \{x \in \mathbf{Z} \mid 3 \mid (x - a)\}$$
$$= \{x \subset \mathbf{Z} \mid x - a = 3k, \text{ for some integer } k\}.$$

Therefore,

$$[a] = \{x \in \mathbf{Z} \mid x = 3k + a, \text{ for some integer } k\}.$$

In particular,

$$[0] = \{x \in \mathbf{Z} \mid x = 3k + 0, \text{ for some integer } k\}$$
$$= \{x \in \mathbf{Z} \mid x = 3k, \text{ for some integer } k\}$$
$$= \{\ldots -9, -6, -3, 0, 3, 6, 9, \ldots\},$$
$$[1] = \{x \in \mathbf{Z} \mid x = 3k + 1, \text{ for some integer } k\}$$
$$= \{\ldots -8, -5, -2, 1, 4, 7, 10, \ldots\},$$
$$[2] = \{x \in \mathbf{Z} \mid x = 3k + 2, \text{ for some integer } k\}$$
$$= \{\ldots -7, -4, -1, 2, 5, 8, 11, \ldots\}.$$

Now since $3 \, R \, 0$, then by Lemma 8.3.2,

$$[3] = [0].$$

More generally, by the same reasoning,

$$[0] = [3] = [-3] = [6] = \lfloor -6 \rfloor = \ldots, \text{ and so on.}$$

Similarly,

$$[1] = [4] = [-2] = [7] = [-5] = \ldots, \text{and so on.}$$

And

$$[2] = [5] = \lceil -1 \rceil = [8] = [-4] = \ldots, \text{and so on.}$$

Notice that every integer is in class [0], [1], or [2]. Hence the distinct equivalence classes are

$$\{x \in \mathbf{Z} \mid x = 3k, \text{for some integer } k\},$$
$$\{x \in \mathbf{Z} \mid x = 3k + 1, \text{for some integer } k\}, \quad \text{and}$$
$$\{x \in \mathbf{Z} \mid x = 3k + 2, \text{for some integer } k\}.$$

In words, the three classes of congruence modulo 3 are (1) the set of all integers that are divisible by 3, (2) the set of all integers that leave a remainder of 1 when divided by 3, and (3) the set of all integers that leave a remainder of 2 when divided by 3. ■

Example 8.3.7 illustrates a very important property of equivalence classes, namely that an equivalence class may have many different names. In Example 8.3.7, for instance, the class of 0, [0], may also be *called* the class of 3, [3], or the class of −6, [−6]. But what the class *is* is the set

$$\{x \in \mathbf{Z} \mid x = 3k, \text{for some integers } k\}.$$

(The quote at the beginning of this section refers in a humorous way to the philosophically interesting distinction between what things are *called* and what they *are*.)

• Definition

Suppose R is an equivalence relation on a set A and S is an equivalence class of R. A **representative** of the class S is any element a such that $[a] = S$.

In exercises 30–35 at the end of this section, you are asked to show in effect, that if a is any element of an equivalence class S, then $S = [a]$. Hence *any* element of an equivalence class is a representative of that class.

Example 8.3.8 Rational Numbers Are Really Equivalence Classes

For a moment, forget what you know about fractional arithmetic and look at the numbers

$$\frac{1}{3} \quad \text{and} \quad \frac{2}{6}$$

as *symbols*. Considered as symbolic expressions, these *appear* quite different. In fact, if they were written as ordered pairs

$$(1, 3) \quad \text{and} \quad (2, 6)$$

they would *be* different. The fact that we regard them as "the same" is a specific instance of our general agreement to regard any two numbers

$$\frac{a}{b} \quad \text{and} \quad \frac{c}{d}$$

as equal provided the *cross products* are equal: $ad = bc$. This can be formalized as follows, using the language of equivalence relations.

Let A be the set of all ordered pairs of integers for which the second element of the pair is nonzero. Symbolically,

$$A = \mathbf{Z} \times (\mathbf{Z} - \{0\}).$$

Define a relation R on A as follows: For all $(a, b), (c, d) \in A$,

$$(a, b)\, R\, (c, d) \quad \Leftrightarrow \quad ad = bc.$$

The fact is that R is an equivalence relation.

a. Prove that R is transitive. (Proofs that R is reflexive and symmetric are left to the exercises.)

b. Describe the distinct equivalence classes of R.

Solution

a. *[We must show that for all $(a, b), (c, d), (e, f) \in A$, if $(a, b)\, R\, (c, d)$ and $(c, d)\,R\, (e, f)$, then $(a, b)\, R\, (e, f)$.]* Suppose $(a, b), (c, d)$, and (e, f) are particular but arbitrarily chosen elements of A such that $(a, b)\, R\, (c, d)$ and $(c, d)\, R\, (e, f)$. *[We must show that $(a, b)\, R\, (e, f)$.]* By definition of R,

$$(1)\ ad = bc \quad \text{and} \quad (2)\ cf = de.$$

Since the second elements of all ordered pairs in A are nonzero, $b \neq 0, d \neq 0$, and $f \neq 0$. Multiply both sides of equation (1) by f and both sides of equation (2) by b to obtain

$$(1')\ adf = bcf \quad \text{and} \quad (2')\ bcf = bde.$$

Thus

$$adf = bde$$

and, since $d \neq 0$, it follows from the cancellation law for multiplication (T7 in Appendix A) that

$$af = be.$$

It follows, by definition of R, that $(a, b)\, R\, (e, f)$ *[as was to be shown]*.

b. There is one equivalence class for each distinct rational number. Each equivalence class consists of all ordered pairs (a, b) that, if written as fractions a/b, would equal each other. The reason for this is that the condition for two rational numbers to be equal is the same as the condition for two ordered pairs to be related. For instance, the class of $(1, 2)$ is

$$[(1, 2)] = \{(1, 2), (-1, -2), (2, 4), (-2, -4), (3, 6), (-3, -6), \ldots\}$$

since $\dfrac{1}{2} = \dfrac{-1}{-2} = \dfrac{2}{4} = \dfrac{-2}{-4} = \dfrac{3}{6} = \dfrac{-3}{-6}$ and so forth. ∎

It is possible to expand the result of Example 8.3.8 to define operations of addition and multiplication on the equivalence classes of R that satisfy all the same properties as the addition and multiplication of rational numbers. (See exercise 37.) It follows that the rational numbers can be defined as equivalence classes of ordered pairs of integers. Similarly (see exercise 38), it can be shown that all integers, negative and zero included, can be defined as equivalence classes of ordered pairs of positive integers. But in the late nineteenth century, F. L. G. Frege and Giuseppe Peano showed that the positive integers can be defined entirely in terms of sets. And just a little earlier, Richard Dedekind (1848–1916) showed that all real numbers can be defined as sets of rational numbers. All together, these results show that the real numbers can be defined using logic and set theory alone.

Test Yourself

1. For a relation on a set to be an equivalence relation, it must be _____.

2. The notation $m \equiv n \pmod{d}$ is read "_____" and means that _____.

3. Given an equivalence relation R on a set A and given an element a in A, the equivalence class of a is denoted _____ and is defined to be _____.

4. If A is a set, R is an equivalence relation on A, and a and b are elements of A, then either $[a] = [b]$ or _____.

5. If A is a set and R is an equivalence relation on A, then the distinct equivalence classes of R form _____.

6. Let $A = \mathbf{Z} \times (\mathbf{Z} - \{0\})$, and define a relation R on A by specifying that for all (a, b) and (c, d) in A, $(a, b) \, R \, (c, d)$ if, and only if, $ad = bc$. Then there is exactly one equivalence class of R for each _____.

Exercise Set 8.3

1. Suppose that $S = \{a, b, c, d, e\}$ and R is a relation on S such that $a \, R \, b$, $b \, R \, c$, and $d \, R \, e$. List all of the following that must be true if R is (a) reflexive (but not symmetric or transitive), (b) symmetric (but not reflexive or transitive), (c) transitive (but not reflexive or symmetric), and (d) an equivalence relation.

   $c \, R \, b \quad c \, R \, c \quad a \, R \, c \quad b \, R \, a \quad a \, R \, d \quad e \, R \, a \quad e \, R \, d \quad c \, R \, a$

2. Each of the following partitions of $\{0, 1, 2, 3, 4\}$ induces a relation R on $\{0, 1, 2, 3, 4\}$. In each case, find the ordered pairs in R.
   a. $\{0, 2\}, \{1\}, \{3, 4\}$ b. $\{0\}, \{1, 3, 4\}, \{2\}$
   c. $\{0\}, \{1, 2, 3, 4\}$

In each of 3–12, the relation R is an equivalence relation on the set A. Find the distinct equivalence classes of R.

3. $A = \{0, 1, 2, 3, 4\}$
   $R = \{(0, 0), (0, 4), (1, 1), (1, 3), (2, 2), (3, 1), (3, 3), (4, 0), (4, 4)\}$

4. $A = \{a, b, c, d\}$
   $R = \{(a, a), (b, b), (b, d), (c, c), (d, b), (d, d)\}$

5. $A = \{1, 2, 3, 4, \ldots, 20\}$. R is defined on A as follows:

   For all $x, y \in A$, $\quad x \, R \, y \quad \Leftrightarrow \quad 4 \,|\, (x - y)$.

6. $A = \{-4, -3, -2, -1, 0, 1, 2, 3, 4, 5\}$. R is defined on A as follows:

   For all $x, y \in A$, $\quad x \, R \, y \quad \Leftrightarrow \quad 3 \,|\, (x - y)$.

7. $A = \{(1, 3), (2, 4), (-4, -8), (3, 9), (1, 5), (3, 6)\}$. R is defined on A as follows: For all $(a, b), (c, d) \in A$,

   $$(a, b) \, R \, (c, d) \quad \Leftrightarrow \quad ad = bc.$$

8. $X = \{a, b, c\}$ and $A = \mathscr{P}(X)$. R is defined on A as follows: For all sets U and V in $\mathscr{P}(X)$,

   $$U \, R \, V \quad \Leftrightarrow \quad N(U) = N(V).$$

   (That is, the number of elements in U equals the number of elements in V.)

9. $X = \{-1, 0, 1\}$ and $A = \mathscr{P}(X)$. R is defined on $\mathscr{P}(X)$ as follows: For all sets S and T in $\mathscr{P}(X)$,

   $S \, R \, T \quad \Leftrightarrow \quad$ the sum of the elements in S equals the sum of the elements in T.

10. $A = \{-5, -4, -3, -2, -1, 0, 1, 2, 3, 4, 5\}$. R is defined on A as follows: For all $m, n \in \mathbf{Z}$,

    $$m \, R \, n \quad \Leftrightarrow \quad 3 \,|\, (m^2 - n^2).$$

11. $A = \{-4, -3, -2, -1, 0, 1, 2, 3, 4\}$. R is defined on A as follows: For all $(m, n) \in A$,

    $$m \, R \, n \Leftrightarrow 4 \,|\, (m^2 - n^2).$$

12. $A = \{-4, -3, -2, -1, 0, 1, 2, 3, 4\}$. R is defined on A as follows: For all $(m, n) \in A$,

    $$m \, R \, n \Leftrightarrow 5 \,|\, (m^2 - n^2).$$

13. Determine which of the following congruence relations are true and which are false.
    a. $17 \equiv 2 \pmod{5}$ b. $4 \equiv -5 \pmod{7}$
    c. $-2 \equiv -8 \pmod{3}$ d. $-6 \equiv 22 \pmod{2}$

14. **a.** Let R be the relation of congruence modulo 3. Which of the following equivalence classes are equal?

$$[7], [-4], [-6], [17], [4], [27], [19]$$

b. Let R be the relation of congruence modulo 7. Which of the following equivalence classes are equal?

$$[35], [3], [-7], [12], [0], [-2], [17]$$

15. **a.** Prove that for all integers m and $n, m \equiv n \pmod{3}$ if, and only if, $m \bmod 3 = n \bmod 3$.
    b. Prove that for all integers m and n and any positive integer $d, m \equiv n \pmod{d}$ if, and only if, $m \bmod d = n \bmod d$.

16. **a.** Give an example of two sets that are distinct but not disjoint.
    b. Find sets A_1 and A_2 and elements x, y and z such that x and y are in A_1 and y and z are in A_2 but x and z are not both in either of the sets A_1 or A_2.

In 17–27, (1) prove that the relation is an equivalence relation, and (2) describe the distinct equivalence classes of each relation.

17. A is the set of all students at your college.
    a. R is the relation defined on A as follows: For all x and y in A,

    $$x \, R \, y \quad \Leftrightarrow \quad x \text{ has the same major (or double major) as } y.$$

    (Assume "undeclared" is a major.)
    b. S is the relation defined on A as follows: For all $x, y \in A$,

    $$x \, S \, y \quad \Leftrightarrow \quad x \text{ is the same age as } y.$$

18. E is the relation defined on $\mathbf{Z}$ as follows:

    For all $m, n \in \mathbf{Z}, \quad m \, E \, n \quad \Leftrightarrow \quad 2 \mid (m - n).$

19. F is the relation defined on $\mathbf{Z}$ as follows:

    For all $m, n \in \mathbf{Z}, \quad m \, F \, n \quad \Leftrightarrow \quad 4 \mid (m - n).$

20. Let A be the set of all statement forms in three variables $p, q,$ and r. $\mathbf{R}$ is the relation defined on A as follows: For all P and Q in A,

    $$\text{P} \, \mathbf{R} \, \text{Q} \quad \Leftrightarrow \quad \text{P and Q have the same truth table.}$$

21. A is the "absolute value" relation defined on $\mathbf{R}$ as follows:

    For all $x, y \in \mathbf{R}, \quad x \, A \, y \quad \Leftrightarrow \quad |x| = |y|.$

H 22. D is the relation defined on $\mathbf{Z}$ as follows: For all $m, n \in \mathbf{Z}$,

    $$m \, D \, n \quad \Leftrightarrow \quad 3 \mid (m^2 - n^2).$$

23. R is the relation defined on $\mathbf{Z}$ as follows: For all $(m, n) \in \mathbf{Z}$,

    $$m \, R \, n \quad \Leftrightarrow \quad 4 \mid (m^2 - n^2).$$

24. I is the relation defined on $\mathbf{R}$ as follows:

    For all $x, y \in \mathbf{R}, \quad x \, I \, y \quad \Leftrightarrow \quad x - y \text{ is an integer.}$

25. Define P on the set $\mathbf{R} \times \mathbf{R}$ of ordered pairs of real numbers as follows: For all $(w, x), (y, z) \in \mathbf{R} \times \mathbf{R}$,

    $$(w, x) \, P \, (y, z) \quad \Leftrightarrow \quad w = y.$$

26. Define Q on the set $\mathbf{R} \times \mathbf{R}$ as follows: For all $(w, x), (y, z) \in \mathbf{R} \times \mathbf{R}$,

    $$(w, x) \, Q \, (y, z) \quad \Leftrightarrow \quad x = z.$$

27. Let P be the set of all points in the Cartesian plane except the origin. R is the relation defined on P as follows: For all p_1 and p_2 in P,

    $$p_1 \, R \, p_2 \quad \Leftrightarrow \quad \begin{array}{l} p_1 \text{ and } p_2 \text{ lie on the same half-line} \\ \text{emanating from the origin.} \end{array}$$

28. Let A be the set of all straight lines in the Cartesian plane. Define a relation $\|$ on A as follows:

    For all l_1 and l_2 in $A, \quad l_1 \| l_2 \quad \Leftrightarrow \quad l_1$ is parallel to l_2.

    Then $\|$ is an equivalence relation on A. Describe the equivalence classes of this relation.

29. Let A be the set of points in the rectangle with x and y coordinates between 0 and 1. That is,

    $$A = \{(x, y) \in \mathbf{R} \times \mathbf{R} \mid 0 \le x \le 1 \quad \text{and} \quad 0 \le y \le 1\}.$$

    Define a relation R on A as follows: For all (x_1, y_1) and (x_2, y_2) in A,

    $(x_1, y_1) \, R \, (x_2, y_2) \Leftrightarrow$
    $\qquad (x_1, y_1) = (x_2, y_2); \quad$ or
    $\qquad x_1 = 0 \quad$ and $\quad x_2 = 1 \quad$ and $\quad y_1 = y_2; \quad$ or
    $\qquad x_1 = 1 \quad$ and $\quad x_2 = 0 \quad$ and $\quad y_1 = y_2; \quad$ or
    $\qquad y_1 = 0 \quad$ and $\quad y_2 = 1 \quad$ and $\quad x_1 = x_2; \quad$ or
    $\qquad y_1 = 1 \quad$ and $\quad y_2 = 0 \quad$ and $\quad x_1 = x_2.$

    In other words, all points along the top edge of the rectangle are related to the points along the bottom edge directly beneath them, and all points directly opposite each other along the left and right edges are related to each other. The points in the interior of the rectangle are not related to anything other than themselves. Then R is an equivalence relation on A. Imagine gluing together all the points that are in the same equivalence class. Describe the resulting figure.

Let R be an equivalence relation on a set A. Prove each of the statements in 30–35 directly from the definitions of equivalence relation and equivalence class without using the results of Lemma 8.3.2, Lemma 8.3.3, or Theorem 8.3.4.

30. For all a in $A, a \in [a]$.

31. For all a and b in A, if $b \in [a]$ then $a \, R \, b$.

32. For all a, b and c in A, if $b \, R \, c$ and $c \in [a]$ then $b \subset [a]$.

33. For all a and b in A, if $[a] = [b]$ then $a \, R \, b$.

34. For all a, b, and x in A, if $a\ R\ b$ and $x \in [a]$, then $x \in [b]$.

H 35. For all a and b in A, if $a \in [b]$ then $[a] = [b]$.

36. Let R be the relation defined in Example 8.3.8.
 a. Prove that R is reflexive.
 b. Prove that R is symmetric.
 c. List four distinct elements in $[(1, 3)]$.
 d. List four distinct elements in $[(2, 5)]$.

★ 37. In Example 8.3.8, define operations of addition ($+$) and multiplication ($\cdot$) as follows: For all $(a, b), (c, d) \in A$,

$$[(a, b)] + [(c, d)] = [(ad + bc, bd)]$$
$$[(a, b)] \cdot [(c, d)] = [(ac, bd)].$$

 a. Prove that this addition is well defined. That is, show that if $[(a, b)] = [(a', b')]$ and $[(c, d)] = [(c', d')]$, then $[(ad + bc, bd)] = [(a'd' + b'c', b'd')]$.
 b. Prove that this multiplication is well defined. That is, show that if $[(a, b)] = [(a', b')]$ and $[(c, d)] = [(c', d')]$, then $[(ac, bd)] = [(a'c', b'd')]$.
 c. Show that $[(0, 1)]$ is an identity element for addition. That is, show that for any $(a, b) \in A$,

$$[(a, b)] + [(0, 1)] = [(0, 1)] + [(a, b)] = [(a, b)].$$

 d. Find an identity element for multiplication. That is, find (i, j) in A so that for all (a, b) in A. $[(a, b)] \cdot [(i, j)] = [(i, j)] \cdot [(a, b)] = [(a, b)]$.
 e. For any $(a, b) \in A$, show that $[(-a, b)]$ is an inverse for $[(a, b)]$ for addition. That is, show that $[(-a, b)] + [(a, b)] = [(a, b)] + [(-a, b)] = [(0, 1)]$.
 f. Given any $(a, b) \in A$ with $a \neq 0$, find an inverse for $[(a, b)]$ for multiplication. That is, find (c, d) in A so that $[(a, b)] \cdot [(c, d)] = [(c, d)] \cdot [(a, b)] = [(i, j)]$, where $[(i, j)]$ is the identity element you found in part (d).

38. Let $A = \mathbf{Z}^+ \times \mathbf{Z}^+$. Define a relation R on A as follows: For all (a, b) and (c, d) in A,

$$(a, b)\ R\ (c, d) \quad \Leftrightarrow \quad a + d = c + b.$$

 a. Prove that R is reflexive.
 b. Prove that R is symmetric.
 H c. Prove that R is transitive.
 d. List five elements in $[(1, 1)]$.
 e. List five elements in $[(3, 1)]$.
 f. List five elements in $[(1, 2)]$.
 g. Describe the distinct equivalence classes of R.

39. The following argument claims to prove that the requirement that an equivalence relation be reflexive is redundant. In other words, it claims to show that if a relation is symmetric and transitive, then it is reflexive. Find the mistake in the argument.

"**Proof:** Let R be a relation on a set A and suppose R is symmetric and transitive. For any two elements x and y in A, if $x\ R\ y$ then $y\ R\ x$ since R is symmetric. But then it follows by transitivity that $x\ R\ x$. Hence R is reflexive."

40. Let R be a relation on a set A and suppose R is symmetric and transitive. Prove the following: If for every x in A there is a y in A such that $x\ R\ y$, then R is an equivalence relation.

41. Refer to the quote at the beginning of this section to answer the following questions.
 a. What is the name of the Knight's song called?
 b. What is the name of the Knight's song?
 c. What is the Knight's song called?
 d. What *is* the Knight's song?
 e. What is your (full, legal) name?
 f. What are you called?
 g. What *are* you? (Do not answer this on paper; just think about it.)

Answers for Test Yourself

1. reflexive, symmetric, and transitive   2. m is congruent to n modulo d; d divides $m - n$ 3. $[a]$; the set of all x in A such that $x R a$
4. $[a] \cap [b] = \emptyset$ 5. a partition of A 6. rational number

8.4 Modular Arithmetic and Z_n

The 'real' mathematics of the 'real' mathematicians, the mathematics of Fermat and Euler and Gauss and Abel and Riemann, is almost wholly 'useless' . . . It is not possible to justify the life of any genuine professional mathematician on the ground of the 'utility' of his work. — G. H. Hardy, *A Mathematician's Apology*, 1941

There is no branch of mathematics, however abstract, which may not some day be applied to phenomena of the real world. —Nikolai Lobatchevsky (1792–1856)

The contradictory quotations at the beginning of this section represent divergent views about the nature of mathematics. G.H. Hardy certainly knew that much of what he called "real" mathematics had essential applications in science and technology, but he believed that the beauty of the mathematics itself would have been sufficient justification for creating it. Hardy thought that number theory especially would not be shown to have practical utility in the future. In that respect Nikolai Lobatchevsky was more prescient. Applications of number theory – mostly through mathematics based on properties of congruence – range from the encryption of material sent over the Internet, such as bank records, PIN numbers, and cable television access codes, to the use of check digits to prevent errors in barcodes such as universal product codes and tracking numbers, to the optimization of communication networks, to the solution of problems in quantum computation – and indeed to an increasing number of aspects of our lives.

In this section we delve more deeply into number theory by developing properties of congruence modulo n and showing how to perform arithmetic modulo n. We then explain how to extend this arithmetic to add and multiply equivalence classes of integers modulo n, and we develop properties of, $\mathbf{Z}_n$, the set of all such classes.

The example of congruence and modular arithmetic with which most people are familiar comes from experience telling time with an 12-hour analog clock.

If it is 5 o'clock and the time moves forward 4 hours, the hands will show $5 + 4 = 9$ o'clock. But if, starting at 5 o'clock, the time moves forward 10 hours, the clock will not show $5 + 10 = 15$ o'clock; it will show 3 o'clock. Observe that 3 is the remainder obtained when $5 + 10$ is divided by 12:

$$3 = (5 + 10) \bmod 12 = 15 \bmod 12.$$

In general, if the clock shows time h hours, where h is an integer from 1 through 12, and the time moves forward x hours, where x is an integer, then the hands will show time y where

$$y = (h + x) \bmod 12$$

For example, if it is 5 o'clock and the time moves forward 30 hours, then the clock hands will show $(5 + 30) \bmod 12 = 35 \bmod 12 = 11$ o'clock. To check this observe that $30 = 6 + 2 \cdot 12$, so, starting from 5 o'clock, the clock hands would move forward 6 hours to 11 o'clock and then make two complete revolutions ending up back at 11 o'clock.

Properties of Congruence Modulo n

The first theorem in this section brings together a variety of equivalent ways of expressing the same basic arithmetic fact. Sometimes one way is most convenient; sometimes another way is best. You need to be comfortable moving from one to another, depending on the nature of the problem you are trying to solve.

Theorem 8.4.1 Modular Equivalences

Let a, b, and n be any integers and suppose $n > 1$. The following statements are all equivalent:

1. $n \mid (a - b)$

2. $a \equiv b \pmod{n}$

3. $a = b + kn$ for some integer k

4. a and b have the same (nonnegative) remainder when divided by n

5. $a \bmod n = b \bmod n$

Proof:

We will show that $(1) \Rightarrow (2) \Rightarrow (3) \Rightarrow (4) \Rightarrow (5) \Rightarrow (1)$. It will follow by the transitivity of if-then that all five statements are equivalent.

So let a, b, and n be any integers with $n > 1$.

***Proof that* (1) $\Rightarrow$ (2):** Suppose that $n \mid (a - b)$. By definition of congruence modulo n, we can immediately conclude that $a \equiv b \pmod{n}$.

***Proof that* (2) $\Rightarrow$ (3):** Suppose that $a \equiv b \pmod{n}$. By definition of congruence modulo n, $n \mid (a - b)$. Thus, by definition of divisibility, $a - b = kn$, for some integer k. Adding b to both sides gives that $a = b + kn$.

***Proof that* (3) $\Rightarrow$ (4):** Suppose that $a = b + kn$, for some integer k. Use the quotient-remainder theorem to divide a by n to obtain

$$a = qn + r \quad \text{where } q \text{ and } r \text{ are integers and } 0 \leq r < n.$$

Substituting $b + kn$ for a in this equation gives that

$$b + kn = qn + r$$

and subtracting kn from both sides and factoring out n yields

$$b = (q - k)n + r.$$

But since $0 \leq r < n$, the uniqueness property of the quotient-remainder theorem guarantees that r is also the remainder obtained when b is divided by n. Thus a and b have the same remainder when divided by n.

***Proof that* (4) $\Rightarrow$ (5):** Suppose that a and b have the same remainder when divided by n. It follows immediately from the definition of the *mod* function that $a \bmod n = b \bmod n$.

***Proof that* (5) $\Rightarrow$ (1):** Suppose that $a \bmod n = b \bmod n$. By definition of the *mod* function, a and b have the same remainder when divided by n. Thus, by the quotient-remainder theorem, we can write

$$a = q_1 n + r \quad \text{and} \quad b = q_2 n + r \quad \text{where } q_1, q_2, \text{ and } r \text{ are integers and } 0 \leq r < n.$$

It follows that

$$a - b = (q_1 n + r) - (q_2 n + r) = (q_1 - q_2)n.$$

Therefore, since $q_1 - q_2$ is an integer, $n \mid (a - b)$.

Example 8.4.1 Verifying a Modular Congruence

Verify that $235 \equiv 53 \pmod 7$ and find an integer k such that $235 = 53 + 7k$.

Solution

$$235 - 53 = 182 \quad \text{and} \quad 182 = 7 \cdot 26.$$

Thus $7 \mid (235 - 53)$ and the congruence is verified. Putting the two equations together gives that

$$235 - 53 = 7 \cdot 26, \quad \text{and so} \quad 235 = 53 + 7 \cdot 26.$$

Thus $k = 26$. ∎

Another consequence of the quotient-remainder theorem is this: When an integer a is divided by a positive integer n, a unique quotient q and remainder r are obtained with the property that $a = nq + r$ and $0 \leq r < n$. Because there are exactly n integers that satisfy the inequality $0 \leq r < n$ (the numbers from 0 through $n - 1$), there are exactly n possible remainders that can occur. These are called the *least nonnegative residues modulo n* or simply the *residues modulo n*.

> • **Definition**
>
> Given integers a and n with $n > 1$, **the residue of *a* modulo *n*** is $a \bmod n$, the nonnegative remainder obtained when a is divided by n. The numbers $0, 1, 2, \ldots, n - 1$ form a **complete set of residues modulo *n***. To **reduce a number modulo *n*** means to replace it by its residue modulo n. If a modulus $n > 1$ is fixed throughout a discussion and an integer a is given, the words "modulo n" are often dropped and we simply speak of **the residue of *a***.

The following theorem generalizes several examples from Section 8.3.

> **Theorem 8.4.2 Congruence Modulo *n* Is an Equivalence Relation**
>
> If n is any integer with $n > 1$, congruence modulo n is an equivalence relation on the set of all integers. The distinct equivalence classes of the relation are the sets $[0], [1], [2], \ldots, [n - 1]$, where for each $a = 0, 1, 2, \ldots, n - 1$,
>
> $$[a] = \{m \in Z \mid m \equiv a \pmod n\},$$
>
> or, equivalently,
>
> $$[a] = \{m \in Z \mid m = a + kn \text{ for some integer } k\}.$$

Proof:

Suppose n is any integer with $n > 1$. We must show that congruence modulo n is reflexive, symmetric, and transitive.

Proof of reflexivity: Suppose a is any integer. To show that $a \equiv a \pmod n$, we must show that $n \mid (a - a)$. But $a - a = 0$, and $n \mid 0$ because $0 = n \cdot 0$. Therefore $a \equiv a \pmod n$.

continued on page 378

Proof of symmetry: Suppose a and b are any integers such that $a \equiv b \pmod{n}$. We must show that $b \equiv a \pmod{n}$. But since $a \equiv b \pmod{n}$, then $n \mid (a - b)$. Thus, by definition of divisibility, $a - b = nk$, for some integer k. Multiply both sides of this equation by -1 to obtain

$$-(a - b) = -nk,$$

or, equivalently,

$$b - a = n(-k).$$

Thus, by definition of divisibility $n \mid (b - a)$, and so, by definition of congruence modulo n, $b \equiv a \pmod{n}$.

Proof of transitivity: This is left as exercise 5 at the end of the section.

Proof that the distinct equivalence classes are $[0], [1], [2], \ldots, [n - 1]$: This is left as exercise 6 at the end of the section.

Observe that there is a one-to-one correspondence between the distinct equivalence classes for congruence modulo n and the elements of a complete set of residues modulo n.

Modular Arithmetic

A fundamental fact about congruence modulo n is that if you first perform an addition, subtraction, or multiplication on integers and then reduce the result modulo n, you will obtain the same answer as if you had first reduced each of the numbers modulo n, performed the operation, and then reduced the result modulo n. For instance,

since $\qquad\qquad 5 \equiv 2 \pmod 3 \quad$ and $\qquad 7 \equiv 1 \pmod 3$

then $\qquad\qquad 5 + 7 \equiv 2 + 1 \pmod 3 \quad$ and $\qquad 5 \cdot 7 \equiv 2 \cdot 1 \pmod 3.$

To check that this is true, note that $5 + 7 = 12$ and

$$12 \equiv 3 \pmod 3 \quad \text{because} \quad 3 \mid (12 - 3),$$

and $5 \cdot 7 = 35$ and $\qquad 35 \equiv 2 \pmod 3 \quad \text{because} \quad 3 \mid (35 - 2).$

The fact that this process works is a result of the following theorem.

Theorem 8.4.3 Modular Arithmetic

Let $a, b, c, d,$ and n be integers with $n > 1$, and suppose

$$a \equiv c \pmod n \text{ and } b \equiv d \pmod n.$$

Then

1. $(a + b) \equiv (c + d) \pmod n$

2. $(a - b) \equiv (c - d) \pmod n$

3. $ab \equiv cd \pmod n$

4. $a^m \equiv c^m \pmod n$ for all positive integers m.

Proof:

We prove part 3 of the theorem, here and leave the proofs of the remaining parts of the theorem to exercises 9–11 at the end of the section.

***Proof of Part 3*:** Suppose $a, b, c, d,$ and n are integers with $n > 1$, and suppose $a \equiv c \pmod{n}$ and $b \equiv d \pmod{n}$. By Theorem 8.4.1, there exist integers s and t such that

$$a = c + sn \quad \text{and} \quad b = d + tn.$$

Then

$$
\begin{aligned}
ab &= (c + sn)(d + tn) &&\text{by substitution} \\
&= cd + ctn + snd + sntn \\
&= cd + n(ct + sd + stn) &&\text{by algebra.}
\end{aligned}
$$

Let $k = ct + sd + stn$. Then k is an integer and $ab = cd + nk$. Thus by Theorem 8.4.1, $ab \equiv cd \pmod{n}$.

Example 8.4.2 Getting Started with Modular Arithmetic

The most practical use of modular arithmetic is to reduce computations involving large integers to computations involving smaller ones. For instance, note that $55 \equiv 3 \pmod{4}$ because $55 - 3 = 52$, which is divisible by 4, and $26 \equiv 2 \pmod{4}$ because $26 - 2 = 24$, which is also divisible by 4. Verify the following statements.

a. $55 + 26 \equiv (3 + 2) \pmod{4}$ b. $55 - 26 \equiv (3 - 2) \pmod{4}$

c. $55 \cdot 26 \equiv (3 \cdot 2) \pmod{4}$ d. $55^2 \equiv 3^2 \pmod{4}$

Solution

a. To show that $81 \equiv 5 \pmod{4}$, you need to show that $4 \mid (81 - 5)$. But this is true because $81 - 5 = 76$, and $4 \mid 76$ since $76 = 4 \cdot 19$.

b. To show that $29 \equiv 1 \pmod{4}$, you need to show that $4 \mid (29 - 1)$. But this is true because $29 - 1 = 28$, and $4 \mid 28$ since $28 = 4 \cdot 7$.

c. To show that $1430 \equiv 6 \pmod{4}$, you need to show that $4 \mid (1430 - 6)$. But this is true because $1430 - 6 = 1424$, and $4 \mid 1424$ since $1424 = 4 \cdot 356$.

d. To show that $3025 \equiv 9 \pmod{4}$, you need to show that $4 \mid (3025 - 9)$. But this is true because $3025 - 9 = 3016$, and $4 \mid 3016$ since $3016 = 4 \cdot 754$.   ■

When an integer is written in ordinary decimal notation, its **units digit** is the digit on its extreme right. For example, the units digit of 247 is 7. The reason 7 is called the "units digit" of 247 is that when 247 is written in expanded form, it becomes

$$247 = 2 \cdot 100 + 4 \cdot 10 + 7 \cdot 1.$$

In other words, 247 equals 2 hundreds plus 4 tens plus 7 ones. So 7 is the number of ones, or single units, in the number. Observe that the units digit of a number is the remainder obtained when the number is divided by 10. For example,

$$247 = 2 \cdot 100 + 4 \cdot 10 + 7 \cdot 1 = 20 \cdot 10 + 4 \cdot 10 + 7 \cdot 1 = 24 \cdot 10 + 7.$$

In general, given any integer n, by the quotient-remainder theorem with $d = 10$ there exist integers q and r with

$$n = 10q + r \text{ and } 0 \le r < 10.$$

Thus the units digit of n is the remainder r of the division by 10, or, in other words,

the units digit of $n = n \bmod 10$.

Example 8.4.3 Applying Modular Arithmetic to Find a Units Digit

a. What is the units digit of 3^{1000}?

b. What is the units digit of 4^{50}?

Solution

Note In Section 9.4, we will show that a repetition of the units digit is guaranteed to occur eventually when the quantities 3^k are computed for $k = 1, 2, 3, \ldots.$

a. Begin by computing some powers of 3 starting with 3^0 and ending as soon as a units digit is repeated or a nonzero power equals 1: $3^0 = 1, 3^1 = 3, 3^2 = 9, 3^3 = 27,$ $3^4 = 81.$ These calculations show that

$$3^4 \equiv 1 \ (\text{mod } 10).$$

By the laws of exponents and part (d) of Theorem 8.4.3,

$$3^{1000} = (3^4)^{250} \equiv 1^{250} \equiv 1^{250} \ (\text{mod } 10) \equiv 1 \ (\text{mod } 10).$$

So the remainder obtained when 3^{1000} is divided by 10 is 1, and thus the units digit of 3^{1000} is 1.

b. Again begin by computing powers and ending when a units digit is repeated: $4^0 = 1,$ $4^1 = 4, 4^2 = 16, 4^3 = 64.$ These calculations show that

$$4^3 \equiv 4 \ (\text{mod } 10).$$

By the laws of exponents and parts (c) and (d) of Theorem 8.4.3,

$$4^{50} = 4^{48} \cdot 4^2 = (4^3)^{16} \cdot 4^2 \equiv 4^{16} \cdot 4^2 \equiv 4^{18} \equiv (4^3)^6 \equiv 4^6 \equiv (4^3)^2$$
$$\equiv 4^2 \equiv 16 \equiv 6 \ (\text{mod } 10).$$

So the remainder obtained when 4^{50} is divided by 10 is 6, and thus the units digit of 4^{50} is 6. ∎

Modular arithmetic can also be used in connection with the problems of finding integer solutions for equations.

Example 8.4.4 Applying Modular Arithmetic to Equation Solving

Reduce each of the following equations modulo 3 to explore the question of whether there are integers x and y that simultaneously satisfy them:

$$16x + 12y = 32$$
$$40x - 9y = 7$$

Solution   Observe that $16 \equiv 1 \ (\text{mod } 3), 12 \equiv 0 \ (\text{mod } 3), 32 \equiv 2 \ (\text{mod } 3), 40 \equiv 1 \ (\text{mod } 3),$ $-9 \equiv 0 \ (\text{mod } 3),$ and $7 \equiv 1 \ (\text{mod } 3)$ (because $16 = 3 \cdot 5 + 1, 12 = 3 \cdot 4 + 0, 32 = 3 \cdot 10 + 2, 40 = 3 \cdot 13 + 1, -9 = 3 \cdot (-3) + 0,$ and $7 = 3 \cdot 2 + 1$). Thus, when reduced modulo 3, the equations become

$$1 \cdot x + 0 \cdot y \equiv 2 \ (\text{mod } 3) \quad \text{and} \quad 1 \cdot x + 0 \cdot y \equiv 1 \ (\text{mod } 3),$$

which are equivalent to

$$x \equiv 2 \ (\text{mod } 3) \quad \text{and} \quad x \equiv 1 \ (\text{mod } 3).$$

However these two congruences are contradictory, and so the given equations do not have a simultaneous solution in the set of integers. ∎

Check digits are used to reduce errors in numbers found alongside barcodes, such as those used as universal product codes, tracking numbers for shipping operations, and book identification numbers (ISBNs). Check digits also occur in the numbers used to identify individual vehicles, individual providers for the U.S. healthcare industry, and many other items.

As an example for how they work, we examine a universal product code, which is a 12-digit number. The check digit, located in position 12, is calculated as follows:

1. Add the digits in positions 1, 3, 5, 7, 9, and 11, and multiply by 3.

2. Add the result of step 1 to the sum of the digits in positions 2, 4, 6, 8, and 10.

3. The check digit is the units digit obtained when the result of step 2 is subtracted from a larger number that is a multiple of 10.

If the first eleven digits of a UPC are $a_1a_2a_3a_4a_5a_6a_7a_8a_9a_{10}a_{11}$, then, because the maximum number obtained in step 2 is 207, the check digit, a_{12}, can be computed using the following formula:

$$a_{12} = \{210 - [3(a_1 + a_3 + a_5 + a_7 + a_9 + a_{11}) + (a_2 + a_4 + a_6 + a_8 + a_{10})]\} \bmod 10.$$

Example 8.4.5 A Check Digit for a Universal Product Code (UPC)

The first eleven digits of the UPC for a package of ink cartridges are 88442334010. What is the check digit?

Solution

$$\{210 - [3(8 + 4 + 2 + 3 + 0 + 0) + (8 + 4 + 3 + 4 + 1)]\} \bmod 10$$
$$= [210 - (3 \cdot 17 + 20)] \bmod 10$$
$$= [210 - 71] \bmod 10$$
$$= 139 \bmod 10$$
$$= 9.$$

So the check digit is 9. ■

Z_n: The Integers Modulo n

We now introduce a notation for the set of distinct equivalence classes for congruence modulo n and define operations on this set.

> **• Notation**
>
> For each integer n greater than 1, $\mathbf{Z}_n$ is the set of distinct equivalence classes for congruence modulo n:
>
> $$\mathbf{Z}_n = \{[0], [1], [2], \ldots, [n - 1]\}.$$

Addition and multiplication of real numbers are examples of *binary operations*. Each involves inputting two numbers and outputting a third. We will define binary operations on $\mathbf{Z}_n$, and, following common usage, we will refer to them as addition and multiplication even though they are not the same as addition and multiplication of real numbers and even though different values of n produce different sums and products for $\mathbf{Z}_n$.

> **• Definition**
>
> Given any set S, a **binary operation** $\star$ on S is a function from $S \times S$ to S. Given x and y in S, we use the notation
>
> $$x \star y$$
>
> to mean the value of the function $\star$ for the input (x, y). A subset A of S is said to be **closed under the operation** $\star$ if, and only if, for all a and b in A, $a \star b$ is in A.

Example 8.4.6 Determining Whether a Set Is Closed under an Operation

    a. Is the set $\mathbf{Z}$ closed under the operation of subtraction?

    b. Is the set $\mathbf{Z}^+$ closed under the operation of subtraction?

Solution

    a. Yes. The difference of any two integers is an integer.

    b. No. For example, $4 - 7$ is a difference of two positive integers, but $4 - 7 = -3$, which is not a positive integer. ∎

> **• Definition**
>
> The following two binary operations are defined on the set $\mathbf{Z}_n$. We call one of them **addition** and denote it by $+$, and we call the other **multiplication** and denote it by $\cdot$. Given any equivalence classes $[a]$, $[b]$, $[c]$, and $[d]$ in $\mathbf{Z}_n$, where a, b, c, and d are integers, $[a] + [b]$ is defined to be the equivalence class $[a + b]$ and $[a] \cdot [b]$ is defined to be the equivalence class $[ab]$. In symbols:
>
> $$[a] + [b] = [a + b] \quad \text{and} \quad [a] \cdot [b] = [ab].$$

    A potential problem with this definition is that it may not be "well-defined," in the sense that there might be integers a, b, c and d with the property that

$$[a] + [b] \neq [c] + [d] \quad \text{even though} \quad [a] = [c] \text{ and } [b] = [d]$$

or that

$$[a] \cdot [b] \neq [c] \cdot [d] \quad \text{even though} \quad [a] = [c] \text{ and } [b] = [d].$$

The following theorem says that you do not have to worry about this problem. For example, in $\mathbf{Z}_5$, $[17] = [52]$ and $[14] = [-6]$. To check that $[17] + [14] = [52] + [-6]$, observe that

$$[17] + [14] = [17 + 14] = [31] = [1] \quad \text{because} \quad 5 \mid (31 - 1)$$

and

$$[52] + [-6] = [52 + (-6)] = [46] = [1] \quad \text{because} \quad 5 \mid (46 - 1).$$

Thus $[17] + [14] = [52] + [-6]$ because both sums equal $[1]$.

Theorem 8.4.4 Addition and Multiplication on $\mathbf{Z}_n$ Are Well-Defined

Given any equivalence classes $[a]$, $[b]$, $[c]$, and $[d]$ in $\mathbf{Z}_n$, where a, b, c, and d are integers. If $[a] = [c]$ and $[b] = [d]$, then

$$(1)\ [a] + [b] = [c] + [d] \quad \text{and} \quad (2)\ [a] \cdot [b] = [c] \cdot [d].$$

Proof of (1):

Suppose $[a]$, $[b]$, $[c]$, and $[d]$ are any equivalence classes in $\mathbf{Z}_n$, where a, b, c, and d are integers, and suppose that $[a] = [c]$ and $[b] = [d]$. *[We must show that $[a] + [b] = [c] + [d]$.]* By definition of equivalence class,

since $[a] = [c]$, then $a \equiv c \pmod n$ and since $[b] = [d]$, then $b \equiv d \pmod n$.

Hence, by Theorem 8.4.3(a),

$$a + b \equiv (c + d) \pmod n$$

Applying the definition of equivalence class to this congruence gives that

$$[a + b] = [c + d].$$

But, according to the definition of addition in $\mathbf{Z}_n$,

$$[a] + [b] = [a + b] \quad \text{and} \quad [c] + [d] = [c + d],$$

and, since $[a + c] = [b + d]$, we conclude that

$$[a] + [b] = [c] + [d]$$

[as was to be shown].

Proof of (2): This is left as exercise 29 at the end of the section.

In Section 6.4 we defined an abstract algebraic structure known as a Boolean algebra, and we observed two examples of Boolean algebras: a set of sets with the operations of union and intersection and a set of statement forms with the operations of disjunction and conjunction. We now define a general algebraic structure known as a *commutative ring*.

● **Definition**

Let S be a set with two binary operations called addition, $+$, and multiplication, $\cdot$. S is called a **commutative ring** if, and only if, the following properties hold:

For all elements a, b, and c in S,

1. **commutative properties:** $a + b = b + a$ and $a \cdot b = b \cdot a$

2. **associative properties:** $(a + b) + c = a + (b + c)$ and $(a \cdot b) \cdot c = a \cdot (b \cdot c)$

3. **distributive property:** $a \cdot (b + c) = a \cdot b + a \cdot c$

4. S contains distinct elements, which we will denote here by 0 and 1, such that $a + 0 = a$ and $a \cdot 1 = a$. We call 0 the **identity for addition** and 1 the **identity for multiplication**.

5. For all elements a in S, there exists an element b in S such that $a + b = 0$. The element b is called the **additive inverse** of a.

Some familiar examples of commutative rings are the sets of integers, rational numbers, and real numbers, with the usual addition and multiplication. Addition and multiplication are binary operations on the set of integers because sums and products of integers are integers, and addition and multiplication are binary operations on the set of rational numbers because sums and products of rational numbers are rational numbers. Both the set of integers and the set of rational numbers inherit the commutative, associative, and distributive properties and the existence of identities for addition and multiplication from the set of real numbers, and both sets have an additive inverse for each of their elements. However, the set of rational numbers and the set of real numbers satisfy a significant additional property beyond those needed to make them commutative rings, namely that every nonzero element has a multiplicative inverse.

Theorem 8.4.5 says that each of the sets $\mathbf{Z}_n$, with the addition and multiplication defined previously, is an example of a commutative ring.

Theorem 8.4.5 $\mathbf{Z}_n$ is a commutative ring

Let n be any integer that is greater than 1, and let operations of addition and multiplication be defined on $\mathbf{Z}_n$ as follows: Given any equivalence classes $[a]$, $[b]$, $[c]$, and $[d]$ in $\mathbf{Z}_n$, where a, b, c, and d are integers,

$$[a] + [b] = [a + b] \quad \text{and} \quad [a] \cdot [b] = [ab].$$

Together with these operations, $\mathbf{Z}_n$ is a commutative ring.

Proof:

We show that $\mathbf{Z}_n$ satisfies parts of properties 4 and 5 of the definition of commutative ring and leave the rest of the proof to the exercises.

Proof of property 4: We claim that $[0]$ is an identity for addition and $[1]$ is an identity for multiplication. To verify the first claim, observe that for all equivalence classes $[a]$ in $\mathbf{Z}_n$, where a is an integer,

$$[a] + [0] = [a + 0] \qquad \text{by definition of addition in } \mathbf{Z}_n$$

$$= [a] \qquad \text{because the set of all integers is closed under addition.}$$

The proof of the second claim, that $[1]$ is the identity for multiplication, is left to exercise 30f at the end of this section.

Proof of property 5: Suppose $[a]$ is any element of $\mathbf{Z}_n$, where a is an integer. Then

$$[a] + [-a] = [0],$$

and so $[-a]$ is an additive inverse for $[a]$.

Note that if $0 < a < n$, then

$[-a] = [n - a]$ because $(n - a) - (-a) = n - a + a = n$, which is divisible by n.

Moreover, multiplying all parts of $0 < a < n$ by -1 gives that

$$0 > -a > -n,$$

and adding n to all parts of this yields

$$n > n - a > 0.$$

Hence if a is a residue modulo n, then so is $n - a$, and $[n - a]$ is another way to write the additive inverse of $[a]$.

Example 8.4.7 Addition and Multiplication in $\mathbf{Z}_4$

a. In $\mathbf{Z}_4$ find the values of $[1] + [3]$, $[2] + [3]$, $[2] \cdot [3]$, and $[2] \cdot [2]$.

b. Make tables showing the values of $[a] + [b]$ and $[a] \cdot [b]$ for all equivalence classes $[a]$ and $[b]$ in $\mathbf{Z}_4$.

Solution

a. $[1] + [3] = [1 + 3] = [4] = [0]$ $[2] + [3] = [2 + 3] = [5] = [1]$
    $[2] \cdot [3] = [2 \cdot 3] = [6] = [2]$ $[2] \cdot [2] = [2 \cdot 2] = [4] = [0]$

b.

$+$	$[0]$	$[1]$	$[2]$	$[3]$
$[0]$	$[0]$	$[1]$	$[2]$	$[3]$
$[1]$	$[1]$	$[2]$	$[3]$	$[0]$
$[2]$	$[2]$	$[3]$	$[0]$	$[1]$
$[3]$	$[3]$	$[0]$	$[1]$	$[2]$

$\cdot$	$[0]$	$[1]$	$[2]$	$[3]$
$[0]$	$[0]$	$[0]$	$[0]$	$[0]$
$[1]$	$[0]$	$[1]$	$[2]$	$[3]$
$[2]$	$[0]$	$[2]$	$[0]$	$[2]$
$[3]$	$[0]$	$[3]$	$[2]$	$[1]$

In the multiplication table for Example 8.4.7(b), observe that two nonzero elements of $\mathbf{Z}_4$ have a zero product: $[2] \cdot [2] = [0]$. Nonzero elements of a ring that multiply out to zero are called **zero divisors**. It turns out that whenever n is composite, $\mathbf{Z}_n$ has zero divisors, and whenever n is prime, $\mathbf{Z}_n$ does not have zero divisors. In fact, when n is prime, each nonzero element of $\mathbf{Z}_n$ has a multiplicative inverse. We prove this result in Section 8.5.

Test Yourself

1. If a, b, and n are integers with $n > 1$, all of the following are different ways to express the fact that $n \mid (a - b)$: _____, _____, _____, _____.

2. If a, b, c, d, m, and n are integers with $n > 1$ and if $a \equiv c \pmod{n}$ and $b \equiv d \pmod{n}$, then $a + b \equiv$ _____, $a - b \equiv$ _____, $ab \equiv$ _____, and $a^m \equiv$ _____.

3. If n is an integer with $n > 1$, the elements of $\mathbf{Z}_n$ can be written using residues modulo n as _____.

4. If $\star$ is a binary operation on a set S and A is a subset of S, then A is closed under $\star$ if, and only if, _____.

5. If n is an integer with $n > 1$ and $[a]$, $[b]$, $[c]$ and $[d]$ are equivalence classes in $\mathbf{Z}_n$ with $[a] = [b]$ and $[c] = [d]$ then $[a] + [c] =$ _____ and $[a] \cdot [c] =$ _____.

6. In a commutative ring S, addition and multiplication satisfy the _____ and _____ properties, the distributive property says that _____, there are identities for _____ and _____, and every element has an additive _____.

Exercise Set 8.4

1. Let $a = 25$ and $b = 19$.
   a. Use the definition of congruence modulo 3 to explain why $25 \equiv 19 \pmod{3}$.
   b. What value of k has the property that $25 = 19 + 3k$?
   c. What is the (nonnegative) remainder obtained when 25 is divided by 3? When 19 is divided by 3?

2. Let $a = 68$ and $b = 33$.
   a. Use the definition of congruence modulo 3 to explain why $68 \equiv 33 \pmod{7}$.
   b. What value of k has the property that $68 = 33 + 7k$?
   c. What is the (nonnegative) remainder obtained when 68 is divided by 7? When 33 is divided by 7?

3. Which of the following alleged congruences are true and which are false?
   a. $58 \equiv 14 \pmod{11}$
   b. $46 \equiv 89 \pmod{13}$
   c. $674 \equiv 558 \pmod{56}$
   d. $432 \equiv 981 \pmod{9}$

4. Which of the following alleged congruences are true and which are false?
   a. $91 \equiv 53 \pmod{14}$
   b. $46 \equiv 123 \pmod{11}$
   c. $283 \equiv 168 \pmod{15}$
   d. $589 \equiv 328 \pmod{29}$

5. Prove the transitivity of modular congruence. That is, prove that for all integers $a, b, c,$ and n with $n > 1$, if $a \equiv b \pmod{n}$ and $b \equiv c \pmod{n}$ then $a \equiv c \pmod{n}$.

H 6. Prove that the distinct equivalence classes of the relation of congruence modulo n are the sets $[0], [1], [2], \ldots, [n-1]$, where for each $a = 0, 1, 2, \ldots, n-1$,

$$[a] = \{m \in Z \mid m \equiv a \pmod{n}\}.$$

7. Verify the following statements.
   a. $128 \equiv 2 \pmod{7}$ and $61 \equiv 5 \pmod{7}$
   b. $(128 + 61) \equiv (2 + 5) \pmod{7}$
   c. $(128 - 61) \equiv (2 - 5) \pmod{7}$
   d. $(128 \cdot 61) \equiv (2 \cdot 5) \pmod{7}$
   e. $128^2 \equiv 2^2 \pmod{7}$

8. Verify the following statements.
   a. $45 \equiv 3 \pmod{6}$ and $104 \equiv 2 \pmod{6}$
   b. $(45 + 104) \equiv (3 + 2) \pmod{6}$
   c. $(45 - 104) \equiv (3 - 2) \pmod{6}$
   d. $(45 \cdot 104) \equiv (3 \cdot 2) \pmod{6}$
   e. $45^2 \equiv 3^2 \pmod{6}$

In 9–11, prove each of the given statements, assuming that $a, b, c, d,$ and n are integers with $n > 1$ and that $a \equiv c \pmod{n}$ and $b \equiv d \pmod{n}$.

9. **a.** $(a + b) \equiv (c + d) \pmod{n}$
   b. $(a - b) \equiv (c - d) \pmod{n}$

10. $a^2 \equiv c^2 \pmod{n}$

11. $a^m \equiv c^m \pmod{n}$ for all integers $m \geq 1$ (Use mathematical induction on m.)

12. **a.** Prove that for all integers $n \geq 0, 10^n \equiv 1 \pmod{9}$.
   b. Use part (a) to prove that a positive integer is divisible by 9 if, and only if, the sum of its digits is divisible by 9.

13. a. Prove that for all integers $n \geq 1, 10^n \equiv (-1)^n \pmod{11}$.
   b. Use part (a) to prove that a positive integer is divisible by 11 if, and only if, the alternating sum of its digits is divisible by 11. (For instance, the alternating sum of the digits of 82,379 is $8 - 2 + 3 - 7 + 9 = 11$ and $82,379 = 11 \cdot 7489$.)

14. What is the units digit of 7^{1066}?

15. What is the units digit of 8^{100}?

16. What is the units digit of 3^{1789}?

For each pair of equations in 17–19, reduce the equations by the given modulus to show that they do not have a simultaneous solution in integers.

17. modulo 7: $\begin{aligned} 36x + 42y &= 84 \\ -6x + 63y &= 76 \end{aligned}$

18. modulo 4: $\begin{aligned} 56x + 37y &= 145 \\ 92x - 7y &= 38 \end{aligned}$

19. modulo 6: $\begin{aligned} 43x + 24y &= 39 \\ -11x + 48y &= 53 \end{aligned}$

20. The first eleven digits for the UPC of an external hard drive are 76364900842. Calculate the check digit.

21. The first eleven digits for the UPC of a box of tissues are 04400000057. Calculate the check digit.

22. The first eleven digits for the UPC of an energy bar are 75004900040. Calculate the check digit.

In 23–28, indicate whether the given subset of the set of real numbers is a commutative ring. Justify your answers.

23. The set of all even integers

24. The set of all nonnegative real numbers

25. The set of all rational numbers

26. The set of all real numbers of the form $a + b\sqrt{2}$, where a and b are rational numbers

27. The set of all real numbers of the form $a + b\sqrt{3}$, where a and b are rational numbers

H 28. The set of all integers of the form $3a + 2b$ where a and b are integers

29. Prove that for any integer $n > 1$, multiplication on $\mathbf{Z}_n$ is well-defined. (This is part 2 of Theorem 8.4.4.)

30. Complete the proof of Theorem 8.4.5. Let n be any integer that is greater than 1, and let operations on $\mathbf{Z}_n$ be defined as in Theorem 8.4.5. Suppose that $[a], [b], [c],$ and $[d]$, where $a, b, c,$ and d are integers, are any equivalence classes in in $\mathbf{Z}_n$. Prove each of the following:
   a. $[a] + [b] = [b] + [a]$
   b. $[a] \cdot [b] = [b] \cdot [a]$
   c. $[a] + ([b] + [c]) = ([a] + [b]) + [c]$
   d. $[a] \cdot ([b] \cdot [c]) = ([a] \cdot [b]) \cdot [c]$
   e. $[a] \cdot ([b] + [c]) = [a] \cdot [b] + [a] \cdot [c]$
   f. $[1]$ is a multiplicative identity for $\mathbf{Z}_n$

31. Prove that in any commutative ring R, there is only one additive identity. In other words, prove that in any commutative ring R, if $z + c = c$ for all elements c in R, then $z = 0$.

32. Prove that in any commutative ring R, there is only one multiplicative identity. In other words, prove that in any commutative ring R, if $ec = c$ for all elements c in R, then $e = 1$.

33. Prove that given any element a in any commutative ring R, there is only one additive inverse for a. In other words, prove that if $a + b = 0$ and $a + c = 0$, then $b = c$.

In 34–36, construct addition and multiplication tables for the given $\mathbf{Z}_n$.

34. $\mathbf{Z}_2$ 35. $\mathbf{Z}_3$ 36. $\mathbf{Z}_6$

Answers for Test Yourself

1. $a \equiv b \pmod{n}$; $a = b + kn$ for some integer k; a and b have the same nonnegative remainder when divided by n; $a \bmod n = b \bmod n$ 2. $(c + d) \pmod{n}$; $(c - d) \pmod{n}$; $(cd) \pmod{n}$; $c^m \pmod{n}$ 3. $[0], [1], [2], \ldots, [n - 1]$ 4. for all a and b in A, $a \star b$ is also in A 5. $[b] + [d]$; $[b] \cdot [d]$ 6. commutative; associative; for all a, b, and c in S, $a(b + c) = ab + ac$; addition; multiplication; inverse

8.5 The Euclidean Algorithm and Applications

Don't just read it; fight it! Ask your own questions, look for your own examples, discover your own proofs. Is the hypothesis necessary? Is the converse true? What happens in the classical special case? What about the degenerate cases? Where does the proof use the hypothesis?—Paul R. Halmos, 1985

In this section we describe the Euclidean algorithm for computing the greatest common divisor of two integers, and we develop an extension of the algorithm to express the result as a linear combination of the integers. One consequence is Euclid's lemma, which is used to complete the final step of the proof of uniqueness for factoring integers into products of prime numbers. Two additional applications are discussed: a method for finding integer solutions for linear equations in two variables and a proof that if p is prime then every nonzero element in $\mathbf{Z}_p$ has a multiplicative inverse.

The Euclidean Algorithm

The greatest common divisor of two integers a and b is the largest integer that divides both a and b. For example, the greatest common divisor of 12 and 30 is 6. The Euclidean algorithm provides a very efficient way to compute the greatest common divisor of two integers.

• Definition

Let a and b be integers that are not both zero. The **greatest common divisor** of a and b, denoted **gcd(a, b)**, is the integer d with the following properties:

1. d is a common divisor of both a and b. In other words,

$$d \mid a \quad \text{and} \quad d \mid b.$$

2. For all integers c, if c is a common divisor of both a and b, then c is less than or equal to d. In other words,

$$\text{for all integers } c, \text{ if } c \mid a \text{ and } c \mid b, \text{ then } c \le d.$$

Integers a and b are called **relatively prime** if, and only if, their greatest common divisor is 1.

Example 8.5.1 Calculating Some gcd's

a. Find gcd(72, 63).

b. Find gcd($10^{20}, 6^{30}$).

c. In the definition of greatest common divisor, gcd$(0, 0)$ is not allowed. Why not? What would gcd$(0, 0)$ equal if it were found in the same way as the greatest common divisors for other pairs of numbers?

d. Are 28 and 15 relatively prime?

Solution

a. $72 = 9 \cdot 8$ and $63 = 9 \cdot 7$. So $9 \mid 72$ and $9 \mid 63$, and no integer larger than 9 divides both 72 and 63. Hence gcd$(72, 63) = 9$.

b. By the laws of exponents, $10^{20} = 2^{20} \cdot 5^{20}$ and $6^{30} = 2^{30} \cdot 3^{30} = 2^{20} \cdot 2^{10} \cdot 3^{30}$. It follows that

$$2^{20} \mid 10^{20} \quad \text{and} \quad 2^{20} \mid 6^{30},$$

and by the unique factorization of integers theorem, no integer larger than 2^{20} divides both 10^{20} and 6^{30} (because no more than twenty 2's divide 10^{20}, no 3's divide 10^{20}, and no 5's divide 6^{30}). Hence gcd$(10^{20}, 6^{30}) = 2^{20}$.

c. Suppose gcd$(0, 0)$ were defined to be the largest common factor that divides 0 and 0. The problem is that *every* positive integer divides 0 and there is no largest integer. So there is no largest common divisor!

d. Yes. The only positive divisors of 28 are 1, 2, 4, 7, and 28, and the only positive divisors of 15 are 1, 3, 5, and 15. So the greatest common divisor of 28 and 15 is 1, and hence 28 and 15 are relatively prime. ∎

Calculating gcd's using the approach illustrated in Example 8.5.1 works only when the numbers can be factored completely. By the unique factorization of integers theorem, all numbers can, in principle, be factored completely. But, in practice, even using the highest-speed computers, the process is unfeasibly long for very large integers. Over 2,000 years ago, Euclid devised a method for finding greatest common divisors that is easy to use and is much more efficient than either factoring the numbers or repeatedly testing both numbers for divisibility by successively larger integers.

The Euclidean algorithm is based on the following two facts, which are stated as lemmas.

Lemma 8.5.1

If r is a positive integer, then gcd$(r, 0) = r$.

Proof:

Suppose r is a positive integer. *[We must show that the greatest common divisor of both r and 0 is r.]* Certainly, r is a common divisor of both r and 0 because r divides itself and also r divides 0 (since every positive integer divides 0). Also no integer larger than r can be a common divisor of r and 0 (since no integer larger than r can divide r). Hence r is the greatest common divisor of r and 0.

The proof of the second lemma is based on a clever pattern of argument that is used in many different areas of mathematics: To prove that $A = B$, prove that $A \leq B$ and that $B \leq A$.

Lemma 8.5.2

If a and b are any integers not both zero, and if q and r are any integers such that

$$a = bq + r,$$

then

$$\gcd(a, b) = \gcd(b, r).$$

Proof:

[The proof is divided into two sections: (1) proof that $\gcd(a, b) \leq \gcd(b, r)$, and (2) proof that $\gcd(b, r) \leq \gcd(a, b)$. Since each $\gcd$ is less than or equal to the other, the two must be equal.]

1. **gcd (a, b) $\leq$ gcd (b, r):**

   a. *[We will first show that any common divisor of a and b is also a common divisor of b and r.]*

      Let a and b be integers, not both zero, and let c be a common divisor of a and b. Then $c \mid a$ and $c \mid b$, and so, by definition of divisibility, $a = nc$ and $b = mc$, for some integers n and m. Now substitute into the equation

      $$a = bq + r$$

      to obtain

      $$nc = (mc)q + r.$$

      Then solve for r:

      $$r = nc - (mc)q = (n - mq)c.$$

      But $n - mq$ is an integer, and so, by definition of divisibility, $c \mid r$. Because we already know that $c \mid b$, we can conclude that c is a common divisor of b and r *[as was to be shown]*.

   b. *[Next we show that $\gcd(a, b) \leq \gcd(b, r)$.]*

      By part (a), every common divisor of a and b is a common divisor of b and r. It follows that the greatest common divisor of a and b is defined because a and b are not both zero, and it is a common divisor of b and r. But then $\gcd(a, b)$ (being one of the common divisors of b and r) is less than or equal to the greatest common divisor of b and r:

      $$\gcd(a, b) \leq \gcd(b, r).$$

2. **gcd (b, r) $\leq$ gcd (a, b):**

   The second part of the proof is very similar to the first part. It is left as exercise 10 at the end of the section.

The Euclidean algorithm can be described as follows:

Euclidean Algorithm

1. Let A and B be integers with $A > B \geq 0$.

2. To find the greatest common divisor of A and B, first check whether $B = 0$. If it is, then $\gcd(A, B) = A$ by Lemma 8.5.1. If it isn't, then $B > 0$ and the

continued on page 390

quotient-remainder theorem can be used to divide A by B to obtain a quotient q and a remainder r:

$$A = Bq + r \quad \text{where } 0 \leq r < B.$$

By Lemma 8.5.2, $\gcd(A, B) = \gcd(B, r)$. Thus the problem of finding the greatest common divisor of A and B is reduced to the problem of finding the greatest common divisor of B and r.

What makes this piece of information useful is that B and r are smaller numbers than A and B. To see this, recall that we assumed

$$A > B \geq 0.$$

Also the r found by the quotient-remainder theorem satisfies

$$0 \leq r < B.$$

Putting these two inequalities together gives

$$0 \leq r < B < A.$$

So the larger number of the pair (B, r) is smaller than the larger number of the pair (A, B).

3. Now just repeat the process, starting again at (2), but use B instead of A and r instead of B. The repetitions are guaranteed to terminate eventually with $r = 0$ because each new remainder is less than the preceding one and all are nonnegative. This conclusion follows from the well-ordering principle for the integers.

By the way, it is always the case that the number of steps required in the Euclidean algorithm is at most five times the number of digits in the smaller integer. This was proved by the French mathematician Gabriel Lamé (1795–1870).

The following example illustrates how to use the Euclidean algorithm.

Example 8.5.2 Using the Euclidean Algorithm to Find a gcd

Use the Euclidean algorithm to find $\gcd(330, 156)$.

Solution

1. Divide 330 by 156:

$$
\begin{array}{r}
2 \leftarrow \text{quotient} \\
156\overline{)330} \\
\underline{312} \\
18 \leftarrow \text{remainder}
\end{array}
$$

Thus $330 = 156 \cdot 2 + 18$ and hence $\gcd(330, 156) = \gcd(156, 18)$ by Lemma 8.5.2.

2. Divide 156 by 18:

$$
\begin{array}{r}
8 \leftarrow \text{quotient} \\
18\overline{)156} \\
\underline{144} \\
12 \leftarrow \text{remainder}
\end{array}
$$

Thus $156 = 18 \cdot 8 + 12$ and hence $\gcd(156, 18) = \gcd(18, 12)$ by Lemma 8.5.2.

3. Divide 18 by 12:

$$\begin{array}{r} 1 \leftarrow \text{quotient} \\ 12\overline{\big)\,18} \\ \underline{12} \\ 6 \leftarrow \text{remainder} \end{array}$$

Thus $18 = 12 \cdot 1 + 6$ and hence $\gcd(18, 12) = \gcd(12, 6)$ by Lemma 8.5.2.

4. Divide 12 by 6:

$$\begin{array}{r} 2 \leftarrow \text{quotient} \\ 6\overline{\big)\,12} \\ \underline{12} \\ 0 \leftarrow \text{remainder} \end{array}$$

Thus $12 = 6 \cdot 2 + 0$ and hence $\gcd(12, 6) = \gcd(6, 0)$ by Lemma 8.5.2.

Putting all the previous equations together gives

$$\begin{aligned} \gcd(330, 156) &= \gcd(156, 18) \\ &= \gcd(18, 12) \\ &= \gcd(12, 6) \\ &= \gcd(6, 0) \\ &= 6 \qquad\qquad \text{by Lemma 8.5.1.} \end{aligned}$$

Therefore, $\gcd(330, 156) = 6$. ■

Extending the Euclidean Algorithm

An extended version of the Euclidean algorithm can be used to find a concrete expression for the greatest common divisor of integers a and b.

• Definition

An integer d is said to be a **linear combination of integers** a and b if, and only if, there exist integers s and t such that $as + bt = d$.

We begin by showing that the greatest common divisor of any two integers, not both zero, can be written as a linear combination of the integers.

Theorem 8.5.3 Writing a Greatest Common Divisor as a Linear Combination

For all integers a and b, not both zero, if $d = \gcd(a, b)$, then there exist integers s and t such that $as + bt = d$. In particular, if a and b are relatively prime, then there are integers s and t with $as + bt = 1$.

Proof:

Given integers a and b, not both zero, and given $d = \gcd(a, b)$, let

$S = \{x \mid x \text{ is a positive integer and } x = as + bt \text{ for some integers } s \text{ and } t\}.$

Note that S is a nonempty set because

(1) if $a > 0$ then $1 \cdot a + 0 \cdot b \in S$,
(2) if $a < 0$ then $(-1) \cdot a + 0 \cdot b \in S$, and
(3) if $a = 0$, then by assumption $b \neq 0$, and hence

$$0 \cdot a + 1 \cdot b \in S \quad \text{or} \quad 0 \cdot a + (-1) \cdot b \in S.$$

continued on page 392

Thus, because S is a nonempty subset of positive integers, by the well-ordering principle for the integers there is a least element c in S. By definition of S,

$$c = as + bt \quad \text{for some integers } s \text{ and } t. \hspace{2em} \text{8.5.1}$$

We will show that (1) $c \geq d$, and (2) $c \leq d$, and we will therefore be able to conclude that $c = d = \gcd(a, b)$.

(1) *Proof that $c \geq d$:*
[In this part of the proof, we show that d is a divisor of c and thus that $d \leq c$.] Because $d = \gcd(a, b)$, by definition of greatest common divisor, $d \mid a$ and $d \mid b$. Hence $a = dx$ and $b = dy$ for some integers x and y. Then

$$
\begin{aligned}
c &= as + bt & \text{by (8.5.1)} \\
&= (dx)s + (dy)t & \text{by substitution} \\
&= d(xs + yt) & \text{by factoring out the } d.
\end{aligned}
$$

But $xs + yt$ is an integer because it is a sum of products of integers. Thus, by definition of divisibility, $d \mid c$. Both c and d are positive, and hence, by Theorem 4.3.1, $c \geq d$.

(2) *Proof that $c \leq d$:*
[In this part of the proof, we show that c is a divisor of both a and b and therefore that c is less than or equal to the greatest common divisor of a and b, which is d.] Apply the quotient-remainder theorem to the division of a by c to obtain

$$a = cq + r \quad \text{for some integers } q \text{ and } r \text{ with } 0 \leq r < c. \hspace{2em} \text{8.5.2}$$

Thus for some integers q and r with $0 \leq r < c$,

$$r = a - cq$$

Now $c = as + bt$. Therefore, for some integers q and r with $0 \leq r < c$,

$$
\begin{aligned}
r &= a - (as + bt)q & \text{by substitution} \\
&= a(1 - sq) - btq.
\end{aligned}
$$

It follows that r is a linear combination of a and b. If $r > 0$, then r would be in S, and so r would be a smaller element of S than c, which would contradict the fact that c is the least element of S. Hence $r = 0$. By substitution into (8.5.2),

$$a = cq$$

and therefore $c \mid a$.

An almost identical argument establishes that $c \mid b$ and is left as exercise 20 at the end of the section. Because $c \mid a$ and $c \mid b$, c is a common divisor of a and b. Hence it is less than or equal to the greatest common divisor of a and b. In other words, $c \leq d$.

From (1) and (2), we conclude that $c = d$. It follows that d, the greatest common divisor of a and b, is equal to $as + bt$.

The following example shows how to extend the Euclidean algorithm to express the greatest common divisor of two integers as a linear combination of the two. It works by starting with the greatest common divisor found by applying the Euclidean algorithm and then substituting back the expressions for the remainders that were found in each of the previous steps.

Example 8.5.3 Expressing a Greatest Common Divisor as a Linear Combination

In Example 8.5.2 we showed how to use the Euclidean algorithm to find that the greatest common divisor of 330 and 156 is 6. Use the results of those calculations to express $\gcd(330, 156)$ as a linear combination of 330 and 156.

Solution   The first four steps of the solution restate and extend results from Example 8.5.2. The fifth step shows how to find the coefficients of the linear combination by substituting back through the results of the previous steps.

Step 1: $330 = 156 \cdot 2 + 18$, which implies that $18 = 330 - 156 \cdot 2$.

Step 2: $156 = 18 \cdot 8 + 12$, which implies that $12 = 156 - 18 \cdot 8$.

Step 3: $18 = 12 \cdot 1 + 6$, which implies that $6 = 18 - 12 \cdot 1$.

Step 4: $12 = 6 \cdot 2 + 0$, which implies that $\gcd(330, 156) = 6$.

Step 5: By substituting back through steps 3 to 1:

$$
\begin{aligned}
6 &= 18 - 12 \cdot 1 && \text{from step 3} \\
&= 18 - (156 - 8 \cdot 18) \cdot 1 && \text{by substitution from step 2} \\
&= 9 \cdot 18 + (-1) \cdot 156 && \text{by algebra} \\
&= 9 \cdot (330 - 156 \cdot 2) + (-1) \cdot 156 && \text{by substitution from step 1} \\
&= 9 \cdot 330 + (-19) \cdot 156 && \text{by algebra.}
\end{aligned}
$$

Thus $\gcd(330, 156) = 9 \cdot 330 + (-19) \cdot 156$. (It is always a good idea to check the result of a calculation like this to be sure you did not make a mistake. In this case, you find that $9 \cdot 330 + (-19) \cdot 156$ does indeed equal 6.) ∎

Euclid's Lemma

One consequence of Theorem 8.5.3 is known as *Euclid's lemma*. It is the crucial fact behind the proof of the uniqueness part of the unique factorization of integers theorem and is also of great importance in many other parts of number theory.

Theorem 8.5.4 Euclid's Lemma

For all integers a, b, and c, if a and c are relatively prime and $a \mid bc$, then $a \mid b$.

Proof:

Suppose a, b and c are integers, a and c are relatively prime, and $a \mid bc$. [*We must show that $a \mid b$.*] By Theorem 8.5.3, there exist integers s and t so that

$$as + ct = 1.$$

Multiply both sides of this equation by b to obtain

$$bas + bct = b. \qquad 8.5.3$$

Since $a \mid bc$, by definition of divisibility there exists an integer k such that

$$bc = ak. \qquad 8.5.4$$

Substituting (8.5.4) into (8.5.3), rewriting, and factoring out an a gives that

$$b = bas + (ak)t = a(bs + kt).$$

Let $r = bs + kt$. Then r is an integer (because b, s, k, and t are all integers), and $b = ar$. Thus $a \mid b$ by definition of divisibility.

The unique factorization of integers theorem states that any integer greater than 1 has a unique representation as a product of prime numbers, except possibly for the order in which the numbers are written. The hint for exercise 13 of Section 5.4 outlined a proof of the existence part of the proof; the uniqueness of the representation follows quickly from Euclid's lemma. In exercise 20 at the end of this section, we outline a proof for you to complete.

The Diophantine Equation ax + by = c

Equations for which only integer solutions are sought are called *Diophantine equations*. They are named after the mathematician Diophantus, who wrote in Greek and lived in about the year 250 in Alexandria, Egypt. Diophantus's great work is the *Arithmetica*, which is probably the first major mathematical work on the subject of algebra. Many of the problems discussed in the *Arithmetica* concern finding integer or rational number solutions for equations.

> **• Definition**
>
> Let a, b, and c be integers with $a \neq 0$ and $b \neq 0$. A **linear Diophantine equation in two variables** is an equation of the form $ax + by = c$ for which only integer solutions are allowed.

Here is an example of the kind of problem that gives rise to a Diophantine equation:

A university department has a budget of $18,500 to purchase certain desktop and laptop computers. The desktop computers cost $740 each and the laptop computers cost $1295 each. What are the the possible purchases the department can make if it spends all the budgeted money.

To answer this question, let x be the number of desktop computers and y be the number of laptop computers. Then

$$740x + 1295y = 18,500.$$

Because x and y must be integers, the equation is Diophantine. We use this example to illustrate the procedure for using the extended Euclidean algorithm to solve a Diophantine equation. When we are done, you will be able to answer the question about the department's purchases.

Example 8.5.4 Solving a Diophantine Equation

Find all nonnegative integer values for x and y that satisfy the equation $740x + 1295y = 18,500$, and determine the possible computer purchases for the department.

Solution   Start by using the extended Euclidean algorithm to find the greatest common divisor of 740 and 1295 (the coefficients of x and y) and express it as a linear combination of 740 and 1295.

Step 1: Divide 1295 by 740 to obtain $1295 = 740 \cdot 1 + 555$, which implies that

$$555 = 1295 - 740 \cdot 1.$$

Step 2: Divide 740 by 555 to obtain $740 = 555 \cdot 1 + 185$, which implies that

$$185 = 740 - 555 \cdot 1.$$

Step 3: Divide 555 by 185 to obtain $555 = 185 \cdot 3 + 0$, which implies that

$$\gcd(740, 1295) = 185.$$

To express 185 as a linear combination of 740 and 1295, substitute back through steps 2 to 1:

$$
\begin{aligned}
185 &= 740 - 555 \cdot 1 && \text{by step 2} \\
&= 740 - (1295 - 740 \cdot 1) \cdot 1 && \text{by substitution from step 1} \\
&= 740 - 1295 \cdot 1 + 740 \cdot 1 \\
&= 740 \cdot 2 + 1295 \cdot (-1). && \text{by algebra.}
\end{aligned}
$$

Thus

$$
740 \cdot 2 + 1295 \cdot (-1) = 185.
$$

Multiply both sides of the equation by 100 to obtain

$$
740 \cdot 200 + 1295 \cdot (-100) = 18{,}500.
$$

This shows that one solution in integers for the given equation is $x = 200$ and $y = -100$, which seems to imply that the budgeted money could be spent by purchasing 200 laptops and -100 desktop computers – a highly unlikely event in the universe we inhabit! To find a realistic solution, consider the quantities

$$
x = 200 + \frac{1295}{185}t \quad \text{and} \quad y = -100 - \frac{740}{185}t,
$$

where $185 = \gcd(1295, 740)$ and t is any integer. The fact is that these quantities satisfy the original equation. To see why, first note that each is an integer because 185 is a common divisor of 1295 and 740. Moreover,

$$
\begin{aligned}
740x + 1295y &= 740 \cdot \left(200 + \frac{1295}{185}t\right) + 1295 \cdot \left(-100 - \frac{740}{185}t\right) && \text{by substitution} \\[2mm]
&= 740 \cdot 200 + 740 \cdot \left(\frac{1295}{185}t\right) + 1295 \cdot (-100) + 1295 \cdot \left(-\frac{740}{185}t\right) && \text{by multiplying out} \\[2mm]
&= 740 \cdot 200 + 1295 \cdot (-100) + 740 \cdot \left(\frac{1295}{185}t\right) + 1295 \cdot \left(-\frac{740}{185}t\right) && \text{by regrouping} \\[2mm]
&= 740 \cdot 200 + 1295 \cdot (-100) + \left(\frac{740 \cdot 1295}{185}t\right) + \left(-\frac{740 \cdot 1295}{185}t\right) && \text{by properties of fractions} \\[2mm]
&= 740 \cdot 200 + 1295 \cdot (-100) && \text{because } \left(\frac{740 \cdot 1295}{185}t\right) + \left(-\frac{740 \cdot 1295}{185}t\right) = 0 \\[2mm]
&= 18{,}500.
\end{aligned}
$$

Thus for any integer t,

$$
x = 200 + \frac{1295}{185}t \quad \text{and} \quad y = -100 - \frac{740}{185}t
$$

are integer solutions for the given equation. Now

$$
\frac{1295}{185} = 7 \quad \text{and} \quad \frac{740}{185} = 4,
$$

and so we can write

$$
x = 200 + 7t \quad \text{and} \quad y = -100 - 4t.
$$

Determining all the nonnegative solutions requires finding the values of t that make $x \geq 0$ and $y \geq 0$. This means finding the integer values of t that satisfy the following inequalities:

$$200 + 7t \geq 0$$
$$-100 - 4t \geq 0.$$

But

$$\left\{ \begin{array}{l} 200 + 7t \geq 0 \\ -100 - 4t \geq 0 \end{array} \right\} \Leftrightarrow \left\{ \begin{array}{l} 7t \geq -200 \\ -100 \geq 4t \end{array} \right\} \Leftrightarrow \left\{ \begin{array}{l} t \geq -28\frac{4}{7} \\ -25 \geq t \end{array} \right\}$$

$$\Leftrightarrow \left\{ \begin{array}{l} t \geq -28 \\ -25 \geq t \end{array} \right\} \Leftrightarrow -28 \leq t \leq -25.$$

Note that because $t \geq -28\frac{4}{7}$ and t is an integer, we can conclude that $t \geq -28$. Thus there are four possible sets of values for x and y:

When $t = -28$, $x = 200 + 7 \cdot (-28) = 4$ and $y = -100 - 4 \cdot (-28) = 12$.
When $t = -27$, $x = 200 + 7 \cdot (-27) = 11$ and $y = -100 - 4 \cdot (-27) = 8$.
When $t = -26$, $x = 200 + 7 \cdot (-26) = 18$ and $y = -100 - 4 \cdot (-26) = 4$.
When $t = -25$, $x = 200 + 7 \cdot (-25) = 25$ and $y = -100 - 4 \cdot (-25) = 0$.

It follows that the possible purchases the department could make if it spent all the budgeted money are, therefore, 4 desktops and 12 laptops, 11 desktops and 8 laptops, 18 desktops and 4 laptops, or 25 desktops and no laptops. ∎

The reason it was possible to find a solution to the problem in the previous example was that the greatest common divisor of the coefficients of x and y was also a divisor of the constant term 18,500. The proof of the next theorem shows that had this not been the case, the problem would not have had a solution. It generalizes the steps of Example 8.5.4 to describe a general method for solving a linear Diophantine equation in two variables.

Theorem 8.5.5 Solving a Diophantine Equation of the Form $ax + by = c$

Consider the equation $ax + by = c$, where a, b, and c are integers and a and b are both nonzero.

1. If the greatest common divisor of a and b is a divisor of c, then there are integers x_1 and y_1 so that the pair (x_1, y_1) is a solution for $ax + by = c$.

2. If $ax + by = c$ has a solution in integers, then the greatest common divisor of a and b is a divisor of c.

3. If the pair (x_0, y_0) is a particular solution for $ax + by = c$ and if t is any integer, then the pair (x_1, y_1) is also a solution, where

$$x_1 = x_0 + \frac{b}{\gcd(a, b)}t \quad \text{and} \quad y_1 = y_0 - \frac{a}{\gcd(a, b)}t.$$

4. If the pair (x_0, y_0) is a particular solution for $ax + by = c$ and (x_1, y_1) is any solution, then there exists an integer t so that

$$x_1 = x_0 + \frac{b}{\gcd(a, b)}t \quad \text{and} \quad y_1 = y_0 - \frac{a}{\gcd(a, b)}t.$$

Proof:

Let a, b, and c be integers, where a and b are both nonzero, and consider the equation $ax + by = c$.

Proof of 1: Let $d = \gcd(a, \ b)$ and suppose that d divides c. *[We will show that $ax + by = c$ has a solution in integers.]* Because d divides c, there exists an integer k so that $c = dk$, and by Theorem 8.5.3, there exist integers r and s such that

$$ar + bs = d.$$

Multiply both sides of this equation by k to obtain

$$ark + bsk = dk = c.$$

Because products of integers are integers, $x = rk$ and $y = sk$ is a solution in integers for $ax + by = c$ *[as was to be shown].*

Proof of 2: Suppose $ax + by = c$ has has at least one solution in integers: the pair (x_0, y_0). Then

$$ax_0 + by_0 = c. \tag{8.5.5}$$

Let $d = \gcd(a, \ b)$. *[We will show that d divides c.]* By definition of divisibility, there exist integers r and s so that

$$a = dr \quad \text{and} \quad b = ds.$$

By substitution into (8.5.5),

$$c = drx_0 + dsy_0 = d(rx_0 + sy_0).$$

But $rx_0 + sy_0$ is an integer because it is a sum of products of integers, and so d divides c by definition of divisibility *[as was to be shown].*

Proof of 3: Suppose that the pair (x_0, y_0) is a particular solution for $ax + by = c$. Let t be any integer, and let

$$x_1 = x_0 + \frac{b}{\gcd(a, \ b)}t \quad \text{and} \quad y_1 = y_0 - \frac{a}{\gcd(a, \ b)}t.$$

We claim that x_1 and y_1 are integers and that the pair (x_1, y_1) also satisfies $ax + by = c$. The proof of this claim is left as exercise 32 at the end of this section.

Proof of 4: Suppose that the pair (x_0, y_0) is a particular solution for $ax + by = c$ and (x_1, y_1) is any other solution. *[We must show that there exists an integer t such that $x_1 = x_0 + \frac{b}{d} \cdot t$ and $y_1 = y_0 - \frac{a}{d} \cdot t$.]* Then

$$ax_0 + by_0 = c = ax_1 + by_1$$

and so

$$a(x_1 - x_0) = b(y_0 - y_1). \tag{8.5.6}$$

Let $d = \gcd(a, \ b)$. Then $d \geq 1$ and there exist nonzero integers r and s so that

$$a = dr \quad \text{and} \quad b = ds. \tag{8.5.7}$$

Note that r and s are relatively prime because d is the greatest common divisor of a and b. Substituting $a = dr$ and $b = ds$ into (8.5.6) gives

$$dr(x_1 - x_0) = ds(y_0 - y_1).$$

Dividing through by d yields

$$r(x_1 - x_0) = s(y_0 - y_1). \tag{8.5.8}$$

By definition of divisibility, r divides $s(y_0 - y_1)$, and, since r and s are relatively prime, Euclid's lemma allows us to conclude that r divides $y_0 - y_1$. Therefore, there

continued on page 398

exists an integer t with

$$y_0 - y_1 = rt \quad \text{and so} \quad y_1 = y_0 - rt. \qquad 8.5.9$$

By substituting rt in place of $y_0 - y_1$ in (8.5.8), we have

$$r(x_1 - x_0) = srt,$$

and dividing through by r gives

$$x_1 - x_0 = st \quad \text{and so} \quad x_1 = x_0 + st. \qquad 8.5.10$$

Now substitute st in place of $x_1 - x_0$ in (8.5.8) to obtain

$$rst = s(y_0 - y_1).$$

Solving the equations in (8.5.7) for r and s yields

$$r = \frac{a}{d} \quad \text{and} \quad s = \frac{b}{d},$$

and substituting these expressions into (8.5.9) and (8.5.10) produces the result that was to be shown:

$$x_1 = x_0 + \frac{b}{d} \cdot t \quad \text{and} \quad y_1 = y_0 - \frac{a}{d} \cdot t.$$

Multiplication in Z_n

In Section 8.4 we pointed out that some of the sets $\mathbf{Z}$ have zero divisors. For example, in $\mathbf{Z}_6$ both [2] and [3] are zero divisors because

$$[2] \neq [0] \quad \text{and} \quad [3] \neq [0] \quad \text{but} \quad [2] \cdot [3] = [6] = [0].$$

Exercise 33 at the end of this section asks you to prove that $\mathbf{Z}_n$ has zero divisors whenever n is a composite number. If p is prime, however, $\mathbf{Z}_p$ every nonzero element in $\mathbf{Z}_p$ has a *multiplicative inverse*. A consequence (see exercise 34) is that $\mathbf{Z}_p$ does not have any zero divisors.

• Definition

An element a in a commutative ring R has a **multiplicative inverse** if, and only if, there exists an element b in R such that $a \cdot b = 1$.

Theorem 8.5.6 Multiplicative Inverses in Z_p

For all prime numbers p, if a is any integer with $[a] \neq [0]$, then there exists a nonzero integer b such that $[a] \cdot [b] = [1]$. In other words, $[a]$ has a multiplicative inverse in $\mathbf{Z}_p$.

Proof:

Suppose p is any prime number and a is any integer with $[a] \neq [0]$. By definition of $\mathbf{Z}_p$, p does not divide $a - 0 = a$. Thus, since the only positive divisors of p are 1 and p, the greatest common divisor of a and p is 1.

By Theorem 8.5.3, there exist integers b and k such that

$$ab + (-p)k = 1 \quad \text{which implies that} \quad ab = 1 + pk.$$

It follows from Theorem 8.4.1 that

$$ab \equiv 1 \ (\text{mod } p)$$

and thus b is a solution for the congruence $ax \equiv 1 \ (\text{mod } n)$. Moreover, by definition of multiplication in $\mathbf{Z}_p$,

$$[a] \cdot [b] = [1],$$

and so $[b]$ is a multiplicative inverse for $[a]$ in $\mathbf{Z}_p$.

Example 8.5.5 Multiplication of Nonzero Elements in $\mathbf{Z}_5$

Find the multiplicative inverse for all the nonzero elements in $\mathbf{Z}_5$, and construct the multiplication table for the nonzero elements.

Solution

A process of trial and error quickly gives the following results:

$$[1] \cdot [1] = [1] \quad [2] \cdot [3] = [3] \cdot [2] = [6] = [1] \quad [4] \cdot [4] = [16] = [1]$$

Thus the multiplicative inverse of $[1]$ is $[1]$, of $[2]$ is $[3]$, $[3]$ is $[2]$, and of $[4]$ is $[4]$. The multiplication table is

$\cdot$	[1]	[2]	[3]	[4]
[1]	[1]	[2]	[3]	[4]
[2]	[2]	[4]	[1]	[3]
[3]	[3]	[1]	[4]	[2]
[4]	[4]	[3]	[2]	[1]

Example 8.5.6 Finding a Multiplicative Inverse in $\mathbf{Z}_{29}$

Use the extended Euclidean algorithm to find a multiplicative inverse for $[9]$ in $\mathbf{Z}_{29}$.

Solution

There exists an integer x such that $[9] \cdot [x] = [1]$ if, and only if, $9x \equiv 1 \ (\text{mod } 29)$, or, equivalently, $9x = 1 + 29k$ for some integer k, which means that

$$9x + 29(-k) = 1.$$

Apply the extended Euclidean algorithm as follows:

Step 1: Divide 29 by 9 to obtain $29 = 9 \cdot 3 + 2$, which implies that $2 = 29 - 9 \cdot 3$.

Step 2: Divide 9 by 2 to obtain $9 = 2 \cdot 4 + 1$, which implies that $1 = 9 - 2 \cdot 4$.

Step 3: Divide 2 by 1 to obtain $2 = 2 \cdot 1 + 0$, which implies that $\gcd(9, 29) = 1$.

To express 1 as a linear combination of 9 and 29, substitute back through steps 2 to 1:

$$
\begin{aligned}
1 &= 9 - 2 \cdot 4 && \text{by step 2} \\
  &= 9 - (29 - 9 \cdot 3) \cdot 4 && \text{by substitution from step 1} \\
  &= 9 - 29 \cdot 4 + 9 \cdot 12 && \\
  &= 9 \cdot 13 + 29(-4). && \text{by algebra.}
\end{aligned}
$$

Thus

$$9 \cdot 13 + 29(-4) = 1 \quad \text{and so} \quad 9 \cdot 13 = 1 + 29 \cdot 4.$$

Hence $9 \cdot 13 \equiv 1 \ \text{mod}(29)$, which means that $[13]$ is a multiplicative inverse for $[9]$ in $\mathbf{Z}_{29}$.

> **• Definition**
>
> A commutative ring R in which every nonzero element has a multiplicative inverse is called a **field**.

Some familiar examples of fields are the set of real numbers and the set of rational numbers. The following corollary shows that a finite set can also be a field. A few additional examples of fields are explored in the exercises.

> **• Corollary 8.5.7 $\mathbf{Z}_p$ Is a Field**
>
> If p is any prime number, then $\mathbf{Z}_p$ is a field.
>
> **Proof**: Let p be any prime number. By Theorem 8.4.5, $\mathbf{Z}_p$ is a commutative ring, and by Theorem 8.5.6 every nonzero element of $\mathbf{Z}_p$ has a multiplicative inverse. Therefore, by definition of field, $\mathbf{Z}_p$ is a field.

Abstract algebra is the study of properties of algebraic structures such as commutative rings, fields, and Boolean algebras. Some other important algebraic structures are integral domains (which are commutative rings without zero divisors), noncommutative rings (which satisfy all the properties of commutative rings except that multiplication is not commutative), and division algebras (which are noncommutative rings in which every nonzero element has a multiplicative inverse). A different kind of commutative ring is the set of all real polynomial functions with either integer, rational, or real coefficients. These rings share many properties with the set of integers, including not having zero divisors and satisfying a polynomial version of the quotient-remainder theorem.

Test Yourself

1. The greatest common divisor of two integers a and b is the integer d with the following two properties: _____ and _____.

2. The Euclidean algorithm is a procedure for computing _____.

3. If r is a positive integer, then $\gcd(r, 0) =$ _____.

4. If a and b are any integers with $b \neq 0$ and if q and r are any integers such that $a = bq + r$, then $\gcd(a, b) =$ _____.

5. Given integers a and b not both zero, the extended Euclidean algorithm makes it possible to express $\gcd(a, b)$ as _____.

6. Euclid's lemma says that for all integers a, b, and c, if a and c are relatively prime and $a|bc$, then _____.

7. A linear Diophantine equation in two variables is an equation of the form _____ for which only _____ solutions are sought.

8. Given an equation $ax + by = c$, where a, b, and c are integers and a and b are nonzero, if $\gcd(a, b)$ divides c, then there are integers x_1 and y_1 such that the pair (x_1, y_1) is _____.

9. If p is a prime number, then every nonzero element in $\mathbf{Z}_p$ has _____.

10. A field is a commutative ring in which _____.

Exercise Set 8.5

Find the greatest common divisor of each of the pairs of integers in 1–4. (Use any method you wish.)

1. 27 and 72   **2.** 5 and 9

3. 7 and 21   **4.** 48 and 54

Use the Euclidean algorithm to calculate the greatest common divisors of each of the pairs of integers in 5–8.

5. 1,188 and 385   **6.** 509 and 1,177

7. 832 and 10,933   **8.** 4,131 and 2,431

H **9.** Prove that for all positive integers a and b, $a|b$ if, and only if, $\gcd(a, b) = a$. (Note that to prove "A if, and only if, B," you need to prove "if A then B" and "if B then A.")

10. Complete the proof of Lemma 8.5.2 by proving the following: If a and b are any integers with $b \neq 0$ and q and r are any integers such that

$$a = bq + r.$$

then

$$\gcd(b, r) \leq \gcd(a, b).$$

H **11.** Let F_0, F_1, $F_2, \ldots$ be the Fibonacci sequence defined in Section 5.5. Prove that for all integers $n \geq 0$, $\gcd(F_{n+1}, F_n) = 1$.

Exercises 12–16 refer to the following definition.

Definition: The **least common multiple** of two nonzero integers a and b, denoted **lcm(a, b)**, is the positive integer c such that
a. $a \mid c$ and $b \mid c$
b. for all positive integers m, if $a \mid m$ and $b \mid m$, then $c \leq m$.

---

12. Find
 a. lcm(12, 18)   b. lcm($2^2 \cdot 3 \cdot 5$, $2^3 \cdot 3^2$)
 c. lcm(2800, 6125)

13. Prove that for all positive integers a and b, $\gcd(a, b) = \text{lcm}(a, b)$ if, and only if, $a = b$.

14. Prove that for all positive integers a and b, $a \mid b$ if, and only if, lcm$(a, b) = b$.

15. Prove that for all integers a and b, $\gcd(a, b) \mid \text{lcm}(a, b)$.

H **16.** Prove that for all positive integers a and b,
 $\gcd(a, b) \cdot \text{lcm}(a, b) = ab$.

In 17–19, use the extended Euclidean algorithm to find the greatest common divisor of the given numbers and express it as a linear combination of the two numbers.

17. 6664 and 765   **18.** 4158 and 1568   **19.** 2583 and 349

20. Finish the proof of Theorem 8.5.3 by proving that if a, b and c are as in the proof, then $c \mid b$.

H **21. a.** Use mathematical induction and Euclid's lemma to prove that for all positive integers s, if p and $q_1, q_2, \ldots, q_s$ are prime numbers and $p \mid q_1 q_2 \cdots q_s$, then $p = q_i$ for some i with $1 \leq i \leq s$.
 b. The uniqueness part of the unique factorization of integers theorem says that given any integer n, if

$$n = p_1 p_2 \cdots p_r = q_1 q_2 \cdots q_s$$

for some positive integers r and s and prime numbers $p_1 \leq p_2 \leq \cdots \leq p_r$ and $q_1 \leq q_2 \leq \cdots \leq q_s$, then $r = s$ and $p_i = q_i$ for all integers i with $1 \leq i \leq r$.

  Use the result of part (a) to fill in the details of the following sketch of a proof: Suppose that n is an integer with two different prime factorizations: $n = p_1 p_2 \cdots p_t = q_1 q_2 \cdots q_u$. All the prime factors that appear on both sides can be cancelled (as many times as they appear on both sides) to arrive at the situation where $p_1 p_2 \cdots p_r = q_1 q_2 \cdots q_s$, $p_1 \leq p_2 \leq \cdots \leq p_r$, $q_1 \leq q_2 \leq \cdots \leq q_s$, and $p_i \neq q_j$ for any integers i and j. Then use part (a) to deduce a contradiction, and so the prime factorization of n is unique except, possibly, for the order in which the prime factors are written.

Use the technique illustrated in Example 8.5.4 to find one particular solution and the formulas for the general solutions for the Diophantine equations in 22 and 23.

22. $243x + 702y = 8289$

23. $1456x + 693y = 4760$

Use the technique illustrated in Example 8.5.4 to find all possible solutions for the problems in 24–25.

24. A gym purchased 2-pound weights and 5-pound weights for a total of 40 pounds of weights in all. What are the possibilities for the number of each kind of weight purchased by the gym if at least one of each kind of weight was purchased?

25. An office purchased packages of paper for $11 per package and packages of pens for $19 per package. If the office spent $400 on the paper and pens, what are the possibilities for the number of packages of paper and the number of packages of pens purchased by the office?

H **26.** A candy store purchased boxes of chocolate creams for $24 each and boxes of caramels for $15 each. If a total of $900 was spent on the candy and if the purchase included at least one box of each kind, what are the possibilities for the number of boxes of each kind of candy purchased by the store?

27. a. Write the contrapositive of the following statement: For all integers a, b, and c, where a and c are nonzero, if $ax + by = c$ has a solution in integers, then the greatest common divisor of a and b is a divisor of c.

   b. How does the contrapositive of the statement in part (a) help decide whether a given Diophantine equation has a solution?

H 28. Use Theorem 8.5.5 to prove that for all integers a, c, and n with $n > 1$, if $\gcd(a, n) = d$ and d divides c, then the congruence $ax \equiv c \pmod{n}$ has a solution.

Use the result of exercise 28 to help solve the congruences in 29 and 30.

29. $24x \equiv 12 \pmod{42}$

30. $2184x \equiv 5481 \pmod{286}$

31. $143x \equiv 195 \pmod{78}$

H 32. Use Theorem 8.5.5 to prove that for any nonzero integers a and c and any integer $n > 1$, if $\gcd(a, n)$ divides c, then the equation $[a] \cdot [x] = [c]$ has a solution in $\mathbf{Z}_n$.

33. Finish the proof of Theorem 8.5.5, part (3).

34. Prove that for any composite number n, $\mathbf{Z}_n$ has zero divisors.

H 35. a. Prove that for any prime number p and for all elements $[a]$ and $[c]$ in $\mathbf{Z}_p$ (where a and c are integers), if $[a] \cdot [c] = [0]$ and $[a] \neq [0]$, then $[c] = [0]$.

   b. Deduce from part (a) that if p is prime, then $\mathbf{Z}_p$ does not have any zero divisors.

In 36–38, construct an addition table for the given $\mathbf{Z}_n$ and construct a multiplication table for its nonzero elements.

36. $\mathbf{Z}_7$ 37. $\mathbf{Z}_8$ 38. $\mathbf{Z}_9$

39. Use the extended Euclidean algorithm to find a positive integer n so that $2 \leq n \leq 28$ and $[n]$ is an inverse for $[13]$ in $\mathbf{Z}_{29}$.

40. Use the extended Euclidean algorithm to find a positive integer n so that $2 \leq n \leq 22$ and $[n]$ is an inverse for $[13]$ in $\mathbf{Z}_{23}$.

41. Use the extended Euclidean algorithm to find a positive integer n so that $2 \leq n \leq 30$ and $[n]$ is an inverse for $[13]$ in $\mathbf{Z}_{31}$.

Definition: Given any integers a, b, and n, where a and b are nonzero and $n > 1$, we say that b is **an inverse for a modulo n** if, and only if, $ab \equiv 1 \pmod{n}$.

---

42. Let R be the set of all real numbers of the form $a + b\sqrt{2}$, where a and b are rational numbers. Show that every nonzero element of R has a multiplicative inverse and use the result of exercise 26 in Section 8.4 to conclude that R is a field.

43. Let R be the set of all real numbers of the form $a + b\sqrt{3}$, where a and b are rational numbers. Show that every nonzero element of R has a multiplicative inverse and use the result of exercise 27 in Section 8.4 to conclude that R is a field.

44. Prove that in any field F, every nonzero element has a unique multiplicative inverse. In other words, if a is any nonzero element of F and b and c are any elements of F such that $ab = ac = 1$, then $b = c$. This fact enables us to speak of "the multiplicative inverse" rather than "a multiplicative inverse" of a nonzero element in a field.

Answers for Test Yourself

1. d divides both a and b; if c is any integer that divides both a and b, then $c \leq d$.   2. the greatest common divisor of two integers   3. r 4. $\gcd(b, r)$ 5. a linear combination of a and b 6. $a|b$ 7. $ax + by = c$, where a, b, and c are integers none of which is equal to zero; integer   8. a solution for $ax + by = c$ 9. a multiplicative inverse   10. every nonzero element has a multiplicative inverse

CHAPTER 9

COUNTING AND PROBABILITY

"It's as easy as 1–2–3."

That's the saying. And in certain ways, counting *is* easy. But other aspects of counting aren't so simple. Have you ever agreed to meet a friend "in three days" and then realized that you and your friend might mean different things? For example, on the European continent, to meet in eight days means to meet on the same day as today one week hence; on the other hand, in English-speaking countries, to meet in seven days means to meet one week hence. The difference is that on the continent, all days including the first and the last are counted. In the English-speaking world, it's the number of 24-hour periods that are counted.

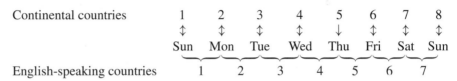

Continental countries	1	2	3	4	5	6	7	8
	Sun	Mon	Tue	Wed	Thu	Fri	Sat	Sun
English-speaking countries	1	2	3	4	5	6	7	

The English convention for counting days follows the almost universal convention for counting hours. If it is 9 A.M. and two people anywhere in the world agree to meet in three hours, they mean that they will get back together again at 12 noon.

Musical intervals, on the other hand, are universally reckoned the way the Continentals count the days of a week. An interval of a third consists of two tones with a single tone in between, and an interval of a second consists of two adjacent tones. (See Figure 9.1.1.)

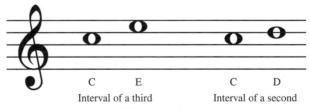

C   E   C   D
Interval of a third   Interval of a second

Figure 9.1.1

Of course, the complicating factor in all these examples is not how to count but rather what to count. And, indeed, in the more complex mathematical counting problems discussed in this chapter, it is what to count that is the central issue. Once one knows exactly what to count, the counting itself is as easy as 1–2–3.

9.1 Introduction

Imagine tossing two coins and observing whether 0, 1, or 2 heads are obtained. It would be natural to guess that each of these events occurs about one-third of the time, but in fact this is not the case. Table 9.1.1 below shows actual data obtained from tossing two quarters 50 times.

Table 9.1.1 Experimental Data Obtained from Tossing Two Quarters 50 Times

Event	Tally	Frequency (Number of times the event occurred)	Relative Frequency (Fraction of times the event occurred)
2 heads obtained	卌 卌 I	11	22%
1 head obtained	卌 卌 卌 卌 卌 II	27	54%
0 heads obtained	卌 卌 II	12	24%

As you can see, the relative frequency of obtaining exactly 1 head was roughly twice as great as that of obtaining either 2 heads or 0 heads. It turns out that the mathematical theory of probability can be used to predict that a result like this will almost always occur. To see how, call the two coins *A* and *B*, and suppose that each is perfectly balanced. Then each has an equal chance of coming up heads or tails, and when the two are tossed together, the four outcomes pictured in Figure 9.1.2 are all equally likely.

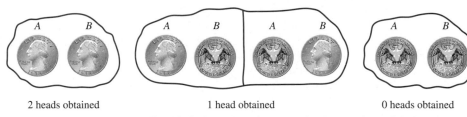

2 heads obtained   1 head obtained   0 heads obtained

Figure 9.1.2 Equally Likely Outcomes from Tossing Two Balanced Coins

Figure 9.1.2 shows that there is a 1 in 4 chance of obtaining two heads and a 1 in 4 chance of obtaining no heads. The chance of obtaining one head, however, is 2 in 4 because either *A* could come up heads and *B* tails or *B* could come up heads and *A* tails. So if you repeatedly toss two balanced coins and record the number of heads, you should expect relative frequencies similar to those shown in Table 9.1.1.

To formalize this analysis and extend it to more complex situations, we introduce the notions of random process, sample space, event and probability. To say that a process is **random** means that when it takes place, one outcome from some set of outcomes is sure to occur, but it is impossible to predict with certainty which outcome that will be. For instance, if an ordinary person performs the experiment of tossing an ordinary coin into the air and allowing it to fall flat on the ground, it can be predicted with certainty that the coin will land either heads up or tails up (so the set of outcomes can be denoted {heads, tails}), but it is not known for sure whether heads or tails will occur. We restricted this experiment to ordinary people because a skilled magician can toss a coin in a way that appears random but is not, and a physicist equipped with first-rate measuring devices may be able to analyze all the forces on the coin and correctly predict its landing position. Just a few of many examples of random processes or experiments are choosing winners in state lotteries, selecting respondents in public opinion polls, and choosing subjects to receive treatments or serve as controls in medical experiments. The set of outcomes that can result from a random process or experiment is called a *sample space*.

> **• Definition**
>
> A **sample space** is the set of all possible outcomes of a random process or experiment. An **event** is a subset of a sample space.

In case an experiment has finitely many outcomes and all outcomes are equally likely to occur, the *probability* of an event (set of outcomes) is just the ratio of the number of outcomes in the event to the total number of outcomes. Strictly speaking, this result can be deduced from a set of axioms for probability formulated in 1933 by the Russian mathematician A. N. Kolmogorov.

Equally Likely Probability Formula

If S is a finite sample space in which all outcomes are equally likely and E is an event in S, then the **probability of E**, denoted $P(E)$, is

$$P(E) = \frac{\text{the number of outcomes in } E}{\text{the total number of outcomes in } S}.$$

> **• Notation**
>
> For any finite set A, $N(A)$ denotes the number of elements in A.

With this notation, the equally likely probability formula becomes

$$P(E) = \frac{N(E)}{N(S)}.$$

Example 9.1.1 Probabilities for a Deck of Cards

An ordinary deck of cards contains 52 cards divided into four *suits*. The *red suits* are diamonds (♦) and hearts (♥) and the *black suits* are clubs (♣) and spades (♠). Each suit contains 13 cards of the following *denominations:* 2, 3, 4, 5, 6, 7, 8, 9, 10, J (jack), Q (queen), K (king), and A (ace). The cards J, Q, and K are called *face cards.*

Mathematician Persi Diaconis, working with David Aldous in 1986 and Dave Bayer in 1992, showed that seven shuffles are needed to "thoroughly mix up" the cards in an ordinary deck. In 2000 mathematician Nick Trefethen, working with his father, Lloyd Trefethen, a mechanical engineer, used a somewhat different definition of "thoroughly mix up" to show that six shuffles will nearly always suffice. Imagine that the cards in a deck have become—by some method—so thoroughly mixed up that if you spread them out face down and pick one at random, you are as likely to get any one card as any other.

a. What is the sample space of outcomes?

b. What is the event that the chosen card is a black face card?

c. What is the probability that the chosen card is a black face card?

Solution

a. The outcomes in the sample space S are the 52 cards in the deck.

b. Let E be the event that a black face card is chosen. The outcomes in E are the jack, queen, and king of clubs and the jack, queen, and king of spades. Symbolically,

$$E = \{J\clubsuit, Q\clubsuit, K\clubsuit, J\spadesuit, Q\spadesuit, K\spadesuit\}.$$

c. By part (b), $N(E) = 6$, and according to the description of the situation, all 52 outcomes in the sample space are equally likely. Therefore, by the equally likely probability formula, the probability that the chosen card is a black face card is

$$P(E) = \frac{N(E)}{N(S)} = \frac{6}{52} \cong 11.5\%. \qquad \blacksquare$$

Example 9.1.2 Rolling a Pair of Dice

A die is one of a pair of dice. It is a cube with six sides, each containing from one to six dots, called *pips*. Suppose a blue die and a gray die are rolled together, and the numbers of dots that occur face up on each are recorded. The possible outcomes can be listed as follows, where in each case the die on the left is blue and the one on the right is gray.

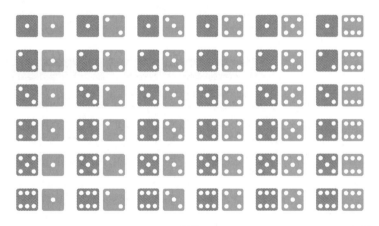

A more compact notation identifies, say, ▨ ▨ with the notation 24, ▨ ▨ with 53, and so forth.

a. Use the compact notation to write the sample space S of possible outcomes.

b. Use set notation to write the event E that the numbers showing face up have a sum of 6 and find the probability of this event.

Solution

a. $S = \{11, 12, 13, 14, 15, 16, 21, 22, 23, 24, 25, 26, 31, 32, 33, 34, 35, 36, 41, 42, 43, 44, 45, 46, 51, 52, 53, 54, 55, 56, 61, 62, 63, 64, 65, 66\}.$

b. $E = \{15, 24, 33, 42, 51\}.$

The probability that the sum of the numbers is $6 = P(E) = \dfrac{N(E)}{N(S)} = \dfrac{5}{36}.$ $\blacksquare$

The next example is called the Monty Hall problem after the host of an old game show, "Let's Make A Deal." When it was originally publicized in a newspaper column and on a radio show, it created tremendous controversy. Many highly educated people, even some with Ph.D.'s, submitted incorrect solutions or argued vociferously against the correct solution. Before you read the answer, think about what your own response to the situation would be.

Example 9.1.3 The Monty Hall Problem

There are three doors on the set for a game show. Let's call them A, B, and C. If you pick the right door you win the prize. You pick door A. The host of the show, Monty Hall, then opens one of the other doors and reveals that there is no prize behind it. Keeping the remaining two doors closed, he asks you whether you want to switch your choice to the other closed door or stay with your original choice of door A. What should you do if you want to maximize your chance of winning the prize: stay with door A or switch—or would the likelihood of winning be the same either way?

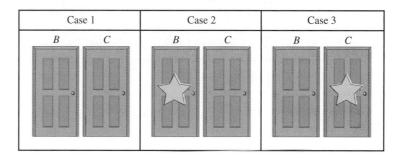

Solution At the point just before the host opens one of the closed doors, there is no information about the location of the prize. Thus there are three equally likely possibilities for what lies behind the doors: (Case 1) the prize is behind A (i.e., it is not behind either B or C), (Case 2) the prize is behind B; (Case 3) the prize is behind C.

Since there is no prize behind the door the host opens, in Case 1 the host could open either door and you would win by staying with your original choice: door A. In Case 2 the host must open door C, and so you would win by switching to door B. In Case 3 the host must open door B, and so you would win by switching to door C. Thus, in two of the three equally likely cases, you would win by switching from A to the other closed door. In only one of the three equally likely cases would you win by staying with your original choice. Therefore, you should switch.

A reality note: The analysis used for this solution applies only if the host *always* opens one of the closed doors and offers the contestant the choice of staying with the original choice or switching. In the original show, Monty Hall made this offer only occasionally—most often when he knew the contestant had already chosen the correct door. ■

Many of the fundamental principles of probability were formulated in the mid-1600s in an exchange of letters between Pierre de Fermat and Blaise Pascal in response to questions posed by a French nobleman interested in games of chance. In 1812, Pierre-Simon Laplace published the first general mathematical treatise on the subject and extended the range of applications to a variety of scientific and practical problems.

Pierre-Simon Laplace (1749–1827)

Bettmann/CORBIS

Counting the Elements of a List

Some counting problems are as simple as counting the elements of a list. For instance, how many integers are there from 5 through 12? To answer this question, imagine going along the list of integers from 5 to 12, counting each in turn.

$$
\begin{array}{lcccccccc}
\text{list:} & 5 & 6 & 7 & 8 & 9 & 10 & 11 & 12 \\
 & \updownarrow & \updownarrow & \updownarrow & \updownarrow & \updownarrow & \updownarrow & \updownarrow & \updownarrow \\
\text{count:} & 1 & 2 & 3 & 4 & 5 & 6 & 7 & 8
\end{array}
$$

So the answer is 8.

More generally, if m and n are integers and $m \leq n$, how many integers are there from m through n? To answer this question, note that $n = m + (n - m)$, where $n - m \geq 0$ [*since* $n \geq m$]. Note also that the element $m + 0$ is the first element of the list, the element $m + 1$ is the second element, the element $m + 2$ is the third, and so forth. In general, the element $m + i$ is the $(i + 1)$st element of the list.

$$\text{list:} \quad m(= m + 0) \quad m + 1 \quad m + 2 \quad \ldots \quad n\,(= m + (n - m))$$
$$\updownarrow \qquad\quad \updownarrow \qquad\quad \updownarrow \qquad\qquad\qquad \updownarrow$$
$$\text{count:} \qquad\qquad\quad 1 \qquad\quad 2 \qquad\quad 3 \quad \ldots \quad (n - m) + 1$$

And so the number of elements in the list is $n - m + 1$.

This general result is important enough to be restated as a theorem, the formal proof of which uses mathematical induction. (See exercise 28 at the end of this section.) The heart of the proof is the observation that if the list $m, m + 1, \ldots, k$ has $k - m + 1$ numbers, then the list $m, m + 1, \ldots, k, k + 1$ has $(k - m + 1) + 1 = (k + 1) - m + 1$ numbers.

Theorem 9.1.1 The Number of Elements in a List

If m and n are integers and $m \leq n$, then there are $n - m + 1$ integers from m to n inclusive.

Example 9.1.4 Counting the Elements of a Sublist

a. How many three-digit integers (integers from 100 to 999 inclusive) are divisible by 5?

b. What is the probability that a randomly chosen three-digit integer is divisible by 5?

Solution

a. Imagine writing the three-digit integers in a row, noting those that are multiples of 5 and drawing arrows between each such integer and its corresponding multiple of 5.

100	101	102	103	104	105	106	107	108	109	110	$\cdots$	994	995	996	997	998	999
$\updownarrow$					$\updownarrow$					$\updownarrow$			$\updownarrow$				
$5 \cdot 20$					$5 \cdot 21$					$5 \cdot 22$		$5 \cdot 199$					

From the sketch it is clear that there are as many three-digit integers that are multiples of 5 as there are integers from 20 to 199 inclusive. By Theorem 9.1.1, there are $199 - 20 + 1$, or 180, such integers. Hence there are 180 three-digit integers that are divisible by 5.

b. By Theorem 9.1.1 the total number of integers from 100 through 999 is $999 - 100 + 1 = 900$. By part (a), 180 of these are divisible by 5. Hence the probability that a randomly chosen three-digit integer is divisible by 5 is $180/900 = 1/5$. ∎

Test Yourself

Answers to Test Yourself questions are located at the end of each section.

1. A sample space of a random process or experiment is _____.

2. An event in a sample space is _____.

3. To compute the probability of an event using the equally likely probability formula, you take the ratio of the _____ to the _____.

4. If $m \leq n$, the number of integers from m to n inclusive is _____.

Exercise Set 9.1*

1. Toss two coins 30 times and make a table showing the relative frequencies of 0, 1, and 2 heads. How do your values compare with those shown in Table 9.1.1?

2. In the example of tossing two quarters, what is the probability that at least one head is obtained? that coin *A* is a head? that coins *A* and *B* are either both heads or both tails?

In 3–6 use the sample space given in Example 9.1.1. Write each event as a set, and compute its probability.

3. The event that the chosen card is red and is not a face card.

4. The event that the chosen card is black and has an even number on it.

5. The event that the denomination of the chosen card is at least 10 (counting aces high).

6. The event that the denomination of the chosen card is at most 4 (counting aces high).

In 7–10, use the sample space given in Example 9.1.2. Write each of the following events as a set and compute its probability.

7. The event that the sum of the numbers showing face up is 8.

8. The event that the numbers showing face up are the same.

9. The event that the sum of the numbers showing face up is at most 6.

10. The event that the sum of the numbers showing face up is at least 9.

11. Suppose that a coin is tossed three times and the side showing face up on each toss is noted. Suppose also that on each toss heads and tails are equally likely. Let *HHT* indicate the outcome heads on the first two tosses and tails on the third, *THT* the outcome tails on the first and third tosses and heads on the second, and so forth.
    a. List the eight elements in the sample space whose outcomes are all the possible head–tail sequences obtained in the three tosses.
    b. Write each of the following events as a set and find its probability:
       (i) The event that exactly one toss results in a head.
       (ii) The event that at least two tosses result in a head.
       (iii) The event that no head is obtained.

12. Suppose that each child born is equally likely to be a boy or a girl. Consider a family with exactly three children. Let *BBG* indicate that the first two children born are boys and the third child is a girl, let *GBG* indicate that the first and third children born are girls and the second is a boy, and so forth.
    a. List the eight elements in the sample space whose outcomes are all possible genders of the three children.
    b. Write each of the events in the next column as a set and find its probability.

    (i) The event that exactly one child is a girl.
    (ii) The event that at least two children are girls.
    (iii) The event that no child is a girl.

13. Suppose that on a true/false exam you have no idea at all about the answers to three questions. You choose answers randomly and therefore have a 50–50 chance of being correct on any one question. Let *CCW* indicate that you were correct on the first two questions and wrong on the third, let *WCW* indicate that you were wrong on the first and third questions and correct on the second, and so forth.
    a. List the elements in the sample space whose outcomes are all possible sequences of correct and incorrect responses on your part.
    b. Write each of the following events as a set and find its probability:
       (i) The event that exactly one answer is correct.
       (ii) The event that at least two answers are correct.
       (iii) The event that no answer is correct.

14. Three people have been exposed to a certain illness. Once exposed, a person has a 50–50 chance of actually becoming ill.
    a. What is the probability that exactly one of the people becomes ill?
    b. What is the probability that at least two of the people become ill?
    c. What is the probability that none of the three people becomes ill?

15. When discussing counting and probability, we often consider situations that may appear frivolous or of little practical value, such as tossing coins, choosing cards, or rolling dice. The reason is that these relatively simple examples serve as models for a wide variety of more complex situations in the real world. In light of this remark, comment on the relationship between your answer to exercise 11 and your answers to exercises 12–14.

16. Two faces of a six-sided die are painted red, two are painted blue, and two are painted yellow. The die is rolled three times, and the colors that appear face up on the first, second, and third rolls are recorded.
    a. Let *BBR* denote the outcome where the color appearing face up on the first and second rolls is blue and the color appearing face up on the third roll is red. Because there are as many faces of one color as of any other, the outcomes of this experiment are equally likely. List all 27 possible outcomes.
    b. Consider the event that all three rolls produce different colors. One outcome in this event is *RBY* and another *RYB*. List all outcomes in the event. What is the probability of the event?

---

*For exercises with blue numbers or letters, solutions are given in Appendix B. The symbol **H** indicates that only a hint or a partial solution is given. The symbol ✶ signals that an exercise is more challenging than usual.

c. Consider the event that two of the colors that appear face up are the same. One outcome in this event is *RRB* and another is *RBR*. List all outcomes in the event. What is the probability of the event?

17. Consider the situation described in exercise 16.
    a. Find the probability of the event that exactly one of the colors that appears face up is red.
    b. Find the probability of the event that at least one of the colors that appears face up is red.

18. An urn contains two blue balls (denoted B_1 and B_2) and one white ball (denoted W). One ball is drawn, its color is recorded, and it is replaced in the urn. Then another ball is drawn, and its color is recorded.
    a. Let $B_1 W$ denote the outcome that the first ball drawn is B_1 and the second ball drawn is W. Because the first ball is replaced before the second ball is drawn, the outcomes of the experiment are equally likely. List all nine possible outcomes of the experiment.
    b. Consider the event that the two balls that are drawn are both blue. List all outcomes in the event. What is the probability of the event?
    c. Consider the event that the two balls that are drawn are of different colors. List all outcomes in the event. What is the probability of the event?

19. An urn contains two blue balls (denoted B_1 and B_2) and three white balls (denoted W_1, W_2, and W_3). One ball is drawn, its color is recorded, and it is replaced in the urn. Then another ball is drawn and its color is recorded.
    a. Let $B_1 W_2$ denote the outcome that the first ball drawn is B_1 and the second ball drawn is W_2. Because the first ball is replaced before the second ball is drawn, the outcomes of the experiment are equally likely. List all 25 possible outcomes of the experiment.
    b. Consider the event that the first ball that is drawn is blue. List all outcomes in the event. What is the probability of the event?
    c. Consider the event that only white balls are drawn. List all outcomes in the event. What is the probability of the event?

20. Refer to Example 9.1.3. Suppose you are appearing on a game show with a prize behind one of five closed doors: A, B, C, D, and E. If you pick the right door, you win the prize. You pick door A. The game show host then opens one of the other doors and reveals that there is no prize behind it. Then the host gives you the option of staying with your original choice of door A or switching to one of the other doors that is still closed.
    a. If you stick with your original choice, what is the probability that you will win the prize?
    b. If you switch to another door, what is the probability that you will win the prize?

21. a. How many positive two-digit integers are multiples of 3?
    b. What is the probability that a randomly chosen positive two-digit integer is a multiple of 3?
    c. What is the probability that a randomly chosen positive two-digit integer is a multiple of 4?

22. a. How many positive three-digit integers are multiples of 6?
    b. What is the probability that a randomly chosen positive three-digit integer is a multiple of 6?
    c. What is the probability that a randomly chosen positive three-digit integer is a multiple of 7?

23. If the largest of 56 consecutive integers is 279, what is the smallest?

24. If the largest of 87 consecutive integers is 326, what is the smallest?

25. How many even integers are between 1 and 1,001?

26. How many integers that are multiples of 3 are between 1 and 1,001?

27. A certain non-leap year has 365 days, and January 1 occurs on a Monday.
    a. How many Sundays are in the year?
    b. How many Mondays are in the year?

✶ 28. Prove Theorem 9.1.1. (Let m be any integer and prove the theorem by mathematical induction on n.)

Answers for Test Yourself

1. the set of all outcomes of the random process or experiment 2. a subset of the sample space 3. number of outcomes in the event; total number of outcomes 4. $n - m + 1$

9.2 Possibility Trees and the Multiplication Rule

Don't believe anything unless you have thought it through for yourself.
— Anna Pell Wheeler, 1883–1966

A tree structure is a useful tool for keeping systematic track of all possibilities in situations in which events happen in order. The following example shows how to use such a structure to count the number of different outcomes of a tournament.

Example 9.2.1 Possibilities for Tournament Play

Teams A and B are to play each other repeatedly until one wins two games in a row or a total of three games. One way in which this tournament can be played is for A to win the first game, B to win the second, and A to win the third and fourth games. Denote this by writing $A–B–A–A$.

a. How many ways can the tournament be played?

b. Assuming that all the ways of playing the tournament are equally likely, what is the probability that five games are needed to determine the tournament winner?

Solution

a. The possible ways for the tournament to be played are represented by the distinct paths from "root" (the start) to "leaf" (a terminal point) in the tree shown sideways in Figure 9.2.1. The label on each branching point indicates the winner of the game. The notations in parentheses indicate the winner of the tournament.

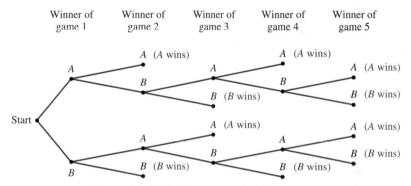

Figure 9.2.1 The Outcomes of a Tournament

The fact that there are ten paths from the root of the tree to its leaves shows that there are ten possible ways for the tournament to be played. They are (moving from the top down): $A–A$, $A–B–A–A$, $A–B–A–B–A$, $A–B–A–B–B$, $A–B–B$, $B–A–A$, $B–A–B–A–A$, $B–A–B–A–B$, $B–A–B–B$, and $B–B$. In five cases A wins, and in the other five B wins. The least number of games that must be played to determine a winner is two, and the most that will need to be played is five.

b. Since all the possible ways of playing the tournament listed in part (a) are assumed to be equally likely, and the listing shows that five games are needed in four different cases ($A–B–A–B–A$, $A–B–A–B–B$, $B–A–B–A–B$, and $B–A–B–A–A$), the probability that five games are needed is $4/10 = 2/5 = 40\%$. ∎

The Multiplication Rule

Consider the following example. Suppose a computer installation has four input/output units (A, B, C, and D) and three central processing units (X, Y, and Z). Any input/output unit can be paired with any central processing unit. How many ways are there to pair an input/output unit with a central processing unit?

To answer this question, imagine the pairing of the two types of units as a two-step operation:

Step 1: Choose the input/output unit.

Step 2: Choose the central processing unit.

The possible outcomes of this operation are illustrated in the possibility tree of Figure 9.2.2.

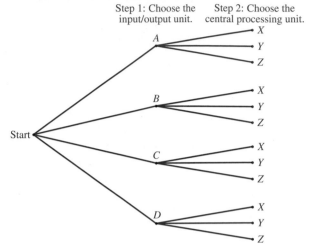

Step 1: Choose the input/output unit.　Step 2: Choose the central processing unit.

Figure 9.2.2 Pairing Objects Using a Possibility Tree

The topmost path from "root" to "leaf" indicates that input/output unit A is to be paired with central processing unit X. The next lower branch indicates that input/output unit A is to be paired with central processing unit Y. And so forth.

Thus the total number of ways to pair the two types of units is the same as the number of branches of the tree, which is

$$3 + 3 + 3 + 3 = 4 \cdot 3 = 12.$$

The idea behind this example can be used to prove the following rule. A formal proof uses mathematical induction and is left to the exercises.

Theorem 9.2.1 The Multiplication Rule

If an operation consists of k steps and

the first step can be performed in n_1 ways,

the second step can be performed in n_2 ways *[regardless of how the first step was performed]*,

$\vdots$

the kth step can be performed in n_k ways *[regardless of how the preceding steps were performed]*,

then the entire operation can be performed in $n_1 n_2 \cdots n_k$ ways.

To apply the multiplication rule, think of the objects you are trying to count as the output of a multistep operation. The possible ways to perform a step may depend on how preceding steps were performed, but the *number* of ways to perform each step must be constant regardless of the action taken in prior steps.

Example 9.2.2 Number of Personal Identification Numbers (PINs)

A typical PIN (personal identification number) is a sequence of any four symbols chosen from the 26 letters in the alphabet and the ten digits, with repetition allowed. How many different PINs are possible?

Solution Typical PINs are CARE, 3387, B32B, and so forth. You can think of forming a PIN as a four-step operation to fill in each of the four symbols in sequence.

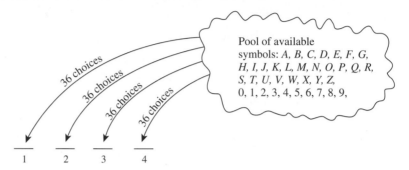

Step 1: Choose the first symbol.

Step 2: Choose the second symbol.

Step 3: Choose the third symbol.

Step 4: Choose the fourth symbol.

There is a fixed number of ways to perform each step, namely 36, regardless of how preceding steps were performed. And so, by the multiplication rule, there are $36 \cdot 36 \cdot 36 \cdot 36 = 36^4 = 1{,}679{,}616$ PINs in all. ∎

Another way to look at the PINs of Example 9.2.2 is as ordered 4-tuples. For example, you can think of the PIN M2ZM as the ordered 4-tuple (M, 2, Z, M). Therefore, the total number of PINs is the same as the total number of ordered 4-tuples whose elements are either letters of the alphabet or digits. One of the most important uses of the multiplication rule is to derive a general formula for the number of elements in any Cartesian product of a finite number of finite sets. In Example 9.2.3, this is done for a Cartesian product of four sets.

Example 9.2.3 The Number of Elements in a Cartesian Product

Suppose A_1, A_2, A_3, and A_4 are sets with n_1, n_2, n_3, and n_4 elements, respectively. Show that the set $A_1 \times A_2 \times A_3 \times A_4$ has $n_1 n_2 n_3 n_4$ elements.

Solution Each element in $A_1 \times A_2 \times A_3 \times A_4$ is an ordered 4-tuple of the form (a_1, a_2, a_3, a_4), where $a_1 \in A_1, a_2 \in A_2, a_3 \in A_3$, and $a_4 \in A_4$. Imagine the process of constructing these ordered tuples as a four-step operation:

Step 1: Choose the first element of the 4-tuple.

Step 2: Choose the second element of the 4-tuple.

Step 3: Choose the third element of the 4-tuple.

Step 4: Choose the fourth element of the 4-tuple.

There are n_1 ways to perform step 1, n_2 ways to perform step 2, n_3 ways to perform step 3, and n_4 ways to perform step 4. Hence, by the multiplication rule, there are $n_1 n_2 n_3 n_4$

ways to perform the entire operation. Therefore, there are $n_1 n_2 n_3 n_4$ distinct 4-tuples in $A_1 \times A_2 \times A_3 \times A_4$. ∎

Example 9.2.4 Number of PINs without Repetition

In Example 9.2.2 we formed PINs using four symbols, either letters of the alphabet or digits, and supposing that letters could be repeated. Now suppose that repetition is not allowed.

a. How many different PINs are there?

b. If all PINs are equally likely, what is the probability that a PIN chosen at random contains no repeated symbol?

Solution

a. Again think of forming a PIN as a four-step operation: Choose the first symbol, then the second, then the third, and then the fourth. There are 36 ways to choose the first symbol, 35 ways to choose the second (since the first symbol cannot be used again), 34 ways to choose the third (since the first two symbols cannot be reused), and 33 ways to choose the fourth (since the first three symbols cannot be reused). Thus, the multiplication rule can be applied to conclude that there are $36 \cdot 35 \cdot 34 \cdot 33 = 1,413,720$ different PINs with no repeated symbol.

b. By part (a) there are 1,413,720 PINs with no repeated symbol, and by Example 9.2.2 there are 1,679,616 PINs in all. Thus the probability that a PIN chosen at random contains no repeated symbol is $\frac{1,413,720}{1,679,616} \cong .8417$. In other words, approximately 84% of PINs have no repeated symbol. ∎

• Definition

Let n be a positive integer. Given a finite set S, **a string of length n over S** is an ordered n-tuple of elements of S written without parentheses or commas. The elements of S are called the **characters** of the string. The **null string** over S is defined to be the "string" with no characters. It is usually denoted ϵ and is said to have length 0. If $S = \{0, 1\}$, then a string over S is called a **bit string**.

Observe that in Examples 9.2.2 and 9.2.4, the set of all PINs of length 4 is the same as the set of all strings of length 4 over the set

$$S = \{x \mid x \text{ is a letter of the alphabet or } x \text{ is a digit}\}.$$

Examples of bit strings are 0111010, 01, 10, 110011, and 0000. Here is a listing of all bit strings of length 3:

$$000, \ 001, \ 010, \ 100, \ 011, \ 101, \ 110, \ 111.$$

When the Multiplication Rule Is Difficult or Impossible to Apply

Consider the following problem:

Three officers—a president, a treasurer, and a secretary—are to be chosen from among four people: Ann, Bob, Cyd, and Dan. Suppose that, for various reasons, Ann cannot be president and either Cyd or Dan must be secretary. How many ways can the officers be chosen?

It is natural to try to solve this problem using the multiplication rule. A person might answer as follows:

> There are three choices for president (all except Ann), three choices for treasurer (all except the one chosen as president), and two choices for secretary (Cyd or Dan). Therefore, by the multiplication rule, there are $3 \cdot 3 \cdot 2 = 18$ choices in all.

Unfortunately, this analysis is incorrect. The number of ways to choose the secretary varies depending on who is chosen for president and treasurer. For instance, if Bob is chosen for president and Ann for treasurer, then there are two choices for secretary: Cyd and Dan. But if Bob is chosen for president and Cyd for treasurer, then there is just one choice for secretary: Dan. The clearest way to see all the possible choices is to construct the possibility tree, as is shown in Figure 9.2.3.

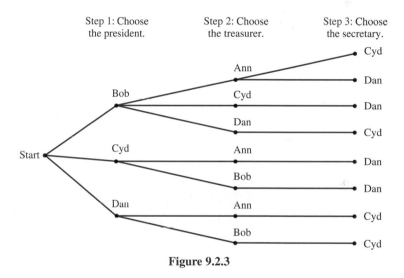

Figure 9.2.3

From the tree it is easy to see that there are only eight ways to choose a president, treasurer, and secretary so as to satisfy the given conditions.

Another way to solve this problem is somewhat surprising. It turns out that the steps can be reordered in a slightly different way so that the number of ways to perform each step is constant regardless of the way previous steps were performed.

Example 9.2.5 A More Subtle Use of the Multiplication Rule

Reorder the steps for choosing the officers in the previous example so that the total number of ways to choose officers can be computed using the multiplication rule.

Solution

Step 1: Choose the secretary.

Step 2: Choose the president.

Step 3: Choose the treasurer.

There are exactly two ways to perform step 1 (either Cyd or Dan may be chosen), two ways to perform step 2 (neither Ann nor the person chosen in step 1 may be chosen but either of the other two may), and two ways to perform step 3 (either of the two people not chosen as secretary or president may be chosen as treasurer). Thus, by the multiplication rule, the total number of ways to choose officers is $2 \cdot 2 \cdot 2 = 8$. A possibility tree illustrating this sequence of choices is shown in Figure 9.2.4. Note how balanced this tree is compared with the one in Figure 9.2.3.

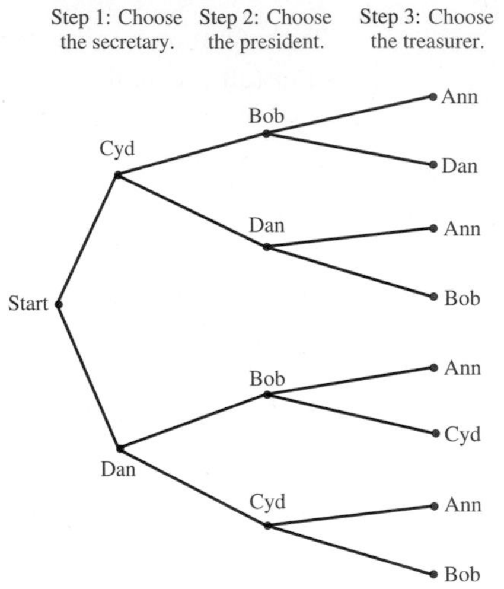

Step 1: Choose   Step 2: Choose   Step 3: Choose
the secretary.   the president.   the treasurer.

Figure 9.2.4

Permutations

A **permutation** of a set of objects is an ordering of the objects in a row. For example, the set of elements a, b, and c has six permutations.

$$abc \quad acb \quad cba \quad bac \quad bca \quad cab$$

In general, given a set of n objects, how many permutations does the set have? Imagine forming a permutation as an n-step operation:

Step 1: Choose an element to write first.

Step 2: Choose an element to write second.

$$\vdots \qquad \vdots$$

Step n: Choose an element to write nth.

Any element of the set can be chosen in step 1, so there are n ways to perform step 1. Any element except that chosen in step 1 can be chosen in step 2, so there are $n - 1$ ways to perform step 2. In general, the number of ways to perform each successive step is one less than the number of ways to perform the preceding step. At the point when the nth element is chosen, there is only one element left, so there is only one way to perform step n. Hence, by the multiplication rule, there are

$$n(n - 1)(n - 2) \cdots 2 \cdot 1 = n!$$

ways to perform the entire operation. In other words, there are $n!$ permutations of a set of n elements. This reasoning is summarized in the following theorem. A formal proof uses mathematical induction and is left as an exercise.

Theorem 9.2.2

For any integer n with $n \geq 1$, the number of permutations of a set with n elements is $n!$.

Example 9.2.6 Permutations of the Letters in a Word

a. How many ways can the letters in the word *COMPUTER* be arranged in a row?

b. How many ways can the letters in the word *COMPUTER* be arranged if the letters *CO* must remain next to each other (in order) as a unit?

c. If letters of the word *COMPUTER* are randomly arranged in a row, what is the probability that the letters *CO* remain next to each other (in order) as a unit?

Solution

a. All the eight letters in the word *COMPUTER* are distinct, so the number of ways in which we can arrange the letters equals the number of permutations of a set of eight elements. This equals $8! = 40,320$.

b. If the letter group *CO* is treated as a unit, then there are effectively only seven objects that are to be arranged in a row.

$$\boxed{CO}\ \boxed{M}\ \boxed{P}\ \boxed{U}\ \boxed{T}\ \boxed{E}\ \boxed{R}$$

Hence there are as many ways to write the letters as there are permutations of a set of seven elements, namely $7! = 5,040$.

c. When the letters are arranged randomly in a row, the total number of arrangements is 40,320 by part (a), and the number of arrangements with the letters *CO* next to each other (in order) as a unit is 5,040. Thus the probability is

$$\frac{5,040}{40,320} = \frac{1}{8} = 12.5\%. \qquad \blacksquare$$

Example 9.2.7 Permutations of Objects Around a Circle

At a meeting of diplomats, the six participants are to be seated around a circular table. Since the table has no ends to confer particular status, it doesn't matter who sits in which chair. But it does matter how the diplomats are seated relative to each other. In other words, two seatings are considered the same if one is a rotation of the other. How many different ways can the diplomats be seated?

Solution   Call the diplomats by the letters $A, B, C, D, E,$ and F. Since only relative position matters, you can start with any diplomat (say A), place that diplomat anywhere (say in the top seat of the diagram shown in Figure 9.2.5), and then consider all arrangements of the other diplomats around that one. B through F can be arranged in the seats around diplomat A in all possible orders. So there are $5! = 120$ ways to seat the group.

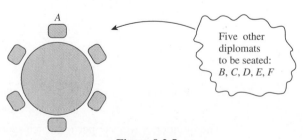

Figure 9.2.5

$\blacksquare$

Permutations of Selected Elements

Given the set $\{a, b, c\}$, there are six ways to select two letters from the set and write them in order.

$$ab \quad ac \quad ba \quad bc \quad ca \quad cb$$

Each such ordering of two elements of $\{a, b, c\}$ is called a 2-*permutation* of $\{a, b, c\}$.

• Definition

An **r-permutation** of a set of n elements is an ordered selection of r elements taken from the set of n elements. The number of r-permutations of a set of n elements is denoted $P(n, r)$.

Theorem 9.2.3

If n and r are integers and $1 \leq r \leq n$, then the number of r-permutations of a set of n elements is given by the formula

$$P(n, r) = n(n - 1)(n - 2) \cdots (n - r + 1) \qquad \text{first version}$$

or, equivalently,

$$P(n, r) = \frac{n!}{(n - r)!} \qquad \text{second version.}$$

A formal proof of this theorem uses mathematical induction and is based on the multiplication rule. The idea of the proof is the following.

Suppose a set of n elements is given. Formation of an r-permutation can be thought of as an r-step process. Step 1 is to choose the element to be first. Since the set has n elements, there are n ways to perform step 1. Step 2 is to choose the element to be second. Since the element chosen in step 1 is no longer available, there are $n - 1$ ways to perform step 2. Step 3 is to choose the element to be third. Since neither of the two elements chosen in the first two steps is available, there are $n - 2$ choices for step 3. This process is repeated r times.

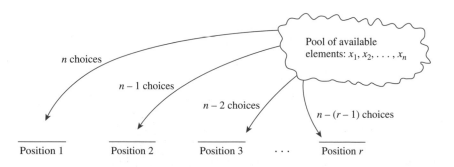

The number of ways to perform each successive step is one less than the number of ways to perform the preceding step. Step r is to choose the element to be rth. At the point just before step r is performed, $r - 1$ elements have already been chosen, and so there are

$$n - (r - 1) = n - r + 1$$

left to choose from. Hence there are $n - r + 1$ ways to perform step r. It follows by the multiplication rule that the number of ways to form an r-permutation is

$$P(n, r) = n(n - 1)(n - 2) \cdots (n - r + 1).$$

Note that

$$\frac{n!}{(n-r)!} = \frac{n(n-1)(n-2)\cdots(n-r+1)(\cancel{n-r})(\cancel{n-r-1})\cdots\cancel{3}\cdot\cancel{2}\cdot\cancel{1}}{(\cancel{n-r})(\cancel{n-r-1})\cdots\cancel{3}\cdot\cancel{2}\cdot\cancel{1}}$$

$$= n(n-1)(n-2)\cdots(n-r+1).$$

Thus the formula can be written as

$$P(n, r) = \frac{n!}{(n-r)!}.$$

The second version of the formula is easier to remember. When you actually use it, however, first substitute the values of n and r and then immediately cancel the numerical value of $(n - r)!$ from the numerator and denominator. Because factorials become so large so fast, direct use of the second version of the formula without cancellation can overload your calculator's capacity for exact arithmetic even when n and r are quite small. For instance, if $n = 15$ and $r = 2$, then

$$\frac{n!}{(n-r)!} = \frac{15!}{13!} = \frac{1,307,674,368,000}{6,227,020,800}.$$

But if you cancel $(n - r)! = 13!$ from numerator and denominator before multiplying out, you obtain

$$\frac{n!}{(n-r)!} = \frac{15!}{13!} = \frac{15 \cdot 14 \cdot \cancel{13!}}{\cancel{13!}} = 15 \cdot 14 = 210.$$

In fact, many scientific calculators allow you to compute $P(n, r)$ simply by entering the values of n and r and pressing a key or making a menu choice. Alternative notations for $P(n, r)$ that you may see in your calculator manual are $_n P_r$, $P_{n,r}$ and $^n P_r$.

Example 9.2.8 Evaluating r-Permutations

    a. Evaluate $P(5, 2)$.

    b. How many 4-permutations are there of a set of seven objects?

    c. How many 5-permutations are there of a set of five objects?

Solution

    a. $P(5, 2) = \dfrac{5!}{(5-2)!} = \dfrac{5 \cdot 4 \cdot \cancel{3} \cdot \cancel{2} \cdot \cancel{1}}{\cancel{3} \cdot \cancel{2} \cdot \cancel{1}} = 20$

    b. The number of 4-permutations of a set of seven objects is

$$P(7, 4) = \frac{7!}{(7-4)!} = \frac{7 \cdot 6 \cdot 5 \cdot 4 \cdot \cancel{3} \cdot \cancel{2} \cdot \cancel{1}}{\cancel{3} \cdot \cancel{2} \cdot \cancel{1}} = 7 \cdot 6 \cdot 5 \cdot 4 = 840.$$

    c. The number of 5-permutations of a set of five objects is

$$P(5, 5) = \frac{5!}{(5-5)!} = \frac{5!}{0!} = \frac{5!}{1} = 5! = 120.$$

Note that the definition of 0! as 1 makes this calculation come out as it should, for the number of 5-permutations of a set of five objects is certainly equal to the number of permutations of the set. ∎

Example 9.2.9 Permutations of Selected Letters of a Word

a. How many different ways can three of the letters of the word *BYTES* be chosen and written in a row?

b. How many different ways can this be done if the first letter must be *B*?

Solution

a. The answer equals the number of 3-permutations of a set of five elements. This equals

$$P(5, 3) = \frac{5!}{(5-3)!} = \frac{5 \cdot 4 \cdot 3 \cdot \cancel{2} \cdot \cancel{1}}{\cancel{2} \cdot \cancel{1}} = 5 \cdot 4 \cdot 3 = 60.$$

b. Since the first letter must be *B*, there are effectively only two letters to be chosen and placed in the other two positions. And since the *B* is used in the first position, there are four letters available to fill the remaining two positions.

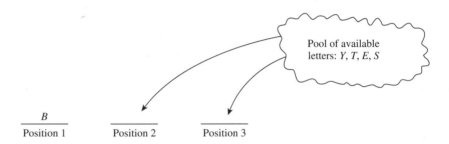

Hence the answer is the number of 2-permutations of a set of four elements, which is

$$P(4, 2) = \frac{4!}{(4-2)!} = \frac{4 \cdot 3 \cdot \cancel{2} \cdot \cancel{1}}{\cancel{2} \cdot \cancel{1}} = 4 \cdot 3 = 12. \qquad \blacksquare$$

In many applications of the mathematics of counting, it is necessary to be skillful in working algebraically with quantities of the form $P(n, r)$. The next example shows a kind of problem that gives practice in developing such skill.

Example 9.2.10 Proving a Property of $P(n, r)$

Prove that for all integers $n \geq 2$,

$$P(n, 2) + P(n, 1) = n^2.$$

Solution   Suppose n is an integer that is greater than or equal to 2. By Theorem 9.2.3,

$$P(n, 2) = \frac{n!}{(n-2)!} = \frac{n(n-1)\cancel{(n-2)!}}{\cancel{(n-2)!}} = n(n-1)$$

and

$$P(n, 1) = \frac{n!}{(n-1)!} = \frac{n \cdot \cancel{(n-1)!}}{\cancel{(n-1)!}} = n.$$

Hence

$$P(n, 2) + P(n, 1) = n \cdot (n-1) + n = n^2 - n + n = n^2,$$

which is what we needed to show. $\qquad \blacksquare$

Test Yourself

1. The multiplication rule says that if an operation can be performed in k steps and, for each i with $1 \leq i \leq k$, the ith step can be performed in n_i ways (regardless of how previous steps were performed), then the operation as a whole can be performed in _____.

2. A permutation of a set of elements is _____.

3. The number of permutations of a set of n elements equals _____.

4. An r-permutation of a set of n elements is _____.

5. The number of r-permutations of a set of n elements is denoted _____.

6. One formula for the number of r-permutations of a set of n elements is _____ and another formula is _____.

Exercise Set 9.2

In 1–4, use the fact that in baseball's World Series, the first team to win four games wins the series.

1. Suppose team A wins the first three games. How many ways can the series be completed? (Draw a tree.)

2. Suppose team A wins the first two games. How many ways can the series be completed? (Draw a tree.)

3. How many ways can a World Series be played if team A wins four games in a row?

4. How many ways can a World Series be played if no team wins two games in a row?

5. In a competition between players X and Y, the first player to win three games in a row or a total of four games wins. How many ways can the competition be played if X wins the first game and Y wins the second and third games? (Draw a tree.)

6. One urn contains two black balls (labeled B_1 and B_2) and one white ball. A second urn contains one black ball and two white balls (labeled W_1 and W_2). Suppose the following experiment is performed: One of the two urns is chosen at random. Next a ball is randomly chosen from the urn. Then a second ball is chosen at random from the same urn without replacing the first ball.
   a. Construct the possibility tree showing all possible outcomes of this experiment.
   b. What is the total number of outcomes of this experiment?
   c. What is the probability that two black balls are chosen?
   d. What is the probability that two balls of opposite color are chosen?

7. One urn contains one blue ball (labeled B_1) and three red balls (labeled R_1, R_2, and R_3). A second urn contains two red balls (R_4 and R_5) and two blue balls (B_2 and B_3). An experiment is performed in which one of the two urns is chosen at random and then two balls are randomly chosen from it, one after the other without replacement.
   a. Construct the possibility tree showing all possible outcomes of this experiment.
   b. What is the total number of outcomes of this experiment?
   c. What is the probability that two red balls are chosen?

8. A person buying a personal computer system is offered a choice of three models of the basic unit, two models of keyboard, and two models of printer. How many distinct systems can be purchased?

9. Suppose there are three roads from city A to city B and five roads from city B to city C.
   a. How many ways is it possible to travel from city A to city C via city B?
   b. How many different round-trip routes are there from city A to B to C to B and back to A?
   c. How many different routes are there from city A to B to C to B and back to A in which no road is traversed twice?

10. Suppose there are three routes from North Point to Boulder Creek, two routes from Boulder Creek to Beaver Dam, two routes from Beaver Dam to Star Lake, and four routes directly from Boulder Creek to Star Lake. (Draw a sketch.)
   a. How many routes from North Point to Star Lake pass through Beaver Dam?
   b. How many routes from North Point to Star Lake bypass Beaver Dam?

11. a. A bit string is a finite sequence of 0's and 1's. How many bit strings have length 8?
   b. How many bit strings of length 8 begin with three 0's?
   c. How many bit strings of length 8 begin and end with a 1?

12. Hexadecimal numbers are made using the sixteen digits 0, 1, 2, 3, 4, 5, 6, 7, 8, 9, A, B, C, D, E, F. They are denoted by the subscript 16. For example, $9A2D_{16}$ and $BC54_{16}$ are hexadecimal numbers.
   a. How many hexadecimal numbers begin with one of the digits 3 through B, end with one of the digits 5 through F, and are 5 digits long?
   b. How many hexadecimal numbers begin with one of the digits 4 through D, end with one of the digits 2 through E, and are 6 digits long?

13. A coin is tossed four times. Each time the result H for heads or T for tails is recorded. An outcome of $HHTT$ means that heads were obtained on the first two tosses and tails on the second two. Assume that heads and tails are equally likely on each toss.
   a. How many distinct outcomes are possible?
   b. What is the probability that exactly two heads occur?
   c. What is the probability that exactly one head occurs?

14. Suppose that in a certain state, all automobile license plates have four letters followed by three digits.
    a. How many different license plates are possible?
    b. How many license plates could begin with *A* and end in 0?
    c. How many license plates could begin with *TGIF*?
    d. How many license plates are possible in which all the letters and digits are distinct?
    e. How many license plates could begin with *AB* and have all letters and digits distinct?

15. A combination lock requires three selections of numbers, each from 1 through 30.
    a. How many different combinations are possible?
    b. Suppose the locks are constructed in such a way that no number may be used twice. How many different combinations are possible?

16. a. How many integers are there from 10 through 99?
    b. How many odd integers are there from 10 through 99?
    c. How many integers from 10 through 99 have distinct digits?
    d. How many odd integers from 10 through 99 have distinct digits?
    e. What is the probability that a randomly chosen two-digit integer has distinct digits? has distinct digits and is odd?

17. a. How many integers are there from 1000 through 9999?
    b. How many odd integers are there from 1000 through 9999?
    c. How many integers from 1000 through 9999 have distinct digits?
    d. How many odd integers from 1000 through 9999 have distinct digits?
    e. What is the probability that a randomly chosen four-digit integer has distinct digits? has distinct digits and is odd?

18. The diagram below shows the keypad for an automatic teller machine. As you can see, the same sequence of keys represents a variety of different PINs. For instance, 2133, AZDE, and BQ3F are all keyed in exactly the same way.

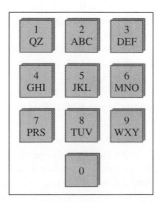

    a. How many different PINs are represented by the same sequence of keys as 2133?

    b. How many different PINs are represented by the same sequence of keys as 5031?
    c. At an automatic teller machine, each PIN corresponds to a four-digit numeric sequence. For instance, TWJM corresponds to 8956. How many such numeric sequences contain no repeated digit?

19. Three officers—a president, a treasurer, and a secretary—are to be chosen from among four people: Ann, Bob, Cyd, and Dan. Suppose that Bob is not qualified to be treasurer and Cyd's other commitments make it impossible for her to be secretary. How many ways can the officers be chosen? Can the multiplication rule be used to solve this problem?

20. Modify Example 9.2.4 by supposing that a PIN must not begin with any of the letters A–M and must end with a digit. Continue to assume that no symbol may be used more than once and that the total number of PINs is to be determined.
    a. Find the error in the following "solution."

    "Constructing a PIN is a four-step process.

    Step 1: Choose the left-most symbol.

    Step 2: Choose the second symbol from the left.

    Step 3: Choose the third symbol from the left.

    Step 4: Choose the right-most symbol.
    Because none of the thirteen letters from A through M may be chosen in step 1, there are $36 - 13 = 23$ ways to perform step 1. There are 35 ways to perform step 2 and 34 ways to perform step 3 because previously used symbols may not be used. Since the symbol chosen in step 4 must be a previously unused digit, there are $10 - 3 = 7$ ways to perform step 4. Thus there are $23 \cdot 35 \cdot 34 \cdot 7 = 191,590$ different PINs that satisfy the given conditions."

    b. Reorder steps 1–4 in part (a) as follows:

    Step 1: Choose the right-most symbol.

    Step 2: Choose the left-most symbol.

    Step 3: Choose the second symbol from the left.

    Step 4: Choose the third symbol from the left.

    Use the multiplication rule to find the number of PINs that satisfy the given conditions.

H 21. Suppose *A* is a set with *m* elements and *B* is a set with *n* elements.
    a. How many relations are there from *A* to *B*? Explain.
    b. How many functions are there from *A* to *B*? Explain.
    c. What fraction of the relations from *A* to *B* are functions?

22. **a.** How many functions are there from a set with three elements to a set with four elements?
    b. How many functions are there from a set with five elements to a set with two elements?
    c. How many functions are there from a set with *m* elements to a set with *n* elements, where *m* and *n* are positive integers?

H ✱ **23.** Consider the numbers 1 through 99,999 in their ordinary decimal representations. How many contain exactly one of each of the digits 2, 3, 4, and 5?

✱ **24.** Let $n = p_1^{k_1} p_2^{k_2} \cdots p_m^{k_m}$ where $p_1, p_2, \ldots, p_m$ are distinct prime numbers and $k_1, k_2, \ldots, k_m$ are positive integers. How many ways can n be written as a product of two positive integers that have no common factors
  a. assuming that order matters (i.e., $8 \cdot 15$ and $15 \cdot 8$ are regarded as different)?
  b. assuming that order does not matter (i.e., $8 \cdot 15$ and $15 \cdot 8$ are regarded as the same)?

✱ **25. a.** If p is a prime number and a is a positive integer, how many distinct positive divisors does p^a have?
  b. If p and q are distinct prime numbers and a and b are positive integers, how many distinct positive divisors does $p^a q^b$ have?
  c. If p, q, and r are distinct prime numbers and a, b, and c are positive integers, how many distinct positive divisors does $p^a q^b r^c$ have?
  d. If $p_1, p_2, \ldots, p_m$ are distinct prime numbers and $a_1, a_2, \ldots, a_m$ are positive integers, how many distinct positive divisors does $p_1^{a_1} p_2^{a_2} \cdots p_m^{a_m}$ have?
  e. What is the smallest positive integer with exactly 12 divisors?

26. a. How many ways can the letters of the word *ALGORITHM* be arranged in a row?
  b. How many ways can the letters of the word *ALGORITHM* be arranged in a row if *A* and *L* must remain together (in order) as a unit?
  c. How many ways can the letters of the word *ALGORITHM* be arranged in a row if the letters *GOR* must remain together (in order) as a unit?

27. Six people attend the theater together and sit in a row with exactly six seats.
  a. How many ways can they be seated together in the row?
  b. Suppose one of the six is a doctor who must sit on the aisle in case she is paged. How many ways can the people be seated together in the row with the doctor in an aisle seat?
  c. Suppose the six people consist of three married couples and each couple wants to sit together with the husband on the left. How many ways can the six be seated together in the row?

28. Five people are to be seated around a circular table. Two seatings are considered the same if one is a rotation of the other. How many different seatings are possible?

29. Write all the 2-permutations of $\{W, X, Y, Z\}$.

30. Write all the 3-permutations of $\{s, t, u, v\}$.

31. Evaluate the following quantities.
  a. $P(6, 4)$ b. $P(6, 6)$ c. $P(6, 3)$ d. $P(6, 1)$

32. a. How many 3-permutations are there of a set of five objects?
  b. How many 2-permutations are there of a set of eight objects?

33. a. How many ways can three of the letters of the word *ALGORITHM* be selected and written in a row?
  b. How many ways can six of the letters of the word *ALGORITHM* be selected and written in a row?
  c. How many ways can six of the letters of the word *ALGORITHM* be selected and written in a row if the first letter must be *A*?
  d. How many ways can six of the letters of the word *ALGORITHM* be selected and written in a row if the first two letters must be *OR*?

34. Prove that for all integers $n \geq 2$, $P(n + 1, 3) = n^3 - n$.

35. Prove that for all integers $n \geq 2$,
$$P(n + 1, 2) - P(n, 2) = 2P(n, 1).$$

36. Prove that for all integers $n \geq 3$,
$$P(n + 1, 3) - P(n, 3) = 3P(n, 2).$$

37. Prove that for all integers $n \geq 2$, $P(n, n) = P(n, n - 1)$.

38. Prove Theorem 9.2.1 by mathematical induction.

H **39.** Prove Theorem 9.2.2 by mathematical induction.

✱ **40.** Prove Theorem 9.2.3 by mathematical induction.

41. A permutation on a set can be regarded as a function from the set to itself. For instance, one permutation of $\{1, 2, 3, 4\}$ is 2341. It can be identified with the function that sends each position number to the number occupying that position. Since position 1 is occupied by 2, 1 is sent to 2 or $1 \to 2$; since position 2 is occupied by 3, 2 is sent to 3 or $2 \to 3$; and so forth. The entire permutation can be written using arrows as follows:

$$
\begin{array}{cccc}
1 & 2 & 3 & 4 \\
\downarrow & \downarrow & \downarrow & \downarrow \\
2 & 3 & 4 & 1
\end{array}
$$

  a. Use arrows to write each of the six permutations of $\{1, 2, 3\}$.
  b. Use arrows to write each of the permutations of $\{1, 2, 3, 4\}$ that keep 2 and 4 fixed.
  c. Which permutations of $\{1, 2, 3\}$ keep no elements fixed?
  d. Use arrows to write all permutations of $\{1, 2, 3, 4\}$ that keep no elements fixed.

Answers for Test Yourself

1. $n_1 n_2 \cdots n_k$ ways   2. an ordering of the elements of the set in a row   3. $n!$ 4. an ordered selection of r of the elements of the set
5. $P(n, r)$ 6. $n(n - 1)(n - 2) \cdots (n - r + 1)$; $\dfrac{n!}{(n-r)!}$

9.3 Counting Elements of Disjoint Sets: The Addition Rule

The whole of science is nothing more than a refinement of everyday thinking.
— Albert Einstein, 1879–1955

In the last section we discussed counting problems that can be solved using possibility trees. In this section we look at counting problems that can be solved by counting the number of elements in the union of two sets, the difference of two sets, or the intersection of two sets.

The basic rule underlying the calculation of the number of elements in a union or difference or intersection is the addition rule. This rule states that the number of elements in a union of mutually disjoint finite sets equals the sum of the number of elements in each of the component sets.

Theorem 9.3.1 The Addition Rule

Suppose a finite set A equals the union of k distinct mutually disjoint subsets A_1, $A_2, \ldots, A_k$. Then

$$N(A) = N(A_1) + N(A_2) + \cdots + N(A_k).$$

A formal proof of this theorem uses mathematical induction and is left to the exercises.

Example 9.3.1 Counting Passwords with Three or Fewer Letters

A computer access password consists of from one to three letters chosen from the 26 in the alphabet with repetitions allowed. How many different passwords are possible?

Solution The set of all passwords can be partitioned into subsets consisting of those of length 1, those of length 2, and those of length 3 as shown in Figure 9.3.1.

Set of All Passwords of Length ≤ 3

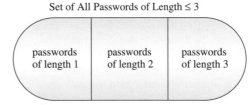

| passwords of length 1 | passwords of length 2 | passwords of length 3 |

Figure 9.3.1

By the addition rule, the total number of passwords equals the number of passwords of length 1, plus the number of passwords of length 2, plus the number of passwords of length 3. Now the

number of passwords of length $1 = 26$ because there are 26 letters in the alphabet

number of passwords of length $2 = 26^2$ because forming such a word can be thought of as a two-step process in which there are 26 ways to perform each step

number of passwords of length $3 = 26^3$ because forming such a word can be thought of as a three-step process in which there are 26 ways to perform each step.

Hence the total number of passwords $= 26 + 26^2 + 26^3 = 18{,}278$. ∎

Example 9.3.2 Counting the Number of Integers Divisible by 5

How many three-digit integers (integers from 100 to 999 inclusive) are divisible by 5?

Solution   One solution to this problem was discussed in Example 9.1.4. Another approach uses the addition rule. Integers that are divisible by 5 end either in 5 or in 0. Thus the set of all three-digit integers that are divisible by 5 can be split into two mutually disjoint subsets A_1 and A_2 as shown in Figure 9.3.2.

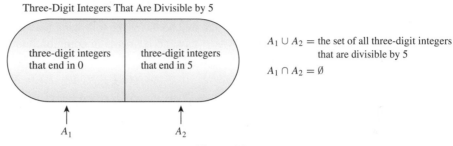

Three-Digit Integers That Are Divisible by 5

three-digit integers that end in 0   three-digit integers that end in 5

A_1 A_2

$A_1 \cup A_2 =$ the set of all three-digit integers that are divisible by 5

$A_1 \cap A_2 = \emptyset$

Figure 9.3.2

Now there are as many three-digit integers that end in 0 as there are possible choices for the left-most and middle digits (because the right-most digit must be a 0). As illustrated below, there are nine choices for the left-most digit (the digits 1 through 9) and ten choices for the middle digit (the digits 0 through 9). Hence $N(A_1) = 9 \cdot 10 = 90$.

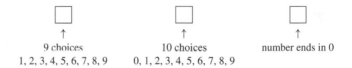

9 choices
1, 2, 3, 4, 5, 6, 7, 8, 9

10 choices
0, 1, 2, 3, 4, 5, 6, 7, 8, 9

number ends in 0

Similar reasoning (using 5 instead of 0) shows that $N(A_2) = 90$ also. So

$$\begin{bmatrix} \text{the number of} \\ \text{three-digit integers} \\ \text{that are divisible by 5} \end{bmatrix} = N(A_1) + N(A_2) = 90 + 90 = 180. \qquad \blacksquare$$

The Difference Rule

An important consequence of the addition rule is the fact that if the number of elements in a set A and the number in a subset B of A are both known, then the number of elements that are in A and not in B can be computed.

Theorem 9.3.2 The Difference Rule

If A is a finite set and B is a subset of A, then

$$N(A - B) = N(A) - N(B).$$

The difference rule is illustrated in Figure 9.3.3.

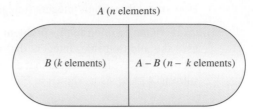

Figure 9.3.3 The Difference Rule

The difference rule holds for the following reason: If B is a subset of A, then the two sets B and $A - B$ have no elements in common and $B \cup (A - B) = A$. Hence, by the addition rule,

$$N(B) + N(A - B) = N(A).$$

Subtracting $N(B)$ from both sides gives the equation

$$N(A - B) = N(A) - N(B).$$

Example 9.3.3 Counting PINs with Repeated Symbols

The PINs discussed in Examples 9.2.2 and 9.2.4 are made from exactly four symbols chosen from the 26 letters of the alphabet and the ten digits, with repetitions allowed.

a. How many PINs contain repeated symbols?

b. If all PINs are equally likely, what is the probability that a randomly chosen PIN contains a repeated symbol?

Solution

a. According to Example 9.2.2, there are $36^4 = 1,679,616$ PINs when repetition is allowed, and by Example 9.2.4, there are 1,413,720 PINs when repetition is not allowed. Thus, by the difference rule, there are

$$1,679,616 - 1,413,720 = 265,896$$

PINs that contain at least one repeated symbol.

b. By Example 9.2.2 there are 1,679,616 PINs in all, and by part (a) 265,896 of these contain at least one repeated symbol. Thus, by the equally likely probability formula, the probability that a randomly chosen PIN contains a repeated symbol is $\frac{265,896}{1,679,616} \cong 0.158 = 15.8\%$. ∎

An alternative solution to Example 9.3.3(b) is based on the observation that if S is the set of all PINs and A is the set of all PINs with no repeated symbol, then $S - A$ is the set of all PINs with at least one repeated symbol. It follows that

$$
\begin{aligned}
P(S - A) &= \frac{N(S - A)}{N(S)} && \text{by definition of probability in the equally likely case} \\[1em]
&= \frac{N(S) - N(A)}{N(S)} && \text{by the difference rule} \\[1em]
&= \frac{N(S)}{N(S)} - \frac{N(A)}{N(S)} && \text{by the laws of fractions} \\[1em]
&= 1 - P(A) && \text{by definition of probability in the equally likely case} \\[0.5em]
&\cong 1 - 0.842 && \text{by Example 9.2.4} \\[0.5em]
&\cong 0.158 = 15.8\%
\end{aligned}
$$

This solution illustrates a more general property of probabilities: that the probability of the complement of an event is obtained by subtracting the probability of the event from the number 1.

Formula for the Probability of the Complement of an Event

If S is a finite sample space and A is an event in S, then

$$P(A^c) = 1 - P(A).$$

The Inclusion/Exclusion Rule

The addition rule says how many elements are in a union of sets if the sets are mutually disjoint. Now consider the question of how to determine the number of elements in a union of sets when some of the sets overlap. For simplicity, begin by looking at a union of two sets A and B, as shown in Figure 9.3.4.

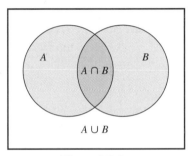

Figure 9.3.4

First observe that the number of elements in $A \cup B$ varies according to the number of elements the two sets have in common. If A and B have no elements in common, then $N(A \cup B) = N(A) + N(B)$. If A and B coincide, then $N(A \cup B) = N(A)$. Thus any general formula for $N(A \cup B)$ must contain a reference to the number of elements the two sets have in common, $N(A \cap B)$, as well as to $N(A)$ and $N(B)$.

The simplest way to derive a formula for $N(A \cup B)$ is to reason as follows: The number $N(A)$ counts the elements that are in A and not in B and also the elements that are in both A and B. Similarly, the number $N(B)$ counts the elements that are in B and not in A and also the elements that are in both A and B. Hence when the two numbers $N(A)$ and $N(B)$ are added, the elements that are in both A and B are counted twice. To get an accurate count of the elements in $A \cup B$, it is necessary to subtract the number of elements that are in both A and B. Because these are the elements in $A \cap B$,

Note An alternative proof is outlined in exercise 46 at the end of this section.

$$N(A \cup B) = N(A) + N(B) - N(A \cap B).$$

A similar analysis gives a formula for the number of elements in a union of three sets, as shown in Theorem 9.3.3.

Theorem 9.3.3 The Inclusion/Exclusion Rule for Two or Three Sets

If A, B, and C are any finite sets, then

$$N(A \cup B) = N(A) + N(B) - N(A \cap B)$$

and

$$N(A \cup B \cup C) = N(A) + N(B) + N(C) - N(A \cap B) - N(A \cap C)$$
$$-N(B \cap C) + N(A \cap B \cap C).$$

It can be shown using mathematical induction (see exercise 46 at the end of this section) that formulas analogous to those of Theorem 9.3.3 hold for unions of any finite number of sets.

Example 9.3.4 Counting Elements of a General Union

a. How many integers from 1 through 1,000 are multiples of 3 or multiples of 5?

b. How many integers from 1 through 1,000 are neither multiples of 3 nor multiples of 5?

Solution

a. Let A = the set of all integers from 1 through 1,000 that are multiples of 3.
Let B = the set of all integers from 1 through 1,000 that are multiples of 5.

Then

$A \cup B$ = the set of all integers from 1 through 1,000 that are multiples of 3
or multiples of 5

and

$A \cap B$ = the set of all integers from 1 through 1,000 that are multiples
of both 3 and 5

= the set of all integers from 1 through 1,000 that are multiples of 15.

[Now calculate $N(A)$, $N(B)$, and $N(A \cap B)$ and use the inclusion/exclusion rule to solve for $N(A \cup B)$.]

Because every third integer from 3 through 999 is a multiple of 3, each can be represented in the form $3k$, for some integer k from 1 through 333. Hence there are 333 multiples of 3 from 1 through 1,000, and so $N(A) = 333$.

$$
\begin{array}{ccccccccccc}
1 & 2 & 3 & 4 & 5 & 6 & \ldots & 996 & 997 & 998 & 999 \\
 & & \updownarrow & & & \updownarrow & & \updownarrow & & & \updownarrow \\
 & & 3 \cdot 1 & & & 3 \cdot 2 & & 3 \cdot 332 & & & 3 \cdot 333
\end{array}
$$

Similarly, each multiple of 5 from 1 through 1,000 has the form $5k$, for some integer k from 1 through 200.

$$
\begin{array}{ccccccccccccccc}
1 & 2 & 3 & 4 & 5 & 6 & 7 & 8 & 9 & 10 & \ldots & 995 & 996 & 997 & 998 & 999 & 1,000 \\
 & & & & \updownarrow & & & & & \updownarrow & & \updownarrow & & & & & \updownarrow \\
 & & & & 5 \cdot 1 & & & & & 5 \cdot 2 & & 5 \cdot 199 & & & & & 5 \cdot 200
\end{array}
$$

Thus there are 200 multiples of 5 from 1 through 1,000 and $N(B) = 200$.
Finally, each multiple of 15 from 1 through 1,000 has the form $15k$, for some integer k from 1 through 66 (since $990 = 66 \cdot 15$).

$$
\begin{array}{ccccccccccccc}
1 & 2 & \ldots & 15 & \ldots & 30 & \ldots & 975 & \ldots & 990 & \ldots & 999 & 1,000 \\
 & & & \updownarrow & & \updownarrow & & \updownarrow & & \updownarrow & & & \\
 & & & 15 \cdot 1 & & 15 \cdot 2 & & 15 \cdot 65 & & 15 \cdot 66 & & &
\end{array}
$$

Hence there are 66 multiples of 15 from 1 through 1,000, and $N(A \cap B) = 66$.

It follows by the inclusion/exclusion rule that

$$N(A \cup B) = N(A) + N(B) - N(A \cap B)$$
$$= 333 + 200 - 66$$
$$= 467.$$

Thus, 467 integers from 1 through 1,000 are multiples of 3 or multiples of 5.

b. There are 1,000 integers from 1 through 1,000, and by part (a), 467 of these are multiples of 3 or multiples of 5. Thus, by the set difference rule, there are $1,000 - 467 = 533$ that are neither multiples of 3 nor multiples of 5. ∎

Note that the solution to part (b) of Example 9.3.4 hid a use of De Morgan's law. The number of elements that are neither in A nor in B is $N(A^c \cap B^c)$, and by De Morgan's law, $A^c \cap B^c = (A \cup B)^c$. So $N((A \cup B)^c)$ was then calculated using the set difference rule: $N((A \cup B)^c) = N(U) - N(A \cup B)$, where the universe U was the set of all integers from 1 through 1,000. Exercises 35–37 at the end of this section explore this technique further.

Example 9.3.5 Counting the Number of Elements in an Intersection

A professor in a discrete mathematics class passes out a form asking students to check all the mathematics and computer science courses they have recently taken. The finding is that out of a total of 50 students in the class,

30 took precalculus;	16 took both precalculus and Java;
18 took calculus;	8 took both calculus and Java;
26 took Java;	47 took at least one of the three courses.
9 took both precalculus and calculus;	

Note that when we write "30 students took precalculus," we mean that the total number of students who took precalculus is 30, and we allow for the possibility that some of these students may have taken one or both of the other courses. If we want to say that 30 students took precalculus *only* (and not either of the other courses), we will say so explicitly.

a. How many students did not take any of the three courses?

b. How many students took all three courses?

c. How many students took precalculus and calculus but not Java? How many students took precalculus but neither calculus nor Java?

Solution

a. By the difference rule, the number of students who did not take any of the three courses equals the number in the class minus the number who took at least one course. Thus the number of students who did not take any of the three courses is

$$50 - 47 = 3.$$

b. Let

$$P = \text{the set of students who took precalculus}$$
$$C = \text{the set of students who took calculus}$$
$$J = \text{the set of students who took Java.}$$

Then, by the inclusion/exclusion rule,

$$N(P \cup C \cup J) = N(P) + N(C) + N(J) - N(P \cap C) - N(P \cap J)$$
$$- N(C \cap J) + N(P \cap C \cap J)$$

Substituting known values, we get

$$47 = 30 + 26 + 18 - 9 - 16 - 8 + N(P \cap C \cap J).$$

Solving for $N(P \cap C \cap J)$ gives

$$N(P \cap C \cap J) = 6.$$

Hence there are six students who took all three courses. In general, if you know any seven of the eight terms in the inclusion/exclusion formula for three sets, you can solve for the eighth term.

c. To answer the questions of part (c), look at the diagram in Figure 9.3.5.

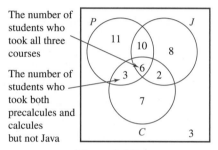

The number of students who took all three courses

The number of students who took both precalcules and calcules but not Java

Figure 9.3.5

Since $N(P \cap C \cap J) = 6$, put the number 6 inside the innermost region. Then work outward to find the numbers of students represented by the other regions of the diagram. For example, since nine students took both precalculus and calculus and six took all three courses, $9 - 6 = 3$ students took precalculus and calculus but not Java. Similarly, since 16 students took precalculus and calculus and six took all three courses, $16 - 6 = 10$ students took precalculus and calculus but not Java. Now the total number of students who took precalculus is 30. Of these 30, three also took calculus but not Java, ten took Java but not calculus, and six took both calculus and Java. That leaves 11 students who took precalculus but neither of the other two courses.

A similar analysis can be used to fill in the numbers for the other regions of the diagram. ∎

Test Yourself

1. The addition rule says that if a finite set A equals the union of k distinct mutually disjoint subsets $A_1, A_2, \ldots, A_k$, then _____.

2. The difference rule says that if A is a finite set and B is a subset of A, then _____.

3. If S is a finite sample space and A is an event in S, then the probability of A^c equals _____.

4. The inclusion/exclusion rule for two sets says that if A and B are any finite sets, then _____.

5. The inclusion/exclusion rule for three sets says that if A, B, and C are any finite sets, then _____.

Exercise Set 9.3

1. a. How many bit strings consist of from one through four digits? (Strings of different lengths are considered distinct. Thus 10 and 0010 are distinct strings.)

b. How many bit strings consist of from five through eight digits?

2. a. How many strings of hexadecimal digits consist of from one through three digits? (Recall that hexadecimal numbers are constructed using the 16 digits 0, 1, 2, 3, 4, 5, 6, 7, 8, 9, A, B, C, D, E, F.)

b. How many strings of hexadecimal digits consist of from two through five digits?

3. a. How many integers from 1 through 999 do not have any repeated digits?

b. How many integers from 1 through 999 have at least one repeated digit?

c. What is the probability that an integer chosen at random from 1 through 999 has at least one repeated digit?

4. How many arrangements in a row of no more than three letters can be formed using the letters of the word *NETWORK* (with no repetitions allowed)?

5. a. How many five-digit integers (integers from 10,000 through 99,999) are divisible by 5?

b. What is the probability that a five-digit integer chosen at random is divisible by 5?

6. In a certain state, license plates consist of from zero to three letters followed by from zero to four digits, with the provision, however, that a blank plate is not allowed.

a. How many different license plates can the state produce?

b. Suppose 85 letter combinations are not allowed because of their potential for giving offense. How many different license plates can the state produce?

***H* 7.** In another state, all license plates consist of from four to six symbols chosen from the 26 letters of the alphabet together with the ten digits 0–9.

a. How many license plates are possible if repetition of symbols is allowed?

b. How many license plates do not contain any repeated symbol?

c. How many license plates have at least one repeated symbol?

d. What is the probability that a license plate chosen at random has a repeated symbol?

8. At a certain company, passwords must be from 3–5 symbols long and composed of the 26 letters of the alphabet, the ten digits 0–9, and the 14 symbols !,@,#,$,%,^,&, *,(,),−,+,{, and }.

a. How many passwords are possible if repetition of symbols is allowed?

b. How many passwords contain no repeated symbols?

c. How many passwords have at least one repeated symbol?

d. What is the probability that a password chosen at random has at least one repeated symbol?

✶ 9. A calculator has an eight-digit display and a decimal point that is located at the extreme right of the number displayed, at the extreme left, or between any pair of digits. The calculator can also display a minus sign at the extreme left of the number. How many distinct numbers can the calculator display? (Note that certain numbers are equal, such as 1.9, 1.90, and 01.900, and should, therefore, not be counted twice.)

10. a. How many ways can the letters of the word *QUICK* be arranged in a row?

b. How many ways can the letters of the word *QUICK* be arranged in a row if the *Q* and the *U* must remain next to each other in the order *QU*?

c. How many ways can the letters of the word *QUICK* be arranged in a row if the letters *QU* must remain together but may be in either the order *QU* or the order *UQ*?

11. a. How many ways can the letters of the word *THEORY* be arranged in a row?

b. How many ways can the letters of the word *THEORY* be arranged in a row if *T* and *H* must remain next to each other as either *TH* or *HT*?

12. A group of eight people are attending the movies together.

a. Two of the eight insist on sitting side-by-side. In how many ways can the eight be seated together in a row?

b. Two of the people do not like each other and do not want to sit side-by-side. Now how many ways can the eight be seated together in a row?

13. An early compiler recognized variable names according to the following rules: Numeric variable names had to begin with a letter, and then the letter could be followed by another letter or a digit or by nothing at all. String variable names had to begin with the symbol $ followed by a letter, which could then be followed by another letter or a digit or by nothing at all. How many distinct variable names were recognized by this compiler?

***H* 14.** Identifiers in a certain database language must begin with a letter, and then the letter may be followed by other characters, which can be letters, digits, or underscores (_). However, 82 keywords (all consisting of 15 or fewer characters) are reserved and cannot be used as identifiers. How many identifiers with 30 or fewer characters are possible? (Write the answer using summation notation and evaluate it using a formula from Section 5.2.)

15. a. If any seven digits could be used to form a telephone number, how many seven-digit telephone numbers would not have any repeated digits?

b. How many seven-digit telephone numbers would have at least one repeated digit?

c. What is the probability that a randomly chosen seven-digit telephone number would have at least one repeated digit?

16. a. How many strings of four hexadecimal digits do not have any repeated digits?
    b. How many strings of four hexadecimal digits have at least one repeated digit?
    c. What is the probability that a randomly chosen string of four hexadecimal digits has at least one repeated digit?

17. Just as the difference rule gives rise to a formula for the probability of the complement of an event, so the addition and inclusion/exclusion rules give rise to formulas for the probability of the union of mutually disjoint events and for a general union of (not necessarily mutually exclusive) events.
    a. Prove that for mutually disjoint events A and B,
    $$P(A \cup B) = P(A) + P(B).$$
    b. Prove that for any events A and B,
    $$P(A \cup B) = P(A) + P(B) - P(A \cap B).$$

***H* 18.** A combination lock requires three selections of numbers, each from 1 through 39. Suppose the lock is constructed in such a way that no number can be used twice in a row but the same number may occur both first and third. For example, 20 13 20 would be acceptable, but 20 20 13 would not. How many different combinations are possible?

✶ 19. a. How many integers from 1 through 100,000 contain the digit 6 exactly once?
    b. How many integers from 1 through 100,000 contain the digit 6 at least once?
    c. If an integer is chosen at random from 1 through 100,000, what is the probability that it contains two or more occurrences of the digit 6?

***H* ✶ 20.** Six new employees, two of whom are married to each other, are to be assigned six desks that are lined up in a row. If the assignment of employees to desks is made randomly, what is the probability that the married couple will have nonadjacent desks? (*Hint:* First find the probability that the couple will have adjacent desks, and then subtract this number from 1.)

✶ 21. Consider strings of length n over the set $\{a, b, c, d\}$.
    a. How many such strings contain at least one pair of adjacent characters that are the same?
    b. If a string of length ten over $\{a, b, c, d\}$ is chosen at random, what is the probability that it contains at least one pair of adjacent characters that are the same?

22. a. How many integers from 1 through 1,000 are multiples of 4 or multiples of 7?
    b. Suppose an integer from 1 through 1,000 is chosen at random. Use the result of part (a) to find the probability that the integer is a multiple of 4 or a multiple of 7.

c. How many integers from 1 through 1,000 are neither multiples of 4 nor multiples of 7?

23. a. How many integers from 1 through 1,000 are multiples of 2 or multiples of 9?
    b. Suppose an integer from 1 through 1,000 is chosen at random. Use the result of part (a) to find the probability that the integer is a multiple of 2 or a multiple of 9.
    c. How many integers from 1 through 1,000 are neither multiples of 2 nor multiples of 9?

24. *Counting Strings:*
    a. Make a list of all bit strings of lengths zero, one, two, three, and four that do not contain the bit pattern 111.
    b. For each integer $n \geq 0$, let d_n = the number of bit strings of length n that do not contain the bit pattern 111. Find d_0, d_1, d_2, d_3, and d_4.
    c. Find a recurrence relation for $d_0, d_1, d_2, \ldots$.
    d. Use the results of parts (b) and (c) to find the number of bit strings of length five that do not contain the pattern 111.

25. *Counting Strings:* Consider the set of all strings of a's, b's, and c's.
    a. Make a list of all of these strings of lengths zero, one, two, and three that do not contain the pattern aa.
    b. For each integer $n \geq 0$, let s_n = the number of strings of a's, b's, and c's of length n that do not contain the pattern aa. Find s_0, s_1, s_2, and s_3.
    ***H* c.** Find a recurrence relation for $s_0, s_1, s_2, \ldots$.
    d. Use the results of parts (b) and (c) to find the number of strings of a's, b's, and c's of length four that do not contain the pattern aa.

26. For each integer $n \geq 0$, let a_k be the number of bit strings of length n that do not contain the pattern 101.
    a. Show that $a_k = a_{k-1} + a_{k-3} + a_{k-4} + \cdots + a_0 + 2$, for all integers $k \geq 3$.
    b. Use the result of part (a) to show that if $k \geq 3$, then $a_k = 2a_{k-1} - a_{k-2} + a_{k-3}$.

✶ 27. For each integer $n \geq 2$ let a_n be the number of permutations of $\{1, 2, 3, \ldots, n\}$ in which no number is more than one place removed from its "natural" position. Thus $a_1 = 1$ since the one permutation of $\{1\}$, namely 1, does not move 1 from its natural position. Also $a_2 = 2$ since neither of the two permutations of $\{1,2\}$, namely 12 and 21, moves either number more than one place from its natural position.
    a. Find a_3.
    b. Find a recurrence relation for $a_1, a_2, a_3, \ldots$.

✶ 28. A row in a classroom has n seats. Let s_n be the number of ways nonempty sets of students can sit in the row so that no student is seated directly adjacent to any other student. (For instance, a row of three seats could contain a single student in any of the seats or a pair of students in the two outer seats. Thus $s_3 = 4$.) Find a recurrence relation for $s_1, s_2, s_3, \ldots$.

29. Assume that birthdays are equally likely to occur in any one of the 12 months of the year.
    a. Given a group of four people, A, B, C, and D, what is the total number of ways in which birth months could be associated with A, B, C, and D? (For instance, A and B might have been born in May, C in September, and D in February. As another example, A might have been born in January, B in June, C in March, and D in October.)
    b. How many ways could birth months be associated with A, B, C, and D so that no two people would share the same birth month?
    c. How many ways could birth months be associated with A, B, C, and D so that at least two people would share the same birth month?
    d. What is the probability that at least two people out of A, B, C, and D share the same birth month?
    e. How large must n be so that in any group of n people, the probability that two or more share the same birth month is at least 50%?

H 30. Assuming that all years have 365 days and all birthdays occur with equal probability, how large must n be so that in any randomly chosen group of n people, the probability that two or more have the same birthday is at least 1/2? (This is called the **birthday problem.** Many people find the answer surprising.)

31. A college conducted a survey to explore the academic interests and achievements of its students. It asked students to place checks beside the numbers of all the statements that were true of them. Statement #1 was "I was on the honor roll last term," statement #2 was "I belong to an academic club, such as the math club or the Spanish club," and statement #3 was "I am majoring in at least two subjects." Out of a sample of 100 students, 28 checked #1, 26 checked #2, and 14 checked #3, 8 checked both #1 and #2, 4 checked both #1 and #3, 3 checked both #2 and #3, and 2 checked all three statements.
    a. How many students checked at least one of the statements?
    b. How many students checked none of the statements?
    c. Let H be the set of students who checked #1, C the set of students who checked #2, and D the set of students who checked #3. Fill in the numbers for all eight regions of the diagram below.

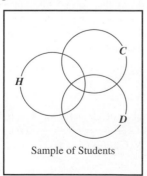

Sample of Students

    d. How many students checked #1 and #2 but not #3?
    e. How many students checked #2 and #3 but not #1?
    f. How many students checked #2 but neither of the other two?

32. A study was done to determine the efficacy of three different drugs—A, B, and C—in relieving headache pain. Over the period covered by the study, 50 subjects were given the chance to use all three drugs. The following results were obtained:

    21 reported relief from drug A.

    21 reported relief from drug B.

    31 reported relief from drug C.

     9 reported relief from both drugs A and B.

    14 reported relief from both drugs A and C.

    15 reported relief from both drugs B and C.

    41 reported relief from at least one of the drugs.

    Note that some of the 21 subjects who reported relief from drug A may also have reported relief from drugs B or C. A similar occurrence may be true for the other data.
    a. How many people got relief from none of the drugs?
    b. How many people got relief from all three drugs?
    c. Let A be the set of all subjects who got relief from drug A, B the set of all subjects who got relief from drug B, and C the set of all subjects who got relief from drug C. Fill in the numbers for all eight regions of the diagram below.

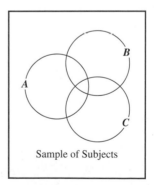

Sample of Subjects

    d. How many subjects got relief from A only?

33. An interesting use of the inclusion/exclusion rule is to check survey numbers for consistency. For example, suppose a public opinion polltaker reports that out of a national sample of 1,200 adults, 675 are married, 682 are from 20 to 30 years old, 684 are female, 195 are married and are from 20 to 30 years old, 467 are married females, 318 are females from 20 to 30 years old, and 165 are married females from 20 to 30 years old. Are the polltaker's figures consistent? Could they have occurred as a result of an actual sample survey?

34. Fill in the reasons for each step below. If A and B are sets in a finite universe U, then

$$N(A \cap B) = N(U) - N((A \cap B)^c) \qquad \underline{\quad\text{(a)}\quad}$$
$$= N(U) - N(A^c \cup B^c) \qquad \underline{\quad\text{(b)}\quad}$$
$$= N(U) - (N(A^c) + N(B^c) - N(A^c \cap B^c)) \quad \underline{\text{(c)}}.$$

For each of exercises 35–37 below, the number of elements in a certain set can be found by computing the number in some larger universe that are not in the set and subtracting this from the total. In each case, as indicated by exercise 34, De Morgan's laws and the inclusion/exclusion rule can be used to compute the number that are not in the set.

35. How many positive integers less than 1,000 have no common factors with 1,000?

✱ 36. How many permutations of *abcde* are there in which the first character is a, b, or c and the last character is c, d, or e?

✱ 37. How many integers from 1 through 999,999 contain each of the digits 1, 2, and 3 at least once? (*Hint:* For each $i = 1, 2$, and 3, let A_i be the set of all integers from 1 through 999,999 that do not contain the digit i.)

For 38 and 39, use the definition of the Euler phi function ϕ on page 304.

H 38. Use the inclusion/exclusion principle to prove the following: If $n = pq$, where p and q are distinct prime numbers, then $\varphi(n) = (p-1)(q-1)$.

39. Use the inclusion/exclusion principle to prove the following: If $n = pqr$, where p, q, and r are distinct prime numbers, then $\varphi(n) = (p-1)(q-1)(r-1)$.

40. A gambler decides to play successive games of blackjack until he loses three times in a row. (Thus the gambler could play five games by losing the first, winning the second, and losing the final three or by winning the first two and losing the final three. These possibilities can be symbolized as *LWLLL* and *WWLLL*.) Let g_n be the number of ways the gambler can play n games.
    a. Find g_3, g_4, and g_5.
    b. Find g_6.
    H c. Find a recurrence relation for g_3, g_4, g_5, ...

✱ 41. A *derangement* of the set $\{1, 2, \ldots, n\}$ is a permutation that moves every element of the set away from its "natural" position. Thus 21 is a derangement of $\{1, 2\}$, and 231 and 312 are derangements of $\{1, 2, 3\}$. For each positive integer n, let d_n be the number of derangements of the set

$\{1, 2, \ldots, n\}$.
    a. Find d_1, d_2, and d_3.
    b. Find d_4.
    H c. Find a recurrence relation for d_1, d_2, d_3, ...

42. Note that a product $x_1 x_2 x_3$ may be parenthesized in two different ways: $(x_1 x_2) x_3$ and $x_1 (x_2 x_3)$. Similarly, there are several different ways to parenthesize $x_1 x_2 x_3 x_4$. Two such ways are $(x_1 x_2)(x_3 x_4)$ and $x_1((x_2 x_3) x_4)$. Let P_n be the number of different ways to parenthesize the product $x_1 x_2 \ldots x_4$. Show that if $P_1 = 1$, then

$$P_n = \sum_{k=1}^{n-1} P_k P_{n-k} \quad \text{for all integers } n \geq 2.$$

(It turns out that the sequence P_1, P_2, P_3, ... is the same as the sequence of Catalan numbers: $P_n = C_{n-1}$ for all integers $n \geq 1$. See Example 5.5.4.)

43. Use mathematical induction to prove Theorem 9.3.1.

44. Prove the inclusion/exclusion rule for two sets A and B by showing that $A \cup B$ can be partitioned into $A \cap B$, $A - (A \cap B)$, and $B - (A \cap B)$, and then using the addition and difference rules.

45. Prove the inclusion/exclusion rule for three sets.

H ✱ 46. Use mathematical induction to prove the general inclusion/exclusion rule:

If $A_1, A_2, \ldots, A_n$ are finite sets, then
$N(A_1 \cup A_2 \cup \cdots \cup A_n)$

$$= \sum_{1 \leq i \leq n} N(A_i) - \sum_{1 \leq i < j \leq n} N(A_i \cap A_j)$$

$$+ \sum_{1 \leq i < j < k \leq n} N(A_i \cap A_j \cap A_k)$$

$$- \cdots + (-1)^{n+1} N(A_1 \cap A_2 \cap \cdots \cap A_n).$$

(The notation $\sum_{1 \leq i < j \leq n} N(A_i \cap A_j)$ means that quantities of the form $N(A_i \cap A_j)$ are to be added together for all integers i and j with $1 \leq i < j \leq n$.)

✱ 47. A circular disk is cut into n distinct sectors, each shaped like a piece of pie and all meeting at the center point of the disk. Each sector is to be painted red, green, yellow, or blue in such a way that no two adjacent sectors are painted the same color. Let S_n be the number of ways to paint the disk. Find a recurrence relation for S_k in terms of S_{k-1} and S_{k-2} for each integer $k \geq 4$.

Answers for Test Yourself

1. the number of elements in A equals $N(A_1) + N(A_2) + \ldots + N(A_n)$ 2. the number of elements in $A - B$ is the difference between the number of elements in A and the number of elements in B, that is, $N(A - B) = N(A) - N(B)$.   3. $1 - P(A)$
4. $N(A \cup B) = N(A) + N(B) - N(A \cap B)$ 5. $N(A \cup B \cup C) = N(A) + N(B) + N(C) - N(A \cap B) - N(A \cap C) - N(B \cap C) + N(A \cap B \cap C)$

9.4 *The Pigeonhole Principle*

*The shrewd guess, the fertile hypothesis, the courageous leap to a tentative
conclusion—these are the most valuable coin of the thinker at work*
— Jerome S. Bruner, 1960

The pigeonhole principle states that if n pigeons fly into m pigeonholes and $n > m$, then at least one hole must contain two or more pigeons. This principle is illustrated in Figure 9.4.1 for $n = 5$ and $m = 4$. Illustration (a) shows the pigeons perched next to their holes, and (b) shows the correspondence from pigeons to pigeonholes. The pigeonhole principle is sometimes called the *Dirichlet box principle* because it was first stated formally by J. P. G. L. Dirichlet (1805–1859).

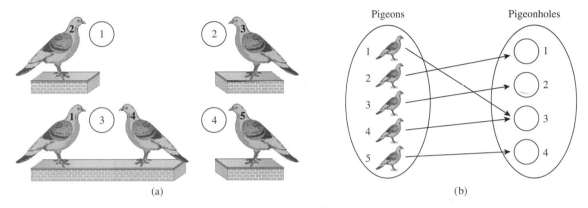

(a)   (b)

Figure 9.4.1

Illustration (b) suggests the following mathematical way to phrase the principle.

Pigeonhole Principle

A function from one finite set to a smaller finite set cannot be one-to-one: There must be a least two elements in the domain that have the same image in the co-domain.

Thus an arrow diagram for a function from a finite set to a smaller finite set must have at least two arrows from the domain that point to the same element of the co-domain. In Figure 9.4.1(b), arrows from pigeons 1 and 4 both point to pigeonhole 3.

Since the truth of the pigeonhole principle is easy to accept on an intuitive basis, we move immediately to applications, leaving a formal proof to the end of the section. Applications of the pigeonhole principle range from the totally obvious to the extremely subtle. A representative sample is given in the examples and exercises that follow.

Example 9.4.1 Applying the Pigeonhole Principle

a. In a group of six people, must there be at least two who were born in the same month? In a group of thirteen people, must there be at least two who were born in the same month? Why?

b. Among the residents of New York City, must there be at least two people with the same number of hairs on their heads? Why?

Solution

a. A group of six people need not contain two who were born in the same month. For instance, the six people could have birthdays in each of the six months January through June.

A group of thirteen people, however, must contain at least two who were born in the same month, for there are only twelve months in a year and $13 > 12$. To get at the essence of this reasoning, think of the thirteen people as the pigeons and the twelve months of the year as the pigeonholes. Denote the thirteen people by the symbols $x_1, x_2, \ldots, x_{13}$ and define a function B from the set of people to the set of twelve months as shown in the following arrow diagram.

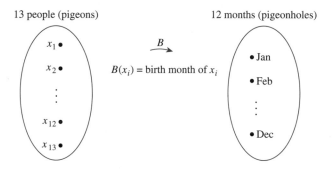

The pigeonhole principle says that no matter what the particular assignment of months to people, there must be at least two arrows pointing to the same month. Thus at least two people must have been born in the same month.

b. The answer is yes. In this example the pigeons are the people of New York City and the pigeonholes are all possible numbers of hairs on any individual's head. Call the population of New York City P. It is known that P is at least 5,000,000. Also the maximum number of hairs on any person's head is known to be no more than 300,000. Define a function H from the set of people in New York City $\{x_1, x_2, \ldots, x_p\}$ to the set $\{0, 1, 2, 3, \ldots, 300\,000\}$, as shown below.

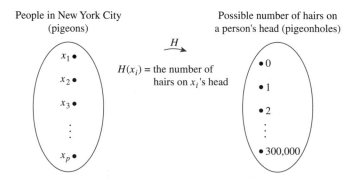

Since the number of people in New York City is larger than the number of possible hairs on their heads, the function H is not one-to-one; at least two arrows point to the same number. But that means that at least two people have the same number of hairs on their heads. ∎

Example 9.4.2 Finding the Number to Pick to Ensure a Result

A drawer contains ten black and ten white socks. You reach in and pull some out without looking at them. What is the *least* number of socks you must pull out to be sure to get a matched pair? Explain how the answer follows from the pigeonhole principle.

Solution   If you pick just two socks, they may have different colors. But when you pick a third sock, it must be the same color as one of the socks already chosen. Hence the answer is three.

This answer could be phrased more formally as follows: Let the socks pulled out be denoted $s_1, s_2, s_3, \ldots, s_n$ and consider the function C that sends each sock to its color, as shown below.

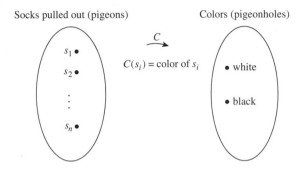

If $n = 2$, C could be a one-to-one correspondence (if the two socks pulled out were of different colors). But if $n > 2$, then the number of elements in the domain of C is larger than the number of elements in the co-domain of C. Thus by the pigeonhole principle, C is not one-to-one: $C(s_i) = C(s_j)$ for some $s_i \neq s_j$. This means that if at least three socks are pulled out, then at least two of them have the same color. ∎

Example 9.4.3 Selecting a Pair of Integers with a Certain Sum

Let $A = \{1, 2, 3, 4, 5, 6, 7, 8\}$.

a. If five integers are selected from A, must at least one pair of the integers have a sum of 9?

b. If four integers are selected from A, must at least one pair of the integers have a sum of 9?

Solution

a. Yes. Partition the set A into the following four disjoint subsets:

$$\{1, 8\}, \quad \{2, 7\}, \quad \{3, 6\}, \quad \text{and} \quad \{4, 5\}$$

Observe that each of the integers in A occurs in exactly one of the four subsets and that the sum of the integers in each subset is 9. Thus if five integers from A are chosen, then by the pigeonhole principle, two must be from the same subset. It follows that the sum of these two integers is 9.

To see precisely how the pigeonhole principle applies, let the pigeons be the five selected integers (call them $a_1, a_2, a_3, a_4,$ and a_5) and let the pigeonholes be the subsets of the partition. The function P from pigeons to pigeonholes is defined by letting $P(a_i)$ be the subset that contains a_i.

The 5 selected integers (pigeons)

The 4 subsets in the partition of A (pigeonholes)

$$
\begin{array}{ccc}
a_1 \bullet & & \bullet\{1, 8\} \\
a_2 \bullet & \xrightarrow{\ P\ } & \bullet\{2, 7\} \\
a_3 \bullet & P(a_i) = \text{the subset that} & \bullet\{3, 6\} \\
a_4 \bullet & \text{contains } a_i & \bullet\{4, 5\} \\
a_5 \bullet & &
\end{array}
$$

The function P is well defined because for each integer a_i in the domain, a_i belongs to one of the subsets (since the union of the subsets is A) and a_i does not belong to more than one subset (since the subsets are disjoint).

Because there are more pigeons than pigeonholes, at least two pigeons must go to the same hole. Thus two distinct integers are sent to the same set. But that implies that those two integers are the two distinct elements of the set, so their sum is 9. More formally, by the pigeonhole principle, since P is not one-to-one, there are integers a_i and a_j such that

$$P(a_i) = P(a_j) \quad \text{and} \quad a_i \neq a_j.$$

But then, by definition of P, a_i and a_j belong to the same subset. Since the elements in each subset add up to 9, $a_i + a_j = 9$.

b. The answer is no. This is a case where the pigeonhole principle does not apply; the number of pigeons is not larger than the number of pigeonholes. For instance, if you select the numbers $1, 2, 3,$ and 4, then since the largest sum of any two of these numbers is 7, no two of them add up to 9. ∎

Application to Decimal Expansions of Fractions

One important consequence of the piegonhole principle is the fact that

the decimal expansion of any rational number either terminates or repeats.

A terminating decimal is one like

$$3.625,$$

and a repeating decimal is one like

$$2.38\overline{246},$$

where the bar over the digits 246 means that these digits are repeated forever.

Note Strictly speaking, a terminating decimal like 3.625 can be regarded as a repeating decimal by adding trailing zeros: $3.625 = 3.625\overline{0}$. This can also be written as $3.624\overline{9}$.

Recall that a rational number is one that can be written as a ratio of integers—in other words, as a fraction. Recall also that the decimal expansion of a fraction is obtained by dividing its numerator by its denominator using long division. For example, the decimal expansion of 4/33 is obtained as follows:

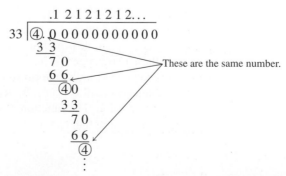

These are the same number.

Because the number 4 reappears as a remainder in the long-division process, the sequence of quotients and remainders that give the digits of the decimal expansion repeats forever; hence the digits of the decimal expansion repeat forever.

In general, when one integer is divided by another, it is the pigeonhole principle (together with the quotient-remainder theorem) that guarantees that such a repetition of remainders and hence decimal digits must always occur. This is explained in the following example. The analysis in the example uses an obvious generalization of the pigeonhole principle, namely that a function from an infinite set to a finite set cannot be one-to-one.

Example 9.4.4 The Decimal Expansion of a Fraction

Consider a fraction a/b, where for simplicity a and b are both assumed to be positive. The decimal expansion of a/b is obtained by dividing the a by the b as illustrated here for $a = 3$ and $b = 14$.

$$
\begin{array}{r}
.2\,1\,4\,2\,8\,5\,7\,1\,4\,2\,8\,5\,7\ldots \\
\hline
14\,\overline{)3.0\,0\,0\,0\,0\,0\,0\,0\,0\,0\,0\,0\,0\,0\,0\,0\,0}
\end{array}
$$

with successive remainders:
$r_0 = 3$
$r_1 = 2$
$r_2 = 6$
$r_3 = 4$
$r_4 = 12$
$r_5 = 8$
$r_6 = 10$
$r_7 = 2 = r_1$
$r_8 = 6 = r_2$
$r_9 = 4 = r_3$

Let $r_0 = a$ and let $r_1, r_2, r_3, \ldots$ be the successive remainders obtained in the long division of a by b. By the quotient-remainder theorem, each remainder must be between 0 and $b - 1$. (In this example, a is 3 and b is 14, and so the remainders are from 0 to 13.) If some remainder $r_i = 0$, then the division terminates and a/b has a terminating decimal expansion. If no $r_i = 0$, then the division process and hence the sequence of remainders continues forever. By the pigeonhole principle, since there are more remainders than values that the remainders can take, some remainder value must repeat: $r_j = r_k$, for some indices j and k with $j < k$. This is illustrated below for $a = 3$ and $b = 14$.

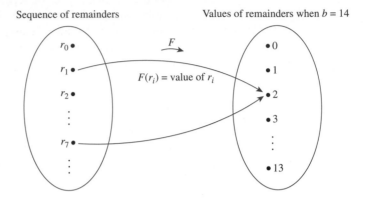

If follows that the decimal digits obtained from the divisions between r_j and r_{k-1} repeat forever. In the case of $3/14$, the repetition begins with $r_7 = 2 = r_1$ and the decimal expansion repeats the quotients obtained from the divisions from r_1 through r_6 forever: $3/14 = 0.2\overline{142857}$. ■

Note that since the decimal expansion of any rational number either terminates or repeats, if a number has a decimal expansion that neither terminates nor repeats, then it cannot be rational. Thus, for example, the following number cannot be rational: $0.01011011101111011111\ldots$ (where each string of 1's is one longer than the previous string).

Generalized Pigeonhole Principle

A generalization of the pigeonhole principle states that if n pigeons fly into m pigeonholes and, for some positive integer k, $k < n/m$, then at least one pigeonhole contains $k + 1$ or more pigeons. This is illustrated in Figure 9.4.2 for $m = 4, n = 9$, and $k = 2$. Since $2 < 9/4 = 2.25$, at least one pigeonhole contains three $(2 + 1)$ or more pigeons. (In this example, pigeonhole 3 contains three pigeons.)

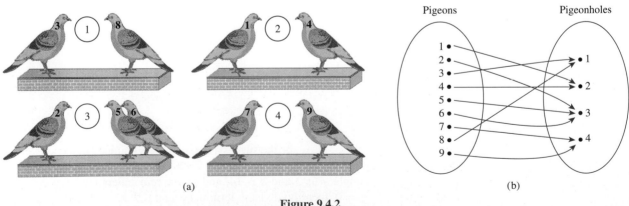

Figure 9.4.2

Generalized Pigeonhole Principle

For any function f from a finite set X with n elements to a finite set Y with m elements and for any positive integer k, if $k < n/m$, then there is some $y \in Y$ such that y is the image of at least $k + 1$ distinct elements of X.

Example 9.4.5 Applying the Generalized Pigeonhole Principle

Show how the generalized pigeonhole principle implies that in a group of 85 people, at least 4 must have the same last initial.

Solution   In this example the pigeons are the 85 people and the pigeonholes are the 26 possible last initials of their names. Note that

$$3 < 85/26 \cong 3.27.$$

Consider the function L from people to initials defined by the following arrow diagram.

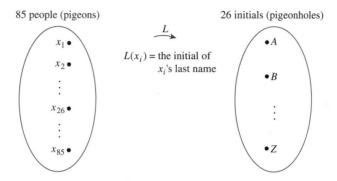

Since $3 < 85/26$, the generalized pigeonhole principle states that some initial must be the image of at least four $(3 + 1)$ people. Thus at least four people have the same last initial. ■

Consider the following contrapositive form of the generalized pigeonhole principle.

Generalized Pigeonhole Principle (Contrapositive Form)

For any function f from a finite set X with n elements to a finite set Y with m elements and for any positive integer k, if for each $y \in Y$, $f^{-1}(y)$ has at most k elements, then X has at most km elements; in other words, $n \leq km$.

You may find it natural to use the contrapositive form of the generalized pigeonhole principle in certain situations. For instance, the result of Example 9.4.5 can be explained as follows:

Suppose no 4 people out of the 85 had the same last initial. Then at most 3 would share any particular one. By the generalized pigeonhole principle (contrapositive form), this would imply that the total number of people is at most $3 \cdot 26 = 78$. But this contradicts the fact that there are 85 people in all. Hence at least 4 people share a last initial.

Example 9.4.6 Using the Contrapositive Form of the Generalized Pigeonhole Principle

There are 42 students who are to share 12 computers. Each student uses exactly 1 computer, and no computer is used by more than 6 students. Show that at least 5 computers are used by 3 or more students.

Solution

a. *Using an Argument by Contradiction:* Suppose not. Suppose that 4 or fewer computers are used by 3 or more students. *[A contradiction will be derived.]* Then 8 or more computers are used by 2 or fewer students. Divide the set of computers into two subsets: C_1 and C_2. Into C_1 place 8 of the computers used by 2 or fewer students; into C_2 place the computers used by 3 or more students plus any remaining computers (to make a total of 4 computers in C_2). (See Figure 9.4.3.)

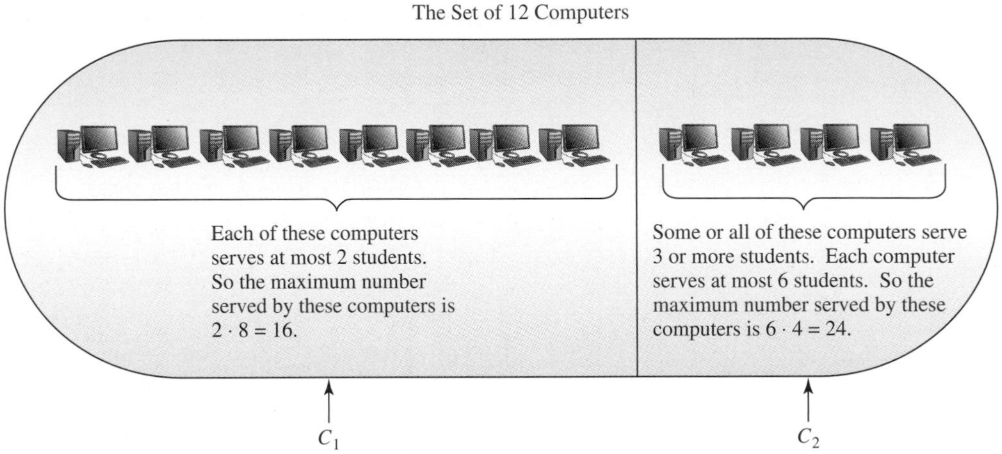

The Set of 12 Computers

Each of these computers serves at most 2 students. So the maximum number served by these computers is $2 \cdot 8 = 16$.

Some or all of these computers serve 3 or more students. Each computer serves at most 6 students. So the maximum number served by these computers is $6 \cdot 4 = 24$.

C_1 C_2

Figure 9.4.3

Since at most 6 students are served by any one computer, by the contrapositive form of the generalized pigeonhole principle, the computers in set C_2 serve at most $6 \cdot 4 = 24$ students. Since at most 2 students are served by any one computer in C_1, by the generalized pigeonhole principle (contrapositive form), the computers in set C_1 serve at most $2 \cdot 8 = 16$ students. Hence the total number of students served by the computers is $24 + 16 = 40$. But this contradicts the fact that each of the 42 students is served by a computer. Therefore, the supposition is false: At least 5 computers are used by 3 or more students.

b. *Using a Direct Argument:* Let k be the number of computers used by 3 or more students. *[We must show that $k \geq 5$.]* Because each computer is used by at most 6 students, these computers are used by at most $6k$ students (by the contrapositive form of the generalized pigeonhole principle). Each of the remaining $12 - k$ computers is used by at most 2 students. Hence, taken together, they are used by at most $2(12 - k) = 24 - 2k$ students (again, by the contrapositive form of the generalized pigeonhole principle). Thus the maximum number of students served by the computers is $6k + (24 - 2k) = 4k + 24$. Because 42 students are served by the computers, $4k + 24 \geq 42$. Solving for k gives that $k \geq 4.5$, and since k is an integer, this implies that $k \geq 5$ *[as was to be shown]*. ∎

Proof of the Pigeonhole Principle

The truth of the pigeonhole principle depends essentially on the sets involved being finite. Recall from Section 7.4 that a set is called **finite** if, and only if, it is the empty set or there is a one-to-one correspondence from $\{1, 2, \ldots, n\}$ to it, where n is a positive integer. In the first case the **number of elements** in the set is said to be 0, and in the second case it is said to be n. A set that is not finite is called **infinite.**

Thus any finite set is either empty or can be written in the form $\{x_1, x_2, \ldots, x_n\}$ where n is a positive integer.

Theorem 9.4.1 The Pigeonhole Principle

For any function f from a finite set X with n elements to a finite set Y with m elements, if $n > m$, then f is not one-to-one.

Proof:

Suppose f is any function from a finite set X with n elements to a finite set Y with m elements where $n > m$. Denote the elements of Y by $y_1, y_2, \ldots, y_m$. Recall that for each y_i in Y, the inverse image set $f^{-1}(y_i) = \{x \in X \mid f(x) = y_i\}$. Now consider the collection of all the inverse image sets for all the elements of Y:

$$f^{-1}(y_1), f^{-1}(y_2), \ldots, f^{-1}(y_m).$$

By definition of function, each element of X is sent by f to some element of Y. Hence each element of X is in one of the inverse image sets, and so the union of all these sets equals X. But also, by definition of function, no element of X is sent by f to more than one element of Y. Thus each element of X is in only one of the inverse image sets, and so the inverse image sets are mutually disjoint. By the addition rule, therefore,

$$N(X) = N(f^{-1}(y_1)) + N(f^{-1}(y_2)) + \cdots + N(f^{-1}(y_m)). \qquad 9.4.1$$

Now suppose that f *is* one-to-one *[which is the opposite of what we want to prove]*. Then each set $f^{-1}(y_i)$ has at most one element, and so

$$N(f^{-1}(y_1)) + N(f^{-1}(y_2)) + \cdots + N(f^{-1}(y_m)) \le \underbrace{1 + 1 + \cdots + 1}_{m \text{ terms}} = m \qquad 9.4.2$$

Putting equations (9.4.1) and (9.4.2) together gives that

$$n = N(X) \le m = N(Y).$$

This contradicts the fact that $n > m$, and so the supposition that f is one-to-one must be false. Hence f is not one-to-one *[as was to be shown]*.

An important theorem that follows from the pigeonhole principle states that a function from one finite set to another finite set of the same size is one-to-one if, and only if, it is onto. As shown in Section 7.4, this result does not hold for infinite sets.

Theorem 9.4.2 One-to-One and Onto for Finite Sets

Let X and Y be finite sets with the same number of elements and suppose f is a function from X to Y. Then f is one-to-one if, and only if, f is onto.

Proof:

Suppose f is a function from X to Y, where X and Y are finite sets each with m elements. Let $X = \{x_1, x_2, \ldots, x_m\}$ and $Y = \{y_1, y_2, \ldots, y_m\}$.

If f is one-to-one, then f is onto: Suppose f is one-to-one. Then $f(x_1), f(x_2), \ldots, f(x_m)$ are all distinct. Consider the set S of all elements of Y that are not the image of any element of X.

(continued on page 444)

Then the sets

$$\{f(x_1)\}, \{f(x_2)\}, \ldots, \{f(x_m)\} \quad \text{and} \quad S$$

are mutually disjoint. By the addition rule,

$$N(Y) = N(\{f(x_1)\}) + N(\{f(x_2)\}) + \cdots + N(\{f(x_m)\}) + N(S)$$

$$= \underbrace{1 + 1 + \cdots + 1}_{m \text{ terms}} + N(S) \quad \text{because each } \{f(x_i)\}$$
$$\text{is a singleton set}$$

$$= m + N(S).$$

Thus

$$m = m + N(S) \qquad \text{because } N(Y) = m,$$
$$\Rightarrow N(S) = 0 \qquad \text{by subtracting } m \text{ from both sides.}$$

Hence S is empty, and so there is no element of Y that is not the image of some element of X. Consequently, f is onto.

If f is onto, then f is one-to-one: Suppose f is onto. Then $f^{-1}(y_i) \neq \emptyset$ and so $N(f^{-1}(y_i)) \geq 1$ for all $i = 1, 2, \ldots, m$. As in the proof of the pigeonhole principle (Theorem 9.4.1), X is the union of the mutually disjoint sets $f^{-1}(y_1), f^{-1}(y_2), \ldots, f^{-1}(y_m)$. By the addition principle,

$$N(X) = \underbrace{N(f^{-1}(y_1)) + N(f^{-1}(y_2)) + \cdots + N(f^{-1}(y_m))}_{m \text{ terms, each} \geq 1} \geq m. \qquad 9.4.3$$

Now if any one of the sets $f^{-1}(y_i)$ has more than one element, then the sum in equation (9.4.3) is greater than m. But we know this is not the case because $N(X) = m$. Hence each set $f^{-1}(y_i)$ has exactly one element, and thus f is one-to-one *[as was to be shown].*

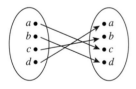

Note that Theorem 9.4.2 applies in particular to the case $X = Y$. Thus a one-to-one function from a finite set to itself is onto, and an onto function from a finite set to itself is one-to-one. Such functions are permutations of the sets on which they are defined. For instance, the function defined by the diagram on the left is another representation for the permutation *cdba* obtained by listing the images of a, b, c, and d in order.

Test Yourself

1. The pigeonhole principle states that _____.

2. The generalized pigeonhole principle states that _____.

3. If X and Y are finite sets and f is a function from X to Y then f is one-to-one if, and only if, _____

Exercise Set 9.4

1. a. If 4 cards are selected from a standard 52-card deck, must at least 2 be of the same suit? Why?

   b. If 5 cards are selected from a standard 52-card deck, must at least 2 be of the same suit? Why?

2. a. If 13 cards are selected from a standard 52-card deck, must at least 2 be of the same denomination? Why?

   b. If 20 cards are selected from a standard 52-card deck, must at least 2 be of the same denomination? Why?

3. A small town has only 500 residents. Must there be 2 residents who have the same birthday? Why?

4. In a group of 700 people, must there be 2 who have the same first and last initials? Why?

5. a. Given any set of four integers, must there be two that have the same remainder when divided by 3? Why?

 b. Given any set of three integers, must there be two that have the same remainder when divided by 3? Why?

6. a. Given any set of seven integers, must there be two that have the same remainder when divided by 6? Why?

 b. Given any set of seven integers, must there be two that have the same remainder when divided by 8? Why?

H **7.** Let $S = \{3, 4, 5, 6, 7, 8, 9, 10, 11, 12\}$. Suppose six integers are chosen from S. Must there be two integers whose sum is 15? Why?

8. Let $T = \{1, 2, 3, 4, 5, 6, 7, 8, 9\}$. Suppose five integers are chosen from T. Must there be two integers whose sum is 10? Why?

9. a. If seven integers are chosen from between 1 and 12 inclusive, must at least one of them be odd? Why?

 b. If ten integers are chosen from between 1 and 20 inclusive, must at least one of them be even? Why?

10. If $n + 1$ integers are chosen from the set

$$\{1, 2, 3, \ldots, 2n\},$$

where n is a positive integer, must at least one of them be odd? Why?

11. If $n + 1$ integers are chosen from the set

$$\{1, 2, 3, \ldots, 2n\},$$

where n is a positive integer, must at least one of them be even? Why?

12. How many cards must you pick from a standard 52-card deck to be sure of getting at least 1 red card? Why?

13. Suppose six pairs of similar-looking boots are thrown together in a pile. How many individual boots must you pick to be sure of getting a matched pair? Why?

14. How many integers from 0 through 60 must you pick in order to be sure of getting at least one that is odd? at least one that is even?

15. If n is a positive integer, how many integers from 0 through $2n$ must you pick in order to be sure of getting at least one that is odd? at least one that is even?

16. How many integers from 1 through 100 must you pick in order to be sure of getting one that is divisible by 5?

17. How many integers must you pick in order to be sure that at least two of them have the same remainder when divided by 7?

18. How many integers must you pick in order to be sure that at least two of them have the same remainder when divided by 15?

19. How many integers from 100 through 999 must you pick in order to be sure that at least two of them have a digit

in common? (For example, 256 and 530 have the common digit 5.)

20. a. If repeated divisions by 20,483 are performed, how many distinct remainders can be obtained?

 b. When 5/20483 is written as a decimal, what is the maximum length of the repeating section of the representation?

21. When 683/1493 is written as a decimal, what is the maximum length of the repeating section of the representation?

22. Is 0.101001000100001000001 ... (where each string of 0's is one longer than the previous one) rational or irrational?

23. Is 56.556655566655556666 ... (where the strings of 5's and 6's become longer in each repetition) rational or irrational?

24. Show that within any set of thirteen integers chosen from 2 through 40, there are at least two integers with a common divisor greater than 1.

25. In a group of 30 people, must at least 3 have been born in the same month? Why?

26. In a group of 30 people, must at least 4 have been born in the same month? Why?

27. In a group of 2,000 people, must at least 5 have the same birthday? Why?

28. A programmer writes 500 lines of computer code in 17 days. Must there have been at least 1 day when the programmer wrote 30 or more lines of code? Why?

29. A certain college class has 40 students. All the students in the class are known to be from 17 through 34 years of age. You want to make a bet that the class contains at least x students of the same age. How large can you make x and yet be sure to win your bet?

30. A penny collection contains twelve 1967 pennies, seven 1968 pennies, and eleven 1971 pennies. If you are to pick some pennies without looking at the dates, how many must you pick to be sure of getting at least five pennies from the same year?

H **31.** A group of 15 executives are to share 5 assistants. Each executive is assigned exactly 1 assistant, and no assistant is assigned to more than 4 executives. Show that at least 3 assistants are assigned to 3 or more executives.

H ✶ **32.** Let A be a set of six positive integers each of which is less than 13. Show that there must be two distinct subsets of A whose elements when added up give the same sum. (For example, if $A = \{5, 12, 10, 1, 3, 4\}$, then the elements of the subsets $S_1 = \{1, 4, 10\}$ and $S_2 = \{5, 10\}$ both add up to 15.)

H **33.** Let A be a set of six positive integers each of which is less than 15. Show that there must be two distinct subsets of A

whose elements when added up give the same sum. (Thanks to Jonathan Goldstine for this problem.)

34. Let S be a set of ten integers chosen from 1 through 50. Show that the set contains at least two different (but not necessarily disjoint) subsets of four integers that add up to the same number. (For instance, if the ten numbers are {3, 8, 9, 18, 24, 34, 35, 41, 44, 50}, the subsets can be taken to be {8, 24, 34, 35} and {9, 18, 24, 50}. The numbers in both of these add up to 101.)

H ✳ **35.** Given a set of 52 distinct integers, show that there must be 2 whose sum or difference is divisible by 100.

H ✳ **36.** Show that if 101 integers are chosen from 1 to 200 inclusive, there must be 2 with the property that one is divisible by the other.

✳ **37.** a. Suppose $a_1, a_2, \ldots, a_n$ is a sequence of n integers none of which is divisible by n. Show that at least one of the differences $a_i - a_j$ (for $i \neq j$) must be divisible by n.

H b. Show that every finite sequence $x_1, x_2, \ldots, x_n$ of n integers has a consecutive subsequence $x_{i+1}, x_{i+2}, \ldots, x_j$ whose sum is divisible by n. (For instance, the sequence 3, 4, 17, 7, 16 has the consecutive subsequence 17, 7, 16 whose sum is divisible by 5.) (From: James E. Schultz and William F. Burger, "An Approach to Problem-Solving Using Equivalence Classes Modulo n," *College Mathematics Journal (15)*, No. 5, 1984, 401–405.)

H ✳ **38.** Observe that the sequence 12, 15, 8, 13, 7, 18, 19, 11, 14, 10 has three increasing subsequences of length four: 12, 15, 18, 19; 12, 13, 18, 19; and 8, 13, 18, 19. It also has one decreasing subsequence of length four: 15, 13, 11, 10. Show that in any sequence of $n^2 + 1$ distinct real numbers, there must be a sequence of length $n + 1$ that is either strictly increasing or strictly decreasing.

✳ **39.** What is the largest number of elements that a set of integers from 1 through 100 can have so that no one element in the set is divisible by another? (*Hint:* Imagine writing all the numbers from 1 through 100 in the form $2^k \cdot m$, where $k \geq 0$ and m is odd.)

Answers for Test Yourself

1. if n pigeons fly into m pigeonholes and $n > m$, then at least two pigeons fly into the same pigeonhole   *Or:* a function from one finite set to a smaller finite set cannot be one-to-one   2. if n pigeons fly into m pigeonholes and, for some positive integer k, $k < n/m$, then at least one pigeonhole contains $k + 1$ or more pigeons   *Or:* for any function f from a finite set X with n elements to a finite set Y with m elements and for any positive integer k, if $k < n/m$, then there is some $y \in Y$ such that y is the image of at least $k + 1$ distinct elements of Y 3. f is onto

9.5 Counting Subsets of a Set: Combinations

"But 'glory' doesn't mean 'a nice knock-down argument,'" Alice objected. "When I use a word," Humpty Dumpty said, in rather a scornful tone, "it means just what I choose it to mean—neither more nor less." —Lewis Carroll, Through the Looking Glass, 1872

Consider the following question:

Suppose five members of a group of twelve are to be chosen to work as a team on a special project. How many distinct five-person teams can be selected?

This question is answered in Example 9.5.4. It is a special case of the following more general question:

Given a set S with n elements, how many subsets of size r can be chosen from S?

The number of subsets of size r that can be chosen from S equals the number of subsets of size r that S has. Each individual subset of size r is called an *r-combination* of the set.

> **• Definition**
>
> Let n and r be nonnegative integers with $r \leq n$. An **r-combination** of a set of n elements is a subset of r of the n elements. As indicated in Section 5.1, the symbol
>
> $$\binom{n}{r},$$
>
> which is read "n choose r," denotes the number of subsets of size r (r-combinations) that can be chosen from a set of n elements.

Recall from Section 5.1 that calculators generally use symbols like $C(n, r)$, $_nC_r$, $C_{n,r}$, or ^nC_r instead of $\binom{n}{r}$.

Example 9.5.1 3-Combinations

Let $S = \{$Ann, Bob, Cyd, Dan$\}$. Each committee consisting of three of the four people in S is a 3-combination of S.

a. List all such 3-combinations of S.   b. What is $\binom{4}{3}$?

Solution

a. Each 3-combination of S is a subset of S of size 3. But each subset of size 3 can be obtained by leaving out one of the elements of S. The 3-combinations are

$$\{Bob, Cyd, Dan\} \quad \text{leave out Ann}$$
$$\{Ann, Cyd, Dan\} \quad \text{leave out Bob}$$
$$\{Ann, Bob, Dan\} \quad \text{leave out Cyd}$$
$$\{Ann, Bob, Cyd\} \quad \text{leave out Dan.}$$

b. Because $\binom{4}{3}$ is the number of 3-combinations of a set with four elements, by part (a),

$$\binom{4}{3} = 4. \qquad\blacksquare$$

There are two distinct methods that can be used to select r objects from a set of n elements. In an **ordered selection,** it is not only what elements are chosen but also the order in which they are chosen that matters. Two ordered selections are said to be the same if the elements chosen are the same and also if the elements are chosen in the same order. An ordered selection of r elements from a set of n elements is an r-permutation of the set.

In an **unordered selection,** on the other hand, it is only the identity of the chosen elements that matters. Two unordered selections are said to be the same if they consist of the same elements, regardless of the order in which the elements are chosen. An unordered selection of r elements from a set of n elements is the same as a subset of size r or an r-combination of the set.

Example 9.5.2 Unordered Selections

How many unordered selections of two elements can be made from the set $\{0, 1, 2, 3\}$?

Solution   An unordered selection of two elements from $\{0, 1, 2, 3\}$ is the same as a 2-combination, or subset of size 2, taken from the set. These can be listed systematically:

$$\{0, 1\}, \{0, 2\}, \{0, 3\} \qquad \text{subsets containing 0}$$
$$\{1, 2\}, \{1, 3\} \qquad \text{subsets containing 1 but not already listed}$$
$$\{2, 3\} \qquad \text{subsets containing 2 but not already listed.}$$

Since this listing exhausts all possibilities, there are six subsets in all. Thus $\binom{4}{2} = 6$, which is the number of unordered selections of two elements from a set of four.   ∎

When the values of n and r are small, it is reasonable to calculate values of $\binom{n}{r}$ using the method of **complete enumeration** (listing all possibilities) illustrated in Examples 9.5.1 and 9.5.2. But when n and r are large, it is not feasible to compute these numbers by listing and counting all possibilities.

The general values of $\binom{n}{r}$ can be found by a somewhat indirect but simple method. An equation is derived that contains $\binom{n}{r}$ as a factor. Then this equation is solved to obtain a formula for $\binom{n}{r}$. The method is illustrated by Example 9.5.3.

Example 9.5.3 Relation between Permutations and Combinations

Write all 2-permutations of the set $\{0, 1, 2, 3\}$. Find an equation relating the number of 2-permutations, $P(4, 2)$, and the number of 2-combinations, $\binom{4}{2}$, and solve this equation for $\binom{4}{2}$.

Solution   According to Theorem 9.2.3, the number of 2-permutations of the set $\{0, 1, 2, 3\}$ is $P(4, 2)$, which equals

$$\frac{4!}{(4 - 2)!} = \frac{4 \cdot 3 \cdot \cancel{2} \cdot \cancel{1}}{\cancel{2} \cdot \cancel{1}} = 12.$$

Now the act of constructing a 2-permutation of $\{0, 1, 2, 3\}$ can be thought of as a two-step process:

Step 1: Choose a subset of two elements from $\{0, 1, 2, 3\}$.

Step 2: Choose an ordering for the two-element subset.

This process can be illustrated by the possibility tree shown in Figure 9.5.1.

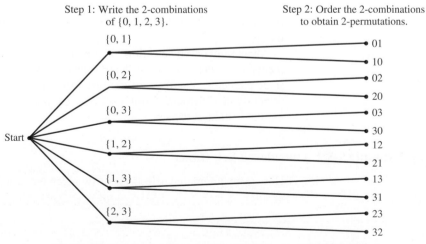

Figure 9.5.1 **Relation between Permutations and Combinations**

The number of ways to perform step 1 is $\binom{4}{2}$, the same as the number of subsets of size 2 that can be chosen from $\{0, 1, 2, 3\}$. The number of ways to perform step 2 is 2!, the number of ways to order the elements in a subset of size 2. Because the number of ways of performing the whole process is the number of 2-permutations of the set $\{0, 1, 2, 3\}$, which equals $P(4, 2)$, it follows from the product rule that

$$P(4, 2) = \binom{4}{2} \cdot 2!. \qquad \text{This is an equation that relates } P(4, 2) \text{ and } \binom{4}{2}.$$

Solving the equation for $\binom{4}{2}$ gives

$$\binom{4}{2} = \frac{P(4, 2)}{2!}$$

Recall that $P(4, 2) = \frac{4!}{(4-2)!}$. Hence, substituting yields

$$\binom{4}{2} = \frac{\dfrac{4!}{(4-2)!}}{2!} = \frac{4!}{2!(4-2)!} = 6. \qquad \blacksquare$$

The reasoning used in Example 9.5.3 applies in the general case as well. To form an r-permutation of a set of n elements, first choose a subset of r of the n elements (there are $\binom{n}{r}$ ways to perform this step), and then choose an ordering for the r elements (there are $r!$ ways to perform this step). Thus the number of r-permutations is

$$P(n, r) = \binom{n}{r} \cdot r!.$$

Now solve for $\binom{n}{r}$ to obtain the formula

$$\binom{n}{r} = \frac{P(n, r)}{r!}.$$

Since $P(n, r) = \frac{n!}{(n-r)!}$, substitution gives

$$\binom{n}{r} = \frac{\dfrac{n!}{(n-r)!}}{r!} = \frac{n!}{r!(n-r)!}.$$

The result of this discussion is summarized and extended in Theorem 9.5.1.

Theorem 9.5.1

The number of subsets of size r (or r-combinations) that can be chosen from a set of n elements, $\binom{n}{r}$, is given by the formula

$$\binom{n}{r} = \frac{P(n, r)}{r!} \qquad \text{first version}$$

or, equivalently,

$$\binom{n}{r} = \frac{n!}{r!(n-r)!} \qquad \text{second version}$$

where n and r are nonnegative integers with $r \le n$.

Note that the analysis presented before the theorem proves the theorem in all cases where n and r are positive. If r is zero and n is any nonnegative integer, then $\binom{n}{0}$ is the number of subsets of size zero of a set with n elements. But you know from Section 6.2 that there is only one set that does not have any elements. Consequently, $\binom{n}{0} = 1$. Also

$$\frac{n!}{0!(n-0)!} = \frac{\not{n}!}{1 \cdot \not{n}!} = 1$$

since $0! = 1$ by definition. (Remember we said that definition would turn out to be convenient!) Hence the formula

$$\binom{n}{0} = \frac{n!}{0!(n-0)!}$$

holds for all integers $n \geq 0$, and so the theorem is true for all nonnegative integers n and r with $r \leq n$.

Example 9.5.4 Calculating the Number of Teams

Consider again the problem of choosing five members from a group of twelve to work as a team on a special project. How many distinct five-person teams can be chosen?

Solution   The number of distinct five-person teams is the same as the number of subsets of size 5 (or 5-combinations) that can be chosen from the set of twelve. This number is $\binom{12}{5}$. By Theorem 9.5.1,

$$\binom{12}{5} = \frac{12!}{5!(12-5)!} = \frac{\not{12} \cdot 11 \cdot \not{10} \cdot 9 \cdot 8 \cdot \not{7}!}{(\not{5} \cdot \not{4} \cdot \not{3} \cdot \not{2} \cdot 1) \cdot \not{7}!} = 11 \cdot 9 \cdot 8 = 792.$$

Thus there are 792 distinct five-person teams.   ■

The formula for the number of r-combinations of a set can be applied in a wide variety of situations. Some of these are illustrated in the following examples.

Example 9.5.5 Teams That Contain Both or Neither

Suppose two members of the group of twelve insist on working as a pair—any team must contain either both or neither. How many five-person teams can be formed?

Solution   Call the two members of the group that insist on working as a pair A and B. Then any team formed must contain both A and B or neither A nor B. The set of all possible teams can be partitioned into two subsets as shown in Figure 9.5.2 on the next page.

Because a team that contains both A and B contains exactly three other people from the remaining ten in the group, there are as many such teams as there are subsets of three people that can be chosen from the remaining ten. By Theorem 9.5.1, this number is

$$\binom{10}{3} = \frac{10!}{3! \cdot 7!} = \frac{10 \cdot \overset{3}{\not{9}} \cdot \overset{4}{\not{8}} \cdot \not{7}!}{\not{3} \cdot \not{2} \cdot 1 \cdot \not{7}!} = 120.$$

Because a team that contains neither A nor B contains exactly five people from the remaining ten, there are as many such teams as there are subsets of five people that can be chosen from the remaining ten. By Theorem 9.5.1, this number is

$$\binom{10}{5} = \frac{10!}{5! \cdot 5!} = \frac{\overset{2}{\not{10}} \cdot 9 \cdot \overset{2}{\not{8}} \cdot 7 \cdot \not{6} \cdot \not{5}!}{\not{5} \cdot \not{4} \cdot \not{3} \cdot \not{2} \cdot 1 \cdot \not{5}!} = 252.$$

Because the set of teams that contain both A and B is disjoint from the set of teams that contain neither A nor B, by the addition rule,

$$\begin{bmatrix} \text{number of teams containing} \\ \text{both } A \text{ and } B \text{ or} \\ \text{neither } A \text{ nor } B \end{bmatrix} = \begin{bmatrix} \text{number of teams} \\ \text{containing} \\ \text{both } A \text{ and } B \end{bmatrix} + \begin{bmatrix} \text{number of teams} \\ \text{containing} \\ \text{neither } A \text{ nor } B \end{bmatrix}$$

$$= 120 + 252 = 372.$$

This reasoning is summarized in Figure 9.5.2.

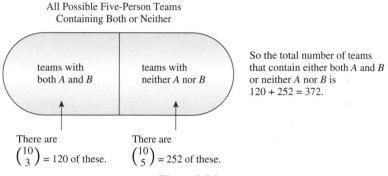

All Possible Five-Person Teams
Containing Both or Neither

teams with both A and B teams with neither A nor B

So the total number of teams that contain either both A and B or neither A nor B is $120 + 252 = 372$.

There are $\binom{10}{3} = 120$ of these.   There are $\binom{10}{5} = 252$ of these.

Figure 9.5.2

Example 9.5.6 Teams That Do Not Contain Both

Suppose two members of the group don't get along and refuse to work together on a team. How many five-person teams can be formed?

Solution   Call the two people who refuse to work together C and D. There are two different ways to answer the given question: One uses the addition rule and the other uses the difference rule.

To use the addition rule, partition the set of all teams that don't contain both C and D into three subsets as shown in Figure 9.5.3 on the next page.

Because any team that contains C but not D contains exactly four other people from the remaining ten in the group, by Theorem 9.5.1 the number of such teams is

$$\binom{10}{4} = \frac{10!}{4!(10-4)!} = \frac{10 \cdot \cancel{9}^{3} \cdot \cancel{8} \cdot 7 \cdot \cancel{6}!}{\cancel{4} \cdot \cancel{3} \cdot \cancel{2} \cdot 1 \cdot \cancel{6}!} = 210.$$

Similarly, there are $\binom{10}{4} = 210$ teams that contain D but not C. Finally, by the same reasoning as in Example 9.5.5, there are 252 teams that contain neither C nor D. Thus, by the addition rule,

$$\begin{bmatrix} \text{number of teams that do} \\ \text{not contain both } C \text{ and } D \end{bmatrix} = 210 + 210 + 252 = 672.$$

This reasoning is summarized in Figure 9.5.3.

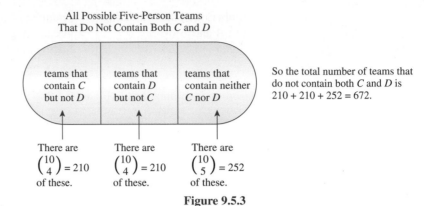

Figure 9.5.3

The alternative solution by the difference rule is based on the following observation: The set of all five-person teams that don't contain both C and D equals the set difference between the set of all five-person teams and the set of all five-person teams that contain both C and D. By Example 9.5.4, the total number of five-person teams is $\binom{12}{5} = 792$. Thus, by the difference rule,

$$\begin{bmatrix} \text{number of teams that don't} \\ \text{contain both } C \text{ and } D \end{bmatrix} = \begin{bmatrix} \text{total number of} \\ \text{teams of five} \end{bmatrix} - \begin{bmatrix} \text{number of teams that} \\ \text{contain both } C \text{ and } D \end{bmatrix}$$

$$= \binom{12}{5} - \binom{10}{3} = 792 - 120 = 672.$$

This reasoning is summarized in Figure 9.5.4. ∎

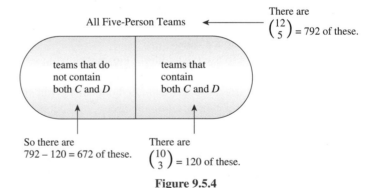

Figure 9.5.4

Before we begin the next example, a remark on the phrases *at least* and *at most* is in order:

> The phrase **at least n** means "n or more."
> The phrase **at most n** means "n or fewer."

For instance, if a set consists of three elements and you are to choose at least two, you will choose two or three; if you are to choose at most two, you will choose none, or one, or two.

Example 9.5.7 Teams with Members of Two Types

Suppose the group of twelve consists of five men and seven women.

a. How many five-person teams can be chosen that consist of three men and two women?

b. How many five-person teams contain at least one man?

c. How many five-person teams contain at most one man?

Solution

a. To answer this question, think of forming a team as a two-step process:

Step 1: Choose the men.

Step 2: Choose the women.

There are $\binom{5}{3}$ ways to choose the three men out of the five and $\binom{7}{2}$ ways to choose the two women out of the seven. Hence, by the product rule,

$$\begin{bmatrix} \text{number of teams of five that} \\ \text{contain three men and two women} \end{bmatrix} = \binom{5}{3}\binom{7}{2} = \frac{5!}{3!2!} \cdot \frac{7!}{2!5!}$$
$$= \frac{7 \cdot 6 \cdot 5 \cdot \cancel{4} \cdot \cancel{3} \cdot \cancel{2} \cdot 1}{\cancel{3} \cdot \cancel{2} \cdot 1 \cdot \cancel{2} \cdot 1 \cdot \cancel{2} \cdot 1}$$
$$= 210.$$

b. This question can also be answered either by the addition rule or by the difference rule. The solution by the difference rule is shorter and is shown first.

Observe that the set of five-person teams containing at least one man equals the set difference between the set of all five-person teams and the set of five-person teams that do not contain any men. See Figure 9.5.5 below.

Now a team with no men consists entirely of five women chosen from the seven women in the group, so there are $\binom{7}{5}$ such teams. Also, by Example 9.5.4, the total number of five-person teams is $\binom{12}{5} = 792$. Hence, by the difference rule,

$$\begin{bmatrix} \text{number of teams} \\ \text{with at least} \\ \text{one man} \end{bmatrix} = \begin{bmatrix} \text{total number} \\ \text{of teams} \\ \text{of five} \end{bmatrix} - \begin{bmatrix} \text{number of teams} \\ \text{of five that do not} \\ \text{contain any men} \end{bmatrix}$$
$$= \binom{12}{5} - \binom{7}{5} = 792 - \frac{7!}{5! \cdot 2!}$$
$$= 792 - \frac{7 \cdot \overset{3}{\cancel{6}} \cdot \cancel{5!}}{\cancel{5!} \cdot \cancel{2} \cdot 1} = 792 - 21 = 771.$$

This reasoning is summarized in Figure 9.5.5.

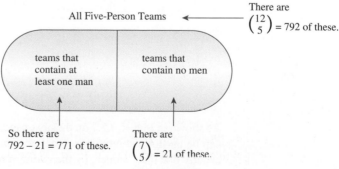

Figure 9.5.5

Alternatively, to use the addition rule, observe that the set of teams containing at least one man can be partitioned as shown in Figure 9.5.6. The number of teams in each subset of the partition is calculated using the method illustrated in part (a). There are

$$\binom{5}{1}\binom{7}{4} \text{ teams with one man and four women}$$

$$\binom{5}{2}\binom{7}{3} \text{ teams with two men and three women}$$

$$\binom{5}{3}\binom{7}{2} \text{ teams with three men and two women}$$

$$\binom{5}{4}\binom{7}{1} \text{ teams with four men and one woman}$$

$$\binom{5}{5}\binom{7}{0} \text{ teams with five men and no women.}$$

Hence, by the addition rule,

$$\left[\begin{array}{l}\text{number of teams with} \\ \text{at least one man}\end{array}\right]$$

$$= \binom{5}{1}\binom{7}{4} + \binom{5}{2}\binom{7}{3} + \binom{5}{3}\binom{7}{2} + \binom{5}{4}\binom{7}{1} + \binom{5}{5}\binom{7}{0}$$

$$= \frac{5!}{1!4!} \cdot \frac{7!}{4!3!} + \frac{5!}{2!3!} \cdot \frac{7!}{3!4!} + \frac{5!}{3!2!} \cdot \frac{7!}{2!5!} + \frac{5!}{4!1!} \cdot \frac{7!}{1!6!} + \frac{5!}{5!0!} \cdot \frac{7!}{0!7!}$$

$$= \frac{5 \cdot \cancel{4!} \cdot 7 \cdot \cancel{6} \cdot 5 \cdot \cancel{4!}}{\cancel{4!} \cdot \cancel{3} \cdot 2 \cdot \cancel{4!}} + \frac{5 \cdot \cancel{4} \cdot \overset{2}{\cancel{3!}} \cdot 7 \cdot \cancel{6} \cdot 5 \cdot \cancel{4!}}{\cancel{3!} \cdot \cancel{2} \cdot \cancel{4!} \cdot \cancel{3} \cdot 2} + \frac{5 \cdot \cancel{4} \cdot \overset{2}{\cancel{3!}} \cdot 7 \cdot \cancel{6} \cdot \overset{3}{\cancel{5!}}}{\cancel{2} \cdot \cancel{3!} \cdot \cancel{5!} \cdot \cancel{2}}$$

$$+ \frac{5 \cdot \cancel{4!} \cdot 7 \cdot \cancel{6!}}{\cancel{4!} \cdot \cancel{6!}} + \frac{\cancel{5!} \cdot \cancel{7!}}{\cancel{5!} \cdot \cancel{7!}}$$

$$= 175 + 350 + 210 + 35 + 1 = 771.$$

This reasoning is summarized in Figure 9.5.6.

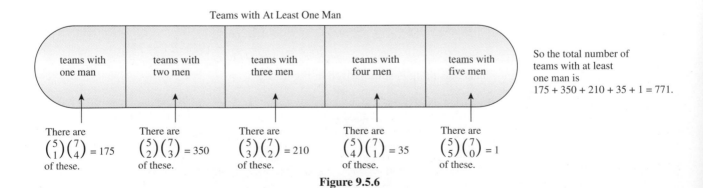

Teams with At Least One Man

| teams with one man | teams with two men | teams with three men | teams with four men | teams with five men |

So the total number of teams with at least one man is $175 + 350 + 210 + 35 + 1 = 771$.

There are $\binom{5}{1}\binom{7}{4} = 175$ of these.　There are $\binom{5}{2}\binom{7}{3} = 350$ of these.　There are $\binom{5}{3}\binom{7}{2} = 210$ of these.　There are $\binom{5}{4}\binom{7}{1} = 35$ of these.　There are $\binom{5}{5}\binom{7}{0} = 1$ of these.

Figure 9.5.6

c. As shown in Figure 9.5.7 on the next page, the set of teams containing at most one man can be partitioned into the set that does not contain any men and the set that contains exactly one man. Hence, by the addition rule,

$$\begin{bmatrix} \text{number of teams} \\ \text{with at} \\ \text{most one man} \end{bmatrix} = \begin{bmatrix} \text{number of} \\ \text{teams without} \\ \text{any men} \end{bmatrix} + \begin{bmatrix} \text{number of} \\ \text{teams with} \\ \text{one man} \end{bmatrix}$$

$$= \binom{5}{0}\binom{7}{5} + \binom{5}{1}\binom{7}{4} = 21 + 175 = 196.$$

This reasoning is summarized in Figure 9.5.7.

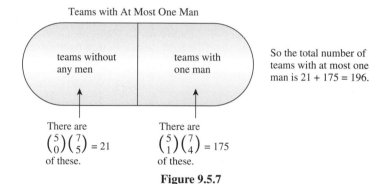

Teams with At Most One Man

teams without any men | teams with one man

So the total number of teams with at most one man is 21 + 175 = 196.

There are $\binom{5}{0}\binom{7}{5} = 21$ of these.

There are $\binom{5}{1}\binom{7}{4} = 175$ of these.

Figure 9.5.7 ■

Example 9.5.8 Poker Hand Problems

The game of poker is played with an ordinary deck of cards (see Example 9.1.1). Various five-card holdings are given special names, and certain holdings beat certain other holdings. The named holdings are listed from highest to lowest below.

Royal flush: 10, J, Q, K, A of the same suit

Straight flush: five adjacent denominations of the same suit but not a royal flush—aces can be high or low, so A, 2, 3, 4, 5 of the same suit is a straight flush.

Four of a kind: four cards of one denomination—the fifth card can be any other in the deck

Full house: three cards of one denomination, two cards of another denomination

Flush: five cards of the same suit but not a straight or a royal flush

Straight: five cards of adjacent denominations but not all of the same suit—aces can be high or low

Three of a kind: three cards of the same denomination and two other cards of different denominations

Two pairs: two cards of one denomination, two cards of a second denomination, and a fifth card of a third denomination

One pair: two cards of one denomination and three other cards all of different denominations

No pairs: all cards of different denominations but not a straight or straight flush or flush

a. How many five-card poker hands contain two pairs?

b. If a five-card hand is dealt at random from an ordinary deck of cards, what is the probability that the hand contains two pairs?

Solution

a. Consider forming a hand with two pairs as a four-step process:

Step 1: Choose the two denominations for the pairs.

Step 2: Choose two cards from the smaller denomination.

Step 3: Choose two cards from the larger denomination.

Step 4: Choose one card from those remaining.

The number of ways to perform step 1 is $\binom{13}{2}$ because there are 13 denominations in all. The number of ways to perform steps 2 and 3 is $\binom{4}{2}$ because there are four cards of each denomination, one in each suit. The number of ways to perform step 4 is $\binom{44}{1}$ because the fifth card is chosen from the eleven denominations not included in the pair and there are four cards of each denomination. Thus

$$\begin{bmatrix} \text{the total number of} \\ \text{hands with two pairs} \end{bmatrix} = \binom{13}{2}\binom{4}{2}\binom{4}{2}\binom{44}{1}$$

$$= \frac{13!}{2!(13-2)!} \cdot \frac{4!}{2!(4-2)!} \cdot \frac{4!}{2!(4-2)!} \cdot \frac{44!}{1!(44-1)!}$$

$$= \frac{13 \cdot 12 \cdot 11!}{(2 \cdot 1) \cdot 11!} \cdot \frac{4 \cdot 3 \cdot 2!}{(2 \cdot 1) \cdot 2!} \cdot \frac{4 \cdot 3 \cdot 2!}{(2 \cdot 1) \cdot 2!} \cdot \frac{44 \cdot 43!}{1 \cdot 43!}$$

$$= 78 \cdot 6 \cdot 6 \cdot 44 = 123{,}552.$$

b. The total number of five-card hands from an ordinary deck of cards is $\binom{52}{5} = 2{,}598{,}960$. Thus if all hands are equally likely, the probability of obtaining a hand with two pairs is $\frac{123{,}552}{2{,}598{,}960} \cong 4.75\%$. ∎

Example 9.5.9 Number of Bit Strings with Fixed Number of 1's

How many eight-bit strings have exactly three 1's?

Solution   To solve this problem, imagine eight empty positions into which the 0's and 1's of the bit string will be placed. In step 1, choose positions for the three 1's, and in step 2, put the 0's into place.

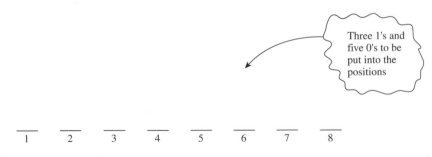

Three 1's and five 0's to be put into the positions

Once a subset of three positions has been chosen from the eight to contain 1's, then the remaining five positions must all contain 0's (since the string is to have exactly three 1's). It follows that the number of ways to construct an eight-bit string with exactly three 1's is the same as the number of subsets of three positions that can be chosen from the eight into which to place the 1's. By Theorem 9.5.1, this equals

$$\binom{8}{3} = \frac{8!}{3! \cdot 5!} = \frac{8 \cdot 7 \cdot 6 \cdot 5!}{3 \cdot 2 \cdot 5!} = 56.$$ ∎

Example 9.5.10 Permutations of a Set with Repeated Elements

Consider various ways of ordering the letters in the word *MISSISSIPPI*:

IIMSSPISSIP, ISSSPMIIPIS, PIMISSSSIIP, and so on.

How many distinguishable orderings are there?

Solution   This example generalizes Example 9.5.9. Imagine placing the 11 letters of *MISSISSIPPI* one after another into 11 positions.

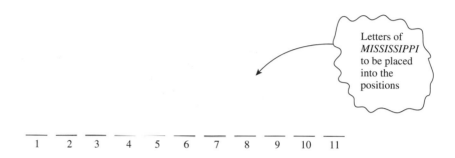

Letters of *MISSISSIPPI* to be placed into the positions

$$\underline{\quad} \;\; \underline{\quad} \;\; \underline{\quad} \;\; \underline{\quad} \;\; \underline{\quad} \;\; \underline{\quad} \;\; \underline{\quad} \;\; \underline{\quad} \;\; \underline{\quad} \;\; \underline{\quad} \;\; \underline{\quad}$$
$$\;\;1\quad 2\quad 3\quad 4\quad 5\quad 6\quad 7\quad 8\quad 9\quad 10\quad 11$$

Because copies of the same letter cannot be distinguished from one another, once the positions for a certain letter are known, then all copies of the letter can go into the positions in any order. It follows that constructing an ordering for the letters can be thought of as a four-step process:

Step 1: Choose a subset of four positions for the *S*'s.

Step 2: Choose a subset of four positions for the *I*'s.

Step 3: Choose a subset of two positions for the *P*'s.

Step 4: Choose a subset of one position for the *M*.

Since there are 11 positions in all, there are $\binom{11}{4}$ subsets of four positions for the *S*'s. Once the four *S*'s are in place, there are seven positions that remain empty, so there are $\binom{7}{4}$ subsets of four positions for the *I*'s. After the *I*'s are in place, there are three positions left empty, so there are $\binom{3}{2}$ subsets of two positions for the *P*'s. That leaves just one position for the *M*. But $1 = \binom{1}{1}$. Hence by the multiplication rule,

$$\begin{bmatrix} \text{number of ways to} \\ \text{position all the letters} \end{bmatrix} = \binom{11}{4}\binom{7}{4}\binom{3}{2}\binom{1}{1}$$

$$= \frac{11!}{4!\,7\!\!\!/\,!} \cdot \frac{7\!\!\!/\,!}{4!\,3\!\!\!/\,!} \cdot \frac{3\!\!\!/\,!}{2!\,1\!\!\!/\,!} \cdot \frac{1\!\!\!/\,!}{1!\,0\!\!\!/\,!}$$

$$= \frac{11!}{4!\cdot 4!\cdot 2!\cdot 1!} = 34{,}650. \qquad \blacksquare$$

In exercise 18 at the end of the section, you are asked to show that changing the order in which the letters are placed into the positions does not change the answer to this example.

The same reasoning used in this example can be used to derive the following general theorem.

Theorem 9.5.2 Permutations with sets of Indistinguishable Objects

Suppose a collection consists of n objects of which

n_1 are of type 1 and are indistinguishable from each other

n_2 are of type 2 and are indistinguishable from each other

$\vdots$

n_k are of type k and are indistinguishable from each other,

and suppose that $n_1 + n_2 + \cdots + n_k = n$. Then the number of distinguishable permutations of the n objects is

$$\binom{n}{n_1}\binom{n-n_1}{n_2}\binom{n-n_1-n_2}{n_3}\cdots\binom{n-n_1-n_2-\cdots-n_{k-1}}{n_k}$$

$$= \frac{n!}{n_1!\,n_2!\,n_3!\cdots n_k!}.$$

Some Advice about Counting

Students learning counting techniques often ask, "How do I know when to add and when to multiply?" Unfortunately, these questions have no easy answers. You need to imagine, as vividly as possible, the objects you are to count. You might even start to list them to get a sense for how to obtain them in a systematic way. You should then construct a model that would allow you to continue counting the objects one by one if you had enough time.

One important principle is only to use the addition rule when you can imagine the items as being broken up into disjoint subsets so that the total number is just the sum of the numbers of the items in each subset. To count the items in the subsets you might use the multiplication rule or r-combinations or a mixture of both. You will use the multiplication rule if you can imagine the items to be counted as being obtained through a multistep process (in which each step is performed in a fixed number of ways regardless of how preceding steps were performed). The number of items will then be the product of the number of ways to perform each step. A rough test to see whether to add or multiply is the "and/or" rule: If the word *and* helps describe the items to be counted, use multiplication (for example, how many teams have four men *and* three women?) whereas if the word *or* helps describe the items to be counted, use addition (for example, how many teams have one man *or* two men?).

The next example explores one of the most common mistakes students make, which is to count certain possibilities more than once.

Example 9.5.11 Double Counting

Consider again the problem of Example 9.5.7(b). A group consists of five men and seven women. How many teams of five contain at least one man?

Incorrect Solution

Imagine constructing the team as a two-step process:

Step 1: Choose a subset of one man from the five men.

Step 2: Choose a subset of four others from the remaining eleven people.

Hence, by the multiplication rule, there are $\binom{5}{1} \cdot \binom{11}{4} = 1,650$ five-person teams that contain at least one man.

Caution! Be careful to avoid counting items twice when using the multiplication rule.

Analysis of the Incorrect Solution The problem with this solution is that some teams are counted more than once. Suppose the men are Anwar, Ben, Carlos, Dwayne, and Ed and the women are Fumiko, Gail, Hui-Fan, Inez, Jill, Kim, and Laura. According to the method described previously, one possible outcome of the two-step process is as follows:

Outcome of step 1: Anwar

Outcome of step 2: Ben, Gail, Inez, and Jill.

In this case the team would be {Anwar, Ben, Gail, Inez, Jill}. But another possible outcome is

Outcome of step 1: Ben

Outcome of step 2: Anwar, Gail, Inez, and Jill,

which also gives the team {Anwar, Ben, Gail, Inez, Jill}. Thus this one team is given by two different branches of the possibility tree, and so it is counted twice. ∎

The best way to avoid mistakes such as the one just described is to visualize the possibility tree that corresponds to any use of the multiplication rule and the set partition that corresponds to a use of the addition rule. Check how your division into steps works by applying it to some actual data—as was done in the analysis above—and try to pick data that are as typical or generic as possible.

It often helps to ask yourself (1) "Am I counting everything?" and (2) "Am I counting anything twice?" When using the multiplication rule, these questions become (1) "Does every outcome appear as some branch of the tree?" and (2) "Does any outcome appear on more than one branch of the tree?" When using the addition rule, the questions become (1) "Does every outcome appear in some subset of the diagram?" and (2) "Do any two subsets in the diagram share common elements?"

Test Yourself

1. The number of subsets of size r that can be formed from a set with n elements is denoted _____, which is read as "_____."

2. The number of r-combinations of a set of n elements is _____.

3. Two unordered selections are said to be the same if the elements chosen are the same, regardless of _____.

4. A formula relating $\binom{n}{r}$ and $P(n, r)$ is _____.

5. The phrase "at least n" means _____, and the phrase "at most n" means _____.

6. Suppose a collection consists of n objects of which, for each i with $1 \leq i \leq k$, n_i are of type i and are indistinguishable from each other. Also suppose that $n = n_1 + n_2 + \cdots + n_k$. Then the number of distinct permutations of the n objects is _____.

Exercise Set 9.5

1. a. List all 2-combinations for the set $\{x_1, x_2, x_3\}$. Deduce the value of $\binom{3}{2}$.
   b. List all unordered selections of four elements from the set $\{a, b, c, d, e\}$. Deduce the value of $\binom{5}{4}$.

2. a. List all 3-combinations for the set $\{x_1, x_2, x_3, x_4, x_5\}$. Deduce the value of $\binom{5}{3}$.

   b. List all unordered selections of two elements from the set $\{x_1, x_2, x_3, x_4, x_5, x_6\}$. Deduce the value of $\binom{6}{2}$.

3. Write an equation relating $P(7, 2)$ and $\binom{7}{2}$.

4. Write an equation relating $P(8, 3)$ and $\binom{8}{3}$.

5. Use Theorem 9.5.1 to compute each of the following.

   a. $\binom{6}{0}$ **b.** $\binom{6}{1}$ c. $\binom{6}{2}$

   d. $\binom{6}{3}$ e. $\binom{6}{4}$ f. $\binom{6}{5}$ g. $\binom{6}{6}$

6. A student council consists of 15 students.

   a. In how many ways can a committee of six be selected from the membership of the council?

   b. Two council members have the same major and are not permitted to serve together on a committee. How many ways can a committee of six be selected from the membership of the council?

   c. Two council members always insist on serving on committees together. If they can't serve together, they won't serve at all. How many ways can a committee of six be selected from the council membership?

   d. Suppose the council contains eight men and seven women.

      (i) How many committees of six contain three men and three women?

      (ii) How many committees of six contain at least one woman?

   e. Suppose the council consists of three freshmen, four sophomores, three juniors, and five seniors. How many committees of eight contain two representatives from each class?

7. A computer programming team has 13 members.

   a. How many ways can a group of seven be chosen to work on a project?

   b. Suppose seven members are women and six are men.

      (i) How many groups of seven can be chosen that contain four women and three men?

      (ii) How many groups of seven can be chosen that contain at least one man?

      (iii) How many groups of seven can be chosen that contain at most three women?

   c. Suppose two team members refuse to work together on projects. How many groups of seven can be chosen to work on a project?

   d. Suppose two team members insist on either working together or not at all on projects. How many groups of seven can be chosen to work on a project?

H 8. An instructor gives an exam with fourteen questions. Students are allowed to choose any ten to answer.

   a. How many different choices of ten questions are there?

   b. Suppose six questions require proof and eight do not.

      (i) How many groups of ten questions contain four that require proof and six that do not?

      (ii) How many groups of ten questions contain at least one that requires proof?

      (iii) How many groups of ten questions contain at most three that require proof?

   c. Suppose the exam instructions specify that at most one of questions 1 and 2 may be included among the ten. How many different choices of ten questions are there?

   d. Suppose the exam instructions specify that either both questions 1 and 2 are to be included among the ten or neither is to be included. How many different choices of ten questions are there?

9. A club is considering changing its bylaws. In an initial straw vote on the issue, 24 of the 40 members of the club favored the change and 16 did not. A committee of six is to be chosen from the 40 club members to devote further study to the issue.

   a. How many committees of six can be formed from the club membership?

   b. How many of the committees will contain at least three club members who, in the preliminary survey, favored the change in the bylaws?

10. Two new drugs are to be tested using a group of 60 laboratory mice, each tagged with a number for identification purposes. Drug *A* is to be given to 22 mice, drug *B* is to be given to another 22 mice, and the remaining 16 mice are to be used as controls. How many ways can the assignment of treatments to mice be made? (A single assignment involves specifying the treatment for each mouse—whether drug *A*, drug *B*, or no drug.)

★ 11. Refer to Example 9.5.8. For each poker holding below, (1) find the number of five-card poker hands with that holding; (2) find the probability that a randomly chosen set of five cards has that holding.

   a. royal flush   b. straight flush   **c.** four of a kind

   d. full house   e. flush   **f.** straight

   g. three of a kind   h. one pair

   i. neither a repeated denomination nor five of the same suit nor five adjacent denominations

12. How many pairs of two distinct integers chosen from the set $\{1, 2, 3, \ldots, 101\}$ have a sum that is even?

13. A coin is tossed ten times. In each case the outcome *H* (for heads) or *T* (for tails) is recorded. (One possible outcome of the ten tossings is denoted $THHTTTHTTH$.)

   a. What is the total number of possible outcomes of the coin-tossing experiment?

   b. In how many of the possible outcomes are exactly five heads obtained?

   c. In how many of the possible outcomes are at least eight heads obtained?

   d. In how many of the possible outcomes is at least one head obtained?

   e. In how many of the possible outcomes is at most one head obtained?

14. a. How many 16-bit strings contain exactly seven 1's?

    b. How many 16-bit strings contain at least thirteen 1's?

    c. How many 16-bit strings contain at least one 1?

    d. How many 16-bit strings contain at most one 1?

15. a. How many even integers are in the set

$$\{1, 2, 3, \ldots, 100\}?$$

    b. How many odd integers are in the set

$$\{1, 2, 3, \ldots, 100\}?$$

c. How many ways can two integers be selected from the set $\{1, 2, 3, \ldots, 100\}$ so that their sum is even?

d. How many ways can two integers be selected from the set $\{1, 2, 3, \ldots, 100\}$ so that their sum is odd?

16. Suppose that three computer boards in a production run of forty are defective. A sample of five is to be selected to be checked for defects.

a. How many different samples can be chosen?

b. How many samples will contain at least one defective board?

c. What is the probability that a randomly chosen sample of five contains at least one defective board?

17. Ten points labeled $A, B, C, D, E, F, G, H, I, J$ are arranged in a plane in such a way that no three lie on the same straight line.

a. How many straight lines are determined by the ten points?

b. How many of these straight lines do not pass through point A?

c. How many triangles have three of the ten points as vertices?

d. How many of these triangles do not have A as a vertex?

18. Suppose that you placed the letters in Example 9.5.10 into positions in the following order: first the M, then the I's, then the S's, and then the P's. Show that you would obtain the same answer for the number of distinguishable orderings.

19. a. How many distinguishable ways can the letters of the word *HULLABALOO* be arranged in order?

b. How many distinguishable orderings of the letters of *HULLABALOO* begin with U and end with L?

c. How many distinguishable orderings of the letters of *HULLABALOO* contain the two letters HU next to each other in order?

20. a. How many distinguishable ways can the letters of the word *MILLIMICRON* be arranged in order?

b. How many distinguishable orderings of the letters of *MILLIMICRON* begin with M and end with N?

c. How many distinguishable orderings of the letters of *MILLIMICRON* contain the letters CR next to each other in order and also the letters ON next to each other in order?

21. In Morse code, symbols are represented by variable-length sequences of dots and dashes. (For example, $A = \cdot -$, $1 = \cdot - - - -$, $? = \cdot \cdot - - \cdot \cdot$.) How many different symbols can be represented by sequences of seven or fewer dots and dashes?

22. Each symbol in the Braille code is represented by a rectangular arrangement of six dots, each of which may be raised or flat against a smooth background. For instance, when the word Braille is spelled out, it looks like this:

Given that at least one of the six dots must be raised, how many symbols can be represented in the Braille code?

23. On an 8×8 chessboard, a rook is allowed to move any number of squares either horizontally or vertically. How many different paths can a rook follow from the bottom-left square of the board to the top-right square of the board if all moves are to the right or upward?

24. The number 42 has the prime factorization $2 \cdot 3 \cdot 7$. Thus 42 can be written in four ways as a product of two positive integer factors (without regard to the order of the factors): $1 \cdot 42, 2 \cdot 21, 3 \cdot 14$, and $6 \cdot 7$. Answer a–d below without regard to the order of the factors.

a. List the distinct ways the number 210 can be written as a product of two positive integer factors.

b. If $n = p_1 p_2 p_3 p_4$, where the p_i are distinct prime numbers, how many ways can n be written as a product of two positive integer factors?

c. If $n = p_1 p_2 p_3 p_4 p_5$, where the p_i are distinct prime numbers, how many ways can n be written as a product of two positive integer factors?

d. If $n = p_1 p_2 \cdots p_k$, where the p_i are distinct prime numbers, how many ways can n be written as a product of two positive integer factors?

25. **a.** How many one-to-one functions are there from a set with three elements to a set with four elements?

b. How many one-to-one functions are there from a set with three elements to a set with two elements?

c. How many one-to-one functions are there from a set with three elements to a set with three elements?

d. How many one-to-one functions are there from a set with three elements to a set with five elements?

H e. How many one-to-one functions are there from a set with m elements to a set with n elements, where $m \leq n$?

26. **a.** How many onto functions are there from a set with three elements to a set with two elements?

b. How many onto functions are there from a set with three elements to a set with five elements?

H c. How many onto functions are there from a set with three elements to a set with three elements?

d. How many onto functions are there from a set with four elements to a set with two elements?

e. How many onto functions are there from a set with four elements to a set with three elements?

H ✶ f. Let $c_{m,n}$ be the number of onto functions from a set of m elements to a set of n elements, where $m \geq n \geq 1$. Find a formula relating $c_{m,n}$ to $c_{m-1,n}$ and $c_{m-1,n-1}$.

27. Let A be a set with eight elements.

a. How many relations are there on A?

b. How many relations on A are reflexive?

c. How many relations on A are symmetric?

d. How many relations on A are both reflexive and symmetric?

H ✶ **28.** A student council consists of three freshmen, four sophomores, four juniors, and five seniors. How many committees of eight members of the council contain at least one member from each class?

★ 29. An alternative way to derive Theorem 9.5.1 uses the following *division rule*: Let n and k be integers so that k divides n. If a set consisting of n elements is divided into subsets that each contain k elements, then the number of such subsets is n/k. Explain how Theorem 9.5.1 can be derived using the division rule.

30. Find the error in the following reasoning: "Consider forming a poker hand with two pairs as a five-step process.

Step 1: Choose the denomination of one of the pairs.
Step 2: Choose the two cards of that denomination.
Step 3: Choose the denomination of the other of the pairs.
Step 4: Choose the two cards of that second denomination.

Step 5: Choose the fifth card from the remaining denominations.

There are $\binom{13}{1}$ ways to perform step 1, $\binom{4}{2}$ ways to perform step 2, $\binom{12}{1}$ ways to perform step 3, $\binom{4}{2}$ ways to perform step 4, and $\binom{44}{1}$ ways to perform step 5. Therefore, the total number of five-card poker hands with two pairs is $13 \cdot 6 \cdot 12 \cdot 6 \cdot 44 = 247{,}104$."

★ 31. Let P_n be the number of partitions of a set with n elements. Show that

$$P_n = \binom{n-1}{0} P_{n-1} + \binom{n-1}{1} P_{n-2} + \cdots + \binom{n-1}{n-1} P_0$$

for all integers $n \geq 1$.

Answers for Test Yourself

1. $\binom{n}{r}$; n choose r 2. $\binom{n}{r}$ (*Or*: n choose r)   3. the order in which they are chosen   4. $\binom{n}{r} = \frac{P(n,r)}{r!}$ 5. n or more; n or fewer
6. $\binom{n}{n_1}\binom{n-n_1}{n_2}\binom{n-n_1-n_2}{n_3} \cdots \binom{n-n_1-n_2-\cdots-n_{k-1}}{n_k}$ $\left(Or: \frac{n!}{n_1!n_2!n_3!\cdots n_k!}\right)$

9.6 *Pascal's Formula and the Binomial Theorem*

I'm very well acquainted, too, with matters mathematical, I understand equations both the simple and quadratical. About binomial theorem I am teaming with a lot of news, With many cheerful facts about the square of the hypotenuse.
—William S. Gilbert, *The Pirates of Penzance*, 1880

In this section we derive several formulas for values of $\binom{n}{r}$. The most important is Pascal's formula, which is the basis for Pascal's triangle and is a crucial component of one of the proofs of the binomial theorem. We offer two distinct proofs for both Pascal's formula and the binomial theorem. One of them is called "algebraic" because it relies to a great extent on algebraic manipulation, and the other is called "combinatorial," because it is based on the kind of counting arguments we have been discussing in this chapter.

Example 9.6.1 Values of $\binom{n}{n}$, $\binom{n}{n-1}$, $\binom{n}{n-2}$

Think of Theorem 9.5.1 as a general template: Regardless of what nonnegative numbers are placed in the boxes, if the number in the lower box is no greater than the number in the top box, then

$$\binom{\square}{\diamond} = \frac{\square!}{\diamond!(\square - \diamond)!}.$$

Use Theorem 9.5.1 to show that for all integers $n \geq 0$,

$$\binom{n}{n} = 1 \tag{9.6.1}$$

$$\binom{n}{n-1} = n, \quad \text{if } n \geq 1 \tag{9.6.2}$$

$$\binom{n}{n-2} = \frac{n(n-1)}{2}, \quad \text{if } n \geq 2. \tag{9.6.3}$$

Solution

$$\binom{n}{n} = \frac{n!}{n!(n-n)!} = \frac{1}{0!} = 1 \qquad \text{since } 0! = 1 \text{ by definition}$$

$$\binom{n}{n-1} = \frac{n!}{(n-1)!(n-(n-1))!}$$

$$= \frac{n \cdot (n-1)!}{(n-1)!(n-n+1)!} = \frac{n}{1} = n$$

$$\binom{n}{n-2} = \frac{n!}{(n-2)!(n-(n-2))!}$$

$$= \frac{n \cdot (n-1) \cdot (n-2)!}{(n-2)!2!} = \frac{n(n-1)}{2} \qquad \blacksquare$$

Note that the result derived algebraically above, that $\binom{n}{n}$ equals 1, agrees with the fact that a set with n elements has just one subset of size n, namely itself. Similarly, exercise 1 at the end of the section asks you to show algebraically that $\binom{n}{0} = 1$, which agrees with the fact that a set with n elements has one subset, the empty set, of size 0. In exercise 2 you are also asked to show algebraically that $\binom{n}{1} = n$. This result agrees with the fact that there are n subsets of size 1 that can be chosen from a set with n elements, namely the subsets consisting of each element taken alone.

Example 9.6.2 $\quad\displaystyle\binom{n}{r} = \binom{n}{n-r}$

In exercise 5 at the end of the section you are asked to verify algebraically that

$$\binom{n}{r} = \binom{n}{n-r}$$

for all nonnegative integers n and r with $r \leq n$.

An alternative way to deduce this formula is to interpret it as saying that a set A with n elements has exactly as many subsets of size r as it has subsets of size $n - r$. Derive the formula using this reasoning.

Solution   Observe that any subset of size r can be specified either by saying which r elements lie in the subset or by saying which $n - r$ elements lie outside the subset.

A, A Set with n Elements

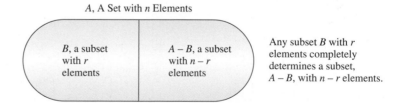

B, a subset with r elements | $A - B$, a subset with $n - r$ elements

Any subset B with r elements completely determines a subset, $A - B$, with $n - r$ elements.

Suppose A has k subsets of size r: $B_1, B_2, \ldots, B_k$. Then each B_i can be paired up with exactly one set of size $n - r$, namely its complement $A - B_i$ as shown below.

Subsets of Size r Subsets of Size $n - r$

$B_1 \longleftrightarrow A - B_1$

$B_2 \longleftrightarrow A - B_2$

$\vdots \qquad\qquad \vdots$

$B_k \longleftrightarrow A - B_k$

All subsets of size r are listed in the left-hand column, and all subsets of size $n - r$ are listed in the right-hand column. The number of subsets of size r equals the number of subsets of size $n - r$, and so $\binom{n}{r} = \binom{n}{n-r}$. ∎

The type of reasoning used in this example is called *combinatorial*, because it is obtained by counting things that are combined in different ways. A number of theorems have both combinatorial proofs and proofs that are purely algebraic.

Pascal's Formula

*Blaise Pascal
(1623–1662)*

Pascal's formula, named after the seventeenth-century French mathematician and philosopher Blaise Pascal, is one of the most famous and useful in combinatorics (which is the formal term for the study of counting and listing problems). It relates the value of $\binom{n+1}{r}$ to the values of $\binom{n}{r-1}$ and $\binom{n}{r}$. Specifically, it says that

$$\binom{n+1}{r} = \binom{n}{r-1} + \binom{n}{r}$$

whenever n and r are positive integers with $r \le n$. This formula makes it easy to compute higher combinations in terms of lower ones: If all the values of $\binom{n}{r}$ are known, then the values of $\binom{n+1}{r}$ can be computed for all r such that $0 < r \le n$.

Pascal's triangle, shown in Table 9.6.1, is a geometric version of Pascal's formula. Sometimes it is simply called the arithmetic triangle because it was used centuries before Pascal by Chinese and Persian mathematicians. But Pascal discovered it independently, and ever since 1654, when he published a treatise that explored many of its features, it has generally been known as Pascal's triangle.

Table 9.6.1 Pascal's Triangle for Values of $\binom{n}{r}$

r \ n	0	1	2	3	4	5	$\cdots$	$r-1$	r	$\cdots$
0	1							.	.	$\cdots$
1	1	1						.	.	$\cdots$
2	1	2	1					.	.	$\cdots$
3	1	3	3	1				.	.	$\cdots$
4	1	4	6 $+$	4	1			.	.	$\cdots$
5	1	5	10 $=$	10	5	1		.	.	$\cdots$
$\vdots$	$\vdots$	$\vdots$	$\vdots$	$\vdots$	$\vdots$	$\vdots$		$\vdots$	$\vdots$	$\vdots\vdots\vdots$
n	$\binom{n}{0}$	$\binom{n}{1}$	$\binom{n}{2}$	$\binom{n}{3}$	$\binom{n}{4}$	$\binom{n}{5}$	$\cdots$	$\binom{n}{r-1}$ $+$	$\binom{n}{r}$	$\cdots$
$n+1$	$\binom{n+1}{0}$	$\binom{n+1}{1}$	$\binom{n+1}{2}$	$\binom{n+1}{3}$	$\binom{n+1}{4}$	$\binom{n+1}{5}$	$\cdots$	$=$	$\binom{n+1}{r}$	$\cdots$
.		.	.	.	.	.		.	.	$\cdots$
.		.	.	.	.	.		.	.	$\cdots$
.		.	.	.	.	.		.	.	$\cdots$

Each entry in the triangle is a value of $\binom{n}{r}$. Pascal's formula translates into the fact that the entry in row $n + 1$, column r equals the sum of the entry in row n, column $r - 1$ plus the entry in row n, column r. That is, the entry in a given interior position equals the sum of the two entries directly above and to the above left. The left-most and right-most entries in each row are 1 because $\binom{n}{n} = 1$ by Example 9.6.1 and $\binom{n}{0} = 1$ by exercise 1 at the end of this section.

Example 9.6.3   Calculating $\binom{n}{r}$ Using Pascal's Triangle

Use Pascal's triangle to compute the values of

$$\binom{6}{2} \quad \text{and} \quad \binom{6}{3}.$$

Solution   By construction, the value in row n, column r of Pascal's triangle is the value of $\binom{n}{r}$, for every pair of positive integers n and r with $r \leq n$. By Pascal's formula, $\binom{n+1}{r}$ can be computed by adding together $\binom{n}{r-1}$ and $\binom{n}{r}$, which are located directly above and above left of $\binom{n+1}{r}$. Thus,

$$\binom{6}{2} = \binom{5}{1} + \binom{5}{2} = 5 + 10 = 15 \quad \text{and}$$

$$\binom{6}{3} = \binom{5}{2} + \binom{5}{3} = 10 + 10 = 20. \qquad \blacksquare$$

Pascal's formula can be derived by two entirely different arguments. One is algebraic; it uses the formula for the number of r-combinations obtained in Theorem 9.5.1. The other is combinatorial; it uses the definition of the number of r-combinations as the number of subsets of size r taken from a set with a certain number of elements. We give both proofs since both approaches have applications in many other situations.

Theorem 9.6.1 Pascal's Formula

Let n and r be positive integers and suppose $r \leq n$. Then

$$\binom{n+1}{r} = \binom{n}{r-1} + \binom{n}{r}.$$

Proof (algebraic version):

Let n and r be positive integers with $r \leq n$. By Theorem 9.5.1,

$$\binom{n}{r-1} + \binom{n}{r} = \frac{n!}{(r-1)!(n-(r-1))!} + \frac{n!}{r!(n-r)!}$$

$$= \frac{n!}{(r-1)!(n-r+1)!} + \frac{n!}{r!(n-r)!}.$$

To add these fractions, a common denominator is needed, so multiply the numerator and denominator of the left-hand fraction by r and multiply the numerator and denominator of the right-hand fraction by $(n-r+1)$. Then

continued on page 466

$$\binom{n}{r-1} + \binom{n}{r} = \frac{n!}{(r-1)!(n-r+1)!} \cdot \frac{r}{r} + \frac{n!}{r!(n-r)!} \cdot \frac{(n-r+1)}{(n-r+1)}$$

$$= \frac{n! \cdot r}{(n-r+1)!r(r-1)!} + \frac{n \cdot n! - n! \cdot r + n!}{(n-r+1)(n-r)!r!}$$

$$= \frac{\cancel{n! \cdot r} + n! \cdot n - \cancel{n! \cdot r} + n!}{(n-r+1)!r!} = \frac{n!(n+1)}{(n+1-r)!r!}$$

$$= \frac{(n+1)!}{((n+1)-r)!r!} = \binom{n+1}{r}.$$

Proof (combinatorial version):

Let n and r be positive integers with $r \le n$. Suppose S is a set with $n + 1$ elements. The number of subsets of S of size r can be calculated by thinking of S as consisting of two pieces: one with n elements $\{x_1, x_2, \ldots, x_n\}$ and the other with one element $\{x_{n+1}\}$.

Any subset of S with r elements either contains x_{n+1} or it does not. If it contains x_{n+1}, then it contains $r - 1$ elements from the set $\{x_1, x_2, \ldots, x_n\}$. If it does not contain x_{n+1}, then it contains r elements from the set $\{x_1, x_2, \ldots, x_n\}$.

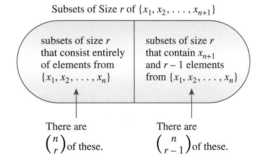

Subsets of Size r of $\{x_1, x_2, \ldots, x_{n+1}\}$

| subsets of size r that consist entirely of elements from $\{x_1, x_2, \ldots, x_n\}$ | subsets of size r that contain x_{n+1} and $r - 1$ elements from $\{x_1, x_2, \ldots, x_n\}$ |

There are $\binom{n}{r}$ of these.　　There are $\binom{n}{r-1}$ of these.

By the addition rule,

$$\begin{bmatrix} \text{number of subsets of} \\ \{x_1, x_2, \ldots, x_n, x_{n+1}\} \\ \text{of size } r \end{bmatrix} = \begin{bmatrix} \text{number of subsets of} \\ \{x_1, x_2, \ldots, x_n\} \\ \text{of size } r - 1 \end{bmatrix} + \begin{bmatrix} \text{number of subsets of} \\ \{x_1, x_2, \ldots, x_n\} \\ \text{of size } r \end{bmatrix}.$$

By Theorem 9.5.1, the set $\{x_1, x_2, \ldots, x_n, x_{n+1}\}$ has $\binom{n+1}{r}$ subsets of size r, the set $\{x_1, x_2, \ldots, x_n\}$ has $\binom{n}{r-1}$ subsets of size $r - 1$, and the set $\{x_1, x_2, \ldots, x_n\}$ has $\binom{n}{r}$ subsets of size r. Thus

$$\binom{n+1}{r} = \binom{n}{r-1} + \binom{n}{r},$$

as was to be shown.

Example 9.6.4 Deriving New Formulas from Pascal's Formula

Use Pascal's formula to derive a formula for $\binom{n+2}{r}$ in terms of values of $\binom{n}{r}$, $\binom{n}{r-1}$, and $\binom{n}{r-2}$. Assume n and r are nonnegative integers and $2 \le r \le n$.

Solution   By Pascal's formula,

$$\binom{n+2}{r} = \binom{n+1}{r-1} + \binom{n+1}{r}.$$

Now apply Pascal's formula to $\binom{n+1}{r-1}$ and $\binom{n+1}{r}$ and substitute into the above to obtain

$$\binom{n+2}{r} = \left[\binom{n}{r-2} + \binom{n}{r-1}\right] + \left[\binom{n}{r-1} + \binom{n}{r}\right].$$

Combining the two middle terms gives

$$\binom{n+2}{r} = \binom{n}{r-2} + 2\binom{n}{r-1} + \binom{n}{r}$$

for all nonnegative integers n and r such that $2 \leq r \leq n$.   ■

The Binomial Theorem

In algebra a sum of two terms, such as $a + b$, is called a **binomial.** The *binomial theorem* gives an expression for the powers of a binomial $(a + b)^n$, for each positive integer n and all real numbers a and b.

Consider what happens when you calculate the first few powers of $a + b$. According to the distributive law of algebra, you take the sum of the products of all combinations of individual terms:

$$(a + b)^2 = (a + b)(a + b) = aa + ab + ba + bb,$$
$$(a + b)^3 = (a + b)(a + b)(a + b)$$
$$= aaa + aab + aba + abb + baa + bab + bba + bbb,$$
$$(a + b)^4 = \underbrace{(a + b)(a + b)(a + b)(a + b)}_{}$$

$$\begin{array}{cccc} \text{1st} & \text{2nd} & \text{3rd} & \text{4th} \\ \text{factor} & \text{factor} & \text{factor} & \text{factor} \end{array}$$

$$= aaaa + aaab + aaba + aabb + abaa + abab + abba + abbb$$
$$+ baaa + baab + baba + babb + bbaa + bbab + bbba + bbbb.$$

Now focus on the expansion of $(a + b)^4$. (It is concrete, and yet it has all the features of the general case.) A typical term of this expansion is obtained by multiplying one of the two terms from the first factor times one of the two terms from the second factor times one of the two terms from the third factor times one of the two terms from the fourth factor. For example, the term *abab* is obtained by multiplying the a's and b's marked with arrows below.

$$\begin{array}{cccc} \downarrow & \quad \downarrow\ \downarrow & \quad \downarrow \end{array}$$
$$(a + b)(a + b)(a + b)(a + b)$$

Since there are two possible values—a or b—for each term selected from one of the four factors, there are $2^4 = 16$ terms in the expansion of $(a + b)^4$.

Now some terms in the expansion are "like terms" and can be combined. Consider all possible orderings of three a's and one b, for example. By the techniques of Section 9.5, there are $\binom{4}{1} = 4$ of them. And each of the four occurs as a term in the expansion of $(a + b)^4$:

$$aaab \quad aaba \quad abaa \quad baaa.$$

By the commutative and associative laws of algebra, each such term equals a^3b, so all four are "like terms." When the like terms are combined, therefore, the coefficient of a^3b equals $\binom{4}{1}$.

Similarly, the expansion of $(a + b)^4$ contains the $\binom{4}{2} = 6$ different orderings of two a's and two b's,

$$aabb \quad abab \quad abba \quad baab \quad baba \quad bbaa,$$

all of which equal a^2b^2, so the coefficient of a^2b^2 equals $\binom{4}{2}$. By a similar analysis, the coefficient of ab^3 equals $\binom{4}{3}$. Also, since there is only one way to order four a's, the coefficient of a^4 is 1 (which equals $\binom{4}{0}$), and since there is only one way to order four b's, the coefficient of b^4 is 1 (which equals $\binom{4}{4}$). Thus, when all of the like terms are combined,

$$(a + b)^4 = \binom{4}{0}a^4 + \binom{4}{1}a^3b + \binom{4}{2}a^2b^2 + \binom{4}{3}ab^3 + \binom{4}{4}b^4$$
$$= a^4 + 4a^3b + 6a^2b^2 + 4ab^3 + b^4.$$

The binomial theorem generalizes this formula to an arbitrary nonnegative integer n.

Theorem 9.6.2 Binomial Theorem

Given any real numbers a and b and any nonnegative integer n,

$$(a + b)^n = \sum_{k=0}^{n} \binom{n}{k} a^{n-k} b^k$$

$$= a^n + \binom{n}{1} a^{n-1} b^1 + \binom{n}{2} a^{n-2} b^2 + \cdots + \binom{n}{n-1} a^1 b^{n-1} + b^n.$$

Note that the second expression equals the first because $\binom{n}{0} = 1$ and $\binom{n}{n} = 1$, for all nonnegative integers n, provided that $b^0 = 1$ and $a^{n-n} = 1$.

It is instructive to see two proofs of the binomial theorem: an algebraic proof and a combinatorial proof. Both require a precise definition of integer power.

• Definition

For any real number a and any nonnegative integer n, the **nonnegative integer powers of a** are defined as follows:

$$a^n = \begin{cases} 1 & \text{if } n = 0 \\ a \cdot a^{n-1} & \text{if } n > 0 \end{cases}$$

Note This is the definition of 0^0 given by Donald E. Knuth in *The Art of Computer Programming, Volume 1: Fundamental Algorithms,* Third Edition (Reading, Mass.: Addison-Wesley, 1997), p. 57.

In some mathematical contexts, 0^0 is left undefined. Defining it to be 1, as is done here, makes it possible to write general formulas such as $\sum_{i=0}^{n} x^i = \frac{1}{1-x}$ without having to exclude values of the variables that result in the expression 0^0.

The algebraic version of the binomial theorem uses mathematical induction and calls upon Pascal's formula at a crucial point.

Proof of the Binomial Theorem (algebraic version):

Suppose a and b are real numbers. We use mathematical induction and let the property $P(n)$ be the equation

$$(a+b)^n = \sum_{k=0}^{n} \binom{n}{k} a^{n-k} b^k. \qquad \leftarrow P(n)$$

Show that P(0) is true: When $n = 0$, the binomial theorem states that:

$$(a+b)^0 = \sum_{k=0}^{0} \binom{0}{k} a^{0-k} b^k. \qquad \leftarrow P(0)$$

But the left-hand side is $(a+b)^0 = 1$ *[by definition of power]*, and the right-hand side is

$$\sum_{k=0}^{0} \binom{0}{k} a^{0-k} b^k = \binom{0}{0} a^{0-0} b^0$$

$$= \frac{0!}{0! \cdot (0-0)!} \cdot 1 \cdot 1 = \frac{1}{1 \cdot 1} = 1$$

also *[since $0! = 1$, $a^0 = 1$, and $b^0 = 1$]*. Hence $P(0)$ is true.

Show that for all integers m≥0, if P(m) is true then P(m+1) is true: Let an integer $m \geq 0$ be given, and suppose $P(m)$ is true. That is, suppose

$$(a+b)^m = \sum_{k=0}^{m} \binom{m}{k} a^{m-k} b^k. \qquad \begin{array}{l} P(m) \\ \text{inductive hypothesis.} \end{array}$$

We need to show that $P(m+1)$ is true:

$$(a+b)^{m+1} = \sum_{k=0}^{m+1} \binom{m+1}{k} a^{(m+1)-k} b^k. \quad P(m+1)$$

Now, by definition of the $(m+1)$st power,
$$(a+b)^{m+1} = (a+b) \cdot (a+b)^m,$$

so by substitution from the inductive hypothesis,

$$(a+b)^{m+1} = (a+b) \cdot \sum_{k=0}^{m} \binom{m}{k} a^{m-k} b^k$$

$$= a \cdot \sum_{k=0}^{m} \binom{m}{k} a^{m-k} b^k + b \cdot \sum_{k=0}^{m} \binom{m}{k} a^{m-k} b^k$$

$$= \sum_{k=0}^{m} \binom{m}{k} a^{m+1-k} b^k + \sum_{k=0}^{m} \binom{m}{k} a^{m-k} b^{k+1}$$

by the generalized distributive law and the facts that $a \cdot a^{m-k} = a^{1+m-k} = a^{m+1-k}$ and $b \cdot b^k = b^{1+k} = b^{k+1}$.

We transform the second summation on the right-hand side by making the change of variable $j = k + 1$. When $k = 0$, then $j = 1$. When $k = m$, then $j = m + 1$. And since $k = j - 1$, the general term is

$$\binom{m}{k} a^{m-k} b^{k+1} = \binom{m}{j-1} a^{m-(j-1)} b^j = \binom{m}{j-1} a^{m+1-j} b^j.$$

(continued on page 470)

Hence the second summation on the right-hand side above is

$$\sum_{j=1}^{m+1} \binom{m}{j-1} a^{m+1-j} b^j.$$

But the j in this summation is a dummy variable; it can be replaced by the letter k, as long as the replacement is made everywhere the j occurs:

$$\sum_{j=1}^{m+1} \binom{m}{j-1} a^{m+1-j} b^j = \sum_{k=1}^{m+1} \binom{m}{k-1} a^{m+1-k} b^k.$$

Substituting back, we get

$$(a+b)^{m+1} = \sum_{k=0}^{m} \binom{m}{k} a^{m+1-k} b^k + \sum_{k=1}^{m+1} \binom{m}{k-1} a^{m+1-k} b^k.$$

[The reason for the above maneuvers was to make the powers of a and b agree so that we can add the summations together term by term, except for the first and the last terms, which we must write separately.]

Thus

$$(a+b)^{m+1} = \binom{m}{0} a^{m+1-0} b^0 + \sum_{k=1}^{m} \left[\binom{m}{k} + \binom{m}{k-1}\right] a^{m+1-k} b^k$$

$$+ \binom{m}{(m+1)-1} a^{m+1-(m+1)} b^{m+1}$$

$$= a^{m+1} + \sum_{k=1}^{m} \left[\binom{m}{k} + \binom{m}{k-1}\right] a^{m+1-k} b^k + b^{m+1}$$

since $a^0 = b^0 = 1$ and $\binom{m}{0} = \binom{m}{m} = 1.$

But

$$\left[\binom{m}{k} + \binom{m}{k-1}\right] = \binom{m+1}{k} \qquad \text{by Pascal's formula.}$$

Hence

$$(a+b)^{m+1} = a^{m+1} + \sum_{k=1}^{m} \binom{m+1}{k} a^{(m+1)-k} b^k + b^{m+1}$$

$$= \sum_{k=0}^{m+1} \binom{m+1}{k} a^{(m+1)-k} b^k \qquad \text{because } \binom{m+1}{0} = \binom{m+1}{m+1} = 1$$

which is what we needed to show.

It is instructive to write out the product $(a+b) \cdot (a+b)^m$ without using the summation notation but using the inductive hypothesis about $(a+b)^m$:

$$(a+b)^{m+1} = (a+b) \cdot \left[a^m + \binom{m}{1} a^{m-1} b + \cdots + \binom{m}{k-1} a^{m-(k-1)} b^{k-1} \right.$$

$$\left. + \binom{m}{k} a^{m-k} b^k + \cdots + \binom{m}{m-1} a b^{m-1} + b^m \right].$$

You will see that the first and last coefficients are clearly 1 and that the term containing $a^{m+1-k}b^k$ is obtained from multiplying $a^{m-k}b^k$ by a and $a^{m-(k-1)}b^{k-1}$ by b [*because* $m + 1 - k = m - (k - 1)$]. Hence the coefficient of $a^{m+1-k}b^k$ equals the sum of $\binom{m}{k}$ and $\binom{m}{k-1}$. This is the crux of the algebraic proof.

If n and r are nonnegative integers and $r \leq n$, then $\binom{n}{r}$ is called a **binomial coefficient** because it is one of the coefficients in the expansion of the binomial expression $(a + b)^n$.

The combinatorial proof of the binomial theorem follows.

Proof of Binomial Theorem (combinatorial version):

[The combinatorial argument used here to prove the binomial theorem works only for $n \geq 1$. If we were giving only this combinatorial proof, we would have to prove the case $n = 0$ separately. Since we have already given a complete algebraic proof that includes the case $n = 0$, we do not prove it again here.]

Let a and b be real numbers and n an integer that is at least 1. The expression $(a + b)^n$ can be expanded into products of n letters, where each letter is either a or b. For each $k = 0, 1, 2, \ldots, n$, the product

$$a^{n-k}b^k = \underbrace{a \cdot a \cdot a \cdots a}_{n - k \text{ factors}} \cdot \underbrace{b \cdot b \cdot b \cdots b}_{k \text{ factors}}$$

occurs as a term in the sum the same number of times as there are orderings of $(n - k)$ a's and k b's. But this number is $\binom{n}{k}$, the number of ways to choose k positions into which to place the b's. *[The other $n - k$ positions will be filled by a's.]* Hence, when like terms are combined, the coefficient of $a^{n-k}b^k$ in the sum is $\binom{n}{k}$. Thus

$$(a + b)^n = \sum_{k=0}^{n} \binom{n}{k} a^{n-k} b^k.$$

This is what was to be proved.

Example 9.6.5 Substituting into the Binomial Theorem

Expand the following expressions using the binomial theorem:

a. $(a + b)^5$ b. $(x - 4y)^4$

Solution

a. $(a + b)^5 = \sum_{k=0}^{5} \binom{5}{k} a^{5-k} b^k$

$= a^5 + \binom{5}{1} a^{5-1}b^1 + \binom{5}{2} a^{5-2}b^2 + \binom{5}{3} a^{5-3}b^3 + \binom{5}{4} a^{5-4}b^4 + b^5$

$= a^5 + 5a^4b + 10a^3b^2 + 10a^2b^3 + 5ab^4 + b^5$

b. Observe that $(x - 4y)^4 = (x + (-4y))^4$. So let $a = x$ and $b = (-4y)$, and substitute into the binomial theorem.

$$(x - 4y)^4 = \sum_{k=0}^{4} \binom{4}{k} x^{4-k}(-4y)^k$$

$$= x^4 + \binom{4}{1}x^{4-1}(-4y)^1 + \binom{4}{2}x^{4-2}(-4y)^2 + \binom{4}{3}x^{4-3}(-4y)^3 + (-4y)^4$$

$$= x^4 + 4x^3(-4y) + 6x^2(16y^2) + 4x^1(-64y^3) + (256y^4)$$

$$= x^4 - 16x^3y + 96x^2y^2 - 256xy^3 + 256y^4 \qquad \blacksquare$$

Example 9.6.6 Deriving Another Combinatorial Identity from the Binomial Theorem

Use the binomial theorem to show that

$$2^n = \sum_{k=0}^{n} \binom{n}{k} = \binom{n}{0} + \binom{n}{1} + \binom{n}{2} + \cdots + \binom{n}{n}$$

for all integers $n \geq 0$.

Solution Since $2 = 1 + 1$, $2^n = (1+1)^n$. Apply the binomial theorem to this expression by letting $a = 1$ and $b = 1$. Then

$$2^n = \sum_{k=0}^{n} \binom{n}{k} \cdot 1^{n-k} \cdot 1^k = \sum_{k=0}^{n} \binom{n}{k} \cdot 1 \cdot 1$$

since $1^{n-k} = 1$ and $1^k = 1$. Consequently,

$$2^n = \sum_{k=0}^{n} \binom{n}{k} = \binom{n}{0} + \binom{n}{1} + \binom{n}{2} + \cdots + \binom{n}{n}. \qquad \blacksquare$$

Example 9.6.7 Using a Combinatorial Argument to Derive the Identity

According to Theorem 6.3.1, a set with n elements has 2^n subsets. Apply this fact to give a combinatorial argument to justify the identity

$$\binom{n}{0} + \binom{n}{1} + \binom{n}{2} + \binom{n}{3} + \cdots + \binom{n}{n} = 2^n.$$

Solution Suppose S is a set with n elements. Then every subset of S has some number of elements k, where k is between 0 and n. It follows that the total number of subsets of S, $N(\mathscr{P}(S))$, can be expressed as the following sum:

$$\begin{bmatrix} \text{number of} \\ \text{subsets} \\ \text{of } S \end{bmatrix} = \begin{bmatrix} \text{number of} \\ \text{subsets of} \\ \text{size 0} \end{bmatrix} + \begin{bmatrix} \text{number of} \\ \text{subsets of} \\ \text{size 1} \end{bmatrix} + \cdots + \begin{bmatrix} \text{number of} \\ \text{subsets of} \\ \text{size } n \end{bmatrix}.$$

Now the number of subsets of size k of a set with n elements is $\binom{n}{k}$. Hence the

$$\text{number of subsets of } S = \binom{n}{0} + \binom{n}{1} + \binom{n}{2} + \cdots + \binom{n}{n}$$

But by Theorem 6.3.1, S has 2^n subsets. Hence

$$\binom{n}{0} + \binom{n}{1} + \binom{n}{2} + \binom{n}{3} + \cdots + \binom{n}{n} = 2^n. \qquad \blacksquare$$

Example 9.6.8 Using the Binomial Theorem to Simplify a Sum

Express the following sum in **closed form** (without using a summation symbol and without using an ellipsis $\cdots$):

$$\sum_{k=0}^{n} \binom{n}{k} 9^k$$

Solution   When the number 1 is raised to any power, the result is still 1. Thus

$$\sum_{k=0}^{n} \binom{n}{k} 9^k = \sum_{k=0}^{n} \binom{n}{k} 1^{n-k} 9^k$$

$$= (1+9)^n \quad \text{by the binomial theorem with } a = 1 \text{ and } b = 9$$

$$= 10^n. \qquad \blacksquare$$

Test Yourself

1. If n and r are nonnegative integers with $r \leq n$, then the relation between $\binom{n}{r}$ and $\binom{n}{n-r}$ is _____.

2. Pascal's formula says that if n and r are positive integers with $r \leq n$, then _____.

3. The crux of the algebraic proof of Pascal's formula is that to add two fractions you need to express both of them with a _____.

4. The crux of the combinatorial proof of Pascal's formula is that the set of subsets of size r of a set $\{x_1, x_2, \ldots, x_{n+1}\}$ can be partitioned into the set of subsets of size r that contain _____ and those that _____.

5. The binomial theorem says that given any real numbers a and b and any nonnegative integer n, _____.

6. The crux of the algebraic proof of the binomial theorem is that, after making a change of variable so that two summations have the same lower and upper limits and the exponents of a and b are the same, you use the fact that $\binom{m}{k} + \binom{m}{k-1} -$ _____.

7. The crux of the combinatorial proof of the binomial theorem is that the number of ways to arrange k b's and $(n-k)$ a's in order is _____.

Exercise Set 9.6

In 1–4, use Theorem 9.5.1 to compute the values of the indicated quantities. (Assume n is an integer.)

1. $\binom{n}{0}$, for $n \geq 0$

2. $\binom{n}{1}$, for $n \geq 1$

3. $\binom{n}{2}$, for $n \geq 2$

4. $\binom{n}{3}$, for $n \geq 3$

5. Use Theorem 9.5.1 to prove algebraically that $\binom{n}{r} = \binom{n}{n-r}$, for integers n and r with $0 \leq r \leq n$. (This can be done by direct calculation; it is not necessary to use mathematical induction.)

Justify the equations in 6–9 either by deriving them from formulas in Example 9.6.1 or by direct computation from Theorem 9.5.1. Assume m, n, k, and r are integers.

6. $\binom{m+k}{m+k-1} = m+k$, for $m+k \geq 1$

7. $\binom{n+3}{n+1} = \dfrac{(n+3)(n+2)}{2}$, for $n \geq -1$

8. $\binom{k-r}{k-r} = 1$, for $k-r \geq 0$

9. $\binom{2(n+1)}{2n} = (n+1)(2n+1)$ for $n \geq 0$

10. a. Use Pascal's triangle given in Table 9.6.1 to compute the values of $\binom{6}{2}$, $\binom{6}{3}$, $\binom{6}{4}$, and $\binom{6}{5}$.
    b. Use the result of part (a) and Pascal's formula to compute $\binom{7}{3}$, $\binom{7}{4}$, and $\binom{7}{5}$.
    c. Complete the row of Pascal's triangle that corresponds to $n = 7$.

11. The row of Pascal's triangle that corresponds to $n = 8$ is as follows:

    1   8   28   56   70   56   28   8   1.

    What is the row that corresponds to $n = 9$?

12. Use Pascal's formula repeatedly to derive a formula for $\binom{n+3}{r}$ in terms of values of $\binom{n}{k}$ with $k \leq r$. (Assume n and r are integers with $n \geq r \geq 3$.)

13. Use Pascal's formula to prove by mathematical induction that if n is an integer and $n \geq 1$, then

$$\sum_{i=2}^{n+1} \binom{i}{2} = \binom{2}{2} + \binom{3}{2} + \cdots + \binom{n+1}{2}$$
$$= \binom{n+2}{3}.$$

H 14. Prove that if n is an integer and $n \geq 1$, then

$$1 \cdot 2 + 2 \cdot 3 + \cdots + n(n+1) = 2 \binom{n+2}{3}.$$

15. Prove the following generalization of exercise 13: Let r be a fixed nonnegative integer. For all integers n with $n \geq r$,

$$\sum_{i=r}^{n} \binom{i}{r} = \binom{n+1}{r+1}.$$

16. Think of a set with $m + n$ elements as composed of two parts, one with m elements and the other with n elements. Give a combinatorial argument to show that

$$\binom{m+n}{r} = \binom{m}{0}\binom{n}{r} + \binom{m}{1}\binom{n}{r-1} + \cdots + \binom{m}{r}\binom{n}{0},$$

where m and n are positive integers and r is an integer that is less than or equal to both m and n.

This identity gives rise to many useful additional identities involving the quantities $\binom{n}{k}$. Because Alexander Vandermonde published an influential article about it in 1772, it is generally called the *Vandermonde convolution*. However, it was known at least in the 1300s in China by Chu Shih-chieh.

H 17. Prove that for all integers $n \geq 0$,

$$\binom{n}{0}^2 + \binom{n}{1}^2 + \cdots + \binom{n}{n}^2 = \binom{2n}{n}.$$

18. Let m be any nonnegative integer. Use mathematical induction and Pascal's formula to prove that for all integers $n \geq 0$,

$$\binom{m}{0} + \binom{m+1}{1} + \cdots + \binom{m+n}{n} = \binom{m+n+1}{n}.$$

Use the binomial theorem to expand the expressions in 19–27.

19. $(1+x)^7$ 20. $(p+q)^6$ 21. $(1-x)^6$

22. $(u-v)^5$ 23. $(p-2q)^4$ 24. $(u^2-3v)^4$

25. $\left(x+\dfrac{1}{x}\right)^5$ 26. $\left(\dfrac{3}{a}-\dfrac{a}{3}\right)^5$ 27. $\left(x^2+\dfrac{1}{x}\right)^5$

28. In Example 9.6.5 it was shown that

$$(a+b)^5 = a^5 + 5a^4b + 10a^3b^2 + 10a^2b^3 + 5ab^4 + b^{35}.$$

Evaluate $(a+b)^6$ by substituting the expression above into the equation

$$(a+b)^6 = (a+b)(a+b)^5$$

and then multiplying out and combining like terms.

In 29–34, find the coefficient of the given term when the expression is expanded by the binomial theorem.

29. x^6y^3 in $(x+y)^9$ 30. x^7 in $(2x+3)^{10}$

31. a^5b^7 in $(a-2b)^{12}$ 32. $u^{16}v^4$ in $(u^2-v^2)^{10}$

33. $p^{16}q^7$ in $(3p^2-2q)^{15}$ 34. x^9y^{10} in $(2x-3y^2)^{14}$

35. As in the proof of the binomial theorem, transform the summation

$$\sum_{k=0}^{n} \binom{m}{k} a^{m-k}b^{k+1}$$

by making the change of variable $j = k + 1$.

Use the binomial theorem to prove each statement in 36–41.

36. For all integers $n \geq 1$,

$$\binom{n}{0} - \binom{n}{1} + \binom{n}{2} - \cdots + (-1)^n \binom{n}{n} = 0.$$

(*Hint:* Use the fact that $1 + (-1) = 0$.)

H 37. For all integers $n \geq 0$,

$$3^n = \binom{n}{0} + 2\binom{n}{1} + 2^2\binom{n}{2} + \cdots + 2^n\binom{n}{n}.$$

38. For all integers $m \geq 0$, $\displaystyle\sum_{i=0}^{m}(-1)^i \binom{m}{i} 2^{m-i} = 1$.

39. For all integers $n \geq 0$, $\displaystyle\sum_{i=0}^{n}(-1)^i \binom{n}{i} 3^{n-i} = 2^n$.

40. For all integers $n \geq 0$ and for all nonnegative real numbers x, $1 + nx \leq (1+x)^n$.

H 41. For all integers $n \geq 1$,

$$\binom{n}{0} - \frac{1}{2}\binom{n}{1} + \frac{1}{2^2}\binom{n}{2} - \frac{1}{2^3}\binom{n}{3}$$
$$+ \cdots + (-1)^{n-1}\frac{1}{2^{n-1}}\binom{n}{n-1} = \begin{cases} 0 & \text{if } n \text{ is even} \\ \dfrac{1}{2^{n-1}} & \text{if } n \text{ is odd} \end{cases}.$$

42. Use mathematical induction to prove that for all integers $n \geq 1$, if S is a set with n elements, then S has the same number of subsets with an even number of elements as with an odd number of elements. Use this fact to give a combinatorial argument to justify the identity of exercise 36.

Express each of the sums in 43–54 in closed form (without using a summation symbol and without using an ellipsis $\cdots$).

43. $\displaystyle\sum_{k=0}^{n} \binom{n}{k} 5^k$ 44. $\displaystyle\sum_{i=0}^{m} \binom{m}{i} 4^i$

45. $\displaystyle\sum_{i=0}^{n} \binom{n}{i} x^i$ 46. $\displaystyle\sum_{k=0}^{m} \binom{m}{k} 2^{m-k}x^k$

47. $\displaystyle\sum_{j=0}^{2n}(-1)^j\binom{2n}{j}x^j$

48. $\displaystyle\sum_{r=0}^{n}\binom{n}{r}x^{2r}$

49. $\displaystyle\sum_{i=0}^{m}\binom{m}{i}p^{m-i}q^{2i}$

50. $\displaystyle\sum_{k=0}^{n}\binom{n}{k}\frac{1}{2^k}$

51. $\displaystyle\sum_{i=0}^{m}(-1)^i\binom{m}{i}\frac{1}{2^i}$

52. $\displaystyle\sum_{k=0}^{n}\binom{n}{k}3^{2n-2k}2^{2k}$

53. $\displaystyle\sum_{i=0}^{n}(-1)^i\binom{n}{i}5^{n-i}2^i$

54. $\displaystyle\sum_{k=0}^{n}(-1)^k\binom{n}{k}3^{2n-2k}2^{2k}$

✱**55.** (For students who have studied calculus)

a. Explain how the equation below follows from the binomial theorem:

$$(1+x)^n = \sum_{k=0}^{n}\binom{n}{k}x^k.$$

b. Write the formula obtained by taking the derivative of both sides of the equation in part (a) with respect to x.

c. Use the result of part (b) to derive the formulas below.

(i) $2^{n-1} = \dfrac{1}{n}\left[\binom{n}{1}+2\binom{n}{2}+3\binom{n}{3}+\cdots+n\binom{n}{n}\right]$

(ii) $\displaystyle\sum_{k=1}^{n}k\binom{n}{k}(-1)^k = 0$

d. Express $\displaystyle\sum_{k=1}^{n}k\binom{n}{k}3^k$ in closed form (without using a summation sign or ellipsis).

Answers for Test Yourself

1. $\binom{n}{r}=\binom{n}{n-r}$ 2. $\binom{n+1}{r}=\binom{n}{r-1}+\binom{n}{r}$ 3. common denominator   4. x_{n+1}; do not contain x_{n+1}

5. $(a+b)^n = \displaystyle\sum_{k=0}^{n}\binom{n}{k}a^{n-k}b^k$ 6. $\binom{m+1}{k}$ 7. $\binom{n}{k}$

CHAPTER 10

GRAPHS AND TREES

Graphs and trees have appeared previously in this book as convenient visualizations. For instance, a possibility tree shows all possible outcomes of a multistep operation with a finite number of outcomes for each step, and the directed graph of a relation on a set shows which elements of the set are related to which.

In this chapter we present an introduction to the mathematics of graphs and trees, discussing concepts such as the degree of a vertex, connectedness, Euler and Hamiltonian circuits, the relation between the number of vertices and the number of edges of a tree, and properties of rooted trees. Applications include uses of graphs and trees in the study of artificial intelligence, chemistry, scheduling problems, and transportation systems.

10.1 Graphs: Definitions and Basic Properties

The whole of mathematics consists in the organization of a series of aids to the imagination in the process of reasoning. — Alfred North Whitehead, 1861–1947

Imagine an organization that wants to set up teams of three to work on some projects. In order to maximize the number of people on each team who had previous experience working together successfully, the director asked the members to provide names of their past partners. This information is displayed below both in a table and in a diagram.

Name	Past Partners
Ana	Dan, Flo
Bev	Cai, Flo, Hal
Cai	Bev, Flo
Dan	Ana, Ed
Ed	Dan, Hal
Flo	Cai, Bev, Ana
Gia	Hal
Hal	Gia, Ed, Bev, Ira
Ira	Hal

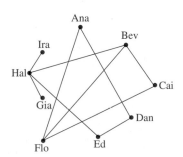

From the diagram, it is easy to see that Bev, Cai, and Flo are a group of three past partners, and so they should form one of these teams. The figure on the next page shows the result when these three names are removed from the diagram.

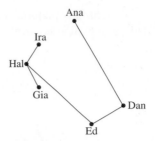

This drawing shows that placing Hal on the same team as Ed would leave Gia and Ira on a team containing no past partners. However, if Hal is placed on a team with Gia and Ira, then the remaining team would consist of Ana, Dan, and Ed, and both teams would contain at least one pair of past partners.

Drawings such as these are illustrations of a structure known as a *graph*. The dots are called *vertices* (plural of *vertex*) and the line segments joining vertices are called *edges*. As you can see from the first drawing, it is possible for two edges to cross at a point that is not a vertex. Note also that the type of graph described here is quite different from the "graph of an equation" or the "graph of a function."

In general, a graph consists of a set of vertices and a set of edges connecting various pairs of vertices. The edges may be straight or curved and should either connect one vertex to another or a vertex to itself, as shown below.

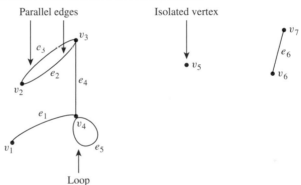

In this drawing, the vertices have been labeled with v's and the edges with e's. When an edge connects a vertex to itself (as e_5 does), it is called a *loop*. When two edges connect the same pair of vertices (as e_2 and e_3 do), they are said to be *parallel*. It is quite possible for a vertex to be unconnected by an edge to any other vertex in the graph (as v_5 is), and in that case the vertex is said to be *isolated*. The formal definition of a graph follows.

> **• Definition**
>
> A **graph** G consists of two finite sets: a nonempty set $V(G)$ of **vertices** and a set $E(G)$ of **edges,** where each edge is associated with a set consisting of either one or two vertices called its **endpoints.** The correspondence from edges to endpoints is called the **edge-endpoint function.**
>
> An edge with just one endpoint is called a **loop,** and two or more distinct edges with the same set of endpoints are said to be **parallel.** An edge is said to **connect** its endpoints; two vertices that are connected by an edge are called **adjacent;** and a vertex that is an endpoint of a loop is said to be **adjacent to itself.**
>
> An edge is said to be **incident on** each of its endpoints, and two edges incident on the same endpoint are called **adjacent.** A vertex on which no edges are incident is called **isolated.**

Graphs have pictorial representations in which the vertices are represented by dots and the edges by line segments. A given pictorial representation uniquely determines a graph.

Example 10.1.1 Terminology

Consider the following graph:

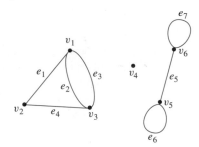

a. Write the vertex set and the edge set, and give a table showing the edge-endpoint function.

b. Find all edges that are incident on v_1, all vertices that are adjacent to v_1, all edges that are adjacent to e_1, all loops, all parallel edges, all vertices that are adjacent to themselves, and all isolated vertices.

Solution

a. vertex set $= \{v_1, v_2, v_3, v_4, v_5, v_6\}$
edge set $= \{e_1, e_2, e_3, e_4, e_5, e_6, e_7\}$
edge-endpoint function:

Edge	Endpoints
e_1	$\{v_1, v_2\}$
e_2	$\{v_1, v_3\}$
e_3	$\{v_1, v_3\}$
e_4	$\{v_2, v_3\}$
e_5	$\{v_5, v_6\}$
e_6	$\{v_5\}$
e_7	$\{v_6\}$

Note that the isolated vertex v_4 does not appear in this table. Although each edge must have either one or two endpoints, a vertex need not be an endpoint of an edge.

b. $e_1, e_2,$ and e_3 are incident on v_1.
v_2 and v_3 are adjacent to v_1.
$e_2, e_3,$ and e_4 are adjacent to e_1.
e_6 and e_7 are loops.
e_2 and e_3 are parallel.
v_5 and v_6 are adjacent to themselves.
v_4 is an isolated vertex. ∎

As noted earlier, a given pictorial representation uniquely determines a graph. However, a given graph may have more than one pictorial representation. Such things as the lengths or curvatures of the edges and the relative position of the vertices on the page may vary from one pictorial representation to another.

Example 10.1.2 Drawing More Than One Picture for a Graph

Consider the graph specified as follows:

$$\text{vertex set} = \{v_1, v_2, v_3, v_4\}$$
$$\text{edge set} = \{e_1, e_2, e_3, e_4\}$$
$$\text{edge-endpoint function:}$$

Edge	Endpoints
e_1	$\{v_1, v_3\}$
e_2	$\{v_2, v_4\}$
e_3	$\{v_2, v_4\}$
e_4	$\{v_3\}$

Both drawings (a) and (b) shown below are pictorial representations of this graph.

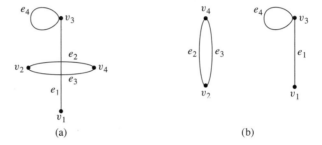

(a)   (b) ∎

Example 10.1.3 Labeling Drawings to Show They Represent the Same Graph

Consider the two drawings shown in Figure 10.1.1. Label vertices and edges in such a way that both drawings represent the same graph.

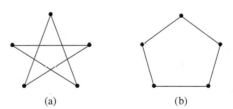

(a)   (b)

Figure 10.1.1

Solution   Imagine putting one end of a piece of string at the top vertex of Figure 10.1.1(a) (call this vertex v_1), then laying the string to the next adjacent vertex on the lower right (call this vertex v_2), then laying it to the next adjacent vertex on the upper left (v_3), and so forth, returning finally to the top vertex v_1. Call the first edge e_1, the second e_2, and so forth, as shown below.

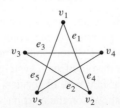

Now imagine picking up the piece of string, together with its labels, and repositioning it as follows:

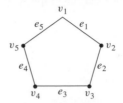

This is the same as Figure 10.1.1(b), so both drawings are representations of the graph with vertex set $\{v_1, v_2, v_3, v_4, v_5\}$, edge set $\{e_1, e_2, e_3, e_4, e_5\}$, and edge-endpoint function as follows:

Edge	Endpoints
e_1	$\{v_1, v_2\}$
e_2	$\{v_2, v_3\}$
e_3	$\{v_3, v_4\}$
e_4	$\{v_4, v_5\}$
e_5	$\{v_5, v_1\}$

■

In Chapter 8 we discussed the directed graph of a binary relation on a set. The general definition of directed graph is similar to the definition of graph, except that one associates an *ordered pair* of vertices with each edge instead of a *set* of vertices. Thus each edge of a directed graph can be drawn as an arrow going from the first vertex to the second vertex of the ordered pair.

• Definition

A **directed graph,** or **digraph,** consists of two finite sets: a nonempty set $V(G)$ of vertices and a set $D(G)$ of directed edges, where each is associated with an ordered pair of vertices called its **endpoints.** If edge e is associated with the pair (v, w) of vertices, then e is said to be the (**directed**) **edge** from v to w.

Note that each directed graph has an associated ordinary (undirected) graph, which is obtained by ignoring the directions of the edges.

Examples of Graphs

Graphs are a powerful problem-solving tool because they enable us to represent a complex situation with a single image that can be analyzed both visually and with the aid of a computer. A few examples follow, and others are included in the exercises.

Example 10.1.4 Using a Graph to Represent a Network

Telephone, electric power, gas pipeline, and air transport systems can all be represented by graphs, as can computer networks—from small local area networks to the global Internet system that connects millions of computers worldwide. Questions that arise in the design of such systems involve choosing connecting edges to minimize cost, optimize a certain type of service, and so forth. A typical network, called a hub and spoke model, is shown on the next page.

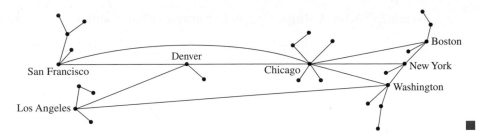

Example 10.1.5 Using a Graph to Represent the World Wide Web

The World Wide Web, or Web, is a system of interlinked documents, or webpages, contained on the Internet. Users employing Web browsers, such as Internet Explorer, Google Chrome, Apple Safari, and Opera, can move quickly from one webpage to another by clicking on hyperlinks, which use versions of software called hypertext transfer protocols (HTTPs). Individuals and individual companies create the pages, which they transmit to servers that contain software capable of delivering them to those who request them through a Web browser. Because the amount of information currently on the Web is so vast, search engines, such as Google, Yahoo, and Bing, have algorithms for finding information very efficiently.

The picture below shows a minute fraction of the hyperlink connections on the Internet that radiate in and out from the Wikipedia main page.

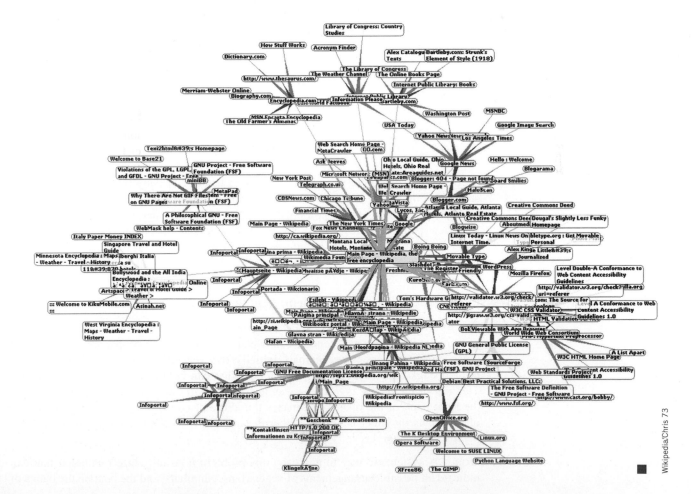

Example 10.1.6 Using a Graph to Represent Knowledge

In many applications of artifical intelligence, a knowledge base of information is collected and represented inside a computer. Because of the way the knowledge is represented and because of the properties that govern the artificial intelligence program, the computer is not limited to retrieving data in the same form as it was entered; it can also derive new facts from the knowledge base by using certain built-in rules of inference. For example, from the knowledge that the *Los Angeles Times* is a big-city daily and that a big-city daily contains national news, an artifical intelligence program could infer that the *Los Angeles Times* contains national news. The directed graph shown in Figure 10.1.2 is a pictorial representation for a simplified knowledge base about periodical publications.

According to this knowledge base, what paper finish does the *New York Times* use?

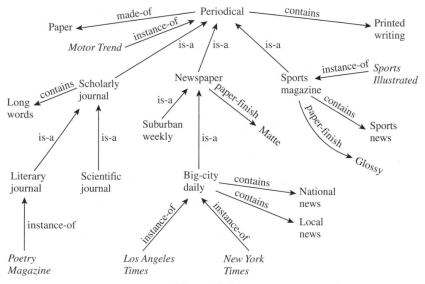

Figure 10.1.2

Solution The arrow going from *New York Times* to big-city daily (labeled "instance-of") shows that the *New York Times* is a big-city daily. The arrow going from big-city daily to newspaper (labeled "is-a") shows that a big-city daily is a newspaper. The arrow going from newspaper to matte (labeled "paper-finish") indicates that the paper finish on a newspaper is matte. Hence it can be inferred that the paper finish on the *New York Times* is matte. ■

Example 10.1.7 Using a Graph to Solve a Problem: Vegetarians and Cannibals

The following is a variation of a famous puzzle often used as an example in the study of artificial intelligence. It concerns an island on which all the people are of one of two types, either vegetarians or cannibals. Initially, two vegetarians and two cannibals are on the left bank of a river. With them is a boat that can hold a maximum of two people. The aim of the puzzle is to find a way to transport all the vegetarians and cannibals to the right bank of the river. What makes this difficult is that at no time can the number of cannibals on either bank outnumber the number of vegetarians. Otherwise, disaster befalls the vegetarians!

Solution A systematic way to approach this problem is to introduce a notation that can indicate all possible arrangements of vegetarians, cannibals, and the boat on the banks of the river. For example, you could write (vvc/Bc) to indicate that there are two vegetarians and one cannibal on the left bank and one cannibal and the boat on the right bank. Then $(vvccB/)$ would indicate the initial position in which both vegetarians, both cannibals, and the boat are on the left bank of the river. The aim of the puzzle is to figure out a sequence of moves to reach the position $(/Bvvcc)$ in which both vegetarians, both cannibals, and the boat are on the right bank of the river.

Construct a graph whose vertices are the various arrangements that can be reached in a sequence of legal moves starting from the initial position. Connect vertex x to vertex y if it is possible to reach vertex y in one legal move from vertex x. For instance, from the initial position there are four legal moves: one vegetarian and one cannibal can take the boat to the right bank; two cannibals can take the boat to the right bank; one cannibal can take the boat to the right bank; or two vegetarians can take the boat to the right bank. You can show these by drawing edges connecting vertex $(vvccB/)$ to vertices (vc/Bvc), (vv/Bcc), $(vvcBc)$, and (cc/Bvv). (It might seem natural to draw directed edges rather than undirected edges from one vertex to another. The rationale for drawing undirected edges is that each legal move is reversible.) From the position (vc/Bvc), the only legal moves are to go back to $(vvccB/)$ or to go to $(vvcB/c)$. You can also show these by drawing in edges. Continue this process until finally you reach $(/Bvvcc)$. From Figure 10.1.3 it is apparent that one successful sequence of moves is $(vvccB/) \rightarrow (vc/Bvc) \rightarrow (vvcB/c) \rightarrow (c/Bvvc) \rightarrow (ccB/vv) \rightarrow (/Bvvcc)$.

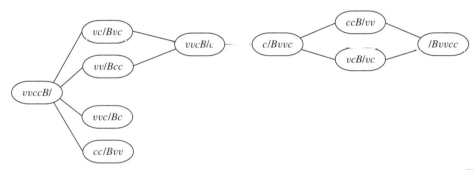

Figure 10.1.3 ∎

Special Graphs

One important class of graphs consists of those that do not have any loops or parallel edges. Such graphs are called *simple*. In a simple graph, no two edges share the same set of endpoints, so specifying two endpoints is sufficient to determine an edge.

> **• Definition and Notation**
>
> A **simple graph** is a graph that does not have any loops or parallel edges. In a simple graph, an edge with endpoints v and w is denoted $\{v, w\}$.

Example 10.1.8 A Simple Graph

Draw all simple graphs with the four vertices $\{u, v, w, x\}$ and two edges, one of which is $\{u, v\}$.

Solution   Each possible edge of a simple graph corresponds to a subset of two vertices. Given four vertices, there are $\binom{4}{2} = 6$ such subsets in all: $\{u, v\}, \{u, w\}, \{u, x\}, \{v, w\}, \{v, x\}$, and $\{w, x\}$. Now one edge of the graph is specified to be $\{u, v\}$, so any of the remaining five from this list can be chosen to be the second edge. The possibilities are shown on the next page.

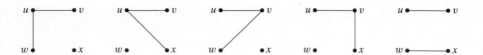

Another important class of graphs consists of those that are "complete" in the sense that all pairs of vertices are connected by edges.

Note The K stands for the German word *komplett*, which means "complete."

> ● **Definition**
>
> Let n be a positive integer. A **complete graph on n vertices,** denoted K_n, is a simple graph with n vertices and exactly one edge connecting each pair of distinct vertices.

Example 10.1.9 Complete Graphs on n Vertices: K_1, K_2, K_3, K_4, K_5

The complete graphs K_1, K_2, K_3, K_4, and K_5 can be drawn as follows:

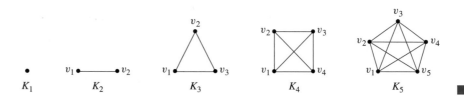

In yet another class of graphs, the vertex set can be separated into two subsets: Each vertex in one of the subsets is connected by exactly one edge to each vertex in the other subset, but not to any vertices in its own subset. Such a graph is called *complete bipartite.*

> ● **Definition**
>
> Let m and n be positive integers. A **complete bipartite graph on (m, n) vertices,** denoted $K_{m,n}$, is a simple graph with distinct vertices $v_1, v_2, \ldots, v_m$ and $w_1, w_2, \ldots, w_n$ that satisfies the following properties: For all $i, k = 1, 2, \ldots, m$ and for all $j, l = 1, 2, \ldots, n$,
>
> 1. There is an edge from each vertex v_i to each vertex w_j.
>
> 2. There is no edge from any vertex v_i to any other vertex v_k.
>
> 3. There is no edge from any vertex w_j to any other vertex w_l.

Example 10.1.10 Complete Bipartite Graphs: $K_{3,2}$ and $K_{3,3}$

The complete bipartite graphs $K_{3,2}$ and $K_{3,3}$ are illustrated below.

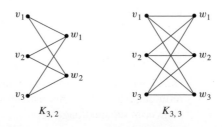

> • **Definition**
>
> A graph H is said to be a **subgraph** of a graph G if, and only if, every vertex in H is also a vertex in G, every edge in H is also an edge in G, and every edge in H has the same endpoints as it has in G.

Example 10.1.11 Subgraphs

List all subgraphs of the graph G with vertex set $\{v_1, v_2\}$ and edge set $\{e_1, e_2, e_3\}$, where the endpoints of e_1 are v_1 and v_2, the endpoints of e_2 are v_1 and v_2, and e_3 is a loop at v_1.

Solution G can be drawn as shown below.

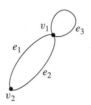

There are 11 subgraphs of G, which can be grouped according to those that do not have any edges, those that have one edge, those that have two edges, and those that have three edges. The 11 subgraphs are shown in Figure 10.1.4.

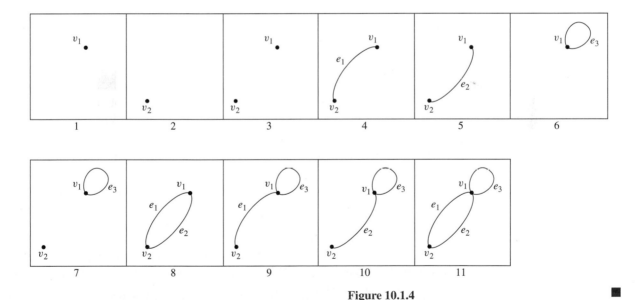

Figure 10.1.4

The Concept of Degree

The *degree of a vertex* is the number of end segments of edges that "stick out of" the vertex. We will show that the sum of the degrees of all the vertices in a graph is twice the number of edges of the graph.

> **• Definition**
>
> Let G be a graph and v a vertex of G. The **degree of v,** denoted **deg(v),** equals the number of edges that are incident on v, with an edge that is a loop counted twice. The **total degree of G** is the sum of the degrees of all the vertices of G.

Since an edge that is a loop is counted twice, the degree of a vertex can be obtained from the drawing of a graph by counting how many end segments of edges are incident on the vertex. This is illustrated below.

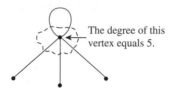

The degree of this vertex equals 5.

Example 10.1.12 Degree of a Vertex and Total Degree of a Graph

Find the degree of each vertex of the graph G shown below. Then find the total degree of G.

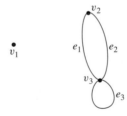

Solution

$\deg(v_1) = 0$ since no edge is incident on v_1 (v_1 is isolated).

$\deg(v_2) = 2$ since both e_1 and e_2 are incident on v_2.

$\deg(v_3) = 4$ since e_1 and e_2 are incident on v_3 and the loop e_3 is also incident on v_3 (and contributes 2 to the degree of v_3).

total degree of $G = \deg(v_1) + \deg(v_2) + \deg(v_3) = 0 + 2 + 4 = 6$. ∎

Note that the total degree of the graph G of Example 10.1.12, which is 6, equals twice the number of edges of G, which is 3. Roughly speaking, this is true because each edge has two end segments, and each end segment is counted once toward the degree of some vertex. This result generalizes to any graph.

In fact, for any graph without loops, the general result can be explained as follows: Imagine a group of people at a party. Depending on how social they are, each person shakes hands with various other people. So each person participates in a certain number of handshakes—perhaps many, perhaps none—but because each handshake is experienced by two different people, if the numbers experienced by each person are added together, the sum will equal twice the total number of handshakes. This is such an attractive way of understanding the situation that the following theorem is often called the *handshake lemma* or the *handshake theorem*. As the proof demonstrates, the conclusion is true even if the graph contains loops.

Theorem 10.1.1 The Handshake Theorem

If G is any graph, then the sum of the degrees of all the vertices of G equals twice the number of edges of G. Specifically, if the vertices of G are $v_1, v_2, \ldots, v_n$, where n is a nonnegative integer, then

$$\text{the total degree of } G = \deg(v_1) + \deg(v_2) + \cdots + \deg(v_n)$$
$$= 2 \cdot (\text{the number of edges of } G).$$

Proof:

Let G be a particular but arbitrarily chosen graph, and suppose that G has n vertices $v_1, v_2, \ldots, v_n$ and m edges, where n is a positive integer and m is a nonnegative integer. We claim that each edge of G contributes 2 to the total degree of G. For suppose e is an arbitrarily chosen edge with endpoints v_i and v_j. This edge contributes 1 to the degree of v_i and 1 to the degree v_j. As shown below, this is true even if $i = j$, because an edge that is a loop is counted twice in computing the degree of the vertex on which it is incident.

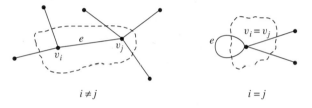

Therefore, e contributes 2 to the total degree of G. Since e was arbitrarily chosen, this shows that *each* edge of G contributes 2 to the total degree of G. Thus

$$\text{the total degree of } G = 2 \cdot (\text{the number of edges of } G).$$

The following corollary is an immediate consequence of Theorem 10.1.1.

Corollary 10.1.2

The total degree of a graph is even.

Proof:

By Theorem 10.1.1 the total degree of G equals 2 times the number of edges, which is an integer, and so the total degree of G is even.

Example 10.1.13 Determining Whether Certain Graphs Exist

Draw a graph with the specified properties or show that no such graph exists.

a. A graph with four vertices of degrees 1, 1, 2, and 3

b. A graph with four vertices of degrees 1, 1, 3, and 3

c. A simple graph with four vertices of degrees 1, 1, 3, and 3

Solution

a. No such graph is possible. By Corollary 10.1.2, the total degree of a graph is even. But a graph with four vertices of degrees 1, 1, 2, and 3 would have a total degree of $1 + 1 + 2 + 3 = 7$, which is odd.

b. Let G be any of the graphs shown below.

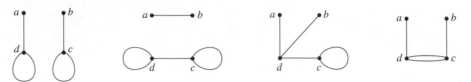

In each case, no matter how the edges are labeled, $\deg(a) = 1$, $\deg(b) = 1$, $\deg(c) = 3$, and $\deg(d) = 3$.

c. There is no simple graph with four vertices of degrees 1, 1, 3, and 3.

Proof (by contradiction):

Suppose there were a simple graph G with four vertices of degrees 1, 1, 3, and 3. Call a and b the vertices of degree 1, and call c and d the vertices of degree 3. Since $\deg(c) = 3$ and G does not have any loops or parallel edges (because it is simple), there must be edges that connect c to a, b, and d.

By the same reasoning, there must be edges connecting d to a, b, and c.

But then $\deg(a) \geq 2$ and $\deg(b) \geq 2$, which contradicts the supposition that these vertices have degree 1. Hence the supposition is false, and consequently there is no simple graph with four vertices of degrees 1, 1, 3, and 3. ∎

Example 10.1.14 Application to an Acquaintance Graph

Is it possible in a group of nine people for each to be friends with exactly five others?

Solution　The answer is no. Imagine constructing an "acquaintance graph" in which each of the nine people represented by a vertex and two vertices are joined by an edge if, and only if, the people they represent are friends. Suppose each of the people were friends with exactly five others. Then the degree of each of the nine vertices of the graph would be five, and so the total degree of the graph would be 45. But this contradicts Corollary 10.1.2, which says that the total degree of a graph is even. This contradiction shows that the supposition is false, and hence it is impossible for each person in a group of nine people to be friends with exactly five others. ∎

The following proposition is easily deduced from Corollary 10.1.2 using properties of even and odd integers.

Proposition 10.1.3

In any graph there are an even number of vertices of odd degree.

Proof:

Suppose G is any graph, and suppose G has n vertices of odd degree and m vertices of even degree, where n is a positive integer and m is a nonnegative integer. *[We must show that n is even.]* Let E be the sum of the degrees of all the vertices of even degree, O the sum of the degrees of all the vertices of odd degree, and T the total degree of G. If $u_1, u_2, \ldots, u_m$ are the vertices of even degree and $v_1, v_2, \ldots, v_n$ are the vertices of odd degree, then

$$E = \deg(u_1) + \deg(u_2) + \cdots + \deg(u_m),$$
$$O = \deg(v_1) + \deg(v_2) + \cdots + \deg(v_n), \quad \text{and}$$
$$T = \deg(u_1) + \cdots + \deg(u_m) + \deg(v_1) + \cdots + \deg(v_n) = E + O.$$

Now T, the total degree of G, is an even integer by Corollary 10.1.2. Also E is even since either E is zero, which is even, or E is a sum of the numbers $\deg(u_i)$, each of which is even. But

$$T = E + O,$$

and therefore

$$O = T - E.$$

Hence O is a difference of two even integers, and so O is even.

By assumption, $\deg(v_i)$ is odd for all $i = 1, 2, \ldots, n$. Thus O, an even integer, is a sum of the n odd integers $\deg(v_1), \deg(v_2), \ldots, \deg(v_n)$. But if a sum of n odd integers is even, then n is even. (See exercise 32 at the end of this section.) Therefore, n is even *[as was to be shown]*.

Example 10.1.15 Applying the Fact That the Number of Vertices with Odd Degree Is Even

Is there a graph with ten vertices of degrees 1, 1, 2, 2, 2, 3, 4, 4, 4, and 6?

Solution No. Such a graph would have three vertices of odd degree, which is impossible by Proposition 10.1.3.

Note that this same result could have been deduced directly from Corollary 10.1.2 by computing the total degree $(1 + 1 + 2 + 2 + 2 + 3 + 4 + 4 + 4 + 6 = 29)$ and noting that it is odd. However, use of Proposition 10.1.3 gives the result without the need to perform this addition. ■

Test Yourself

Answers to Test Yourself questions are located at the end of each section.

1. A graph consists of two finite sets: _____ and _____, where each edge is associated with a set consisting of _____.

2. A loop in a graph is _____.

3. Two distinct edges in a graph are parallel if, and only if, _____.

4. Two vertices are called adjacent if, and only if, _____.

5. An edge is incident on _____.

6. Two edges incident on the same endpoint are _____.

7. A vertex on which no edges are incident is _____.

8. In a directed graph, each edge is associated with _____.

9. A simple graph is _____.

10. A complete graph on n vertices is a _____.

11. A complete bipartite graph on (m, n) vertices is a simple graph whose vertices can be partitioned into two disjoint sets

V_1 and V_2 in such a way that (1) each of the m vertices in V_1 is _____ to each of the n vertices in V_2, no vertex in V_1 is connected to _____, and no vertex in V_2 is connected to _____.

12. A graph H is a subgraph of a graph G if, and only if, (1) _____, (2) _____, and (3) _____.

13. The degree of a vertex in a graph is _____.

14. The total degree of a graph is defined as _____.

15. The handshake theorem says that the total degree of a graph is _____.

16. In any graph the number of vertices of odd degree is _____.

Exercise Set 10.1*

In 1 and 2, graphs are represented by drawings. Define each graph formally by specifying its vertex set, its edge set, and a table giving the edge-endpoint function.

1.

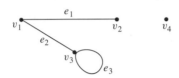

2.

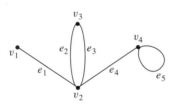

In 3 and 4, draw pictures of the specified graphs.

3. Graph G has vertex set $\{v_1, v_2, v_3, v_4, v_5\}$ and edge set $\{e_1, e_2, e_3, e_4\}$, with edge-endpoint function as follows:

Edge	Endpoints
e_1	$\{v_1, v_2\}$
e_2	$\{v_1, v_2\}$
e_3	$\{v_2, v_3\}$
e_4	$\{v_2\}$

4. Graph H has vertex set $\{v_1, v_2, v_3, v_4, v_5\}$ and edge set $\{e_1, e_2, e_3, e_4\}$ with edge-endpoint function as follows:

Edge	Endpoints
e_1	$\{v_1\}$
e_2	$\{v_2, v_3\}$
e_3	$\{v_2, v_3\}$
e_4	$\{v_1, v_5\}$

In 5–7, show that the two drawings represent the same graph by labeling the vertices and edges of the right-hand drawing to correspond to those of the left-hand drawing.

5.

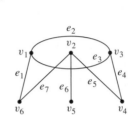

6.

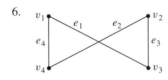

7.

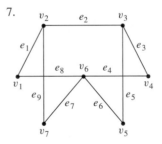

*For exercises with blue numbers or letters, solutions are given in Appendix B. The symbol **H** indicates that only a hint or a partial solution is given. The symbol ✻ signals that an exercise is more challenging than usual.

For each of the graphs in 8 and 9:
  (i) Find all edges that are incident on v_1.
  (ii) Find all vertices that are adjacent to v_3.
  (iii) Find all edges that are adjacent to e_1.
  (iv) Find all loops.
  (v) Find all parallel edges.
  (vi) Find all isolated vertices.
  (vii) Find the degree of v_3.
  (viii) Find the total degree of the graph.

8.

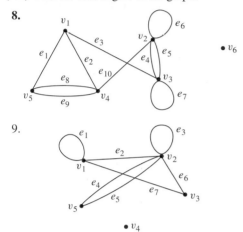

9.

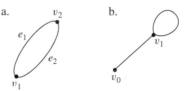

10. Use the graph of Example 10.1.6 to determine
  a. whether *Sports Illustrated* contains printed writing;
  b. whether *Poetry Magazine* contains long words.

11. Find three other winning sequences of moves for the vegetarians and the cannibals in Example 10.1.7.

12. Another famous puzzle used as an example in the study of artificial intelligence seems first to have appeared in a collection of problems, *Problems for the Quickening of the Mind*, which was compiled about A.D. 775. It involves a wolf, a goat, a bag of cabbage, and a ferryman. From an initial position on the left bank of a river, the ferryman is to transport the wolf, the goat, and the cabbage to the right bank. The difficulty is that the ferryman's boat is only big enough for him to transport one object at a time, other than himself. Yet, for obvious reasons, the wolf cannot be left alone with the goat, and the goat cannot be left alone with the cabbage. How should the ferryman proceed?

13. Solve the vegetarians-and-cannibals puzzle for the case where there are three vegetarians and three cannibals to be transported from one side of a river to the other.

H **14.** Two jugs A and B have capacities of 3 quarts and 5 quarts, respectively. Can you use the jugs to measure out exactly 1 quart of water, while obeying the following restrictions? You may fill either jug to capacity from a water tap; you may empty the contents of either jug into a drain; and you may pour water from either jug into the other.

15. A graph has vertices of degrees 0, 2, 2, 3, and 9. How many edges does the graph have?

16. A graph has vertices of degrees 1, 1, 4, 4, and 6. How many edges does the graph have?

In each of 17–25, either draw a graph with the specified properties or explain why no such graph exists.

17. Graph with five vertices of degrees 1, 2, 3, 3, and 5.

18. Graph with four vertices of degrees 1, 2, 3, and 3.

19. Graph with four vertices of degrees 1, 1, 1, and 4.

20. Graph with four vertices of degrees 1, 2, 3, and 4.

21. Simple graph with four vertices of degrees 1, 2, 3, and 4.

22. Simple graph with five vertices of degrees 2, 3, 3, 3, and 5.

23. Simple graph with five vertices of degrees 1, 1, 1, 2, and 3.

24. Simple graph with six edges and all vertices of degree 3.

25. Simple graph with nine edges and all vertices of degree 3.

26. Find all subgraphs of each of the following graphs.

a.

b.

c.

27. **a.** In a group of 15 people, is it possible for each person to have exactly 3 friends? Explain. (Assume that friendship is a symmetric relationship: If x is a friend of y, then y is a friend of x.)
  b. In a group of 4 people, is it possible for each person to have exactly 3 friends? Why?

28. In a group of 25 people, is it possible for each to shake hands with exactly 3 other people? Explain.

29. Is there a simple graph, each of whose vertices has even degree? Explain.

30. Suppose that G is a graph with v vertices and e edges and that the degree of each vertex is at least d_{min} and at most d_{max}. Show that

$$\frac{1}{2}d_{min} \cdot v \leq e \leq \frac{1}{2}d_{max} \cdot v.$$

31. Prove that any sum of an odd number of odd integers is odd.

H **32.** Deduce from exercise 31 that for any positive integer n, if there is a sum of n odd integers that is even, then n is even.

33. Recall that K_n denotes a complete graph on n vertices.
  a. Draw K_6.
  H **b.** Show that for all integers $n \geq 1$, the number of edges of K_n is $\dfrac{n(n-1)}{2}$.

34. Use the result of exercise 33 to show that the number of edges of a simple graph with n vertices is less than or equal to $\dfrac{n(n-1)}{2}$.

35. Is there a simple graph with twice as many edges as vertices? Explain. (You may find it helpful to use the result of exercise 34.)

36. Recall that $K_{m,n}$ denotes a complete bipartite graph on (m, n) vertices.
    a. Draw $K_{4,2}$
    b. Draw $K_{1,3}$
    c. Draw $K_{3,4}$
    d. How many vertices of $K_{m,n}$ have degree m? degree n?
    e. What is the total degree of $K_{m,n}$?
    f. Find a formula in terms of m and n for the number of edges of $K_{m,n}$. Explain.

37. A **bipartite graph** G is a simple graph whose vertex set can be partitioned into two disjoint nonempty subsets V_1 and V_2 such that vertices in V_1 may be connected to vertices in V_2, but no vertices in V_1 are connected to other vertices in V_1 and no vertices in V_2 are connected to other vertices in V_2. For example, the graph G illustrated in (i) can be redrawn as shown in (ii). From the drawing in (ii), you can see that G is bipartite with mutually disjoint vertex sets $V_1 = \{v_1, v_3, v_5\}$ and $V_2 = \{v_2, v_4, v_6\}$.

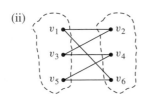

Find which of the following graphs are bipartite. Redraw the bipartite graphs so that their bipartite nature is evident.

a. 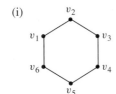 b.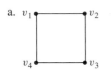

c. 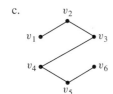 d.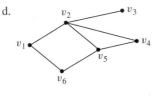

e. 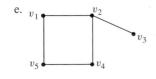 f.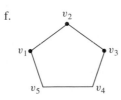

38. Suppose r and s are any positive integers. Does there exist a graph G with the property that G has vertices of degrees r and s and of no other degrees? Explain.

Definition: If G is a simple graph, the **complement of G,** denoted G', is obtained as follows: The vertex set of G' is identical to the vertex set of G. However, two distinct vertices v and w of G' are connected by an edge if, and only if, v and w are not connected by an edge in G. For example, if G is the graph

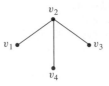

then G' is

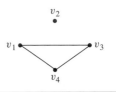

39. Find the complement of each of the following graphs.

a. 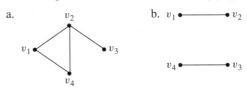 b.

40. a. Find the complement of the graph K_4, the complete graph on four vertices. (See Example 10.1.9.)
    b. Find the complement of the graph $K_{3,2}$, the complete bipartite graph on $(3, 2)$ vertices. (See Example 10.1.10.)

41. Suppose that in a group of five people A, B, C, D, and E the following pairs of people are acquainted with each other:
    A and C, A and D, B and C, C and D, C and E.
    a. Draw a graph to represent this situation.
    b. Draw a graph that illustrates who among these five people are *not* acquainted. That is, draw an edge between two people if, and only if, they are not acquainted.

H 42. Let G be a simple graph with n vertices. What is the relation between the number of edges of G and the number of edges of the complement G'?

43. Show that at a party with at least two people, there are at least two mutual acquaintances or at least two mutual strangers.

44. a. In a simple graph, must every vertex have degree that is less than the number of vertices in the graph? Why?
    b. Can there be a simple graph that has four vertices each of different degrees?
    H ✶ c. Can there be a simple graph that has n vertices all of different degrees?

H ✶ 45. In a group of two or more people, must there always be at least two people who are acquainted with the same number of people within the group? Why?

46. Imagine that the diagram shown below is a map with countries labeled *a–g*. Is it possible to color the map with only three colors so that no two adjacent countries have the same color? To answer this question, draw and analyze a graph in which each country is represented by a vertex and two vertices are connected by an edge if, and only if, the countries share a common border.

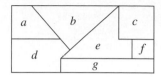

H **47.** In this exercise a graph is used to help solve a scheduling problem. Twelve faculty members in a mathematics department serve on the following committees:

Undergraduate Education: Tenner, Peterson, Kashina, Cohen
Graduate Education: Gatto, Yang, Cohen, Catoiu
Colloquium: Sahin, McMurry, Ash
Library: Cortzen, Tenner, Sahin
Hiring: Gatto, McMurry, Yang, Peterson
Personnel: Yang, Wang, Cortzen

The committees must all meet during the first week of classes, but there are only three time slots available. Find

a schedule that will allow all faculty members to attend the meetings of all committees on which they serve. To do this, represent each committee as the vertex of a graph, and draw an edge between two vertices if the two committees have a common member. Find a way to color the vertices using only three colors so that no two committees have the same color, and explain how to use the result to schedule the meetings.

48. A department wants to schedule final exams so that no student has more than one exam on any given day. The vertices of the graph below show the courses that are being taken by more than one student, with an edge connecting two vertices if there is a student in both courses. Find a way to color the vertices of the graph with only four colors so that no two adjacent vertices have the same color and explain how to use the result to schedule the final exams.

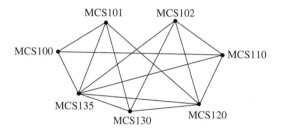

Answers for Test Yourself

1. a finite, nonempty set of vertices; a finite set of edges; one or two vertices called its endpoints   2. an edge with a single endpoint 3. they have the same set of endpoints   4. they are connected by an edge   5. each of its endpoints   6. adjacent   7. isolated   8. an ordered pair of vertices called its endpoints   9. a graph with no loops or parallel edges   10. simple graph with *n* vertices whose set of edges contains exactly one edge for each pair of vertices   11. connected by an edge; any other vertex in V_1; any other vertex in V_2 12. every vertex in *H* is also a vertex in *G*; every edge in *H* is also an edge in *G*; every edge in *H* has the same endpoints as it has in *G*   13. the number of edges that are incident on the vertex, with an edge that is a loop counted twice   14. the sum of the degrees of all the vertices of the graph   15. equal to twice the number of edges of the graph   16. an even number

10.2 Trails, Paths, and Circuits

One can begin to reason only when a clear picture has been formed in the imagination.
— W. W. Sawyer, *Mathematician's Delight*, 1943

The subject of graph theory began in the year 1736 when the great mathematician Leonhard Euler published a paper giving the solution to the following puzzle:

The town of Königsberg in Prussia (now Kaliningrad in Russia) was built at a point where two branches of the Pregel River came together. It consisted of an island and some land along the river banks. These were connected by seven bridges as shown in Figure 10.2.1.

The question is this: Is it possible for a person to take a walk around town, starting and ending at the same location and crossing each of the seven bridges exactly once?*

---

*In his original paper, Euler did not require the walk to start and end at the same point. The analysis of the problem is simplified, however, by adding this condition. Later in the section, we discuss walks that start and end at different points.

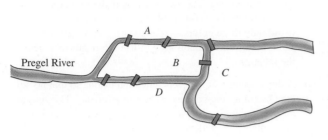

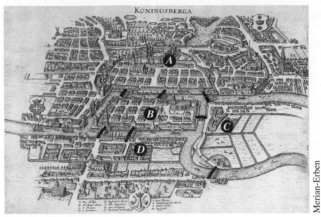

Merian-Erben

Figure 10.2.1 The Seven Bridges of Königsberg

Bettmann/CORBIS

*Leonhard Euler
(1707–1783)*

To solve this puzzle, Euler translated it into a graph theory problem. He noticed that all points of a given land mass can be identified with each other since a person can travel from any one point to any other point of the same land mass without crossing a bridge. Thus for the purpose of solving the puzzle, the map of Königsberg can be identified with the graph shown in Figure 10.2.2, in which the vertices A, B, C, and D represent land masses and the seven edges represent the seven bridges.

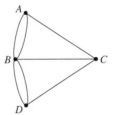

Figure 10.2.2 Graph Version of Königsberg Map

In terms of this graph, the question becomes the following:

> Is it possible to find a route through the graph that starts and ends at some vertex, one of A, B, C, or D, and traverses each edge exactly once?

Equivalently:

> Is it possible to trace this graph, starting and ending at the same point, without ever lifting your pencil from the paper?

Take a few minutes to think about the question yourself. Can you find a route that meets the requirements? Try it!

Looking for a route is frustrating because you continually find yourself at a vertex that does not have an unused edge on which to leave, while elsewhere there are unused edges that must still be traversed. If you start at vertex A, for example, each time you pass through vertex B, C, or D, you use up two edges because you arrive on one edge and depart on a different one. So, if it is possible to find a route that uses all the edges of the graph and starts and ends at A, then the total number of arrivals and departures from each vertex B, C, and D must be a multiple of 2. Or, in other words, the degrees of the vertices B, C, and D must be even. But they are not: $\deg(B) = 5$, $\deg(C) = 3$, and $\deg(D) = 3$. Hence there is no route that solves the puzzle by starting and ending at A. Similar reasoning can be used to show that there are no routes that solve the puzzle by starting and ending at B, C, or D. Therefore, it is impossible to travel all around the city crossing each bridge exactly once.

Definitions

Travel in a graph is accomplished by moving from one vertex to another along a sequence of adjacent edges. In the graph below, for instance, you can go from u_1 to u_4 by taking f_1 to u_2 and then f_7 to u_4. This is represented by writing

$$u_1 f_1 u_2 f_7 u_4.$$

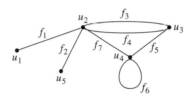

Or you could take the roundabout route

$$u_1 f_1 u_2 f_3 u_3 f_4 u_2 f_3 u_3 f_5 u_4 f_6 u_4 f_7 u_2 f_3 u_3 f_5 u_4.$$

Certain types of sequences of adjacent vertices and edges are of special importance in graph theory: those that do not have a repeated edge, those that do not have a repeated vertex, and those that start and end at the same vertex.

• Definition

Let G be a graph, and let v and w be vertices in G.

A **walk from v to w** is a finite alternating sequence of adjacent vertices and edges of G. Thus a walk has the form

$$v_0 e_1 v_1 e_2 \cdots v_{n-1} e_n v_n,$$

where the v's represent vertices, the e's represent edges, $v_0 = v$, $v_n = w$, and for all $i = 1, 2, \ldots n$, v_{i-1} and v_i are the endpoints of e_i. The **trivial walk from v to v** consists of the single vertex v.

A **trail from v to w** is a walk from v to w that does not contain a repeated edge.

A **path from v to w** is a trail that does not contain a repeated vertex.

A **closed walk** is a walk that starts and ends at the same vertex.

A **circuit** is a closed walk that contains at least one edge and does not contain a repeated edge.

A **simple circuit** is a circuit that does not have any other repeated vertex except the first and last.

For ease of reference, these definitions are summarized in the following table:

	Repeated Edge?	Repeated Vertex?	Starts and Ends at Same Point?	Must Contain at Least One Edge?
Walk	allowed	allowed	allowed	no
Trail	no	allowed	allowed	no
Path	no	no	no	no
Closed walk	allowed	allowed	yes	no
Circuit	no	allowed	yes	yes
Simple circuit	no	first and last only	yes	yes

Often a walk can be specified unambiguously by giving either a sequence of edges or a sequence of vertices. The next two examples show how this is done.

Example 10.2.1 Notation for Walks

a. In the graph below, the notation $e_1e_2e_4e_3$ refers unambiguously to the following walk: $v_1e_1v_2e_2v_3e_4v_3e_3v_2$. On the other hand, the notation e_1 is ambiguous if used to refer to a walk. It could mean either $v_1e_1v_2$ or $v_2e_1v_1$.

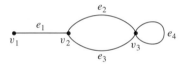

b. In the graph of part (a), the notation v_2v_3 is ambiguous if used to refer to a walk. It could mean $v_2e_2v_3$ or $v_2e_3v_3$. On the other hand, in the graph below, the notation $v_1v_2v_2v_3$ refers unambiguously to the walk $v_1e_1v_2e_2v_2e_3v_3$.

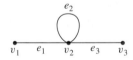

■

Note that if a graph G does not have any parallel edges, then any walk in G is uniquely determined by its sequence of vertices.

Example 10.2.2 Walks, Trails, Paths, and Circuits

In the graph below, determine which of the following walks are trails, paths, circuits, or simple circuits.

a. $v_1e_1v_2e_3v_3e_4v_3e_5v_4$ b. $e_1e_3e_5e_5e_6$ c. $v_2v_3v_4v_5v_3v_6v_2$
d. $v_2v_3v_4v_5v_6v_2$ e. $v_1e_1v_2e_1v_1$ f. v_1

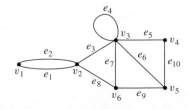

Solution

   a. This walk has a repeated vertex but does not have a repeated edge, so it is a trail from v_1 to v_4 but not a path.

   b. This is just a walk from v_1 to v_5. It is not a trail because it has a repeated edge.

   c. This walk starts and ends at v_2, contains at least one edge, and does not have a repeated edge, so it is a circuit. Since the vertex v_3 is repeated in the middle, it is not a simple circuit.

   d. This walk starts and ends at v_2, contains at least one edge, does not have a repeated edge, and does not have a repeated vertex. Thus it is a simple circuit.

   e. This is just a closed walk starting and ending at v_1. It is not a circuit because edge e_1 is repeated.

   f. The first vertex of this walk is the same as its last vertex, but it does not contain an edge, and so it is not a circuit. It is a closed walk from v_1 to v_1. (It is also a trail from v_1 to v_1.) ∎

Because most of the major developments in graph theory have happened relatively recently and in a variety of different contexts, the terms used in the subject have not been standardized. For example, what this book calls a *graph* is sometimes called a *multigraph*, what this book calls a *simple graph* is sometimes called a *graph*, what this book calls a *vertex* is sometimes called a *node*, and what this book calls an *edge* is sometimes called an *arc*. Similarly, instead of the word *trail*, the word *path* is sometimes used; instead of the word *path*, the words *simple path* are sometimes used; and instead of the words *simple circuit*, the word *cycle* is sometimes used. The terminology in this book is among the most common, but if you consult other sources, be sure to check their definitions.

Connectedness

It is easy to understand the concept of connectedness on an intuitive level. Roughly speaking, a graph is connected if it is possible to travel from any vertex to any other vertex along a sequence of adjacent edges of the graph. The formal definition of connectedness is stated in terms of walks.

> **• Definition**
>
> Let G be a graph. Two **vertices v and w of G are connected** if, and only if, there is a walk from v to w. The **graph G is connected** if, and only if, given *any* two vertices v and w in G, there is a walk from v to w. Symbolically,
>
> G is connected ⇔ ∀ vertices $v, w \in V(G)$, ∃ a walk from v to w.

    If you take the negation of this definition, you will see that a graph G is *not connected* if, and only if, there are two vertices of G that are not connected by any walk.

Example 10.2.3 Connected and Disconnected Graphs

Which of the following graphs are connected?

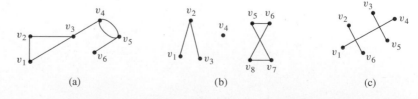

       (a)   (b)   (c)

Solution The graph represented in (a) is connected, whereas those of (b) and (c) are not. To understand why (c) is not connected, recall that in a drawing of a graph, two edges may cross at a point that is not a vertex. Thus the graph in (c) can be redrawn as follows:

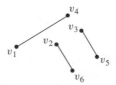

Some useful facts relating circuits and connectedness are collected in the following lemma. Proofs of (a) and (b) are left for the exercises. The proof of (c) is in Section 10.3.

Lemma 10.2.1

Let G be a graph.

a. If G is connected, then any two distinct vertices of G can be connected by a path.

b. If vertices v and w are part of a circuit in G and one edge is removed from the circuit, then there still exists a trail from v to w in G.

c. If G is connected and G contains a circuit, then an edge of the circuit can be removed without disconnecting G.

Look back at Example 10.2.3. The graphs in (b) and (c) are both made up of three pieces, each of which is itself a connected graph. A *connected component* of a graph is a connected subgraph of largest possible size.

• Definition

A graph H is a **connected component** of a graph G if, and only if,

1. H is subgraph of G;

2. H is connected; and

3. no connected subgraph of G has H as a subgraph and contains vertices or edges that are not in H.

The fact is that any graph is a kind of union of its connected components.

Example 10.2.4 Connected Components

Find all connected components of the following graph G.

Solution G has three connected components: H_1, H_2, and H_3 with vertex sets V_1, V_2, and V_3 and edge sets E_1, E_2, and E_3, where

$$V_1 = \{v_1, v_2, v_3\}, \qquad E_1 = \{e_1, e_2\},$$
$$V_2 = \{v_4\}, \qquad\qquad E_2 = \emptyset,$$
$$V_3 = \{v_5, v_6, v_7, v_8\}, \qquad E_3 = \{e_3, e_4, e_5\}. \qquad ■$$

Euler Circuits

Now we return to consider general problems similar to the puzzle of the Königsberg bridges. The following definition is made in honor of Euler.

• Definition

Let G be a graph. An **Euler circuit** for G is a circuit that contains every vertex and every edge of G. That is, an Euler circuit for G is a sequence of adjacent vertices and edges in G that has at least one edge, starts and ends at the same vertex, uses every vertex of G at least once, and uses every edge of G exactly once.

The analysis used earlier to solve the puzzle of the Königsberg bridges generalizes to prove the following theorem:

Theorem 10.2.2

If a graph has an Euler circuit, then every vertex of the graph has positive even degree.

Proof:

Suppose G is a graph that has an Euler circuit. *[We must show that given any vertex v of G, the degree of v is even.]* Let v be any particular but arbitrarily chosen vertex of G. Since the Euler circuit contains every edge of G, it contains all edges incident on v. Now imagine taking a journey that begins in the middle of one of the edges adjacent to the start of the Euler circuit and continues around the Euler circuit to end in the middle of the starting edge. (See Figure 10.2.3. There is such a starting edge because the Euler circuit has at least one edge.) Each time v is entered by traveling along one edge, it is immediately exited by traveling along another edge (since the journey ends in the *middle* of an edge).

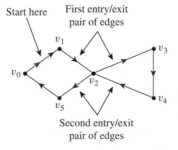

In this example, the Euler circuit is $v_0 v_1 v_2 v_3 v_4 v_2 v_5 v_0$, and v is v_2. Each time v_2 is entered by one edge, it is exited by another edge.

Figure 10.2.3 Example for the Proof of Theorem 10.2.2

(continued on page 500)

> Because the Euler circuit uses every edge of G exactly once, every edge incident on v is traversed exactly once in this process. Hence the edges incident on v occur in entry/exit pairs, and consequently the degree of v must be a positive multiple of 2. But that means that v has positive even degree *[as was to be shown]*.

Recall that the contrapositive of a statement is logically equivalent to the statement. The contrapositive of Theorem 10.2.2 is as follows:

Contrapositive Version of Theorem 10.2.2

If some vertex of a graph has odd degree, then the graph does not have an Euler circuit.

This version of Theorem 10.2.2 is useful for showing that a given graph does *not* have an Euler circuit.

Example 10.2.5 Showing That a Graph Does Not Have an Euler Circuit

Show that the graph below does not have an Euler circuit.

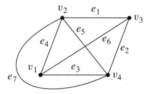

Solution   Vertices v_1 and v_3 both have degree 3, which is odd. Hence by (the contrapositive form of) Theorem 10.2.2, this graph does not have an Euler circuit. ∎

Now consider the converse of Theorem 10.2.2: If every vertex of a graph has even degree, then the graph has an Euler circuit. Is this true? The answer is no. There is a graph G such that every vertex of G has even degree but G does not have an Euler circuit. In fact, there are many such graphs. The illustration below shows one example.

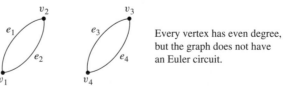

Every vertex has even degree, but the graph does not have an Euler circuit.

Note that the graph in the preceding drawing is not connected. It turns out that although the converse of Theorem 10.2.2 is false, a modified converse is true: If every vertex of a graph has positive even degree *and* if the graph is connected, then the graph has an Euler circuit. The proof of this fact is constructive: It contains an algorithm to find an Euler circuit for any connected graph in which every vertex has even degree.

Theorem 10.2.3

If a graph G is connected and the degree of every vertex of G is a positive even integer, then G has an Euler circuit.

Proof:

Suppose that G is any connected graph and suppose that every vertex of G is a positive even integer. *[We must find an Euler circuit for G.]* Construct a circuit C by the following algorithm:

Step 1: Pick any vertex v of G at which to start.

[This step can be accomplished because the vertex set of G is nonempty by assumption.]

Step 2: Pick any sequence of adjacent vertices and edges, starting and ending at v and never repeating an edge. Call the resulting circuit C.

[This step can be performed for the following reasons: Since the degree of each vertex of G is a positive even integer, as each vertex of G is entered by traveling on one edge, either the vertex is v itself and there is no other unused edge adjacent to v, or the vertex can be exited by traveling on another previously unused edge. Since the number of edges of the graph is finite (by definition of graph), the sequence of distinct edges cannot go on forever. The sequence can eventually return to v because the degree of v is a positive even integer, and so if an edge connects v to another vertex, there must be a different edge that connects back to v.]

Step 3: Check whether C contains every edge and vertex of G. If so, C is an Euler circuit, and we are finished. If not, perform the following steps.

Step 3a: Remove all edges of C from G and also any vertices that become isolated when the edges of C are removed. Call the resulting subgraph G'.

[Note that G' may not be connected (as illustrated in Figure 10.2.4), but every vertex of G' has positive, even degree (since removing the edges of C removes an even number of edges from each vertex, the difference of two even integers is even, and isolated vertices with degree 0 were removed.)]

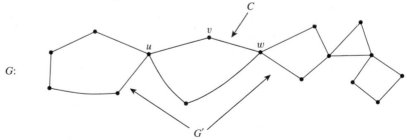

Figure 10.2.4

Step 3b: Pick any vertex w common to both C and G'.

[There must be at least one such vertex since G is connected. (See exercise 44.) (In Figure 10.2.4 there are two such vertices: u and w.)]

Step 3c: Pick any sequence of adjacent vertices and edges of G', starting and ending at w and never repeating an edge. Call the resulting circuit C'.

[This can be done since each vertex of G' has positive, even degree and G' is finite. See the justification for step 2.]

(continued on page 502)

Step 3d: Patch C and C' together to create a new circuit C'' as follows: Start at v and follow C all the way to w. Then follow C' all the way back to w. After that, continue along the untraveled portion of C to return to v. *[The effect of executing steps 3c and 3d for the graph of Figure 10.2.4 is shown in Figure 10.2.5.]*

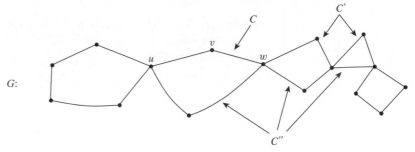

Figure 10.2.5

Step 3e: Let $C = C''$ and go back to step 3.

Since the graph G is finite, execution of the steps outlined in this algorithm must eventually terminate. At that point an Euler circuit for G will have been constructed. (Note that because of the element of choice in steps 1, 2, 3b, and 3c, a variety of different Euler circuits can be produced by using this algorithm.)

Example 10.2.6 Finding an Euler Circuit

Use Theorem 10.2.3 to check that the graph below has an Euler circuit. Then use the algorithm from the proof of the theorem to find an Euler circuit for the graph.

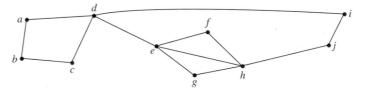

Solution Observe that

$$\deg(a) = \deg(b) = \deg(c) = \deg(f) = \deg(g) = \deg(i) = \deg(j) = 2$$

and that $\deg(d) = \deg(e) = \deg(h) = 4$. Hence all vertices have even degree. Also, the graph is connected. Thus, by Theorem 10.2.3, the graph has an Euler circuit.

To construct an Euler circuit using the algorithm of Theorem 10.2.3, let $v = a$ and let C be

$$C:\ abcda.$$

C is represented by the labeled edges shown below.

Observe that C is not an Euler circuit for the graph but that C intersects the rest of the graph at d. Let C' be

$$C': deghjid.$$

Patch C' into C to obtain

$$C'': abcdeghjida.$$

Set $C = C''$. Then C is represented by the labeled edges shown below.

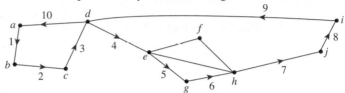

Observe that C is not an Euler circuit for the graph but that it intersects the rest of the graph at e. Let C' be

$$C': efhe.$$

Patch C' into C to obtain

$$C'': abcdefheghjida.$$

Set $C = C''$. Then C is represented by the labeled edges shown below.

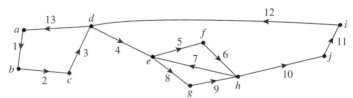

Since C includes every edge of the graph exactly once, C is an Euler circuit for the graph. ∎

In exercise 45 at the end of this section you are asked to show that any graph with an Euler circuit is connected. This result can be combined with Theorems 10.2.2 and 10.2.3 to give a complete characterization of graphs that have Euler circuits, as stated in Theorem 10.2.4.

Theorem 10.2.4

A graph G has an Euler circuit if, and only if, G is connected and every vertex of G has positive even degree.

A corollary to Theorem 10.2.4 gives a criterion for determining when it is possible to find a walk from one vertex of a graph to another, passing through every vertex of the graph at least once and every edge of the graph exactly once.

• Definition

Let G be a graph, and let v and w be two distinct vertices of G. An **Euler trail from v to w** is a sequence of adjacent edges and vertices that starts at v, ends at w, passes through every vertex of G at least once, and traverses every edge of G exactly once.

> **Corollary 10.2.5**
>
> Let G be a graph, and let v and w be two distinct vertices of G. There is an Euler trail from v to w if, and only if, G is connected, v and w have odd degree, and all other vertices of G have positive even degree.

The proof of this corollary is left as an exercise.

Example 10.2.7 Finding an Euler Trail

The floor plan shown below is for a house that is open for public viewing. Is it possible to find a trail that starts in room A, ends in room B, and passes through every interior doorway of the house exactly once? If so, find such a trail.

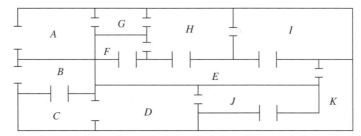

Solution Let the floor plan of the house be represented by the graph below.

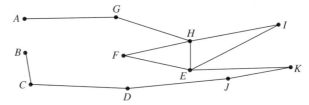

Each vertex of this graph has even degree except for A and B, each of which has degree 1. Hence by Corollary 10.2.5, there is an Euler trail from A to B. One such trail is

$$AGHFEIHEKJDCB.$$

∎

Hamiltonian Circuits

Theorem 10.2.4 completely answers the following question: Given a graph G, is it possible to find a circuit for G in which all the *edges* of G appear exactly once? A related question is this: Given a graph G, is it possible to find a circuit for G in which all the *vertices* of G (except the first and the last) appear exactly once?

In 1859 the Irish mathematician Sir William Rowan Hamilton introduced a puzzle in the shape of a dodecahedron (DOH-dek-a-HEE-dron). (Figure 10.2.6 contains a drawing of a dodecahedron, which is a solid figure with 12 identical pentagonal faces.)

Sir Wm. Hamilton
(1805–1865)

Figure 10.2.6 Dodecahedron

Each vertex was labeled with the name of a city—London, Paris, Hong Kong, New York, and so on. The problem Hamilton posed was to start at one city and tour the world by visiting each other city exactly once and returning to the starting city. One way to solve the puzzle is to imagine the surface of the dodecahedron stretched out and laid flat in the plane, as follows:

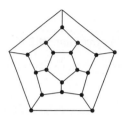

The circuit denoted with black lines is one solution. Note that although every city is visited, many edges are omitted from the circuit. (More difficult versions of the puzzle required that certain cities be visited in a certain order.)

The following definition is made in honor of Hamilton.

• Definition

Given a graph G, a **Hamiltonian circuit** for G is a simple circuit that includes every vertex of G. That is, a Hamiltonian circuit for G is a sequence of adjacent vertices and distinct edges in which every vertex of G appears exactly once, except for the first and the last, which are the same.

Note that although an Euler circuit for a graph G must include every vertex of G, it may visit some vertices more than once and hence may not be a Hamiltonian circuit. On the other hand, a Hamiltonian circuit for G does not need to include all the edges of G and hence may not be an Euler circuit.

Despite the analogous-sounding definitions of Euler and Hamiltonian circuits, the mathematics of the two are very different. Theorem 10.2.4 gives a simple criterion for determining whether a given graph has an Euler circuit. Unfortunately, there is no analogous criterion for determining whether a given graph has a Hamiltonian circuit, nor is there even an efficient algorithm for finding such a circuit. There is, however, a simple technique that can be used in many cases to show that a graph does *not* have a Hamiltonian circuit. This follows from the following considerations:

Suppose a graph G with at least two vertices has a Hamiltonian circuit C given concretely as

$$C: v_0 e_1 v_1 e_2 \cdots v_{n-1} e_n v_n.$$

Since C is a simple circuit, all the e_i are distinct and all the v_j are distinct except that $v_0 = v_n$. Let H be the subgraph of G that is formed using the vertices and edges of C. An example of such an H is shown below.

H is indicated by the black lines.

Note that H has the same number of edges as it has vertices since all its n edges are distinct and so are its n vertices $v_1, v_2, \ldots, v_n$. Also, by definition of Hamiltonian circuit, every vertex of G is a vertex of H, and H is connected since any two of its vertices lie on a circuit. In addition, every vertex of H has degree 2. The reason for this is that there are exactly two edges incident on any vertex. These are e_i and e_{i+1} for any vertex v_i except $v_0 = v_n$, and they are e_1 and e_n for $v_0 (= v_n)$. These observations have established the truth of the following proposition in all cases where G has at least two vertices.

Proposition 10.2.6

If a graph G has a Hamiltonian circuit, then G has a subgraph H with the following properties:

1. H contains every vertex of G.

2. H is connected.

3. H has the same number of edges as vertices.

4. Every vertex of H has degree 2.

Note that if G contains only one vertex and G has a Hamiltonian circuit, then the circuit has the form $v\, e\, v$, where v is the vertex of G and e is an edge incident on v. In this case, the subgraph H consisting of v and e satisfies conditions (1)–(4) of Proposition 10.2.6.

Recall that the contrapositive of a statement is logically equivalent to the statement. The contrapositive of Proposition 10.2.6 says that if a graph G does *not* have a subgraph H with properties (1)–(4), then G does *not* have a Hamiltonian circuit.

Example 10.2.8 Showing That a Graph Does Not Have a Hamiltonian Circuit

Prove that the graph G shown below does not have a Hamiltonian circuit.

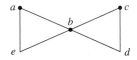

Solution   If G has a Hamiltonian circuit, then by Proposition 10.2.6, G has a subgraph H that (1) contains every vertex of G, (2) is connected, (3) has the same number of edges as vertices, and (4) is such that every vertex has degree 2. Suppose such a subgraph H exists. In other words, suppose there is a connected subgraph H of G such that H has five vertices (a, b, c, d, e) and five edges and such that every vertex of H has degree 2. Since the degree of b in G is 4 and every vertex of H has degree 2, two edges incident on b must be removed from G to create H. Edge $\{a, b\}$ cannot be removed because if it were, vertex a would have degree less than 2 in H. Similar reasoning shows that edges $\{e, b\}$, $\{b, a\}$, and $\{b, d\}$ cannot be removed either. It follows that the degree of b in H must be 4, which contradicts the condition that every vertex in H has degree 2 in H. Hence no such subgraph H exists, and so G does not have a Hamiltonian circuit.   ∎

The next example illustrates a type of problem known as a **traveling salesman problem.** It is a variation of the problem of finding a Hamiltonian circuit for a graph.

Example 10.2.9 A Traveling Salesman Problem

Imagine that the drawing below is a map showing four cities and the distances in kilometers between them. Suppose that a salesman must travel to each city exactly once, starting and ending in city A. Which route from city to city will minimize the total distance that must be traveled?

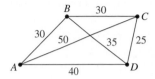

Solution   This problem can be solved by writing all possible Hamiltonian circuits starting and ending at A and calculating the total distance traveled for each.

Route	Total Distance (In Kilometers)
$ABCDA$	$30 + 30 + 25 + 40 = 125$
$ABDCA$	$30 + 35 + 25 + 50 = 140$
$ACBDA$	$50 + 30 + 35 + 40 = 155$
$ACDBA$	140 [$ABDCA$ backwards]
$ADBCA$	155 [$ACBDA$ backwards]
$ADCBA$	125 [$ABCDA$ backwards]

Thus either route $ABCDA$ or $ADCBA$ gives a minimum total distance of 125 kilometers. ∎

The general traveling salesman problem involves finding a Hamiltonian circuit to minimize the total distance traveled for an arbitrary graph with n vertices in which each edge is marked with a distance. One way to solve the general problem is to use the method of Example 10.2.9: Write down all Hamiltonian circuits starting and ending at a particular vertex, compute the total distance for each, and pick one for which this total is minimal. However, even for medium-sized values of n this method is impractical. For a complete graph with 30 vertices, there would be $(29!)/2 \cong 4.42 \times 10^{30}$ Hamiltonian circuits starting and ending at a particular vertex to check. Even if each circuit could be found and its total distance computed in just one nanosecond, it would require approximately 1.4×10^{14} years to finish the computation. At present, there is no known algorithm for solving the general traveling salesman problem that is more efficient. However, there are efficient algorithms that find "pretty good" solutions—that is, circuits that, while not necessarily having the least possible total distances, have smaller total distances than most other Hamiltonian circuits.

Test Yourself

1. Let G be a graph and let v and w be vertices in G.

   (a) A walk from v to w is _____.

   (b) A trail from v to w is _____.

   (c) A path from v to w is _____.

   (d) A closed walk is _____.

   (e) A circuit is _____.

   (f) A simple circuit is _____.

   (g) A trivial walk is _____.

   (h) Vertices v and w are connected if, and only if, _____.

2. A graph is connected if, and only if, _____.

3. Removing an edge from a circuit in a graph does not _____.

4. An Euler circuit in a graph is _____.

5. A graph has an Euler circuit if, and only if, _____.

6. Given vertices v and w in a graph, there is an Euler path from v to w if, and only if, _____.

7. A Hamiltonian circuit in a graph is _____.

8. If a graph G has a Hamiltonian circuit, then G has a subgraph H with the following properties: _____, _____, _____, and _____.

9. A traveling salesman problem involves finding a _____ that minimizes the total distance traveled for a graph in which each edge is marked with a distance.

Exercise Set 10.2

1. In the graph below, determine whether the following walks are trails, paths, closed walks, circuits, simple circuits, or just walks.

   a. $v_0 e_1 v_1 e_{10} v_5 e_9 v_2 e_2 v_1$ b. $v_4 e_7 v_2 e_9 v_5 e_{10} v_1 e_3 v_2 e_9 v_5$

   c. v_2 d. $v_5 v_2 v_3 v_4 v_4 v_5$

   e. $v_2 v_3 v_4 v_5 v_2 v_4 v_3 v_2$ f. $e_5 e_8 e_{10} e_3$

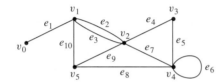

2. In the graph below, determine whether the following walks are trails, paths, closed walks, circuits, simple circuits, or just walks.

   a. $v_1 e_2 v_2 e_3 v_3 e_4 v_4 e_5 v_2 e_2 v_1 e_1 v_0$ b. $v_2 v_3 v_4 v_5 v_2$

   c. $v_4 v_2 v_3 v_4 v_5 v_2 v_4$ d. $v_2 v_1 v_5 v_2 v_3 v_4 v_2$

   e. $v_0 v_5 v_2 v_3 v_4 v_2 v_1$ f. $v_5 v_4 v_2 v_1$

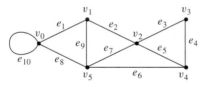

3. Let G be the graph

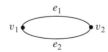

and consider the walk $v_1 e_1 v_2 e_2 v_1$.

   a. Can this walk be written unambiguously as $v_1 v_2 v_1$? Why?

   b. Can this walk be written unambiguously as $e_1 e_2$? Why?

4. Consider the following graph.

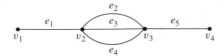

a. How many paths are there from v_1 to v_4?

b. How many trails are there from v_1 to v_4?

c. How many walks are there from v_1 to v_4?

5. Consider the following graph.

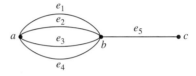

a. How many paths are there from a to c?

b. How many trails are there from a to c?

c. How many walks are there from a to c?

6. An edge whose removal disconnects the graph of which it is a part is called a **bridge.** Find all bridges for each of the following graphs.

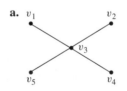

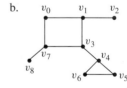

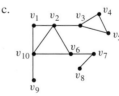

7. Given any positive integer n, (a) find a connected graph with n edges such that removal of just one edge disconnects the graph; (b) find a connected graph with n edges that cannot be disconnected by the removal of any single edge.

8. Find the number of connected components for each of the following graphs.

a.
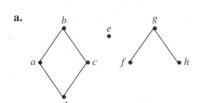

b.
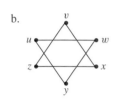

c.
d.
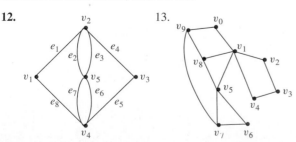

9. Each of (a)–(c) describes a graph. In each case answer *yes, no,* or *not necessarily* to this question: Does the graph have an Euler circuit? Justify your answers.

a. G is a connected graph with five vertices of degrees 2, 2, 3, 3, and 4.

b. G is a connected graph with five vertices of degrees 2, 2, 4, 4, and 6.

c. G is a graph with five vertices of degrees 2, 2, 4, 4, and 6.

10. The solution for Example 10.2.5 shows a graph for which every vertex has even degree but which does not have an Euler circuit. Give another example of a graph satisfying these properties.

11. Is it possible for a citizen of Königsberg to make a tour of the city and cross each bridge exactly twice? (See Figure 10.2.1.) Why?

Determine which of the graphs in 12–17 have Euler circuits. If the graph does not have an Euler circuit, explain why not. If it does have an Euler circuit, describe one.

12.
13.
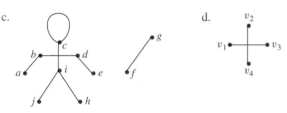

14.
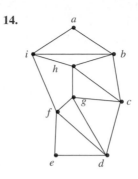

15.
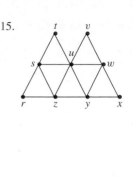

16.
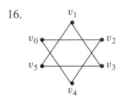

17.
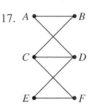

18. Is it possible to take a walk around the city whose map is shown below, starting and ending at the same point and crossing each bridge exactly once? If so, how can this be done?

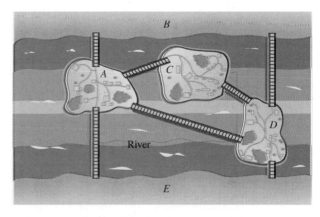

For each of the graphs in 19–21, determine whether there is an Euler trail from u to w. If there is, find such a path.

19.
20.
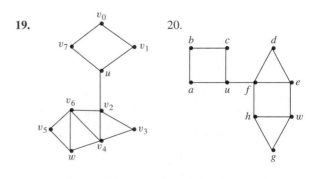

21.

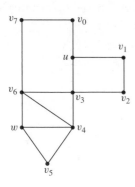

22. The following is a floor plan of a house. Is it possible to enter the house in room *A*, travel through every interior doorway of the house exactly once, and exit out of room *E*? If so, how can this be done?

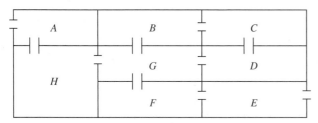

Find Hamiltonian circuits for each of the graphs in 23 and 24.

23.

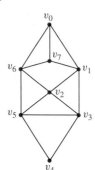

24.

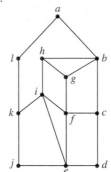

Show that none of the graphs in 25–27 has a Hamiltonian circuit.

H **25.**

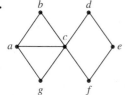

26.

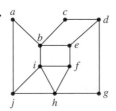

27.

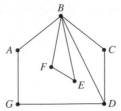

In 28–31 find Hamiltonian circuits for those graphs that have them. Explain why the other graphs do not.

H **28.**

29.

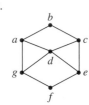

30.

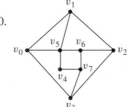

31.

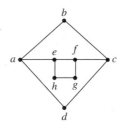

H **32.** Give two examples of graphs that have Euler circuits but not Hamiltonian circuits.

H **33.** Give two examples of graphs that have Hamiltonian circuits but not Euler circuits.

H **34.** Give two examples of graphs that have circuits that are both Euler circuits and Hamiltonian circuits.

H **35.** Give two examples of graphs that have Euler circuits and Hamiltonian circuits that are not the same.

36. A traveler in Europe wants to visit each of the cities shown on the map exactly once, starting and ending in Brussels. The distance (in kilometers) between each pair of cities is given in the table. Find a Hamiltonian circuit that minimizes the total distance traveled. (Use the map to narrow the possible circuits down to just a few. Then use the table to find the total distance for each of those.)

	Berlin	Brussels	Düsseldorf	Luxembourg	Munich
Brussels	783				
Düsseldorf	564	223			
Luxembourg	764	219	224		
Munich	585	771	613	517	
Paris	1,057	308	497	375	832

37. **a.** Prove that if a walk in a graph contains a repeated edge, then the walk contains a repeated vertex.
    b. Explain how it follows from part (a) that any walk with no repeated vertex has no repeated edge.

38. Prove Lemma 10.2.1(a): If G is a connected graph, then any two distinct vertices of G can be connected by a path.

39. Prove Lemma 10.2.1(b): If vertices v and w are part of a circuit in a graph G and one edge is removed from the circuit, then there still exists a trail from v to w in G.

40. Draw a picture to illustrate Lemma 10.2.1(c): If a graph G is connected and G contains a circuit, then an edge of the circuit can be removed without disconnecting G.

41. Prove that if there is a trail in a graph G from a vertex v to a vertex w, then there is a trail from w to v.

H 42. If a graph contains a circuit that starts and ends at a vertex v, does the graph contain a simple circuit that starts and ends at v? Why?

43. Prove that if there is a circuit in a graph that starts and ends at a vertex v and if w is another vertex in the circuit, then there is a circuit in the graph that starts and ends at w.

44. Let G be a connected graph, and let C be any circuit in G that does not contain every vertex of C. Let G' be the subgraph obtained by removing all the edges of C from G and also any vertices that become isolated when the edges of C are removed. Prove that there exists a vertex v such that v is in both C and G'.

45. Prove that any graph with an Euler circuit is connected.

46. Prove Corollary 10.2.5.

47. For what values of n does the complete graph K_n with n vertices have (a) an Euler circuit? (b) a Hamiltonian circuit? Justify your answers.

✱ 48. For what values of m and n does the complete bipartite graph on (m, n) vertices have (a) an Euler circuit? (b) a Hamiltonian circuit? Justify your answers.

✱ 49. What is the maximum number of edges a simple disconnected graph with n vertices can have? Prove your answer.

✱ 50. Show that a graph is bipartite if, and only if, it does not have a circuit with an odd number of edges. (See exercise 37 of Section 10.1 for the definition of bipartite graph.)

Answers for Test Yourself

1. (a) a finite alternating sequence of adjacent vertices and edges of G (b) a walk that does not contain a repeated edge   (c) a trail that does not contain a repeated vertex   (d) a walk that starts and ends at the same vertex   (e) a closed walk that contains at least one edge and does not contain a repeated edge   (f) a circuit that does not have any repeated vertex other than the first and the last   (g) a walk consisting of a single vertex and no edge   (h) there is a walk from v to w 2. given any two vertices in the graph, there is a walk from one to the other   3. disconnect the graph   4. a circuit that contains every vertex and every edge of the graph   5. the graph is connected, and every vertex has positive, even degree   6. the graph is connected, v and w have odd degree, and all other vertices have positive even degree   7. a simple circuit that includes every vertex of the graph   8. H contains every vertex of G; H is connected; H has the same number of edges as vertices; every vertex of H has degree 2   9. Hamiltonian circuit

10.3 Trees

We are not very pleased when we are forced to accept a mathematical truth
by virtue of a complicated chain of formal conclusions and computations, which we
traverse blindly, link by link, feeling our way by touch. We want first an overview of the
aim and of the road; we want to understand the idea of the proof, the deeper context.
— Hermann Weyl, 1885–1955

If a friend asks what you are studying and you answer "trees," your friend is likely to infer you are taking a course in botany. But trees are also a subject for mathematical investigation. In mathematics, a tree is a connected graph that does not contain any circuits. Mathematical trees are similar in certain ways to their botanical namesakes.

• Definition

A graph is said to be **circuit-free** if, and only if, it has no circuits. A graph is called a **tree** if, and only if, it is circuit-free and connected. A **trivial tree** is a graph that consists of a single vertex. A graph is called a **forest** if, and only if, it is circuit-free and not connected.

Example 10.3.1 Trees and Non-Trees

All the graphs shown in Figure 10.3.1 are trees, whereas those in Figure 10.3.2 are not.

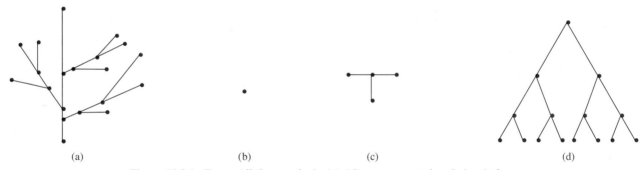

(a) (b) (c) (d)

Figure 10.3.1 Trees. All the graphs in (a)–(d) are connected and circuit-free.

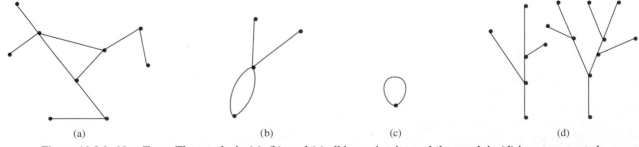

(a) (b) (c) (d)

Figure 10.3.2 Non-Trees. The graphs in (a), (b), and (c) all have circuits, and the graph in (d) is not connected. ∎

Examples of Trees

The following examples illustrate just a few of the many and varied situations in which mathematical trees arise.

Example 10.3.2 A Decision Tree

During orientation week, a college administers an exam to all entering students to determine placement in the mathematics curriculum. The exam consists of two parts, and placement recommendations are made as indicated by the tree shown in Figure 10.3.3. Read the tree from left to right to decide what course should be recommended for a student who scored 9 on part I and 7 on part II.

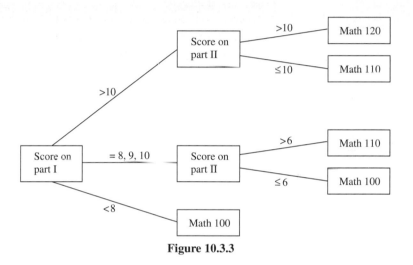

Figure 10.3.3

Solution   Since the student scored 9 on part I, the score on part II is checked. Since it is greater than 6, the student should be advised to take Math 110.   ■

Example 10.3.3 A Parse Tree

In the last 30 years, Noam Chomsky and others have developed new ways to describe the syntax (or grammatical structure) of natural languages such as English. This work has proved useful in constructing compilers for high-level computer languages. In the study of grammars, trees are often used to show the derivation of grammatically correct sentences from certain basic rules. Such trees are called **syntactic derivation trees** or **parse trees.**

A very small subset of English grammar, for example, specifies that

1. a sentence can be produced by writing first a noun phrase and then a verb phrase;

2. a noun phrase can be produced by writing an article and then a noun;

3. a noun phrase can also be produced by writing an article, then an adjective, and then a noun;

4. a verb phrase can be produced by writing a verb and then a noun phrase;

5. one article is "the";

6. one adjective is "young";

7. one verb is "caught";

8. one noun is "man";

9. one (other) noun is "ball."

John Backus
(1924–1998)

The rules of a grammar are called **productions.** It is customary to express them using the shorthand notation illustrated below. This notation, introduced by John Backus in 1959 and modified by Peter Naur in 1960, was used to describe the computer language Algol and is called the **Backus-Naur notation.** In the notation, the symbol | represents the word *or,* and angle brackets ⟨ ⟩ are used to enclose terms to be defined (such as a sentence or noun phrase).

1. ⟨sentence⟩ → ⟨noun phrase⟩⟨verb phrase⟩

2., 3. ⟨noun phrase⟩ → ⟨article⟩⟨noun⟩ | ⟨article⟩⟨adjective⟩⟨noun⟩

4. ⟨verb phrase⟩ → ⟨verb⟩⟨noun phrase⟩

5. ⟨article⟩ → the

6. ⟨adjective⟩ → young

7, 8. ⟨noun⟩ → man | ball

9. ⟨verb⟩ → caught

The derivation of the sentence "The young man caught the ball" from the above rules is described by the tree shown below.

Peter Naur
(born 1928)

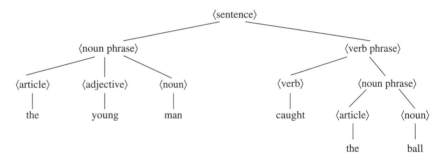

In the study of linguistics, **syntax** refers to the grammatical structure of sentences, and **semantics** refers to the meanings of words and their interrelations. A sentence can be syntactically correct but semantically incorrect, as in the nonsensical sentence "The young ball caught the man," which can be derived from the rules given above. Or a sentence can contain syntactic errors but not semantic ones, as, for instance, when a two-year-old child says, "Me hungry!" ∎

Example 10.3.4 Structure of Hydrocarbon Molecules

The German physicist Gustav Kirchhoff (1824–1887) was the first to analyze the behavior of mathematical trees in connection with the investigation of electrical circuits. Soon after (and independently), the English mathematician Arthur Cayley used the mathematics of trees to enumerate all isomers for certain hydrocarbons. Hydrocarbon molecules are composed of carbon and hydrogen; each carbon atom can form up to four chemical bonds with other atoms, and each hydrogen atom can form one bond with another atom. Thus the structure of hydrocarbon molecules can be represented by graphs such as those shown following, in which the vertices represent atoms of hydrogen and carbon, denoted H and C, and the edges represent the chemical bonds between them.

Arthur Cayley
(1821–1895)

Bettmann/CORBIS

Butane

Isobutane

Note that each of these graphs has four carbon atoms and ten hydrogen atoms, but the two graphs show different configurations of atoms. When two molecules have the same chemical formulae (in this case C_4H_{10}) but different chemical bonds, they are called *isomers*.

Certain *saturated hydrocarbon* molecules contain the maximum number of hydrogen atoms for a given number of carbon atoms. Cayley showed that if such a saturated hydrocarbon molecule has k carbon atoms, then it has $2k + 2$ hydrogen atoms. The first step in doing so is to prove that the graph of such a saturated hydrocarbon molecule is a tree. Prove this using proof by contradiction. (You are asked to finish the derivation of Cayley's result in exercise 4 at the end of this section.)

Solution   Suppose there is a hydrocarbon molecule that contains the maximum number of hydrogen atoms for the number of its carbon atoms and whose graph G is not a tree. *[We must derive a contradiction.]* Since G is not a tree, G is not connected or G has a circuit. But the graph of any molecule is connected (all the atoms in a molecule must be connected to each other), and so G must have a nontrivial circuit. Now the edges of the circuit can link only carbon atoms because every vertex of a circuit has degree at least 2 and a hydrogen atom vertex has degree 1. Delete one edge of the circuit and add two new edges to join each of the newly disconnected carbon atom vertices to a hydrogen atom vertex as shown below.

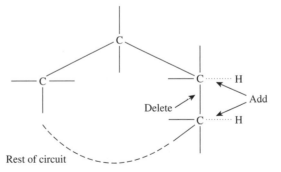

The resulting molecule has two more hydrogen atoms than the given molecule, but the number of carbon atoms is unchanged. This contradicts the supposition that the given molecule has the maximum number of hydrogen atoms for the given number of carbon atoms. Hence the supposition is false, and so G is a tree. ■

Characterizing Trees

There is a somewhat surprising relation between the number of vertices and the number of edges of a tree. It turns out that if n is a positive integer, then any tree with n vertices (no matter what its shape) has $n - 1$ edges. Perhaps even more surprisingly, a partial converse to this fact is also true—namely, any *connected* graph with n vertices and $n - 1$ edges is a tree. It follows from these facts that if even one new edge (but no new vertex) is added to a tree, the resulting graph must contain a circuit. Also, from the fact that removing an edge from a circuit does not disconnect a graph, it can be shown that every connected graph has a subgraph that is a tree. It follows that if n is a positive integer, any graph with n vertices and *fewer* than $n - 1$ edges is not connected.

A small but very important fact necessary to derive the first main theorem about trees is that any nontrivial tree must have at least one vertex of degree 1.

Lemma 10.3.1

Any tree that has more than one vertex has at least one vertex of degree 1.

A constructive way to understand this lemma is to imagine being given a tree T with more than one vertex. You pick a vertex v at random and then search outward along a path from v looking for a vertex of degree 1. As you reach each new vertex, you check whether it has degree 1. If it does, you are finished. If it does not, you exit from the vertex along a different edge from the one you entered on. Because T is circuit-free, the vertices included in the path never repeat. And since the number of vertices of T is finite, the process of building a path must eventually terminate. When that happens, the final vertex v' of the path must have degree 1. This process is illustrated below.

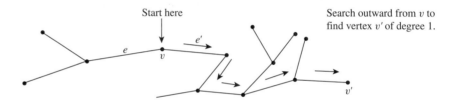

This discussion is made precise in the following proof.

Proof:

Let T be a particular but arbitrarily chosen tree that has more than one vertex, and consider the following algorithm:

Step 1: Pick a vertex v of T and let e be an edge incident on v.

    [If there were no edge incident on v, then v would be an isolated vertex. But this would contradict the assumption that T is connected (since it is a tree) and has at least two vertices.]

Step 2: While $\deg(v) > 1$, repeat steps 2a, 2b, and 2c:

    Step 2a: Choose e' to be an edge incident on v such that $e' \neq e$. *[Such an edge exists because $\deg(v) > 1$ and so there are at least two edges incident on v.]*

    Step 2b: Let v' be the vertex at the other end of e' from v. *[Since T is a tree, e' cannot be a loop and therefore e' has two distinct endpoints.]*

    Step 2c: Let $e = e'$ and $v = v'$. *[This is just a renaming process in preparation for a repetition of step 2.]*

The algorithm just described must eventually terminate because the set of vertices of the tree T is finite and T is circuit-free. When it does, a vertex v of degree 1 will have been found.

Using Lemma 10.3.1 it is not difficult to show that, in fact, any tree that has more than one vertex has at least *two* vertices of degree 1. This extension of Lemma 10.3.1 is left to the exercises at the end of this section.

> **• Definition**
>
> Let T be a tree. If T has only one or two vertices, then each is called a **terminal vertex**. If T has at least three vertices, then a vertex of degree 1 in T is called a **terminal vertex** (or a **leaf**), and a vertex of degree greater than 1 in T is called an **internal vertex** (or a **branch vertex**).

Example 10.3.5 Terminal and Internal Vertices

Find all terminal vertices and all internal vertices in the following tree:

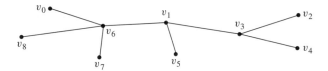

Solution   The terminal vertices are v_0, v_2, v_4, v_5, v_7, and v_8. The internal vertices are v_6, v_1, and v_3. ∎

The following is the first of the two main theorems about trees:

> **Theorem 10.3.2**
>
> For any positive integer n, any tree with n vertices has $n - 1$ edges.

The proof is by mathematical induction. To do the inductive step, you assume the theorem is true for a positive integer k and then show it is true for $k + 1$. Thus you assume you have a tree T with $k + 1$ vertices, and you must show that T has $(k + 1) - 1 = k$ edges. As you do this, you are free to use the inductive hypothesis that *any* tree with k vertices has $k - 1$ edges. To make use of the inductive hypothesis, you need to reduce the tree T with $k + 1$ vertices to a tree with just k vertices. But by Lemma 10.3.1, T has a vertex v of degree 1, and since T is connected, v is attached to the rest of T by a single edge e as sketched below.

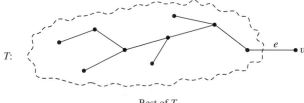

Rest of T

Now if e and v are removed from T, what remains is a tree T' with $(k + 1) - 1 = k$ vertices. By inductive hypothesis, then, T' has $k - 1$ edges. But the original tree T has one more vertex and one more edge than T'. Hence T must have $(k - 1) + 1 = k$ edges, as was to be shown. A formal version of this argument is given on the next page.

Proof (by mathematical induction):

Let the property $P(n)$ be the sentence

Any tree with n vertices has $n - 1$ edges.   ← $P(n)$

We use mathematical induction to show that this property is true for all integers $n \geq 1$.

Show that $P(1)$ is true: Let T be any tree with one vertex. Then T has zero edges (since it contains no loops). But $0 = 1 - 1$, so $P(1)$ is true.

Show that for all integers $k \geq 1$, if $P(k)$ is true then $P(k + 1)$ is true:
Suppose k is any positive integer for which $P(k)$ is true. In other words, suppose that

Any tree with k vertices has k - 1 edges.   ← $P(k)$
                                                                     inductive hypothesis

We must show that $P(k + 1)$ is true. In other words, we must show that

Any tree with $k + 1$ vertices has $(k + 1) - 1 = k$ edges. ← $P(k + 1)$

Let T be a particular but arbitrarily chosen tree with $k + 1$ vertices. *[We must show that T has k edges.]* Since k is a positive integer, $(k + 1) \geq 2$, and so T has more than one vertex. Hence by Lemma 10.3.1, T has a vertex v of degree 1. Also, since T has more than one vertex, there is at least one other vertex in T besides v. Thus there is an edge e connecting v to the rest of T. Define a subgraph T' of T so that

$$V(T') = V(T) - \{v\}$$

Then

$$E(T') = E(T) - \{e\}.$$

1. The number of vertices of T' is $(k + 1) - 1 = k$.

2. T' is circuit-free (since T is circuit-free, and removing an edge and a vertex cannot create a circuit).

3. T' is connected (see exercise 24 at the end of this section).

Hence, by the definition of tree, T' is a tree. Since T' has k vertices, by inductive hypothesis

the number of edges of $T' =$ (the number of vertices of T') $- 1$
$$= k - 1.$$

But then

the number of edges of $T =$ (the number of edges of T') $+ 1$
$$= (k - 1) + 1$$
$$= k.$$

[This is what was to be shown.]

Example 10.3.6 Determining Whether a Graph Is a Tree

A graph G has ten vertices and twelve edges. Is it a tree?

Solution   No. By Theorem 10.3.2, any tree with ten vertices has nine edges, not twelve. ∎

Example 10.3.7 Finding Trees Satisfying Given Conditions

Find all possible combinations of degrees of the vertices for a tree with four vertices, and draw an example for each such tree.

Solution   By Theorem 10.3.2, any tree with four vertices has three edges. Thus the total degree of a tree with four vertices must be 6. Also, every tree with more than one vertex has at least two vertices of degree 1 (see the comment following Lemma 10.3.1 and exercise 29 at the end of this section). Thus the following combinations of degrees for the vertices are the only ones possible:

$$1, 1, 1, 3 \quad \text{and} \quad 1, 1, 2, 2.$$

Two trees corresponding to these possibilities are shown below.

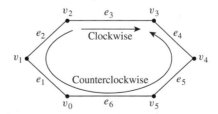

and

To prove the second major theorem about trees, we need another lemma.

Lemma 10.3.3

If G is any connected graph, C is any circuit in G, and any one of the edges of C is removed from G, then the graph that remains is connected.

Essentially, the reason why Lemma 10.3.3 is true is that any two vertices in a circuit are connected by two distinct paths. It is possible to draw the graph so that one of these goes "clockwise" and the other goes "counterclockwise" around the circuit. For example, in the circuit shown below, the clockwise path from v_2 to v_3 is

$$v_2 e_3 v_3$$

and the counterclockwise path from v_2 to v_3 is

$$v_2 e_2 v_1 e_1 v_0 e_6 v_5 e_5 v_4 e_4 v_3.$$

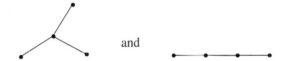

Proof:

Suppose G is a connected graph, C is a circuit in G, and e is an edge of C. Form a subgraph G' of G by removing e from G. Thus

$$V(G') = V(G)$$
$$E(G') = E(G) - \{e\}.$$

(continued on page 520)

We must show that G' is connected. *[To show a graph is connected, we must show that if u and w are any vertices of the graph, then there exists a walk in G' from u to w.]* Suppose u and w are any two vertices of G'. *[We must find a walk from u to w.]* Since the vertex sets of G and G' are the same, u and w are both vertices of G, and since G is connected, there is a walk W in G from u to w.

Case 1 (e is not an edge of W): The only edge in G that is not in G' is e, so in this case W is also a walk in G'. Hence u is connected to w by a walk in G'.

Case 2 (e is an edge of W): In this case the walk W from u to w includes a section of the circuit C that contains e. Let C be denoted as follows:

$$C: v_0 e_1 v_1 e_2 v_2 \cdots e_n v_n \, (= v_0).$$

Now e is one of the edges of C, so, to be specific, let $e = e_k$. Then the walk W contains either the sequence

$$v_{k-1} e_k v_k \quad \text{or} \quad v_k e_k v_{k-1}.$$

If W contains $v_{k-1} e_k v_k$, connect v_{k-1} to v_k by taking the "counterclockwise" walk W' defined as follows:

$$W': v_{k-1} e_{k-1} v_{k-2} \cdots v_0 e_n v_{n-1} \cdots e_{k+1} v_k.$$

An example showing how to go from u to w while avoiding e_k is given in Figure 10.3.4.

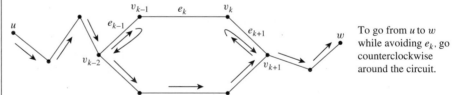

To go from u to w while avoiding e_k, go counterclockwise around the circuit.

Figure 10.3.4 An Example of a Walk from u to w That Does Not Include Edge e_k

If W contains $v_k e_k v_{k-1}$, connect v_k to v_{k-1} by taking the "clockwise" walk W'' defined as follows:

$$W'': v_k e_{k+1} v_{k+1} \cdots v_n e_1 v_1 e_2 \cdots e_{k-1} v_{k-1}.$$

Now patch either W' or W'' into W to form a new walk from u to w. For instance, to patch W' into W, start with the section of W from u to v_{k-1}, then take W' from v_{k-1} to v_k, and finally take the section of W from v_k to w. If this new walk still contains an occurrence of e, just repeat the process described previously until all occurrences are eliminated. *[This must happen eventually since the number of occurrences of e in C is finite.]* The result is a walk from u to w that does not contain e and hence is a walk in G'.

The previous arguments show that both in case 1 and in case 2 there is a walk in G' from u to w. Since the choice of u and w was arbitrary, G' is connected.

The second major theorem about trees is a modified converse to Theorem 10.3.2.

Theorem 10.3.4

For any positive integer n, if G is a connected graph with n vertices and $n-1$ edges, then G is a tree.

Proof:

Let n be a positive integer and suppose G is a particular but arbitrarily chosen graph that is connected and has n vertices and $n-1$ edges. *[We must show that G is a tree. Now a tree is a connected, circuit-free graph. Since we already know G is connected, it suffices to show that G is circuit-free.]* Suppose G is not circuit-free. That is, suppose G has a circuit C. *[We must derive a contradiction.]* By Lemma 10.3.3, an edge of C can be removed from G to obtain a graph G' that is connected. If G' has a circuit, then repeat this process: Remove an edge of the circuit from G' to form a new connected graph. Continue repeating the process of removing edges from circuits until eventually a graph G'' is obtained that is connected and is circuit-free. By definition, G'' is a tree. Since no vertices were removed from G to form G'', G'' has n vertices just as G does. Thus, by Theorem 10.3.2, G'' has $n-1$ edges. But the supposition that G has a circuit implies that at least one edge of G is removed to form G''. Hence G'' has no more than $(n-1)-1=n-2$ edges, which contradicts its having $n-1$ edges. So the supposition is false. Hence G is circuit-free, and therefore G is a tree *[as was to be shown].*

Theorem 10.3.4 is not a full converse of Theorem 10.3.2. Although it is true that every *connected* graph with n vertices and $n-1$ edges (where n is a positive integer) is a tree, it is not true that *every* graph with n vertices and $n-1$ edges is a tree.

Example 10.3.8 A Graph with n Vertices and $n-1$ Edges That Is Not a Tree

Give an example of a graph with five vertices and four edges that is not a tree.

Solution   By Theorem 10.3.4, such a graph cannot be connected. One example of such an unconnected graph is shown below.

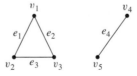

Test Yourself

1. A circuit-free graph is a graph with _____.

2. A forest is a graph that is _____, and a tree is a graph that is _____.

3. A trivial tree is a graph that consists of _____.

4. Any tree with at least two vertices has at least one vertex of degree _____.

5. If a tree T has at least two vertices, then a terminal vertex (or leaf) in T is a vertex of degree _____ and an internal vertex (or branch vertex) in T is a vertex of degree _____.

6. For any positive integer n, any tree with n vertices has _____.

7. For any positive integer n, if G is a connected graph with n vertices and $n-1$ edges then _____.

Exercise Set 10.3

1. Read the tree in Example 10.3.2 from left to right to answer the following questions:
   a. What course should a student who scored 12 on part I and 4 on part II take?
   b. What course should a student who scored 8 on part I and 9 on part II take?

2. Draw trees to show the derivations of the following sentences from the rules given in Example 10.3.3.
   a. The young ball caught the man.
   b. The man caught the young ball.

H 3. What is the total degree of a tree with n vertices? Why?

4. Let G be the graph of a hydrocarbon molecule with the maximum number of hydrogen atoms for the number of its carbon atoms.
   a. Draw the graph of G if G has three carbon atoms and eight hydrogen atoms.
   b. Draw the graphs of three isomers of C_5H_{12}.
   c. Use Example 10.3.4 and exercise 3 to prove that if the vertices of G consist of k carbon atoms and m hydrogen atoms, then G has a total degree of $2k + 2m - 2$.
   H d. Prove that if the vertices of G consist of k carbon atoms and m hydrogen atoms, then G has a total degree of $4k + m$.
   e. Equate the results of (c) and (d) to prove Cayley's result that a saturated hydrocarbon molecule with k carbon atoms and a maximum number of hydrogen atoms has $2k + 2$ hydrogen atoms.

H 5. Extend the argument given in the proof of Lemma 10.3.1 to show that a tree with more than one vertex has at least two vertices of degree 1.

6. If graphs are allowed to have an infinite number of vertices and edges, then Lemma 10.3.1 is false. Give a counterexample that shows this. In other words, give an example of an "infinite tree" (a connected, circuit-free graph with an infinite number of vertices and edges) that has no vertex of degree 1.

7. Find all terminal vertices and all internal vertices for the following trees.

   a.
   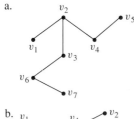

   b.
   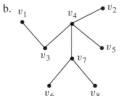

In each of 8–21, either draw a graph with the given specifications or explain why no such graph exists.

8. Tree, nine vertices, nine edges

9. Graph, connected, nine vertices, nine edges

10. Graph, circuit-free, nine vertices, six edges

11. Tree, six vertices, total degree 14

12. Tree, five vertices, total degree 8

13. Graph, connected, six vertices, five edges, has a circuit

14. Graph, two vertices, one edge, not a tree

15. Graph, circuit-free, seven vertices, four edges

16. Tree, twelve vertices, fifteen edges

17. Graph, six vertices, five edges, not a tree

18. Tree, five vertices, total degree 10

19. Graph, connected, ten vertices, nine edges, has a circuit

20. Simple graph, connected, six vertices, six edges

21. Tree, ten vertices, total degree 24

22. A connected graph has twelve vertices and eleven edges. Does it have a vertex of degree 1? Why?

23. A connected graph has nine vertices and twelve edges. Does it have a circuit? Why?

24. Suppose that v is a vertex of degree 1 in a connected graph G and that e is the edge incident on v. Let G' be the subgraph of G obtained by removing v and e from G. Must G' be connected? Why?

25. A graph has eight vertices and six edges. Is it connected? Why?

H 26. If a graph has n vertices and $n - 2$ or fewer edges, can it be connected? Why?

27. A circuit-free graph has ten vertices and nine edges. Is it connected? Why?

H 28. Is a circuit-free graph with n vertices and at least $n - 1$ edges connected? Why?

29. Prove that every nontrivial tree has at least two vertices of degree 1 by filling in the details and completing the following argument: Let T be a nontrivial tree and let S be the set of all paths from one vertex to another of T. Among all the paths in S, choose a path P with the most edges. (Why is it possible to find such a P?) What can you say about the initial and final vertices of P? Why?

30. Find all possible combinations for the degrees of the vertices of a tree with five vertices, and draw an example of each such tree.

Answers for Test Yourself

1. no circuits   2. circuit-free and not connected; connected and circuit-free   3. a single vertex (and no edges)   4. 1   5. 1; greater than 1 (*Or*: at least 2)   6. $n - 1$ edges   7. G is a tree

10.4 Rooted Trees

Let us grant that the pursuit of mathematics is a divine madness of the human spirit, a refuge from the goading urgency of contingent happenings.
— Alfred North Whitehead, 1861–1947

An outdoor tree is rooted and so is the kind of family tree that shows all the descendants of one particular person. The terminology and notation of rooted trees blends the language of botanical trees and that of family trees. In mathematics, a rooted tree is a tree in which one vertex has been distinguished from the others and is designated the *root*. Given any other vertex v in the tree, there is a unique path from the root to v. (After all, if there were two distinct paths, a circuit could be constructed.) The number of edges in such a path is called the *level* of v, and the *height* of the tree is the length of the longest such path. It is traditional in drawing rooted trees to place the root at the top (as is done in family trees) and show the branches descending from it.

> **• Definition**
>
> A **rooted tree** is a tree in which there is one vertex that is distinguished from the others and is called the **root.** The **level** of a vertex is the number of edges along the unique path between it and the root. The **height** of a rooted tree is the maximum level of any vertex of the tree. Given the root or any internal vertex v of a rooted tree, the **children** of v are all those vertices that are adjacent to v and are one level farther away from the root than v. If w is a child of v, then v is called the **parent** of w, and two distinct vertices that are both children of the same parent are called **siblings.** Given two distinct vertices v and w, if v lies on the unique path between w and the root, then v is an **ancestor** of w and w is a **descendant** of v.

These terms are illustrated in Figure 10.4.1.

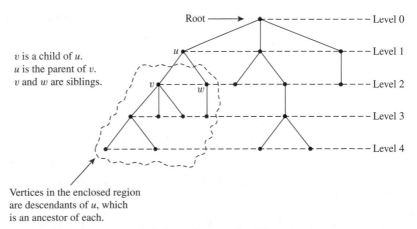

v is a child of *u*.
u is the parent of *v*.
v and *w* are siblings.

Vertices in the enclosed region are descendants of *u*, which is an ancestor of each.

Figure 10.4.1 A Rooted Tree

Example 10.4.1 Rooted Trees

Consider the tree with root v_0 shown below.

a. What is the level of v_5? b. What is the level of v_0?

c. What is the height of this rooted tree? d. What are the children of v_3?

e. What is the parent of v_2? f. What are the siblings of v_8?

g. What are the descendants of v_3?

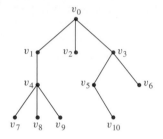

Solution

a. 2 b. 0 c. 3 d. v_5 and v_6 e. v_0 f. v_7 and v_9 g. v_5, v_6, v_{10}

∎

Note that in the tree with root v_0 shown below, v_1 has level 1 and is the child of v_0, and both v_0 and v_1 are terminal vertices.

Binary Trees

When every vertex in a rooted tree has at most two children and each child is designated either the (unique) left child or the (unique) right child, the result is a *binary tree*.

> **• Definition**
>
> A **binary tree** is a rooted tree in which every parent has at most two children. Each child in a binary tree is designated either a **left child** or a **right child** (but not both), and every parent has at most one left child and one right child. A **full binary tree** is a binary tree in which each parent has exactly two children.
>
> Given any parent v in a binary tree T, if v has a left child, then the **left subtree** of v is the binary tree whose root is the left child of v, whose vertices consist of the left child of v and all its descendants, and whose edges consist of all those edges of T that connect the vertices of the left subtree. The **right subtree** of v is defined analogously.

These terms are illustrated in Figure 10.4.2.

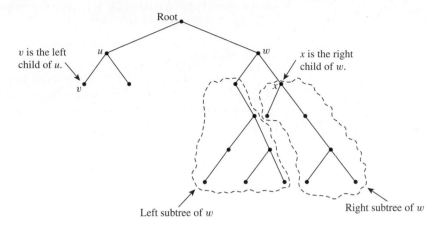

Figure 10.4.2 A Binary Tree

Example 10.4.2 Representation of Algebraic Expressions

Binary trees are used in many ways in computer science. One use is to represent algebraic expressions with arbitrary nesting of balanced parentheses. For instance, the following (labeled) binary tree represents the expression a/b: The operator is at the root and acts on the left and right children of the root in left-right order.

More generally, the binary tree shown below represents the expression $a/(c + d)$. In such a representation, the internal vertices are arithmetic operators, the terminal vertices are variables, and the operator at each vertex acts on its left and right subtrees in left-right order.

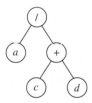

Draw a binary tree to represent the expression $((a - b) \cdot c) + (d/e)$.

Solution

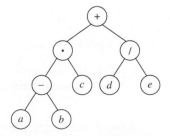

An interesting theorem about binary trees says that if you know the number of internal vertices of a full binary tree, then you can calculate both the total number of vertices and the number of terminal vertices, and conversely. More specifically, a full binary tree with k internal vertices has a total of $2k + 1$ vertices of which $k + 1$ are terminal vertices.

Theorem 10.4.1

If k is a positive integer and T is a full binary tree with k internal vertices, then T has a total of $2k + 1$ vertices and has $k + 1$ terminal vertices.

Proof:

Suppose k is a positive integer and T is a full binary tree with k internal vertices. Observe that the set of all vertices of T can be partitioned into two disjoint subsets: the set of all vertices that have a parent and the set of all vertices that do not have a parent. Now there is just one vertex that does not have a parent, namely the root. Also, since every internal vertex of a full binary tree has exactly two children, the number of vertices that have a parent is twice the number of parents, or $2k$, since each parent is an internal vertex. Hence

$$\begin{bmatrix} \text{the total number} \\ \text{of vertices of } T \end{bmatrix} = \begin{bmatrix} \text{the number of} \\ \text{vertices that} \\ \text{have a parent} \end{bmatrix} + \begin{bmatrix} \text{the number of} \\ \text{vertices that do} \\ \text{not have a parent} \end{bmatrix}$$

$$= \qquad 2k \qquad + \qquad 1.$$

But it is also true that the total number of vertices of T equals the number of internal vertices plus the number of terminal vertices. Thus

$$\begin{bmatrix} \text{the total number} \\ \text{of vertices of } T \end{bmatrix} = \begin{bmatrix} \text{the number of} \\ \text{internal vertices} \end{bmatrix} + \begin{bmatrix} \text{the number of} \\ \text{terminal vertices} \end{bmatrix}$$

$$= \qquad k \qquad + \begin{bmatrix} \text{the number of} \\ \text{terminal vertices} \end{bmatrix}$$

Now equate the two expressions for the total number of vertices of T:

$$2k + 1 = k + \begin{bmatrix} \text{the number of} \\ \text{terminal vertices} \end{bmatrix}$$

Solving this equation gives

$$\begin{bmatrix} \text{the number of} \\ \text{terminal vertices} \end{bmatrix} = (2k + 1) - k = k + 1.$$

Thus the total number of vertices is $2k + 1$ and the number of terminal vertices is $k + 1$ *[as was to be shown].*

Example 10.4.3 Determining Whether a Certain Full Binary Tree Exists

Is there a full binary tree that has 10 internal vertices and 13 terminal vertices?

Solution No. By Theorem 10.4.1, a full binary tree with 10 internal vertices has $10 + 1 = 11$ terminal vertices, not 13. ∎

Another interesting theorem about binary trees specifies the maximum number of terminal vertices of a binary tree of a given height. Specifically, the maximum number of terminal vertices of a binary tree of height h is 2^h. Another way to say this is that a binary tree with t terminal vertices has height of at least $\log_2 t$.

Theorem 10.4.2

For all integers $h \geq 0$, if T is any binary tree with height h and t terminal vertices, then

$$t \leq 2^h.$$

Equivalently, $\log_2 t \leq h.$

Proof (by strong mathematical induction):

Let $P(h)$ be the sentence

> If T is any binary tree of height h, then the number of $\leftarrow P(h)$
> terminal vertices of T is at most 2^h.

Show that $P(0)$ is true: We must show that if T is any binary tree of height 0, then the number of terminal vertices of T is at most 2^0. Suppose T is a tree of height 0. Then T consists of a single vertex, the root. By definition this is a terminal vertex and so the number of terminal vertices is $t = 1 = 2^\circ = 2^h$. Hence $t \leq 2^h$ *[as was to be shown]*.

Show that for all integers $k \geq 0$, if $P(i)$ is true for all integers i from 0 through k, then it is true for $k + 1$:
Let k be any integer with $k \geq 0$, and suppose that

> For all integers i from 0 through k, if T is any
> binary tree of height i, then the number of $\leftarrow$ inductive hypothesis
> terminal vertices of T is at most 2^i.

We must show that

> If T is any binary tree of height $k + 1$, then the number of $\leftarrow P(k + 1)$
> terminal vertices of T is 2^{k+1}.

Let T be a binary tree of height $k + 1$, root v, and t terminal vertices. Because $k \geq 0$, we have that $k + 1 \geq 1$ and so v has at least one child.

Case 1 (v has only one child): In this case we may assume without loss of generality that v's child is a left child and denote it by v_L. Let T_L be the left subtree of v. Then v_L is the root of T_L. (This situation is illustrated in Figure 10.4.3.) Because v has only one child, v is itself a terminal vertex, so the total number of terminal vertices in T equals the number of terminal vertices in T_L plus 1. Thus if t_L is the number of terminal vertices in T_L, then $t = t_L + 1$.

Now by inductive hypothesis, $t_L \leq 2^k$ because the height of T_L is k, one less than the height of T. Also, because v has a child, $k + 1 \geq 1$, and so $2^k \geq 2^0 = 1$. Therefore,

$$t = t_L + 1 \leq 2^k + 1 \leq 2^k + 2^k = 2 \cdot 2^k = 2^{(k+1)}.$$

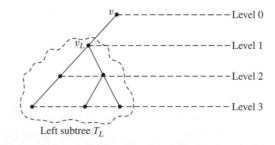

Left subtree T_L

Figure 10.4.3 A Binary Tree Whose Root Has One Child

(continued on page 528)

Case 2 (v has two children): In this case, v has both a left child, v_L, and a right child, v_R, and v_L and v_R are roots of a left subtree T_L and a right subtree T_R. Note that T_L and T_R are binary trees because T is a binary tree. (This situation is illustrated in Figure 10.4.4.)

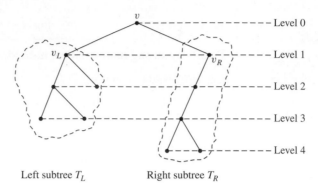

Left subtree T_L Right subtree T_R

Figure 10.4.4 A Binary Tree Whose Root Has Two Children

Now v_L and v_R are the roots of the left and right subtrees of v, denoted T_L and T_R, respectively. Note that T_L and T_R are binary trees because T is a binary tree. Let h_L and h_R be the heights of T_L and T_R, respectively. Then $h_L \leq k$ and $h_R \leq k$ since T is obtained by joining T_L and T_R and adding a level. Let t_L and t_R be the numbers of terminal vertices of T_L and T_R, respectively. Then, since both T_L and T_R have heights less than $k + 1$, by inductive hypothesis

$$t_L \leq 2^{h_L} \quad \text{and} \quad t_R \leq 2^{h_R}.$$

But the terminal vertices of T consist exactly of the terminal vertices of T_L together with the terminal vertices of T_R. Therefore,

$$t = t_L + t_R \leq 2^{h_L} + 2^{h_R} \qquad \text{by inductive hypothesis since } h_L \leq k \text{ and } h_R \leq k$$

Hence,

$$t \leq 2^k + 2^k = 2 \cdot 2^k = 2^{k+1} \qquad \text{by basic algebra.}$$

Thus the number of terminal vertices is at most $2k + 1$ *[as was to be shown]*.

Since both the basis step and the inductive step have been proved, we conclude that for all integers $h \geq 0$, if T is any binary tree with height h and t terminal vertices, then $t \leq 2^h$.

The equivalent inequality $\log_2 t \leq h$ follows from the fact that the logarithmic function with base 2 is increasing. In other words, for all positive real numbers x and y,

$$\text{if } x < y \text{ then } \log_2 x < \log_2 y.$$

Thus if we apply the logarithmic function with base 2 to both sides of

$$t \leq 2^h,$$

we obtain

$$\log_2 t \leq \log_2(2^h).$$

Now by definition of logarithm, $\log_2(2^h) = h$ *[because $\log_2(2^h)$ is the exponent to which 2 must be raised to obtain 2^h]*. Hence

$$\log_2 t \leq h$$

[as was to be shown].

Example 10.4.4 Determining Whether a Certain Binary Tree Exists

Is there a binary tree that has height 5 and 38 terminal vertices?

Solution No. By Theorem 10.4.2, any binary tree T with height 5 has at most $2^5 = 32$ terminal vertices, so such a tree cannot have 38 terminal vertices. ■

Test Yourself

1. A rooted tree is a tree in which _____. The level of a vertex in a rooted tree is _____. The height of a rooted tree is _____.

2. A binary tree is a rooted tree in which _____.

3. A full binary tree is a rooted tree in which _____.

4. If k is a positive integer and T is a full binary tree with k internal vertices, then T has a total of _____ vertices and has _____ terminal vertices.

5. If T is a binary tree that has t terminal vertices and height h, then t and h are related by the inequality _____.

Exercise Set 10.4

1. Consider the tree shown below with root a.
   a. What is the level of n?
   b. What is the level of a?
   c. What is the height of this rooted tree?
   d. What are the children of n?
   e. What is the parent of g?
   f. What are the siblings of j?
   g. What are the descendants of f?

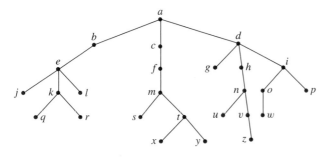

2. Consider the tree shown below with root v_0.
   a. What is the level of v_8?
   b. What is the level of v_0?
   c. What is the height of this rooted tree?
   d. What are the children of v_{10}?
   e. What is the parent of v_5?
   f. What are the siblings of v_1?
   g. What are the descendants of v_{12}?

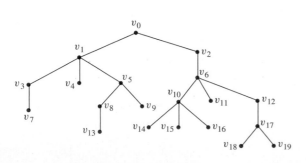

3. Draw binary trees to represent the following expressions:
   a. $a \cdot b - (c/(d + e))$ b. $a/(b - c \cdot d)$

In each of 4–20 either draw a graph with the given specifications or explain why no such graph exists.

4. Full binary tree, five internal vertices

5. Full binary tree, five internal vertices, seven terminal vertices

6. Full binary tree, seven vertices, of which four are internal vertices

7. Full binary tree, twelve vertices

8. Full binary tree, nine vertices

9. Binary tree, height 3, seven terminal vertices

10. Full binary tree, height 3, six terminal vertices

11. Binary tree, height 3, nine terminal vertices

12. Full binary tree, eight internal vertices, seven terminal vertices.

13. Binary tree, height 4, eight terminal vertices

14. Full binary tree, seven vertices

15. Full binary tree, nine vertices, five internal vertices

16. Full binary tree, four internal vertices

17. Binary tree, height 4, eighteen terminal vertices

18. Full binary tree, sixteen vertices

19. Full binary tree, height 3, seven terminal vertices

20. What can you deduce about the height of a binary tree if you know that it has the following properties?
    a. Twenty-five terminal vertices
    b. Forty terminal vertices
    c. Sixty terminal vertices

Answers for Test Yourself

1. one vertex is distinguished from the others and is called the root; the number of edges along the unique path between it and the root; the maximum level of any vertex of the tree   2. every parent has at most two children   3. every parent has exactly two children   4. $2k + 1$; $k + 1$ 5. $t \leq 2^h$, or, equivalently, $\log_2 t \leq h$

PROPERTIES OF THE REAL NUMBERS*

In this text we take the real numbers and their basic properties as our starting point. We give a core set of properties, called axioms, which the real numbers are assumed to satisfy, and we state some useful properties that can be deduced from these axioms.

We assume that there are two binary operations defined on the set of real numbers, called **addition** and **multiplication,** such that if a and b are any two real numbers, the **sum** of a and b, denoted $a + b$, and the **product** of a and b, denoted $a \cdot b$ or ab, are also real numbers. These operations satisfy properties F1–F6, which are called the **field axioms.**

F1. *Commutative Laws*   For all real numbers a and b,

$$a + b = b + a \quad \text{and} \quad ab = ba.$$

F2. *Associative Laws*   For all real numbers a, b, and c,

$$(a + b) + c = a + (b + c) \quad \text{and} \quad (ab)c = a(bc).$$

F3. *Distributive Laws*   For all real numbers a, b, and c,

$$a(b + c) = ab + ac \quad \text{and} \quad (b + c)a = ba + ca.$$

F4. *Existence of Identity Elements*   There exist two distinct real numbers, denoted 0 and 1, such that for every real number a,

$$0 + a = a + 0 = a \quad \text{and} \quad 1 \cdot a = a \cdot 1 = a.$$

F5. *Existence of Additive Inverses*   For every real number a, there is a real number, denoted $-a$ and called the **additive inverse** of a, such that

$$a + (-a) = (-a) + a = 0.$$

F6. *Existence of Reciprocals*   For every real number $a \neq 0$, there is a real number, denoted $1/a$ or a^{-1}, called the **reciprocal** of a, such that

$$a \cdot \left(\frac{1}{a}\right) = \left(\frac{1}{a}\right) \cdot a = 1.$$

All the usual algebraic properties of the real numbers that do not involve order can be derived from the field axioms. The most important are collected as theorems T1–T16 as follows. In all these theorems the symbols a, b, c, and d represent arbitrary real numbers.

*Adapted from Tom M. Apostol, *Calculus, Volume I* (New York: Blaisdell, 1961), pp. 13–19.

T1. *Cancellation Law for Addition* If $a + b = a + c$, then $b = c$. (In particular, this shows that the number 0 of Axiom F4 is unique.)

T2. *Possibility of Subtraction* Given a and b, there is exactly one x such that $a + x = b$. This x is denoted by $b - a$. In particular, $0 - a$ is the additive inverse of a, $-a$.

T3. $b - a = b + (-a)$.

T4. $-(-a) = a$.

T5. $a(b - c) = ab - ac$.

T6. $0 \cdot a = a \cdot 0 = 0$.

T7. *Cancellation Law for Multiplication* If $ab = ac$ and $a \neq 0$, then $b = c$. (In particular, this shows that the number 1 of Axiom F4 is unique.)

T8. *Possibility of Division* Given a and b with $a \neq 0$, there is exactly one x such that $ax = b$. This x is denoted by b/a and is called the **quotient** of b and a. In particular, $1/a$ is the reciprocal of a.

T9. If $a \neq 0$, then $b/a = b \cdot a^{-1}$.

T10. If $a \neq 0$, then $(a^{-1})^{-1} = a$.

T11. *Zero Product Property* If $ab = 0$, then $a = 0$ or $b = 0$.

T12. *Rule for Multiplication with Negative Signs*

$$(-a)b = a(-b) = -(ab), \quad (-a)(-b) = ab,$$

and

$$-\frac{a}{b} = \frac{-a}{b} = \frac{a}{-b}.$$

T13. *Equivalent Fractions Property*

$$\frac{a}{b} = \frac{ac}{bc}, \quad \text{if } b \neq 0 \text{ and } c \neq 0.$$

T14. *Rule for Addition of Fractions*

$$\frac{a}{b} + \frac{c}{d} = \frac{ad + bc}{bd}, \quad \text{if } b \neq 0 \text{ and } d \neq 0.$$

T15. *Rule for Multiplication of Fractions*

$$\frac{a}{b} \cdot \frac{c}{d} = \frac{ac}{bd}, \quad \text{if } b \neq 0 \text{ and } d \neq 0.$$

T16. *Rule for Division of Fractions*

$$\frac{\dfrac{a}{b}}{\dfrac{c}{d}} = \frac{ad}{bc}, \quad \text{if } b \neq 0, c \neq 0, \text{ and } d \neq 0.$$

The real numbers also satisfy the following axioms, called the **order axioms.** It is assumed that among all real numbers there are certain ones, called the **positive real numbers,** that satisfy properties Ord1–Ord3.

Ord1. For any real numbers a and b, if a and b are positive, so are $a + b$ and ab.

Ord2. For every real number $a \neq 0$, either a is positive or $-a$ is positive but not both.

Ord3. The number 0 is not positive.

The symbols $<, >, \leq$, and $\geq$, and negative numbers are defined in terms of positive numbers.

• Definition

Given real numbers a and b,

$a < b$ means $b + (-a)$ is positive. $b > a$ means $a < b$.
$a \leq b$ means $a < b$ or $a = b$. $b \geq a$ means $a \leq b$.
If $a < 0$, we say that a is **negative.**   If $a \geq 0$, we say that a is **nonnegative.**

From the order axioms Ord1–Ord3 and the above definition, all the usual rules for calculating with inequalities can be derived. The most important are collected as theorems T17–T27 as follows. In all these theorems the symbols a, b, c, and d represent arbitrary real numbers.

T17. *Trichotomy Law*   For arbitrary real numbers a and b, exactly one of the three relations $a < b, b < a$, or $a = b$ holds.

T18. *Transitive Law*   If $a < b$ and $b < c$, then $a < c$.

T19. If $a < b$, then $a + c < b + c$.

T20. If $a < b$ and $c > 0$, then $ac < bc$.

T21. If $a \neq 0$, then $a^2 > 0$.

T22. $1 > 0$.

T23. If $a < b$ and $c < 0$, then $ac > bc$.

T24. If $a < b$, then $-a > -b$. In particular, if $a < 0$, then $-a > 0$.

T25. If $ab > 0$, then both a and b are positive or both are negative.

T26. If $a < c$ and $b < d$, then $a + b < c + d$.

T27. If $0 < a < c$ and $0 < b < d$, then $0 < ab < cd$.

One final axiom distinguishes the set of real numbers from the set of rational numbers. It is called the **least upper bound axiom.**

LUB. Any nonempty set S of real numbers that is bounded above has a least upper bound. That is, if B is the set of all real numbers x such that $x \geq s$ for all s in S and if B has at least one element, then B has a smallest element. This element is called the **least upper bound** of S.

The least upper bound axiom holds for the set of real numbers but not for the set of rational numbers. For example, the set of all rational numbers that are less than $\sqrt{2}$ has upper bounds but not a least upper bound within the set of rational numbers.

SOLUTIONS AND HINTS TO SELECTED EXERCISES

Section 1.1

1. a. $x^2 = -1$ (*Or*: the square of x is -1)

   b. A real number x

3. a. Between a and b

   b. Real numbers a and b; there is a real number c

5. a. r is positive

   b. Positive; the reciprocal of r is positive (*Or*: positive; $1/r$ is positive)

   c. Is positive; $1/r$ is positive (*Or*: is positive; the reciprocal of r is positive)

7. a. There are real numbers whose sum is less than their difference.

     True. For example, $1 + (-1) = 0$, $1 - (-1) = 1 + 1 = 2$, and $0 < 2$.

   c. The square of any positive integer is greater than the integer.

     True. If n is any positive integer, then $n \geq 1$. Multiplying both sides by the positive number n does not change the direction of the inequality (see Appendix A, T20), and so $n^2 \geq n$.

8. a. Have four sides

   b. Has four sides

   c. Has four sides

   d. Is a square; J has four sides

   e. J has four sides

10. a. Have a reciprocal

   b. A reciprocal

   c. s is a reciprocal for r

12. a. Real number; product with every number leaves the number unchanged

   b. With every number leaves the number unchanged

   c. $rs = s$

Section 1.2

1. $A = C$ and $B = D$

2. a. The set of all positive real numbers x such that 0 is less than x and x is less than 1

   c. The set of all integers n such that n is a factor of 6

3. a. No, $\{4\}$ is a set with one element, namely 4, whereas 4 is just a symbol that represents the number 4

   b. Three: the elements of the set are 3, 4, and 5.

   c. Three: the elements are the symbol 1, the set $\{1\}$, and the set $\{1, \{1\}\}$

5. *Hint:* $\mathbf{R}$ is the set of all real numbers, $\mathbf{Z}$ is the set of all integers, and $\mathbf{Z}^+$ is the set of all positive integers

6. *Hint:* T_0 and T_1 do not have the same number of elements as T_2 and T_{-3}.

7. a. $\{1, -1\}$

   c. $\emptyset$ (the set has no elements)

   d. $\mathbf{Z}$ (every integer is in the set)

8. a. No, $B \nsubseteq A \therefore j \in B$ and $j \notin A$

   d. Yes, C is a proper subset of A. Both elements of C are in A, but A contains elements (namely c and f) that are not in C.

9. a. Yes

   b. No

   f. No

   i. Yes

10. a. No. Observe that $(-2)^2 = (-2)(-2) = 4$ whereas $-2^2 = -(2^2) = -4$. So $((-2)^2, -2^2) = (4, -4)$, $(-2^2, (-2)^2) = (-4, 4)$, and $(4, -4) \neq (-4, 4)$ because $-4 \neq 4$.

   c. Yes. Note that $8 - 9 = -1$ and $\sqrt[3]{-1} = -1$, and so $(8 - 9, \sqrt[3]{-1}) = (-1, -1)$.

11. a. $\{(w, a), (w, b), (x, a), (x, b), (y, a), (y, b), (z, a),$
$(z, b)\}$ $A \times B$ has $4 \cdot 2 = 8$ elements.

b. $\{(a, w), (b, w), (a, x), (b, x), (a, y), (b, y), (a, z),$
$(b, z)\}$ $B \times A$ has $4 \cdot 2 = 8$ elements.

c. $\{(w, w), (w, x), (w, y), (w, z), (x, w), (x, x), (x, y),$
$(x, z), (y, w), (y, x), (y, y), (y, z), (z, w), (z, x),$
$(z, y), (z, z)\}$ $A \times A$ has $4 \cdot 4 = 16$ elements.

d. $\{(a, a), (a, b), (b, a), (b, b)\}$ $B \times B$ has $2 \cdot 2 = 4$
elements.

Section 1.3

1. a. No. Yes. No. Yes.

b. $R = \{(2, 6), (2, 8), (2, 10), (3, 6), (4, 8)\}$

c. Domain of $R = A = \{2, 3, 4\}$, co-domain of $R = B = \{6, 8, 10\}$

d.
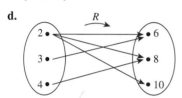

3. a. $3 \, T \, 0$ because $\frac{3-0}{3} = \frac{3}{3} = 1$, which is an integer.

$1 \, T \, (-1)$ because $\frac{1-(-1)}{3} = \frac{2}{3}$, which is not an integer.

$(2, -1) \in T$ because $\frac{2-(-1)}{3} = \frac{3}{3} = 1$, which is an
integer.

$(3, -2) \notin T$ because $\frac{3-(-2)}{3} = \frac{5}{3}$, which is not an
integer.

b. $T = \{(1, -2), (2, -1), (3, 0)\}$

c. Domain of $T = E = \{1, 2, 3\}$, co-domain of $T = F = \{-2, -1, 0\}$

d.
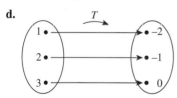

5. a. $(2, 1) \in S$ because $2 \geq 1$. $(2, 2) \in S$ because $2 \geq 2$.
$2 \, \$ \, 3$ because $2 \not\geq 3$. $(-1) \, S \, (-2)$ because $-1 \geq -2$.

b.
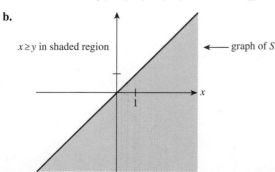

$x \geq y$ in shaded region ← graph of S

7. a.
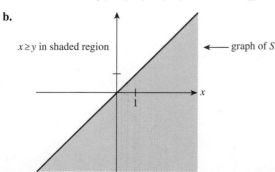

b. R is not a function because it satisfies neither property
(1) nor property (2) of the definition. It fails property (1)
because $(4, y) \notin R$, for any y in B. It fails property (2)
because $(6, 5) \in R$ and $(6, 6) \in R$ and $5 \neq 6$.

   S is not a function because $(5, 5) \in S$ and $(5, 7) \in S$
and $5 \neq 7$. So S does not satisfy property (2) of the definition of function.

   T is not a function both because $(5, x) \notin T$ for any x
in B and because $(6, 5) \in T$ and $(6, 7) \in T$ and $5 \neq 7$.
So T does not satisfy either property (1) or property (2)
of the definition of function.

9. a. $\emptyset, \{(0, 1)\}, \{(1, 1)\}, \{(0, 1), (1, 1)\}$

b. $\{(0, 1), (1, 1)\}$

c. $1/4$

11. No, P is not a function because, for example, $(4, 2) \in P$
and $(4, -2) \in P$ but $2 \neq -2$.

13. a. Domain $= A = \{-1, 0, 1\}$, co-domain $= B = \{t, u, v, w\}$

b. $F(-1) = u$, $F(0) = w$, $F(1) = u$

15. a. This diagram does not determine a function because 2 is
related to both 2 and 6.

b. This diagram does not determine a function because 5 is
in the domain but it is not related to any element in the
co-domain.

16. $f(-1) = (-1)^2 = 1$, $f(0) = 0^2 = 0$, $f\left(\frac{1}{2}\right) = \left(\frac{1}{2}\right)^2 = \frac{1}{4}$.

19. For all $x \in \mathbf{R}$, $g(x) = \frac{2x^3 + 2x}{x^2 + 1} = \frac{2x(x^2 + 1)}{x^2 + 1} = 2x = f(x)$.
Therefore, by definition of equality of functions, $f = g$.

Section 2.1

1. Common form: If p then q.
$$p.$$
Therefore, q.
$(a + 2b)(a^2 - b)$ can be written in prefix notation.
All algebraic expressions can be written in prefix notation.

3. Common form: $p \vee q$.
$$\sim p.$$
Therefore, q.
My mind is shot. Logic is confusing.

5. a. It is a statement because it is a true sentence. 1,024
is a perfect square because $1,024 = 32^2$, and the next
smaller perfect square is $31^2 = 961$, which has less than
four digits.

6. a. $s \wedge i$ **b.** $\sim s \wedge \sim i$

8. a. $(h \wedge w) \wedge \sim s$ **d.** $(\sim w \wedge \sim s) \wedge h$

10. a. $p \wedge q \wedge r$ **c.** $p \wedge (\sim q \vee \sim r)$

11. Inclusive or. For instance, a team could win the playoff by winning games 1, 3, and 4 and losing game 2. Such an outcome would satisfy both conditions.

12.

p	q	$\sim p$	$\sim p \wedge q$
T	T	F	F
T	F	F	F
F	T	T	T
F	F	T	F

14.

p	q	r	$q \wedge r$	$p \wedge (q \wedge r)$
T	T	T	T	T
T	T	F	F	F
T	F	T	F	F
T	F	F	F	F
F	T	T	T	F
F	T	F	F	F
F	F	T	F	F
F	F	F	F	F

16.

p	q	$p \wedge q$	$p \vee (p \wedge q)$	p
T	T	T	T	T
T	F	F	T	T
F	T	F	F	F
F	F	F	F	F

$p \vee (p \wedge q)$ and p always have the same truth values, so they are logically equivalent. (This proves one of the absorption laws.)

18.

p	t	$p \vee$ t
T	T	T
F	T	T

$p \vee$ t and t always have the same truth values, so they are logically equivalent. (This proves one of the universal bound laws.)

21.

p	q	r	$p \wedge q$	$q \wedge r$	$(p \wedge q) \wedge r$	$p \wedge (q \wedge r)$
T	T	T	T	T	T	T
T	T	F	T	F	F	F
T	F	T	F	F	F	F
T	F	F	F	F	F	F
F	T	T	F	T	F	F
F	T	F	F	F	F	F
F	F	T	F	F	F	F
F	F	F	F	F	F	F

$(p \wedge q) \wedge r$ and $p \wedge (q \wedge r)$ always have the same truth values, so they are logically equivalent. (This proves the associative law for $\wedge$.)

23.

p	q	r	$p \wedge q$	$q \vee r$	$(p \wedge q) \vee r$	$p \wedge (q \vee r)$
T	T	T	T	T	T	T
T	T	F	T	T	T	T
T	F	T	F	T	T	T
T	F	F	F	F	F	F
F	T	T	F	T	T	F
F	T	F	F	T	F	F
F	F	T	F	T	T	F
F	F	F	F	F	F	F

$(p \wedge q) \vee r$ and $p \wedge (q \vee r)$ have different truth values in the fifth and seventh rows, so they are not logically equivalent. (This proves that parentheses are needed with $\wedge$ and $\vee$.)

25. Hal is not a math major or Hal's sister is not a computer science major.

27. The connector is not loose and the machine is not unplugged.

32. $-2 \geq x$ or $x \geq 7$

34. $2 \leq x \leq 5$

36. $1 \leq x$ or $x < -3$

38. This statement's logical form is $(p \wedge q) \vee r$, so its negation has the form $\sim((p \wedge q) \vee r) \equiv \sim(p \wedge q) \wedge \sim r \equiv (\sim p \vee \sim q) \wedge \sim r$. Thus a negation for the statement is (*num_orders* ≤ 100 or *num_instock* > 500) and *num_instock* ≥ 200.

40.

p	q	$\sim p$	$\sim q$	$p \wedge q$	$p \wedge \sim q$	$\sim p \vee (p \wedge \sim q)$	$(p \wedge q) \vee (\sim p \vee (p \wedge \sim q))$
T	T	F	F	T	F	F	T
T	F	F	T	F	T	T	T
F	T	T	F	F	F	T	T
F	F	T	T	F	F	T	T

↑

Its truth values are all T's, so $(p \wedge q) \vee (\sim p \vee (p \wedge \sim q))$ is a tautology.

41.

p	q	$\sim p$	$\sim q$	$p \wedge \sim q$	$\sim p \vee q$	$(p \wedge \sim q) \wedge (\sim p \vee q)$
T	T	F	F	F	T	F
T	F	F	T	T	F	F
F	T	T	F	F	T	F
F	F	T	T	F	T	F

↑

Its truth values are all F's, so $(p \wedge \sim q) \wedge (\sim p \vee q)$ is a contradiction.

44. Let p be '$x < 2$', q be '$1 < x$', and r be '$x < 3$'. Then the sentences in (a) and (b) are symbolized as $p \vee \sim(q \wedge r)$ and $\sim q \vee (p \vee \sim r)$, respectively.

p	q	r	$\sim q$	$\sim r$	$q \wedge r$	$\sim(q \wedge r)$	$p \vee \sim r$	$p \vee \sim(q \wedge r)$	$\sim q \vee (p \vee \sim r)$
T	T	T	F	F	T	F	T	T	T
T	T	F	F	T	F	T	T	T	T
T	F	T	T	F	F	T	T	T	T
T	F	F	T	T	F	T	T	T	T
F	T	T	F	F	T	F	F	F	F
F	T	F	F	T	F	T	T	T	T
F	F	T	T	F	F	T	F	T	T
F	F	F	T	T	F	T	T	T	T

↑   ↑

The statement forms $p \vee \sim(q \wedge r)$ and $\sim q \vee (p \vee \sim r)$ always have the same truth values, so they are logically equivalent.

Therefore the statements in (a) and (b) are logically equivalent.

46. a. *Solution 1:* Construct a truth table for $p \oplus p$ using the truth values for *exclusive or*.

p	$p \oplus p$
T	F
F	F

because an *exclusive or* statement is false when both components are true and when both components are false.

Since all its truth values are false, $p \oplus p \equiv \mathbf{c}$, a contradiction.

Solution 2: Replace q by p in the logical equivalence $p \oplus q \equiv (p \vee q) \wedge \sim(p \wedge q)$, and simplify the result.

$$p \oplus p \equiv (p \vee q) \wedge \sim(p \wedge p) \quad \text{by defintion of } \oplus$$
$$\equiv p \wedge \sim p \quad \text{by the identity laws}$$
$$\equiv \mathbf{c} \quad \text{by the negation law for } \wedge$$

47. There is a famous story about a philosopher who once gave a talk in which he observed that whereas in English and many other languages a double negative is equivalent to a positive, there is no language in which a double positive is equivalent to a negative. To this, another philosopher, Sidney Morgenbesser, responded sarcastically, "Yeah, yeah."

[Strictly speaking, sarcasm functions like negation. When spoken sarcastically, the words "Yeah, yeah" are not a true double positive; they just mean "no."]

Section 2.2

1. If this loop does not contain a **stop** or a **go to**, then it will repeat exactly N times.

3. If you do not freeze, then I'll shoot.

5.

p	q	$\sim p$	$\sim q$	conclusion $\sim p \vee q$	hypothesis $\sim p \vee q \rightarrow \sim q$
T	T	F	F	T	F
T	F	F	T	F	T
F	T	T	F	T	F
F	F	T	T	T	T

7.

| | | | | conclusion | hypothesis | |
| | | | | | | |

p	q	r	$\sim q$	$p \wedge \sim q$	$p \wedge \sim q \to r$
T	T	T	F	F	T
T	T	F	F	F	T
T	F	T	T	T	T
T	F	F	T	T	F
F	T	T	F	F	T
F	T	F	F	F	T
F	F	T	T	F	T
F	F	F	T	F	T

9.

p	q	r	$\sim r$	$p \wedge \sim r$	$q \vee r$	$p \wedge \sim r \leftrightarrow q \vee r$
T	T	T	F	F	T	F
T	T	F	T	T	T	T
T	F	T	F	F	T	F
T	F	F	T	T	F	F
F	T	T	F	F	T	F
F	T	F	T	F	T	F
F	F	T	F	F	T	F
F	F	F	T	F	F	T

12. If $x > 2$ then $x^2 > 4$, and if $x < -2$ then $x^2 > 4$.

13. a.

p	q	$\sim p$	$p \to q$	$\sim p \vee q$
T	T	F	T	T
T	F	F	F	F
F	T	T	T	T
F	F	T	T	T

$p \to q$ and $\sim p \vee q$ always have the same truth values, so they are logically equivalent.

14. a. *Hint:* $p \to q \vee r$ is true in all cases except when p is true and both q and r are false.

16. Let p represent "You paid full price" and q represent "You didn't buy it at Crown Books." Thus, "If you paid full price, you didn't buy it at Crown Books" has the form $p \to q$. And "You didn't buy it at Crown Books or you paid full price" has the form $q \vee p$.

p	q	$p \to q$	$q \vee p$
T	T	T	T
T	F	F	T
F	T	T	T
F	F	T	F

These two statements are not logically equivalent because their forms have different truth values in rows 2 and 4.

(An alternative representation for the forms of the two statements is $p \to \sim q$ and $\sim q \vee p$. In this case, the truth values differ in rows 1 and 3.)

19. False. The negation of an if-then statement is not an if-then statement. It is an *and* statement.

20. a. P is a square and P is not a rectangle.

 d. n is prime and both n is not odd and n is not 2.

 Or: n is prime and n is neither odd nor 2.

 f. Tom is Ann's father and either Jim is not her uncle or Sue is not her aunt.

21. a. Because $p \to q$ is false, p is true and q is false. Hence $\sim p$ is false, and so $\sim p \to q$ is true.

22. a. If P is not a rectangle, then P is not a square.

 d. If n is not odd and n is not 2, then n is not prime.

 f. If either Jim is not Ann's uncle or Sue is not her aunt, then Tom is not her father.

23. a. *Converse:* If P is a rectangle, then P is a square.
 Inverse: If P is not a square, then P is not a rectangle.

 d. *Converse:* If n is odd or n is 2, then n is prime.
 Inverse: If n is not prime, then n is not odd and n is not 2.

 f. *Converse:* If Jim is Ann's uncle and Sue is her aunt, then Tom is her father.
 Inverse: If Tom is not Ann's father, then Jim is not her uncle or Sue is not her aunt.

24.

p	q	$p \to q$	$q \to p$
T	T	T	T
T	F	F	T
F	T	T	F
F	F	T	T

$p \to q$ and $q \to p$ have different truth values in the second and third rows, so they are not logically equivalent.

26.

p	q	$\sim q$	$\sim p$	$\sim q \to \sim p$	$p \to q$
T	T	F	F	T	T
T	F	T	F	F	F
F	T	F	T	T	T
F	F	T	T	T	T

$\sim q \to \sim p$ and $p \to q$ always have the same truth values, so they are logically equivalent.

28. *Hint:* A person who says "I mean what I say" claims to speak sincerely. A person who says "I say what I mean" claims to speak with precision.

29. $(p \rightarrow (q \vee r)) \leftrightarrow ((p \wedge \sim q) \rightarrow r)$

p	q	r	$\sim q$	$q \vee r$	$p \wedge \sim q$	$p \rightarrow (q \vee r)$	$p \wedge \sim q \rightarrow r$	$(p \rightarrow (q \vee r)) \leftrightarrow ((p \wedge \sim q) \rightarrow r)$
T	T	T	F	T	F	T	T	T
T	T	F	F	T	F	T	T	T
T	F	T	T	T	T	T	T	T
T	F	F	T	F	T	F	F	T
F	T	T	F	T	F	T	T	T
F	T	F	F	T	F	T	T	T
F	F	T	T	T	F	T	T	T
F	F	F	T	F	F	T	T	T

$\uparrow$

$(p \rightarrow (q \vee r)) \leftrightarrow ((p \wedge \sim q) \rightarrow r)$ is a tautology because all of its truth values are T.

32. If this quadratic equation has two distinct real roots, then its discriminant is greater than zero, and if the discriminant of this quadratic equation is greater than zero, then the equation has two real roots.

34. If the Cubs do not win tomorrow's game, then they will not win the pennant.

If the Cubs win the pennant, then they will have won tomorrow's game.

37. If a new hearing is not granted, payment will be made on the fifth.

40. If I catch the 8:05 bus, then I am on time for work.

42. If this number is not divisible by 3, then it is not divisible by 9.

If this number is divisible by 9, then it is divisible by 3.

44. If Jon's team wins the rest of its games, then it will win the championship.

46. a. This statement is the converse of the given statement, and so it is not necessarily true. For instance, if the actual boiling point of compound X were 200°C, then the given statement would be true but this statement would be false.

b. This statement must be true. It is the contrapositive of the given statement.

Section 2.3

1. $\sqrt{2}$ is not rational.   **3.** Logic is not easy.

6.

		premises		conclusion
p	q	$p \rightarrow q$	$q \rightarrow p$	$p \vee q$
T	T	T	T	T
T	F	F	T	
F	T	T	F	
F	F	T	T	F

$\leftarrow$

This row shows that it is possible for an argument of this form to have true premises and a false conclusion. Thus this argument form is invalid.

7.

					premises		conclusion
p	q	r	$\sim q$	p	$p \rightarrow q$	$\sim q \vee r$	r
T	T	T	F	T	T	T	T
T	T	F	F	T	T	F	
T	F	T	T	T	F	T	
T	F	F	T	T	F	T	
F	T	T	F	F	T	T	
F	T	F	F	F	T	F	
F	F	T	T	F	T	T	
F	F	F	T	F	T	T	

$\leftarrow$

This row describes the only situation in which all the premises are true. Because the conclusion is also true here, the argument form is valid.

8.

p	q	r	~q	p ∨ q	p → ~q	p → r	r
T	T	T	F	T	F	T	
T	T	F	F	T	F	F	
T	F	T	T	T	T	T	T
T	F	F	T	T	T	F	
F	T	T	F	T	T	T	T
F	T	F	F	T	T	T	F
F	F	T	T	F	T	T	
F	F	F	T	F	T	T	

with header spanning: premises (~q, p ∨ q, p → ~q, p → r) and conclusion (r).

This row shows that it is possible for an argument of this form to have true premises and a false conclusion. Thus this argument form is invalid.

12. a.

p	q	p → q	q	p
T	T	T	T	T
T	F	F	F	
F	T	T	T	F
F	F	T	F	

with header: premises (p → q, q) and conclusion (p).

This row shows that it is possible for an argument of this form to have true premises and a false conclusion. Thus this argument form is invalid.

14.

p	q	p	p ∨ q
T	T	T	T
T	F	T	T
F	T	F	
F	F	F	

with header: premise (p) and conclusion (p ∨ q).

These two rows show that in all situations where the premise is true, the conclusion is also true. Thus the argument form is valid.

18.

p	q	p ∨ q	~q	p
T	T	T	F	
T	F	T	T	T
F	T	T	F	
F	F	F	T	

with header: premises (p ∨ q, ~q) and conclusion (p).

This row represents the only situation in which both premises are true. Because the conclusion is also true here the argument form is valid.

22. Let p represent "Tom is on team A" and q represent "Hua is on team B." Then the argument has the form

$$\sim p \to q$$
$$\sim q \to p$$
$$\therefore \sim p \lor \sim q$$

p	q	~p	~q	~p → q	~q → p	~p ∨ ~q
T	T	F	F	T	T	F
T	F	F	T	T	T	T
F	T	T	F	T	T	T
F	F	T	T	F	F	

with header: premises (~p → q, ~q → p) and conclusion (~p ∨ ~q).

This row shows that it is possible for an argument of this form to have true premises and a false conclusion. Thus this argument form is invalid.

24.
$$p \to q$$
$$q$$
$$\therefore p \qquad \text{invalid: converse error}$$

25.
$$p \lor q$$
$$\sim p$$
$$\therefore q \qquad \text{valid: elimination}$$

26.
$$p \to q$$
$$q \to r$$
$$\therefore p \to r \qquad \text{valid: transitivity}$$

27.
$$p \to q$$
$$\sim p$$
$$\therefore \sim q \qquad \text{invalid: inverse error}$$

36. The program contains an undeclared variable.
One explanation:
1. There is not a missing semicolon and there is not a misspelled variable name. *(by (c) and (d) and definition of ∧)*
2. It is not the case that there is a missing semicolon or a misspelled variable name. *(by (1) and De Morgan's laws)*
3. There is not a syntax error in the first five lines. *(by (b) and (2) and modus tollens)*
4. There is an undeclared variable. *(by (a) and (3) and elimination)*

37. The treasure is buried under the flagpole.
One explanation:
1. The treasure is not in the kitchen. *(by (c) and (a) and modus ponens)*
2. The tree in the front yard is not an elm. *(by (b) and (1) and modus tollens)*
3. The treasure is buried under the flagpole. *(by (d) and (2) and elimination)*

38. a. *A* is a knave and *B* is a knight.
One explanation:
1. Suppose *A* is a knight.
2. ∴ What *A* says is true. *(by definition of knight)*

3. ∴ *B* is a knight also. *(That's what A said.)*

4. ∴ What *B* says is true. *(by definition of knight)*

5. ∴ *A* is a knave. *(That's what B said.)*

6. ∴ We have a contradiction: *A* is a knight and a knave. *(by (1) and (5))*

7. ∴ The supposition that *A* is a knight is false. *(by the contradiction rule)*

8. ∴ *A* is a knave. *(negation of supposition)*

9. ∴ What *B* says is true. *(B said A was a knave, which we now know to be true.)*

10. ∴ *B* is a knight. *(by definition of knight)*

d. *Hint:* *W* and *Y* are knights; the rest are knaves.

39. The chauffeur killed Lord Hazelton.

One explanation:

1. Suppose the cook was in the kitchen at the time of the murder.

2. ∴ The butler killed Lord Hazelton with strychnine. *(by (c) and (1) and modus ponens)*

3. ∴ We have a contradiction: Lord Hazelton was killed by strychnine and a blow on the head. *(by (2) and (a))*

4. ∴ The supposition that the cook was in the kitchen is false. *(by the contradiction rule)*

5. ∴ The cook was not in the kitchen at the time of the murder. *(negation of supposition)*

6. ∴ Sara was not in the dining room when the murder was committed. *(by (e) and (5) and modus ponens)*

7. ∴ Lady Hazelton was in the dining room when the murder was committed. *(by (b) and (6) and elimination)*

8. ∴ The chauffeur killed Lord Hazelton. *(by (d) and (7) and modus ponens)*

41. (1) $p \to t$ by premise (d)

      $\sim t$ by premise (c)

      $\therefore \sim p$ by modus tollens

(2) $\sim p$ by (1)

      $\therefore \sim p \vee q$ by generalization

(3) $\sim p \vee q \to r$ by premise (a)

      $\sim p \vee q$ by (2)

      $\therefore r$ by modus ponens

(4) $\sim p$ by (1)

      r by (3)

      $\therefore \sim p \wedge r$ by conjunction

(5) $\sim p \wedge r \to \sim s$ by premise (e)

      $\sim p \wedge r$ by (4)

      $\therefore \sim s$ by modus ponens

(6) $s \vee \sim q$ by premise (b)

      $\sim s$ by (5)

      $\therefore \sim q$ by elimination

43. (1) $\sim w$ by premise (d)

      $u \vee w$ by premise (e)

      $\therefore u$ by elimination

(2) $u \to \sim p$ by premise (c)

      u by (1)

      $\therefore \sim p$ by modus ponens

(3) $\sim p \to r \wedge \sim s$ by premise (a)

      $\sim p$ by (2)

      $\therefore r \wedge \sim s$ by modus ponens

(4) $r \wedge \sim s$ by (3)

      $\therefore \sim s$ by specialization

(5) $t \to s$ by premise (b)

      $\sim s$ by (4)

      $\therefore \sim t$ by modus tollens

Section 3.1

1. a. False   **b.** True

2. a. The statement is true. The integers correspond to certain of the points on a number line, and the real numbers correspond to all the points on the number line.

   b. The statement is false; 0 is neither positive nor negative.

   c. The statement is false. For instance, let $r = -2$. Then $-r = -(-2) = 2$, which is positive.

   d. The statement is false. For instance, the number $\frac{1}{2}$ is a real number, but it is not an integer.

3. a. $P(2)$ is "$2 > \frac{1}{2}$," which is true.

$P\left(\frac{1}{2}\right)$ is "$\frac{1}{2} > \frac{1}{\frac{1}{2}}$." This is false because $\frac{1}{\frac{1}{2}} = 2$, and $\frac{1}{2} \not> 2$.

$P(-1)$ is "$-1 > \frac{1}{-1}$." This is false because $\frac{1}{-1} = -1$, and $-1 \not> -1$.

$P\left(-\frac{1}{2}\right)$ is "$-\frac{1}{2} > \frac{1}{-\frac{1}{2}}$." This is true because $\frac{1}{-\frac{1}{2}} = -2$ and $-\frac{1}{2} > -2$.

$P(-8)$ is "$-8 > \frac{1}{-8}$." This is false because $\frac{1}{-8} = -\frac{1}{8}$ and $-8 \not> -\frac{1}{8}$.

   b. If the domain of $P(x)$ is the set of all real numbers, then its truth set is the set of all real numbers x for which either $x > 1$ or $-1 < x < 0$.

   c. If the domain of $P(x)$ is the set of all positive real numbers, then its truth set is the set of all real numbers x for which $x > 1$.

4. b. If the domain of $Q(n)$ is the set of all integers, then its truth set is $\{-5, -4, -3, -2, -1, 0, 1, 2, 3, 4, 5\}$.

5. a. $Q(-2, 1)$ is the statement "If $-2 < 1$ then $(-2)^2 < 1^2$." The hypothesis of this statement is $-2 < 1$, which is true. The conclusion is $(-2)^2 < 1^2$, which is false because $(-2)^2 = 4$ and $1^2 = 1$ and $4 \not< 1$. Thus $Q(-2, 1)$ is a conditional statement with a true hypothesis and a false conclusion. So $Q(-2, 1)$ is false.

   c. $Q(3, 8)$ is the statement "If $3 < 8$ then $3^2 < 8^2$." The hypothesis of this statement is $3 < 8$, which is true. The conclusion is $3^2 < 8^2$, which is also true because $3^2 = 9$ and $8^2 = 64$ and $9 < 64$. Thus $Q(3, 8)$ is a conditional statement with a true hypothesis and a true conclusion. So $Q(3, 8)$ is true.

7. a. The truth set is the set of all integers d such that $6/d$ is an integer, so the truth set is $\{-6, -3, -2, -1, 1, 2, 3, 6\}$.

c. The truth set is the set of all real numbers x with the property that $1 \leq x^2 \leq 4$, so the truth set is $\{x \in \mathbf{R} \mid -2 \leq x \leq -1 \text{ or } 1 \leq x \leq 2\}$. In other words, the truth set is the set of all real numbers between -2 and -1 inclusive together with those between 1 and 2 inclusive.

8. a. $\{-9, -8, -7, -6, -5, -4, -3, -2, -1, 0, 1, 2, 3, 4, 5, 6, 7, 8, 9\}$

9. Counterexample: Let $x = 1 : 1 \not> \frac{1}{1}$. (*This is one counterexample among many.*)

11. Counterexample: Let $m = 1$ and $n = 1$. Then $m \cdot n = 1 \cdot 1 = 1$ and $m + n = 1 + 1 = 2$. But $1 \not> 2$, and so $m \cdot n \not> m + n$. (*This is one counterexample among many.*)

13. (a), (e), (f)   **14.** (b), (c), (e), (f)

15. a. *Partial answer:* Every rectangle is a quadrilateral.
   b. *Partial answer:* At least one set has 16 subsets.

16. a. $\forall$ dinosaurs x, x is extinct.
   c. $\forall$ irrational numbers x, x is not an integer.
   e. $\forall$ integers x, x^2 does not equal $2, 147, 581, 953$.

17. a. $\exists$ an exercise x such that x has an answer.

18. a. $\exists s \in D$ such that $E(s)$ and $M(s)$. (Or: $\exists s \in D$ such that $E(s) \wedge M(s)$.)
   b. $\forall s \in D$, if $C(s)$ then $E(s)$. (Or: $\forall s \in D$, $C(s) \rightarrow E(s)$.)
   e. $(\exists s \in D$ such that $C(s) \wedge E(s)) \wedge (\exists s \in D$ such that $C(s) \wedge {\sim}E(s))$

19. (b), (d), (e)

20. *Partial answer:* The square root of a positive real number is positive.

21. a. The total degree of G is even, for any graph G.
   c. p is even, for some prime number p.

22. a. $\forall x$, if x is a Java program, then x has at least 5 lines.

23. a. $\forall x$ if x is an equilateral triangle, then x is isosceles.
   $\forall$ equilateral triangles x, x is isosceles.

24. a. $\exists$ a hatter x such that x is mad.
   $\exists x$ such that x is a hatter and x is mad.

25. a. $\forall$ nonzero fractions x, the reciprocal of x is a fraction.
   $\forall x$, if x is a nonzero fraction, then the reciprocal of x is a fraction.
   c. $\forall$ triangles x, the sum of the angles of x is $180°$.
   $\forall x$, if x is a triangle, then the sum of the angles of x is $180°$.
   e. $\forall$ even integers x and y, the sum of x and y is even.
   $\forall x$ and y, if x and y are even integers, then the sum of x and y is even.

26. b. $\forall x (\text{Int}(x) \longrightarrow \text{Ratl}(x)) \wedge \exists x (\text{Ratl}(x) \wedge {\sim}\text{Int}(x))$

27. a. False. Figure b is a circle that is not gray.
   b. True. All the gray figures are circles.

28. b. *One answer among many:* If a real number is negative, then when its opposite is computed, the result is a positive real number.
   This statement is true because for all real numbers x, $-(-|x|) = |x|$ (and any negative real number can be represented as $-|x|$, for some real number x).

d. *One answer among many:* There is a real number that is not an integer. This statement is true. For instance, $\frac{1}{2}$ is a real number that is not an integer.

30. b. *One answer among many:* If an integer is prime, then it is not a perfect square.
   This statement is true because a prime number is an integer greater than 1 that is not a product of two smaller positive integers. So a prime number cannot be a perfect square because if it were, it would be a product of two smaller positive integers.

31. *Hint:* Your answer should have the appearance shown in the following made-up example:
   Statement: "If a function is differentiable, then it is continuous."
   Formal version: $\forall$ functions f, if f is differentiable, then f is continuous.
   Citation: Calculus by D. R. Mathematician, Best Publishing Company, 2004, page 263.

32. a. True: Any real number that is greater than 2 is greater than 1.
   c. False: $(-3)^2 > 4$ but $-3 \not> 2$.

33. a. True. Whenever both a and b are positive, so is their product.
   b. False. Let $a = -2$ and $b = -3$. Then $ab = 6$, which is not less than zero.

Section 3.2

1. (a) and (e) are negations.

3. a. $\exists$ a fish x such that x does not have gills.
   c. $\forall$ movies m, m is less than or equal to 6 hours long. (*Or:* $\forall$ movies m, m is no more than 6 hours long.)

In 4–6 there are other correct answers in addition to those shown.

4. a. Some dogs are unfriendly. (*Or:* There is at least one unfriendly dog.)
   c. All suspicions were unsubstantiated. (*Or:* No suspicions were substantiated.)

5. a. There is a valid argument that does not have a true conclusion. (*Or:* At least one valid argument does not have a true conclusion.)

6. a. Sets A and B have at least one point in common.

7. The statement is not existential.
   Informal negation: There is at least one order from store A for item B.
   Formal version of statement: $\forall$ orders x, if x is from store A, then x is not for item B.

9. $\exists$ a real number x such that $x > 3$ and $x^2 \leq 9$.

11. The proposed negation is not correct. Consider the given statement: "The sum of any two irrational numbers is irrational." For this to be false means that it is possible to find at least one pair of irrational numbers whose sum is rational. On the other hand, the negation proposed in the exercise ("The sum of any two irrational numbers is rational") means that given *any* two irrational numbers, their sum is rational. This is a much stronger statement than the actual negation: The truth of this statement implies the truth of the negation (assuming that there are at least two irrational numbers), but the negation can be true without having this statement be true.

Correct negation: There are at least two irrational numbers whose sum is rational.

Or: The sum of some two irrational numbers is rational.

13. The proposed negation is not correct. There are two mistakes: The negation of a "for all" statement is not a "for all" statement; and the negation of an if-then statement is not an if-then statement.

Correct negation: There exists an integer n such that n^2 is even and n is not even.

15. **a.** True: All the odd numbers in D are positive.

c. False: $x = 16$, $x = 26$, $x = 32$, and $x = 36$ are all counterexamples.

16. $\exists$ a real number x such that $x^2 \geq 1$ and $x \not> 0$. In other words, $\exists$ a real number x such that $x^2 \geq 1$ and $x \leq 0$.

There is a real number whose square is at least 1 but that is not greater than 0.

Some real numbers that are less than or equal to zero have squares that are greater than or equal to one.

18. $\exists$ a real number x such that $x(x+1) > 0$ and both $x \leq 0$ and $x \geq -1$.

20. $\exists$ integers a, b, and c such that $a - b$ is even and $b - c$ is even and $a - c$ is not even.

22. There is an integer such that the square of the integer is odd but the integer is not odd. (*Or:* At least one integer has an odd square but is not itself odd.)

24. **a.** If a person is a child in Tom's family, then the person is female.

If a person is a female in Tom's family, then the person is a child.

The second statement is the converse of the first.

25. **a.** *Converse:* If $n + 1$ is an even integer, then n is a prime number that is greater than 2.

Counterexample: Let $n = 15$. Then $n + 1 = 16$, which is even but n is not a prime number that is greater than 2.

26. *Statement:* $\forall$ real numbers x, if $x^2 \geq 1$ then $x > 0$.

Contrapositive: $\forall$ real numbers x, if $x \leq 0$ then $x^2 < 1$.

Converse: $\forall$ real numbers x, if $x > 0$ then $x^2 \geq 1$.

Inverse: $\forall$ real numbers x, if $x^2 < 1$ then $x \leq 0$.

The statement and its contrapositive are false. As a counterexample, let $x = -2$. Then $x^2 = (-2)^2 = 4$, and so $x^2 \geq 1$. However $x \not> 0$.

The converse and the inverse are also false. As a counterexample, let $x = 1/2$. Then $x^2 = 1/4$, and so $x > 0$ but $x^2 \not\geq 1$.

28. *Statement:* $\forall x \in \mathbf{R}$, if $x(x+1) > 0$ then $x > 0$ or $x < -1$.

Contrapositive: $\forall x \in \mathbf{R}$, if $x \leq 0$ and $x \geq -1$, then $x(x+1) \leq 0$.

Converse: $\forall x \in \mathbf{R}$, if $x > 0$ or $x < -1$ then $x(x+1) > 0$.

Inverse: $\forall x \in \mathbf{R}$, if $x(x+1) \leq 0$ then $x \leq 0$ and $x \geq -1$.

The statement, its contrapositive, its converse, and its inverse are all true.

30. *Statement:* $\forall$ integers a, b, and c, if $a - b$ is even and $b - c$ is even, then $a - c$ is even.

Contrapositive: $\forall$ integers a, b, and c, if $a - c$ is not even, then $a - b$ is not even or $b - c$ is not even.

Converse: $\forall$ integers a, b and c, if $a - c$ is even then $a - b$ is even and $b - c$ is even.

Inverse: $\forall$ integers a, b, and c, if $a - b$ is not even or $b - c$ is not even, then $a - c$ is not even.

The statement is true, but its converse and inverse are false. As a counterexample, let $a = 3$, $b = 2$, and $c = 1$. Then $a - c = 2$, which is even, but $a - b = 1$ and $b - c = 1$, so it is not the case that both $a - b$ and $b - c$ are even.

32. *Statement:* If the square of an integer is odd, then the integer is odd.

Contrapositive: If an integer is not odd, then the square of the integer is not odd.

Converse: If an integer is odd, then the square of the integer is odd.

Inverse: If the square of an integer is not odd, then the integer is not odd.

The statement, its contrapositive, its converse, and its inverse are all true.

34. **a.** If n is divisible by some prime number between 1 and $\sqrt{n}$ inclusive, then n is not prime.

36. **a.** *One possible answer:* Let $P(x)$ be "$2x \neq 1$." The statement "$\forall x \in \mathbf{Z}, 2x \neq 1$" is true, but the statements "$\forall x \in \mathbf{Q}, 2x \neq 1$" and "$\forall x \in \mathbf{R}, 2x \neq 1$" are both false.

37. The claim is "$\forall x$, if $x = 1$ and x is in the sequence 0204, then x is to the left of all the 0's in the sequence."

The negation is "$\exists x$ such that $x = 1$ and x is in the sequence 0204, and x is not to the left of all the 0's in the sequence." The negation is false because the sequence does not contain the character 1. So the claim is vacuously true (or true by default).

39. If a person earns a grade of C^- in this course, then the course counts toward graduation.

41. If a person is not on time each day, then the person will not keep this job.

43. It is not the case that if a number is divisible by 4, then that number is divisible by 8. In other words, there is a number that is divisible by 4 and is not divisible by 8.

45. It is not the case that if a person has a large income, then that person is happy. In other words, there is a person who has a large income and is not happy.

48. No. Interpreted formally, the statement says, "If carriers do not offer the same lowest fare, then you may not select among them," or, equivalently, "If you may select among carriers, then they offer the same lowest fare."

Section 3.3

1. **a.** True: Tokyo is the capital of Japan.

b. False: Athens is not the capital of Egypt.

2. a. True: $2^2 > 3$ **b.** False: $1^2 \not> 1$

3. a. $y = \frac{1}{2}$ **b.** $y = -1$

4. a. Let $n = 16$. Then $n > x$ because $16 > 15.83$.

5. The statement says that no matter what circle anyone might give you, you can find a square of the same color. This is true because the only circles are $a, c,$ and $b,$ and given a or $c,$ which are blue, square j is also blue, and given $b,$ which is gray, squares g and h are also gray.

7. This is true because triangle d is above every square.

9. a. There are five elements in D. For each, an element in E must be found so that the sum of the two equals 0. So: if $x = -2,$ take $y = 2;$ if $x = -1,$ take $y = 1;$ if $x = 0,$ take $y = 0;$ if $x = 1,$ take $y = -1;$ if $x = 2,$ take $y = -2.$

Alternatively, note that for each integer x in $D,$ the integer $-x$ is also in $D,$ including 0 (because $-0 = 0$), and for all integers $x,$ $x + (-x) = 0.$

10. a. True. Every student chose at least one dessert: Uta chose pie, Tim chose both pie and cake, and Yuen chose pie.

   c. This statement says that some particular dessert was chosen by every student. This is true: Every student chose pie.

11. a. There is a student who has seen *Casablanca*.

   c. Every student has seen at least one movie.

   d. There is a movie that has been seen by every student. (There are many other acceptable ways to state these answers.)

12. a. Negation: $\exists x$ in D such that $\forall y$ in $E,$ $x + y \neq 1.$
   The negation is true. When $x = -2,$ the only number y with the property that $x + y = 1$ is $y = 3,$ and 3 is not in $E.$

   b. *Negation*: $\forall x$ in $D,$ $\exists y$ in E such that $x + y \neq -y.$
   The negation is true and the original statement is false. To see that the original statement is false, take any x in D and choose y to be any number in E with $y \neq -\frac{x}{2}.$ Then $2y \neq -x,$ and adding x and subtracting y from both sides gives $x + y \neq -y.$

In 13–19 there are other correct answers in addition to those shown.

13. a. *Statement:* For every color, there is an animal of that color.
   There are animals of every color.

   b. *Negation:* $\exists$ a color C such that $\forall$ animals $A,$ A is not colored $C.$
   For some color, there is no animal of that color.

14. *Statement*: There is a book that all people have read.
   Negation: There is no book that all people have read.
   (*Or:* $\forall$ books $b,$ $\exists$ a person p such that p has not read $b.$)

15. a. *Statement:* For every odd integer $n,$ there is an integer k such that $n = 2k + 1.$
   Given any odd integer, there is another integer for which the given integer equals twice the other integer plus 1.
   Given any odd integer $n,$ we can find another integer k so that $n = 2k + 1.$

An odd integer is equal to twice some other integer plus 1.
Every odd integer has the form $2k + 1$ for some integer $k.$

   b. *Negation:* $\exists$ an odd integer n such that $\forall$ integers $k,$ $n \neq 2k + 1.$
   There is an odd integer that is not equal to $2k + 1$ for any integer $k.$
   Some odd integer does not have the form $2k + 1$ for any integer $k.$

18. a. *Statement:* $\forall x \in \mathbf{R}^+,$ $\exists y \in \mathbf{R}^+$ such that $y > x.$
   Given any positive real number, it is possible to find a larger positive real number.
   For any positive real number, it is possible to find a positive real number that is larger.
   There is no largest positive real number.

   b. *Negation*: $\exists x \in \mathbf{R}^+$ such that $\forall y \in \mathbf{R}^+,$ $y \leq x.$
   There is a positive real number that is greater than or equal to every positive real number.
   There is a positive real number with the property that all positive real numbers are less than or equal to it.
   Some positive real number is greater than or equal to every positive real number.
   There is a largest positive real number.

20. Statement (1) says that no matter what square anyone might give you, you can find a triangle of a different color. This is true because the only squares are $e, g, h,$ and $j,$ and given squares g and $h,$ which are gray, you could take triangle $d,$ which is black; given square $e,$ which is black, you could take either triangle f or $i,$ which are gray; and given square $j,$ which is blue, you could take either triangle f or $h,$ which are gray, or triangle $d,$ which is black.

21. a. (1) The statement "$\forall$ real numbers $x,$ $\exists$ a real number y such that $2x + y = 7$" is true.
   (2) The statement "$\exists$ a real number x such that $\forall$ real numbers $y,$ $2x + y = 7$" is false.

   b. Both statements (1) "$\forall$ real numbers $x,$ $\exists$ a real number y such that $x + y = y + x$" and (2) "$\exists$ a real number x such that $\forall$ real numbers $y,$ $x + y = y + x$" are true.

22. a. Given any real number, you can find a real number so that the sum of the two is zero. In other words, every real number has an additive inverse. This statement is true.

   b. There is a real number with the following property: No mattter what real number is added to it, the sum of the two will be zero. In other words, there is one particular real number whose sum with any real number is zero. This statement is false; no one number will work for all numbers. For instance, if $x + 0 = 0,$ then $x = 0,$ but in that case $x + 1 = 1 \neq 0.$

24. a. $\sim(\forall x \in D(\forall y \in E(P(x, y))))$

$$\equiv \exists x \in D(\sim(\forall y \in E(P(x, y))))$$
$$\equiv \exists x \in D(\exists y \in E(\sim P(x, y)))$$

25. This statement says that all of the circles are above all of the squares. This statement is true because the circles are $a, b,$ and $c,$ and the squares are $e, g, h,$ and $j,$ and all of $a, b,$ and c lie above all of $e, g, h,$ and $j.$

Negation: There is a circle x and a square y such that x is not above y. In other words, at least one of the circles is not above at least one of the squares.

27. The statement says that there are a circle and a square with the property that the circle is above the square and has a different color from the square. This statement is true. For example, circle a lies above square e and is differently colored from e. (Several other examples could also be given.)

29. a. *Version with interchanged quantifiers:* $\exists x \in \mathbf{R}$ such that $\forall y \in \mathbf{R}, x < y$.

   b. The given statement says that for any real number x, there is a real number y that is greater than x. This is true: For any real number x, let $y = x + 1$. Then $x < y$. The version with interchanged quantifiers says that there is a real number that is less than every other real number. This is false.

31. $\forall$ people x, $\exists$ a person y such that x is older than y.

32. $\exists$ a person x such that $\forall$ people y, x is older than y.

33. a. *Formal version:* $\forall$ people x, $\exists$ a person y such that x loves y.

   b. *Negation:* $\exists$ a person x such that $\forall$ people y, x does not love y. In other words, there is someone who does not love anyone.

34. a. *Formal version:* $\exists$ a person x such that $\forall$ people y, x loves y.

   b. *Negation:* $\forall$ people x, $\exists$ a person y such that x does not love y. In other words, everyone has someone whom they do not love.

37. a. *Statement:* $\forall$ even integers n, $\exists$ an integer k such that $n = 2k$.

   b. *Negation:* $\exists$ an even integer n such that $\forall$ integers k, $n \neq 2k$.
   There is some even integer that is not equal to twice any other integer.

39. a. *Statement:* $\exists$ a program P such that $\forall$ questions Q posed to P, P gives the correct answer to Q.

   b. *Negation:* $\forall$ programs P, there is a question Q that can be posed to P such that P does not give the correct answer to Q.

40. a. $\forall$ minutes m, $\exists$ a sucker s such that s was born in minute m.

41. a. This statement says that given any positive integer, there is a positive integer such that the first integer is one more than the second integer. This is false. Given the positive integer $x = 1$, the only integer with the property that $x = y + 1$ is $y = 0$, and 0 is not a positive integer.

   b. This statement says that given any integer, there is an integer such that the first integer is one more than the second integer. This is true. Given any integer x, take $y = x - 1$. Then y is an integer, and $y + 1 = (x - 1) + 1 = x$.

   e. This statement says that given any real number, there is a real number such that the product of the two is equal to 1. This is false because $0 \cdot y = 0 \neq 1$ for every number

y. So when $x = 0$, there is no real number y with the property that $xy = 1$.

42. $\exists \varepsilon > 0$ such that $\forall$ integers N, $\exists$ an integer n such that $n > N$ and either $L - \varepsilon \geq a_n$ or $a_n \geq L + \varepsilon$. In other words, there is a positive number ε such that for all integers N, it is possible to find an integer n that is greater than N and has the property that a_n does not lie between $L - \varepsilon$ and $L + \varepsilon$.

44. a. This statement is true. The unique real number with the given property is 1. Note that

$$1 \cdot y = y \quad \text{for all real numbers } y,$$

and if x is any real number such that for instance, $x \cdot 2 = 2$, then dividing both sides by 2 gives $x = 2/2 = 1$.

46. True. Both triangles a and c lie above all the squares.

48. False. There is no square to the right of circle k.

51. False. There is no object that has a different color from every other object.

53. True. Circle b and squares h and j have the same color.

Section 3.4

1. b. $(f_i + f_j)^2 = f_i^2 + 2f_if_j + f_j^2$

   c. $(3u + 5v)^2 = (3u)^2 + 2(3u)(5v) + (5v)^2$
   $(= 9u^2 + 30uv + 25v^2)$

   d. $(g(r) + g(s))^2 = (g(r))^2 + 2g(r)g(s) + (g(s))^2$

2. 0 is even.

3. $\frac{2}{3} + \frac{4}{5} = \frac{(2 \cdot 5 + 3 \cdot 4)}{(3 \cdot 5)} \left(= \frac{22}{15} \right)$

5. $\frac{1}{0}$ is not an irrational number.

7. Invalid; converse error

8. Valid by universal modus ponens (or universal instantiation)

9. Invalid; inverse error

10. Valid by universal modus tollens

16. Invalid; converse error

19. $\forall x$, if x is a good car, then x is not cheap.

   a. Valid, universal modus ponens (or universal instantiation)

   b. Invalid, converse error

21. Valid. (A valid argument can have false premises and a true conclusion!)

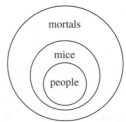

The major premise says the set of people is included in the set of mice. The minor premise says the set of mice is included in the set of mortals. Assuming both of these premises are true, it must follow that the set of people is included in the set of mortals. Since it is impossible for the conclusion to be false if the premises are true, the argument is valid.

23. Valid. The major and minor premises can be diagrammed as follows:

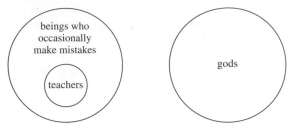

According to the diagram, the set of teachers and the set of gods can have no common elements. Hence, if the premises are true, then the conclusion must also be true, and so the argument is valid.

25. Invalid. Let C represent the set of all college cafeteria food, G the set of all good food, and W the set of all wasted food. Then any one of the following diagrams could represent the given premises.

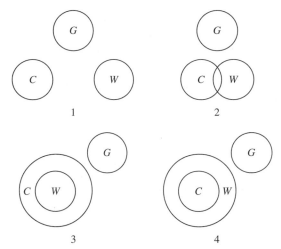

Only in drawing (1) is the conclusion true. Hence it is possible for the premises to be true while the conclusion is false, and so the argument is invalid.

28. (3) *Contrapositive form:* If an object is gray, then it is a circle.
(2) If an object is a circle, then it is to the right of all the blue objects.
(1) If an object is to right of all the blue objects, then it is above all the triangles.
∴ If an object is gray, then it is above all the triangles.

31. 4. If an animal is in the yard, then it is mine.
1. If an animal belongs to me, then I trust it.
5. If I trust an animal, then I admit it into my study.
3. If I admit an animal into my study, then it will beg when told to do so.
6. If an animal begs when told to do so, then that animal is a dog.
2. If an animal is a dog, then that animal gnaws bones.

∴ If an animal is in the yard, then that animal gnaws bones; that is, all the animals in the yard gnaw bones.

33. 2. If a bird is in this aviary, then it belongs to me.
4. If a bird belongs to me, then it is at least 9 feet high.
1. If a bird is at least 9 feet high, then it is an ostrich.
3. If a bird lives on mince pies, then it is not an ostrich.
Contrapositive: If a bird is an ostrich, then it does not live on mince pies.
∴ If a bird is in this aviary, then it does not live on mince pies; that is, no bird in this aviary lives on mince pies.

Section 4.1

1. a. Yes: $-17 = 2(-9) + 1$
b. Yes: $0 = 2 \cdot 0$
c. Yes: $2k - 1 = 2(k - 1) + 1$ and $k - 1$ is an integer because it is a difference of integers.

2. a. Yes: $6m + 8n = 2(3m + 4n)$ and $(3m + 4n)$ is an integer because 3, 4, m, and n are integers, and products and sums of integers are integers.
b. Yes: $10mn + 7 = 2(5mn + 3) + 1$ and $5mn + 3$ is an integer because 3, 5, m, and n are integers, and products and sums of integers are integers.
c. Not necessarily. For instance, if $m = 3$ and $n = 2$, then $m^2 - n^2 = 9 - 4 = 5$, which is prime. (Note that $m^2 - n^2$ is composite for many values of m and n because of the identity $m^2 - n^2 = (m - n)(m + n)$.)

4. For example, let $m = n = 2$. Then m and n are integers such that $m > 1$ and $n > 1$ and $\frac{1}{m} + \frac{1}{n} = \frac{1}{2} + \frac{1}{2} = 1$, which is an integer.

7. For example, let $n = 7$. Then n is an integer such that $n > 5$ and $2^n - 1 = 127$, which is prime.

9. For example, 25, 9, and 16 are all perfect squares, because $25 = 5^2$, $9 = 3^2$, and $16 = 4^2$, and $25 = 9 + 16$. Thus 25 is a perfect square that can be written as a sum of two other perfect squares.

11. Counterexample: Let $a = -2$ and $b = -1$. Then $a < b$ because $-2 < -1$, but $a^2 \not< b^2$ because $(-2)^2 = 4$ and $(-1)^2 = 1$ and $4 \not< 1$. *[So the hypothesis of the statement is true but its conclusion is false.]*

14. This property is true for some integers and false for other integers. For instance, if $a = 0$ and $b = 1$, the property is true because $(0 + 1)^2 = 0^2 + 1^2$, but if $a = 1$ and $b = 1$, the property is false because $(1 + 1)^2 = 4$ and $1^2 + 1^2 = 2$ and $4 \neq 2$.

15. *Hint:* This property is true for some integers and false for other integers. To justify this answer you need to find examples of both.

17. $2 = 1^2 + 1^2, \quad 4 = 2^2, \quad 6 = 2^2 + 1^2 + 1^2,$

$8 = 2^2 + 2^2, \quad 10 = 3^2 + 1^2, \quad 12 = 2^2 + 2^2 + 2^2,$

$14 = 3^2 + 2^2 + 1^2, \quad 16 = 4^2,$

$18 = 3^2 + 3^2 = 4^2 + 1^2 + 1^2, \quad 20 = 4^2 + 2^2,$

$22 = 3^2 + 3^2 + 2^2, \quad 24 = 4^2 + 2^2 + 2^2$

19. a. ∀ integers m and n, if m is even and n is odd, then $m + n$ is odd.

∀ even integers m and odd integers n, $m + n$ is odd.

If m is any even integer and n is any odd integer, then $m + n$ is odd.

b. **(a)** any odd integer **(b)** integer r
(c) $2r + (2s + 1)$ **(d)** $m + n$ is odd

20. a. If an integer is greater than 1, then its reciprocal is between 0 and 1.

b. *Start of proof*: Suppose m is any integer such that $m > 1$. *Conclusion to be shown*: $0 < 1/m < 1$.

22. a. If the product of two integers is 1, then either both are 1 or both are -1.

b. *Start of proof*: Suppose m and n are any integers with $mn = 1$.
Conclusion to be shown: $m = n = 1$ or $m = n = -1$.

24. *Two versions of a correct proof are given below to illustrate some of the variety that is possible.*

Proof 1: Suppose n is any *[particular but arbitrarily chosen]* even integer. *[We must show that $-n$ is even.]* By definition of even, $n = 2k$ for some integer k. Multiplying both side by -1 gives that

$$-n = -(2k) = 2(-k).$$

Let $r = -k$. Then r is an integer because $r = -k = (-1)k$, -1 and k are integers, and a product of two integers is an integer. Hence, $-n = 2r$ for some integer r, and so $-n$ is even *[as was to be shown]*.

Proof 2: Suppose n is any even integer. By definition of even, $n = 2k$ for some integer k. Then

$$-n = -2k = 2(-k).$$

But $-k$ is an integer because it is a product of integers -1 and k. Thus $-n$ equals twice some integer, and so $-n$ is even by definition of even.

25. Proof: Suppose a is any even integer and b is any odd integer. *[We must show that $a - b$ is odd.]* By definition of even and odd, $a = 2r$ and $b = 2s + 1$ for some integers r and s. By substitution and algebra,

$$a - b = 2r - (2s + 1) = 2r - 2s - 1 = 2(r - s - 1) + 1.$$

Let $t = r - s - 1$. Then t is an integer because differences of integers are integers. Thus $a - b = 2t + 1$, where t is an integer, and so, by definition of odd, $a - b$ is odd *[as was to be shown]*.

26. *Hint:* The conclusion to be shown is that a certain quantity is odd. To show this, you need to show that the quantity equals twice some integer plus one.

29. Proof: Suppose n is any *[particular but arbitrarily chosen]* odd integer. *[We must show that $3n + 5$ is even.]* By definition of odd, there is an integer r such that $n = 2r + 1$. Then

$$3n + 5 = 3(2r + 1) + 5 \qquad \text{by substitution}$$
$$= 6r + 3 + 5$$
$$= 6r + 8$$
$$= 2(3r + 4) \qquad \text{by algebra.}$$

Let $t = 3r + 4$. Then t is an integer because products and sums of integers are integers. Hence, $3n + 5 = 2t$, where t is an integer, and so, by definition of even, $3n + 5$ is even *[as was to be shown]*.

31. Proof: Suppose k is any *[particular but arbitrarily chosen]* odd integer and m is any even integer. *[We must show that $k^2 + m^2$ is odd.]* By definition of odd and even, $k = 2a + 1$ and $m = 2b$ for some integers a and b. Then

$$k^2 + m^2 = (2a + 1)^2 + (2b)^2 \qquad \text{by substitution}$$
$$= 4a^2 + 4a + 1 + 4b^2$$
$$= 4(a^2 + a + b^2) + 1$$
$$= 2(2a^2 + 2a + 2b^2) + 1 \quad \text{by algebra.}$$

But $2a^2 + 2a + 2b^2$ is an integer because it is a sum of products of integers. Thus $k^2 + m^2$ is twice an integer plus 1, and so $k^2 + m^2$ is odd *[as was to be shown]*.

33. Proof: Suppose n is any even integer. Then $n = 2k$ for some integer k. Hence

$$(-1)^n = (-1)^{2k} \qquad \text{by substitution}$$
$$= ((-1)^2)^k \qquad \text{by a law of exponents}$$
$$= 1^k \qquad \text{because } (-1)^2 = 1$$
$$= 1 \qquad \text{because 1 to any power equals 1.}$$

This is what was to be shown.

35. The negation of the statement is "For all integers $m \geq 3$, $m^2 - 1$ is not prime."
Proof of the negation: Suppose m is any integer with $m \geq 3$. By basic algebra, $m^2 - 1 = (m - 1)(m + 1)$. Because $m \geq 3$, both $m - 1$ and $m + 1$ are positive integers greater than 1, and each is smaller than $m^2 - 1$. So $m^2 - 1$ is a product of two smaller positive integers, each greater than 1, and hence $m^2 - 1$ is not prime.

38. The incorrect proof just shows the theorem to be true in the one case where $k = 2$. A real proof must show that it is true for *all* integers $k > 0$.

39. The mistake in the "proof" is that the same symbol, k, is used to represent two different quantities. By setting $m = 2k$ and $n = 2k + 1$, the proof implies that $n = m + 1$, and thus it deduces the conclusion only for this one situation. When $m = 4$ and $n = 17$, for instance, the computations in the proof indicate that $n - m = 1$, but actually $n - m = 13$. In other words, the proof does not deduce the conclusion for an arbitrarily chosen even integer m and odd integer n, and hence it is invalid.

40. This incorrect proof exhibits circular reasoning. The word *since* in the third sentence is completely unjustified. The second sentence tells only what happens *if* $k^2 + 2k + 1$ is composite. But at that point in the proof, it has not been established that $k^2 + 2k + 1$ *is* composite. In fact, that is exactly what is to be proved.

43. True. <u>Proof:</u> Suppose m and n are any odd integers. *[We must show that mn is odd.]* By definition of odd, $n = 2r + 1$ and $m = 2s + 1$ for some integers r and s. Then

$$\begin{aligned} mn &= (2r + 1)(2s + 1) \quad \text{by substitution} \\ &= 4rs + 2r + 2s + 1 \\ &= 2(2rs + r + s) + 1 \quad \text{by algebra.} \end{aligned}$$

Now $2rs + r + s$ is an integer because products and sums of integers are integers and $2, r$, and s are all integers. Hence $mn = 2 \cdot (\text{some integer}) + 1$, and so, by definition of odd, mn is odd.

44. True. <u>Proof:</u> Suppose n is any odd integer. *[We must show that $-n$ is odd.]* By definition of odd, $n = 2k + 1$ for some integer k. By substitution and algebra,

$$-n = -(2k + 1) = -2k - 1 = 2(-k - 1) + 1.$$

Let $t = -k - 1$. Then t is an integer because differences of integers are integers. Thus $-n = 2t + 1$, where t is an integer, and so, by definition of odd, $-n$ is odd *[as was to be shown]*.

45. False. Counterexample: Both 3 and 1 are odd, but their difference is $3 - 1 = 2$, which is even.

47. False. Counterexample: Let $m = 1$ and $n = 3$. Then $m + n = 4$ is even, but neither summand m nor summand n is even.

54. <u>Proof:</u> Suppose n is any integer. Then

$$\begin{aligned} 4(n^2 + n + 1) - 3n^2 &= 4n^2 + 4n + 4 - 3n^2 \\ &= n^2 + 4n + 4 = (n + 2)^2 \end{aligned}$$

(by algebra). But $(n + 2)^2$ is a perfect square because $n + 2$ is an integer (being a sum of n and 2). Hence $4(n^2 + n + 1) - 3n^2$ is a perfect square, as was to be shown.

56. *Hint:* This is true.

62. *Hint:* The answer is no.

Section 4.2

1. $\frac{-35}{6} = \frac{-35}{6}$

3. $\frac{4}{5} + \frac{2}{9} = \frac{4 \cdot 9 + 2 \cdot 5}{45} = \frac{46}{45}$

4. Let $x = 0.3737373737\ldots$.
Then $100x = 37.37373737\ldots$, and so
$100x - x = 37.37373737\ldots - 0.3737373737\ldots$.
Thus $99x = 37$, and hence $x = \frac{37}{99}$.

6. Let $x = 320.5492492492\ldots$.
Then $10000x = 3205492.492492\ldots$, and
$10x = 3205.492492492\ldots$, and so
$10000x - 10x = 3205492 - 3205$.
Thus $9990x = 3202287$, and hence $x = \frac{3202287}{9990}$.

8. b. $\forall$ real numbers x and y, if $x \neq 0$ and $y \neq 0$ then $xy \neq 0$.

9. Because a and b are integers, $b - a$ and ab^2 are both integers (since differences and products of integers are inte-

gers). Also, by the zero product property, $ab^2 \neq 0$ because neither a nor b is zero. Hence $(b - a)/ab^2$ is a quotient of two integers with nonzero denominator, and so it is rational.

11. <u>Proof:</u> Suppose n is any *[particular but arbitrarily chosen]* integer. Then $n = n \cdot 1$, and so $n = n/1$ by by dividing both sides by 1. Now n and 1 are both integers, and $1 \neq 0$. Hence n can be written as a quotient of integers with a nonzero denominator, and so n is rational.

12. (a) any *[particular but arbitrarily chosen]* rational number **(b)** integers a and b **(c)** $(a/b)^2$ **(d)** b^2 **(e)** zero product property **(f)** r^2 is rational

13. a. $\forall$ real numbers r, if r is rational then $-r$ is rational. *Or:* $\forall r$, if r is a rational number then $-r$ is rational. *Or:* $\forall$ rational numbers r, $-r$ is rational.

b. The statement is true. <u>Proof:</u> Suppose r is a *[particular but arbitrarily chosen]* rational number. *[We must show that $-r$ is rational.]* By definition of rational, $r = a/b$ for some integers a and b with $b \neq 0$. Then

$$\begin{aligned} -r &= -\frac{a}{b} \quad \text{by substitution} \\ &= \frac{-a}{b} \quad \text{by algebra.} \end{aligned}$$

But since a is an integer, so is $-a$ (being the product of -1 and a). Hence $-r$ is a quotient of integers with a nonzero denominator, and so $-r$ is rational *[as was to be shown]*.

15. <u>Proof:</u> Suppose r and s are rational numbers. By definition of rational, $r = a/b$ and $s = c/d$ for some integers a, b, c, and d with $b \neq 0$ and $d \neq 0$. Then

$$\begin{aligned} rs &= \frac{a}{b} \cdot \frac{c}{d} \quad \text{by substitution} \\ &= \frac{ac}{bd} \quad \text{by the rules of algebra for multiplying fractions.} \end{aligned}$$

Now ac and bd are both integers (being products of integers) and $bd \neq 0$ (by the zero product property). Hence rs is a quotient of integers with a nonzero denominator, and so, by definition of rational, rs is rational.

16. *Hint:* Counterexample: Let r be any rational number and $s = 0$. Then r and s are both rational, but the quotient of r divided by s is undefined and therefore is not a rational number.
Revised statement to be proved: For all rational numbers r and s, if $s \neq 0$ then r/s is rational.

17. *Hint:* The conclusion to be shown is that a certain quantity (the difference of two rational numbers) is rational. To show this, you need to show that the quantity can be expressed as a ratio of two integers with a nonzero denominator.

18. *Hint:* $\dfrac{a/b + c/d}{2} = \dfrac{(ad + bc)/(bd)}{2} = \dfrac{ad + bc}{2bd}$

19. *Hint:* If $a < b$ then $a + a < a + b$ (by T19 of Appendix A), or equivalently $2a < a + b$. Thus $a < \frac{a+b}{2}$ (by T20 Appendix A).

21. True. <u>Proof:</u> Suppose m is any even integer and n is any odd integer. *[We must show that $m^2 + 3n$ is odd.]* By properties

1 and 3 of Example 4.2.3, m^2 is even (because $m^2 = m \cdot m$) and $3n$ is odd (because both 3 and n are odd). It follows from property 5 *[and the commutative law for addition]* that $m^2 + 3n$ is odd *[as was to be shown]*.

24. Proof: Suppose r and s are any rational numbers. By Theorem 4.2.1, both 2 and 3 are rational, and so, by exercise 15, both $2r$ and $3s$ are rational. Hence, by Theorem 4.2.2, $2r + 3s$ is rational.

27. Let

$$x = \frac{1 - \dfrac{1}{2^{n+1}}}{1 - \dfrac{1}{2}} = \frac{1 - \dfrac{1}{2^{n+1}}}{\dfrac{1}{2}} = \frac{1 - \dfrac{1}{2^{n+1}}}{\dfrac{1}{2}} \cdot \frac{2^{n+1}}{2^{n+1}} = \frac{2^{n+1} - 1}{2^n}.$$

But $2^{n+1} - 1$ and 2^n are both integers (since n is a nonnegative integer) and $2^n \neq 0$ by the zero product property. Therefore, x is rational.

31. Proof: Suppose c is a real number such that

$$r_3 c^3 + r_2 c^2 + r_1 c + r_0 = 0,$$

where r_0, r_1, r_2, and r_3 are rational numbers. By definition of rational, $r_0 = a_0/b_0$, $r_1 = a_1/b_1, r_2 = a_2/b_2$, and $r_3 = a_3/b_3$ for some integers, a_0, a_1, a_2, a_3, and nonzero integers b_0, b_1, b_2, and b_3. By substitution,

$$r_3 c^3 + r_2 c^2 + r_1 c + r_0$$

$$= \frac{a_3}{b_3} c^3 + \frac{a_2}{b_2} c^2 + \frac{a_1}{b_1} c + \frac{a_0}{b_0}$$

$$= \frac{b_0 b_1 b_2 a_3}{b_0 b_1 b_2 b_3} c^3 + \frac{b_0 b_1 b_3 a_2}{b_0 b_1 b_2 b_3} c^2 + \frac{b_0 b_2 b_3 a_1}{b_0 b_1 b_2 b_3} c + \frac{b_1 b_2 b_3 a_0}{b_0 b_1 b_2 b_3}$$

$$= 0.$$

Multiplying both sides by $b_0 b_1 b_2 b_3$ gives

$$b_0 b_1 b_2 a_3 \cdot c^3 + b_0 b_1 b_3 a_2 \cdot c^2 + b_0 b_2 b_3 a_1 \cdot c + b_1 b_2 b_3 a_0 = 0.$$

Let $n_3 = b_0 b_1 b_3 a_3$, $n_2 = b_0 b_1 b_3 a_2$, $n_1 = b_0 b_2 b_3 a_1$, and $n_0 = b_1 b_2 b_3 a_0$. Then n_0, n_1, n_2, and n_3 are all integers (being products of integers). Hence c satisfies the equation

$$n_3 c^3 + n_2 c^2 + n_1 c + n_0 = 0.$$

where n_0, n_1, n_2, and n_3 are all integers. This is what was to be shown.

33. a. *Hint:* Note that $(x - r)(x - s) = x^2 - (r + s)x + rs$. If both r and s are odd, then $r + s$ is even and rs is odd. So the coefficient of x^2 is 1 (odd), the coefficient of x is even, and the constant coefficient, rs, is odd.

35. This "proof" assumes what is to be proved.

37. By setting both r and s equal to a/b, this incorrect proof violates the requirement that r and s be arbitrarily chosen rational numbers. If both r and s equal a/b, then $r = s$.

Section 4.3

1. Yes, $52 = 13 \cdot 4$ **2.** Yes, $56 = 7 \cdot 8$

4. Yes, $(3k + 1)(3k + 2)(3k + 3) =$ $3[(3k + 1)(3k + 2)(k + 1)]$, and $(3k + 1)(3k + 2)(k + 1)$ is an integer because k is an integer and sums and products of integers are integers.

6. No, $29/3 \cong 9.67$, which is not an integer.

7. Yes, $66 = (-3)(-22)$.

8. Yes, $6a(a + b) = 3a[2(a + b)]$, and $2(a + b)$ is an integer because a and b are integers and sums and products of integers are integers.

10. No, $34/7 \cong 4.86$, which is not an integer.

12. Yes, $n^2 - 1 = (4k + 1)^2 - 1 = (16k^2 + 8k + 1) - 1 = 16k^2 + 8k = 8(2k^2 + k)$, and $2k^2 + k$ is an integer because k is an integer and sums and products of integers are integers.

14. (a) $a \mid b$ **(b)** $b = a \cdot r$ **(c)** $-r$ **(d)** $a \mid (-b)$

15. Proof: Suppose a, b, and c are any integers such that $a \mid b$ and $a \mid c$. *[We must show that $a \mid (b + c)$.]* By definition of divides, $b = ar$ and $c = as$ for some integers r and s. Then

$$b + c = ar + as = a(r + s) \quad \text{by algebra.}$$

Let $t = r + s$. Then t is an integer (being a sum of integers), and thus $b + c = at$ where t is an integer. By definition of divides, then, $a \mid (b + c)$ *[as was to be shown]*.

16. *Hint:* The conclusion to be shown is that a certain quantity is divisible by a. To show this, you need to show that the quantity equals a times some integer.

17. a. $\forall$ integers n if n is a multiple of 3 then $-n$ is a multiple of 3.

b. The statement is true. Proof: Suppose n is any integer that is a multiple of 3. *[We must show that $-n$ is a multiple of 3.]* By definition of multiple, $n = 3k$ for some integer k. Then

$$-n = -(3k) \qquad \text{by substitution}$$
$$= 3(-k) \qquad \text{by algebra.}$$

Hence, by definition of multiple, $-n$ is a multiple of 3 *[as was to be shown]*.

18. Counterexample: Let $a = 2$ and $b = 1$. Then $a + b = 2 + 1 = 3$, and so $3 \mid (a + b)$ because $3 = 3 \cdot 1$. On the other hand, $a - b = 2 - 1 = 1$, and $3 \nmid 1$ because $1/3$ is not an integer. Thus $3 \nmid (a - b)$. *[So the hypothesis of the statement is true but its conclusion is false.]*

19. *Start of proof:* Suppose a, b, and c are any integers such that a divides b. *[We must show that a divides bc.]*

22. *Hint:* The given statement can be rewritten formally as "$\forall$ integers n, if n is divisible by 6, then n is divisible by 2." This statement is true.

24. The statement is true. Proof: Suppose a, b, and c are any integers such that $a \mid b$ and $a \mid c$. *[We must show that $a \mid (2b - 3c)$.]* By definition of divisibility, we know that $b = am$ and $c = an$ for some integers m and n. It follows that $2b - 3c = 2(am) - 3(an)$ *(by substitution)* $= a(2m - 3n)$ *(by basic algebra)*. Let $t = 2m - 3n$. Then t is an integer because it is a difference of products of integers. Hence

$2b - 3c = at$, where t is an integer, and so $a \mid (2b - 3c)$ by definition of divisibility *[as was to be shown]*.

25. The statement is false. <u>Counterexample:</u> Let $a = 2$, $b = 3$, and $c = 8$. Then $a \mid c$ because 2 divides 8, but $ab \nmid c$ because $ab = 6$ and 6 does not divide 8.

26. *Hint:* The statement is true.

27. *Hint:* The statement is false.

32. No. Each of these numbers is divisible by 3, and so their sum is also divisible by 3. But 100 is not divisible by 3. Thus the sum cannot equal $100.

36. **a.** The sum of the digits is 54, which is divisible by 9. Therefore, $637,425,403,705,125$ is divisible by 9 and hence also divisible by 3 (by transitivity of divisibility). Because the rightmost digit is 5, then $637,425,403,705,125$ is divisible by 5. And because the two rightmost digits are 25, which is not divisible by 4, then $637,425,403,705,125$ is not divisible by 4.

37. **a.** $1176 = 2^3 \cdot 3 \cdot 7^2$

38. **a.** $p_1^{2e_1} p_2^{2e_2} \dots p_k^{2e_k}$
    b. $n = 42$, $2^5 \cdot 3 \cdot 5^2 \cdot 7^3 \cdot n = 5880^2$

40. **a.** Because $12a = 25b$, the unique factorization theorem guarantees that the standard factored forms of $12a$ and $25b$ must be the same. Thus $25b$ contains the factors $2^2 \cdot 3 (= 12)$. But since neither 2 nor 3 divide 25, the factors $2^2 \cdot 3$ must all occur in b, and hence $12 \mid b$. Similarly, $12a$ contains the factors $5^2 = 25$, and since 5 is not a factor of 12, the factors 5^2 must occur in a. So $25 \mid a$.

41. *Hint:* $45^8 \cdot 88^5 = (3^2 \cdot 5)^8 \cdot (2^3 \cdot 11)^5 = 3^{16} \cdot 5^8 \cdot 2^{15} \cdot 11^5$. How many factors of 10 does this number contain?

42. **a.** $6! = 6 \cdot 5 \cdot 4 \cdot 3 \cdot 2 \cdot 1 = 2 \cdot 3 \cdot 5 \cdot 2 \cdot 2 \cdot 3 \cdot 2 = 2^4 \cdot 3^2 \cdot 5$

44. <u>Proof:</u> Suppose n is a nonnegative integer whose decimal representation ends in 0. Then $n = 10m + 0 = 10m$ for some integer m. Factoring out a 5 yields $n = 10m = 5(2m)$, and $2m$ is an integer since m is an integer. Hence $10m$ is divisible by 5, which is what was to be shown.

47. *Hint:* You may take it as a fact that for any positive integer k,

$$10^k = \underbrace{99 \dots 9}_{k \text{ of these}} + 1; \text{ that is,}$$

$$10^k = 9 \cdot 10^{k-1} + 9 \cdot 10^{k-2} + \dots + 9 \cdot 10^1 + 9 \cdot 10^0 + 1.$$

Section 4.4

1. $q = 7, r = 7$ 3. $q = 0, r = 36$

5. $q = -5, r = 10$ 7. **a.** 4   **b.** 7

11. **a.** When today is Saturday, 15 days from today is two weeks (which is Saturday) plus one day (which is Sunday). Hence $DayN$ should be 0. According to the formula, when today is Saturday, $DayT = 6$, and so when $N = 15$,

$$DayN = (DayT + N) \bmod 7$$
$$= (6 + 15) \bmod 7$$
$$= 21 \bmod 7 = 0, \text{ which agrees.}$$

13. *Solution 1:* $30 = 4 \cdot 7 + 2$. Hence the answer is two days after Monday, which is Wednesday.
    Solution 2: By the formula, the answer is $(1 + 30) \bmod 7 = 31 \bmod 7 = 3$, which is Wednesday.

14. *Hint:* There are two ways to solve this problem. One is to find that $1,000 = 7 \cdot 142 + 6$ and note that if today is Tuesday, then 1,000 days from today is 142 weeks plus 6 days from today. The other way is to use the formula $DayN = (DayT + N) \bmod 7$, with $DayT = 2$ (Tuesday) and $N = 1000$.

16. Because $d \mid n$, $n = dq + 0$ for some integer q. Thus the remainder is 0.

18. <u>Proof:</u> Suppose n is any odd integer. By definition of odd, $n = 2q + 1$ for some integer q. Then $n^2 = (2q + 1)^2 = 4q^2 + 4q + 1 = 4(q^2 + q) + 1 = 4q(q + 1) + 1$. By the result of exercise 17, the product $q(q + 1)$ is even, so $q(q + 1) = 2m$ for some integer m. Then, by substitution, $n^2 = 4 \cdot 2m + 1 = 8m + 1$.

20. Because $a \bmod 7 = 4$, the remainder obtained when a is divided by 7 is 4, and so $a = 7q + 4$ for some integer q. Multiplying this equation through by 5 gives that $5a = 35q + 20 = 35q + 14 + 6 = 7(5q + 2) + 6$. Because q is an integer, $5q + 2$ is also an integer, and so $5a = 7 \cdot (\text{an integer}) + 6$. Thus, because $0 \leq 6 < 7$, the remainder obtained when $5a$ is divided by 7 is 6, and so $5a \bmod 7 = 6$.

23. <u>Proof:</u> Suppose n is any *[particular but arbitrarily chosen]* integer such that $n \bmod 5 = 3$. Then the remainder obtained when n is divided by 5 is 3, and so $n = 5q + 3$ for some integer q. By substitution,

$$n^2 = (5q + 3)^2 = 25q^2 + 30q + 9$$
$$= 25q^2 + 30q + 5 + 4 = 5(5q^2 + 6q + 1) + 4.$$

Because products and sums of integers are integers, $5q^2 + 6q + 1$ is an integer, and hence $n^2 = 5 \cdot (\text{an integer}) + 4$. Thus, since $0 \leq 4 < 5$, the remainder obtained when n^2 is divided by 5 is 4, and so $n^2 \bmod 5 = 4$.

26. *Hint:* You need to show that (1) for all nonnegative integers n and positive integers d, if n is divisible by d then $n \bmod d = 0$; and (2) for all nonnegative integers n and positive integers d, if $n \bmod d = 0$ then n is divisible by d.

27. <u>Proof:</u> Suppose n is any integer. By the quotient-remainder theorem with $d = 3$, there exist integers q and r such that $n = 3q + r$ and $0 \leq r < 3$. But the only nonnegative integers r that are less than 3 are 0, 1, and 2. Therefore, $n = 3q + 0 = 3q$, or $n = 3q + 1$, or $n = 3q + 2$ for some integer q.

28. **a.** <u>Proof:</u> Suppose n, $n + 1$, and $n + 2$ are any three consecutive integers. *[We must show that $n(n + 1)(n + 2)$ is*

divisible by 3.] By the quotient-remainder theorem, n can be written in one of the three forms, $3q, 3q + 1$, or $3q + 2$ for some integer q. We divide into cases accordingly.

Case 1 ($n = 3q$ for some integer q): In this case,

$n(n + 1)(n + 2)$

$\quad = 3q(3q + 1)(3q + 2) \qquad$ by substitution

$\quad = 3 \cdot [q(3q + 1)(3q + 2)] \qquad$ by factoring out a 3.

Let $m = q(3q + 1)(3q + 2)$. Then m is an integer because q is an integer, and sums and products of integers are integers. By substitution,

$\quad n(n + 1)(n + 2) = 3m \quad$ where m is an integer.

And so, by definition of divisible, $n(n + 1)(n + 2)$ is divisible by 3.

Case 2 ($n = 3q + 1$ for some integer q): In this case,

$n(n + 1)(n + 2)$

$\quad = (3q + 1)((3q + 1) + 1)((3q + 1) + 2)$

$\qquad\qquad\qquad\qquad\qquad\qquad$ by substitution

$\quad = (3q + 1)(3q + 2)(3q + 3)$

$\quad = (3q + 1)(3q + 2)3(q + 1)$

$\quad = 3 \cdot [(3q + 1)(3q + 2)(q + 1)] \quad$ by algebra.

Let $m = (3q + 1)(3q + 2)(q + 1)$. Then m is an integer because q is an integer, and sums and products of integers are integers. By substitution,

$\quad n(n + 1)(n + 2) = 3m \quad$ where m is an integer.

And so, by definition of divisible, $n(n + 1)(n + 2)$ is divisible by 3.

Case 3 ($n = 3q + 2$ for some integer q): In this case,

$n(n + 1)(n + 2)$

$\quad = (3q + 2)((3q + 2) + 1)((3q + 2) + 2)$

$\qquad\qquad\qquad\qquad\qquad\qquad$ by substitution

$\quad = (3q + 2)(3q + 3)(3q + 4)$

$\quad = (3q + 2)3(q + 1)(3q + 4)$

$\quad = 3 \cdot [(3q + 2)(q + 1)(3q + 4)] \quad$ by algebra

Let $m = (3q + 2)(q + 1)(3q + 4)$. Then m is an integer because q is an integer, and sums and products of integers are integers. By substitution,

$\quad n(n + 1)(n + 2) = 3m \quad$ where m is an integer.

And so, by definition of divisible, $n(n + 1)(n + 2)$ is divisible by 3.

In each of the three cases, $n(n + 1)(n + 2)$ was seen to be divisible by 3. But by the quotient-remainder theorem, one of these cases must occur. Therefore, the product of *any* three consecutive integers is divisible by 3.

b. For all integers n, $n(n + 1)(n + 2) \bmod 3 = 0$.

29. a. *Hint:* Given any integer n, begin by using the quotient-remainder theorem to say that n can be written in one of the three forms: $n = 3q$, or $n = 3q + 1$, or

$n = 3q + 2$ for some integer q. Then divide into three cases according to these three possibilities. Show that in each case either $n^2 = 3k$ for some integer k, or $n^2 = 3k + 1$ for some integer k. For instance, when $n = 3q + 2$, then $n^2 = (3q + 2)^2 = 9q^2 + 12q + 4 = 3(3q^2 + 4q + 1) + 1$, and $3q^2 + 4q + 1$ is an integer because it is a sum of products of integers.

31. b. If $m^2 - n^2 = 56$, then $56 = (m + n)(m - n)$. Now $56 = 2^3 \cdot 7$, and by the unique factorization theorem, this factorization is unique. Hence the only representations of 56 as a product of two positive integers are $56 = 7 \cdot 8 = 14 \cdot 4 = 28 \cdot 2 = 56 \cdot 1$. By part (a), m and n must both be odd or both be even. Thus the only solutions are either $m + n = 14$ and $m - n = 4$ or $m + n = 28$ and $m - n = 2$. This gives either $m = 9$ and $n = 5$ or $m = 15$ and $n = 13$ as the only solutions.

32. For any integers a, b, and c, if $a - b$ and $b - c$ are even, then $2a - (b + c)$ is even.

Proof: Suppose a, b, and c are any integers such that $a - b$ is even and $b - c$ is even. *[We must show that $2a - (b + c)$ is even.]* Note first that $2a - (b + c) = (a - b) + (a - c)$. Also note that $(a - b) + (b - c)$ is a sum of two even integers and hence is even by Example 4.2.3 #1. But $(a - b) + (b - c) = a - c$, and so $a - c$ is even. Hence $2a - (b + c)$ is a sum of two even integers, and thus it is even *[as was to be shown]*.

34. *Hint:* Express n using the quotient-remainder theorem with $d = 3$.

36. *Hint:* Use the quotient-remainder theorem (as in Example 3.4.5) to say that $n = 4q, n = 4q + 1, n = 4q + 2$, or $n = 4q + 3$ and divide into cases accordingly.

38. *Hint:* Given any integer n, consider the two cases where n is even and where n is odd.

39. *Hint:* Given any integer n, analyze the sum $n + (n + 1) + (n + 2) + (n + 3)$.

42. *Hint:* Use the quotient-remainder theorem to say that n must have one of the forms $6q, 6q + 1, 6q + 2, 6q + 3, 6q + 4$, or $6q + 5$ for some integer q.

44. *Hint:* There are four cases: Either x and y are both positive, or x is positive and y is negative, or x is negative and y is positive, or both x and y are negative.

Section 4.5

1. (a) A contradiction
(b) A positive real number
(c) x
(d) Both sides by 2
(e) Contradiction

3. Proof: Suppose not. That is, suppose there is an integer n such that $3n + 2$ is divisible by 3. *[We must derive a contradiction.]* By definition of divisibility, $3n + 2 = 3k$ for some integer k. Subtracting $3n$ from both sides gives that $2 = 3k - 3n = 3(k - n)$. So, by definition of divisibility, $3 \mid 2$. But by Theorem 4.3.1 this implies that $3 \le 2$, which contradicts the fact that $3 > 2$. *[Thus for all integers n, $3n + 2$ is not divisible by 3.]*

5. *Negation of statement*: There is a greatest even integer.
Proof of statement: Suppose not. That is, suppose there is a greatest even integer; call it N. Then N is an even integer, and $N \geq n$ for every even integer n. *[We must deduce a contradiction.]* Let $M = N + 2$. Then M is an even integer since it is a sum of even integers, and $M > N$ since $M = N + 2$. This contradicts the supposition that $N \geq n$ for *every* even integer n. *[Hence the supposition is false and the statement is true.]*

8. (a) a rational number
  (b) an irrational number
  (c) $\frac{a}{b}$
  (d) $\frac{c}{d}$
  (e) $\frac{a}{b} - \frac{c}{d}$
  (f) integers
  (g) integers
  (h) zero product property
  (i) rational

9. a. The mistake in this proof occurs in the second sentence where the negation written by the student is incorrect: Instead of being existential, it is universal. The problem is that if the student proceeds in a logically correct manner, all that is needed to reach a contradiction is one example of a rational and an irrational number whose difference is irrational. To prove the given statement, however, it is necessary to show that there is *no* rational number and *no* irrational number whose difference is rational.

10. Proof by contradiction: Suppose not. That is, suppose there is an irrational number x such that the square root of x is rational. *[We must derive a contradiction.]* By definition of rational, $\sqrt{x} = \frac{a}{b}$ for some integers a and b with $b \neq 0$. By substitution,

$$(\sqrt{x})^2 = \left(\frac{a}{b}\right)^2,$$

and so, by algebra,

$$x = \frac{a^2}{b^2}.$$

But a^2 and b^2 are both products of integers and thus are integers, and b^2 is nonzero by the zero product property. Thus $\frac{a^2}{b^2}$ is rational. It follows that x is both irrational and rational, which is a contradiction. *[This is what was to be shown.]*

11. Proof: Suppose not. That is, suppose $\exists$ a nonzero rational number x and an irrational number y such that xy is rational. *[We must derive a contradiction.]* By definition of rational, $x = a/b$ and $xy = c/d$ for some integers a, b, c, and d with $b \neq 0$ and $d \neq 0$. Also $a \neq 0$ because x is nonzero. By substitution, $xy = (a/b)y = c/d$. Solving for y gives $y = bc/ad$. Now bc and ad are integers (being products of integers) and $ad \neq 0$ (by the zero product property). Thus,

by definition of rational, y is rational, which contradicts the supposition that y is irrational. *[Hence the supposition is false and the statement is true.]*

13. *Hint*: Suppose $n^2 - 2$ is divisible by 4, and consider the two cases where n is even and n is odd. (An alternative solution uses Proposition 4.5.4.)

14. *Hint*: $a^2 = c^2 - b^2 = (c - b)(c + b)$

15. *Hint*: (1) For any integer c, if 2 divides c, then 4 divides c^2. (2) The result of exercise 13 may be helpful.

16. *Hint*: Suppose a, b, and c are odd integers, z is a solution to $ax^2 + bx + c = 0$, and z is rational. Then $z = p/q$ for some integers p and q with $q \neq 0$. We may assume p and q have no common factor. (Why? If p and q do have a common factor, we can divide out their greatest common factor to obtain two integers p' and q' that (1) have no common factor and (2) satisfy the equation $z = p'/q'$. Then we can redefine $q = q'$ and $p = p'$.) Note that because p and q have no common factor, they are not both even. Substitute p/q into $ax^2 + bx + c = 0$, and multiply through by q^2. Show that (1) the assumption that p is even leads to a contradiction, (2) the assumption that q is even leads to a contradiction, and (3) the assumption that both p and q are odd leads to a contradiction. The only remaining possibility is that both p and q are even, which has been ruled out.

18. a. $5 \mid n$ **b.** $5 \mid n^2$ **c.** $5k$ **d.** $(5k)^2$ **e.** $5 \mid n^2$

19. Proof (by contraposition): *[To go by contraposition, we must prove that $\forall$ positive real numbers, r and s, if $r \leq 10$ and $s \leq 10$, then $rs \leq 100$.]* Suppose r and s are positive real numbers and $r \leq 10$ and $s \leq 10$. By the algebra of inequalities, $rs \leq 100$. *[To derive this fact, multiply both sides of $r \leq 10$ by s to obtatin $rs \leq 10s$. And multiply both sides of $s \leq 10$ by 10 to obtain $10s \leq 10 \cdot 10 = 100$. By transitivity of $\leq$, then, $rs \leq 100$.]* But this is what was to be shown.

21. a. Proof by contradiction: Suppose not. That is, suppose there is an integer n such that n^2 is odd and n is even. Show that this supposition leads logically to a contradiction.
  b. Proof by contraposition: Suppose n is any integer such that n is not odd. Show that n^2 is not odd.

23. a. The contrapositive is the statement "$\forall$ real numbers x, if $-x$ is not irrational, then x is not irrational." Equivalently (because $-(-x) = x$), "$\forall$ real numbers x, if x is rational then $-x$ is rational."
  Proof by contraposition: Suppose x is any rational number. *[We must show that $-x$ is also rational.]* By definition of rational, $x = a/b$ for some integers a and b with $b \neq 0$. Then $x = -(a/b) = (-a)/b$. Since both $-a$ and b are integers and $b \neq 0$, $-x$ is rational *[as was to be shown]*.
  b. Proof by contradiction: Suppose not. *[We take the negation and suppose it to be true.]* That is, suppose $\exists$ an irrational number x such that $-x$ is rational. *[We must derive a contradiction.]* By definition of rational, $-x = a/b$ for some integers a and b with $b \neq 0$. Multiplying

both sides by -1 gives $x = -(a/b) = -a/b$. But $-a$ and b are integers (since a and b are) and $b \neq 0$. Thus x is a ratio of the two integers $-a$ and b with $b \neq 0$. Hence x is rational (by definition of rational), which is a contradiction. *[This contradiction shows that the supposition is false, and so the given statement is true.]*

25. *Hints:* See the answer to exercise 21 and look carefully at the two proofs for Proposition 4.5.4.

26. **a.** Proof by contraposition: Suppose $a, b,$ and c are any *[particular but arbitrarily chosen]* integers such that $a \mid b$. *[We must show that $a \mid bc$.]* By definition of divides, $b = ak$ for some integer k. Then $bc = (ak)c = a(kc)$. But kc is an integer (because it is a product of the integers k and c). Hence $a \mid bc$ by definition of divisibility *[as was to be shown]*.

 b. Proof by contradiction: Suppose not. *[We take the negation and suppose it to be true.]* Suppose $\exists$ integers $a, b,$ and c such that $a \nmid bc$ and $a \mid b$. Since $a \mid b$, there exists an integer k such that $b = ak$ by definition of divides. Then $bc = (ak)c = a(kc)$ *[by the associative law of algebra]*. But kc is an integer (being a product of integers), and so $a \mid bc$ by definition of divides. Thus $a \nmid bc$ and $a \mid bc$, which is a contradiction. *[This contradiction shows that the supposition is false, and hence the given statement is true.]*

27. **a.** *Hint:* The contrapositive is "For all integers m and n, if m and n are not both even and m and n are not both odd, then $m + n$ is not even." Equivalently: "For all integers m and n, if one of m and n is even and the other is odd, then $m + n$ is odd."

 b. *Hint:* The negation of the given statement is the following: $\exists$ integers m and n such that $m + n$ is even, and either m is even and n is odd, or m is odd and n is even.

30. The negation of "Every integer is rational" is "There is at least one integer that is irrational" not "Every integer is irrational." Deriving a contradiction from an incorrect negation of a statement does not prove the statement is true.

31. **a.** Proof: Suppose $r, s,$ and n are integers and $r > \sqrt{n}$ and $s > \sqrt{n}$. Note that r and s are both positive because $\sqrt{n}$ cannot be negative. By multiplying both sides of the first inequality by s and both sides of the second inequality by $\sqrt{n}$ (Appendix A, T20), we have that $rs > \sqrt{n}s$ and $\sqrt{n}s > \sqrt{n}\sqrt{n} = n$. Thus, by the transitive law for inequality (Appendix A, T18), $rs > n$.

32. **a.** $\sqrt{667} \cong 25.8$, and so the possible prime factors to be checked are 2, 3, 5, 7, 11, 13, 17, 19, and 23. Testing each in turn shows that 667 is not prime because $667 = 23 \cdot 29$.

 b. $\sqrt{557} \cong 23.6$, and so the possible prime factors to be checked are 2, 3, 5, 7, 11, 13, 17, 19, and 23. Testing each in turn shows that none divides 557. Therefore, 557 is prime.

34. **a.** $\sqrt{9269} \cong 96.3$, and so the possible prime factors to be checked are all among those you found for exercise 33. Testing each in turn shows that 9,269 is not prime because $9{,}269 = 13 \cdot 713$.

 b. $\sqrt{9103} \cong 95.4$, and so the possible prime factors to be checked are all among those you found for exercise 33. Testing each in turn shows that none divides 9,103. Therefore, 9,103 is prime.

35. *Hint:* Is it possible for all three of $n - 4, n - 6,$ and $n - 8$ to be prime?

Section 4.6

1. The value of $\sqrt{2}$ given by a calculator is an approximation. Calculators can give exact values only for numbers that can be represented using at most the number of decimal digits in the calculator display. In particular, every number in a calculator display is rational, but even many rational numbers cannot be represented exactly. For instance, consider the number formed by writing a decimal point and following it with the first 1 million digits of $\sqrt{2}$. By the discussion in Section 4.2, this number is rational, but you could not infer this from the calculator display.

3. Proof by contradiction: Suppose not. That is, suppose $6 - 7\sqrt{2}$ is rational. *[We must prove a contradiction.]* By definition of rational, there exist integers a and $b \neq 0$ with

$$6 - 7\sqrt{2} = \frac{a}{b}.$$

Then $\sqrt{2} = \frac{1}{-7}\left(\frac{a}{b} - 6\right)$ by subtracting 6 from both sides and dividing both sides by -7

and so $\sqrt{2} = \dfrac{a - 6b}{-7b}$ by the rules of algebra.

But $a - 6b$ and $-7b$ are both integers (since a and b are integers and products and difference of integers are integers), and $-7b \neq 0$ by the zero product property. Hence $\sqrt{2}$ is a ratio of the two integers $a - 6b$ and $-7b$ with $-7b \neq 0$, so $\sqrt{2}$ is a rational number (by definition of rational). This contradicts the fact that $\sqrt{2}$ is irrational, and so the supposition is false and $6 - 7\sqrt{2}$ is irrational.

5. This is false. $\sqrt{4} = 2 = 2/1$, which is rational.

7. Counterexample: Let $x = \sqrt{2}$ and let $y = -\sqrt{2}$. Then x and y are irrational, but $x + y = 0 = 0/1$, which is rational.

9. True.
 Formal version of the statement: $\forall$ positive real numbers r, if r is irrational, then $\sqrt{r}$ is irrational.
 Proof by contraposition: Suppose r is any positive real number such that $\sqrt{r}$ is rational. *[We must show that r is rational.]* By definition of rational, $\sqrt{r} = \frac{a}{b}$ for some integers a and b with $b \neq 0$. Then $r = \left(\sqrt{r}\right)^2 = \left(\frac{a}{b}\right)^2 = \frac{a^2}{b^2}$. But both a^2 and b^2 are integers because they are products of integers, and $b^2 \neq 0$ by the zero product property. Thus r is rational *[as was to be shown]*.
 (The statement may also be proved by contradiction.)

13. *Hint:* Can you think of any "nice" integers x and y that are greater than 1 and have the property that $x^2 = y^3$?

16. a. Proof by contradiction: Suppose not. That is, suppose there is an integer n such that $n = 3q_1 + r_1 = 3q_2 + r_2$, where $q_1, q_2, r_1,$ and r_2 are integers, $0 \le r_1 < 3$, $0 \le r_2 < 3$, and $r_1 \ne r_2$. By interchanging the labels for r_1 and r_2 if necessary, we may assume that $r_2 > r_1$. Then $3(q_1 - q_2) = r_2 - r_1 > 0$, and because both r_1 and r_2 are less than 3, either $r_2 - r_1 = 1$ or $r_2 - r_1 = 2$. So either $3(q_1 - q_2) = 1$ or $3(q_1 - q_2) = 2$. The first case implies that $3 \mid 1$, and hence, by Theorem 4.3.1, that $3 \le 1$, and the second case implies that $3 \mid 2$, and hence, by Theorem 4.3.1, that $3 \le 2$. These results contradict the fact that 3 is greater than both 1 and 2. Thus in either case we have reached a contradiction, which shows that the supposition is false and the given statement is true.

b. Proof by contradiction: Suppose not. That is, suppose there is an integer n such that n^2 is divisible by 3 and n is not divisible by 3. *[We must deduce a contradiction.]* By definition of divisible, $n^2 = 3q$ for some integer q, and by the quotient-remainder theorem and part (a), $n = 3k + 1$ or $n = 3k + 2$ some integer k.

Case 1 ($n = 3k + 1$ for some integer k): In this case

$$n^2 = (3k + 1)^2 = 9k^2 + 6k + 1 = 3(3k^2 + 2k) + 1.$$

Let $s = 3k^2 + 2k$. Then $n^2 = 3s + 1$, and s is an integer because it is a sum of products of integers. It follows that $n^2 = 3q = 3s + 1$ for some integers q and s, which contradicts the result of part (a).

Case 2 ($n = 3k + 2$ for some integer k): In this case

$$n^2 = (3k + 2)^2 = 9k^2 + 12k + 4 = 3(3k^2 + 6k + 1) + 1.$$

Let $t = 3k^2 + 6k + 1$. Then $n^2 = 3t + 1$, and t is an integer because it is a sum of products of integers. It follows that $n^2 = 3q = 3t + 1$ for some integers q and t, which contradicts the result of part (a).

Thus in either case, a contradiction is reached, which shows that the supposition is false and the given statement is true.

c. Proof by contradiction: Suppose not. That is, suppose $\sqrt{3}$ is rational. By definition of rational, $\sqrt{3} = \frac{a}{b}$ for some integers a and b with $b \ne 0$. Without loss of generality, assume that a and b have no common factor. (If not, divide both a and b by their greatest common factor to obtain integers a' and b' with the property that a' and b' have no common factor and $\sqrt{3} = \frac{a'}{b'}$. Then redefine $a = a'$ and $b = b'$.) Squaring both sides of $\sqrt{3} = \frac{a}{b}$ gives $3 = \frac{a^2}{b^2}$, and multiplying both sides by b^2 gives

$$3b^2 = a^2 \quad (*).$$

Thus a^2 is divisible by 3, and so, by part (b), a is also divisible by 3. By definition of divisibility, then, $a = 3k$ for some integer k, and so

$$a^2 = 9k^2 \quad (**).$$

Substituting equation (**) into equation (*) gives $3b^2 = 9k^2$, and dividing both sides by 3 yields

$$b^2 = 3k^2.$$

Hence b^2 is divisible by 3, and so, by part (b), b is also divisible by 3. Consequently, both a and b are divisible by 3, which contradicts the assumption that a and b have no common factor. Thus the supposition is false, and so $\sqrt{3}$ is irrational.

18. *Hint:* The proof is a generalization of the one given in the solution for exercise 16(a).

19. *Hint:* First prove that for all integers a, if 5 divides a squared then 5 divides a. The rest of the proof is similar to the solution for exercises 16(c).

20. *Hint:* This statement is true. If $a^2 - 3 = 9b$, then $a^2 = 9b + 3 = 3(3b + 1)$, and so a^2 is divisible by 3. Hence, by exercise 16(b), a is divisible by 3. Thus $a^2 = (3c)^2$ for some integer c.

21. Proof by contradiction: Suppose not. That is, suppose $\sqrt{2}$ is rational. *[We will show that this supposition leads to a contradiction.]* By definition of rational, we may write $\sqrt{2} = a/b$ for some integers a and b with $b \ne 0$. Then $2 = a^2/b^2$, and so $a^2 = 2b^2$. Consider the prime factorizations for a^2 and for $2b^2$. By the unique factorization of integers theorem, these factorizations are unique except for the order in which the factors are written. Now because every prime factor of a occurs twice in the prime factorization of a^2, the prime factorization of a^2 contains an even number of 2's. (If 2 is a factor of a, then this even number is positive, and if 2 is not a factor of a, then this even number is 0.) On the other hand, because every prime factor of b occurs twice in the prime factorization of b^2, the prime factorization of $2b^2$ contains an odd number of 2's. Therefore, the equation $a^2 = 2b^2$ cannot be true. So the supposition is false, and hence $\sqrt{2}$ is irrational.

23. *Hint:* One solution uses only Theorem 4.7.1. Another uses the result of exercise 22 that $\sqrt{6}$ is irrational.

25. *Hint:* $\dfrac{2 \cdot 3 \cdot 5 \cdot 7 + 1}{2} = 3 \cdot 5 \cdot 7 + \dfrac{1}{2}$ and

$\dfrac{2 \cdot 3 \cdot 5 \cdot 7 + 1}{3} = 2 \cdot 5 \cdot 7 + \dfrac{1}{3}$.

26. *Hint:* You can deduce that $p = 3$.

27. a. *Hint:* For example, $N_4 = 2 \cdot 3 \cdot 5 \cdot 7 + 1 = 211$.

29. *Hint:* By Theorem 4.3.4 (divisibility by a prime) there is a prime number p such that $p \mid (n! - 1)$. Show that the supposition that $p \le n$ leads to a contradiction. It will then follow that $n < p < n!$.

30. *Hint:* Every odd integer can be written as $4k + 1$ or as $4k + 3$ for some integer k. (Why?) If $p_1 p_2 \ldots p_n + 1 = 4k + 1$, then $4 \mid p_1 p_2 \ldots p_n$. Is this possible?

31. a. *Hint:* Prove the contrapositive: If for some integer $n > 2$ that is not a power of 2, $x^n + y^n = z^n$ has a positive integer solution, then for some prime number

$p > 2, x^p + y^p = z^p$ has a positive integer solution. Note that if $n = kp$, then $x^n = x^{kp} = (x^k)^p$.

32. Existence proof: When $n = 2$, then $n^2 - 1 = 3$, which is prime. Hence there exists a prime number of the form $n^2 - 1$, where n is an integer and $n \geq 2$.

Uniqueness proof (by contradiction): Suppose to the contrary that m is another integer satisfying the given conditions. That is, $m > 2$ and $m^2 - 1$ is prime. *[We must derive a contradiction.]* Factor $m^2 - 1$ to obtain $m^2 - 1 = (m - 1)(m + 1)$. But $m > 2$, and so $m - 1 > 1$ and $m + 1 > 1$. Hence $m^2 - 1$ is not prime, which is a contradiction. *[This contradiction shows that the supposition is false, and so there is no other integer $m > 2$ such that $n^2 - 1$ is prime.]*

Uniqueness proof (direct): Suppose m is any integer such that $m \geq 2$ and $m^2 - 1$ is prime. *[We must show that $m = 2$.]* By factoring, $m^2 - 1 = (m - 1)(m + 1)$. Since $m^2 - 1$ is prime, either $m - 1 = 1$ or $m + 1 = 1$. But $m + 1 \geq 2 + 1 = 3$. Hence, by elimination, $m - 1 = 1$, and so $m = 2$.

34. Proof (by contradiction): Suppose not. That is, suppose there are two distinct real numbers a_1 and a_2 such that for all real numbers r,

$$(1) \ a_1 + r = r \quad \text{and} \quad (2) \ a_2 + r = r$$

Then

$$a_1 + a_2 = a_2 \quad \text{by (1) with} \quad r = a_2$$

and

$$a_2 + a_1 = a_1 \quad \text{by (2) with} \quad r = a_1.$$

It follows that

$$a_2 = a_1 + a_2 = a_2 + a_1 = a_1$$

which implies that $a_2 = a_1$. But this contradicts the supposition that a_1 and a_2 are distinct. *[Thus the supposition is false and there is at most one real number a such that $a + r = r$ for all real numbers r.]*

Proof (direct): Suppose a_1 and a_2 are real numbers such that for all real numbers r,

$$(1) \ a_1 + r = r \quad \text{and} \quad (2) \ a_2 + r = r$$

Then

$$a_1 + a_2 = a_2 \quad \text{by (1) with} \quad r = a_2$$

and

$$a_2 + a_1 = a_1 \quad \text{by (2) with} \quad r = a_1.$$

It follows that

$$a_2 = a_1 + a_2 = a_2 + a_1 = a_1.$$

Hence $a_2 = a_1$. *[Thus there is at most one real number a such that $a + r = r$ for all real numbers r.]*

Section 5.1

1. $\dfrac{1}{11}, \dfrac{2}{12}, \dfrac{3}{13}, \dfrac{4}{14}$ **3.** $1, -\dfrac{1}{3}, \dfrac{1}{9}, -\dfrac{1}{27}$

5. $a_0 = 2 \cdot 0 + 1 = 1, a_1 = 2 \cdot 1 + 1 = 3, a_2 = 2 \cdot 2 + 1 = 5$,
$a_3 = 2 \cdot 3 + 1 = 7, \quad b_0 = (0 - 1)^3 + 0 + 2 = 1$,
$b_1 = (1 - 1)^3 + 1 + 2 = 3, b_2 = (2 - 1)^3 + 2 + 2 = 5$,
$b_3 = (3 - 1)^3 + 3 + 2 = 13$
So $a_0 = b_0$, $a_1 = b_1$, and $a_2 = b_2$, but $a_3 \neq b_3$.

Exercises 6–12 have more than one correct answer.

6. $a_n = (-1)^n$, where n is an integer and $n \geq 1$.

7. $a_n = (n - 1)(-1)^n$, where n is an integer and $n \geq 1$.

8. $a_n = \dfrac{n}{(n + 1)^2}$, where n is an integer and $n \geq 1$

10. $a_n = \dfrac{n^2}{3^n}$, where n is an integer and $n \geq 1$

13. a. $2 + 3 + (-2) + 1 + 0 + (-1) + (-2) = 1$
  b. $a_0 = 2$
  c. $a_2 + a_4 + a_6 = -2 + 0 + (-2) = -4$
  d. $2 \cdot 3 \cdot (-2) \cdot 1 \cdot 0 \cdot (-1) \cdot (-2) = 0$

14. $2 + 3 + 4 + 5 + 6 = 20$ **20.** $2^2 \cdot 3^2 \cdot 4^2 = 576$

18. $1(1 + 1) = 2$

22. $\left(\dfrac{1}{1} - \dfrac{1}{2}\right) + \left(\dfrac{1}{2} - \dfrac{1}{3}\right) + \left(\dfrac{1}{3} - \dfrac{1}{4}\right) + \left(\dfrac{1}{4} - \dfrac{1}{5}\right)$

$+ \left(\dfrac{1}{5} - \dfrac{1}{6}\right) + \left(\dfrac{1}{6} - \dfrac{1}{7}\right) + \left(\dfrac{1}{7} - \dfrac{1}{8}\right) + \left(\dfrac{1}{8} - \dfrac{1}{9}\right)$

$+ \left(\dfrac{1}{9} - \dfrac{1}{10}\right) + \left(\dfrac{1}{10} - \dfrac{1}{11}\right) = 1 - \dfrac{1}{11} = \dfrac{10}{11}$

24. $(-2)^1 + (-2)^2 + (-2)^3 + \cdots + (-2)^n$
$$= -2 + 2^2 - 2^3 + \cdots + (-1)^n 2^n$$

26. $\displaystyle\sum_{k=0}^{n+1} \dfrac{1}{k!} = \dfrac{1}{0!} + \dfrac{1}{1!} + \dfrac{1}{2!} + \cdots + \dfrac{1}{(n+1)!}$

28. $\dfrac{1}{1^2} = 1$

30. $\left(\dfrac{1}{1+1}\right)\left(\dfrac{2}{2+1}\right)\left(\dfrac{3}{3+1}\right) = \left(\dfrac{1}{2}\right)\left(\dfrac{2}{3}\right)\left(\dfrac{3}{4}\right) = \dfrac{1}{4}$

32. $\displaystyle\sum_{i=1}^{k+1} i(i!) = \sum_{i=1}^{k} i(i!) + (k + 1)(k + 1)!$

35. $\displaystyle\sum_{i=1}^{k} i^3 + (k + 1)^3 = \sum_{i=1}^{k+1} i^3$

Exercises 38–47 have more than one correct answer.

38. $\displaystyle\sum_{k=1}^{7} (-1)^{k+1} k^2 \quad \text{or} \quad \sum_{k=0}^{6} (-1)^k (k + 1)^2$

41. $\displaystyle\sum_{j=2}^{6} \dfrac{(-1)^j j}{(j + 1)(j + 2)} \quad \text{or} \quad \sum_{k=3}^{7} \dfrac{(-1)^{k+1}(k - 1)}{k(k + 1)}$

42. $\displaystyle\sum_{i=0}^{5} (-1)^i r^i$ **44.** $\displaystyle\sum_{k=1}^{n} k^3$

46. $\displaystyle\sum_{i=0}^{n-1} (n - i)$

48. When $k = 0$, then $i = 1$. When $k = 5$, then $i = 6$. Since $i = k + 1$, then $k = i - 1$. Thus,

$$k(k - 1) = (i - 1)((i - 1) - 1) = (i - 1)(i - 2),$$

and so

$$\sum_{k=0}^{5} k(k - 1) = \sum_{i=1}^{6} (i - 1)(i - 2)$$

50. When $i = 1$, then $j = 0$. When $i = n + 1$, then $j = n$. Since $j = i - 1$, then $i = j + 1$. Thus,

$$\frac{(i - 1)^2}{i \cdot n} = \frac{((j + 1) - 1)^2}{(j + 1) \cdot n} = \frac{j^2}{jn + n}.$$

(Note that n is constant as far as the sum is concerned.)

So $\sum_{i=1}^{n+1} \frac{(i - 1)^2}{i \cdot n} = \sum_{j=0}^{n} \frac{j^2}{jn + n}.$

51. When $i = 3$, then $j = 2$. When $i = n$ then $j = n - 1$. Since $j = i - 1$, then $i = j + 1$. Thus,

$$\sum_{i=3}^{n} \frac{i}{i + n - 1} = \sum_{j=2}^{n-1} \frac{j + 1}{(j + 1) + n - 1}$$

$$= \sum_{j=2}^{n-1} \frac{j + 1}{j + n}.$$

54. $\sum_{k=1}^{n} [3(2k - 3) + (4 - 5k)]$

$$= \sum_{k=1}^{n} [(6k - 9) + (4 - 5k)] = \sum_{k=1}^{n} (k - 5)$$

57. $\dfrac{4 \cdot 3 \cdot \cancel{2 \cdot 1}}{3 \cdot \cancel{2 \cdot 1}} = 4$

60. $\dfrac{n\cancel{(n - 1)(n - 2) \cdots 3 \cdot 2 \cdot 1}}{\cancel{(n - 1)(n - 2) \cdots 3 \cdot 2 \cdot 1}} = n$

61. $\dfrac{\cancel{(n - 1)(n - 2) \cdots 3 \cdot 2 \cdot 1}}{(n + 1)n\cancel{(n - 1)(n - 2) \cdots 3 \cdot 2 \cdot 1}} = \dfrac{1}{n(n + 1)}$

63. $\dfrac{[(n + 1)n(n - 1)(n - 2) \cdots 3 \cdot 2 \cdot 1]^2}{[n(n - 1)(n - 2) \cdots 3 \cdot 2 \cdot 1]^2} = (n + 1)^2$

64.
$$\dfrac{n(n - 1)(n - 2) \cdots (n - k + 1)\cancel{(n - k)(n - k - 1) \cdots 2 \cdot 1}}{\cancel{(n - k)(n - k - 1) \cdots 2 \cdot 1}}$$

$$= n(n - 1)(n - 2) \cdots (n - k + 1)$$

66. a. Proof: Let n be an integer such that $n \geq 2$. By definition of factorial,

$$n! = \begin{cases} 2 \cdot 1 & \text{if } n = 2 \\ 3 \cdot 2 \cdot 1 & \text{if } n = 3 \\ n \cdot (n - 1) \cdots 2 \cdot 1 & \text{if } n > 3. \end{cases}$$

In each case, $n!$ has a factor of 2, and so $n! = 2k$ for some integer k. Then

$$n! + 2 = 2k + 2 \qquad \text{by substitution}$$

$$= 2(k + 1) \qquad \text{by factoring out the 2.}$$

Since $k + 1$ is an integer, $n! + 2$ is divisible by 2 [as was to be shown].

c. *Hint:* Consider the sequence $m! + 2, m! + 3, m! + 4,$ $\ldots, m! + m.$

67. $\dbinom{5}{3} = \dfrac{5!}{(3!)(5-3)!} = \dfrac{5 \cdot 4 \cdot \cancel{3 \cdot 2 \cdot 1}}{(\cancel{3 \cdot 2 \cdot 1})(2 \cdot 1)} = 10$

69. $\dbinom{3}{0} = \dfrac{3!}{(0!)(3-0)!} = \dfrac{\cancel{3!}}{(1)(\cancel{3!})} = 1$

71. $\dbinom{n}{n - 1} = \dfrac{n!}{(n-1)!(n-(n-1))!} = \dfrac{n(\cancel{n-1})!}{(\cancel{n-1})!(n-n+1)!} = \dfrac{n}{1} = n$

73. Proof: Suppose n and r are nonnegative integers with $r + 1 \leq n$. The right-hand side of the equation to be shown is

$$\frac{n - r}{r + 1} \cdot \binom{n}{r} = \frac{n - r}{r + 1} \cdot \frac{n!}{r!(n - r)!}$$

$$= \frac{n - r}{r + 1} \cdot \frac{n!}{r!(n - r) \cdot (n - r - 1)!}$$

$$= \frac{n!}{(r + 1)! \cdot (n - r - 1)!}$$

$$= \frac{n!}{(r + 1)! \cdot (n - (r + 1))!}$$

$$= \binom{n}{r + 1},$$

which is the left-hand side of the equation to be shown.

Section 5.2

1. Proof: Let $P(n)$ be the property "n cents can be obtained by using 3-cent and 8-cent coins."

Show that P(14) is true:

Fourteen cents can be obtained by using two 3-cent coins and one 8-cent coin.

Show that for all integers $k \geq 14$, if P(k) is true, then P(k + 1) is true:

Suppose k cents (where $k \geq 14$) can be obtained using 3-cent and 8-cent coins. *[Inductive hypothesis]* We must show that $k + 1$ cents can be obtained using 3-cent and 8-cent coins. If the k cents includes an 8-cent coin, replace it by three 3-cent coins to obtain a total of $k + 1$ cents. Otherwise the k cents consists of 3-cent coins exclusively, and so there must be least five 3-cent coins (since the total amount is at least 14 cents). In this case, replace five of the 3-cent coins by two 8-cent coins to obtain a total of $k + 1$ cents. Thus, in either case, $k + 1$ cents can be obtained using 3-cent and 8-cent coins. *[This is what we needed to show.]*

[Since we have proved the basis step and the inductive step, we conclude that the given statement is true for all integers $n \geq 14$.]

3. a. $P(1)$ is "$1^2 = \dfrac{1 \cdot (1+1) \cdot (2 \cdot 1 + 1)}{6}$." $P(1)$ is true because

$$1^2 = 1 \text{ and } \frac{1 \cdot (1+1) \cdot (2 \cdot 1 + 1)}{6} = \frac{2 \cdot 3}{6} = 1 \text{ also.}$$

b. $P(k)$ is "$1^2 + 2^2 + \cdots + k^2 = \frac{k(k+1)(2k+1)}{6}$."

c. $P(k+1)$ is "$1^2 + 2^2 + \cdots + (k+1)^2$
$$= \frac{(k+1)((k+1)+1)(2 \cdot (k+1)+1)}{6}."$$

d. *Must show:* If for some integer $k \geq 1$,
$$1^2 + 2^2 + \cdots + k^2 = \frac{k(k+1)(2k+1)}{6}, \text{ then}$$
$$1^2 + 2^2 + \cdots + (k+1)^2$$
$$= \frac{(k+1)[(k+1)+1][(2(k+1)+1)]}{6}.$$

5. a. 1^2 **b.** k^2

c. $1 + 3 + 5 + \cdots + [2(k+1) - 1]$

d. $(k+1)^2$

e. the odd integer just before $2k+1$ is $2k - 1$

f. inductive hypothesis

6. Proof: For the given statement, the property $P(n)$ is the equation
$$2 + 4 + 6 + \cdots + 2n = n^2 + n. \quad \leftarrow P(n)$$

Show that P(1) is true:
To prove $P(1)$, we must show that when 1 is substituted into the equation in place of n, the left-hand side equals the right-hand side. But when 1 is substituted for n, the left-hand side is the sum of all the even integers from 2 to $2 \cdot 1$, which is just 2, and the right-hand side is $1^2 + 1$, which also equals 2. Thus $P(1)$ is true.

Show that for all integers $k \geq 1$, if P(k) is true then P(k+1) is true:
Let k be any integer with $k \geq 1$, and suppose $P(k)$ is true. That is, suppose
$$2 + 4 + 6 + \cdots + 2k = k^2 + k. \quad \leftarrow P(k)$$
$$\text{inductive hypothesis}$$

We must show that $P(k+1)$ is true. That is, we must show that
$$2 + 4 + 6 + \cdots + 2(k+1) = (k+1)^2 + (k+1).$$

Because $(k+1)^2 + (k+1) = k^2 + 2k + 1 + k + 1 = k^2 + 3k + 2$, this is equivalent to showing that
$$2 + 4 + 6 + \cdots + 2(k+1) = k^2 + 3k + 2. \leftarrow P(k+1)$$

But the left-hand side of $P(k+1)$ is

$2 + 4 + 6 + \cdots + 2(k+1)$
$$= 2 + 4 + 6 + \cdots + 2k + 2(k+1)$$
$$\text{by making the next-to-last term explicit}$$
$$= (k^2 + k) + 2(k+1) \quad \text{by substitution from the inductive hypothesis}$$
$$= k^2 + 3k + 2, \quad \text{by algebra,}$$

and this is the right-hand side of $P(k+1)$. Hence $P(k+1)$ is true.

[Since both the basis step and the inductive step have been proved, $P(n)$ is true for all integers $n \geq 1$.]

8. Proof: For the given statement, the property $P(n)$ is the equation
$$1 + 2 + 2^2 + \cdots + 2^n = 2^{n+1} - 1. \quad \leftarrow P(n)$$

Show that P(0) is true:
The left-hand side of $P(0)$ is 1, and the right-hand side is $2^{0+1} - 1 = 2 - 1 = 1$ also. Thus $P(0)$ is true.

Show that for all integers $k \geq 0$, if P(k) is true then P(k+1) is true:
Let k be any integer with $k \geq 0$, and suppose $P(k)$ is true. That is, suppose
$$1 + 2 + 2^2 + \cdots + 2^k = 2^{k+1} - 1. \leftarrow P(k) \text{ inductive hypothesis}$$

We must show that $P(k+1)$ is true. That is, we must show that
$$1 + 2 + 2^2 + \cdots + 2^{k+1} = 2^{(k+1)+1} - 1,$$
or, equivalently,
$$1 + 2 + 2^2 + \cdots + 2^{k+1} = 2^{k+2} - 1. \leftarrow P(k+1)$$

But the left-hand side of $P(k+1)$ is

$1 + 2 + 2^2 + \cdots + 2^{k+1}$
$$= 1 + 2 + 2^2 + \cdots + 2^k + 2^{k+1}$$
$$\text{by making the next-to-last term explicit}$$
$$= (2^{k+1} - 1) + 2^{k+1} \quad \text{by substitution from the inductive hypothesis}$$
$$= 2 \cdot 2^{k+1} - 1 \quad \text{by combining like terms}$$
$$= 2^{k+2} - 1, \quad \text{by the laws of exponents,}$$

and this is the right-hand side of $P(k+1)$. Hence the property is true for $n = k+1$.

[Since both the basis step and the inductive step have been proved, $P(n)$ is true for all integers $n \geq 0$.]

10. Proof: For the given statement, the property is the equation
$$1^2 + 2^2 + 3^2 + \cdots + n^2$$
$$= \frac{n(n+1)(2n+1)}{6}. \quad \leftarrow P(n)$$

Show that P(1) is true:
The left-hand side of $P(1)$ is $1^2 = 1$, and the right-hand side is $\frac{1(1+1)(2 \cdot 1+1)}{6} = \frac{2 \cdot 3}{6} = 1$ also. Thus $P(1)$ is true.

Show that for all integers $k \geq 1$, if P(k) is true then P(k+1) is true:
Let k be any integer with $k \geq 1$, and suppose $P(k)$ is true. That is, suppose
$$1^2 + 2^2 + 3^2 + \cdots + k^2$$
$$= \frac{k(k+1)(2k+1)}{6}. \quad \leftarrow P(k) \text{ inductive hypothesis}$$

We must show that $P(k+1)$ is true. That is, we must show that

$$1^2 + 2^2 + 3^2 + \cdots + (k+1)^2$$
$$= \frac{(k+1)((k+1)+1)(2(k+1)+1)}{6},$$

or, equivalently,

$$1^2 + 2^2 + 3^2 + \cdots + (k+1)^2$$
$$= \frac{(k+1)(k+2)(2k+3)}{6}. \qquad \leftarrow P(k+1)$$

But the left-hand side of $P(k+1)$ is

$$1^2 + 2^2 + 3^2 + \cdots + (k+1)^2$$

$= 1^2 + 2^2 + 3^2 + \cdots + k^2 + (k+1)^2$ by making the next-to-last term explicit

by substitution from the inductive hypothesis

$= \dfrac{k(k+1)(2k+1)}{6} + (k+1)^2$

$= \dfrac{k(k+1)(2k+1)}{6} + \dfrac{6(k+1)^2}{6}$ because $\frac{6}{6} = 1$

$= \dfrac{k(k+1)(2k+1) + 6(k+1)^2}{6}$ by adding fractions

$= \dfrac{(k+1)[k(2k+1) + 6(k+1)]}{6}$ by factoring out $(k+1)$

$= \dfrac{(k+1)(2k^2 + 7k + 6)}{6}$ by multiplying out and combining like terms

$= \dfrac{(k+1)(k+2)(2k+3)}{6}$ because $(k+2)$ $(2k+3) = 2k^2 + 7k + 6$,

and this is the right-hand side of $P(k+1)$. Hence the property is true for $n = k + 1$.

[Since both the basis step and the inductive step have been proved, $P(n)$ is true for all integers $n \geq 1$.]

13. <u>Proof</u>: For the given statement, the property $P(n)$ is the equation

$$\sum_{i=1}^{n-1} i(i+1) = \frac{n(n-1)(n+1)}{3}. \qquad \leftarrow P(n)$$

Show that $P(2)$ is true:
The left-hand side of $P(2)$ is $\sum_{i=1}^{1} i(i+1) = 1 \cdot (1+1) = 2$, and the right-hand side is $\frac{2(2-1)(2+1)}{3} = \frac{6}{3} = 2$ also. Thus $P(2)$ is true.

Show that for all integers $k \geq 2$, if $P(k)$ is true then $P(k+1)$ is true:
Let k be any integer with $k \geq 2$, and suppose $P(k)$ is true. That is, suppose

$$\sum_{i=1}^{k-1} i(i+1) = \frac{k(k-1)(k+1)}{3} \qquad \begin{array}{l} \leftarrow P(k) \\ \text{inductive hypothesis} \end{array}$$

We must show that $P(k+1)$ is true. That is, we must show that

$$\sum_{i=1}^{(k+1)-1} i(i+1) = \frac{(k+1)((k+1)-1)((k+1)+1)}{3},$$

or, equivalently,

$$\sum_{i=1}^{k} i(i+1) = \frac{(k+1)k(k+2)}{3}. \qquad \leftarrow P(k+1)$$

But the left-hand side of $P(k+1)$ is

$$\sum_{i=1}^{k} i(i+1)$$

$= \displaystyle\sum_{i=1}^{k-1} i(i+1) + k(k+1)$ by writing the last term separately

$= \dfrac{k(k-1)(k+1)}{3} + k(k+1)$ by substitution from the inductive hypothesis

$= \dfrac{k(k-1)(k+1)}{3} + \dfrac{3k(k+1)}{3}$ because $\frac{3}{3} = 1$

$= \dfrac{k(k-1)(k+1) + 3k(k+1)}{3}$ by adding the fractions

$= \dfrac{k(k+1)[(k-1) + 3]}{3}$ by factoring out $k(k+1)$

$= \dfrac{k(k+1)(k+2)}{3}$,   by algebra,

and this is the right-hand side of $P(k+1)$. Hence $P(k+1)$ is true.

[Since both the basis step and the inductive step have been proved, $P(n)$ is true for all integers $n \geq 2$.]

15. *Hint:* To prove the basis step, show that $\sum_{i=1}^{1} i(i!) = (1+1)! - 1$. To prove the inductive step, suppose that $\sum_{i=1}^{k} i(i!) = (k+1)! - 1$ for some integer $k \geq 1$ and show that $\sum_{i=1}^{k+1} i(i!) = (k+2)! - 1$. Note that $[(k+1)! - 1] + (k+1)[(k+1)!] = (k+1)![1 + (k+1)] - 1$.

18. *Hints:* $\sin^2 x + \cos^2 x = 1$, $\cos(2x) = \cos^2 x - \sin^2 x = 1 - 2\sin^2 x$, $\sin(a+b) = \sin a \cos b + \cos a \sin b$, $\sin(2x) = 2\sin x \cos x$, $\cos(a+b) = \cos a \cos b - \sin a \sin b$.

20. $4 + 8 + 12 + 16 + \cdots + 200 = 4(1 + 2 + 3 + \cdots + 50)$
$= 4\left(\dfrac{50 \cdot 51}{2}\right) = 5100$

22. $3 + 4 + 5 + 6 + \cdots + 1000 = (1 + 2 + 3 + 4 + \cdots + 1000) - (1 + 2) = \left(\dfrac{1000 \cdot 1001}{2}\right) - 3 = 500{,}497$

24. $\dfrac{(k-1)((k-1)+1)}{2} = \dfrac{k(k-1)}{2}$

25. **a.** $\dfrac{2^{26} - 1}{2 - 1} = 2^{26} - 1 = 67{,}108{,}863$

b. $2 + 2^2 + 2^3 + \cdots + 2^{26}$

$= 2(1 + 2 + 2^2 + \cdots + 2^{25})$

$= 2 \cdot (67{,}108{,}863) \quad$ by part (a)

$= 134{,}217{,}726$

28. $\dfrac{\left(\frac{1}{2}\right)^{n+1} - 1}{\frac{1}{2} - 1} = \dfrac{\frac{1}{2^{n+1}} - 1}{-\frac{1}{2}} = \left(\dfrac{1}{2^{n+1}} - 1\right)(-2)$

$= -\dfrac{2}{2^{n+1}} + 2 = 2 - \dfrac{1}{2^n}$

30. *Hint:* $c + (c+d) + (c+2d) + \cdots + (c+nd)$

$= (n+1)\, c + d \cdot \dfrac{n(n+1)}{2}.$

33. In the inductive step, both the inductive hypothesis and what is to be shown are wrong. The inductive hypothesis should be

Suppose that for some integer $k \geq 1$,

$$1^2 + 2^2 + \cdots + k^2 = \frac{k(k+1)(2k+1)}{6}.$$

And what is to be shown should be

$1^2 + 2^2 + \cdots + (k+1)^2$

$$= \frac{(k+1)((k+1)+1)(2(k+1)+1)}{6}.$$

34. *Hint:* See the Caution note for Example 5.1.8.

35. *Hint:* See the subsection Proving an Equality on page 254.

37. *Hint:* Form the sum $n^2 + (n+1)^2 + (n+2)^2 + \cdots + (n+(p-1))^2$, and show that it equals

$pn^2 + 2n(1 + 2 + 3 + \cdots + (p-1))$

$+ (1 + 4 + 9 + 16 + \cdots + (p-1)^2).$

Section 5.3

1. *General formula:* $\prod_{i=2}^{n}\left(1 - \frac{1}{i}\right) = \frac{1}{n}$ for all integers $n \geq 2$.
Proof (by mathematical induction): Let the property $P(n)$ be the equation

$$\prod_{i=2}^{n}\left(1 - \frac{1}{i}\right) = \frac{1}{n}. \qquad \leftarrow P(n)$$

Show that $P(2)$ is true:
The left-hand side of $P(2)$ is $\prod_{i=2}^{2}\left(1 - \frac{1}{i}\right) = 1 - \frac{1}{2} = \frac{1}{2}$, which equals the right-hand side.

Show that for all integers $k \geq 2$, if $P(k)$ is true then $P(k+1)$ is also true:

Suppose that k is any integer with $k \geq 2$ such that

$$\prod_{i=2}^{k}\left(1 - \frac{1}{i}\right) = \frac{1}{k}. \qquad \begin{array}{l} \leftarrow P(k) \\ \text{Inductive hypothesis} \end{array}$$

We must show that

$$\prod_{i=2}^{k+1}\left(1 - \frac{1}{i}\right) = \frac{1}{k+1}. \qquad \leftarrow P(k+1)$$

But by the laws of algebra and substitution from the inductive hypothesis, the left-hand side of $P(k+1)$ is

$$\prod_{i=2}^{k+1}\left(1 - \frac{1}{i}\right)$$

$$= \prod_{i=2}^{k}\left(1 - \frac{1}{i}\right)\left(1 - \frac{1}{k+1}\right)$$

$$= \left(\frac{1}{k}\right)\left(1 - \frac{1}{k+1}\right) = \left(\frac{1}{k}\right)\left(\frac{(k+1) - 1}{k+1}\right)$$

$$= \frac{1}{k+1} \text{ which is the right-hand side of } P(k+1)$$

[as was to be shown].

3. *General formula:* $\dfrac{1}{1\cdot3} + \dfrac{1}{3\cdot5} + \cdots + \dfrac{1}{(2n-1)(2n+1)} = \dfrac{n}{2n+1}$ for all integers $n \geq 1$.
Proof (by mathematical induction): Let the property $P(n)$ be the equation

$$\frac{1}{1\cdot3} + \frac{1}{3\cdot5} + \cdots + \frac{1}{(2n-1)(2n+1)} = \frac{n}{2n+1}.$$

Show that $P(1)$ is true:
The left-hand side of $P(1)$ equals $\frac{1}{1\cdot3}$, and the right-hand side equals $\frac{1}{2\cdot1+1}$. But both of these equal $\frac{1}{3}$, so $P(1)$ is true.

Show that for any integer $k \geq 1$, if $P(k)$ is true then $P(k+1)$ is true:

Suppose that k is any integer with $k \geq 1$, and

$$\frac{1}{1\cdot3} + \frac{1}{3\cdot5} + \cdots + \frac{1}{(2k-1)(2k+1)} = \frac{k}{2k+1}$$

$$\uparrow P(k) \text{ inductive hypothesis}$$

We must show that

$$\frac{1}{1\cdot3} + \frac{1}{3\cdot5} + \cdots + \frac{1}{(2(k+1)-1)(2(k+1)+1)}$$

$$= \frac{k+1}{2(k+1)+1}.$$

or, equivalently,

$$\frac{1}{1\cdot3} + \frac{1}{3\cdot5} + \cdots + \frac{1}{(2k+1)(2k+3)} = \frac{k+1}{2k+3}.$$

$$\uparrow P(k+1)$$

But the left-hand side of $P(k + 1)$ is

$$\frac{1}{1 \cdot 3} + \frac{1}{3 \cdot 5} + \cdots + \frac{1}{(2k + 1)(2k + 3)}$$

$$= \frac{1}{1 \cdot 3} + \frac{1}{3 \cdot 5} + \cdots + \frac{1}{(2k - 1)(2k + 1)}$$

$$+ \frac{1}{(2k + 1)(2k + 3)}$$

$$= \frac{k}{2k + 1} + \frac{1}{(2k + 1)(2k + 3)} \quad \text{by inductive hypothesis}$$

$$= \frac{k(2k + 3)}{(2k + 1)(2k + 3)} + \frac{1}{(2k + 1)(2k + 3)}$$

$$= \frac{2k^2 + 3k + 1}{(2k + 1)(2k + 3)}$$

$$= \frac{(2k + 1)(k + 1)}{(2k + 1)(2k + 3)}$$

$$= \frac{k + 1}{2k + 3} \quad \text{by algebra,}$$

and this is the right-hand side of $P(k + 1)$ *[as was to be shown]*.

4. *Hint 1:* The general formula is

$$1 - 4 + 9 - 16 + \cdots + (-1)^{n-1} n^2$$

$$= (-1)^{n-1} (1 + 2 + 3 + \cdots + n) \quad \text{in expanded form}$$

$$Or: \sum_{i=1}^{n} (-1)^{i-1} i^2 = (-1)^{n-1} \left(\sum_{i=1}^{n} i \right) \quad \begin{array}{l} \text{in summation} \\ \text{notation.} \end{array}$$

Hint 2: In the proof, use the fact that

$$1 + 2 + 3 + \cdots + n = \sum_{i=1}^{n} i = \frac{n(n + 1)}{2}.$$

6. a. $P(0)$ is "$5^0 - 1$ is divisible by 4." $P(0)$ is true because $5^0 - 1 = 0$, which is divisible by 4.
  b. $P(k)$ is "$5^k - 1$ is divisible by 4."
  c. $P(k + 1)$ is "$5^{k+1} - 1$ is divisible by 4."
  d. *Must show:* If for some integer $k \geq 0$, $5^k - 1$ is divisible by 4, then $5^{k+1} - 1$ is divisible by 4.

8. Proof (by mathematical induction): For the given statement, the property is the sentence "$5^n - 1$ is divisible by 4."

Show that $P(0)$ is true:

$P(0)$ is the sentence "$5^0 - 1$ is divisible by 4." But $5^0 - 1 = 1 - 1 = 0$, and 0 is divisible by 4 because $0 = 4 \cdot 0$. Thus $P(0)$ is true.

Show that for all integers $k \geq 0$, if $P(k)$ is true then $P(k + 1)$ is true:

Let k be any integer with $k \geq 0$, and suppose $P(k)$ is true. That is, suppose $5^k - 1$ is divisible by 4. *[This is the inductive hypothesis.]* We must show that $P(k + 1)$ is true. That is, we must show that $5^{k+1} - 1$ is divisible by 4. Now

$$5^{k+1} - 1 = 5^k \cdot 5 - 1$$

$$= 5^k \cdot (4 + 1) - 1 = 5^k \cdot 4 + (5^k - 1). \quad (*)$$

By the inductive hypothesis $5^k - 1$ is divisible by 4, and so $5^k - 1 = 4r$ for some integer r. By substitution into equation $(*)$,

$$5^{k+1} - 1 = 5^k \cdot 4 + 4r = 4(5^k + r).$$

But $5^k + r$ is an integer because k and r are integers. Hence, by definition of divisibility, $5^{k+1} - 1$ is divisible by 4 *[as was to be shown]*.

An alternative proof of the inductive step goes as follows: Suppose that for some integer $k \geq 0$, $5^k - 1$ is divisible by 4. Then $5^k - 1 = 4r$ for some integer r, and hence $5^k = 4r + 1$.
It follows that $5^{k+1} = 5^k \cdot 5 = (4r + 1) \cdot 5 = 20r + 5$. Subtracting 1 from both sides gives that $5^{k+1} - 1 = 20r + 4 = 4(5r + 1)$. But $5r + 1$ is an integer, and so, by definition of divisibility, $5^{k+1} - 1$ is divisible by 4.

11. Proof (by mathematical induction): For the given statement, the property $P(n)$ is the sentence "$3^{2n} - 1$ is divisible by 8."

Show that $P(0)$ is true:

$P(0)$ is the sentence "$3^{2 \cdot 0} - 1$ is divisible by 8." But $3^{2 \cdot 0} - 1 = 1 - 1 = 0$, and 0 is divisible by 8 because $0 = 8 \cdot 0$. Thus $P(0)$ is true.

Show that for all integers $k \geq 0$, if $P(k)$ is true then $P(k + 1)$ is true:

Let k be any integer with $k \geq 0$, and suppose $P(k)$ is true. That is, suppose $3^{2k} - 1$ is divisible by 8. *[This is the inductive hypothesis.]* We must show that $P(k + 1)$ is true. That is, we must show that $3^{2(k+1)} - 1$ is divisible by 8, or equivalently, $3^{2k+2} - 1$ is divisible by 8. Now

$$3^{2k+2} - 1 = 3^{2k} \cdot 3^2 - 1 = 3^{2k} \cdot 9 - 1$$

$$= 3^{2k} \cdot (8 + 1) - 1 = 3^{2k} \cdot 8 + (3^{2k} - 1) \cdot \quad (*)$$

By the inductive hypothesis $3^{2k} - 1$ is divisible by 8, and so $3^{2k} - 1 = 8r$ for some integer r. By substitution into equation $(*)$,

$$3^{2k+2} - 1 = 3^{2k} \cdot 8 + 8r = 8(3^{2k} + r).$$

But $3^{2k} + r$ is an integer because k and r are integers. Hence, by definition of divisibility, $3^{2k+2} - 1$ is divisible by 8 *[as was to be shown]*.

13. *Hint:* $x^{k+1} - y^{k+1} = x^{k+1} - x \cdot y^k + x \cdot y^k - y^{k+1}$
$$= x \cdot (x^k - y^k) + y^k \cdot (x - y)$$

14. *Hint 1:* $(k + 1)^3 - (k + 1) = k^3 + 3k^2 + 3k + 1 - k - 1$
$$= (k^3 - k) + 3k^2 + 3k$$
$$= (k^3 - k) + 3k(k + 1)$$

Hint 2: $k(k + 1)$ is a product of two consecutive integers. By Theorem 4.4.3, one of these must be even.

16. Proof (by mathematical induction): For the given statement, let the property $P(n)$ be the inequality $2^n < (n + 1)!$.

Show that $P(2)$ is true:

$P(2)$ says that $2^2 < (2 + 1)!$. The left-hand side is $2^2 = 4$ and the right-hand side is $3! = 6$. So, because $4 < 6$, $P(2)$ is true.

Show that for all integers $k \geq 2$, if $P(k)$ is true then $P(k + 1)$ is true:

Let k be any integer with $k \geq 2$, and suppose $P(k)$ is true. That is, suppose $2^k < (k + 1)!$. *[This is the inductive hypothesis.]* We must show that $P(k + 1)$ is true. That is, we must show that $2^{k+1} < ((k + 1) + 1)$, or, equivalently, $2^{k+1} < (k + 2)!$. By the laws of exponents and the inductive hypothesis,

$$2^{k+1} = 2 \cdot 2^k < 2(k + 1)!. \qquad (*)$$

Since $k \geq 2$, then $2 < k + 2$, and so

$$2(k + 1)! < (k + 2)(k + 1)! = (k + 2)!. \qquad (**)$$

Combining inequalities (*) and (**) gives

$$2^{k+1} < (k + 2)!$$

[as was to be shown].

19. Proof (by mathematical induction): For the given statement, let the property $P(n)$ be the inequality $n^2 < 2^n$.

Show that $P(5)$ is true:

$P(5)$ says that $5^2 < 2^5$. But $5^2 = 25$ and $2^5 = 32$, and $25 < 32$. Hence $P(5)$ is true.

Show that for any integer $k \geq 5$, if $P(k)$ is true then $P(k + 1)$ is true:

Let k be any integer with $k \geq 5$, and suppose $P(k)$ is true. That is, suppose $k^2 < 2^k$. *[This is the inductive hypothesis.]* We must show that $P(k + 1)$ is true. That is, we must show that $(k + 1)^2 < 2^{k+1}$. But

$$(k + 1)^2 = k^2 + 2k + 1 < 2^k + 2k + 1$$

by inductive hypothesis
Also, by Proposition 5.3.2,

$$2k + 1 < 2^k \qquad \text{Prop. 5.3.2 applies since } k \geq 5 \geq 3.$$

Putting these inequalities together gives

$$(k + 1)^2 < 2^k + 2k + 1 < 2^k + 2^k = 2^{k+1}$$

[as was to be shown].

24. Proof (by mathematical induction): For the given statement, let the property $P(n)$ be the equation $a_n = 3 \cdot 7^{n-1}$.

Show that $P(1)$ is true:

The left-hand side of $P(1)$ is a_1, which equals 3 by definition of the sequence. The right-hand side is $3 \cdot 7^{1-1} = 3$ also. Thus $P(1)$ is true.

Show that for all integers $k \geq 1$, if $P(k)$ is true then $P(k + 1)$ is true:

Let k be any integer with $k \geq 1$, and suppose $P(k)$ is true. That is, suppose $a_k = 3 \cdot 7^{k-1}$. *[This is the inductive*

hypothesis.] We must show that $P(k + 1)$ is true. That is, we must show that $a_{k+1} = 3 \cdot 7^{(k+1)-1}$, or, equivalently, $a_{k+1} = 3 \cdot 7^k$. But the left-hand side of $P(k + 1)$ is

$$
\begin{aligned}
a_{k+1} &= 7a_k && \text{by definition of the sequence} \\
&&& a_1, a_2, a_3, \ldots \\
&= 7(3 \cdot 7^{k-1}) && \text{by inductive hypothesis} \\
&= 3 \cdot 7^k && \text{by the laws of exponents,}
\end{aligned}
$$

and this is the right-hand side of $P(k + 1)$ *[as was to be shown].*

30. The inductive step fails for going from $n = 1$ to $n = 2$, because when $k = 1$,

$$A = \{a_1, a_2\} \quad \text{and} \quad B = \{a_1\},$$

and no set C can be defined to have the properties claimed for the C in the proof. The reason is that $C = \{a_1\} = B$, and so an element of A, namely a_2, is not in either B or C.

Since the inductive step fails for going from $n = 1$ to $n = 2$, the truth of the following statement is never proved: "All the numbers in a set of two numbers are equal to each other." This breaks the sequence of inductive steps, and so none of the statements for $n > 2$ is proved true either.

Here is an explanation for what happens in terms of the domino analogy. The first domino *is* tipped backward (the basis step is proved). Also, if any domino from the second onward tips backward, then it tips the one behind it backward (the inductive step works for $n \geq 2$). However, when the first domino is tipped backward, it does *not* tip the second one backward. So only the first domino falls down; the rest remain standing.

31. *Hint:* Is the basis step true?

32. *Hint:* Consider the problem of trying to cover a 3×3 checkerboard with trominoes. Place a checkmark in certain squares as shown in the following figure.

Observe that no two squares containing checkmarks can be covered by the same tromino. Since there are four checkmarks, four tromiones would be needed to cover these squares. But, since each tromino covers three squares, four trominoes would cover twelve squares, not the nine squares in this checkerboard. It follows that such a covering is impossible.

34. a. *Hint:* For the inductive step, note that a $2 \times 3(k + 1)$ checkerboard can be split into a $2 \times 3k$ checkerboard and a 2×3 checkerboard.

35. b. *Hint:* Consider a 3×5 checkerboard, and refer to the hint for exercise 32. Figure out a way to place six checkmarks in squares so that no two of the squares that contain checkmarks can be covered by the same tromino.

37. *Hint:* Use proof by contradiction. If the statement is false, then there exists some ordering of the integers from 1 to 30, say $x_1, x_2, \ldots, x_{30}$, such that $x_1 + x_2 + x_3 < 45$,

$x_2 + x_3 + x_4 < 45, \ldots,$ and $x_{30} + x_1 + x_2 < 45$. Evaluate the sum of all these inequalities using the fact that $\sum_{i=1}^{30} x_i = \sum_{i=1}^{30} i$ and Theorem 5.2.2.

38. *Hint:* Given $k + 1$ a's and $k + 1$ b's arrayed around the outside of the circle, there has to be at least one location where an a is followed by a b as one travels in the clockwise direction. In the inductive step, temporarily remove such an a and the b that follows it, and apply the inductive hypothesis.

40. b. *Hint:* In the inductive step, imagine dividing a $2(k + 1) \times 2(k + 1)$ checkerboard into two sections: a center checkerboard of dimensions $2k \times 2k$ and an outer perimeter of single, adjacent squares. Then examine three cases: case 1 is where both removed squares are in the central $2k \times 2k$ checkerboard, case 2 is where one removed square is in the central $2k \times 2k$ checkerboard and the other is on the perimeter, and case 3 is where both removed squares are on the perimeter.

Section 5.4

1. Proof (by strong mathematical induction): Let the property $P(n)$ be the sentence "a_n is odd."

Show that $P(1)$ and $P(2)$ are true:

Observe that $a_1 = 1$ and $a_2 = 3$ and both 1 and 3 are odd. Thus $P(1)$ and $P(2)$ are true.

Show that for any integer $k \geq 2$, if $P(i)$ is true for all integers i with $1 \leq i \leq k$, then $P(k + 1)$ is true:

Let $k \geq 2$ be any integer, and suppose a_i is odd for all integers i with $1 \leq i \leq k$. *[This is the inductive hypothesis.]* We must show that a_{k+1} is odd. We know that $a_{k+1} = a_{k-1} + 2a_k$ by definition of $a_1, a_2, a_3, \ldots$. Moreover, $k - 1$ is less than $k + 1$ and is greater than or equal to 1 (because $k \geq 2$). Thus, by inductive hypothesis, a_{k-1} is odd. Also, every term of the sequence is an integer (being a sum of products of integers), and so $2a_k$ is even by definition of even. Hence a_{k+1} is the sum of an odd integer and an even integer and hence is odd (by exercise 19, in Section 4.1). *[This is what was to be shown.]*

4. Proof (by strong mathematical induction): Let the property $P(n)$ be the inequality $d_n \leq 1$.

Show that $P(1)$ and $P(2)$ are true:

Observe that $d_1 = \frac{9}{10}$ and $d_2 = \frac{10}{11}$ and both $\frac{9}{10} \leq 1$ and $\frac{10}{11} \leq 1$. Thus $P(1)$ and $P(2)$ are true.

Show that for any integer $k \geq 2$, if $P(i)$ is true for all integers i with $1 \leq i \leq k$, then $P(k + 1)$ is true:

Let $k \geq 2$ be any integer, and suppose $d_i \leq 1$ for all integers i with $1 \leq i \leq k$. *[This is the inductive hypothesis.]* We must show that $d_{k+1} \leq 1$. But, by definition of $d_1, d_2, d_3, \ldots, d_{k+1} = d_k \cdot d_{k-1}$. Now $d_k \leq 1$ and $d_{k-1} \leq 1$ by inductive hypothesis *[since $1 \leq k < k + 1$ and $1 \leq k - 1 < k + 1$ because $k \geq 2$.]*. Consequently, $d_{k+1} = d_k \cdot d_{k-1} \leq 1$ because if two positive numbers are each less than or equal to 1, then their product is less than or equal to 1. *[If $0 < a \leq 1$ and $0 < b \leq 1$, then multiplying $a \leq 1$ by b*

gives $ab \leq b$, and since $b \leq 1$, then by transitivity of order, $ab \leq 1$.] This is what was to be shown. *[Since we have proved both the basis step and the inductive step, we conclude that $d_n \leq 1$ for all integers $n \geq 1$.]*

5. Proof (by strong mathematical induction): Let the property $P(n)$ be the equation $e_n = 5 \cdot 3^n + 7 \cdot 2^n$.

Show that $P(0)$ and $P(1)$ are true.

We must show that $e_0 = 5 \cdot 3^0 + 7 \cdot 2^0$ and $e_1 = 5 \cdot 3^1 + 7 \cdot 2^1$. The left-hand side of the first equation is 12 (by definition of $e_0, e_1, e_2, \ldots$), and its right-hand side is $5 \cdot 1 + 7 \cdot 1 = 12$ also. The left-hand side of the second equation is 29 (by definition of $e_0, e_1, e_2, \ldots$), and its right-hand side is $5 \cdot 3 + 7 \cdot 2 = 29$ also. Thus $P(0)$ and $P(1)$ are true.

Show that for any integer $k \geq 1$, if $P(i)$ is true for all integers i with $0 \leq i \leq k$, then $P(k + 1)$ is true:

Let $k \geq 1$ be an integer, and suppose $e_i = 5 \cdot 3^i + 7 \cdot 2^i$ for all integers i with $0 \leq i \leq k$. *[Inductive hypothesis]* We must show that $e_{k+1} = 5 \cdot 3^{k+1} + 7 \cdot 2^{k+1}$. But

$$
\begin{aligned}
e_{k+1} &= 5e_k - 6e_{k-1} && \text{by definition of } e_0, e_1, e_2, \ldots \\
&= 5(5 \cdot 3^k + 7 \cdot 2^k) - 6(5 \cdot 3^{k-1} + 7 \cdot 2^{k-1}) \\
& && \text{by inductive hypothesis} \\
&= 25 \cdot 3^k + 35 \cdot 2^k - 30 \cdot 3^{k-1} - 42 \cdot 2^{k-1} \\
&= 25 \cdot 3^k + 35 \cdot 2^k - 10 \cdot 3 \cdot 3^{k-1} - 21 \cdot 2 \cdot 2^{k-1} \\
&= 25 \cdot 3^k + 35 \cdot 2^k - 10 \cdot 3^k - 21 \cdot 2^k \\
&= (25 - 10) \cdot 3^k + (35 - 21) \cdot 2^k \\
&= 15 \cdot 3^k + 14 \cdot 2^k \\
&= 5 \cdot 3 \cdot 3^k + 7 \cdot 2 \cdot 2^k \\
&= 5 \cdot 3^{k+1} + 7 \cdot 2^{k+1} && \text{by algebra.}
\end{aligned}
$$

[This is what was to be shown.]

10. *Hint:* In the basis step, show that $P(14)$, $P(15)$, and $P(16)$ are all true. For the inductive step, note that $k + 1 = [(k + 1) - 3] + 3$, and if $k \geq 16$, then $(k + 1) - 3 \geq 14$.

11. Proof (by strong mathematical induction): Let the property $P(n)$ be the sentence

"*A jigsaw puzzle consisting of n pieces takes $n - 1$ steps to put together.*"

Show that $P(1)$ is true:

A jigsaw puzzle consisting of just one piece does not take any steps to put together. Hence it is correct to say that it takes zero steps to put together.

Show that for any integer $k \geq 1$, if $P(i)$ is true for all integers i with $1 \leq i \leq k$ then $P(k + 1)$ is true:

Let $k \geq 1$ be an integer and suppose that for all integers i with $1 \leq i \leq k$, a jigsaw puzzle consisting of i pieces takes $i - 1$ steps to put together. *[This is the inductive hypothesis.]* We must show that a jigsaw puzzle consisting of $k + 1$ pieces takes k steps to put together. Consider assembling a jigsaw puzzle consisting of $k + 1$ pieces. The last step involves fitting together two blocks. Suppose one of the blocks consists of r pieces and the other consists of

s pieces. Then $r + s = k + 1$, and $1 \leq r \leq k$ and $1 \leq s \leq k$. Thus by inductive hypothesis, the numbers of steps required to assemble the blocks are $r - 1$ and $s - 1$, respectively. Then the total number of steps required to assemble the puzzle is $(r - 1) + (s - 1) + 1 = (r + s) - 1 = (k + 1) - 1 = k$ [as was to be shown].

12. *Hint:* For any collection of cans, at least one must contain enough gasoline to enable the car to get to the next can. (Why?) Imagine taking all the gasoline from that can and pouring it into the can that immediately precedes it in the direction of travel around the track.

13. *Sketch of proof:* Given any integer $k > 1$, either k is prime or k is a product of two smaller positive integers, each greater than 1. In the former case, the property is true. In the latter case, the inductive hypothesis ensures that both factors of k are products of primes and hence that k is also a product of primes.

14. Proof (by strong mathematical induction): Let the property $P(n)$ be the sentence "Any product of n odd integers is odd."

Show that $P(2)$ is true:

We must show that any product of two odd integers is odd. But this was established in Chapter 4 (exercise 43 of Section 4.1).

Show that for any integer $k \geq 2$, if $P(i)$ is true for all integers i with $2 \leq i \leq k$ then $P(k + 1)$ is true:

Let k be any integer with $k \geq 2$, and suppose that for all integers i with $2 \leq i \leq k$, any product of i odd integers is odd. *[Inductive hypothesis]* Consider any product M of $k + 1$ odd integers. Some multiplication is the final one that is used to obtain M. Thus there are integers A and B such that $M = AB$, and each of A and B is a product of between 1 and k odd integers. (For instance, if $M = ((a_1 a_2) a_3) a_4$, then $A = (a_1 a_2) a_3$ and $B = a_4$.) By inductive hypothesis, each of A and B is odd, and, as in the basis step, we know that any product of two odd integers is odd. Hence $M = AB$ is odd.

16. *Hint:* Let the property $P(n)$ be the sentence "If n is even, then any sum of n odd integers is even, and if n is odd, then any sum of n odd integers is odd." For the inductive step, consider any sum S of $k + 1$ odd integers. Some addition is the final one that is used to obtain S. Thus there are integers A and B such that $S = A + B$, and A is a sum of r odd integers and B is a sum of $(k + 1) - r$ odd integers. Consider the two cases where $k + 1$ is even and $k + 1$ is odd, and for each case consider the two subcases where r is even and where r is odd.

17. $4^1 = 4$, $4^2 = 16$, $4^3 = 64$, $4^4 = 256$, $4^5 = 1024$, $4^6 = 4096$, $4^7 = 16384$, and $4^8 = 65536$.

Conjecture: The units digit of 4^n equals 4 if n is odd and equals 6 if n is even.

Proof by strong mathematical induction: Let the property $P(n)$ be the sentence "The units digit of 4^n equals 4 if n is odd and equals 6 if n is even."

Show that $P(1)$ and $P(2)$ are true:

When $n = 1$, $4^n = 4^1 = 4$, and the units digit is 4. When $n = 2$, then $4^n = 4^2 = 16$, and the units digits is 6. Thus $P(1)$ and $P(2)$ are true.

Show that for any integer $k \geq 2$, if the property is true for all integers i with $1 \leq i \leq k$ then it is true for $k+1$:

Let k by any integer with $k \geq 2$, and suppose that for all integers i with $0 \leq i \leq k$, the units digit of 4^i equals 4 if i is odd and equals 6 if i is even. *[Inductive hypothesis]* We must show that the units digit of 4^{k+1} equals 4 if $k + 1$ is odd and equals 6 if $k + 1$ is even.

Case 1 ($k + 1$ is odd): In this case, k is even, and so, by inductive hypothesis, the units digits of 4^k is 6. Thus $4^k = 10q + 6$ for some nonnegative integer q. It follows that $4^{k+1} = 4^k \cdot 4 = (10q + 6) \cdot 4 = 40q + 24 = 10(4q + 2) + 4$. Thus the units digit of 4^{k+1} is 4 [as was to be shown].

Case 2 ($k + 1$ is even): In this case, k is odd, and so, by inductive hypothesis, the units digit of 4^k is 4. Thus $4^k = 10q + 4$ for some nonnegative integer q. It follows that $4^{k+1} = 4^k \cdot 4 = (10q + 4) \cdot 4 = 40q + 16 = 10(4q + 1) + 6$. Thus the units digit of 4^{k+1} is 6 [as was to be shown].

20. Proof: Let n be any integer greater than 1. Consider the set S of all positive integers other than 1 that divide n. Since $n \mid n$ and $n > 1$, there is at least one element in S. Hence, by the well-ordering principle for the integers, S has a smallest element; call it p. We claim that p is prime. For suppose p is not prime. Then there are integers a and b with $1 < a < p$, $1 < b < p$, and $p = ab$. By definition of divides, $a \mid p$. Also $p \mid n$ because p is in S and every element in S divides n. Therefore, $a \mid p$ and $p \mid n$, and so, by transitivity of divisibility, $a \mid n$. Consequently, $a \in S$. But this contradicts the fact that $a < p$, and p is the smallest element of S. *[This contradiction shows that the supposition that p is not prime is false.]* Hence p is prime, and we have shown the existence of a prime number that divides n.

22. **a.** Proof: Suppose r is any rational number. *[We need to show that there is an integer n such that $r < n$.]*

Case 1 ($r \leq 0$): In this case, take $n = 1$. Then $r < n$.

Case 2 ($r > 0$): In this case, $r = \frac{a}{b}$ for some positive integers a and b (by definition of rational and because r is positive). Note that $r = \frac{a}{b} < n$ if, and only if, $a < nb$. Let $n = 2a$. Multiply both sides of the inequality $1 < 2$ by a to obtain $a < 2a$, and multiply both sides of the inequality $1 < b$ by $2a$ to obtain $2a < 2ab = nb$. Thus $a < 2a < nb$, and so, by transitivity of order, $a < nb$. Dividing both sides by b gives that $\frac{a}{b} < n$, or, equivalently, that $r < n$.

Hence, in both cases, $r < n$ [as was to be shown].

23. *Hint:* If r is any rational number, let S be the set of all integers n such that $r < n$. Use the results of exercises 22(a), 22(c), and the well-ordering principle for the integers to show that S has a least element, say v, and then show that $v - 1 \leq r < v$.

24. Proof: Let S be the set of all integers r such that $n = 2^i \cdot r$ for some integer i. Then $n \in S$ because $n = 2^0 \cdot n$, and so

$S \neq \emptyset$. Also, since $n \geq 1$, each r in S is positive, and so, by the well-ordering principle, S has a least element m. This means that $n = 2^k \cdot m$ (*) for some nonnegative integer k and $m \leq r$ for every r in S. We claim that m is odd. The reason is that if m were even, then $m = 2p$ for some integer p. Substituting into equation (*) gives

$$n = 2^k \cdot m = 2^k \cdot 2p = (2^k \cdot 2)p = 2^{k+1} \cdot p.$$

It follows that $p \in S$ and $p < m$, which contradicts the fact that m is the *least* element of S. Hence m is odd, and so $n = m \cdot 2^k$ for some odd integer m and nonnegative integer k.

29. a. $1110_2 = 1 \cdot 2^3 + 1 \cdot 2^2 + 1 \cdot 2^1 + 0 \cdot 2^0$
$= 8 + 4 + 2 = 14_{10}$

b. $10111_2 = 1 \cdot 2^4 + 0 \cdot 2^3 + 1 \cdot 2^2 + 1 \cdot 2^1 + 1 \cdot 2^0 =$
$16 + 4 + 2 + 1 = 23_{10}$

30. *Hint:* In the inductive step, divide into cases depending upon whether k can be written as $k = 3x$ or $k = 3x + 1$ or $k = 3x + 2$ for some integer x.

31. *Hint:* In the inductive step, let an integer $k \geq 0$ be given and suppose that there exist integers q' and r' such that $k = dq' + r'$ and $0 \leq r' < d$. You must show that there exist integers q and r such that

$$k + 1 = dq + r \quad \text{and} \quad 0 \leq r < d.$$

To do this, consider the two cases $r' < d - 1$ and $r' = d - 1$.

32. *Hint:* Given a predicate $P(n)$ that satisfies conditions (1) and (2) of the principle of mathematical induction, let S be the set of all integers greater than or equal to a for which $P(n)$ is false. Suppose that S has one or more elements, and use the well-ordering principle for the integers to derive a contradiction.

33. *Hint:* Suppose S is a set containing one or more integers, all of which are greater than or equal to some integer a, and suppose that S does not have a least element. Let the property $P(n)$ be the sentence "$i \notin S$ for any integer i with $a \leq i \leq n$." Use mathematical induction to prove that $P(n)$ is true for all integers $n \geq a$, and explain how this result contradicts the supposition that S does not have a least element.

Section 5.5

1. $a_1 = 1, a_2 = 2a_1 + 2 = 2 \cdot 1 + 2 = 4,$
$a_3 = 2a_2 + 3 = 2 \cdot 4 + 3 = 11,$
$a_4 = 2a_3 + 4 = 2 \cdot 11 + 4 = 26$

3. $c_0 = 1, c_1 = 1 \cdot (c_0)^2 = 1 \cdot (1)^2 = 1,$
$c_2 = 2(c_1)^2 = 2 \cdot (1)^2 = 2,$
$c_3 = 3(c_2)^2 = 3 \cdot (2)^2 = 12$

5. $s_0 = 1, s_1 = 1, s_2 = s_1 + 2s_0 = 1 + 2 \cdot 1 = 3,$
$s_3 = s_2 + 2s_1 = 3 + 2 \cdot 1 = 5$

7. $u_1 = 1, u_2 = 1, u_3 = 3u_2 - u_1 = 3 \cdot 1 - 1 = 2,$
$u_4 = 4u_3 - u_2 = 4 \cdot 2 - 1 = 7$

9. By definition of $a_0, a_1, a_2, \ldots$, for each integer $k \geq 1$,

(*) $\qquad a_k = 3k + 1 \quad$ and
(**) $\qquad a_{k-1} = 3(k-1) + 1.$

Then $a_{k-1} + 3$

$= 3(k-1) + 1 + 3$
$= 3k - 3 + 1 + 3$
$= 3k + 1$
$= a_k$

11. By definition of $c_0, c_1, c_2, \ldots, c_n = 2^n - 1$, for each integer $n \geq 0$. Substitute k and $k - 1$ in place of n to get

(*) $\qquad c_k = 2^k - 1 \quad$ and
(**) $\qquad c_{k-1} = 2^{k-1} - 1$

for all integers $k \geq 1$. Then

$2c_{k-1} + 1 = 2(2^{k-1} - 1) + 1 \quad$ by substitution from (**)
$\qquad = 2^k - 2 + 1$
$\qquad = 2^k - 1 \qquad$ by basic algebra
$\qquad = c_k \qquad$ by substitution from (*)

13. By definition of $t_0, t_1, t_2, \ldots, t_n = 2 + n$, for each integer $n \geq 0$. Substitute $k, k - 1$, and $k - 2$ in place of n to get

(*) $\qquad t_k = 2 + k,$
(**) $\qquad t_{k-1} = 2 + (k - 1), \quad$ and
(***) $\qquad t_{k-2} = 2 + (k - 2)$

for each integer $k \geq 2$. Then

$2t_{k-1} - t_{k-2}$
$\qquad = 2(2 + (k-1)) - (2 + (k-2)) \quad$ by substitution from (**) and (***)
$\qquad = 2(k + 1) - k$
$\qquad = 2 + k \qquad$ by basic algebra
$\qquad = t_k \qquad$ by substitution from (*).

15. *Hint:* Mathematical induction is not needed for the proof. Start with the right-hand side of the equation and use algebra to transform it into the left-hand side of the equation.

17. a. $a_1 = 2$

$a_2 = 2$ (moves to move the top disk from pole A to pole C)

$\qquad + 1$ (move to move the bottom disk from pole A to pole B)

$\qquad + 2$ (moves to move the top disk from pole C to pole A)

$\qquad + 1$ (move to move the bottom disk from pole B to pole C)

$\qquad + 2$ (moves to move top disk from pole A to pole C)

$\qquad = 8$

$a_3 = 8 + 1 + 8 + 1 + 8 = 26$

c. For all integers $k \geq 2$.

$a_k = a_{k-1}$ (moves to move the top $k - 1$ disks from pole A to pole C)

$+ 1$ (move to move the bottom disk from pole A to pole B)

$+ a_{k-1}$ (moves to move the top disk from pole C to pole A)

$+ 1$ (move to move the bottom disks from pole B to pole C)

$+ a_{k-1}$ (moves to move the top disks from pole A to pole C)

$= 3a_{k-1} + 2$.

18. b. $b_4 = 40$

e. *Hint:* One solution is to use mathematical induction and apply the formula from part (c). Another solution is to prove by mathematical induction that when a most efficient transfer of n disks from one end pole to the other end pole is performed, at some point all the disks are on the middle pole.

19. a. $s_1 = 1$, $s_2 = 1 + 1 + 1 = 3$,
$s_3 = s_1 + (1 + 1 + 1) + s_1 = 5$
b. $s_4 = s_2 + (1 + 1 + 1) + s_2 = 9$

20. b. Call the poles A, B, and C. Compute c_2 by using the following sequence of steps to transfer two disks from A to B:

1 (move to move the top disk for A to B)
$+1$ (move to move the top disk from B to C)
$+1$ (move to move the bottom disk from A to B)
$+1$ (move to move the top disk from C to A)
$+1$ (move to move the top disk from A to B)

This sequence of steps is the least possible, and so $c_2 = 5$.

A tower of 3 disks can be transferred from A to B by using the following sequence of steps:

1 (move to move the top disk from A to B)
$+1$ (move to move the top disk from B to C)
$+1$ (move to move the middle disk from A to B)
$+1$ (move to move the top disk from C to A)
$+1$ (move to move the middle disk from B to C)
$+1$ (move to move the top disk from A to B)
$+1$ (move to move the top disk from B to C).

After these 7 steps have been completed, the bottom disk can be moved from A to B. At that point the top two disks are on C, and a modified version of the initial seven steps can be used to move them from C to B. Thus the total number of steps is $7 + 1 + 7 = 15$, and $15 < 21 = 4c_2 + 1$.

21. b. $t_3 = 14$

22. b. $r_0 = 1$, $r_1 = 1$, $r_2 = 1 + 4 \cdot 1 = 5$, $r_3 = 5 + 4 \cdot 1 = 9$,
$r_4 = 9 + 4 \cdot 5 = 29$, $r_5 = 29 + 4 \cdot 9 = 65$,
$r_6 = 65 + 4 \cdot 29 = 181$

23. c. There are 904 rabbit pairs, or 1,808 rabbits, after 12 months.

25. a. Each term of the Fibonacci sequence beyond the second equals the sum of the previous two. For any integer $k \geq 1$, the two terms previous to F_{k+1} are F_k and F_{k-1}. Hence, for all integers $k \geq 1$, $F_{k+1} = F_k + F_{k-1}$.

26. By repeated use of definition of the Fibonacci sequence, for all integers $k \geq 4$,

$$F_k = F_{k-1} + F_{k-2} = (F_{k-2} + F_{k-3}) + (F_{k-3} + F_{k-4})$$
$$= ((F_{k-3} + F_{k-4}) + F_{k-3}) + (F_{k-3} + F_{k-4})$$
$$= 3F_{k-3} + 2F_{k-4}.$$

27. For all integers $k \geq 1$,

$$F_k^2 - F_{k-1}^2$$
$$= (F_k - F_{k-1})(F_k + F_{k-1}) \qquad \text{by basic algebra (difference of two squares)}$$
$$= (F_k - F_{k-1})F_{k+1} \qquad \text{by definition of the Fibonacci sequence}$$
$$= F_k F_{k+1} - F_{k-1}F_{k+1}$$

33. *Hint:* Let $L = \lim_{n \to \infty} \dfrac{F_{n+1}}{F_n}$ and show that $L = \dfrac{1}{L} + 1$. Deduce that $L = \dfrac{1 + \sqrt{5}}{2}$.

34. *Hint:* Use the result of exercise 30 to prove that the infinite sequence $\dfrac{F_0}{F_1}, \dfrac{F_2}{F_3}, \dfrac{F_4}{F_5}, \ldots$ is strictly decreasing and that the infinite sequence $\dfrac{F_1}{F_2}, \dfrac{F_3}{F_4}, \dfrac{F_5}{F_6}, \ldots$ is strictly increasing. The first sequence is bounded below by 0, and the second sequence is bounded above by 1. Deduce that the limits of both sequences exist, and show that they are equal.

36. a. Because the 4% annual interest is compounded quarterly, the quarterly interest rate is $(4\%)/4 = 1\%$. Then $R_k = R_{k-1} + 0.01R_{k-1} = 1.01R_{k-1}$.
b. Because one year equals four quarters, the amount on deposit at the end of one year is $R_4 = \$5203.02$ (rounded to the nearest cent).
c. The annual percentage rate (APR) for the account is $\dfrac{\$5203.02 - \$5000.00}{\$5000.00} = 4.0604\%$.

38. When one is climbing a staircase consisting of n stairs, the last step taken is either a single stair or two stairs together. The number of ways to climb the staircase and have the final step be a single stair is c_{n-1}; the number of ways to climb the staircase and have the final step be two stairs is c_{n-2}. Therefore, $c_n = c_{n-1} + c_{n-2}$. Note also that $c_1 = 1$ and $c_2 = 2$ *[because either the two stairs can be climbed one by one or they can be climbed as a unit]*.

40. Proof (by mathematical induction): Let the property, $P(n)$, be the equation $\sum_{i=1}^{n} ca_i = c \sum_{i=1}^{n} a_i$, where $a_1, a_2, a_3, \ldots, a_n$ and c are any real numbers.

Show that $P(1)$ is true:

Let a_1 and c be any real numbers. By the recursive definition of sum, $\sum_{i=1}^{1} (ca_i) = ca_1$ and $\sum_{i=1}^{1} a_i = a_1$. Therefore, $\sum_{i=1}^{1} (ca_i) = c \sum_{i=1}^{1} a_i$, and so $P(1)$ is true.

Show that for all integers $k \geq 1$, if $P(k)$ is true, then $P(k + 1)$ is true:

Let k be any integer with $k \geq 1$. Suppose that for any real numbers $a_1, a_2, a_3, \ldots, a_k$ and c, $\sum_{i=1}^{k}(ca_i) = c\sum_{i=1}^{k} a_i$. *[This is the inductive hypothesis]. [We must show that for any real numbers $a_1, a_2, a_3, \ldots a_{k+1}$ and c, $\sum_{i=1}^{k+1}(ca_i) = c\sum_{i=1}^{k+1} a_i$.]*

Let $a_1, a_2, a_3, \ldots, a_{k+1}$ and c be any real numbers. Then

$$\sum_{i=1}^{k+1} ca_i = \sum_{i=1}^{k} ca_i + ca_{k+1} \quad \text{by the recursive definition of } \Sigma$$

$$= c\sum_{i=1}^{k} a_i + ca_{k+1} \quad \text{by inductive hypothesis}$$

$$= c\left(\sum_{i=1}^{k} a_i + a_{k+1}\right) \quad \begin{array}{l}\text{by the distributive law} \\ \text{for the real numbers}\end{array}$$

$$= c\sum_{i=1}^{k+1} a_i \quad \text{by the recursive definition of } \Sigma.$$

43. *Hint:* Let the property be the inequality

$$\left|\sum_{i=1}^{n} a_i\right| \leq \sum_{i=1}^{n} |a_i|.$$

To prove the inductive step, note that because $\left|\sum_{i=1}^{k+1} a_i\right| = \left|\sum_{i=1}^{k} a_i + a_{k+1}\right|$, you can use the triangle inequality for absolute value (Theorem 4.4.6) to deduce $\left|\sum_{i=1}^{k} a_i + a_{k+1}\right| \leq \left|\sum_{i=1}^{k} a_i\right| + |a_{k+1}|$.

Section 5.6

1. a. $1 + 2 + 3 + \cdots + (k - 1)$

$$= \frac{(k-1)((k-1)+1)}{2} = \frac{(k-1)k}{2}$$

b. $3 + 2 + 4 + 6 + 8 + \cdots + 2n$

$$= 3 + 2(1 + 2 + 3 + \cdots + n)$$

$$= 3 + 2\frac{n(n+1)}{2} = 3 + n(n+1)$$

$$= n^2 + n + 3$$

2. a. $1 + 2 + 2^2 + \cdots + 2^{i-1} = \dfrac{2^{(i-1)+1} - 1}{2 - 1} = 2^i - 1$

c. $2^n + 2^{n-2} \cdot 3 + 2^{n-3} \cdot 3 + \cdots + 2^2 \cdot 3 + 2 \cdot 3 + 3$

$$= 2^n + 3(2^{n-2} + 2^{n-3} + \cdots + 2^2 + 2 + 1)$$

$$= 2^n + 3(1 + 2 + 2^2 + \cdots + 2^{n-3} + 2^{n-2})$$

$$= 2^n + 3\left(\frac{2^{(n-2)+1} - 1}{2 - 1}\right)$$

$$= 2^n + 3(2^{n-1} - 1)$$

$$= 2 \cdot 2^{n-1} + 3 \cdot 2^{n-1} - 3$$

$$= 5 \cdot 2^{n-1} - 3$$

3. $a_0 = 1$

$a_1 = 1 \cdot a_0 = 1 \cdot 1 = 1$

$a_2 = 2a_1 = 2 \cdot 1$

$a_3 = 3a_2 = 3 \cdot 2 \cdot 1$

$a_4 = 4a_3 = 4 \cdot 3 \cdot 2 \cdot 1$

$\vdots$

Guess:

$a_n = n(n - 1) \cdots 3 \cdot 2 \cdot 1 = n!$

5. $c_1 = 1$

$c_2 = 3c_1 + 1 = 3 \cdot 1 + 1 = 3 + 1$

$c_3 = 3c_2 + 1 = 3 \cdot (3 + 1) + 1 = 3^2 + 3 + 1$

$c_4 = 3c_3 + 1 = 3 \cdot (3^2 + 3 + 1) + 1$

$\quad = 3^3 + 3^2 + 3 + 1$

$\vdots$

Guess:

$c_n = 3^{n-1} + 3^{n-2} + \cdots + 3^3 + 3^2 + 3 + 1$

$\quad = \dfrac{3^n - 1}{3 - 1} \quad \text{by Theorem 5.2.3 with } r = 3$

$\quad = \dfrac{3^n - 1}{2}$

6. *Hint:*

$d_n = 2^n + 2^{n-2} \cdot 3 + 2^{n-3} \cdot 3 + \cdots + 2^2 \cdot 3 + 2 \cdot 3 + 3$

$\quad = 5 \cdot 2^{n-1} - 3$ for all integers $n \geq 1$

9. *Hint:* For any positive real numbers a and b,

$$\frac{\frac{a}{b}}{\frac{a}{b} + 2} = \frac{\frac{a}{b}}{\frac{a}{b} + 2} \cdot \frac{b}{b} = \frac{a}{a + 2b}.$$

10. $h_0 = 1$

$h_1 = 2^1 - h_0 = 2^1 - 1$

$h_2 = 2^2 - h_1 = 2^2 - (2^1 - 1) = 2^2 - 2^1 + 1$

$h_3 = 2^3 - h_2 = 2^3 - (2^2 - 2^1 + 1)$

$\quad = 2^3 - 2^2 + 2^1 - 1$

$h_4 = 2^4 - h_3 = 2^4 - (2^3 - 2^2 + 2^2 - 1)$

$\quad = 2^4 - 2^3 + 2^2 - 2^1 + 1$

$\vdots$

Guess:

$h_n = 2^n - 2^{n-1} + \cdots + (-1)^n \cdot 1$

$\quad = (-1)^n[1 - 2 + 2^2 - \cdots + (-1)^n \cdot 2^n]$

$\quad = (-1)^n[1 + (-2)$

$\qquad + (-2)^2 - \cdots + (-2)^n] \quad \text{by basic algebra}$

$\quad = (-1)^n\left[\dfrac{(-2)^{n+1} - 1}{(-2) - 1}\right] \quad \text{by Theorem 5.2.3}$

$\quad = \dfrac{(-1)^{n+1} \cdot [(-2)^{n+1} - 1]}{(-1) \cdot (-3)}$

$\quad = \dfrac{2^{n+1} - (-1)^{n+1}}{3} \quad \text{by basic algebra}$

12. $s_0 = 3$

$s_1 = s_0 + 2 \cdot 1 = 3 + 2 \cdot 1$

$s_2 = s_1 + 2 \cdot 2 = [3 + 2 \cdot 1] + 2 \cdot 2$

$\quad = 3 + 2 \cdot (1 + 2)$

$s_3 = s_2 + 2 \cdot 3 = [3 + 2 \cdot (1 + 2)] + 2 \cdot 3$

$\quad = 3 + 2 \cdot (1 + 2 + 3)$

$s_4 = s_3 + 2 \cdot 4 = [3 + 2 \cdot (1 + 2 + 3)] + 2 \cdot 4$

$\quad = 3 + 2 \cdot (1 + 2 + 3 + 4)$

$\vdots$

Guess:

$s_n = 3 + 2 \cdot (1 + 2 + 3 + \cdots + (n-1) + n)$

$\quad = 3 + 2 \cdot \dfrac{n(n+1)}{2} \qquad$ by Theorem 5.2.2

$\quad = 3 + n(n+1) \qquad\qquad$ by basic algebra

14. $x_1 = 1$

$x_2 = 3x_1 + 2 = 3 + 2$

$x_3 = 3x_2 + 3 = 3(3 + 2) + 3 = 3^2 + 3 \cdot 2 + 3$

$x_4 = 3x_3 + 4 = 3(3^2 + 3 \cdot 2 + 3) + 4$

$\quad = 3^3 + 3^2 \cdot 2 + 3 \cdot 3 + 4$

$x_5 = 3x_4 + 5 = 3(3^3 + 3^2 \cdot 2 + 3 \cdot 3 + 4) + 5$

$\quad = 3^4 + 3^3 \cdot 2 + 3^2 \cdot 3 + 3 \cdot 4 + 5$

$x_6 = 3x_5 + 6$

$\quad = 3(3^4 + 3^3 \cdot 2 + 3^2 \cdot 3 + 4 \cdot 3 + 5) + 6$

$\quad = 3^5 + 3^4 \cdot 2 + 3^3 \cdot 3 + 3^2 \cdot 4 + 3 \cdot 5 + 6$

$\vdots$

Guess:

$x_n = 3^{n-1} + 3^{n-2} \cdot 2 + 3^{n-3} \cdot 3 + \cdots + 3(n-1) + n$

$\quad = 3^{n-1} + \underbrace{3^{n-2} + 3^{n-2}}_{2 \text{ times}} + \underbrace{3^{n-3} + 3^{n-3} + 3^{n-3}}_{3 \text{ times}} +$

$\qquad\qquad + \underbrace{3 + 3 + \cdots + 3}_{(n-1) \text{ times}} + \underbrace{1 + 1 + \cdots + 1}_{n \text{ times}}$

$\quad = (3^{n-1} + 3^{n-2} + \cdots + 3^2 + 3 + 1)$

$\qquad\quad + (3^{n-2} + 3^{n-3} + \cdots + 3^2 + 3 + 1) + \cdots$

$\qquad\qquad\qquad + (3^2 + 3 + 1) + (3 + 1) + 1$

$\quad = \dfrac{3^n - 1}{2} + \dfrac{3^{n-1} - 1}{2} + \cdots + \dfrac{3^3 - 1}{2}$

$\qquad\qquad\qquad\qquad + \dfrac{3^2 - 1}{2} + \dfrac{3 - 1}{2}$

$\quad = \frac{1}{2}[(3^n + 3^{n-1} + \cdots + 3^2 + 3) - n]$

$\quad = \frac{1}{2}[3(3^{n-1} + 3^{n-2} + \cdots + 3 + 1) - n]$

$\quad = \frac{1}{2}\left(3\left(\dfrac{3^n - 1}{3 - 1}\right) - n\right)$

$\quad = \frac{1}{4}(3^{n+1} - 3 - 2n)$

18. Proof: Let d be any fixed constant, and let $a_0, a_1, a_2, \ldots$ be the sequence defined recursively by $a_k = a_{k-1} + d$ for all integers $k \geq 1$. The property $P(n)$ is the equation $a_n = a_0 + nd$. We show by mathematical induction that $P(n)$ is true for all integers $n \geq 0$.

Show that P(0) is true:

When $n = 0$, the left-hand side of the equation is a_0, and the right-hand side is $a_0 + 0 \cdot d = a_0$, which equals the left-hand side. Thus $P(0)$ is true.

Show that for all integers $k \geq 0$, if $P(k)$ is true, then $P(k+1)$ is true:

Suppose

$a_k = a_0 + kd, \;$ for some integer $k \geq 0$.

$\qquad\qquad\qquad$ [*This is the inductive hypothesis.*]

We must show that $a_{k+1} = a_0 + (k+1)d$. But

$a_{k+1} = a_k + d \qquad$ by definition of $a_0, a_1, a_2, \ldots$

$\quad = [a_0 + kd] + d \quad$ by substitution from the inductive hypothesis

$\quad = a_0 + (k+1)d \quad$ by basic algebra

[*as was to be shown*].

19. Let U_n = the number of units produced on day n. Then

$U_k = U_{k-1} + 2 \quad$ for all integers $k \geq 1$,

$U_0 = 170.$

Hence $U_0, U_1, U_2, \ldots$ is an arithmetic sequence with fixed constant 2. It follows that when $n = 30$,

$U_n = U_0 + n \cdot 2 = 170 + 2n = 170 + 2 \cdot 30$

$\quad = 230$ units.

Thus the worker must produce 230 units on day 30.

24. $\displaystyle\sum_{k=0}^{20} 5^k = \dfrac{5^{21} - 1}{4} \cong 1.192 \times 10^{14} \cong$

$119,200,000,000,000 \cong 119$ trillion people (This is about 20,000 times the current population of the earth!)

26. b. *Hint:* Before simplification,

$A_n = 1000(1.0025)^n + 200[(1.0025)^{n-1} + (1.0025)^{n-1} + \cdots + (1.0025)^2 + 1.0025 + 1].$

d. $A_{240} \cong \$67,481.15, \; A_{480} \cong \$188,527.05$

e. *Hint:* Use logarithms to solve the equation $A_n = 10,000$, where A_n is the expression found (after simplification) in part (b).

27. a. *Hint:* APR $\cong 19.6\%$

c. *Hint:* approximately two years

28. Proof: Let $a_0, a_1, a_2, \ldots$ be the sequence defined recursively by $a_0 = 1$ and $a_k = ka_{k-1}$ for all integers $k \geq 1$. Let the property $P(n)$ be the equation $a_n = n!$. We show by mathematical induction that $P(n)$ is true for all integers $n \geq 0$.

Show that P(0) is true:

When $n = 0$, the right-hand side of the equation is $0! = 1$, and by definition of $a_0, a_1, a_2, \ldots$, the left-hand side of the equation, a_0, is also 1. Thus the property is true for $n = 0$.

Show that for all integers $k \geq 0$, if $P(k)$ is true, then $P(k + 1)$ is true:

Suppose

$$a_k = k! \quad \text{for some integer } k \geq 0.$$

[This is the inductive hypothesis.]

We must show that $a_{k+1} = (k + 1)!$. But

$$
\begin{aligned}
a_{k+1} &= (k + 1) \cdot a_k & \text{by definition of } a_0, a_1, a_2, \ldots \\
&= (k + 1) \cdot k! & \text{by substitution from the inductive hypotheses} \\
&= (k + 1)! & \text{by definition of factorial.}
\end{aligned}
$$

[Hence if $P(k)$ is true, then $P(k + 1)$ is true.]

30. Proof: Let $c_1, c_2, c_3, \ldots$ be the sequence defined recursively by $c_1 = 1$ and $c_k = 3c_{k-1} + 1$ for all integers $k \geq 2$. Let the property $P(n)$ be the equation $c_n = \dfrac{3^n - 1}{2}$. We show by mathematical induction that $P(n)$ is true for all integers $n \geq 1$.

Show that $P(1)$ is true:

When $n = 1$, the right-hand side of the equation is $\dfrac{3^1 - 1}{2} = \dfrac{3-1}{2} = 1$, and by definition of $c_1, c_2, c_3, \ldots$, the left-hand side of the equation, c_1, is also 1. Thus the property is true for $n = 1$.

Show that for all integers $k \geq 1$, if $P(k)$ is true, then $P(k + 1)$ is true:

Suppose that

$$c_k = \frac{3^k - 1}{2} \quad \text{for some integer } k \geq 1.$$

[This is the inductive hypothesis.]

We must show that $c_{k+1} = \dfrac{3^{k+1} - 1}{2}$. But

$$
\begin{aligned}
c_{k+1} &= 3c_k + 1 & \text{by definition of } c_1, c_2, c_3, \ldots \\
&= 3\left(\frac{3^k - 1}{2}\right) + 1 & \text{by substitution from the inductive hypothesis} \\
&= \frac{3^{k+1} - 3}{2} + \frac{2}{2} & \\
&= \frac{3^{k+1} - 1}{2} & \text{by basic algebra.}
\end{aligned}
$$

35. *Hint:*

$$
\begin{aligned}
2^{k+1} &- \frac{2^{k+1} - (-1)^{k+1}}{3} \\
&= \frac{3 \cdot 2^{k+1}}{3} - \frac{2^{k+1} - (-1)^{k+1}}{3} \\
&= \frac{2 \cdot 2^{k+1} + (-1)^{k+1}}{3} = \frac{2^{k+2} - (-1)^{k+2}}{3}
\end{aligned}
$$

37. *Hint:*

$$
\begin{aligned}
[3 &+ k(k + 1)] + 2(k + 1) \\
&= 3 + k^2 + k + 2k + 2 = 3 + [k^2 + 3k + 2] \\
&= 3 + (k + 1)(k + 2) \\
&= 3 + (k + 1)[(k + 1) + 1]
\end{aligned}
$$

39. Proof: Let $x_1, x_2, x_3, \ldots$ be the sequence defined recursively by $x_1 = 1$ and $x_k = 3x_{k-1} + k$ for all integers $k \geq 2$. Let the property, $P(n)$, be the equation $x_n = \dfrac{3^{n+1} - 2n - 3}{4}$. We show by mathematical induction that $P(n)$ is true for all integers $n \geq 1$.

Show that $P(1)$ is true:

When $n = 1$, the right-hand side of the equation is $\dfrac{3^{1+1} - 2 \cdot 1 - 3}{4} = \dfrac{3^2 - 2 - 3}{4} = 1$, and by definition of $x_1, x_2, x_3, \ldots$, the left-hand side of the equation, x_1, is also 1. Thus $P(1)$ is true.

Show that for all integers $k \geq 1$, if $P(k)$ is true for, then $P(k + 1)$ is true.

Suppose that for some integer $k \geq 0$, $x_k = \dfrac{3^{k+1} - 2k - 3}{4}$. *[Inductive hypothesis]* We must show that

$$x_{k+1} = \frac{3^{(k+1)+1} - 2(k + 1) - 3}{4}, \text{ or, equivalently,}$$

$$x_{k+1} = \frac{3^{k+2} - 2k - 5}{4}. \text{ But}$$

$$
\begin{aligned}
x_{k+1} &= 3x_k + k & \text{by definition of } x_1, x_2, x_3, \\
&= 3\left(\frac{3^{k+1} - 2k - 3}{4}\right) + k + 1 & \text{by inductive hypothesis} \\
&= \frac{3 \cdot 3^{k+1} - 3 \cdot 2k - 3 \cdot 3}{4} + \frac{4(k + 1)}{4} & \\
&= \frac{3^{k+2} - 6k - 9 + 4k + 4}{4} & \\
&= \frac{3^{k+2} - 2k - 5}{4} & \text{by algebra.}
\end{aligned}
$$

[This is what was to be shown.]

43. a. $a_0 = 2$

$$a_1 = \frac{a_0}{2a_0 - 1} = \frac{2}{2 \cdot 2 - 1} = \frac{2}{3}$$

$$a_2 = \frac{a_1}{2a_1 - 1} = \frac{\frac{2}{3}}{2 \cdot \frac{2}{3} - \frac{3}{3}} = \frac{\frac{2}{3}}{\frac{1}{3}} = 2$$

$$a_3 = \frac{a_2}{2a_2 - 1} = \frac{2}{2 \cdot 2 - 1} = \frac{2}{3}$$

$$a_4 = \frac{a_3}{2a_3 - 1} = \frac{\frac{2}{3}}{2 \cdot \frac{2}{3} - \frac{3}{3}} = \frac{\frac{2}{3}}{\frac{1}{3}} = 2$$

$$\text{Guess: } a_n = \begin{cases} 2 & \text{if } n \text{ is even} \\ \frac{2}{3} & \text{if } n \text{ is odd} \end{cases}.$$

b. <u>Proof:</u> Let $a_0, a_1, a_2, \ldots$ be the sequence defined recursively by $x_0 = 2$ and $a_k = \frac{a_{k-1}}{2a_{k-1}-1}$ for all integers $k \geq 1$. Let the property, $P(n)$, be the equation

$$a_n = \begin{cases} 2 & \text{if } n \text{ is even} \\ \frac{2}{3} & \text{if } n \text{ is odd} \end{cases}.$$

We show by strong mathematical induction that $P(n)$ is true for all integers $n \geq 1$.

Show that $P(0)$ and $P(1)$ are true:
The results of part (a) show that $P(0)$ and $P(1)$ are true.

Show that for all integers $k \geq 0$, if $P(k)$ is true for all integers i with $0 \leq i \leq k$, then $P(k+1)$ is true:
Let k be any integer with $k \geq 0$, and suppose that for all integers i with $0 \leq i \leq k$,

$$a_i = \begin{cases} 2 & \text{if } i \text{ is even} \\ \frac{2}{3} & \text{if } i \text{ is odd} \end{cases}. \quad \textit{[Inductive hypothesis]}$$

We must show that

$$a_{k+1} = \begin{cases} 2 & \text{if } k \text{ is even} \\ \frac{2}{3} & \text{if } k \text{ is odd} \end{cases}.$$

But

$$a_{k+1} = \frac{a_k}{2a_k - 1} \qquad \text{by definition of } a_0, a_1, a_2, \ldots$$

$$= \begin{cases} \dfrac{2}{2 \cdot 2 - 1} & \text{if } k \text{ is even} \\[2mm] \dfrac{\frac{2}{3}}{2 \cdot \frac{2}{3} - 1} & \text{if } k \text{ is odd} \end{cases} \qquad \text{by inductive hypothesis}$$

$$= \begin{cases} \dfrac{2}{3} & \text{if } k \text{ is even} \\[2mm] \dfrac{\frac{2}{3}}{\frac{1}{3}} & \text{if } k \text{ is odd} \end{cases}$$

$$= \begin{cases} \dfrac{2}{3} & \text{if } k+1 \text{ is odd} & \text{because } k+1 \text{ is odd when } k \text{ is even} \\[2mm] 2 & \text{if } k+1 \text{ is even} & \text{and } k+1 \text{ is even when } k \text{ is odd.} \end{cases}$$

[This is what was to be shown.]

45. $v_1 = 1$

$v_2 = v_{\lfloor 2/2 \rfloor} + v_{\lfloor 3/2 \rfloor} + 2 = v_1 + v_1 + 2$
$\quad = 1 + 1 + 2$

$v_3 = v_{\lfloor 3/2 \rfloor} + v_{\lfloor 4/2 \rfloor} + 2 = v_1 + v_2 + 2$
$\quad = 1 + (1 + 1 + 2) + 2 = 3 + 2 \cdot 2$

$v_4 = v_{\lfloor 4/2 \rfloor} + v_{\lfloor 5/2 \rfloor} + 2 = v_2 + v_2 + 2$
$\quad = (1 + 1 + 2) + (1 + 1 + 2) + 2$
$\quad = 4 + 3 \cdot 2$

$v_5 = v_{\lfloor 5/2 \rfloor} + v_{\lfloor 6/2 \rfloor} + 2 = v_2 + v_3 + 2$
$\quad = (3 + 2 \cdot 2) + (1 + 1 + 2) + 2$
$\quad = 5 + 4 \cdot 2$

$v_6 = v_{\lfloor 6/2 \rfloor} + v_{\lfloor 7/2 \rfloor} + 2 = v_3 + v_3 + 2$
$\quad = (3 + 2 \cdot 2) + (3 + 2 \cdot 2) + 2$
$\quad = 6 + 5 \cdot 2$
$\quad \vdots$

Guess:
$v_n = n + 2(n-1) = 3n - 2$ for all integers $n \geq 1$

b. <u>Proof:</u> Let $v_1, v_2, v_3, \ldots$ be the sequence defined recursively by $v_1 = 1$ and $v_k = v_{\lfloor k/2 \rfloor} + v_{\lfloor (k+1)/2 \rfloor} + 2$ for all integers $k \geq 1$. Let the property, $P(n)$, be the equation

$$v_n = 3n - 2.$$

We show by strong mathematical induction that $P(n)$ is true for all integers $n \geq 1$.

Show that $P(1)$ is true:

When $n = 1$, the right-hand side of the equation is $3 \cdot 1 - 2 = 1$, which equals v_1 by definition of $v_1, v_2, v_3, \ldots$. Thus $P(1)$ is true.

Show that for all integers $k \geq 1$, if $P(i)$ is true for all integers i with $0 \leq i \leq k$, then $P(k+1)$ is true:

Let k be any integer with $k \geq 1$, and suppose that for all integers i with $1 \leq i \leq k$, $v_i = 3i - 2$.

[This is the inductive hypothesis.] We must show that $v_{k+1} = 3(k+1) - 2 = 3k + 1$.

$$v_{k+1} = v_{\lfloor (k+1)/2 \rfloor} + v_{\lfloor (k+2)/2 \rfloor} + 2 \qquad \text{by definition of } v_1, v_2, v_3, \ldots$$

$$= \left(3 \left\lfloor \frac{k+1}{2} \right\rfloor - 2\right) + \left(3 \left\lfloor \frac{k+2}{2} \right\rfloor - 2\right) + 2$$

$$= 3 \left(\left\lfloor \frac{k+1}{2} \right\rfloor + \left\lfloor \frac{k+2}{2} \right\rfloor\right) - 2$$

$$= \begin{cases} 3 \left(\frac{k}{2} + \frac{k+2}{2}\right) - 2 & \text{if } k \text{ is even} \\[2mm] 3 \left(\frac{k+1}{2} + \frac{k+1}{2}\right) - 2 & \text{if } k \text{ is odd} \end{cases}$$

$$= 3 \left(\frac{2k+2}{2}\right) - 2$$

$$= 3(k+1) - 2$$

$$= 3k + 1 \qquad \text{by the laws of algebra.}$$

[This is what was to be shown.]

46. *Hint:* Show that for all integers $n \geq 0$, $s_{2n} = 2^n$ and $s_{2n+1} = 2^{n+1}$. Then combine these formulas using the ceiling function to obtain $s_n = 2^{\lceil n/2 \rceil}$.

48. a. *Hint:* $w_n = \begin{cases} \left(\frac{n+1}{2}\right)^2 & \text{if } n \text{ is odd} \\[2mm] \frac{n}{2}\left(\frac{n}{2} + 1\right) & \text{if } n \text{ is even} \end{cases}$

49. a. *Hint:* Express the answer using the Fibonacci sequence.

50. The sequence does not satisfy the formula. According to the formula, $a_4 = (4-1)^2 = 9$. But by definition of the sequence, $a_1 = 0$, $a_2 = 2 \cdot 0 + (2 + 1) = 1$, $a_3 = 2 \cdot 1 + (3 - 1) = 4$, and so $a_4 = 2 \cdot 4 + (4 - 1) = 11$. Hence the sequence does not satisfy the formula for $n = 4$.

52. a. *Hint:* The maximum number of regions is obtained when each additional line crosses all the previous lines, but not at any point that is already the intersection of two lines. When a new line is added, it divides each region through which it passes into two pieces. The number of regions a newly added line passes through is one more than the number of lines it crosses.

53. *Hint:* The answer involves the Fibonacci numbers!

Section 6.1

1. a. $A = \{2, \{2\}, (\sqrt{2})^2\} = \{2, \{2\}, 2\} = \{2, \{2\}\}$ and $B = \{2, \{2\}, \{\{2\}\}\}$. So $A \subseteq B$ because every element in A is in B, but $B \nsubseteq A$ because $\{\{2\}\} \in B$ and $\{\{2\}\} \notin A$. Also A is a proper subset of B because $\{\{2\}\}$ is in B but not A.

  c. $A = \{\{1, 2\}, \{2, 3\}\}$ and $B = \{1, 2, 3\}$. So $A \nsubseteq B$ because $\{1, 2\} \in A$ and $\{1, 2\} \notin B$. Also $B \nsubseteq A$ because $1 \in B$ and $1 \notin A$.

  e. $A = \left\{ \sqrt{16}, \{4\} \right\} = \{4, \{4\}\}$ and $B = \{4\}$. Then $B \subseteq A$ because the only element in B is 4 and 4 is in A, but $A \nsubseteq B$ because $\{4\} \in A$ and $\{4\} \notin B$. Also B is a proper subset of A because $\{4\}$ is in A but not B.

2. *Proof That $B \subseteq A$:*
Suppose x is a particular but arbitrarily chosen element of B.
  [We must show that $x \in A$. By definition of A, this means we must show that $x = 2 \cdot$ (some integer).]
By definition of B, there is an integer b such that $x = 2b - 2$.
  [Given that $x = 2b - 2$, can x also be expressed as $2 \cdot$ (some integer)? I.e., is there an integer, say a, such that $2b - 2 = 2a$? Solve for a to obtain $a = b - 1$. Check to see if this works.]
Let $a = b - 1$.
  [First check that a is an integer.]
Then a is an integer because it is a difference of integers.
  [Then check that $x = 2a$.]
Also $2a = 2(b - 1) = 2b - 2 = x$,
Thus, by definition of A, x is an element of A,
  [which is what was to be shown].

3. a. No. $R \nsubseteq T$ because there are elements in R that are not in T. For example, the number 2 is in R but 2 is not in T since 2 is not divisible by 6.

  b. Yes. $T \subseteq R$ because every element in T is in R since every integer divisible by 6 is divisible by 2. To see why this is so, suppose n is any integer that is divisible by 6. Then $n = 6m$ for some integer m. Since $6m = 2(3m)$ and since $3m$ is an integer (being a product of integers), it follows that $n = 2 \cdot$(*some integer*), and, hence, that n is divisible by 2.

5. a. $C \subseteq D$ <u>Proof:</u> *[We will show that every element of C is in D.]* Suppose n is any element of C. Then $n = 6r - 5$ for some integer r. Let $s = 2r - 2$. Then s is an integer (because products and differences of integers are integers), and

$$3s + 1 = 3(2r - 2) + 1 = 6r - 6 + 1 = 6r - 5,$$

which equals n. Thus n satisfies the condition for being in D. Hence, every element in C is in D.

  b. $D \nsubseteq C$ because there are elements of D that are not in C. For example, 4 is in D because $4 = 3 \cdot 1 + 1$. But 4 is not in C because if it were, then $4 = 6r - 5$ for some integer r, which would imply that $9 = 6r$, or, equivalently, that $r = 3/2$, and this contradicts the fact that r is an integer.

6. c. *Sketch of proof that $B \subseteq C$:* If r is any element of B then there is an integer b such that $r = 10b - 3$. To show that r is in C, you must show that there is an integer c such that $r = 10c + 7$. In scratch work, assume that c exists and use the information that $10b - 3$ would have to equal $10c + 7$ to deduce the only possible value for $\dot{c}$. Then show that this value is (1) an integer and (2) satisfies the equation $r = 10c + 7$, which will allow you to conclude that r is an element of C.

  Sketch of proof that $C \subseteq B$: If s is any element of C then there is an integer c such that $s = 10c + 7$. To show that s is in B, you must show that there is an integer b such that $s = 10c - 3$. In scratch work, assume that b exists and use the information that $10c + 7$ would have to equal $10b - 3$ to deduce the only possible value for b. Then show that this value is (1) an integer and (2) satisfies the equation $s = 10b - 3$, which will allow you to conclude that s is an element of B.

8. a. The set of all x in U such that x is in A and x is in B. The shorthand notation is $A \cap B$.

9. a. $x \notin A$ and $x \notin B$

10. a. $\{1, 3, 5, 6, 7, 9\}$ **b.** $\{3, 9\}$
  c. $\{1, 2, 3, 4, 5, 6, 7, 8, 9\}$ **d.** $\varnothing$ **e.** $\{1, 5, 7\}$

11. a. $A \cup B = \{x \in \mathbf{R} \mid 0 < x < 4\}$
  b. $A \cap B = \{x \in \mathbf{R} \mid 1 \leq x \leq 2\}$
  c. $A^c = \{x \in \mathbf{R} \mid x \leq 0 \text{ or } x > 2\}$
  d. $A \cup C = \{x \in \mathbf{R} \mid 0 < x \leq 2 \text{ or } 3 \leq x < 9\}$
  e. $A \cap C = \varnothing$
  f. $B^c = \{x \in \mathbf{R} \mid x < 1 \text{ or } x \geq 4\}$
  g. $A^c \cap B^c = \{x \in \mathbf{R} \mid x \leq 0 \text{ or } x \geq 4\}$
  h. $A^c \cup B^c = \{x \in \mathbf{R} \mid x < 1 \text{ or } x > 2\}$
  i. $(A \cap B)^c = \{x \in \mathbf{R} \mid x < 1 \text{ or } x > 2\}$
  j. $(A \cup B)^c = \{x \in \mathbf{R} \mid x \leq 0 \text{ or } x \geq 4\}$

13. b. False. Many negative real numbers are not rational. For example, $-\sqrt{2} \in \mathbf{R}$ but $-\sqrt{2} \notin \mathbf{Q}$.
  d. False. $0 \in \mathbf{Z}$ but $0 \notin \mathbf{Z}^- \cup \mathbf{Z}^+$.

14. a.

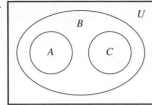

15. a.

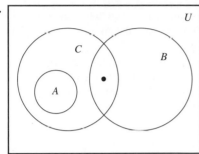

16. a. $A \cup (B \cap C) = \{a, b, c\}$, $(A \cup B) \cap C = \{b, c\}$, and
$(A \cup B) \cap (A \cup C) = \{a, b, c, d\} \cap \{a, b, c, e\} = \{a, b, c\}$.
Hence $A \cup (B \cap C) = (A \cup B) \cap (A \cup C)$.

17. a.

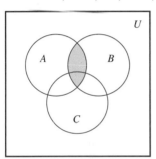

18. a. The number 0 is not in $\emptyset$ because $\emptyset$ has no elements.
 b. No. The left-hand set is the empty set; it does not have any elements. The right-hand set is a set with one element, namely $\emptyset$.

19. $A_1 = \{1, 1^2\} = \{1\}$, $A_2 = \{2, 2^2\} = \{2, 4\}$,
$A_3 = \{3, 3^2\} = \{3, 9\}$, $A_4 = \{4, 4^2\} = \{4, 16\}$
 a. $A_1 \cup A_2 \cup A_3 \cup A_4 = \{1\} \cup \{2, 4\} \cup \{3, 9\} \cup \{4, 16\}$
$= \{1, 2, 3, 4, 9, 16\}$
 b. $A_1 \cap A_2 \cap A_3 \cap A_4 = \{1\} \cap \{2, 4\} \cap \{3, 9\} \cap \{4, 16\}$
$= \emptyset$
 c. A_1, A_2, A_3, and A_4 are not mutually disjoint because $A_2 \cap A_4 = \{4\} \neq \emptyset$.

21. $C_0 = \{0, -0\} = \{0\}$, $C_1 = \{1, -1\}$, $C_1 = \{2, -2\}$,
$C_1 = \{3, -3\}$, $C_1 = \{4, -4\}$
 a. $\bigcup_{i=0}^{4} C_i = \{0\} \cup \{1, -1\} \cup \{2, -2\} \cup \{3, -3\} \cup \{4, -4\} = \{-4, -3, -2, -1, 0, 1, 2, 3, 4\}$
 b. $\bigcap_{i=0}^{4} C_i = \{0\} \cap \{1, -1\} \cap \{2, -2\} \cap \{3, -3\} \cap \{4, -4\}$
$= \emptyset$
 c. C_0, C_1, C_2, ... are mutually disjoint because no two of the sets have any elements in common.

d. $\bigcup_{i=0}^{n} C_i = \{-n, -(n-1), \ldots, -2, -1, 0, 1, 2, \ldots, (n-1), n\}$
 e. $\bigcap_{i=0}^{n} C_i = \emptyset$
 f. $\bigcup_{i=0}^{\infty} C_i = \mathbf{Z}$, the set of all integers
 g. $\bigcap_{i=0}^{\infty} C_i = \emptyset$

22. $D_0 = [-0, 0] = \{0\}$, $D_1 = [-1, 1]$, $D_2 = [-2, 2]$,
$D_3 = [-3, 3]$, $D_4 = [-4, 4]$
 a. $\bigcup_{i=0}^{4} D_i = \{0\} \cup [-1, 1] \cup [-2, 2] \cup [-3, 3] \cup [-4, 4]$
$= [-4, 4]$
 b. $\bigcap_{i=0}^{4} D_i = \{0\} \cap [-1, 1] \cap [-2, 2] \cap [-3, 3] \cap [-4, 4]$
$= \{0\}$
 c. D_0, D_1, D_2, ... are not mutually disjoint. In fact, each $D_k \subseteq D_{k+1}$.
 d. $\bigcup_{i=0}^{n} D_i = [-n, n]$
 e. $\bigcap_{i=0}^{n} D_i = \{0\}$
 f. $\bigcup_{i=0}^{\infty} D_i = \mathbf{R}$, the set of all real numbers
 g. $\bigcap_{i=0}^{\infty} D_i = \{0\}$

24. $W_0 = (0, \infty)$, $W_1 = (1, \infty)$, $W_2 = (2, \infty)$,
$W_3 = (3, \infty)$, $W_4 = (4, \infty)$
 a. $\bigcup_{i=0}^{4} W_i = (0, \infty) \cup (1, \infty) \cup (2, \infty) \cup (3, \infty) \cup (4, \infty) = (0, \infty)$
 b. $\bigcap_{i=0}^{4} W_i = (0, \infty) \cap (1, \infty) \cap (2, \infty) \cap (3, \infty) \cap (4, \infty) = (4, \infty)$
 c. W_0, W_1, W_2, ... are not mutually disjoint. In fact, $W_{k+1} \subseteq W_k$ for all integers $k \geq 0$.
 d. $\bigcup_{i=0}^{n} W_i = (0, \infty)$
 e. $\bigcap_{i=0}^{n} W_i = (n, \infty)$
 f. $\bigcup_{i=0}^{\infty} W_i = (0, \infty)$
 g. $\bigcap_{i=0}^{\infty} W_i = \emptyset$

27. a. No. The element d is in two of the sets.
 b. No. None of the sets contains 6.

28. Yes. Every integer is either even or odd, and no integer is both even and odd.

31. a. $A \cap B = \{2\}$, so $\mathscr{P}(A \cap B) = \{\emptyset, \{2\}\}$.
 b. $A = \{1, 2\}$, so $\mathscr{P}(A) = \{\emptyset, \{1\}, \{2\}, \{1, 2\}\}$.
 c. $A \cup B = \{1, 2, 3\}$, so $\mathscr{P}(A \cup B) = \{\emptyset, \{1\}, \{2\}, \{3\}, \{1, 2\}, \{1, 3\}, \{2, 3\}, \{1, 2, 3\}\}$.
 d. $A \times B = \{(1, 2), (1, 3), (2, 2), (2, 3)\}$, so
$\mathscr{P}(A \times B) = \{\emptyset, \{(1, 2)\}, \{(1, 3)\}, \{(2, 2)\}, \{(2, 3)\}, \{(1, 2), (1, 3)\}, \{(1, 2), (2, 2)\}, \{(1, 2), (2, 3)\}, \{(1, 3), (2, 2)\}, \{(1, 3), (2, 3)\},$

$\{(2, 2), (2, 3)\}, \{(1, 2), (1, 3), (2, 2)\},$
$\{(1, 2), (1, 3), (2, 3)\},$
$\{(1, 2), (2, 2), (2, 3)\}, \{(1, 3), (2, 2), (2, 3)\},$
$\{(1, 2), (1, 3), (2, 2), (2, 3)\}\}.$

32. a. $\mathscr{P}(A \times B) = \{\emptyset, \{(1, u)\}, \{(1, v)\}, \{(1, u), (1, v)\}\}$

33. b. $\mathscr{P}(\mathscr{P}(\emptyset)) = \mathscr{P}(\{\emptyset\}) = \{\emptyset, \{\emptyset\}\}$

34. a. $A_1 \times (A_2 \times A_3) = \{(1, (u, m)), (2, (u, m)),$
$(3, (u, m)), (1, (u, n)), (2, (u, n)), (3, (u, n)),$
$(1, (v, m)), (2, (v, m)), (3, (v, m)), (1, (v, n)),$
$(2, (v, n)), (3, (v, n))\}$

35. a. $A \times (B \cup C) = \{a, b\} \times \{1, 2, 3\}$
$= \{(a, 1), (a, 2), (a, 3), (b, 1), (b, 2), (b, 3)\}$
   b. $(A \times B) \cup (A \times C) = \{(a, 1), (a, 2), (b, 1), (b, 2),$
$(a, 2), (a, 3), (b, 2), (b, 3)\}$
$= \{(a, 1), (a, 2), (b, 1), (b, 2),$
$(a, 3), (b, 3)\}$

Section 6.2

1. a. (1) A (2) $B \cup C$
   b. (1) $A \cap B$ (2) C

2. a. (1) $A - B$ (2) A (3) A (4) B
   b. (1) $x \in A$ (2) A (3) B (4) A

3. (a.) A **(b)** C **(c)** B **(d)** C **(e)** $B \subseteq C$

5. Proof: Suppose A and B are sets.
   $\boldsymbol{B - A \subseteq B \cap A^c}$: Suppose $x \in B - A$. By definition of set difference, $x \in B$ and $x \notin A$. But then by definition of complement, $x \in B$ and $x \in A^c$, and so by definition of intersection, $x \in B \cap A^c$. *[Thus $B - A \subseteq B \cap A^c$ by definition of subset].*
   $\boldsymbol{B \cap A^c \subseteq B - A}$: Suppose $x \in B \cap A^c$. By definition of intersection, $x \in B$ and $x \in A^c$. But then by definition of complement, $x \in B$ and $x \notin A$, and so by definition of set difference, $x \in B - A$. *[Thus $B \cap A^c \subseteq B - A$ by definition of subset.]*
   [Since both set containments have been proved, $B - A = B \cap A^c$ by definition of set equality.]

6. *Partial answers*
   (1) a. $(A \cap B) \cup (A \cap C)$ **b.** A **c.** $B \cup C$
   d. $x \in C$ **e.** $A \cap B$ **f.** by definition of intersection, $x \in A \cap C$, and so by definition of union, $x \in (A \cap B) \cup (A \cap C)$.

7. *Hint:* This is somewhat similar to the proof in Example 6.2.3.

8. Proof: Suppose A and B are any sets.
   Proof that $(A \cap B) \cup (A \cap B^c) \subseteq A$: Suppose $x \in (A \cap B) \cup (A \cap B^c)$. *[We must show that $x \in A$.]* By definition of union, $x \in A \cap B$ or $x \in (A \cap B^c)$.
   Case 1 ($\mathbf{x \in A \cap B}$): In this case x is in A and x is in B, and so, in particular, $x \in A$.
   Case 2 ($\mathbf{x \in A \cap B^c}$): In this case x is in A and x is not in B, and so, in particular, $x \in A$.
   Thus, in either case, $x \in A$ *[as was to be shown]. [Thus $(A \cap B) \cup (A \cap B^c) \subseteq A$ by definition of subset.]*
   Proof that $A \subseteq (A \cap B) \cup (A \cap B^c)$: Suppose $x \in A$. *[We must show that $x \in (A \cap B) \cup (A \cap B^c)$.]* Either $x \in B$ or $x \notin B$.

Case 1 ($\mathbf{x \in B}$): In this case we know that x is in A and we are also assuming that x is in B. Hence, by definition of intersection, $x \in A \cap B$.
Case 2 ($\mathbf{x \in A \cap B^c}$): In this case we know that x is in A and we are also assuming that x is in B^c. Hence, by definition of intersection, $x \in A \cap B^c$.
Thus, $x \in A \cap B$ or $x \in A \cap B^c$, and so, by definition of union, $x \in (A \cap B) \cup (A \cap B^c)$ *[as was to be shown. Thus $A \subseteq (A \cap B) \cup (A \cap B^c)$ by definition of subset.]*
Conclusion: Since both set containments have been proved, it follows by definition of set equality that $(A \cap B) \cup (A \cap B^c) = A$.

9. Partial proof: Suppose A, B, and C are any sets. To show that $(A - B) \cup (C - B) = (A \cup C) - B$, we must show that $(A - B) \cup (C - B) \subseteq (A \cup C) - B$ and that $(A \cup C) - B \subseteq (A - B) \cup (C - B)$.
Proof that $(A - B) \cup (C - B) \subseteq (A \cup C) - B$: Suppose that x is any element in $(A - B) \cup (C - B)$. *[We must show that $x \in (A \cup C) - B$.]* By definition of union, $x \in A - B$ or $x \in C - B$.
Case 1 ($x \in A - B$): Then, by definition of set difference, $x \in A$ and $x \notin B$. But because $x \in A$, we have that $x \in A \cup C$ by definition of union. Hence $x \in A \cup C$ and $x \notin B$, and so, by definition of set difference, $x \in (A \cup C) - B$.
Case 2 ($x \in C - B$): Then, by definition of set difference, $x \in C$ and $x \notin B$. But because $x \in C$, we have that $x \in A \cup C$ by definition of union. Hence $x \in A \cup C$ and $x \notin B$, and so, by definition of set difference, $x \in (A \cup C) - B$.
Thus, in both cases, $x \in (A \cup C) - B$ *[as was to be shown]*. So $(A - B) \cup (C - B) \subseteq (A \cup C) - B$.
To complete the proof that $(A - B) \cup (C - B) = (A \cup C) - B$, you must show that $(A \cup C) - B \subseteq (A - B) \cup (C - B)$.

11. Partial proof: Suppose A and B are any sets. We will show that $A \cup (A \cap B) \subseteq A$. Suppose x is any element in $A \cup (A \cap B)$. *[We must show that $x \in A$.]* By definition of union, $x \in A$ or $x \in A \cap B$. In the case where $x \in A$, clearly $x \in A$. In the case where $x \in A \cap B$, $x \in A$ and $x \in B$ (by definition of intersection). Thus, in particular, $x \in A$. Hence, in both cases $x \in A$ *[as was to be shown]*.
To complete the proof that $A \cup (A \cap B) = A$, you must show that $A \subseteq A \cup (B \cap A)$.

12. Proof: Let A be a set. *[We must show that $A \cup \emptyset = A$.]*
$A \cup \emptyset \subseteq A$: Suppose $x \in A \cup \emptyset$. Then $x \in A$ or $x \in \emptyset$ by definition of union. But $x \notin \emptyset$ since $\emptyset$ has no elements. Hence $x \in A$.

$A \subseteq A \cup \emptyset$: Suppose $x \in A$. Then the statement "$x \in A$ or $x \in \emptyset$" is true. Hence $x \in A \cup \emptyset$ by definition of union. *[Alternatively, $A \subseteq A \cup \emptyset$ by the inclusion in union property.]* Since $A \cup \emptyset \subseteq A$ and $A \subseteq A \cup \emptyset$, then $A \cup \emptyset = A$ by definition of set equality.

13. Proof: Suppose A, B, and C are sets and $A \subseteq B$. Let $x \in A \cap C$. By definition of intersection, $x \in A$ and $x \in C$. But since $A \subseteq B$ and $x \in A$, then $x \in B$. Hence $x \in B$ and $x \in C$, and so, by definition of intersection, $x \in B \cap C$. *[Thus $A \cap C \subseteq B \cap C$ by definition of subset.]*

16. *Hint:* The proof has the following outline:

Suppose A, B, and C are any sets such that $A \subseteq B$ and $A \subseteq C$.

$$\vdots$$

Therefore, $A \subseteq B \cap C$.

18. <u>Proof</u>: Suppose A, B, and C are arbitrarily chosen sets.

$A \times (B \cup C) \subseteq (A \times B) \cup (A \times C)$: Suppose $(x, y) \in A \times (B \cup C)$. *[We must show that $(x, y) \in (A \times B) \cup (A \times C)$.]* Then $x \in A$ and $y \in B \cup C$. By definition of union, this means that $y \in B$ or $y \in C$.

Case 1 ($y \in B$): Then, since $x \in A$, $(x, y) \in A \times B$ by definition of Cartesian product. Hence $(x, y) \in (A \times B) \cup (A \times C)$ by the inclusion in union property.

Case 2 ($y \in C$): Then, since $x \in A$, $(x, y) \in A \times C$ by definition of Cartesian product. Hence $(x, y) \in (A \times B) \cup (A \times C)$ by the inclusion in union property.

Hence, in either case, $(x, y) \in (A \times B) \cup (A \times C)$ *[as was to be shown].*

Thus $A \times (B \cup C) \subseteq (A \times B) \cup (A \times C)$ by definition of subset.

$(A \times B) \cup (A \times C) \subseteq A \times (B \cup C)$: Suppose $(x, y) \in (A \times B) \cup (A \times C)$. Then $(x, y) \in A \times B$ or $(x, y) \in A \times C$.

Case 1 ($(x, y) \in A \times B$): In this case, $x \in A$ and $y \in B$. By definition of union, since $y \in B$, then $y \in B \cup C$. Hence $x \in A$ and $y \in B \cup C$, and so, by definition of Cartesian product, $(x, y) \in A \times (B \cup C)$.

Case 2 ($(x, y) \in A \times C$): In this case, $x \in A$ and $y \in C$. By definition of union, since $y \in C$, then $y \in B \cup C$. Hence $x \in A$ and $y \in B \cup C$, and so, by definition of Cartesian product, $(x, y) \in A \times (B \cup C)$.

Thus, in either case, $(x, y) \in A \times (B \cup C)$. *[Hence, by definition of subset, $(A \times B) \cup (A \times C) \subseteq A \times (B \cup C)$.]*

[Since both subset relations have been proved, we can conclude that $A \times (B \cup C) = (A \times B) \cup (A \times C)$ by definition of set equality.]

20. There is more than one error in this "proof." The most serious is the misuse of the definition of subset. To say that A is a subset of B means that for all x, **if** $x \in A$ **then** $x \in B$. It does not mean that there exists an element of A that is also an element of B. The second error in the proof occurs in the last sentence. Just because there is an element in A that is in B and an element in B that is in C, it does not follow that there is an element in A that is in C. For instance, suppose $A = \{1, 2\}$, $B = \{2, 3\}$, and $C = \{3, 4\}$. Then there is an element in A that is in B (namely 2) and there is an element in B that is in C (namely 3), but there is no element in A that is in C.

21. *Hint:* The statement "since $x \notin A$ or $x \notin B$, $x \notin A \cup B$" is fallacious. Try to think of an example of sets A and B and an element x such that the statement "$x \notin A$ or $x \notin B$" is true and the statement "$x \notin A \cup B$" is false.

23. a.

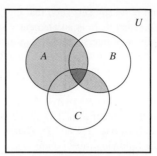

Entire shaded region is $A \cup (B \cap C)$.

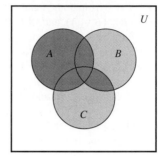

Darkly shaded region is $(A \cup B) \cap (A \cup C)$.

24. **(a)** $(A - B) \cap (B - A)$ **(b)** intersection **(c)** $B - A$ **(d)** B **(e)** A **(f)** A **(g)** $(A - B) \cap (B - A) = \emptyset$

25. Proof by contradiction: Suppose not. That is, suppose there exist sets A and B such that $(A \cap B) \cap (A \cap B^c) \neq \emptyset$. Then there is an element x in $(A \cap B) \cap (A \cap B^c)$. By definition of intersection, $x \in (A \cap B)$ and $x \in (A \cap B^c)$. Applying the definition of intersection again, we have that since $x \in (A \cap B)$, $x \in A$ and $x \in B$, and since $x \in (A \cap B^c)$, $x \in A$ and $x \notin B$. Thus, in particular, $x \in B$ and $x \notin B$, which is a contradiction. It follows that the supposition is false, and so $(A \cap B) \cap (A \cap B^c) = \emptyset$.

27. <u>Proof</u>: Let A be a subset of a universal set U. Suppose $A \cap A^c \neq \emptyset$, that is, suppose there is an element x such that $x \in A \cap A^c$. Then by definition of intersection, $x \in A$ and $x \in A^c$, and so by definition of complement, $x \in A$ and $x \notin A$. This is a contradiction. *[Hence the supposition is false, and we conclude that $A \cap A^c = \emptyset$.]*

29. <u>Proof</u>: Let A be a set. Suppose $A \times \emptyset \neq \emptyset$. Then there would be an element (x, y) in $A \times \emptyset$. By definition of Cartesian product, $x \in A$ and $y \in \emptyset$. But there are no elements y such that $y \in \emptyset$. Hence there are no elements (x, y) such that $x \in A$ and $y \in \emptyset$. Consequently, $(x, y) \notin A \times \emptyset$. *[Thus the supposition is false, and so $A \times \emptyset = \emptyset$.]*

30. <u>Proof</u>: Let A and B be sets such that $A \subseteq B$. *[We must show that $A \cap B^c = \emptyset$.]* Suppose $A \cap B^c \neq \emptyset$; that is, suppose there were an element x such that $x \in A \cap B^c$. Then $x \in A$ and $x \in B^c$ by definition of intersection. So $x \in A$ and $x \notin B$ by definition of complement. But $A \subseteq B$ by hypothesis. So since $x \in A$, $x \in B$ by definition of subset. Thus $x \notin B$ and also $x \in B$, which is a contradiction. Hence the supposition that $A \cap B^c \neq \emptyset$ is false, and so $A \cap B^c = \emptyset$.

33. <u>Proof</u>: Let A, B, and C be any sets such that $C \subseteq B - A$. Suppose $A \cap C \neq \emptyset$. Then there is an element x such that $x \in A \cap C$. By definition of intersection, $x \in A$ and $x \in C$. Since $C \subseteq B - A$, then $x \in B$ and $x \notin A$. So $x \in A$ and $x \notin A$, which is a contradiction. Hence the supposition is false, and thus $A \cap C = \emptyset$.

36. a. *Start of proof that* $A \cup B \subseteq (A - B) \cup (B - A) \cup$ $(A \cap B)$: Given any element x in $A \cup B$, by definition of union x is in at least one of A and B. Thus x satisfies exactly one of the following three conditions:
(1) $x \in A$ and $x \notin B$ (x is in A only)
(2) $x \in B$ and $x \notin A$ (x is in B only)
(3) $x \in A$ and $x \in B$ (x is in both A and B)

b. To show that $(A - B)$, $(B - A)$, and $(A \cap B)$ are mutually disjoint, we must show that the intersection of any two of them is the empty set. But, by definition of set difference and set intersection, saying that $x \in A - B$ means that (1) $x \in A$ and $x \notin B$, saying that $x \in B - A$ means that (2) $x \in B$ and $x \notin A$, and saying that $x \in A \cap B$ means that (3) $x \in A$ and $x \in B$. Conditions (1)–(3) are mutually exclusive, and so no two of them can be satisfied at the same time. Thus no element can be in the intersection of any two of the sets, and, therefore, the intersection of any two of the sets is the empty set. Hence, $(A - B)$, $(B - A)$, and $(A \cap B)$ are mutually disjoint.

37. Suppose A and $B_1, B_2, B_3, \ldots, B_n$ are any sets.

Proof that $A \cap \left(\bigcup_{i=1}^{n} B_i \right) \subseteq \bigcup_{i=1}^{n} (A \cap B_i)$:

Suppose x is any element in $A \cap \left(\bigcup_{i=1}^{n} B_i \right)$. *[We must show that $x \in \bigcup_{i=1}^{n} (A \cap B_i)$.]* By definition of intersection, $x \in A$ and $x \in \bigcup_{i=1}^{n} B_i$. Since $x \in \bigcup_{i=1}^{n} B_i$, the definition of general union implies that $x \in B_i$ for some $i = 1, 2, \ldots, n$, and so, since $x \in A$, the definition of intersection implies that $x \in A \cap B_i$. Thus, by definition of general union, $x \in \bigcup_{i=1}^{n} (A \cap B_i)$ *[as was to be shown].*

Proof that $\bigcup_{i=1}^{n} (A \cap B_i) \subseteq A \cap \left(\bigcup_{i=1}^{n} B_i \right)$:

Suppose x is any element in $\bigcup_{i=1}^{n} (A \cap B_i)$. *[We must show that $x \in A \cap \left(\bigcup_{i=1}^{n} B_i \right)$.]* By definition of general union, $x \in A \cap B_i$ for some $i = 1, 2, \ldots, n$. Thus, by definition of intersection, $x \in A$ and $x \in B_i$. Since $x \in B_i$ for some $i = 1, 2, \ldots, n$, by definition of general union, $x \in \bigcup_{i=1}^{n} B_i$.

Thus we have that $x \in A$ and $x \in \bigcup_{i=1}^{n} B_i$, and so, by definition of intersection, $x \in A \cap \left(\bigcup_{i=1}^{n} B_i \right)$ *[as was to be shown].*

Conclusion: Since both set containments have been proved, it follows by definition of set equality that $A \cap \left(\bigcup_{i=1}^{n} B_i \right) = \bigcup_{i=1}^{n} (A \cap B_i)$.

38. *Proof sketch*: If $x \in \bigcup_{i=1}^{n} (A_i - B)$, then $x \in A_i - B$ for some $i = 1, 2, \ldots, n$, and so, (1) for some $i = 1, 2, \ldots, n$, $x \in A_i$ (which implies that $x \in \left(\bigcup_{i=1}^{n} A_i \right)$) and (2) $x \notin B$.

Conversely, if $x \in \left(\bigcup_{i=1}^{n} A_i \right) - B$, then $x \in \bigcup_{i=1}^{n} A_i$ and $x \notin B$, and so, by definition of general union, $x \in A_i$ for some $i = 1, 2, \ldots, n$, $x \in A_i$ and $x \notin B$. This implies that there is an integer i such that $x \in A_i - B$, and thus that $x \in \bigcup_{i=1}^{n} (A_i - B)$.

40. Suppose A and $B_1, B_2, B_3, \ldots, B_n$ are any sets.

Proof that $\bigcup_{i=1}^{n} (A \times B_i) \subseteq A \times \left(\bigcup_{i=1}^{n} B_i \right)$:

Suppose (x, y) is any element in $\bigcup_{i=1}^{n} (A \times B_i)$. *[We must show that $(x, y) \in A \times \left(\bigcup_{i=1}^{n} B_i \right)$.]* By definition of general union, $(x, y) \in A \times B_i$ for some $i = 1, 2, \ldots, n$. By definition of Cartesian product, this implies that (1) $x \in A$ and (2) $y \in B_i$ for some $i = 1, 2, \ldots, n$. By definition of general union, (2) implies that $y \in \bigcup_{i=1}^{n} B_i$. Thus $x \in A$ and $y \in \bigcup_{i=1}^{n} B_i$, and so by definition of Cartesian product, $(x, y) \in A \times \left(\bigcup_{i=1}^{n} B_i \right)$ *[as was to be shown].*

Proof that $A \times \left(\bigcup_{i=1}^{n} B_i \right) \subseteq \bigcup_{i=1}^{n} (A \times B_i)$:

Suppose (x, y) is any element in $A \times \left(\bigcup_{i=1}^{n} B_i \right)$. *[We must show that $(x, y) \in \bigcup_{i=1}^{n} (A \times B_i)$.]* By definition of Cartesian product, (1) $x \in A$ and (2) $y \in \bigcup_{i=1}^{n} B_i$. By definition of general union, (2) implies that $y \in B_i$ for some $i = 1, 2, \ldots, n$. Thus $x \in A$ and $y \in B_i$ for some $i = 1, 2, \ldots, n$, and so, by definition of Cartesian product, $(x, y) \in A \times B_i$ for some $i = 1, 2, \ldots, n$. It follows from the definition of general union that $(x, y) \in \bigcup_{i=1}^{n} (A \times B_i)$ *[as was to be shown].*

Conclusion: Since both set containments have been proved, it follows by definition of set equality that $\bigcup_{i=1}^{n} (A \times B_i) = A \times \left(\bigcup_{i=1}^{n} B_i \right)$.

Section 6.3

1. Counterexample: Any sets A, B, and C where C contains elements that are not in A will serve as a counterexample. For instance, let $A = \{1, 3\}$, $B = \{2, 3\}$, and $C = \{4\}$. Then $(A \cap B) \cup C = \{3\} \cup \{4\} = \{3, 4\}$, whereas $A \cap (B \cup C) = \{1, 3\} \cap \{2, 3, 4\} = \{3\}$. Since $\{3, 4\} \neq \{3\}$, $(A \cap B) \cup C \neq A \cap (B \cup C)$.

3. Counterexample: Any sets, A, B, and C where $A \subseteq C$ and B contains at least one element that is not in either A or C will serve as a counterexample. For instance, let $A = \{1\}$, $B = \{2\}$, and $C = \{1, 3\}$. Then $A \nsubseteq B$ and $B \nsubseteq C$ but $A \subseteq C$.

5. False. Counterexample: Any sets A, B, and C where A and C have elements in common that are not in B will serve as a counterexample. For instance, let $A = \{1, 2, 3\}$, $B = \{2, 3\}$, and $C = \{3\}$. Then $B - C = \{2\}$, and so $A - (B - C) = \{1, 2, 3\} - \{2\} = \{1, 3\}$. On the other hand $A - B = \{1, 2, 3\} - \{2, 3\} = \{1\}$, and so $(A - B) - C = \{1\} - \{3\} = \{1\}$. Since $\{1, 3\} \neq \{1\}$, $A - (B - C) \neq (A - B) - C$.

6. True. Proof: Let A and B be any sets.

$A \cap (A \cup B) \subseteq A$: Suppose $x \in A \cap (A \cup B)$. By definition of intersection, $x \in A$ and $x \in A \cup B$. In particular $x \in A$. Thus, by definition of subset, $A \cap (A \cup B) \subseteq A$.

$A \subseteq A \cap (A \cup B)$: Suppose $x \in A$. Then by definition of union, $x \in A \cup B$. Hence $x \in A$ and $x \in A \cup B$, and so, by definition of intersection $x \in A \cap (A \cup B)$. Thus, by definition of subset, $A \subseteq A \cap (A \cup B)$.

Because both $A \cap (A \cup B) \subseteq A$ and $A \subseteq A \cap (A \cup B)$ have been proved, we conclude that $A \cap (A \cup B) = A$.

9. True. Proof: Suppose A, B, and C are sets and $A \subseteq C$ and $B \subseteq C$. Let $x \in A \cup B$. By definition of union, $x \in A$ or $x \in B$. But if $x \in A$ then $x \in C$ (because $A \subseteq C$), and if $x \in B$ then $x \in C$ (because $B \subseteq C$). Hence, in either case, $x \subset C$. [So, by definition of subset, $A \cup B \subseteq C$.]

11. Hint: The statement is false. Consider sets U, A, B, and C as follows: $U = \{1, 2, 3, 4\}$, $A = \{1, 2\}$, $B = \{1, 2, 3\}$, and $C = \{2\}$.

12. Hint: The statement is true. *Sketch of proof*: If $x \in A \cap (B - C)$, then $x \in A$ and $x \in B$ and $x \notin C$. So it is true that $x \in A$ and $x \in B$ and that $x \in A$ and $x \notin C$. Conversely, if $x \in (A \cap B) - (A \cap C)$, then $x \in A$ and $x \in B$, but $x \notin A \cap C$, and so $x \notin C$.

14. Hint: The statement is true.

15. Hint: The statement is true. *Sketch of part of proof*: Suppose $x \in A$. [We must show that $x \in B$.] Either $x \in C$ or $x \notin C$. In case $x \in C$, make use of the fact that $A \cap C \subseteq B \cap C$ to show that $x \in B$. In case $x \notin C$, make use of the fact that $A \cup C \subseteq B \cup C$ to show that $x \in B$.

17. True. Proof: Suppose A and B are any sets with $A \subseteq B$. [We must show that $\mathscr{P}(A) \subseteq \mathscr{P}(B)$.] So suppose $X \in \mathscr{P}(A)$. Then $X \subseteq A$ by definition of power set. But because $A \subseteq B$, we also have that $X \subseteq B$ by the transitive property for subsets, and thus, by definition of power set,

$X \in \mathscr{P}(B)$. This proves that for all X, if $X \in \mathscr{P}(A)$ then $X \in \mathscr{P}(B)$, and so $\mathscr{P}(A) \subseteq \mathscr{P}(B)$ [as was to be shown].

18. False. Counterexample: For any sets A and B, $\mathscr{P}(A) \cup \mathscr{P}(B)$ contains only sets that are subsets of either A or B, whereas the sets in $\mathscr{P}(A \cup B)$ can contain elements of both A and B. Thus, if at least one of A or B contains elements that are not in the other set, $\mathscr{P}(A) \cup \mathscr{P}(B)$ and $\mathscr{P}(A \cup B)$ will not be equal. For instance, let $A = \{1\}$ and $B = \{2\}$. Then $\{1, 2\} \in \mathscr{P}(A \cup B)$ but $\{1, 2\} \notin \mathscr{P}(A) \cup \mathscr{P}(B)$.

19. Hint: The statement is true. To prove it, suppose A and B are any sets, and suppose $X \in \mathscr{P}(A) \cup \mathscr{P}(B)$. Show that $X \subseteq A \cup B$, and deduce the conclusion from this result.

22. **a.** *Statement:* $\forall$ sets S, $\exists$ a set T such that $S \cap T = \varnothing$.
   Negation: $\exists$ a set S such that $\forall$ sets T, $S \cap T \neq \varnothing$.
   The statement is true. Given any set S, take $T = S^c$. Then $S \cap T = S \cap S^c = \varnothing$ by the complement law for $\cap$. Alternatively, T could be taken to be $\varnothing$.

23. Hint: $S_0 = \{\varnothing\}$, $S_1 = \{\{a\}, \{b\}, \{c\}\}$

25. **a.** $S_1 = \{\varnothing, \{t\}, \{u\}, \{v\}, \{t, u\}, \{t, v\}, \{u, v\}, \{t, u, v\}\}$
   b. $S_2 = \{\{w\}, \{t, w\}, \{u, w\}, \{v, w\}, \{t, u, w\}, \{t, v, w\}, \{u, v, w\}, \{t, u, v, w\}\}$
   c. Yes

26. Hint: Use mathematical induction. In the inductive step, you will consider the set of all nonempty subsets of $\{2, \ldots, k\}$ and the set of all nonempty subsets of $\{2, \ldots, k + 1\}$. Any subset of $\{2, \ldots, k + 1\}$ either contains $k + 1$ or does not contain $k + 1$. Thus

$$\begin{bmatrix} \text{the sum of all products} \\ \text{of elements of nonempty} \\ \text{subsets of } \{2, \ldots, k + 1\} \end{bmatrix}$$

$$= \begin{bmatrix} \text{the sum of all products} \\ \text{of elements of nonempty} \\ \text{subsets of } \{2, \ldots, k + 1\} \\ \text{that do not contain } k + 1 \end{bmatrix} + \begin{bmatrix} \text{the sum of all products} \\ \text{of elements of nonempty} \\ \text{subsets of } \{2, \ldots, k + 1\} \\ \text{that contain } k + 1 \end{bmatrix}$$

But any subset of $\{2, \ldots, k + 1\}$ that does not contain $k + 1$ is a subset of $\{2, \ldots, k\}$. And any subset of $\{2, \ldots, k + 1\}$ that contains $k + 1$ is the union of a subset of $\{2, \ldots, k\}$ and $\{k + 1\}$.

27. **a.** commutative law for $\cap$
   b. distributive law
   c. commutative law for $\cap$

28. Partial answer:
   a. set difference law
   b. set difference law
   c. commutative law for $\cap$
   d. De Morgan's law

29. Hint: Remember to use the properties in Theorem 6.2.2 exactly as they are written. For example, the distributive law does not state that for all sets A, B, and C, $(A \cup B) \cap C = (A \cap C) \cup (B \cap C)$.

30. Proof: Let sets A, B, and C be given. Then

$(A \cap B) \cup C$

$= C \cup (A \cap B)$	by the commutative law for $\cup$
$= (C \cup A) \cap (C \cup B)$	by the distributive law
$= (A \cup C) \cap (B \cup C)$	by the commutative law for $\cup$.

31. Proof: Suppose A and B are sets. Then

$A \cup (B - A)$

$= A \cup (B \cap A^c)$	by the set difference law
$= (A \cup B) \cap (A \cup A^c)$	by the distributive law
$= (A \cup B) \cap U$	by the complement law for $\cup$
$= A \cup B$	by the identity law for $\cap$.

36. Proof: Let A and B, be any sets. Then

$((A^c \cup B^c) - A)^c$

$= ((A^c \cup B^c) \cap A^c)^c$	by the set difference law
$= (A^c \cup B^c)^c \cup (A^c)^c$	by De Morgan's law
$= ((A^c)^c \cap (B^c)^c) \cup (A^c)^c$	by De Morgan's law
$= (A \cap B) \cup A$	by the double complement law
$= A \cup (A \cap B)$	by the commutative law for $\cup$
$= A$	by the absorption law

39. Partial proof: Let A and B be any sets. Then

$(A - B) \cup (B - A)$

$= (A \cap B^c) \cup (B \cap A^c)$	by the set difference law
$= [(A \cap B^c) \cup B] \cap [(A \cap B^c) \cup A^c)]$	
	by the distributive law
$= [(B \cup (A \cap B^c)] \cap [A^c \cup (A \cap B^c)]$	
	by the commutative law for $\cup$
$= [(B \cup A) \cap (B \cup B^c)] \cap [(A^c \cup A) \cap (A^c \cup B^c)]$	
	by the distributive law
$= [(A \cup B) \cap (B \cup B^c)] \cap [(A \cup A^c) \cap (A^c \cup B^c)]$	
	by the commutative law for $\cup$

41. Hint: The answer is $\emptyset$.

44. a. Proof: Suppose not. That is, suppose there exist sets A and B such that $A - B$ and B are not disjoint. *[We must derive a contradiction.]* Then $(A - B) \cap B \neq \emptyset$, and so there is an element x in $(A - B) \cap B$. By definition of intersection, $x \in A - B$ and $x \in B$, and by definition of difference, $x \in A$ and $x \notin B$. Hence $x \in B$ and also $x \notin B$, which is a contradiction. Thus the supposition is false, and we conclude that $A - B$ and B are disjoint.

b. Let A and B be any sets. Then

$(A - B) \cap B$

$= (A \cap B^c) \cap B$	by the set difference law
$= A \cap (B^c \cap B)$	by the associative law for $\cap$
$= A \cap (B \cap B^c)$	by the commutative law for $\cap$
$= A \cap \emptyset$	by the complement law for $\cap$
$= \emptyset$	by the universal bound law for $\cap$.

46. a. $A \triangle B = (A - B) \cup (B - A) = \{1, 2\} \cup \{5, 6\} = \{1, 2, 5, 6\}$

47. Proof: Let A and B be any subsets of a universal set. By definition of $\triangle$, showing that $A \triangle B = B \triangle A$ is equivalent to showing that $(A - B) \cup (B - A) = (B - A) \cup (A - B)$. But this follows immediately from the commutative law for $\cup$.

48. Proof: Let A be any subset of a universal set. Then

$A \triangle \emptyset$

$= (A - \emptyset) \cup (\emptyset - A)$	by definition of $\triangle$
$= (A \cap \emptyset^c) \cup (\emptyset \cap A^c)$	by the set difference law
$= (A \cap U) \cup (A^c \cap \emptyset)$	by the complement of U law and the commutative law for $\cap$
$= A \cup \emptyset$	by the identity law for $\cap$ and the universal bound law for $\cap$
$= A$.	by the identity law for $\cup$

51. Hint: First show that for any sets A and B and for any element x,

$x \in A \triangle B \Leftrightarrow (x \in A \text{ and } x \notin B) \text{ or } (x \in B \text{ and } x \notin A),$

and

$x \notin A \triangle B \Leftrightarrow (x \notin A \text{ and } x \notin B) \text{ or } (x \in B \text{ and } x \in A).$

52. Same hint as for exercise 51.

53. *Start of proof*: Suppose A and B are any subsets of a universal set U. By the universal bound law for union, $B \cup U = U$, and so, by the commutative law for union, $U \cup B = U$. Take the intersection of both sides of the equation with A.

Section 6.4

1. a. because 1 is an identity for $\cdot$
b. by the complement law for $+$
c. by the distributive law for $+$ over $\cdot$
d. by the complement law for $\cdot$
e. because 0 is an identity for $+$

4. Proof: For all elements a in B,

$a \cdot 0 = a \cdot (a \cdot \overline{a})$	by the complement law for $\cdot$
$= (a \cdot a) \cdot \overline{a}$	by the associative law for $\cdot$
$= a \cdot \overline{a}$	by exercise 1
$= 0$.	by the complement law for $\cdot$

6. a. Proof: $0 \cdot 1 = 0$ because 1 is an identity for $\cdot$, and $0 + 1 = 1 + 0 = 1$ because $+$ is commutative and 0 is an identity for $+$. Thus, by the uniqueness of the complement law, $\overline{0} = 1$.

7. a. Proof: Suppose 0 and $0'$ are elements of B both of which are identities for $+$. Then both 0 and $0'$ satisfy the identity, complement, and universal bound laws. *[We will show that $0 = 0'$.]* By the identity law for $+$, for all $a \in B$,

$a + 0 = a \quad \text{and} \quad a + 0' = a.$

It follows that

$$
\begin{array}{rrcll}
\Rightarrow & a + 0 &=& a + 0' & \text{because both quantities equals } a \\
\Rightarrow & \bar{a} \cdot (a + 0) &=& \bar{a} \cdot (a + 0') & \text{by "multiplying" both sides by } \bar{a} \\
\Rightarrow & (\bar{a} \cdot a) + (\bar{a} \cdot 0) &=& (\bar{a} \cdot a) + (\bar{a} \cdot 0') & \text{by the distributive law} \\
\Rightarrow & (a \cdot \bar{a}) + 0 &=& (a \cdot \bar{a}) + 0' & \text{by the universal bound law for } \cdot \\
\Rightarrow & 0 \cdot 0 &=& 0' \cdot 0' & \text{by the complement law for } \cdot \\
\Rightarrow & 0 &=& 0' & \text{by the universal bound law for } \cdot
\end{array}
$$

[This is what was to be shown.]

b. *Hint:* Suppose 1 and $1'$ are elements of B both of which are identities for $\cdot$. Then for all $a \in B$, by the identity law for $\cdot$, $a \cdot 1 = a$ and $a \cdot 1' = a$. It follows that $a \cdot 1 = a \cdot 1'$, and $\bar{a} + a \cdot 1 = \bar{a} + a \cdot 1'$. Etc.

8. <u>Proof:</u> Suppose B is a Boolean algebra and a and b are any elements of B. We first prove that $(a \cdot b) + (\bar{a} + \bar{b}) = 1$.

$a \cdot b + (\bar{a} + \bar{b})$

$\quad = (\bar{a} + \bar{b}) + (a \cdot b)$

$\qquad$ by the commutative law for $+$

$\quad = ((\bar{a} + \bar{b}) + a) \cdot ((\bar{a} + \bar{b}) + b)$

$\qquad$ by the distributive law of $+$ over $\cdot$

$\quad = ((\bar{b} + \bar{a}) + a) \cdot (\bar{a} + (\bar{b} + b))$

$\qquad$ by the commutative and associative laws for $+$

$\quad = (\bar{b} + (\bar{a} + a)) \cdot (\bar{a} + (b + \bar{b}))$

$\qquad$ by the associative and commutative laws for $+$

$\quad = (\bar{b} + (a + \bar{a})) \cdot (\bar{a} + 1)$

$\qquad$ by the commutative and complement laws for $+$

$\quad = (\bar{b} + 1) \cdot 1$

$\qquad$ by the complement and universal bound laws for $+$

$\quad = 1 \cdot 1$

$\qquad$ by the universal bound law for $+$

$\quad = 1$

$\qquad$ by the identity law for $\cdot$.

Next we prove that $(a \cdot b) \cdot (\bar{a} + \bar{b}) = 0$.

$(a \cdot b) \cdot (\bar{a} + \bar{b})$

$\quad = ((a \cdot b) \cdot \bar{a}) + (((a \cdot b) \cdot \bar{b})$

$\qquad$ by the distributive law of $\cdot$ over $+$

$\quad = ((b \cdot a) \cdot \bar{a}) + ((a \cdot (b \cdot \bar{b}))$

$\qquad$ by the commutative and associative laws for $\cdot$

$\quad = (b \cdot (a \cdot \bar{a})) + (a \cdot 0)$

$\qquad$ by the associative and complement laws for $\cdot$

$\quad = (b \cdot 0) + 0$

$\qquad$ by the complement and universal bound laws for $\cdot$

$\quad = 0 + 0 \quad$ by the universal bound law for $\cdot$

$\quad = 0 \qquad$ by the identity law for $+$.

Because both $(a \cdot b) + (\bar{a} + \bar{b}) = 1$ and $(a \cdot b) \cdot (\bar{a} + \bar{b}) = 0$, it follows, by the uniqueness of the complement law, that $\overline{a \cdot b} = \bar{a} + \bar{b}$.

10. *Hint:* One way to prove the statement is to use the result of exercise 3. Some stages in the proof are the following:

$$y = (y + x) \cdot y = (x \cdot y) + (z \cdot y) = z \cdot (x + y) = z.$$

11. a. (i) Because S has only two distinct elements, 0 and 1, we only need to check that $0 + 1 = 1 + 0$. But this is true because both sums equal 1.

(v) *Partial answer:*

$0 + (0 \cdot 0) = 0 + 0 = 0$ and $(0 + 0) \cdot (0 + 0) = 0 \cdot 0 = 0$ also

$0 + (0 \cdot 1) = 0 + 0 = 0$ and $(0 + 0) \cdot (0 + 1) = 0 \cdot 1 = 0$ also

$0 + (1 \cdot 0) = 0 + 0 = 0$ and $(0 + 1) \cdot (0 + 0) = 1 \cdot 0 = 0$ also

$0 + (1 \cdot 1) = 0 + 1 = 1$ and $(0 + 1) \cdot (0 + 1) = 1 \cdot 1 = 1$ also

b. *Hint:* Verify that $0 + x = x$ and that $1 \cdot x = x$ for all $x \in S$.

12. *Hints:* (1) The universal bound law $a + 1 = 1$ can be derived without using the associative law by using $(a + 1) \cdot (a + \bar{a}) = a + 1 \cdot \bar{a}$. Similarly, $a \cdot 0 = 0$ can be derived using $(a \cdot 1) + (a \cdot \bar{a}) = a \cdot (1 + \bar{a})$. To derive the absorption laws without using the associative law, note that $a + a \cdot b = a \cdot 1 + a \cdot b = a \cdot (1 + b) = a \cdot (b + 1) = a \cdot 1 = a$. The other absorption law can be derived using the same sequence of steps but changing each $+$ to $\cdot$ and each $\cdot$ to $+$.

(2) Show that for all x, y, and z in B, $x \cdot (x + (y + z)) \cdot x = x$ and $((x + y) + z)) \cdot x = x$.

(3) Show that for all a, b, and c in B, both $a + (b + c)$ and $(a + b) + c$ equal $((a + b) + c) \cdot (a + (b + c))$.

(4) The other associative law can be derived using the hints (2) and (3) but changing each $+$ to $\cdot$ and each $\cdot$ to $+$.

13. The sentence is not a statement because it is neither true nor false. If the sentence were true, then because it declares itself to be false, the sentence would be false. Therefore, the sentence is not true. On the other hand, if the sentence were false, then it would be false that "This sentence is false," and so the sentence would be true. Consequently, the sentence is not false.

14. This sentence is a statement because it is true. Recall that the only way for an if-then statement to be false is for the hypothesis to be true and the conclusion false. In this case the hypothesis is not true. So regardless of what the conclusion states, the sentence is true. (This is an example of a statement that is vacuously true, or true by default.)

17. This sentence is not a statement because it is neither true nor false. If the sentence were true, then either the sentence is false or $1 + 1 = 3$. But $1 + 1 \neq 3$, and so the sentence is false. Therefore, the sentence is not true. On the other hand, if the sentence were false, then it would be true that "This sentence is false or $1 + 1 = 3$," and so the sentence would be true. Consequently, the sentence is not false.

20. *Hint:* Suppose that apart from statement (ii), all of Nixon's other assertions about Watergate are evenly split between true and false.

21. No. Suppose there were a computer program P that had as output a list of all computer programs that do not list themselves in their output. If P lists itself as output, then it would be on the output list of P, which consists of all computer programs that do not list themselves in their output. Hence P would not list itself as output. But if P does not list itself as output, then P would be a member of the list of all computer programs that do not list themselves in their output, and this list is exactly the output of P. Hence P would list itself as output. This analysis shows that the assumption of the existence of such a program P is contradictory, and so no such program exists.

Section 7.1

1. a. domain of $f = \{1, 3, 5\}$, co-domain of $f = \{s, t, u, v\}$
   b. $f(1) = v$, $f(3) = s$, $f(5) = v$
   c. range of $f = \{s, v\}$
   d. yes, no
   e. inverse image of $s = \{3\}$, inverse image of $u = \emptyset$, inverse image of $v = \{1, 5\}$
   f. $\{(1, v), (3, s), (5, v)\}$

3. a. True. The definition of function says that for any input there is one and only one output, so if two inputs are equal, their outputs must also be equal.
   c. True. The definition of function does not prohibit this occurrence.

4. a. There are four functions from X to Y as shown below.

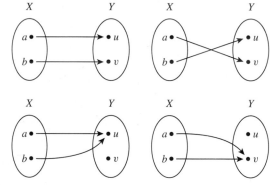

5. a. $I_{\mathbf{Z}}(e) = e$
   b. $I_{\mathbf{Z}}\left(b_i^{jk}\right) = b_i^{jk}$

6. a. The sequence is given by the function $f: \mathbf{Z}^{nonneg} \to \mathbf{R}$ defined by the rule
$$f(n) = \frac{(-1)^n}{2n + 1} \quad \text{for all nonnegative integers } n.$$

7. a. 1 *[because there is an odd number of elements in $\{1, 3, 4\}$]*
   c. 0 *[because there is an even number of elements in $\{2, 3\}$]*

8. a. $F(0) = (0^3 + 2 \cdot 0 + 4) \bmod 5 = 4 \bmod 5 = 4$
   b. $F(1) = (1^3 + 2 \cdot 1 + 4) \bmod 5 = 7 \bmod 5 = 2$

9. a. $S(1) = 1$ **b.** $S(15) = 1 + 3 + 5 + 15 = 24$
   c. $S(17) = 1 + 17 = 18$

10. a. $T(1) = \{1\}$ **b.** $T(15) = \{1, 3, 5, 15\}$
   c. $T(17) = \{1, 17\}$

11. a. $F(4, 4) = (2 \cdot 4 + 1, 3 \cdot 4 - 2) = (9, 10)$
   b. $F(2, 1) = (2 \cdot 2 + 1, 3 \cdot 1 - 2) = (5, 1)$

12. a. $G(4, 4) = ((2 \cdot 4 + 1) \bmod 5, (3 \cdot 4 - 2) \bmod 5) =$
   $(9 \bmod 5, 10 \bmod 5) = (4, 0)$
   b. $G(2, 1) = ((2 \cdot 2 + 1) \bmod 5, (3 \cdot 1 - 2) \bmod 5) =$
   $(5 \bmod 5, 1 \bmod 5) = (0, 1)$

13.

x	$f(x)$	$g(x)$
0	$4^2 \bmod 5 = 1$	$(0^2 + 3 \cdot 0 + 1) \bmod 5 = 1$
1	$5^2 \bmod 5 = 0$	$(1^2 + 3 \cdot 1 + 1) \bmod 5 = 0$
2	$6^2 \bmod 5 = 1$	$(2^2 + 3 \cdot 2 + 1) \bmod 5 = 1$
3	$7^2 \bmod 5 = 4$	$(3^2 + 3 \cdot 3 + 1) \bmod 5 = 4$
4	$8^2 \bmod 5 = 4$	$(4^2 + 3 \cdot 4 + 1) \bmod 5 = 4$

The table shows that $f(x) = g(x)$ for all x in J_5. Thus, by definition of equality of functions, $f = g$.

15. $F \cdot G$ and $G \cdot F$ are equal because for all real numbers x,
$(F \cdot G)(x) = F(x) \cdot G(x)$ by definition of $F \cdot G$
$\qquad\qquad = G(x) \cdot F(x)$ by the commutative law for multiplication of real numbers
$\qquad\qquad = (G \cdot F)(x)$ by definition of $G \cdot F$.

17. a. $2^3 = 8$ **c.** $4^1 = 4$

18. a. $\log_3 81 = 4$ because $3^4 = 81$
   c. $\log_3 \left(\frac{1}{27}\right) = -3$ because $3^{-3} = \frac{1}{27}$

19. Let b be any positive real number with $b \neq 1$. Since $b^1 = b$, by definition of logarithm, $\log_b b = 1$.

21. Proof: Suppose b and u are any positive real numbers. *[We must show that $\log_b \left(\frac{1}{u}\right) = - \log_b(u)$.]* Let $v = \log_b \left(\frac{1}{u}\right)$. By definition of logarithm, $b^v = \frac{1}{u}$. Multiplying both sides by u and dividing by b^v gives $u = b^{-v}$, and thus, by definition of logarithm, $-v = \log_b(u)$. Now multiply both sides of this equation by -1 to obtain $v = - \log_b(u)$. Therefore, $\log_b \left(\frac{1}{u}\right) = - \log_b(u)$ because both expressions equal v. *[This is what was to be shown.]*

22. *Hint:* Use a proof by contradiction. Suppose $\log_3 7$ is rational. Then $\log_3 7 = \frac{a}{b}$ for some integers a and b with $b \neq 0$. Apply the definition of logarithm to rewrite $\log_3 7 = \frac{a}{b}$ in exponential form.

23. Suppose b and y are positive real numbers with $\log_b y = 3$. By definition of logarithm, this implies that $b^3 = y$. Then
$$y = b^3 = \frac{1}{\frac{1}{b^3}} = \frac{1}{\left(\frac{1}{b}\right)^3} = \left(\frac{1}{b}\right)^{-3}.$$
Thus, by definition of logarithm (with base $1/b$), $\log_{1/b}(y) = -3$.

25. a. $p_1(2, y) = 2$, $p_1(5, x) = 5$, range of $p_1 = \{2, 3, 5\}$

26. a. $mod(67, 10) = 7$ and $div(67, 10) = 6$ since $67 = 10 \cdot 6 + 7$.

27. If g were well defined, then $g(1/2) = g(2/4)$ because $1/2 = 2/4$. However, $g(1/2) = 1 - 2 = -1$ and $g(2/4) = 2 - 4 = -2$. Since $-1 \neq -2$, $g(1/2) \neq g(2/4)$. Thus g is not well defined.

29. Student B is correct. If R were well defined, then $R(3)$ would have a uniquely determined value. However, on the one hand, $R(3) = 2$ because $(3 \cdot 2) \bmod 5 = 1$, and, on the other hand, $R(3) = 7$ because $(3 \cdot 7) \bmod 5 = 1$. Hence $R(3)$ does not have a uniquely determined value, and so R is not well defined.

31. a.

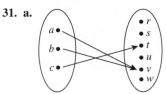

b. $f(A) = \{v\}$, $\quad f(X) = \{t, v\}$, $\quad f^{-1}(C) = \{c\}$, $\quad f^{-1}(D) = \{a, b\}$, $\quad f^{-1}(E) = \emptyset$ $\quad f^{-1}(Y) = \{a, b, c\} = X$

33. *Partial answer:* **(a)** $y \in F(A)$ or $y \in F(B)$, **(b)** some, **(c)** $A \cup B$, **(d)** $F(A \cup B)$

34. The statement is true. Proof: Let F be a function from X to Y, and suppose $A \subseteq X$, $B \subseteq X$, and $A \subseteq B$. To show that $F(A) \subseteq F(B)$, let $y \in F(A)$. *[We must show that $y \in F(B)$.]* Then, by definition of image of a set, $y = F(x)$ for some $x \in A$. Since $A \subseteq B$, $x \in B$, and so $y = F(x)$ for some $x \in B$. Hence $y \in F(B)$ *[as was to be shown].* Thus $F(A) \subseteq F(B)$.

36. The statement is false. Counterexample: Let $X = \{1, 2, 3\}$, let $Y = \{a, b\}$, and define a function $F: X \to Y$ by the arrow diagram shown below.

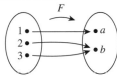

Let $A = \{1, 2\}$ and $B = \{1, 3\}$. Then $F(A) = \{a, b\} = F(B)$, and so $F(A) \cap F(B) = \{a, b\}$. But $F(A \cap B) = F(\{1\}) = \{a\} \neq \{a, b\}$. And so $F(A) \cap F(B) \not\subseteq F(A \cap B)$. (This is just one of many possible counterexamples.)

38. The statement is true. Proof: Let F be a function from a set X to a set Y, and suppose $C \subseteq Y$, $D \subseteq Y$, and $C \subseteq D$. *[We must show that $F^{-1}(C) \subseteq F^{-1}(D)$.]* Suppose $x \in F^{-1}(C)$. Then $F(x) \in C$. Since $C \subseteq D$, $F(x) \in D$ also. Hence by definition of inverse image, $x \in F^{-1}(D)$. *[So every element in $F^{-1}(C)$ is in $F^{-1}(D)$, and thus $F^{-1}(C) \subseteq F^{-1}(D)$.]*

39. *Hint:* $x \in F^{-1}(C \cup D) \Leftrightarrow F(x) \in C \cup D \Leftrightarrow F(x) \in C$ or $F(x) \in D$

44. a. $\phi(15) = 8$ *[because 1, 2, 4, 7, 8, 11, 13, and 14 have no common factors with 15 other than ± 1]*

b. $\phi(2) = 1$ *[because the only positive integer less than or equal to 2 having no common factors with 2 other than ± 1 is 1]*

c. $\phi(5) = 4$ *[because 1, 2, 3, and 4 have no common factors with 5 other than ± 1]*

45. Proof: Let p be any prime number and n any integer with $n \geq 1$. There are p^{n-1} positive integers less than or equal to p^n that have a common factor other than ± 1 with p^n,

namely $p, 2p, 3p, \ldots, (p^{n-1})p$. Hence, by the difference rule, there are $p^n - p^{n-1}$ positive integers less than or equal to p^n that have no common factor with p^n except ± 1.

46. *Hint:* Use the result of exercise 52 with $p = 2$.

Section 7.2

1. The second statement is the contrapositive of the first.

2. a. most **b.** least

3. *Hint:* One counterexample is given and explained below. Give a different counterexample and accompany it with an explanation. Counterexample: Consider the function defined by the following arrow diagram:

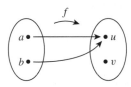

Observe that a is sent to exactly one element of Y, namely, u, and b is also sent to exactly one element of Y, namely, u also. So it is true that every element of X is sent to exactly one element of Y. But f is not one-to-one because $f(a) = f(b)$ but $a \neq b$. *[Note that to say, "Every element of X is sent to exactly one element of Y" is just another way of saying that in the arrow diagram for the function there is only one arrow coming out of each element of X. But this statement is part of the definition of any function, not just of a one-to-one function.]*

4. *Hint:* The statement is true.

5. *Hint:* One of the incorrect ways is (b).

6. a. f is not one-to-one because $f(1) = 4 = f(9)$ and $1 \neq 9$. f is not onto because $f(x) \neq 3$ for any x in X.

b. g is one-to-one because $g(1) \neq g(5)$, $g(1) \neq g(9)$, and $g(5) \neq g(9)$. g is onto because each element of Y is the image of some element of X: $3 = g(5)$, $4 = g(9)$, and $7 = g(1)$.

7. a. F is not one-to-one because $F(c) = x = F(d)$ and $c \neq d$. F is onto because each element of Y is the image of some element of X: $x = F(c) = F(d)$, $y = F(a)$, and $z = F(b)$.

9. a. One example of many is the following:

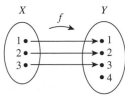

10. a. (i) f *is one-to-one:* Suppose $f(n_1) = f(n_2)$ for some integers n_1 and n_2. *[We must show that $n_1 = n_2$.]* By definition of f, $2n_1 = 2n_2$, and dividing both sides by 2 gives $n_1 = n_2$, as was to be shown.

(ii) f *is not onto:* Consider $1 \in \mathbf{Z}$. We claim that $1 \neq f(n)$, for any integer n, because if there were an integer n such that $1 = f(n)$, then, by definition of f, $1 = 2n$. Dividing both sides by 2 would give $n = 1/2$. But $1/2$ is not an integer. Hence $1 \neq f(n)$ for any integer n, and so f is not onto.

b. h *is onto:* Suppose $m \in 2\mathbf{Z}$. *[We must show that there exists an integer n such that $h(n) = m$.] Since $m \in 2\mathbf{Z}$, $m = 2k$ for some integer k. Let $n = k$. Then $h(n) = 2n = 2k = m$. Hence there exists an integer (namely, n) such that $h(n) = m$. This is what was to be shown.*

11. *Hints:* **a.** (i) g is one-to-one   (ii) g is not onto
   b. G is onto. Proof: Suppose y is any element of $\mathbf{R}$. *[We must show that there is an element x in $\mathbf{R}$ such that $G(x) = y$. What would x be if it exists? Scratch work shows that x would have to equal $(y + 5)/4$. The proof must then show that x has the necessary properties.]* Let $x = (y + 5)/4$. Then (1) $x \in \mathbf{R}$, and (2) $G(x) = G((y + 5)/4) = 4[(y + 5)/4] - 5 = (y + 5) - 5 = y$ *[as was to be shown]*.

13. a. (i) H *is not one-to-one:* $H(1) = 1 = H(-1)$ but $1 \neq -1$.
   (ii) H *is not onto:* $H(x) \neq -1$ for any real number x (since no real numbers have negative squares).

14. The "proof" claims that f is one-to-one because for each integer n there is only one possible value for $f(n)$. But to say that for each integer n there is only one possible value for $f(n)$ is just another way of saying that f satisfies one of the conditions necessary for it to be a function. To show that f is one-to-one, one must show that any integer n has a *different* function value from that of the integer m whenever $n \neq m$.

15. f is one-to-one. Proof: Suppose $f(x_1) = f(x_2)$ where x_1 and x_2 are nonzero real numbers. *[We must show that $x_1 = x_2$.]* By definition of f,

$$\frac{x_1 + 1}{x_1} = \frac{x_2 + 1}{x_2}$$

cross-multiplying gives

$$x_1 x_2 + x_2 = x_1 x_2 + x_1,$$

and so

$$x_1 = x_2 \quad \text{by subtracting } x_1 x_2 \text{ from both sides}$$

[This is what was to be shown.]

16. f is not one-to-one. Note that

$$\frac{x_1}{x_1^2 + 1} = \frac{x_2}{x_2^2 + 1} \Rightarrow x_1 x_2^2 + x_1 = x_2 x_1^2 + x_2$$
$$\Rightarrow x_1 x_2^2 - x_2 x_1^2 = x_2 - x_1$$
$$\Rightarrow x_1 x_2 (x_2 - x_1) = x_2 - x_1$$
$$\Rightarrow x_1 = x_2 \text{ or } x_1 x_2 = 1.$$

Thus for a counterexample take any x_1 and x_2 with $x_1 \neq x_2$ but $x_1 x_2 = 1$. For instance, take $x_1 = 2$ and $x_2 = 1/2$. Then $f(x_1) = f(2) = 2/5$ and $f(x_2) = f(1/2) = 2/5$, but $2 \neq 1/2$.

19. a. F *is not one-to-one:* Let $A = \{a\}$ and $B = \{b\}$. Then $F(A) = F(B) = 1$ but $A \neq B$.

20. S is not one-to-one. Counterexample: $S(6) = 1 + 2 + 3 + 6 = 12$ and $S(11) = 1 + 11 = 12$. So $S(6) = S(11)$ but $6 \neq 11$.
S is not onto. Counterexample: In order for there to be a positive integer n such that $S(n) = 5$, n would have to be less than 5. But $S(1) = 1$, $S(2) = 3$, $S(3) = 4$, and $S(4) = 7$. Hence there is no positive integer n such that $S(n) = 5$.

21. *Hint:* **a.** T is one-to-one.   **b.** T is not onto.

22. a. G is one-to-one. Proof: Suppose (x_1, y_1) and (x_2, y_2) are any elements of $\mathbf{R} \times \mathbf{R}$ such that $G(x_1, y_1) = G(x_2, y_2)$. *[We must show that $(x_1, y_1) = (x_2, y_2)$.]* Then, by definition of G, $(2y_1, -x_1) = (2y_2, -x_2)$, and, by definition of ordered pair,

$$2y_1 = 2y_2 \quad \text{and} \quad -x_1 = -x_1.$$

Dividing both sides of the left equation by 2 and both sides of the right equation by -1 gives that

$$y_1 = y_2 \quad \text{and} \quad x_1 = x_2,$$

and so, by definition of ordered pair, $(x_1, y_1) = (x_2, y_2)$ *[as was to be shown]*.

b. G is onto. Proof: Suppose (u, v) is any element of $\mathbf{R} \times \mathbf{R}$. *[We must show that there is an element (x, y) in $\mathbf{R} \times \mathbf{R}$ such that $G(x, y) = (u, v)$.]* Let $(x, y) = (-v, u/2)$. Then (1) $(x, y) \in \mathbf{R} \times \mathbf{R}$ and (2) $G(x, y) = (2y, -x) = (2(u/2), -(-v)) = (u, v)$ *[as was to be shown.]*

25. a. *Hint:* F is one-to-one. Use the unique factorization of integers theorem in the proof.

26. a. Let $x = \log_8 27$ and $y = \log_2 3$. *[The question is: Is $x = y$?]* By definition of logarithm, both of these equations can be written in exponential form as

$$8^x = 27 \quad \text{and} \quad 2^y = 3.$$

Now $8 = 2^3$. So

$$8^x = (2^3)^x = 2^{3x}.$$

Also $27 = 3^3$ and $3 = 2^y$. So

$$27 = 3^3 = (2^y)^3 = 2^{3y}.$$

Hence, since $8^x = 27$,

$$2^{3x} = 2^{3y}.$$

By (7.2.5), then,

$$3x = 3y,$$

and so

$$x = y.$$

But $x = \log_8 27$ and $y = \log_2 3$, and so $\log_8 27 = y = \log_2 3$ and the answer to the question is yes.

27. Proof: Suppose that b, x, and y are positive real numbers and $b \neq 1$. Let $u = \log_b(x)$ and $v = \log_b(y)$. By definition of logarithm, $b^u = x$ and $b^v = y$. By substitution, $\frac{x}{y} = \frac{b^u}{b^v} = b^{u-v}$ *[by (7.2.3) and the fact that $b^{-v} = \frac{1}{b^v}$]*. Translating $\frac{x}{y} = b^{u-v}$ into logarithmic form gives

$\log_b\left(\frac{x}{y}\right) = u - v$, and so, by substitution, $\log_b\left(\frac{x}{y}\right) = \log_b(x) - \log_b(y)$ *[as was to be shown]*.

29. *Start of Proof:* Suppose a, b, and x are *[particular but arbitrarily chosen]* real numbers such that b and x are positive and $b \neq 1$. *[We must show that $\log_b(x^a) = a\,\log_b x$.]* Let

$$r = \log_b(x^a) \text{ and } s = \log_b x.$$

30. No. Counterexample: Define $f: \mathbf{R} \to \mathbf{R}$ and $g: \mathbf{R} \to \mathbf{R}$ as follows: $f(x) = x$ and $g(x) = -x$ for all real numbers x. Then f and g are both one-to-one *[because for all real number x_1 and x_2, if $f(x_1) = f(x_2)$ then $x_1 = x_2$, and if $g(x_1) = g(x_2)$ then $-x_1 = -x_2$ and so $x_1 = x_2$ also]*, but $f + g$ is not one-to-one *[because $f + g$ satisfies the equation $(f + g)(x) = x + (-x) = 0$ for all real numbers x, and so, for instance, $(f + g)(1) = (f + g)(2)$ but $1 \neq 2$]*.

32. Yes. Proof: Let f be a one-to-one function from $\mathbf{R}$ to $\mathbf{R}$, and let c be any nonzero real number. Suppose $(cf)(x_1) = (cf)(x_2)$. *[We must show that $x_1 = x_2$.]* It follows by definition of cf that $cf(x_1) = cf(x_2)$. Since $c \neq 0$, we may divide both sides of the equation by c to obtain $f(x_1) = f(x_2)$. But since f is one-to-one, this implies that $x_1 = x_2$ *[as was to be shown]*.

34. a. *Hint:* The assumption that F is one-to-one is needed in the proof that $F^{-1}(F(A)) \subseteq A$. If $F(r) \in F(A)$, the definition of image of a set implies that there is an element x in A such that $F(r) = F(x)$.

b. *Hint:* The assumption that F is one-to-one is needed in the proof that $F(A_1) \cap F(A_2) \subseteq F(A_1 \cap A_2)$. If $u \in F(A_1)$ and $u \in F(A_2)$, then the definition of image of a set implies that there are elements x_1 in A_1 and x_2 in A_2 such that $F(x_1) = u$ and $F(x_2) = u$ and, thus, that $F(x_1) = F(x_2)$.

36.

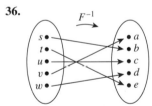

38. The function is not onto. Hence it is not a one-to-one correspondence.

39. The answer to exercise 10(b) shows that h is onto. To show that h is one-to-one, suppose $h(n_1) = h(n_2)$. By definition of h, this implies that $2n_1 = 2n_2$. Dividing both sides by 2 gives $n_1 = n_2$. Hence h is one-to-one. Given any even integer m, if $m = h(n)$, then by definition of h, $m = 2n$, and so $n = m/2$. Thus

$$h^{-1}(m) = \frac{m}{2} \text{ for all } m \in 2\mathbf{Z}.$$

40. The function g is not a one-to-one correspondence because it is not onto. For instance, if $m = 2$, it is impossible to find an integer n such that $g(n) = m$. (This is because if $g(n) = m$, then $4n - 5 = 2$, which implies that $n = 7/4$. Thus the

only number n with the property that $g(n) = m$ is 7/4. But 7/4 is not an integer.)

41. The answer to exercise 11b shows that G is onto. In addition, G is one-to-one. To prove this, suppose $G(x_1) = G(x_2)$ for some x_1 and x_2 in $\mathbf{R}$. *[We must show that $x_1 = x_2$.]* By definition of G, $4x_1 - 5 = 4x_2 - 5$. Add 5 to both sides of this equation and divide both sides by 4 to obtain $x_1 = x_2$ *[as was to be shown]*. We claim that $G^{-1}(y) = (y + 5)/4$. By definition of inverse function, this is true if, and only if, $G((y + 5)/4) = y$. But $G((y + 5)/4) = 4((y + 5)/4) - 5 = (y + 5) - 5 = y$, so it is the case that $G^{-1}(y) = (y + 5)/4$.

44. The solution to exercise 22 showed that G is one-to-one and onto, and so G is a one-to-one correspondence. Given any element (u, v) in $\mathbf{R} \times \mathbf{R}$ (the co-domain of G), $G^{-1}(u, v) = (-v, \frac{u}{2})$ because (1) both $-v$ and $\frac{u}{2}$ are real numbers, and (2) by definition of G,

$$G\left(-v, \frac{u}{2}\right) = (2\left(\frac{u}{2}\right), -(-v)) = (u, v).$$

46. The answer to exercise 15 shows that f is one-to-one, and if the co-domain is taken to be the set of all real numbers not equal to 1, then f is also onto. The reason is that given any real number $y \neq 1$, if we take $x = \frac{1}{y-1}$, then

$$f(x) = f\left(\frac{1}{y - 1}\right) = \frac{\frac{1}{y-1} + 1}{\frac{1}{y-1}} = \frac{1 + (y - 1)}{1} = y.$$

Thus $f^{-1}(y) = \frac{1}{y - 1}$ for each real number $y \neq 1$.

47. *Hint:* Is there a real number x such that $f(x) = 1$?

Section 7.3

1. $g \circ f$ is defined as follows:

$$(g \circ f)(1) = g(f(1)) = g(5) = 1,$$
$$(g \circ f)(3) = g(f(3)) = g(3) = 5,$$
$$(g \circ f)(5) = g(f(5)) = g(1) = 3.$$

$f \circ g$ is defined as follows:

$$(f \circ g)(1) = f(g(1)) = f(3) = 3,$$
$$(f \circ g)(3) = f(g(3)) = f(5) = 1,$$
$$(f \circ g)(5) = f(g(5)) = f(1) = 5.$$

Then $g \circ f \neq f \circ g$ because, for example, $(g \circ f)(1) \neq (f \circ g)(1)$.

3. $(G \circ F)(x) = G(F(x)) = G(x^3) = x^3 - 1$ for all real numbers x.
$(F \circ G)(x) = F(G(x)) = F(x - 1) = (x - 1)^3$ for all real numbers x.
$G \circ F \neq F \circ G$ because, for instance, $(G \circ F)(2) = 2^3 - 1 = 7$, whereas $(F \circ G)(2) = (2 - 1)^3 = 1$.

6. $(G \circ F)(0) = G(F(0)) = G(7.0) = G(0) = 0 \bmod 5 = 0$
$(G \circ F)(1) = G(F(1)) = G(7.1) = G(7) = 7 \bmod 5 = 2$
$(G \circ F)(2) = G(F(2)) = G(7.2) = G(14) = 14 \bmod 5 = 4$
$(G \circ F)(3) = G(F(3)) = G(7.3) = G(21) = 21 \bmod 5 = 1$
$(G \circ F)(4) = G(F(4)) = G(7.4) = G(28) = 28 \bmod 5 = 3$

8. a. $(L \circ M)(12) = L(M(12)) = L(12 \bmod 5) = L(2)$
$= 2^2 = 4$

$(M \circ L)(12) = M(L(12)) = M(12^2) = M(144)$
$= 144 \bmod 5 = 4$

$(L \circ M)(9) = L(M(9)) = L(9 \bmod 5) = L(4)$
$= 4^2 = 16$

$(M \circ L)(9) = M(L(9)) = M(9^2) = M(81)$
$= 81 \bmod 5 = 1$

9. $(F^{-1} \circ F)(x) = F^{-1}(F(x)) = F^{-1}(3x + 2)$
$= \dfrac{(3x + 2) - 2}{3} = \dfrac{3x}{3} = x = I_{\mathbf{R}}(x)$

for all x in $\mathbf{R}$. Hence $F^{-1} \circ F = I_{\mathbf{R}}$ by definition of equality of functions.

$(F \circ F^{-1})(y) = F(F^{-1}(y)) = F\left(\dfrac{y-2}{3}\right) = 3\left(\dfrac{y-2}{3}\right) + 2$
$= (y - 2) + 2 = y = I_{\mathbf{R}}(y)$

for all y in $\mathbf{R}$. Hence $F \circ F^{-1} = I_{\mathbf{R}}$ by definition of equality of functions.

12. a. By definition of logarithm with base b, for each real number x, $\log_b(b^x)$ is the exponent to which b must be raised to obtain b^x. But this exponent is just x. So $\log_b(b^x) = x$.

13. *Hint:* Suppose f is any function from a set X to a set Y, and show that for all x in X, $(I_Y \circ f)(x) = f(x)$.

15. a. $s_k = s_m$

16. No. Counterexample: Define f and g by the arrow diagrams below.

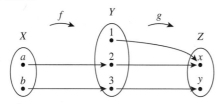

Then $g \circ f$ is one-to-one but g is not one-to-one. (So it is false that *both* f and g are one-to-one by De Morgan's law!) (This is one counterexample among many. Can you construct a different one?)

18. *Hint:* Suppose $f: X \to Y$ and $g: Y \to Z$ are functions and $g \circ f$ is one-to-one. Given x_1 and x_2 in X, if $f(x_1) = f(x_2)$ then $(g \circ f)(x_1) = (g \circ f)(x_2)$. (Why?) Then use the fact that $g \circ f$ is one-to-one.

19. *Hint:* Suppose $f: X \to Y$ and $g: Y \to Z$ are functions and $g \circ f$ is onto. Given $z \in Z$, there is an element x in X such that $(g \circ f)(x) = z$. (Why?) Let $y = f(x)$. Then $g(y) = z$.

21. True. <u>Proof:</u> Suppose X is any set and f, g, and h are functions from X to X such that h is one-to-one and $h \circ f = h \circ g$. *[We must show that for all x in X, $f(x) = g(x)$.]* Suppose x is any element in X. Because $h \circ f = h \circ g$, we have that $(h \circ f)(x) = (h \circ g)(x)$ by definition of equality of functions. Then, by definition of composition of functions, $h(f(x)) = h(g(x))$. But since h is one-to-one, this implies that $f(x) = g(x)$ *[as was to be shown].*

23.

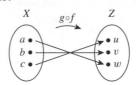 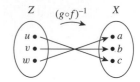

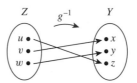 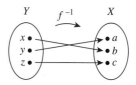

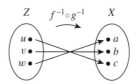

The functions $(g \circ f)^{-1}$ and $f^{-1} \circ g^{-1}$ are equal.

26. *Hints:* (1) Theorems 7.3.3 and 7.3.4 taken together insure that $g \circ f$ is one-to-one and onto. (2) Use the inverse function property: $F^{-1}(b) = a \Leftrightarrow F(a) = b$, for all a in the domain of F and b in the domain of F^{-1}.

Section 7.4

1. The student should have replied that for A to have the same cardinality as B means that there is a function from A to B that is one-to-one and onto. A set cannot have the property of being one-to-one or onto another set; only a function can have these properties.

2. Define a function $f: \mathbf{Z}^+ \to S$ as follows: For all positive integers k, $f(k) = k^2$.
f is *one-to-one:* *[We must show that for all $k_1, k_2 \in \mathbf{Z}^+$, if $f(k_1) = f(k_2)$ then $k_1 = k_2$.]* Suppose k_1 and k_2 are positive integers and $f(k_1) = f(k_2)$. By definition of f, $(k_1)^2 = (k_2)^2$, so $k_1 = \pm k_2$. But k_1 and k_2 are *positive.* Hence $k_1 = k_2$.
f is *onto:* *[We must show that for all $n \in S$, there exists $k \in \mathbf{Z}^+$ such that $n = f(k)$.]* Suppose $n \in S$. By definition of S, $n = k^2$ for some positive integer k. But then by definition of f, $n = f(k)$.
Since there is a one-to-one, onto function (namely, f) from $\mathbf{Z}^+$ to S, the two sets have the same cardinality.

3. Define $f: \mathbf{Z} \to 3\mathbf{Z}$ by the rule $f(n) = 3n$ for all integers n. The function f is one-to-one because for any integers n_1

and n_2, if $f(n_1) = f(n_2)$ then $3n_1 = 3n_2$ and so $n_1 = n_2$. Also f is onto because if m is any element in $3\mathbf{Z}$, then $m = 3k$ for some integer k. But then $f(k) = 3k = m$ by definition of f. Thus, since there is a function $f: \mathbf{Z} \to 3\mathbf{Z}$ that is one-to-one and onto, $\mathbf{Z}$ has the same cardinality as $3\mathbf{Z}$.

7. *Hint:* If $m \in \mathbf{Z}^+$, show that $j(m) = j(m+1) = m$.

8. It was shown in Example 7.4.2 that $\mathbf{Z}$ is countably infinite, which means that $\mathbf{Z}^+$ has the same cardinality as $\mathbf{Z}$. By exercise 3, $\mathbf{Z}$ has the same cardinality as $3\mathbf{Z}$. It follows by the transitive property of cardinality (Theorem 7.4.1 (c)) that $\mathbf{Z}^+$ has the same cardinality as $3\mathbf{Z}$. Thus $3\mathbf{Z}$ is countably infinite *[by definition of countably infinite]*, and hence $3\mathbf{Z}$ is countable *[by definition of countable]*.

10. <u>Proof:</u> Define $f: S \to U$ by the rule $f(x) = 2x$ for all real numbers x in S. Then f is one-to-one by the same argument as in exercise 10a of Section 7.2 with $\mathbf{R}$ in place of $\mathbf{Z}$. Furthermore, f is onto because if y is any element in U, then $0 < y < 2$ and so $0 < y/2 < 1$. Consequently, $y/2 \in S$ and $f(y/2) = 2(y/2) = y$. Hence f is a one-to-one correspondence, and so S and U have the same cardinality.

11. *Hint:* Define $h: S \to V$ as follows: $h(x) = 3x + 2$, for all $x \in S$.

13.

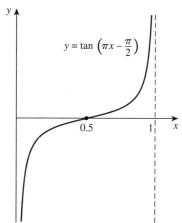

$$y = \tan\left(\pi x - \frac{\pi}{2}\right)$$

It is clear from the graph that f is one-to-one (since it is increasing) and that the image of f is all of $\mathbf{R}$ (since the lines $x = 0$ and $x = 1$ are vertical asymptotes). Thus S and $\mathbf{R}$ have the same cardinality.

16. In Example 7.4.4 it was shown that there is a one-to-one correspondence from $\mathbf{Z}^+$ to $\mathbf{Q}^+$. This implies that the positive rational numbers can be written as an infinite sequence: $r_1, r_2, r_3, r_4, \ldots$. Now the set $\mathbf{Q}$ of all rational numbers consists of the numbers in this sequence together with 0 and the negative rational numbers: $-r_1, -r_2, -r_3, -r_4, \ldots$. Let $r_0 = 0$. Then the elements of the set of all rational numbers can be "counted" as follows:

$$r_0, r_1, -r_1, r_2, -r_2, r_3, -r_3, r_4, -r_4, \ldots.$$

In other words, we can define a one-to-one correspondence

$$G(n) = \begin{cases} r_{n/2} & \text{if } n \text{ is even} \\ -r_{(n-1)/2} & \text{if } n \text{ is odd} \end{cases} \quad \text{for all integers } n \geq 1.$$

Therefore, $\mathbf{Q}$ is countably infinite and hence countable.

18. *Hint:* No.

19. *Hint:* Suppose r and s are real numbers with $s > r > 0$. Let n be an integer such that $n > \dfrac{\sqrt{2}}{s-r}$. Then $s - r > \dfrac{\sqrt{2}}{n}$. Let $m = \left\lfloor \dfrac{nr}{\sqrt{2}} \right\rfloor + 1$. Then $m > \dfrac{nr}{\sqrt{2}} \geq m - 1$. Use the fact that $s = r + (s - r)$ to show that $r < \dfrac{\sqrt{2}m}{n} < s$.

22. *Hint:* Use the unique factorization of integers theorem (Theorem 4.3.5) and Theorem 7.4.3.

23. a. Define a function $G: \mathbf{Z}^{nonneg} \to \mathbf{Z}^{nonneg} \times \mathbf{Z}^{nonneg}$ as follows: Let $G(0) = (0, 0)$, and then follow the arrows in the diagram, letting each successive ordered pair of integers be the value of G for the next successive integer. Thus, for instance, $G(1) = (1, 0)$, $G(2) = (0, 1)$, $G(3) = (2, 0)$, $G(4) = (1, 1)$, $G(5) = (0, 2)$, $G(6) = (3, 0), G(7) = (2, 1), G(8) = (1, 2)$, and so forth.

b. *Hint:* Observe that if the top ordered pair of any given diagonal is $(k, 0)$, the entire diagonal (moving from top to bottom) consists of $(k, 0)$, $(k - 1, 1)$, $(k - 2, 2), \ldots$, $(2, k - 2)$, $(1, k - 1)$, $(0, k)$. Thus for all the ordered pairs (m, n) within any given diagonal, the value of $m + n$ is constant, and as you move down the ordered pairs in the diagonal, starting at the top, the value of the second element of the pair keeps increasing by 1.

25. *Hint:* There are at least two different approaches to this problem. One is to use the method discussed in Section 4.2. Another is to suppose that $1.999999\ldots < 2$ and derive a contradiction. (Show that the difference between 2 and $1.999999\ldots$ can be made smaller than any given positive number.)

26. <u>Proof:</u> Let A be an infinite set. Construct a countably infinite subset $a_1, a_2, a_3, \ldots$ of A, by letting a_1 be any element of A, letting a_2 be any element of A other than a_1, letting a_3 be any element of A other than a_1 or a_2, and so forth. This process never stops (and hence $a_1, a_2, a_3, \ldots$ is an infinite sequence) because A is an infinite set. More formally,

1. Let a_1 be any element of A.
2. For each integer $n \geq 2$, let a_n be any element of $A - \{a_1, a_2, a_3, \ldots, a_{n-1}\}$. Such an element exists, for otherwise $A - \{a_1, a_2, a_3, \ldots, a_{n-1}\}$ would be empty and A would be finite.

27. <u>Proof:</u> Suppose A is any countably infinite set, B is any set, and $g: A \to B$ is onto. Since A is countably infinite, there is a one-to-one correspondence $f: \mathbf{Z}^+ \to A$. Then, in particular, f is onto, and so by Theorem 7.3.4, $g \circ f$ is an onto function from $\mathbf{Z}^+$ to B. Define a function $h: B \to \mathbf{Z}^+$ as follows: Suppose x is any element of B. Since $g \circ f$ is onto, $\{m \in \mathbf{Z}^+ \mid (g \circ f)(m) = x\} \neq \emptyset$. Thus, by the well-ordering principle for the integers, this set has a least element. In other words, there is a least positive integer n with $(g \circ f)(n) = x$. Let $h(x)$ be this integer.

We claim that h is a one-to-one. For suppose $h(x_1) = h(x_2) = n$. By definition of h, n is the least positive integer with $(g \circ f)(n) = x_1$. But also by definition of h, n is the least positive integer with $(g \circ f)(n) = x_2$. Hence $x_1 = (g \circ f)(n) = x_2$.

Thus h is a one-to-one correspondence between B and a subset S of positive integers (the range of h). Since any subset of a countable set is countable (Theorem 7.4.3), S is countable, and so there is a one-to-one correspondence between B and a countable set. Hence, by the transitive property of cardinality, B is countable.

29. *Hint:* Suppose A and B are any two countably infinite sets. Then there are one-to-one correspondences $f : \mathbf{Z}^+ \to A$ and $g : \mathbf{Z}^+ \to B$.
    Case 1 ($A \cap B = \emptyset$): In this case define $h : \mathbf{Z}^+ \to A \cup B$ as follows: For all integers $n \geq 1$,

$$h(n) \begin{cases} f(n/2) & \text{if } n \text{ is even} \\ g((n+1)/2) & \text{if } n \text{ is odd.} \end{cases}$$

Show that h is one-to-one and onto.
    Case 2 ($A \cap B \neq \emptyset$): In this case let $C = B - A$. Then $A \cup B = A \cup C$ and $A \cap C = \emptyset$. If C is countably infinite, use the result of case 1 to complete the proof. If C is finite, use the result of exercise 28 to complete the proof.

30. *Hint:* Use proof by contradiction and the fact that the set of all real numbers is uncountable.

31. *Hint:* Consider the following cases: (1) A and B are both finite, (2) at least one of A or B is infinite and $A \cap B = \emptyset$, (3) at least one of A or B is infinite and $A \cap B \neq \emptyset$. In case 3 use the fact that $A \cup B = (A - B) \cup (B - A) \cup (A \cap B)$ and that the sets $(A - B)$, $(B - A)$, and $(A \cap B)$ are mutually disjoint.

32. *Hint:* Use the one-to-one correspondence $F : \mathbf{Z}^+ \to \mathbf{Z}$ of Example 7.4.2 to define a function $G : \mathbf{Z}^+ \times \mathbf{Z}^+ \to \mathbf{Z} \times \mathbf{Z}$ by the formula $G(m, n) = (F(m), F(n))$. Show that G is a one-to-one correspondence, and use the result of exercise 22 and the transitive property of cardinality.

34. *Hint for Solution 1:* Define a function $f : \mathscr{P}(S) \to T$ as follows: For each subset A of S, let $f(A) = \chi_A$, the *characteristic function* of A, where $\chi_A : S \to \{0, 1\}$ is defined by the rule

$$\chi_A(x) = \begin{cases} 1 & \text{if } x \in A \\ 0 & \text{if } x \notin A \text{ for all } x \in S \end{cases}.$$

Show that f is one-to-one (for all $A_1, A_2 \subseteq S$, if $\chi_{A_1} = \chi_{A_2}$ then $A_1 = A_2$) and that f is onto (given any function $g : S \to \{0, 1\}$, there is a subset A of S such that $g = \chi_A$).
    Hint for Solution 2: Define $H : T \to \mathscr{P}(S)$ by letting $H(f) = \{x \in S \mid f(x) = 1\}$. Show that H is a one-to-one correspondence?

35. Partial proof (by contradiction): Suppose not. Suppose there is a one-to-one, onto function $f : S \to \mathscr{P}(S)$. Let

$$A = \{x \in S \mid x \notin f(x)\}.$$

Then $A \in \mathscr{P}(S)$ and since f is onto, there is a $z \in S$ such that $A = f(z)$. *[Now derive a contradiction!]*

37. *Hint:* Since A and B are countable, their elements can be listed as

$$A: a_1, a_2, a_3, \ldots \quad \text{and} \quad B: b_1, b_2, b_3, \ldots$$

Represent the elements of $A \times B$ in a grid:

$$\begin{array}{lll} (a_1, b_1) & (a_1, b_2) & (a_1, b_3) \cdots \\ (a_2, b_1) & (a_2, b_2) & (a_2, b_3) \cdots \\ (a_3, b_1) & (a_3, b_2) & (a_3, b_3) \cdots \\ \vdots & \vdots & \vdots \end{array}$$

Now use a counting method similar to that of Example 7.4.4.

Section 8.1

1. **a.** $0\ E\ 0$ because $0 - 0 = 0 = 2 \cdot 0$, so $2 \mid (0 - 0)$.
    $5\ \not{E}\ 2$ because $5 - 2 = 3$ and $3 \neq 2k$ for any integer k so $2 \nmid (5 - 2)$.
    $(6, 6) \in E$ because $6 - 6 = 0 = 2 \cdot 0$, so $2 \mid (6 - 6)$.
    $(-1, 7) \in E$ because $-1 - 7 = -8 = 2 \cdot (-4)$, so $2 \mid (-1 - 7)$.

2. *Hint:* To show a statement of the form $p \leftrightarrow (q \vee r)$, you need to show $p \to (q \vee r)$ and $(q \vee r) \to p$. To show a statement of the form $p \to (q \vee r)$, you can show $(p \wedge \sim q) \to r$ (since these two statement forms are logically equivalent). To show a statement of the form $(q \vee r) \to p$, you can show $(q \to p) \wedge (r \to p)$ (since these two statement forms are logically equivalent). In this case, suppose m and n are any integers, and let p be "$m - n$ is even," let q be "m and n are both even," and let r be "$m - n$ is even," let q be "m and n are both even," and let r be "m and n are both odd."

3. **a.** $10\ T\ 1$ because $10 - 1 = 9 = 3 \cdot 3$, so $3 \mid (10 - 1)$.
    $1\ T\ 10$ because $1 - 10 = -9 = 3 \cdot (-3)$, so $3 \mid (1 - 10)$.
    $2\ T\ 2$ because $2 - 2 = 0 = 3 \cdot 0$, so $3 \mid (2 - 2)$.
    $8\ \not{T}\ 1$ because $8 - 1 = 7 \neq 3k$, for any integer k. So $3 \nmid (8 - 1)$.
    b. *One possible answer:* $3, 6, 9, -3, -6$
    e. *Hint:* All integers of the form $3k + 1$, for some integer k, are related by T to 1.

4. **a.** Yes, because 15 and 25 are both divisible by 5, which is prime.
    b. No, because 22 and 27 have no common prime factor.

5. **a.** Yes, because both $\{a, b\}$ and $\{b, c\}$ have two elements.

6. **a.** No, because $\{a\} \cap \{c\} = \emptyset$.

7. **a.** Yes. $1\ R(-9) \Leftrightarrow 5 \mid (1^2 - (-9)^2)$. But $1^2 - (-9)^2 = 1 - 81 = -80$, and $5 \mid (-80)$ because $-80 = 5 \cdot (-16)$.

8. **a.** Yes, because both *abaa* and *abba* have the same first two characters *ab*.
    b. No, because the first two characters of *aabb* are different from the first two characters of *bbaa*.

9. a. Yes, because the sum of the characters in 0121 is 4 and the sum of the characters in 2200 is also 4.

   b. No, because the sum of the characters in 1011 is 3 whereas the sum of the characters in 2101 is 4.

10. $R = \{(3, 4), (3, 5), (3, 6), (4, 5), (4, 6), (5, 6)\}$

   $R^{-1} = \{(4, 3), (5, 3), (6, 3), (5, 4), (6, 4), (6, 5)\}$

12. a. No. If $F: X \to Y$ is not onto, then F^{-1} is not defined on all of Y. In other words, there is an element y in Y such that $(y, x) \notin F^{-1}$ for any $x \in X$. Consequently, F^{-1} does not satisfy property (1) of the definition of function.

13.   **15.**

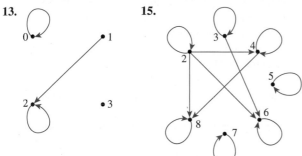

16. *Hint:* See Example 8.1.6.

19. $A \times B = \{(2, 6), (2, 8), (2, 10), (4, 6), (4, 8), (4, 10)\}$

   $R = \{(2, 6), (2, 8), (2, 10), (4, 8)\}$

   $S = \{(2, 6), (4, 8)\}$

   $R \cup S = R, R \cap S = S$

21.

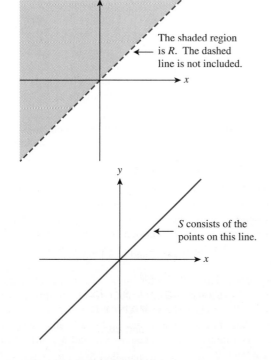

The shaded region is R. The dashed line is not included.

S consists of the points on this line.

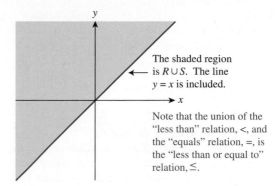

The shaded region is $R \cup S$. The line $y = x$ is included.

Note that the union of the "less than" relation, $<$, and the "equals" relation, $=$, is the "less than or equal to" relation, $\leq$.

The graph of the intersection of R and S is obtained by finding the set of all points common to both graphs. But there are no points for which both $x < y$ and $x = y$. Hence $R \cap S = \emptyset$ and the graph consists of no points at all.

Section 8.2

1. R_1:

a.

b. R_1 is not reflexive: $2 \ \not{R_1} \ 2$.

c. R_1 is not symmetric: $2 \ R_1 \ 3 \ but \ 3 \ \not{R_1} \ 2$.

d. R_1 is not transitive: $1 \ R_1 \ 0$ and $0 \ R_1 \ 3$ but $1 \ \not{R_1} \ 3$.

3. R_3:

   a. $0 \bullet \qquad\qquad \bullet 1$

   $2 \bullet \rightleftharpoons \bullet 3$

   b. R_3 is not reflexive: $(0, 0) \notin R_3$

   c. R_3 is symmetric. (If R_3 were not symmetric, there would be elements x and y in $A = \{0, 1, 2, 3\}$ such that $(x, y) \in R_3$ but $(y, x) \notin R_3$. It is clear by inspection that no such elements exist.)

   d. R_3 is not transitive: $(2, 3) \in R_3$ and $(3, 2) \in R_3$ but $(2, 2) \notin R_3$

6. R_6:

a.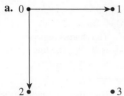

b. R_6 is not reflexive: $(0, 0) \notin R_6$

c. R_6 is not symmetric: $(0, 1) \in R_6$ but $(1, 0) \notin R_6$.

d. R_6 is transitive. (If R_6 were not transitive, there would be elements x, y, and z in $\{0, 1, 2, 3\}$ such that $(x, y) \in R_6$ and $(y, z) \in R_6$ and $(x, z) \notin R_6$. It is clear by inspection that no such elements exist.)

9. *R is reflexive:* R is reflexive $\Leftrightarrow$ for all real numbers x, $x\ R\ x$. By definition of R, this means that for all real numbers x, $x \geq x$. In other words, for all real numbers x, $x > x$ or $x = x$. But this is true.

R is not symmetric: R is symmetric $\Leftrightarrow$ for all real numbers x and y, if $x\ R\ y$ then $y\ R\ x$. By definition of R, this means that for all real numbers x and y, if $x \geq y$ then $y \geq x$. But this is false. As a counterexample, take $x = 1$ and $y = 0$. Then $x \geq y$ but $y \not\geq x$ because $1 \geq 0$ but $0 \not\geq 1$.

R is transitive: R is transitive $\Leftrightarrow$ for all real numbers x, y, and z, if $x\ R\ y$ and $y\ R\ z$ then $x\ R\ z$. By definition of R, this means that for all real numbers x, y and z, if $x \geq y$ and $y \geq z$ then $x \geq z$. But this is true by definition of $\geq$ and the transitive property of order for the real numbers. (See Appendix A, T18.)

11. *D is reflexive:* For D to be reflexive means that for all real numbers x, $x\ D\ x$. But by definition of D, this means that for all real numbers x, $xx = x^2 \geq 0$, which is true.

D is symmetric: For D to be symmetric means that for all real numbers x and y, if $x\ D\ y$ then $y\ D\ x$. But by definition of D, this means that for all real numbers x and y, if $xy \geq 0$ then $yx \geq 0$, which is true by the commutative law of multiplication.

D is not transitive: For D to be transitive means that for all real numbers x, y, and z, if $x\ D\ y$ and $y\ D\ z$ then $x\ D\ z$. By definition of D, this means that for all real numbers x, y, and z, if $xy \geq 0$ and $yz \geq 0$ then $xz \geq 0$. But this is false: there exist real numbers x, y, and z such that $xy \geq 0$ and $yz \geq 0$ but $xz \not\geq 0$. As a counterexample, let $x = 1$, $y = 0$, and $z = -1$. Then $x\ D\ y$ and $y\ D\ z$ because $1 \cdot 0 \geq 0$ and $0 \cdot (-1) \geq 0$. But $x\ \not\!D\ z$ because $1 \cdot (-1) \not\geq 0$.

12. *E is reflexive:* [We must show that for all integers m, $m\ E\ m$.] Suppose m is any integer. Since $m - m = 0$ and $2 \mid 0$, we have that $2 \mid (m - m)$. Consequently, $m\ E\ m$ by definition of E.

E is symmetric: [We must show that for all integers m and n, if $m\ E\ n$ then $n\ E\ m$.] Suppose m and n are any integers such that $m\ E\ n$. By definition of E, this means that $2 \mid (m - n)$, and so, by definition of divisibility, $m - n = 2k$ for some integer k. Now $n - m = -(m - n)$. Hence, by substitution, $n - m = -(2k) = 2(-k)$. It follows that $2 \mid (n - m)$

by definition of divisibility (since $-k$ is an integer), and thus $n\ E\ m$ by definition of E.

E is transitive: [We must show that for all integers m, n and p if $m\ E\ n$ and $n\ E\ p$ then $m\ E\ p$.] Suppose m, n, and p are any integers such that $m\ E\ n$ and $n\ E\ p$. By definition of E this means that $2 \mid (m - n)$ and $2 \mid (n - p)$, and so, by definition of divisibility, $m - n = 2k$ for some integer k and $n - p = 2l$ for some integer l. Now $m - p = (m - n) + (n - p)$. Hence, by substitution, $m - p = 2k + 2l = 2(k + l)$. It follows that $2 \mid (m - p)$ by definition of divisibility (since $k + l$ is an integer), and thus $m\ E\ p$ by definition of E.

15. *D is reflexive:* [We must show that for all positive integers m, $m\ D\ m$.] Suppose m is any positive integer. Since $m = m \cdot 1$, by definition of divisibility $m \mid m$. Hence $m\ D\ m$ by definition of D.

D is not symmetric: For D to be symmetric would mean that for all positive integers m and n, if $m\ D\ n$ then $n\ D\ m$. By definition of divisibility, this would mean that for all positive integers m and n, if $m \mid n$ then $n \mid m$. But this is false. As a counterexample, take $m = 2$ and $n = 4$. Then $m \mid n$ because $2 \mid 4$ but $n \nmid m$ because $4 \nmid 2$.

D is transitive: To prove transitivity of D, we must show that for all positive integers m, n, and p, if $m\ D\ n$ and $n\ D\ p$ then $m\ D\ p$. By definition of D, this means that for all positive integers m, n, and p, if $m \mid n$ and $n \mid p$ then $m \mid p$. But this is true by Theorem 4.3.3 (the transitivity of divisbility).

18. *Hint:* Q is reflexive, symmetric, and transitive.

20. *E is reflexive:* $\mathbf{E}$ is reflexive $\Leftrightarrow$ for all subsets A of X, A $\mathbf{E}$ A. By definition of $\mathbf{E}$, this means that for all subsets A of X, A has the same number of elements as A. But this is true.

E is symmetric: $\mathbf{E}$ is symmetric $\Leftrightarrow$ for all subsets A and B of X, if A $\mathbf{E}$ B then B $\mathbf{E}$ A. By definition of $\mathbf{E}$, this means that if A has the same number of elements as B, then B has the same number of elements as A. But this is true.

E is transitive: $\mathbf{E}$ is transitive $\Leftrightarrow$ for all subsets A, B, and C of X, if A $\mathbf{E}$ B and B $\mathbf{E}$ C, then A $\mathbf{E}$ C. By definition of $\mathbf{E}$, this means that for all subsets, A, B, and C of X, if A has the same number of elements as B and B has the number of elements as C, then A has the same number of elements as C. But this is true.

23. *S is reflexive:* $\mathbf{S}$ is reflexive $\Leftrightarrow$ for all subsets A of X, A$\mathbf{S}$A. By definition of $\mathbf{S}$, this means that for all subsets A of X, A $\subseteq$ A. But this is true because every set is a subset of itself.

S is not symmetric: $\mathbf{S}$ is symmetric $\Leftrightarrow$ for all subsets A and B of X, if A$\mathbf{S}$B then B$\mathbf{S}$A. By definition of $\mathbf{S}$, this means that for all subsets A and B of X, if A $\subseteq$ B then B $\subseteq$ A. But this is false because $X \neq \emptyset$ and so there is an element a in X. As a counterexample, take A $= \emptyset$, and B $= \{a\}$. Then A $\subseteq$ B but B $\not\subseteq$ A.

S is transitive: $\mathbf{S}$ is transitive $\Leftrightarrow$ for all subsets A, B, and C of X, if A$\mathbf{S}$B and B$\mathbf{S}$C, then A $\mathbf{S}$ C. By definition of $\mathbf{S}$, this means that for all subsets A, B, and C of X, if $A \subseteq B$ and $B \subseteq C$ then $A \subseteq C$. But this is true by the transitive property of subsets (Theorem 6.2.1 (3)).

25. *R is reflexive:* Suppose s is any string in A. Then $s\ R\ s$ because s has the same first two characters as s.

R is symmetric: Suppose s and t are any strings in A such that $s\ R\ t$. By definition of R, s has the same first two characters as t. It follows that t has the same first two characters as s, and so $t\ R\ s$.

R is transitive: Suppose s, t, and u, are any strings in A such that $s\ R\ t$ and $t\ R\ u$. By definition of R, s has the same first two characters as t and t has the same first two characters as u. It follows that s has the same two characters as u, and so $s\ R\ u$.

27. *I is reflexive: [We must show that for all statements p, p I p.]* Suppose p is a statement. The only way a conditional statement can be false is for its hypothesis to be true and its conclusion false. Consider the statement $p \rightarrow p$. Both the hypothesis and the conclusion have the same truth value. Thus it is impossible for $p \rightarrow p$ to be false, and so $p \rightarrow p$ must be true.

I is not symmetric: I is symmetric $\Leftrightarrow$ for all statements p and q, if p I q then q I p. By definition of I, this means that for all statements p and q, if $p \rightarrow q$ then $q \rightarrow p$. But this false. As a counterexample, let p be the statement "10 is divisible by 4" and let q be "10 is divisible by 2." Then $p \rightarrow q$ is the statement "If 10 is divisible by 4, then 10 is divisible by 2." This is true because its hypothesis, p, is false. On the other hand, $q \rightarrow p$ is the statement "If 10 is divisible by 2, then 10 is divisible by 4." This is false because its hypothesis, q, is true and its conclusion, p, is false.

I is transitive: [We must show that for all statements p, q, and r, if p I q and q I r then p I r.] Suppose p, q, and r are statements such that p I q and q I r. By definition of I, this means that $p \rightarrow q$ and $q \rightarrow r$ are both true. By transitivity of if-then (Example 2.3.6 and exercise 20 of Section 2.3), we can conclude that $p \rightarrow r$ is true. Hence, by definition of I, p I r.

28. *F is reflexive:* **F** is reflexive $\Leftrightarrow$ for all elements (x, y) in **R** $\times$ **R**, (x, y) **F** (x, y). By definition of **F**, this means that for all elements (x, y) in **R** $\times$ **R**, $x = x$. But this is true.

*F is symmetric: [We must show that for all elements (x_1, y_1) and (x_2, y_2) in **R** $\times$ **R**, if (x_1, y_1)**F**(x_2, y_2) then (x_2, y_2)**F**(x_1, y_1).]* Suppose (x_1, y_1) and (x_2, y_2) are elements of **R** $\times$ **R** such that (x_1, y_1), **F**(x_2, y_2). By definition of **F**, this means that $x_1 = x_2$. By symmetry of equality, $x_2 = x_1$. Thus, by definition of **F**, (x_2, y_2)**F**(x_1, y_1).

*F is transitive: [We must show that for all elements (x_1, y_1), (x_2, y_2) and (x_3, y_3) in **R** $\times$ **R**, if (x_1, y_1)**F**(x_2, y_2) and (x_2, y_2)**F**(x_3, y_3) then (x_1, y_1)**F**(x_3, y_3).]* Suppose (x_1, y_1), (x_2, y_2), and (x_3, y_3) are elements of **R** $\times$ **R** such that (x_1, y_1)**F**(x_2, y_2) and (x_2, y_2)**F**(x_3, y_3). By definition of **F**, this means that $x_1 = x_2$ and $x_2 = x_3$. By transitivity of equality, $x_1 = x_3$. Hence, by definition of **F**, (x_1, y_1)**F**(x_3, y_3).

31. *R is reflexive:* R is reflexive $\Leftrightarrow$ for all people p in A, $p\ R\ p$. By definition of R, this means that for all people p living in the world today, p lives within 100 miles of p. But this is true.

R is symmetric: [We must show that for all people p and q in A, if $p\ R\ q$ then $q\ R\ p$.] Suppose p and q are people in A such that $p\ R\ q$. By definition of R, this means that p lives within 100 miles of q. But this implies that q lives within 100 miles of p. So, by definition of R, $q\ R\ p$.

R is not transitive: R is transitive $\Leftrightarrow$ for all people p, q and r, if $p\ R\ q$ and $q\ R\ r$ then $p\ R\ r$. But this is false. As a counterexample, take p to be an inhabitant of Chicago, Illinois, q an inhabitant of Kankakee, Illinois, and r an inhabitant of Champaign, Illinois. Then $p\ R\ q$ because Chicago is less then 100 miles from Kankakee, and $q\ R\ r$ because Kankakee is less than 100 miles from Champaign, but $p\ \not{R}\ r$ because Chicago is not less than 100 miles from Champaign.

34. Proof: Suppose R is any reflexive relation on a set A. *[We must show that R^{-1} is reflexive. To show this, we must show that for all x in A, $x\ R^{-1}\ x$.]* Given any element x in A, since R is reflexive, $x\ R\ x$, and by definition of relation, this means that $(x, x) \in R$. It follows, by definition of the inverse of a relation, that $(x, x) \in R^{-1}$, and so, by definition of relation, $x\ R^{-1}\ x$ *[as was to be shown].*

37. a. *R $\cap$ S is reflexive:* Suppose R and S are reflexive. *[To show that $R \cap S$ is reflexive, we must show that $\forall x \in A$, $(x, x) \in R \cap S$.]* So suppose $x \in A$. Since R is reflexive, $(x, x) \in R$, and since S is reflexive, $(x, x) \in S$. Thus, by definition of intersection, $(x, x) \in R \cap S$ *[as was to be shown].*

38. *Hint:* The answer is yes.

41. Yes. To prove this we must show that for all x and y in A, if $(x, y) \in R \cup S$ then $(y, x) \in R \cup S$. So suppose (x, y) is a particular but arbitrarily chosen element in $R \cup S$. *[We must show that $(y, x) \in R \cup S$.]* By definition of union, $(x, y) \in R$ or $(x, y) \in S$. If $(x, y) \in R$, then $(y, x) \in R$ because R is symmetric. Hence $(y, x) \in R \cup S$ by definition of union. But also, if $(x, y) \in S$ then $(y, x) \in S$ because S is symmetric. Hence $(y, x) \in R \cup S$ by definition of union. Thus, in either case, $(y, x) \in R \cup S$ *[as was to be shown].*

43. R_1 is not irreflexive because $(0, 0) \in R_1$. R_1 is not asymmetric because $(0, 1) \in R_1$ and $(1, 0) \in R_1$. R_1 is not intransitive because $(0, 1) \in R_1$ and $(1, 0) \in R_1$ and $(0, 0) \in R_1$.

45. R_3 is irreflexive. R_3 is not asymmetric because $(2, 3) \in R_3$ and $(3, 2) \in R_3$. R_3 is intransitive.

48. R_6 is irreflexive. R_6 is asymmetric. R_6 is intransitive (by default).

Section 8.3

1. a. cRc **b.** bRa, cRb, eRd **c.** aRc
d. $cRc, bRa, cRb, eRd, aRc, cRa$

2. a. $R = \{(0,0), (0,2), (1,1), (2,0), (2,2), (3,3), (3,4),$
$(4,3), (4,4)\}$

3. $\{0,4\}, \{1,3\}, \{2\}$

5. $\{1,5,9,13,17\}, \{2,6,10,14,18\}, \{3,7,11,15,19\},$
$\{4,8,12,16,20\}$

7. $\{(1,3),(3,9)\}, \{(2,4),(-4,-8),(3,6)\}, \{(1,5)\}$

8. $\{\emptyset\}, \{\{a\},\{b\},\{c\}\}, \{\{a,b\},\{a,c\},\{b,c\}\}, \{\{a,b,c\}\}$

11. $[0] = \{x \in A \mid 4 \mid (x^2 - 0)\} = \{x \in A \mid 4 \mid x^2\} =$
$\{-4, -2, 0, 2, 4\}$ $[1] = \{x \in A \mid 4 \mid (x^2 - 1^2)\} =$
$\{x \in A \mid 4 \mid (x^2 - 1)\} = \{-3, -1, 1, 3\}$

13. a. True. $17 - 2 = 15$ and $5 \mid 15$.

14. a. $[7] = [4] = [19], [-4] = [17], [-6] = [27]$

15. a. <u>Proof</u>: Suppose that m and n are integers such that
$m \equiv n \pmod 3$. *[We must show that m mod 3 = n mod 3.]*
By definition of congruence, $3 \mid (m - n)$, and so by
definition of divisibility, $m - n = 3k$ for some integer
k. Let $m \bmod 3 r =$. Then $m = 3l + r$ for some integer l. Since $m - n = 3k$, then by substitution, $(3l +
r) - n = 3k$, or, equivalently, $n = 3(l - k) + r$. Since
$l - k$ is an integer and $0 \le r < 3$, it follows, by definition of *mod*, that $n \bmod 3 = r$ also. So $m \bmod 3 =
n \bmod 3$.
Suppose that m and n are integers such that
$m \bmod 3 = n \bmod 3$. *[We must show that m ≡ n (mod 3).]*
Let $r = m \bmod 3 = n \bmod 3$. Then, by definition of
mod, $m = 3p + r$ and $n = 3q + r$ for some integers
p and q. By substitution, $m - n = (3p + r) - (3q +
r) = 3(p - q)$. Since $p - q$ is an integer, it follows that
$3 \mid (m - n)$, and so, by definition of congruence, $m \equiv n
\pmod 3$.

16. a. For example, let $A = \{1, 2\}$ and $B = \{2, 3\}$. Then $A \ne
B$, so A and B are distinct. But A and B are not disjoint
since $2 \in A \cap B$.

17. a. (1) <u>Proof</u>: R is reflexive because it is true that for each
student x at a college, x has the same major (or double
major) as x.
R is symmetric because it is true that for all students
x and y at a college, if x has the same major (or double
major) as y, then y has the same major (or double major)
as x.
R is transitive because it is true that for all students x, y,
and z at a college, if x has the same major (or double
major) as y and y has the same major (or double major)
as z, then x has the same major (or double major) as z.
R is an equivalence relation because it is reflexive, symmetric, and transitive.

(2) There is one equivalence class for each major and
double major at the college. Each class consists of all
students with that major (or double major).

18. (1) See the solution to exercise 12 in Section 8.2.
(2) Two distinct classes: $\{x \in \mathbf{Z} \mid x = 2k,$ for some integer $k\}$
and $\{x \in \mathbf{Z} \mid x = 2k + 1,$ for some integer $k\}$.

21. (1) <u>Proof</u>: A is reflexive because each real number has the
same absolute value as itself.

A is symmetric because for all real numbers x and y, if
$|x| = |y|$ then $|y| = |x|$.
A is transitive because for all real numbers x, y, and z, if
$|x| = |y|$ and $|y| = |z|$ then $|x| = |z|$.
A is an equivalence relation because it is reflexive, symmetric, and transitive.

(2) The distinct classes are all sets of the form $\{x, -x\}$,
where x is a real number.

22. *Hints*: (1) D is reflexive, symmetric, and transitive. The
proofs are very similar to the proofs in Example 8.2.4.

(2) There are two distinct equivalence classes. Note that
$m^2 - n^2 = (m - n)(m + n)$ for all integers m and n. In
addition, $3 \mid (m - n)$ or $3 \mid (m + n) \Leftrightarrow$ either $m - n = 3r$
or $m + n = 3r$, for some integer r

24. (1) <u>Proof</u>: I is reflexive because the difference
between each real number and itself is 0, which is an
integer.

I is symmetric because for all real numbers x and y, if
$x - y$ is in integer, then $y - x = (-1)(x - y)$, which is
also an integer.

I is transitive because for all real numbers x, y, and z, if
$x - y$ is an integer and $y - z$ is an integer, then $x - z =
(x - y) + (y - z)$ is the sum of two integers and thus an
integer.
I is an equivalence relation because it is reflexive, symmetric, and transitive.

(2) There is one class for each real number x with $0 \le x <
1$. The distinct classes are all sets of the form $\{y \in \mathbf{R} \mid y =
n + x,$ for some integer $n\}$, where x is a real number such
that $0 \le x < 1$.

25. (1) <u>Proof</u>: P is reflexive because each ordered pair of real
numbers has the same first element as itself.

P is symmetric for the following reason: Suppose (w, x)
and (y, z) are ordered pairs of real numbers such that
$(w, x)P(y, z)$. Then, by definition of P, $w = y$. But by the
symmetric property of equality, this implies that $y = w$,
and so, by definition of P, $(y, z)P(w, x)$.
P is transitive for the following reason: Suppose
(u, v), (w, x), and (y, z) are ordered pairs of real numbers
such that $(u, v)P(w, x)$ and $(w, x)P(y, z)$. Then, by definition of P, $u = w$ and $w = y$. But by the transitive property
of equality, this implies that $u = w$, and so, by definition of
P, $(u, v)P(w, x)$.
P is an equivalence relation because it is reflexive, symmetric, and transitive.

(2) There is one equivalence class for each real number.
The distinct equivalence classes are all sets of ordered pairs
$\{(x, y) \in \mathbf{R} \times \mathbf{R} \mid x = a\}$, for each real number a. Equivalently, the equivalence classes consist of all vertical lines in
the Cartesian plane.

28. *Solution*: There is one equivalence class for each real number t such that $0 \le t < \pi$. One line in each class goes
through the origin, and that line makes an angle of t with
the positive horizontal axis.

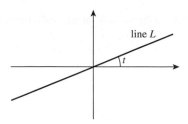

Alternatively, there is one equivalence class for every possible slope: all real numbers plus "undefined."

30. Proof: Suppose R is an equivalence relation on a set A and $a \in A$. Because R is an equivalence relation, R is reflexive, and because R is reflexive, each element of A is related to itself by R. In particular, $a\ R\ a$. Hence by definition of equivalence class, $a \in [a]$.

32. Proof: Suppose R is an equivalence relation on a set A and a, b, and c are elements of A with $b\ R\ c$ and $c \in [a]$. Since $c \in [a]$, then $c\ R\ a$ by definition of equivalence class. But R is transitive since R is an equivalence relation. Thus since $b\ R\ c$ and $c\ R\ a$, then $b\ R\ a$. It follows that $b \in [a]$ by definition of class.

34. Proof: Suppose a, b and x are in A, $a\ R\ b$, and $x \in [a]$. By definition of equivalence class, $x\ R\ a$. So $x\ R\ a$ and $a\ R\ b$, and thus, by transitivity, $x\ R\ b$. Hence $x \in [b]$.

35. *Hint:* To show that $[a] = [b]$, show that $[a] \subseteq [b]$ and $[b] \subseteq [a]$. To show that $[a] \subseteq [b]$, show that for all x in A, if $x \in [a]$ then $x \in [b]$.

36. c. For example $(2, 6)$, $(-2, -6)$, $(3, 9)$, $(-3, -9)$.

37. a. Suppose that (a, b), (a', b'), (c, d) and (c', d') are any elements of A such that $[(a, b)] = [(a', b')]$ and $[(c, d)] = [(c', d')]$. By definition of the relation, $ab' = ba'$ (*) and $cd' = dc'$ (**). We must show that $[(a, b)] + [(c, d)] = [(a', b')] + [(c', d')]$. By definition of the addition, this equation is true if, and only if,

$$[(ad + bc, bd)] - [(a'd' + b'c', b'd')].$$

And, by definition of the relation, this equation is true if, and only if,

$$(ad + bc)b'd' = bd(a'd' + b'c'),$$

which is equivalent to

$$adb'd' + bcb'd' = bda'd' + bdb'c', \quad \text{by multiplying out.}$$

But this equation is equivalent to

$$(ab')(dd') + (cd')(bb')$$
$$= (ba')(dd') + (dc')(bb') \quad \text{by regrouping}$$

and, by substitution from (*) and (**), this last equation is true.

c. Suppose that (a, b) is any element of A. We must show that $[(a, b)] + [(0, 1)] = [(0, 1)] + [(a, b)] = [(a, b)]$. By definition of the addition, these equations are true if, and only if,

$$[(a \cdot 1 + b \cdot 0, b \cdot 1)] = [(0 \cdot b + 1 \cdot a), 1 \cdot b)] = [(a, b)].$$

But this last equation is true because $a \cdot 1 + b \cdot 0 = 0 \cdot b + 1 \cdot a = a$ and $b \cdot 1 = 1 \cdot b = b$.

e. Suppose that (a, b) is any element of A. We must show that $[(a, b)] + [(-a, b)] = [(-a, b)] + [(a, b)] = [(0, 1)]$. By definition of the addition, this equation is true if, and only if,

$$[(ab + b(-a), bb)] = [(0, 1)],$$

or, equivalently,

$$[(0, bb)] = [(0, 1)]$$

By definition of the relation, this last equation is true if, and only if, $0 \cdot 1 = bb \cdot 0$, which is true.

38. a. Let (a, b) be any element of $\mathbf{Z}^+ \times \mathbf{Z}^+$. We must show that $(a, b)R(a, b)$. By definition of R, this relationship holds if, and only if, $a + b = b + a$. But this equation is true by the commutative law of addition for real numbers. Hence R is reflexive.

c. *Hint:* You will need to show that for any positive integers a, b, c, and d, if $a + d = c + b$ and $c + f = d + e$, then $a + f = b + e$.

d. *One possible answer:* $(1, 1)$, $(2, 2)$, $(3, 3)$, $(4, 4)$, $(5, 5)$

g. Observe that for any positive integers a and b, the equivalence class of (a, b) consists of all ordered pairs in $\mathbf{Z}^+ \times \mathbf{Z}^+$ for which the difference between the first and second coordinates equals $a - b$. Thus there is one equivalence class for each integer: positive, negative, and zero. Each positive integer n corresponds to the class of $(n + 1, 1)$; each negative integer $-n$ corresponds to the class of $(1, n + 1)$; and zero corresponds to the class $(1, 1)$.

41. c. "Ways and Means"

Section 8.4

1. a. By definition of congruence modulo 3, $25 \equiv 19 \pmod 3$ because $25 - 19 = 6$ and 3 divides 6 since $6 = 3 \cdot 2$.

b. From part (a), $25 - 19 = 6 = 3 \cdot 2$, and so $25 = 19 + 3 \cdot 2$. Thus $k = 2$ satisfies the equation $25 = 19 + 3k$.

c. When 25 is divided by 3, the remainder is 1 because $25 = 3 \cdot 8 + 1$. When 19 is divided by 3, the remainder is also 1 because $19 = 3 \cdot 6 + 1$. Thus 25 and 19 have the same remainder when divided by 3.

3. a. True because $58 - 14 = 11 \cdot 4$.

b. False because $46 - 89 = -43$, and $\dfrac{-43}{13} \cong -3.3$ is not an integer.

c. False because $674 - 558 = 116$, and $\dfrac{116}{56} \cong 2.1$ is not an integer.

d. True because $432 - 981 = 9 \cdot (-61)$.

6. *Hints:* (1) Use the quotient-remainder theorem and Theorem 8.4.1 to show that given any integer a, a is in one of the classes $[0]$, $[1]$, $[2]$, $\ldots [n - 1]$. (2) Use Theorem 4.3.1 to prove that if $0 \le a < n$, $0 \le b < n$, and $a \equiv b \pmod n$, then $a = b$.

7. a. $128 \equiv 2 \pmod 7$ because $128 - 2 = 126 = 7 \cdot 18$, and $61 \equiv 5 \pmod 7$ because $61 - 5 = 56 = 7 \cdot 8$

b. $128 + 61 \equiv (2 + 5) \pmod 7$ because $128 + 61 = 189$, $2 + 5 = 7$, and $189 - 7 = 182 = 7 \cdot 26$

c. $128 - 61 \equiv (2 - 5) \pmod 7$ because $128 - 61 = 67$, $2 - 5 = -3$, and $67 - (-3) = 70 = 7 \cdot 10$

d. $128 \cdot 61 \equiv (2 \cdot 5) \pmod 7$ because $128 \cdot 61 = 7808$, $2 \cdot 5 = 10$, and $7808 - (10) = 7798 = 7 \cdot 1114$

e. $128^2 \equiv 2^2 \pmod 7$ because $128^2 = 16384$, $2^2 = 4$, and $16384 - 4 = 16380 = 7 \cdot 2340$.

9. a. Proof: Suppose a, b, c, d, and n are integers with $n > 1$, $a \equiv c \pmod n$, and $b \equiv d \pmod n$. By Theorem 8.4.1, $a - c = nr$ and $b - d = ns$ for some integers r and s. Then

$$(a + b) - (c + d) = (a - c) + (b - d) = nr + ns$$
$$= n(r + s).$$

But $r + s$ is an integer, and so, by Theorem 8.4.1, $a + b \equiv (c + d) \pmod n$.

12. a. Proof (by mathematical induction): Let the property $P(n)$ be the congruence $10^n \equiv 1 \pmod 9$.

Show that P(0) is true:

When $n = 0$, the left-hand side of the congruence is $10^0 = 1$ and the right-hand side is also 1.

Show that for all integers $k \geq 0$, if $P(k)$ is true, then $P(k + 1)$ is true.

Let k be any integer with $k \geq 0$, and suppose $P(k)$ is true. That is, suppose $10^k \equiv 1 \pmod 9$. (*) *[This is the inductive hypothesis.]* By Theorem 8.4.1, $10 \equiv 1 \pmod 9$ (**) because $10 - 1 = 9 = 9 \cdot 1$. And by Theorem 8.4.3, we can multiply the left- and right-hand sides of (*) and (**) to obtain $10^k \cdot 10 \equiv 1 \cdot 1 \pmod 9$, or, equivalently, $10^{k+1} \equiv 1 \pmod 9$. Hence $P(k + 1)$ is true.

Alternative Proof: Note that $10 \equiv 1 \pmod 9$ because $10 - 1 = 9$ and $9|9$. Thus by Theorem 8.4.3(4), $10^n \equiv 1^n \equiv 1 \pmod 9$.

14. Note that $7^2 = 49$, $7^3 = 343$, and $7^4 = 2401$. Thus $7^4 \equiv 1 \pmod{10}$. Note also that $49 \equiv 9 \pmod{10}$. Now $1066 = 4 \cdot 266 + 2$, and so, by the laws of exponents and Theorem 8.4.3,

$$7^{1066} = 7^{4 \cdot 266 + 2} = 7^{4 \cdot 266} \cdot 7^2 = (7^4)^{266} \cdot 49 \equiv 1^{266} \cdot 9$$
$$\equiv 9 \pmod{10}.$$

17. Observe that $36 \equiv 1 \pmod 7$, $42 \equiv 0 \pmod 7$, $84 \equiv 0 \pmod 7$, $-6 \equiv 1 \pmod 7$, $63 \equiv 0 \pmod 7$, and $76 \equiv 6 \pmod 7$ *[because $36 = 7 \cdot 5 + 1$, $42 = 7 \cdot 6 + 0$, $84 = 7 \cdot 12 + 0$, $-6 = 7 \cdot (-1) + 1$, $63 = 7 \cdot 9 + 0$, and $76 = 7 \cdot 10 + 6$].* Thus, when reduced modulo 7, the equations become

$$1 \cdot x + 0 \cdot y \equiv 0 \pmod 7 \text{ and } 1 \cdot x + 0 \cdot y \equiv 6 \pmod 7,$$

which are equivalent to

$$x \equiv 0 \pmod 7 \text{ and } x \equiv 6 \pmod 7.$$

However these two congruences are contradictory, and so the given equations do not have a simultaneous solution in the set of integers.

20. According to the formula on page 381, the check digit is

$$\{210 - [3(7 + 3 + 4 + 0 + 8 + 2) + (6 + 6 + 9 + 0 + 4)]\} mod \, 10$$
$$= [210 - (3 \cdot 24 + 25)] \, mod \, 10$$
$$= [210 - (3 \cdot 24 + 25)] \, mod \, 10$$
$$= 113 \, mod \, 10$$
$$= 3$$

23. The set of all even integers is not a commutative ring because it does not contain an identity for multiplication.

25. The set of all rational numbers is a commutative ring. To see why, first note that, by Theorem 4.2.2 and exercise 15 in Section 4.2, the sum and product of any two rational numbers is rational. Moreover, because all real numbers satisfy the commutative and associative properties for addition and multiplication and the distributive property that links addition and multiplication, these properties are satisfied for all rational numbers. In addition, the real numbers 1 and 0 are rational and act as identities for multiplication and addition respectively. Finally, by exercise 13 in Section 4.2, the negative of any rational number is rational, and so every rational number has an additive inverse.

26. Let S be the set of all real numbers of the form $a + b\sqrt{2}$ where a and b are rational numbers. Then S is a commutative ring. To see why, first note that for all rational numbers a, b, c, and d,

$$(a + b\sqrt{2}) + (c + d\sqrt{2}) = (a + c) + (b + d)\sqrt{2}$$

and

$$(a + b\sqrt{2}) \cdot (c + d\sqrt{2}) = (ac + 2bd) + (ad + bc)\sqrt{2}.$$

Thus the sum of any two elements of S is in S and the product of any two elements of S are in S because products and sums of rational numbers are rational. Moreover, because all real numbers satisfy the commutative and associative properties for addition and multiplication and the distributive property that links addition and multiplication, these properties are satisfied for all the elements of S. In addition, $1 = 1 + 0 \cdot \sqrt{2}$ and $0 = 0 + 0 \cdot \sqrt{2}$, and so S contains both a multiplicative identity and an additive identity because 1 and 0 are multiplicative and additive identities, respectively, for all real numbers. Finally, given any element $a + b\sqrt{2}$ in S (where a and b are integers), $(-a) + (-b)\sqrt{2}$ is also in S (because the negative of any rational number is rational), and

$$(a + b\sqrt{2}) + ((-a) + (-b)\sqrt{2})$$
$$= (a + (-a)) + (b + (-b))\sqrt{2} = 0 + 0\sqrt{2} = 0.$$

Thus every element in S has an additive inverse.

28. *Hint*: Show that 1 is in the set and deduce that every integer is in the set.

31. Proof: Let R be any commutative ring with additive identity 0, and suppose that z is an element in R such that $z + c = c$ for all elements of R. *[We will show that $z = 0$.]* Because $z + c = c$ for every element of R, in particular, $z + 0 = 0$. On the other hand, because 0 is an additive identity for R, $z + 0 = z$. Hence, $z = 0$ because both are equal to $z + 0$. *[This is what was to be shown.]*

34. $\mathbf{Z}_2 = \{[0], [1]\}$

+	[0]	[1]
[0]	[0]	[1]
[1]	[1]	[0]

·	[0]	[1]
[0]	[0]	[0]
[1]	[0]	[1]

Section 8.5

1. $\gcd(27, 72) = 9$ **2.** $\gcd(5, 9) = 1$

5. Divide the larger number, 1,188, by the smaller, 385, to obtain a quotient of 3 and a remainder of 33. Next divide 385 by 33 to obtain a quotient of 11 and a remainder of 22. Then divide 33 by 22 to obtain a quotient of 1 and a remainder of 11. Finally, divide 22 by 11 to obtain a quotient of 2 and a remainder of 0. Thus, by Lemma 8.5.2, $\gcd(1188, 385) = \gcd(385, 33) = \gcd(33, 22) = \gcd(22, 11) = \gcd(11, 0)$, and by Lemma 8.5.1, $\gcd(11, 0) = 11$. So gcd $(1188, 385) = 11$.

6. Divide the larger number, 1,177, by the smaller, 509, to obtain a quotient of 2 and a remainder of 159. Next divide 509 by 159 to obtain a quotient of 3 and a remainder of 32. Next divide 159 by 32 to obtain a quotient of 4 and a remainder of 31. Then divide 32 by 31 to obtain a quotient of 1 and a remainder of 1. Finally, divide 31 by 1 to obtain a quotient of 31 and a remainder of 0. Thus, by Lemma 8.5.2, $\gcd(1177, 509) = \gcd(509, 159) = \gcd(159, 32) = \gcd(32, 31) = \gcd(31, 1) = \gcd(1, 0)$, and by Lemma 8.5.1, $\gcd(1, 0) = 1$. So $\gcd(1177, 509) = 1$.

9. *Hint:* Divide the proof into two parts. In part 1 suppose a and b are any positive integers such that $a \mid b$, and derive the conclusion that $\gcd(a, b) = a$. To do this, note that because it is also the case that $a \mid a$, a is a common divisor of a and b. Thus, by definition of greatest common divisor, a is less than or equal to the greatest common divisor of a and b. In symbols, $a \le \gcd(a, b)$. Then show that $a \ge \gcd(a, b)$ by using Theorem 4.3.1. In part 2 of the proof, suppose a and b are any positive integers such that $\gcd(a, b) = a$, and deduce that $a \mid b$.

11. *Hint:* Use mathematical induction. In the inductive step, use Lemma 8.5.2 and the fact that $F_{k+2} = F_{k+1} + F_k$ to deduce that

$$gcd(F_{k+2}, F_{k+1}) = gcd(F_{k+1}, F_k).$$

12. a. $\mathrm{lcm}(12, 18) = 36$

13. Proof: *Part 1:* Let a and b be positive integers, and suppose $d = \gcd(a, b) = \mathrm{lcm}(a, b)$. By definition of greatest common divisor and least common multiple, $d > 0$, $d \mid a$, $d \mid b$, $a \mid d$, and $b \mid d$. Thus, in particular, $a = dm$ and $d = an$ for some integers m and n. By substitution, $a = dm = (an)m = anm$. Dividing both sides by a gives $1 = nm$. But the only divisors of 1 are 1 and -1 (Theorem 4.3.2), and so $m = n = \pm 1$. Since both a and d are positive, $m = n = 1$, and hence $a = d$. Similar reasoning shows that $b = d$ also, and so $a = b$.
Part 2: Given any positive integers a and b such that $a = b$, we have $\gcd(a, b) = \gcd(a, a) = a$ and $\mathrm{lcm}(a, b) = \mathrm{lcm}(a, a) = a$, and hence $\gcd(a, b) = \mathrm{lcm}(a, b)$.

16. *Hint:* Divide the proof into two parts. In part 1, suppose a and b are any positive integers, and deduce that

$$\gcd(a, b) \cdot \mathrm{lcm}(a, b) \le ab.$$

Derive this result by showing that $\mathrm{lcm}\,(a, b) \le \dfrac{ab}{\gcd(a,b)}$. To do this, show that $\dfrac{ab}{\gcd(a,b)}$ is a multiple of both a and b. For instance, to see that $\dfrac{ab}{\gcd(a,b)}$ is a multiple of b, note that $\dfrac{ab}{\gcd(a,b)} = \dfrac{a}{\gcd(a,b)} \cdot b$. But since $\gcd(a,b)$ divides a, $\dfrac{a}{\gcd(a,b)}$ is an integer, and thus $\dfrac{ab}{\gcd(a,b)} = (\text{an integer}) \cdot b$. The argument that $\dfrac{ab}{\gcd(a,b)}$ is a multiple of a is almost identical. In part 2 of the proof, use the definition of least common multiple to show that $\dfrac{ab}{\mathrm{lcm}(a,b)} \mid a$ and $\dfrac{ab}{\mathrm{lcm}(a,b)} \mid b$. Conclude that $\dfrac{ab}{\mathrm{lcm}(a,b)} \le \gcd(a, b)$ and hence that $ab \le \gcd(a, b) \cdot \mathrm{lcm}(a, b)$.

17. Step 1: $6664 = 765 \cdot 8 + 544$, and so $544 = 6664 - 765 \cdot 8$
Step 2: $765 = 544 \cdot 1 + 221$, and so $221 = 765 - 544$
Step 3: $544 = 221 \cdot 2 + 102$, and so $102 = 544 - 221 \cdot 2$
Step 4: $221 = 102 \cdot 2 + 17$, and so $17 = 221 - 102 \cdot 2$
Step 5: $102 = 17 \cdot 6 + 0$

Thus $\gcd(6664, 765) = 17$ (which is the remainder obtained just before the final division). Substitute back through steps 4–1 to express 17 as a linear combination of 6664 and 765:

$$17 = 221 - 102 \cdot 2$$
$$= 221 - (544 - 221 \cdot 2) = 221 \cdot 5 - 544 \cdot 2$$
$$= (765 - 544) \cdot 5 - 544 \cdot 2 = 765 \cdot 5 - 544 \cdot 7$$
$$= 765 \cdot 5 - (6664 - 765 \cdot 8) \cdot 7 = (-7) \cdot 6664 + 61 \cdot 765.$$

(When you have finished this final step, it is wise to verify that you have not made a mistake by checking that the final expression really does equal the greatest common divisor.)

21. a. *Hint:* For the inductive step, assume $p \mid q_1 q_2 \ldots q_{s+1}$ and let $a = q_1 q_2 \ldots q_s$. Then $p \mid aq_{s+1}$, and either $p = q_{s+1}$ or Euclid's lemma and the inductive hypothesis can be applied.

22. First use the extended Euclidean algorithm to find the greatest common divisor of 243 and 702 and express it as a linear combination of 243 and 702. Successive divisions give the following results:

Step 1: $702 = 243 \cdot 2 + 216$, and so $216 = 702 - 243 \cdot 2$
Step 2: $243 = 216 \cdot 1 + 27$, and so $27 = 243 - 216 \cdot 1$
Step 3: $216 = 27 \cdot 8 + 0$, and so $\gcd(243, 702) = 27$

Substituting back through steps 2 and 1 gives

$$27 = 243 - 216 \cdot 1$$
$$= 243 - (702 - 243 \cdot 2) \cdot 1$$
$$= 243 - 702 + 243 \cdot 2$$
$$= 243 \cdot 3 + 702 \cdot (-1).$$

Hence

$$243 \cdot 3 + 702 \cdot (-1) = 27. \text{(*)}$$

Now $8289 = 27 \cdot 307$, and multiplying both sides of equation (*) by 307 gives

$$243 \cdot (3 \cdot 307) + 702 \cdot [(-1) \cdot 307] = 27 \cdot 307,$$

or, equivalently,

$$243 \cdot 921 + 702 \cdot (-307) = 8289.$$

Thus one integer solution to the equation is $x = 921$ and $y = 307$. By Theorem 8.5.5, the general solution in integers is

$$x = 921 + \frac{702}{27}t \text{ and } y = -307 - \frac{243}{27}t$$

where t is any integer.

24. Let x be the number of 2-pound weights and y be the number of 5-pound weights. Then

$$2x + 5y = 40.$$

Because 2 and 5 are small numbers, you may see immediately that $\gcd(2,5) = 1$ and that $1 = 5 - 2 \cdot 2 = 2 \cdot (-2) + 5 \cdot 1$. The extended Euclidean algorithm can also be used to derive this result by finding the greatest common divisor of 2 and 5 and expressing it as a linear combination of 2 and 5:

Step 1: $5 = 2 \cdot 2 + 1$, and so $1 = 5 - 2 \cdot 2$
Step 2: $2 = 2 \cdot 1 + 0$, and so $\gcd(2, 5) = 1$

Rewriting the result of step 1 gives

$$2 \cdot (-2) + 5 \cdot 1 = 1,$$

and multiplying both sides of this equation by 40 yields

$$2 \cdot (-80) + 5 \cdot 40 = 40.$$

Thus one integer solution to the equation is $x = -80$ and $y = 40$. This solution is not a realistic answer to the question because the number of 2-pound weights cannot be negative. However, realistic answers can be obtained using Theorem 8.5.5, which gives the following general solution in integers:

$$x = -80 + \frac{5}{1}t = -80 + 5t \text{ and } y = 40 - \frac{2}{1}t = 40 - 2t$$

where t is any integer. Because at least one of each kind of weight was purchased, a realistic solution requires that both x and y be greater than zero:

$$-80 + 5t > 0$$
$$40 - 2t > 0.$$

But

$$\begin{Bmatrix} -80 + 5t > 0 \\ 40 - 2t > 0 \end{Bmatrix} \Leftrightarrow \begin{Bmatrix} 5t > 80 \\ 40 > 2t \end{Bmatrix} \Leftrightarrow \begin{Bmatrix} t > 16 \\ 20 > t \end{Bmatrix}$$

$$\Leftrightarrow 16 > t > 20.$$

Thus t= 17, 18, or 19, and there are three possible sets of values for x and y:

When $t = 17$, $x = -80 + 5 \cdot 17 = 5$ and

$$y = 40 - 2 \cdot 17 = 6.$$

When $t = 18$, $x = -80 + 5 \cdot 18 = 10$ and

$$y = 40 - 2 \cdot 18 = 4.$$

When $t = 19$, $x = -80 + 5 \cdot 19 = 15$ and

$$y = 40 - 2 \cdot 19 = 2.$$

The possible numbers of weights of each kind are, therefore, five 2-pound and six 5-pound, or ten 2-pound and four 5-pound, or fifteen 2-pound and two 5-pound.

26. *Hint*: The general solution is $x = 600 - 5t$ and $y = -900 + 8t$, and there are seven possible sets of values for x and y.

27. b. The contrapositive of the statement in part (a) is "For all integers a, b, and c, where a and c are nonzero, if the greatest common divisor of a and b is not a divisor of c, then $ax + by = c$ does not have a solution in integers." This statement provides the following test for a solution to a Diophantine equation: Given an equation $ax + by = c$, where a and c are nonzero integers, check to see whether $\gcd(a, b)$ divides c. If not, then the equation has no integer solution.

28. *Hint:* There exists an integer x such that $ax \equiv c \pmod{n}$ $\Leftrightarrow n$ divides $ax - c \Leftrightarrow$ there exists an integer k such that $ax - c = kn \Leftrightarrow ax + n(-k) = c$.

29. In this case, $\gcd(24, 42) = 6$, and since 6 divides 12, the congruence $24x \equiv 12 \pmod{42}$ has a solution. Now $24x \equiv 12 \pmod{42} \Leftrightarrow$ there exists an integer k such that $24x - 12 = 42k \Leftrightarrow$ there exists an integer k such that

$$24x + (-42)k = 12.$$

Because 6 divides 12, 24, and 42, we may divide both sides of the equation by this number to obtain the equivalent (Diophantine) equation

$$4x + (-7)k = 2.$$

Either use the extended Euclidean algorithm to find that $\gcd(4, -7) = 1$ and express it as a linear combination of 4 and -7, or simply observe that

$$4 \cdot 2 + (-7) \cdot 1 = 1.$$

Multiply both sides by 2 to obtain

$$4 \cdot (2 \cdot 2) + (-7) \cdot (1 \cdot 2) = 2, \text{ or, equivalently,}$$
$$4 \cdot 4 + (-7) \cdot 2 = 2,$$

and so $x = 4$ and $k = 2$ is a solution for $4x + (-7)k = 2$. It follows that $x = 4$ is a solution for the given congruence. To check that this solution is correct, note that

$$24 \cdot 4 \equiv 12 \pmod{42} \Leftrightarrow 96 \equiv 12 \pmod{42}$$
$$\Leftrightarrow 42 \mid (96 - 12) \Leftrightarrow 42 \mid 84,$$

which is true because $84 = 42 \cdot 2$.

32. *Hint:* There is an integer x such that in $\mathbf{Z}_n$ $[a] \cdot [x] = [c]$ if, and only if, there is an integer x such that $ax \equiv c \pmod{n}$. To prove this congruence, see the hint for exercise 28.

35. a. *Hint:* Suppose p is any prime number and $[a]$ and $[c]$ are any elements in $\mathbf{Z}_p$ *[where a and c are integers]* such that $[a] \cdot [c] = [0]$ and $[a] \neq [0]$. Deduce that $\gcd(p, a) = 1$ and apply Euclid's lemma to help conclude that $[c] = [0]$.

b. *Hint:* See exercise 14 in Section 2.2.

36. To fill in the tables, note that in $\mathbf{Z}_7$, $[0] = [7]$, $[1] = [8] = [15] = [36]$, $[2] = [9] = [16] = [30]$, $[3] = [10] = [24]$, $[4] = [11] = [18] = [25]$, $[5] = [12]$, and $[6] = [20]$.

+	[0]	[1]	[2]	[3]	[4]	[5]	[6]
[0]	[0]	[1]	[2]	[3]	[4]	[5]	[6]
[1]	[1]	[2]	[3]	[4]	[5]	[6]	[0]
[2]	[2]	[3]	[4]	[5]	[6]	[0]	[1]
[3]	[3]	[4]	[5]	[6]	[0]	[1]	[2]
[4]	[4]	[5]	[6]	[0]	[1]	[2]	[3]
[5]	[5]	[6]	[0]	[1]	[2]	[3]	[4]
[6]	[6]	[0]	[1]	[2]	[3]	[4]	[5]

·	[1]	[2]	[3]	[4]	[5]	[6]
[1]	[1]	[2]	[3]	[4]	[5]	[6]
[2]	[2]	[4]	[6]	[1]	[3]	[5]
[3]	[3]	[6]	[2]	[5]	[1]	[4]
[4]	[4]	[1]	[5]	[2]	[6]	[3]
[5]	[5]	[3]	[1]	[6]	[4]	[2]
[6]	[6]	[5]	[4]	[3]	[2]	[1]

42. Proof: Suppose R is the set of all real numbers of the form $a + b\sqrt{2}$, where a and b are rational numbers, and suppose that r is any nonzero element of R. *[We will show that r has a multiplicative inverse in R.]* Then $r = a + b\sqrt{2}$ for some rational numbers a and b, where at least one of a or b is nonzero. To show that $a + b\sqrt{2}$ has a multiplicative inverse, we must find rational numbers c and d such that $(a + b\sqrt{2})(c + d\sqrt{2}) = 1$.

Observe that $(a + b\sqrt{2})(a - b\sqrt{2}) = (a^2 - 2b^2) + 0 \cdot \sqrt{2} = a^2 - 2b^2$, and note that, because $\sqrt{2}$ is irrational and at least one of a or b is nonzero, it is impossible for $a^2 - 2b^2$ to equal zero. *[Otherwise, $a^2 = 2b^2$, and so $\sqrt{2}a = b$, which would imply that $\sqrt{2}$ is rational.]* So let

$$c = \frac{a}{a^2 - 2b^2} \text{ and } d = \frac{-b}{a^2 - 2b^2}.$$

Then c and d are rational *[because for rational numbers, products, differences, and quotients with nonzero denominators are rational]*, and

$$(a + b\sqrt{2})(c + d\sqrt{2})$$
$$= (a + b\sqrt{2})\left(\frac{a}{a^2 - 2b^2} + \left(\frac{-b}{a^2 - 2b^2}\right)\sqrt{2}\right)$$
$$= \left(\frac{a^2}{a^2 - 2b^2} - \frac{2b^2}{a^2 - 2b^2}\right)$$
$$+ \left(\frac{-ab}{a^2 - 2b^2} + \frac{ab}{a^2 - 2b^2}\right)\sqrt{2}$$
$$= \frac{a^2 - 2b^2}{a^2 - 2b^2}$$
$$= 1.$$

Since the choice of r was arbitrary, we conclude that every nonzero element of R has a multiplicative inverse.

By exercise 26 in Section 8.4, R is a commutative ring, and, because it has now been shown that every nonzero element of R has a multiplicative inverse we conclude that, by definition of field, R is a field.

Section 9.1

2. 3/4, 1/2, 1/2

3. $\{1\spadesuit, 2\spadesuit, 3\spadesuit, 4\spadesuit, 5\spadesuit, 6\spadesuit, 7\spadesuit, 8\spadesuit, 9\spadesuit, 10\spadesuit, 1\heartsuit, 2\heartsuit, 3\heartsuit, 4\heartsuit, 5\heartsuit, 6\heartsuit, 7\heartsuit, 8\heartsuit, 9\heartsuit, 10\heartsuit\}$, probability $= 20/52 \cong 38.5\%$

5. $\{10\clubsuit, J\clubsuit, Q\clubsuit, K\clubsuit, A\clubsuit, 10\spadesuit, J\spadesuit, Q\spadesuit, K\spadesuit, A\spadesuit, 10\heartsuit, J\heartsuit, Q\heartsuit, K\heartsuit, A\heartsuit, 10\spadesuit, J\spadesuit, Q\spadesuit, K\spadesuit, K\spadesuit, A\spadesuit\}$ probability $= 20/52 = 5/13 \cong 38.5\%$.

7. $\{26, 35, 44, 53, 62\}$, probability $= 5/36 \cong 13.9\%$

9. $\{11, 12, 13, 14, 15, 21, 22, 23, 24, 31, 32, 33, 41, 42, 51\}$ probability $= 15/36 = 41\frac{2}{3}\%$

11. a. $\{HHH, HHT, HTH, HTT, THH, THT, TTH, TTT\}$

b. (i) $\{HTT, THT, TTH\}$, probability $= 3/8 = 37.5\%$

12. a. $\{BBB, BBG, BGB, BGG, GBB, GBG, GGB, GGG\}$

b. (i) $\{GBB, BGB, BBG\}$ probability $= 3/8 = 37.5\%$

13. a. $\{CCC, CCW, CWC, CWW, WCC, WCW, WWC, WWW\}$

b. (i) $\{CWW, WCW, WWC\}$, probability $= 3/8 = 37.5\%$

14. a. probability $= 3/8 = 37.5\%$

16. a. $\{RRR, RRB, RRY, RBR, RBB, RBY, RYR, RYB, RYY, BRR, BRB, BRY, BBR, BBB, BBY, BYR, BYB, BYY, YRR, YRB, YRY, YBR, YBB, YBY, YYR, YYB, YYY\}$

b. $\{RBY, RYB, YBR, BRY, BYR, YRB\}$, probability $= 6/27 = 2/9 \cong 22.2\%$

c. $\{RRR, RBR, BRR, RRY, RYR, YRR, BBR, BRB, RBB, BBY, BYB, YBB, YYR, YRY, RYY, YYB, YBY, BYY\}$ probability $= 18/27 = 2/3 = 66\frac{2}{3}\%$

18. a. $\{B_1B_1, B_1B_2, B_1W, B_2B_1, B_2B_2, B_2W, WB_1, WB_2, WW\}$

b. $\{B_1B_1, B_1B_2, B_2B_1, B_2B_2\}$ probability $= 4/9 \cong 44.4\%$

c. $\{B_1W, B_2W, WB_1, WB_2\}$ probability $= 4/9 \cong 44.4\%$

21. a.

10	11	12	13	14	15	16	17	18	...	96	97	98	99
				$\updownarrow$			$\updownarrow$			$\updownarrow$	$\updownarrow$		$\updownarrow$
				$3 \cdot 4$			$3 \cdot 5$			$3 \cdot 6$	$3 \cdot 32$		$3 \cdot 33$

The above diagram shows that there are as many positive two-digit integers that are multiples of 3 as there are integers from 4 to 33 inclusive. By Theorem 9.1.1, there are $33 - 4 + 1$, or 30, such integers.

b. There are $99 - 10 + 1 = 90$ positive two-digit integers in all, and by part (a), 30 of these are multiples of 3. So the probability that a randomly chosen positive two-digit integer is a multiple of 3 is $30/90 = 1/3 = 33\frac{1}{3}\%$.

c. Of the integers from 10 through 99 that are multiples of 4, the smallest is $12 (= 4 \cdot 3)$ and the largest is $96 (= 4 \cdot 24)$. Thus there are $24 - 3 + 1 = 22$ two-digit integers that are multiples of 4. Hence the probability that a randomly chosen two-digit integer is a multiple of 4 is $22/90 = 36\frac{2}{3}\%$.

23. Let m be the smallest of the integers. By Theorem 9.1.1, $279 - m + 1 = 56$, and so $m = 279 - 56 + 1 = 224$. Thus the smallest of the integers is 224.

26. 1 2 3 4 5 6 7 8 9 ... 999 1000 1001

   $\updownarrow$ $\updownarrow$ $\updownarrow$ $\updownarrow$

   $3 \cdot 1$ $3 \cdot 2$ $3 \cdot 3$ $3 \cdot 333$

Thus there are 333 multiples of 3 between 1 and 1001.

27. a. M Tu W Th F Sa Su M Tu W Th F Sa Su $\cdots$ F Sa Su M
     1   2   3   4   5   6   7   8   9 10 11 12 13 14   362 363 364 365

   $\updownarrow$ $\updownarrow$ $\updownarrow$

   $7 \cdot 1$ $7 \cdot 2$ $7 \cdot 52$

Sundays occur on the 7th day of the year, the 14th day of the year, and in fact on all days that are multiples of 7. There are 52 multiples of 7 between 1 and 365, and so there are 52 Sundays in the year.

Section 9.2

1.

Game 4   Game 5   Game 6   Game 7

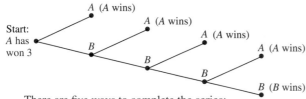

A (A wins)

Start: A has won 3

A (A wins)

B

A (A wins)

B

A (A wins)

B

B (B wins)

There are five ways to complete the series:
$A, B–A, B–B–A, B–B–B–A,$ and $B–B–B–B.$

3. Four ways: $A–A–A–A,$ $B–A–A–A–A,$ $B–B–A–A–A–A,$ and $B–B–B–A–A–A–A.$

4. Two ways: $A–B–A–B–A–B–A$ and $B–A–B–A–B–A–B$

6. a.

Step 1: Choose urn.   Step 2: Choose ball 1.   Step 3: Choose ball 2.

B_1 → B_2, W
B_2 → B_1, W
W → B_1, B_2

Urn 1

Start

Urn 2

B → W_1, W_2
W_1 → B, W_2
W_2 → B, W_1

b. There are 12 equally likely outcomes of the experiment.

c. $2/12 = 1/6 = 16\frac{2}{3}\%$ **d.** $8/12 = 2/3 = 66\frac{2}{3}\%$

8. By the multiplication rule, the answer is $3 \cdot 2 \cdot 2 = 12$.

9. a. In going from city A to city B, one may take any of the 3 roads. In going from city B to city C, one may take any of the 5 roads. So, by the multiplication rule, there are $3 \cdot 5 = 15$ ways to travel from city A to city C via city B.

b. A round-trip journey can be thought of as a four-step operation:

Step 1: Go from A to B.
Step 3: Go from B to C.
Step 2: Go from C to B.
Step 4: Go from B to A.

Since there are 3 ways to perform step 1, 5 ways to perform step 2, 5 ways to perform step 3, and 3 ways to perform step 4, by the multiplication rule, there are $3 \cdot 5 \cdot 5 \cdot 3 = 225$ round-trip routes.

c. In this case the steps for making a round-trip journey are the same as in part (b), but since no route segment may be repeated, there are only 4 ways to perform step 3 and only 2 ways to perform step 4. So, by the multiplication rule, there are $3 \cdot 5 \cdot 4 \cdot 2 = 120$ round-trip routes in which no road is traversed twice.

11. a. Imagine constructing a bit string of length 8 as an eight-step process:

Step 1: Choose either a 0 or a 1 for the left-most position,

Step 2: Choose either a 0 or a 1 for the next position to the right.

Step 3: Choose either a 0 or a 1 for the next position to the right.

Since there are 2 ways to perform each step, the total number of ways to accomplish the entire operation, which is the number of different bit strings of length 8, is $2 \cdot 2 \cdot 2 \cdot 2 \cdot 2 \cdot 2 \cdot 2 \cdot 2 = 2^8 = 256$.

b. Imagine that there are three 0's in the three left-most positions, and imagine filling in the remaining 5 positions as a 5-step process, where step i is to fill in the $(i + 3)$rd position. Since there are 2 ways to perform each of the 5 steps, there are 2^5 ways to perform the entire operation. So there are 2^5, or 32, 8-bit strings that begin with three 0's.

12. a. There are 9 hexadecimal digits from 3 through B and 11 hexadecimal digits from 5 through F. Thus the answer is $9 \cdot 16 \cdot 16 \cdot 11 = 405{,}504$.

13. a. In each of the four tosses there are two possible results: Either a head (H) or a tail (T) is obtained. Thus, by the multiplication rule, the number of outcomes is $2 \cdot 2 \cdot 2 \cdot 2 = 2^4 = 16$.

b. There are six outcomes with two heads:
$HHTT, HTHT, HTTH, THHT, THTH, TTHH$.
Thus the probability of obtaining exactly two heads is $6/16 = 3/8$.

14. a. Let each of steps 1–4 be to choose a letter of the alphabet to put in positions 1–4, and let each of steps 5–7 be to choose a digit to put in positions 5–7. Since there are 26 letters and 10 digits (0–9), the number of license plates is

$$26 \cdot 26 \cdot 26 \cdot 26 \cdot 10 \cdot 10 \cdot 10 = 456{,}976{,}000.$$

b. In this case there is only one way to perform step 1 (because the first letter must be an A) and only one way to perform step 7 (because the last digit must be a 0). Therefore, the number of license plates is $26 \cdot 26 \cdot 26 \cdot 10 \cdot 10 = 17{,}576{,}000$.

d. In this case there are 26 ways to perform step 1, 25 ways to perform step 2, 24 ways to perform step 3, 10 ways to perform step 4, 9 ways to perform step 5, and 8 ways to perform step 6, so the number of license plates is $26 \cdot 25 \cdot 24 \cdot 23 \cdot 10 \cdot 9 \cdot 8 = 258{,}336{,}000$.

16. a. Two solutions:

  (i) number of integers

$$= \begin{bmatrix} \text{number of} \\ \text{ways to pick} \\ \text{first digit} \end{bmatrix} \begin{bmatrix} \text{number of} \\ \text{ways to pick} \\ \text{second digit} \end{bmatrix} = 9 \cdot 10 = 90$$

  (ii) Using Theorem 9.1.1, number of integers $= 99 - 10 + 1 = 90$.

b. Odd integers end in 1, 3, 5, 7, or 9.
number of odd integers

$$= \begin{bmatrix} \text{number of} \\ \text{ways to pick} \\ \text{first digit} \end{bmatrix} \begin{bmatrix} \text{number of} \\ \text{ways to pick} \\ \text{second digit} \end{bmatrix} = 9 \cdot 5 = 45$$

Alternative solution: Use the listing method shown in the solution for Example 9.1.4.

c. $\begin{bmatrix} \text{number of integers} \\ \text{with distinct digits} \end{bmatrix}$

$$= \begin{bmatrix} \text{number of} \\ \text{ways to pick} \\ \text{first digit} \end{bmatrix} \begin{bmatrix} \text{number of} \\ \text{ways to pick} \\ \text{second digit} \end{bmatrix}$$
$$= 9 \cdot 9 = 81$$

d. $\begin{bmatrix} \text{number of odd integers} \\ \text{with distinct digits} \end{bmatrix}$

$$= \begin{bmatrix} \text{number of} \\ \text{ways to pick} \\ \text{second digit} \end{bmatrix} \begin{bmatrix} \text{number of} \\ \text{ways to pick} \\ \text{first digit} \end{bmatrix}$$
$$= 5 \cdot 8 = 40 \qquad \text{because the first digit}$$

because the first digit can't equal 0, nor can it equal the second digit

e. $81/90 = 9/10,\ 40/90 = 4/9$

18. a. Let step 1 be to choose either the number 2 or one of the letters corresponding to the number 2 on the keypad, let step 2 be to choose either the number 1 or one of the letters corresponding to the number 1 on the keypad, and let steps 3 and 4 be to choose either the number 3 or one of the letters corresponding to the number 3 on the keypad. There are 4 ways to perform step 1, 3 ways to perform step 2, and 4 ways to perform each of steps 3 and 4. So by the multiplication rule, there are $4 \cdot 3 \cdot 4 \cdot 4 = 192$ ways to perform the entire operation. Thus there are 192 different PINs that are keyed the same as 2133. Note that on a computer keyboard, these PINs would not be keyed the same way.

19.

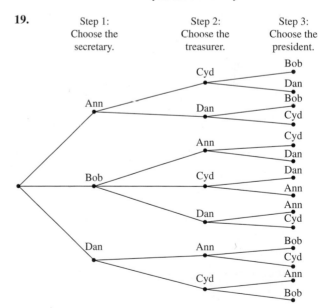

Step 1: Choose the secretary.	Step 2: Choose the treasurer.	Step 3: Choose the president.

There are 14 different paths from "root" to "leaf" of this possibility tree, and so there are 14 ways the officers can be chosen. Because $14 = 2 \cdot 7$, reordering the steps will not make it possible to use the multiplication rule alone to solve this problem.

20. a. The number of ways to perform step 4 is not constant; it depends on how the previous steps were performed. For instance, if 3 digits had been chosen in steps 1–3, then there would be $10 - 3 = 7$ ways to perform step 4, but if 3 letters had been chosen in steps 1–3, then there would be 10 ways to perform step 4.

21. *Hint:*
   a. The answer is 2^{mn}.   **b.** The answer is n^m.

22. a. The answer is $4 \cdot 4 \cdot 4 = 4^3 = 64$. Imagine creating a function from a 3-element set to a 4-element set as a three-step process: Step 1 is to send the first element of the 3-element set to an element of the 4-element set (there are four ways to perform this step); step 2 is to send the second element of the 3-element set to an element of the 4-element set (there are also four ways to perform this step); and step 3 is to send the third element of the 3-element set to an element of the 4-element set (there are four ways to perform this step). Thus the entire process can be performed in $4 \cdot 4 \cdot 4$ different ways.

23. *Hints:* One solution is to add leading zeros as needed to make each number five digits long. For instance, write 1 as 00001. Let some of the steps be to choose positions for the given digits. The answer is 720. Another solution is to consider separately the cases of four-digit and five-digit numbers.

25. a. There are $a + 1$ divisors: $1, p, p^2, \ldots, p^a$.
   b. A divisor is a product of any one of the $a + 1$ numbers listed in part (a) times any one of the $b + 1$ numbers $1, q, q^2, \ldots, q^b$. So, by the multiplication rule, there are $(a + 1)(b + 1)$ divisors in all.

26. a. Since the nine letters of the word $ALGORITHM$ are all distinct, there are as many arrangements of these letters in a row as there are permutations of a set with nine elements: $9! = 362,880$.
   b. In this case there are effectively eight symbols to be permuted (because $\boxed{AL}$ may be regarded as a single symbol). So the number of arrangements is $8! = 40,320$.

28. The same reasoning as in Example 9.2.9 gives an answer of $4! = 24$.

29. $WX, WY, WZ, XW, XY, XZ, YW, YX, YZ, ZW, ZX, ZY$

31. a. $P(6, 4) = \dfrac{6!}{(6 - 4)!} = \dfrac{6 \cdot 5 \cdot 4 \cdot 3 \cdot \cancel{2 \cdot 1}}{\cancel{2 \cdot 1}} = 360$

32. a. $P(5, 3) = \dfrac{5 \cdot 4 \cdot 3 \cdot \cancel{2!}}{\cancel{2!}} = 60$

33. a. $P(9, 3) = \dfrac{9 \cdot 8 \cdot 7 \cdot \cancel{6!}}{\cancel{6!}} = 504$

   c. $P(8, 5) = \dfrac{8 \cdot 7 \cdot 6 \cdot 5 \cdot 4 \cdot \cancel{3!}}{\cancel{3!}} = 6,720$

35. <u>Proof:</u> Let n be an integer and $n \geq 2$. Then

$$P(n + 1, 2) - P(n, 2)$$

$$= \frac{(n + 1)!}{[(n + 1) - 2]!} - \frac{n!}{(n - 2)!} = \frac{(n + 1)!}{(n - 1)!} - \frac{n!}{(n - 2)!}$$

$$= \frac{(n + 1) \cdot n \cdot \cancel{(n - 1)!}}{\cancel{(n - 1)!}} - \frac{n \cdot (n - 1) \cdot \cancel{(n - 2)!}}{\cancel{(n - 2)!}}$$

$$= n^2 + n - (n^2 - n) = 2n$$

$$= 2 \cdot \frac{n \cdot (n - 1)!}{(n - 1)!}$$

$$= 2 \cdot \frac{n!}{(n - 1)!} = 2P(n, 1).$$

This is what was to be proved.

39. *Hint:* In the inductive step, suppose there exist $k!$ permutations of a set with k elements. Let X be a set with $k + 1$ elements. The process of forming a permutation of the elements of X can be considered a two-step operation where step 1 is to choose the element to write first. Step 2 is to write the remaining elements of X in some order.

41. a.

1 2 3	1 2 3	1 2 3
↓ ↓ ↓	↓ ↓ ↓	↓ ↓ ↓
1 2 3	2 1 3	3 2 1

1 2 3	1 2 3	1 2 3
↓ ↓ ↓	↓ ↓ ↓	↓ ↓ ↓
1 3 2	2 3 1	3 1 2

   c.

1 2 3	1 2 3
↓ ↓ ↓	↓ ↓ ↓
2 3 1	3 1 2

Section 9.3

1. a. Set of Bit Strings Consisting of from 1 through 4 Bits

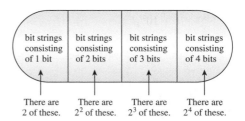

| bit strings consisting of 1 bit | bit strings consisting of 2 bits | bit strings consisting of 3 bits | bit strings consisting of 4 bits |

There are 2 of these. There are 2^2 of these. There are 2^3 of these. There are 2^4 of these.

Applying the addition rule to the figure above shows that there are $2 + 2^2 + 2^3 + 2^4 = 30$ bit strings consisting of from one through four bits.
   b. By reasoning similar to that of part (a), there are $2^5 + 2^6 + 2^7 + 2^8 = 480$ bit strings of from five through eight bits.

3. a. $\begin{bmatrix} \text{number of integers from 1 through 999} \\ \text{with no repeated digits} \end{bmatrix}$

$= \begin{bmatrix} \text{number of integers} \\ \text{from 1 through 9} \\ \text{with no repeated digits} \end{bmatrix} + \begin{bmatrix} \text{number of integers} \\ \text{from 10 through 99} \\ \text{with no repeated digits} \end{bmatrix}$

$\quad + \begin{bmatrix} \text{number of integers from} \\ \text{100 through 999 with} \\ \text{no repeated digits} \end{bmatrix}$

$= 9 + 9 \cdot 9 + 9 \cdot 9 \cdot 8 = 738$

b. $\begin{bmatrix} \text{number of integers from 1 through 999} \\ \text{with at least one repeated digit} \end{bmatrix}$

$= \begin{bmatrix} \text{total number of} \\ \text{integers from} \\ \text{1 through 999} \end{bmatrix} - \begin{bmatrix} \text{number of integers} \\ \text{from 1 through 999} \\ \text{with no repeated digits} \end{bmatrix}$

$= 999 - 738 = 261$

c. The probability that an integer chosen at random has at least one repeated digit is $261/999 \cong 26.1\%$.

4.

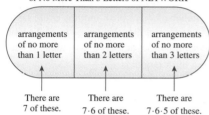

Set of Arrangements (without repetition)
or No More Than 3 Letters of NETWORK

arrangements of no more than 1 letter	arrangements of no more than 2 letters	arrangements of no more than 3 letters
There are 7 of these.	There are $7 \cdot 6$ of these.	There are $7 \cdot 6 \cdot 5$ of these.

Applying the addition rule to the figure above shows that there are $7 + 7 \cdot 6 + 7 \cdot 6 \cdot 5 = 259$ arrangements of three letters of the word *NETWORK* if repetition of letters is not permitted.

6. a. There are $1 + 26 + 26^2 + 26^3$ arrangements of from 0 through 3 letters of the alphabet. Any of these may be paired with all but one arrangement of from 0 through 4 digits, and there are $1 + 10 + 10^2 + 10^3 + 10^4$ arrangements of from 0 through 4 digits. So, by the multiplication rule and the difference rule, the number of license plates is

$(1 + 26 + 26^2 + 26^3)$
$\quad \cdot (1 + 10 + 10^2 + 10^3 + 10^4) - 1 = 203{,}097{,}968$
$\qquad\qquad\qquad \uparrow$
$\qquad\quad$ the blank plate

b. $(1 + 26 + 26^2 + 26^3 - 85)$
$\quad \cdot (1 + 10 + 10^2 + 10^3 + 10^4) - 1 = 202{,}153{,}533$

7. c. *Hint:* See the solutions to exercises 9.2.2, 9.2.4, and 9.3.3. The answers are as follows: (a) 2,238,928,128, (b) 1,449,063,000, (c) 789,865,128, (d) approximately 35.3%.

8. a. $50^3 + 50^4 + 50^5 = 318{,}875{,}000$

10. a. The answer is the number of permutations of the five letters in *QUICK*, which equals $5! = 120$.

b. Because *QU* (in order) is to be considered as a single unit, the answer is the number of permutations of the four symbols $\boxed{QU}, I, C, K$. This is $4! = 24$.

c. By part (b), there are 4! arrangements of $\boxed{QU}, I, C, K$. Similarly, there are 4! arrangements of $\boxed{UQ}, I, C, K$. Therefore, by the addition rule, there are $4! + 4! = 48$ arrangements in all.

12. a. $\begin{bmatrix} \text{number of ways to place eight people} \\ \text{in a row keeping } A \text{ and } B \text{ together} \end{bmatrix}$

$= \begin{bmatrix} \text{number of ways to arrange} \\ \boxed{AB}\, CDEFGH \end{bmatrix}$

$\quad + \begin{bmatrix} \text{number of ways to arrange} \\ \boxed{BA}\, CDEFGH \end{bmatrix}$

$= 7! + 7! = 5{,}040 + 5{,}040 = 10{,}080$

b. $\begin{bmatrix} \text{number of ways to arrange the eight} \\ \text{people in a row keeping } A \text{ and } B \text{ apart} \end{bmatrix}$

$= \begin{bmatrix} \text{total number of ways} \\ \text{to place the eight} \\ \text{people in a row} \end{bmatrix} - \begin{bmatrix} \text{number of ways} \\ \text{to place the eight} \\ \text{people in a row} \\ \text{keeping } A \text{ and } B \\ \text{together} \end{bmatrix}$

$= 8! - 10{,}080 = 40{,}320 - 10{,}080$
$= 30{,}240$

13. number of variable names

$= \begin{bmatrix} \text{number of} \\ \text{numeric} \\ \text{variable names} \end{bmatrix} + \begin{bmatrix} \text{number of} \\ \text{string} \\ \text{variable names} \end{bmatrix}$

$= (26 + 26 \cdot 36) + (26 + 26 \cdot 36) = 1{,}924$

14. *Hint:* In exercise 13 note that

$$26 + 26 \cdot 36 = 26 \sum_{k=0}^{1} 36^k.$$

Generalize this idea here. Use Theorem 5.2.3 to evaluate the expression you obtain.

15. a. $10 \cdot 9 \cdot 8 \cdot 7 \cdot 6 \cdot 5 \cdot 4 = 604{,}800$

b. $\begin{bmatrix} \text{number of phone numbers with} \\ \text{at least one repeated digit} \end{bmatrix}$

$= \begin{bmatrix} \text{total number of} \\ \text{phone numbers} \end{bmatrix} - \begin{bmatrix} \text{number of phone numbers} \\ \text{with no repeated digits} \end{bmatrix}$

$= 10^7 - 604{,}800 = 9{,}395{,}200$

c. $9{,}395{,}200/10^7 \cong 93.95\%$

17. a. Proof: Let A and B be mutually disjoint events in a sample space S. By the addition rule, $N(A \cup B) = N(A) + N(B)$. Therefore, by the equally likely probability formula,

$$P(A \cup B) = \frac{N(A \cup B)}{N(S)} = \frac{N(A) + N(B)}{N(S)}$$

$$= \frac{N(A)}{N(S)} + \frac{N(B)}{N(S)} = P(A) + P(B).$$

18. *Hint:* Justify the following answer: $39 \cdot 38 \cdot 38$.

19. a. Identify the integers from 1 to 100,000 that contain the digit 6 exactly once with strings of five digits. Thus, for example, 306 would be identified with 00306. It is not necessary to use strings of six digits, because 100,000 does not contain the digit 6. Imagine the process of constructing a five-digit string that contains the digit 6 exactly once as a five-step operation that consists of filling in the five digit positions $\underline{}_1 \ \underline{}_2 \ \underline{}_3 \ \underline{}_4 \ \underline{}_5$.

Step 1: Choose one of the five positions for the 6.

Step 2: Choose a digit for the left-most remaining position.

Step 3: Choose a digit for the next remaining position to the right.

Step 4: Choose a digit for the next remaining position to the right.

Step 5: Choose a digit for the right-most position.

Since there are 5 choices for step 1 (any one of the five positions) and 9 choices for each of steps 2–5 (any digit except 6), by the multiplication rule, the number of ways to perform this operation is $5 \cdot 9 \cdot 9 \cdot 9 \cdot 9 = 32,805$. Hence there are 32,805 integers from 1 to 100,000 that contain the digit 6 exactly once.

20. *Hint:* The answer is 2/3.

22. a. Let A = the set of integers that are multiples of 4 and B = the set of integers that are multiples of 7. Then $A \cap B$ = the set of integers that are multiples of 28.

But $n(A) = 250$ since 1 2 3 4 5 6 7 8 ... 999 1000,

$$\updownarrow \qquad \updownarrow \qquad\quad \updownarrow$$
$$4 \cdot 1 \quad\ 4 \cdot 2 \ldots \quad 4 \cdot 250$$

or, equivalently, since $1,000 = 4 \cdot 250$.

Also $n(B) = 142$ since 1 2 3 4 5 6 7 ... 14 ...   994 995 ... 1000

$$\updownarrow \quad\ \updownarrow \qquad\quad \updownarrow$$
$$7 \cdot 1 \ \ 7 \cdot 2 \ldots \ 7 \cdot 142$$

or, equivalently, since $1,000 = 7 \cdot 142 + 6$.

and $n(A \cap B) = 35$ since 1 2 3 ... 28 ... 56 ...   980 ... 1000,

$$\updownarrow \qquad \updownarrow \qquad\quad \updownarrow$$
$$28 \cdot 1 \ \ 28 \cdot 2 \ldots \ 28 \cdot 35$$

or, equivalently, since $1,000 = 28 \cdot 35 + 20$.

So $n(A \cup B) = 250 + 142 - 35 = 357$.

24. a.   Length 0: ϵ

Length 1: 0, 1

Length 2: 00, 01, 10, 11

Length 3: 000, 001, 010, 011, 100, 101, 110

Length 4: 0000, 0001, 0010, 0011, 0100, 0101, 0110, 1000, 1001, 1010, 1011, 1100, 1101

b. By part (a), $d_0 = 1$, $d_1 = 2$, $d_2 = 4$, $d_3 = 7$, and $d_4 = 13$.

c. Let k be an integer with $k \geq 3$. Any string of length k that does not contain the bit pattern 111 starts either with a 0 or with a 1. If it starts with a 0, this can be followed by any string of $k - 1$ bits that does not contain the pattern 111. There are d_{k-1} of these. If the string starts with a 1, then the first two bits are 10 or 11. If the first two bits are 10, then these can be followed by

any string of $k - 2$ bits that does not contain the pattern 111. There are d_{k-2} of these. If the string starts with a 11, then the third bit must be 0 (because the string does not contain 111), and these three bits can be followed by any string of $k - 3$ bits that does not contain the pattern 111. There are d_{k-3} of these. Therefore, for all integers $k \geq 3$, $d_k = d_{k-1} + d_{k-2} + d_{k-3}$.

d. By parts (b) and (c), $d_5 = d_4 + d_3 + d_2 = 13 + 7 + 4 = 24$. This is the number of bit strings of length five that do not contain the pattern 111.

25. c. *Hint:* $s_k = 2s_{k-1} + 2s_{k-2}$.

27. a. $a_3 = 3$ (The three permutations that do not move more than one place from their "natural" positions are 213, 132, and 123.)

29. a. There are 12 possible birth months for A, 12 for B, 12 for C, and 12 for D, so the total is $12^4 = 20,736$.

b. If no two people share the same birth month, there are 12 possible birth months for A, 11 for B, 10 for C, and 9 for D. Thus the total is $12 \cdot 11 \cdot 10 \cdot 9 = 11,880$.

c. If at least two people share the same birth month, the total number of ways birth months could be associated with A, B, C, and D is $20,736 - 11,880 = 8,856$.

d. The probability that at least two of the four people share the same birth month is $\frac{8856}{20736} \cong 42.7\%$.

e. When there are five people, the probability that at least two share the same birth month is $\frac{12^5 - 12 \cdot 11 \cdot 10 \cdot 9 \cdot 8}{12^5}$ $\cong 61.8\%$, and when there are more than five people, the probability is even greater. Thus, since the probability for four people is less than 50%, the group must contain five or more people for the propability to be at least 50% that two or more share the same birth month.

30. *Hint:* Analyze the solution to exercise 29.

31. a. The number of students who checked at least one of the statements is $N(H) + N(C) + N(D) - N(H \cap C) - N(H \cap D) - N(C \cap D) + N(H \cap C \cap D) = 28 + 26 + 14 - 8 - 4 - 3 + 2 = 55$

b. By the difference rule, the number of students who checked none of the statements is the total number of students minus the number who checked at least one statement. This is $100 - 55 = 45$.

d. The number of students who checked #1 and #2 but not #3 is $N(H \cap D) - N(H \cap C \cap D) = 8 - 2 = 6$.

33. Let

M = the set of married people in the sample,
Y = the set of people between 20 and 30 in the sample, and
F = the set of females in the sample.

Then the number of people in the set $M \cup Y \cup F$ is less than or equal to the size of the sample. And so

$$1,200 \geq N(M \cup Y \cup F)$$
$$= N(M) + N(Y) + N(F) - N(M \cap Y)$$
$$- N(M \cap F) - N(Y \cap F) + N(M \cap Y \cap F)$$
$$= 675 + 682 + 684 - 195 - 467 - 318 + 165$$
$$= 1,226.$$

This is impossible since $1,200 < 1,226$, so the polltaker's figures are inconsistent. They could not have occurred as a result of an actual sample survey.

35. Let A be the set of all positive integers less than $1,000$ that are not multiples of 2, and let B be the set of all positive integers less than $1,000$ that are not multiples of 5. Since the only prime factors of $1,000$ are 2 and 5, the number of positive integers that have no common factors with $1,000$ is $N(A \cap B)$. Let the universe U be the set of all positive integers less than $1,000$. Then A^c is the set of positive integers less than $1,000$ that are multiples of 2, B^c is the set of positive integers less than $1,000$ that are multiples of 5, and $A^c \cap B^c$ is the set of positive integers less than $1,000$ that are multiples of 10. By one of the procedures discussed in Section 9.1 or 9.2, it is easily found that $N(A^c) = 499$, $N(B^c) = 199$, and $N(A^c \cap B^c) = 99$. Thus, by the inclusion/exclusion rule,

$$N(A^c \cup B^c) = N(A^c) + N(B^c) - N(A^c \cap B^c)$$
$$= 499 + 199 - 99 = 599.$$

But by De Morgan's law, $N(A^c \cup B^c) = N((A \cap B)^c)$, and so

$$N((A \cap B)^c) = 599. \qquad (*)$$

Now since $(A \cap B)^c = U - (A \cap B)$, by the difference rule we have

$$N((A \cap B)^c) = N(U) - N(A \cap B). \qquad (**)$$

Equating the right-hand sides of $(*)$ and $(**)$ gives $N(U) - N(A \cap B) = 599$. And because $N(U) = 999$, we conclude that $999 - N(A \cap B) = 599$, or, equivalently, $N(A \cap B) = 999 - 599 = 400$. So there are 400 positive integers less than $1,000$ that have no common factor with $1,000$.

38. *Hint:* Let A and B be the sets of all positive integers less than or equal to n that are divisible by p and q, respectively. Then $\phi(n) = n - (N(A \cup B))$.

40. c. *Hint:* If $k \geq 6$, any sequence of k games must begin with W, LW, or LLW, where L stands for "lose" and W stands for "win."

41. c. *Hint:* Divide the set of all derangements into two subsets: one subset consists of all derangements in which the number 1 changes places with another number, and the other subset consists of all derangements in which the number 1 goes to position $i \neq 1$ but i does not go to position 1. The answer is $d_k = (k-1)d_{k-1} + (k-1)d_{k-2}$. Can you justify it?

46. *Hint:* Use the associative law for sets and the generalized distributive law for sets from exercise 37, Section 6.2.

Section 9.4

1. a. No. For instance, the aces of the four different suits could be selected.

b. Yes. Let x_1, x_2, x_3, x_4, x_5 be the five cards. Consider the function S that sends each card to its suit.

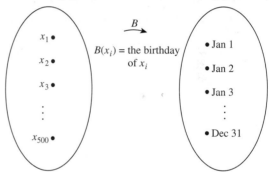

By the pigeonhole principle, S is not one-to-one: $S(x_i) = S(x_j)$ for some two cards x_i and x_j. Hence at least two cards have the same suit.

3. Yes. Denote the residents by $x_1, x_2, \ldots, x_{500}$. Consider the function B from residents to birthdays that sends each resident to his or her birthday:

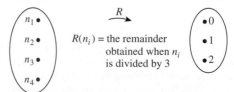

By the pigeonhole principle, B is not one-to-one: $B(x_i) = B(x_j)$ for some two residents x_i and x_j. Hence at least two residents have the same birthday.

5. a. Yes. There are only three possible remainders that can be obtained when an integer is divided by 3: 0, 1, and 2. Thus, by the pigeonhole principle, if four integers are each divided by 3, then at least two of them must have the same remainder.

More formally, call the integers n_1, n_2, n_3, and n_4, and consider the function R that sends each integer to the remainder obtained when that integer is divided by 3:

4 integers (pigeons)   3 remainders (pigeonholes)

$n_1 \bullet$

$n_2 \bullet$ $R(n_i) =$ the remainder $\bullet 0$

$n_3 \bullet$ obtained when n_i $\bullet 1$

$n_4 \bullet$ is divided by 3 $\bullet 2$

By the pigeonhole principle, R is not one-to-one, $R(n_i) = R(n_j)$ for some two integers n_i and n_j. Hence at least two integers must have the same remainder.

b. No. For instance, $\{0, 1, 2\}$ is a set of three integers no two of which have the same remainder when divided by 3.

7. *Hint:* Look at Example 9.4.3.

9. a. Yes.

Solution 1: Only six of the numbers from 1 to 12 are even (namely, 2, 4, 6, 8, 10, 12), so at most six even numbers can be chosen from between 1 and 12 inclusive. Hence if seven numbers are chosen, at least one must be odd.

Solution 2: Partition the set of all integers from 1 through 12 into six subsets (the pigeonholes), each consisting of an odd and an even number: {1, 2}, {3, 4}, {5, 6}, {7, 8}, {9, 10}, {11, 12}. If seven integers (the pigeons) are chosen from among 1 through 12, then, by the pigeonhole principle, at least two must be from the same subset. But each subset contains one odd and one even number. Hence at least one of the seven numbers is odd.

Solution 3: Let $S = \{x_1, x_2, x_3, x_4, x_5, x_6, x_7\}$ be a set of seven numbers chosen from the set $T = \{1, 2, 3, 4, 5, 6, 7, 8, 9, 10, 11, 12\}$, and let P be the following partition of T: {1, 2}, {3, 4}, {5, 6}, {7, 8}, {9, 10}, and {11, 12}. Since each element of S lies in exactly one subset of the partition, we can define a function F from S to P by letting $F(x_i)$ be the subset that contains x_i.

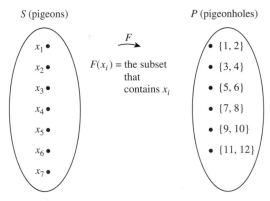

Since S has 7 elements and P has 6 elements, by the pigeonhole principle, F is not one-to-one. Thus two distinct numbers of the seven are sent to the same subset, which implies that these two numbers are the two distinct elements of the subset. Therefore, since each pair consists of one odd and one even integer, one of the seven numbers is odd.

b. No. For instance, none of the 10 numbers 1, 3, 5, 7, 9, 11, 13, 15, 17, 19 is even.

10. Yes. There are n even integers in the set $\{1, 2, 3, \ldots, 2n\}$, namely $2(= 2 \cdot \underline{1}), 4(= 2 \cdot \underline{2}), 6(= 2 \cdot \underline{3}), \ldots, 2\underline{n}(= 2 \cdot n)$. So the maximum number of even integers that can be chosen is n. Thus if $n + 1$ integers are chosen, at least one of them must be odd.

12. The answer is 27. There are only 26 black cards in a standard 52-card deck, so at most 26 black cards can be chosen. Hence if 27 are taken, at least one must be red.

14. There are 61 integers from 0 to 60 inclusive. Of these, 31 are even $(0 = 2 \cdot \underline{0}, 2 = 2 \cdot \underline{1}, 4 = 2 \cdot \underline{2}, \ldots, 60 = 2 \cdot \underline{30})$ and so 30 are odd. Hence if 32 integers are chosen, at least one

must be odd, and if 31 integers are chosen, at least one must be even.

17. The answer is 8. (There are only seven possible remainders for division by 7: 0, 1, 2, 3, 4, 5, 6.)

20. a. The answer is 20,483 *[namely, 0, 1, 2, ..., 20482].*

b. The length of the repeating section of the decimal representation of 5/20483 is less than or equal to 20,482. The reason is that 20,482 is the number of nonzero remainders that can be obtained when an integer is divided by 20,483. Thus, in the long-division process of dividing 5 by 20,483, either some remainder is 0 and the decimal expansion terminates, or only nonzero remainders are obtained and at some point within the first 20,482 successive divisions, a nonzero remainder is repeated. At that point the digits in the developing decimal expansion begin to repeat because the sequence of successive remainders repeats those previously obtained.

22. This number is irrational; the decimal expansion neither terminates nor repeats.

24. Let A be the set of the thirteen chosen numbers, and let B be the set of all prime numbers between 1 and 40. Note that $B = \{2, 3, 5, 7, 11, 13, 17, 19, 23, 29, 31, 37\}$. For each x in A, let $F(x)$ be the smallest prime number that divides x. Since A has 13 elements and B has 12 elements, by the pigeonhole principle F is not one-to-one. Thus $F(x_1) = F(x_2)$ for some $x_1 \neq x_2$ in A. By definition of F, this means that the smallest prime number that divides x_1 equals the smallest prime number that divides x_2. Therefore, two numbers in A, namely x_1 and x_2, have a common divisor greater than 1.

25. Yes. This follows from the generalized pigeonhole principle with 30 pigeons, 12 pigeonholes, and $k = 2$, using the fact that $30 > 2 \cdot 12$.

26. No. For instance, the birthdays of the 30 people could be distributed as follows: three birthdays in each of the six months January through June and two birthdays in each of the six months July through December.

29. The answer is $x = 3$. There are 18 years from 17 through 34. Now $40 > 18 \cdot 2$, so by the generalized pigeonhole principle, you can be sure that there are at least $x = 3$ students of the same age. However, since $18 \cdot 3 > 40$, you cannot be sure of having more than three students with the same age. (For instance, three students could be each of the ages 17 through 20, and two could be each of the ages from 21 through 34.) So x cannot be taken to be greater than 3.

31. *Hint:* Use the same type of reasoning as in Example 9.4.6.

32. *Hints:* (1) The number of subsets of the six integers is $2^6 = 64$. (2) Since each integer is less than 13, the largest possible sum is 57. (Why? What gives this sum?)

33. *Hint:* The power set of A has $2^6 = 64$ elements, and so there are 63 nonempty subsets of A. Let k be the smallest number in the set A. Then the sums over the elements in the nonempty subsets of A lie in the range from k through $k + 10 + 11 + 12 + 13 + 14 = k + 60$. How many numbers are in this range?

35. *Hint:* Let X be the set consisting of the given 52 positive integers, and let Y be the set containing the following elements: $\{00\}$, $\{50\}$, $\{01, 99\}$, $\{02, 98\}$, $\{03, 97\}$, ..., $\{48, 52\}$, $\{49, 51\}$. Define a function F from X to Y by the rule $F(x) =$ the set containing the last two digits of x. Use the pigeonhole principle to argue that F is not one-to-one, and show how the desired conclusion follows.

36. *Hint:* Represent each of the 101 integers x_i as $a_i 2^{k_i}$ where a_i is odd and $k_i \geq 0$. Now $1 \leq x_i \leq 200$, and so $1 \leq a_i \leq 199$ for all i. There are only 100 odd integers from 1 to 199 inclusive.

37. b. *Hint:* For each $k = 1, 2, \ldots, n$, let $a_k = x_1 + x_2 + \cdots + x_k$. If some a_k is divisible by n, then the problem is solved: the consecutive subsequence is $x_1, x_2, \ldots, x_k$. If no a_k is divisible by n, then $a_1, a_2, a_3, \ldots, a_n$ satisfies the hypothesis of part (a). Hence $a_j - a_i$ is divisible by n for some integers i and j with $j > i$. Write $a_j - a_i$ in terms of the x_i's to derive the given conclusion.

38. *Hint:* Let $a_1, a_2, \ldots, a_{n^2+1}$ be any sequence of $n^2 + 1$ distinct real numbers, and suppose that this sequence contains neither a strictly increasing subsequence of length $n + 1$ nor a strictly decreasing subsequence of length $n + 1$. Let S be the set of all ordered pairs of integers (i, d), where $1 \leq i \leq n$ and $1 \leq d \leq n$. For each term a_k in the sequence, let $F(a_k) = (i_k, d_k)$, where i_k is the length of the longest increasing sequence starting at a_k, and d_k is the length of the longest decreasing sequence starting at a_k. Suppose that F is one-to-one and derive a contradiction.

Section 9.5

1. a. 2-combinations: $\{x_1, x_2\}$, $\{x_1, x_3\}$, $\{x_2, x_3\}$.

Hence, $\dbinom{3}{2} = 3$.

b. Unordered selections: $\{a, b, c, d\}$, $\{a, b, c, e\}$, $\{a, b, d, e\}$, $\{a, c, d, e\}$, $\{b, c, d, e\}$.

Hence, $\dbinom{5}{4} = 5$.

3. $P(7, 2) = \dbinom{7}{2} \cdot 2!$

5. a. $\dbinom{6}{0} = \dfrac{6!}{0!(6-0)!} = \dfrac{\not{6}!}{1 \cdot \not{6}!} = 1$

b. $\dbinom{6}{1} = \dfrac{6!}{1!(6-1)!} = \dfrac{6 \cdot \not{5}!}{1 \cdot \not{5}!} = 6$

6. a. number of committees of 6

$$= \binom{15}{6} = \frac{15!}{(15-6)!6!}$$

$$= \frac{\overset{7}{\not{15}} \cdot \not{14} \cdot 13 \cdot \not{12} \cdot 11 \cdot \overset{5}{\not{10}} \cdot \not{9}!}{\not{9}! \cdot \not{6} \cdot \not{5} \cdot \not{4} \cdot \not{3} \cdot \not{2}} = 5{,}005$$

b. $\begin{bmatrix} \text{number of committees} \\ \text{that don't contain } A \\ \text{and } B \text{ together} \end{bmatrix}$

$$= \begin{bmatrix} \text{number of} \\ \text{committees with } A \\ \text{and five others—} \\ \text{none of them } B \end{bmatrix} + \begin{bmatrix} \text{number of} \\ \text{committees with } B \\ \text{and five others—} \\ \text{none of them } A \end{bmatrix}$$

$$+ \begin{bmatrix} \text{number of committees} \\ \text{with neither } A \text{ nor } B \end{bmatrix}$$

$$= \binom{13}{5} + \binom{13}{5} + \binom{13}{6}$$

$$= 1{,}287 + 1{,}287 + 1{,}716 = 4{,}290$$

Alternative solution:

$$\begin{bmatrix} \text{number of committees} \\ \text{that don't contain } A \\ \text{and } B \text{ together} \end{bmatrix}$$

$$= \begin{bmatrix} \text{total number} \\ \text{of committees} \end{bmatrix} - \begin{bmatrix} \text{number of committees} \\ \text{that contain both } A \text{ and } B \end{bmatrix}$$

$$= \binom{15}{6} - \binom{13}{4}$$

$$= 5{,}005 - 715 = 4{,}290$$

c. $\begin{bmatrix} \text{number of} \\ \text{committees with} \\ \text{both } A \text{ and } B \end{bmatrix} + \begin{bmatrix} \text{number of} \\ \text{committees with} \\ \text{neither } A \text{ and } B \end{bmatrix}$

$$= \binom{13}{4} + \binom{13}{6} = 715 + 1{,}716 = 2{,}431$$

d. (i) $\begin{bmatrix} \text{number of subsets} \\ \text{of three men} \\ \text{chosen from eight} \end{bmatrix} \cdot \begin{bmatrix} \text{number of subsets} \\ \text{of three women} \\ \text{chosen from seven} \end{bmatrix}$

$$= \binom{8}{3}\binom{7}{3} = 56 \cdot 35 = 1{,}960$$

(ii) $\begin{bmatrix} \text{number of committees} \\ \text{with at least one woman} \end{bmatrix}$

$$= \begin{bmatrix} \text{total number of} \\ \text{committees} \end{bmatrix} - \begin{bmatrix} \text{number of all-male} \\ \text{committees} \end{bmatrix}$$

$$= \binom{15}{6} - \binom{8}{6} = 5{,}005 - 28$$

$$= 4{,}977$$

e. $\begin{bmatrix} \text{number of} \\ \text{ways to choose} \\ \text{two freshmen} \end{bmatrix} \cdot \begin{bmatrix} \text{number of} \\ \text{ways to choose two} \\ \text{sophomores} \end{bmatrix}$

$$\cdot \begin{bmatrix} \text{number of ways} \\ \text{to choose two juniors} \end{bmatrix} \cdot \begin{bmatrix} \text{number of ways} \\ \text{to choose two seniors} \end{bmatrix}$$

$$= \binom{3}{2}\binom{4}{2}\binom{3}{2}\binom{5}{2}$$

$$= 540$$

8. *Hint:* The answers are **a.** 1001, **b.** (i) 420, (ii) all 1001 require proof, (iii) 175, **c.** 506, **d.** 561

9. b. $\binom{24}{3}\binom{16}{3} + \binom{24}{4}\binom{16}{2} + \binom{24}{5}\binom{16}{1} + \binom{24}{6}\binom{16}{0} = 3{,}223{,}220$

11. a.(1) 4 (because there are as many royal flushes as there are suits)

(2) $\dfrac{4}{\binom{52}{5}} = \dfrac{4}{2{,}598{,}960} \cong 0.0000015$

  c.(1) $13 \cdot \binom{48}{1} = 624$ (because one can first choose the denomination of the four-of-a-kind and then choose one additional card from the 48 remaining)

(2) $\dfrac{624}{\binom{52}{5}} = \dfrac{624}{2{,}598{,}960} = 0.00024$

  f.(1) Imagine constructing a straight (including a straight flush and a royal flush) as a six-step process: step 1 is to choose the lowest denomination of any card of the five (which can be any one of $A, 2, \ldots, 10$), step 2 is to choose a card of that denomination, step 3 is to choose a card of the next higher denomination, and so forth until all five cards have been selected. By the multiplication rule, the number of ways to perform this process is

$$10 \cdot \binom{4}{1}\binom{4}{1}\binom{4}{1}\binom{4}{1}\binom{4}{1} = 10 \cdot 4^5 = 10{,}240.$$

By parts (a) and (b), 40 of these numbers represent royal or straight flushes, so there are $10{,}240 - 40 = 10{,}200$ straights in all.

(2) $\dfrac{10{,}200}{\binom{52}{5}} = \dfrac{10{,}200}{2{,}598{,}960} \cong 0.0039$

13. a. $2^{10} = 1{,}024$

  d. $\begin{bmatrix} \text{number of outcomes} \\ \text{with at least one head} \end{bmatrix}$

$= \begin{bmatrix} \text{total number} \\ \text{of outcomes} \end{bmatrix} - \begin{bmatrix} \text{number of outcomes} \\ \text{with no heads} \end{bmatrix}$

$= 1{,}024 - 1 = 1{,}023$

15. a. 50 **b.** 50

  c. To get an even sum, both numbers must be even or both must be odd. Hence

$\begin{bmatrix} \text{number of subsets of two integers from} \\ \text{1 to 100 inclusive whose sum is even} \end{bmatrix}$

$= \begin{bmatrix} \text{number of subsets} \\ \text{of two even} \\ \text{integers chosen from} \\ \text{the 50 possible} \end{bmatrix} + \begin{bmatrix} \text{number of subsets} \\ \text{of two odd} \\ \text{integers chosen from} \\ \text{the 50 possible} \end{bmatrix}$

$= \binom{50}{2} + \binom{50}{2} = 2{,}450.$

  d. To obtain an odd sum, one of the numbers must be even and the other odd. Hence the answer is $\binom{50}{1} \cdot \binom{50}{1} = 2{,}500$. Alternatively, note that the answer equals the total number of subsets of two integers chosen from 1 through 100 minus the number of such subsets for which the sum of the elements is even. Thus the answer is $\binom{100}{2} - 2{,}450 = 2{,}500.$

17. a. Two points determine a line. Hence

$\begin{bmatrix} \text{number of straight lines} \\ \text{determined by the} \\ \text{ten points} \end{bmatrix} = \begin{bmatrix} \text{number of subsets} \\ \text{of two points chosen} \\ \text{from ten} \end{bmatrix}$

$= \binom{10}{2} = 45.$

19. a. $\dfrac{10!}{2!1!1!3!2!1!} = 151{,}200$ since there are 2 A's, 1 B, 1 H, 3 L's, 2 O's, and 1 U

  b. $\dfrac{8!}{2!1!1!2!2!} = 5{,}040$ **c.** $\dfrac{9!}{1!2!1!3!2!} = 15{,}120$

23. Rook must move seven squares to the right and seven squares up, so

$\begin{bmatrix} \text{the number of} \\ \text{paths the rook} \\ \text{can take} \end{bmatrix} = \begin{bmatrix} \text{the number} \\ \text{of orderings} \\ \text{of seven R's} \\ \text{and seven U's} \end{bmatrix}$ where R stands for "right" and U stands for "up"

$= \dfrac{14!}{7!7!} = 3{,}432.$

24. **b.** *Solution 1:* One factor can be 1, and the other factor can be the product of all the primes. (This gives 1 factorization.) One factor can be one of the primes, and the other factor can be the product of the other three. (This gives $\binom{4}{1} = 4$ factorizations.) One factor can be a product of two of the primes, and the other factor can be a product of the two other primes. The number $\binom{4}{2} = 6$ counts all possible sets of two primes chosen from the four primes, and each set of two primes corresponds to a factorization. Note, however, that the set $\{p_1, p_2\}$ corresponds to the same factorization as the set $\{p_3, p_4\}$, namely, $p_1 p_2 p_3 p_4$ (just written in a different order). In general, each choice of two primes corresponds to the same factorization as one other choice of two primes. Thus the number of factorizations in which each factor is a product of two primes is $\dfrac{\binom{4}{2}}{2} = 3$. (This gives 3 factorizations.) The foregoing cases account for all the possibilities, so the answer is $4 + 3 + 1 = 8$.

Solution 2: Let $S = \{p_1, p_2, p_3, p_4\}$. Let $p_1 p_2 p_3 p_4 = P$, and let $f_1 f_2$ be any factorization of P. The product of the numbers in any subset $A \subseteq S$ can be used for f_1, with the product of the numbers in A^c being f_2. There are as many ways to write $f_1 f_2$ as there are subsets of S, namely $2^4 = 16$ (by Theorem 6.3.1). But given any factors f_1 and f_2, $f_1 f_2 = f_2 f_1$. Thus counting the number of ways to write $f_1 f_2$ counts each factorization twice, so the answer is $\frac{16}{2} = 8$.

25. **a.** There are four choices for where to send the first element of the domain (any element of the co-domain may be chosen), three choices for where to send the second (since the function is one-to-one, the second element of the domain must go to a different element of the co-domain from the one to which the first element went), and two choices for where to send the third element (again since the function is one-to-one). Thus the answer is $4 \cdot 3 \cdot 2 = 24$.

   b. none

   e. *Hint:* The answer is $n(n - 1) \cdots (n - m + 1)$.

26. **a.** Let the elements of the domain be called a, b, and c and the elements of the co-domain be called u and v. In order for a function from $\{a, b, c\}$ to $\{u, v\}$ to be onto, two elements of the domain must be sent to u and one to v, or two elements must be sent to v and one to u. There are as many ways to send two elements of the domain to u and one to v as there are ways to choose which elements of $\{a, b, c\}$ to send to u, namely, $\binom{3}{2} = 3$. Similarly, there are $\binom{3}{2} = 3$ ways to send two elements of the domain to v and one to u. Therefore, there are $3 + 3 = 6$ onto functions from a set with three elements to a set with two elements.

   c. *Hint:* The answer is 6.

   d. Consider functions from a set with four elements to a set with two elements. Denote the set of four elements by $X = \{a, b, c, d\}$ and the set of two elements by $Y = \{u, v\}$. Divide the set of all onto functions from X to Y into two categories. The first category consists of all those that send the three elements in $\{a, b, c\}$ onto $\{u, v\}$ and that send d to either u or v. The functions in this category can be defined by the following two-step process:

   Step 1: Construct an onto function from $\{a, b, c\}$ to $\{u, v\}$.

   Step 2: Choose whether to send d to u or to v.

   By part (a), there are six ways to perform step 1, and, because there are two choices for where to send d, there are two ways to perform step 2. Thus, by the multiplication rule, there are $6 \cdot 2 = 12$ ways to define the functions in the first category.

   The second category consists of all those onto functions from X to Y that send all three elements in $\{a, b, c\}$ to either u or v and that send d to whichever of u or v is not the image of the others. Because there are only two choices for where to send the elements in $\{a, b, c\}$, and because d is simply sent to wherever the others do not go, there are just two functions in the second category.

   Every onto function from X to Y either sends at least two elements of X to $f(d)$ or it does not. If it sends at least two elements of X to $f(d)$ then it is in the second category. If it does not, then the image of $\{a, b, c\}$ is $\{u, v\}$ and so the "restriction" of the function to $\{a, b, c\}$ is onto. Therefore, the function is one of those included in the first category. Thus all onto functions from X to Y are in one of the two categories and no function is in both categories, and so the total number of onto functions is $12 + 2 = 14$.

   Hints: **a.** (i) g is one-to-one   (ii) g is not onto

   b. G is onto. <u>Proof:</u> Suppose y is any element of $\mathbf{R}$. *[We must show that there is an element x in $\mathbf{R}$ such that $G(x) = y$. Use of scratch work to determine what x would have to be if it exists shows that x would have to equal $(y + 5)/4$. The proof must then show that x has the necessary properties.]* Let $x = (y + 5)/4$. Then (1) $x \in \mathbf{R}$, and (2) $G(x) = G((y + 5)/4) = 4[(y + 5)/4] - 5 = (y + 5) - 5 = y$ *[as was to be shown].*

27. **a.** A relation on A is any subset of $A \times A$, and $A \times A$ has $8^2 = 64$ elements. So there are 2^{64} binary relations on A.

   c. Form a symmetric relation by a two-step process: (1) pick a set of elements of the form (a, a) (there are eight such elements, so 2^8 sets); (2) pick a set of pairs of elements of the form (a, b) and (b, a) where $a \neq b$ (there are $(64 - 8)/2 = 28$ such pairs, so 2^{28} such sets). The answer is therefore $2^8 \cdot 2^{28} = 2^{36}$.

28. *Hint:* Use the difference rule and the generalization of the inclusion/exclusion rule for 4 sets. (See exercise 48 in Section 9.3.)

31. Call the set X, and suppose that $X = \{x_1, x_2, \ldots, x_n\}$. For each integer $i = 0, 1, 2, \ldots, n - 1$, we can consider the set of all partitions of X (let's call them *partitions of type i*) where one of the subsets of the partition is an $(i + 1)$-element set that contains x_n and i elements chosen from $\{x_1, \ldots, x_{n-1}\}$. The remaining subsets of the partition will be a partition of the remaining $(n - 1) - i$ elements of $\{x_1, \ldots, x_{n-1}\}$. For instance, if $X = \{x_1, x_2, x_3\}$, there are five partitions of the various types, namely,

Type 0: two partitions where one set is a 1-element set containing x_3: $[\{x_3\}, \{x_1\}, \{x_2\}]$, $[\{x_3\}, \{x_1, x_2\}]$

Type 1: two partitions where one set is a 2-element set containing x_3: $[\{x_1, x_3\}, \{x_2\}]$, $[\{x_2, x_3\}, \{x_1\}]$

Type 2: one partition where one set is a 3-element set containing x_3: $\{x_1, x_2, x_3\}$

In general, we can imagine constructing a partition of type i as a two-step process:

Step 1: Select out the i elements of $\{x_1, \ldots, x_{n-1}\}$ to put together with x_n,

Step 2: Choose any partition of the remaining $(n - 1) - i$ elements of $\{x_1, \ldots, x_{n-1}\}$ to put with the set formed in step 1.

There are $\binom{n-1}{i}$ ways to perform step 1 and $P_{(n-1)-i}$ ways to perform step 2. Therefore, by the multiplication rule, there are $\binom{n-1}{i} \cdot P_{(n-1)-i}$ partitions of type i. Because any partition of X is of type i for some $i = 0, 1, 2, \ldots, n-1$, it follows from the addition rule that the total number of partitions is

$$\binom{n-1}{0} P_{n-1} + \binom{n-1}{1} P_{n-2}$$
$$+ \binom{n-1}{2} P_{n-3} + \cdots + \binom{n-1}{n-1} P_0.$$

Section 9.6

1. $\binom{n}{0} = \dfrac{n!}{0!(n-0)!} = \dfrac{\cancel{n!}}{1 \cdot \cancel{n!}} = 1$

3. $\binom{n}{2} = \dfrac{n!}{(n-2)! \cdot 2!} = \dfrac{n \cdot (n-1) \cdot \cancel{(n-2)!}}{\cancel{(n-2)!} \cdot 2!}$

$= \dfrac{n(n-1)}{2}$

5. Proof: Suppose n and r are nonnegative integers and $r \leq n$. Then

$$\binom{n}{r} = \frac{n!}{r!(n-r)!} \qquad \text{by Theorem 9.5.1}$$

$$= \frac{n!}{(n-(n-r))!(n-r)!} \qquad \begin{array}{l}\text{since } n-(n-r) = \\ n-n+r = r\end{array}$$

$$= \frac{n!}{(n-r)!(n-(n-r))!} \qquad \begin{array}{l}\text{by interchanging the} \\ \text{factors in the denominator}\end{array}$$

$$= \binom{n}{n-r} \qquad \text{by Theorem 9.5.1.}$$

6. *Solution 1*: Apply formula (9.7.2) with $m + k$ in place of n. This is legal because $m + k \geq 1$.
Solution 2:

$$\binom{m+k}{m+k-1} = \frac{(m+k)!}{(m+k-1)![(m+k)-(m+k-1)]!}$$

$$= \frac{(m+k) \cdot (m+k-1)!}{(m+k-1)!(m+k-m-k+1)!}$$

$$= \frac{(m+k) \cdot (m+k-1)!}{(m+k-1)! \cdot 1!} = m+k$$

10. a. $\binom{6}{2} = \binom{5}{2} + \binom{5}{1} = 10 + 5 = 15,$

$\binom{6}{3} = \binom{5}{3} + \binom{5}{2} = 10 + 10 = 20$

b. $\binom{6}{4} = \binom{5}{4} + \binom{5}{3} = 5 + 10 = 15,$

$\binom{6}{5} = \binom{5}{5} + \binom{5}{4} = 1 + 5 = 6,$

$\binom{7}{3} = \binom{6}{3} + \binom{6}{2} = 20 + 15 = 35,$

$\binom{7}{4} = \binom{6}{4} + \binom{6}{3} = 15 + 20 = 35,$

$\binom{7}{5} = \binom{6}{5} + \binom{6}{4} = 6 + 15 = 21$

c. Row for $n = 7$:   1   7   21   35   35   21   7   1

13. Proof by mathematical induction: Let the property $P(n)$ be the formula

$$\sum_{i=2}^{n+1} \binom{i}{2} = \binom{n+2}{3}. \qquad \leftarrow P(n)$$

Show that $P(1)$ is true:
To prove $P(1)$ we must show that

$$\sum_{i=2}^{1+1} \binom{i}{2} = \binom{1+2}{3}. \qquad \leftarrow P(1)$$

But

$$\sum_{i=2}^{1+1} \binom{i}{2} = \sum_{i=2}^{2} \binom{i}{2} = \binom{2}{2} = 1$$

$$= \binom{3}{3} = \binom{1+2}{3},$$

so $P(1)$ is true.

Show that for all integers $k \geq 1$, $P(k)$ is true, then $P(k+1)$ is true:
Let k be any integer with $k \geq 1$, and suppose that

$$\sum_{i=2}^{k+1} \binom{i}{2} = \binom{k+2}{3} \qquad \begin{array}{l}\leftarrow P(k) \\ \text{inductive hypothesis}\end{array}$$

We must show that

$$\sum_{i=2}^{(k+1)+1} \binom{i}{2} = \binom{(k+1)+2}{3},$$

or, equivalently,

$$\sum_{i=2}^{k+2} \binom{i}{2} = \binom{k+3}{3}. \qquad \leftarrow P(k+1)$$

But the left-hand side of $P(k+1)$ is

$$\sum_{i=2}^{k+2} \binom{i}{2} = \sum_{i=1}^{k+1} \binom{i}{2} + \binom{k+2}{2} \qquad \begin{array}{l}\text{by writing the last} \\ \text{term separately}\end{array}$$

$$= \binom{k+2}{3} + \binom{k+2}{2} \qquad \text{by inductive hypothesis}$$

$$= \binom{(k+2)+1}{3} \qquad \text{by Pascal's formula}$$

$$= \binom{k+3}{3},$$

which is the right-hand side of $P(k+1)$ *[as was to be shown].*
[Since we have proved the basis step and the inductive step, we conclude that $P(n)$ is true for all $n \geq 1$.]

14. *Hint:* Use the results of exercises 3 and 13.

17. *Hint:* This follows by letting $m = n = r$ in exercise 16 and using the result of Example 9.7.2.

19. $1 + 7x + \binom{7}{2}x^2 + \binom{7}{3}x^3 + \binom{7}{4}x^4 + \binom{7}{5}x^5 + \binom{7}{6}x^6 + x^7 = 1 + 7x + 21x^2 + 35x^3 + 35x^4 + 21x^5 + 7x^6 + x^7$

21. $1 + 6(-x) + \binom{6}{2}(-x)^2 + \binom{6}{3}(-x)^3 + \binom{6}{4}(-x)^4 + \binom{6}{5}(-x)^5 + (-x)^6 = 1 - 6x + 15x^2 - 20x^3 + 15x^4 - 6x^5 + x^6$

23. $(p - 2q)^4 = \sum_{k=0}^{4} \binom{4}{k} p^{4-k}(-2q)^k$

$= \binom{4}{0} p^4(-2q)^0 + \binom{4}{1} p^3(-2q)^1$

$\quad + \binom{4}{2} p^2(-2q)^2 + \binom{4}{3} p^1(-2q)^3$

$\quad + \binom{4}{4} p^0(-2q)^4$

$= p^4 - 8p^3q + 24p^2q^2 - 32pq^3 + 16q^4$

25. $\left(x + \dfrac{1}{x}\right)^5 = \sum_{k=0}^{5} \binom{5}{k} x^{5-k} \left(\dfrac{1}{x}\right)^k$

$= \binom{5}{0} x^5 \left(\dfrac{1}{x}\right)^0 + \binom{5}{1} x^4 \left(\dfrac{1}{x}\right)^1$

$\quad + \binom{5}{2} x^3 \left(\dfrac{1}{x}\right)^2 + \binom{5}{3} x^2 \left(\dfrac{1}{x}\right)^3$

$\quad + \binom{5}{4} x^1 \left(\dfrac{1}{x}\right)^4 + \binom{5}{5} x^0 \left(\dfrac{1}{x}\right)^5$

$= x^5 + 5x^3 + 10x + \dfrac{10}{x} + \dfrac{5}{x^3} + \dfrac{1}{x^5}$

29. The term is $\binom{9}{3} x^6 y^3 = 84 x^6 y^3$, so the coefficient is 84.

31. The term is $\binom{12}{7} a^5(-2b)^7 = 792a^5(-128)b^7 = -101{,}376 a^5 b^7$, so the coefficient is $-101{,}376$.

33. The term is $\binom{15}{8}(3p^2)^8(-2q)^7 = \binom{15}{8} 3^8 (-2)^7 p^{16} q^7$, so the coefficient is $\binom{15}{8} 3^8 (-2)^7 = -5{,}404{,}164{,}480$.

36. <u>Proof:</u> Let $a = 1$, let $b = -1$, and let n be a positive integer. Substitute into the binomial theorem to obtain

$(1 + (-1))^n = \sum_{k=0}^{n} \binom{n}{k} \cdot 1^{n-k} \cdot (-1)^k$

$= \sum_{k=0}^{n} \binom{n}{k} (-1)^k \quad$ since $1^{n-k} = 1$.

But $(1 + (-1))^n = 0^n = 0$, so

$0 = \sum_{k=0}^{n} \binom{n}{k}(-1)^k$

$= \binom{n}{0} - \binom{n}{1} + \binom{n}{2} - \binom{n}{3} + \cdots + (-1)^n \binom{n}{n}.$

37. *Hint:* $3 = 1 + 2$

38. <u>Proof:</u> Let m be any integer with $m \geq 0$, and apply the binomial theorem with $a = 2$ and $b = -1$. The result is

$1 = 1^m = (2 + (-1))^m = \sum_{i=0}^{m} \binom{m}{i} 2^{m-i}(-1)^i$

$= \sum_{i=0}^{m} (-1)^i \binom{m}{i} 2^{m-i}.$

41. *Hint:* Apply the binomial theorem with $a = 1$ and $b = -\left(\frac{1}{2}\right)$, and analyze the resulting equation when n is even and when n is odd.

43. $\sum_{k=0}^{n} \binom{n}{k} 5^k = \sum_{k=0}^{n} \binom{n}{k} 1^{n-k} 5^k = (1 + 5)^n = 6^n$

45. $\sum_{i=0}^{n} \binom{n}{i} x^i = \sum_{i=0}^{n} \binom{n}{i} 1^{n-i} x^i = (1 + x)^n$

47. $\sum_{j=0}^{2n} (-1)^j \binom{2n}{j} x^j = \sum_{j=0}^{2n} \binom{2n}{j} 1^{2n-j}(-x)^j = (1 - x)^{2n}$

51. $\sum_{i=0}^{m} (-1)^i \binom{m}{i} \dfrac{1}{2^i} = \sum_{i=0}^{m} \binom{m}{i} 1^{m-i} \left(-\dfrac{1}{2}\right)^i$

$= \left(1 - \dfrac{1}{2}\right)^m = \dfrac{1}{2^m}$

53. $\sum_{i=0}^{n} (-1)^i \binom{n}{i} 5^{n-i} 2^i = \sum_{i=0}^{n} \binom{n}{i} 5^{n-i}(-2)^i = (5 - 2)^n = 3^n$

55. b. $n(1 + x)^{n-1} = \sum_{k=1}^{n} \binom{n}{k} k x^{k-1}.$

[The term corresponding to $k = 0$ is zero because

$\dfrac{d}{dx}(x^0) = 0.]$

c. (i) Substitute $x = 1$ in part (b) above to obtain

$n(1 + 1)^{n-1} = \sum_{k=1}^{n} \binom{n}{k} k \cdot 1^{k-1} = \sum_{k=1}^{n} \binom{n}{k} k$

$= \binom{n}{1} \cdot 1 + \binom{n}{2} \cdot 2 + \binom{n}{3} \cdot 3 + \cdots + \binom{n}{n} n.$

Dividing both sides by n and simplifying gives

$2^{n-1} = \dfrac{1}{n}\left[\binom{n}{1} + 2\binom{n}{2} + 3\binom{n}{3} + \cdots + n\binom{n}{n}\right].$

Section 10.1

1. $V(G) = \{v_1, v_2, v_3, v_4\}$, $E(G) = \{e_1, e_2, e_3\}$
Edge-endpoint function:

Edge	Endpoints
e_1	$\{v_1, v_2\}$
e_2	$\{v_1, v_3\}$
e_3	$\{v_3\}$

3.

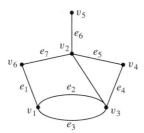

5. Imagine that the edges are strings and the vertices are knots. You can pick up the left-hand figure and lay it down again to form the right-hand figure as shown below.

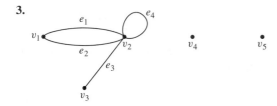

8. (i) e_1, e_2, and e_3 are incident on v_1.
(ii) v_1, v_2, and v_3 are adjacent to v_3.
(iii) e_2, e_8, e_9, and e_3 are adjacent to e_1.
(iv) Loops are e_6 and e_7.
(v) e_8 and e_9 are parallel; e_4 and e_5 are parallel.
(vi) v_6 is an isolated vertex.
(vii) degree of $v_3 = 5$
(viii) total degree $= 20$

10. a. Yes. According to the graph, *Sports Illustrated* is an instance of a sports magazine, a sports magazine is a periodical, and a periodical contains printed writing.

12. To solve this puzzle using a graph, introduce a notation in which, for example, wc/fg means that the wolf and the cabbage are on the left bank of the river and the ferryman and the goat are on the right bank. Then draw those arrangements of wolf, cabbage, goat, and ferryman that can be reached from the initial arrangement $(wgcf/)$ and that are not arrangements to be avoided (such as (wg/fc)). At each stage ask yourself, "Where can I go from here?" and draw lines or arrows pointing to those arrangements. This method gives the graph shown at the top of the next column.

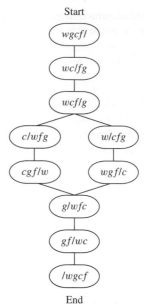

Examination of the diagram shows the solutions

$$(wgcf/) \to (wc/gf) \to (wcf/g) \to (w/gcf) \to$$
$$(wgf/c) \to (g/wcf) \to (gf/wc) \to (/wgcf)$$

and

$$(wgcf/) \to (wc/gf) \to (wcf/g) \to (c/wgf) \to$$
$$(gcf/w) \to (g/wcf) \to (gf/wc) \to (/wgcf)$$

14. *Hint:* The answer is yes. Represent possible amounts of water in jugs A and B by ordered pairs. For instance, the ordered pair $(1, 3)$ would indicate that there is one quart of water in jug A and three quarts in jug B. Starting with $(0, 0)$, draw arrows from one ordered pair to another if it is possible to go from the situation represented by one pair to that represented by the other by either filling a jug, emptying a jug, or transferring water from one jug to another. You need only draw arrows from states that have arrows pointing to them; the other states cannot be reached. Then find a directed path (sequence of directed edges) from the initial state $(0, 0)$ to a final state $(1, 0)$ or $(0, 1)$.

15. The total degree of the graph is $0 + 2 + 2 + 3 + 9 = 16$, so by Theorem 10.1.1, the number of edges is $16/2 = 8$.

17. One such graph is

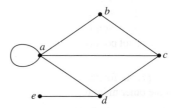

18. If there were a graph with four vertices of degrees 1, 2, 3, and 3, then its total degree would be 9, which is odd. But by Corollary 10.1.2, the total degree of the graph must be even. *[This is a contradiction.]* Hence there is no such graph. (Alternatively, if there were such a graph, it would have an odd number of vertices of odd degree. But by Proposition 10.1.3 this is impossible.)

21. Suppose there were a simple graph with four vertices of degrees 1, 2, 3, and 4. Then the vertex of degree 4 would have to be connected by edges to four distinct vertices other than itself because of the assumption that the graph is simple (and hence has no loops or parallel edges.) This contradicts the assumption that the graph has four vertices in total. Hence there is no simple graph with four vertices of degrees 1, 2, 3, and 4.

24.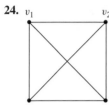

26. **a.** The nonempty subgraphs are as follows:

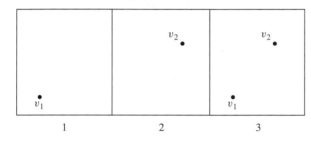

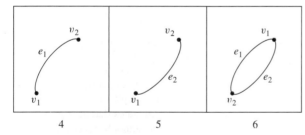

27. **a.** Suppose that, in a group of 15 people, each person had exactly three friends. Then you could draw a graph representing each person by a vertex and connecting two vertices by an edge if the corresponding people were friends. But such a graph would have 15 vertices, each of degree 3, for a total degree of 45. This would contradict the fact that the total degree of any graph is even. Hence the supposition must be false, and in a group of 15 people it is not possible for each to have exactly three friends.

31. We give two proofs for the following statement, one less formal and the other more formal.

For all integers $n \geq 0$, if $a_1, a_2, a_3, \ldots, a_{2n+1}$ are odd integers, then $\sum_{i=1}^{2n+1} a_i$ is odd.

Proof 1 (by mathematical induction): It is certainly true that the "sum" of one odd integer is odd. Suppose that for a certain positive odd integer r, the sum of r odd integers

is odd. We must show that the sum of $r + 2$ odd integers is odd (because $r + 2$ is the next odd integer after r). But any sum of $r + 2$ odd integers equals a sum of r odd integers (which is odd by inductive hypothesis) plus a sum of two more odd integers (which is even). Thus the total sum is an odd integer plus an even integer, which is odd. *[This is what was to be shown.]*

Proof 2 (by mathematical induction): Let the property $P(n)$ be the following sentence: "If $a_1, a_2, a_3, \ldots, a_{2n+1}$ are odd integers, then $\sum_{i=1}^{2n+1} a_i$ is odd.

Show that $P(0)$ is true:

Suppose a_1 is an odd integer. Then $\sum_{i=1}^{2 \cdot 0 + 1} a_i = \sum_{i=1}^{1} a_i = a_1$, which is odd.

Show that for all integers $k \geq 0$, if $P(k)$ is true then $P(k + 1)$ is true:

Let k be an integer with $k \geq 0$, and suppose that

if $a_1, a_2, \ldots, a_{2k+1}$ are odd integers, then $\sum_{i=1}^{2k+1} a_i$ is odd.

[This is the inductive hypothesis $P(k)$.]

Suppose $a_1, a_2, a_3, \ldots, a_{2(k+1)+1}$ are odd integers. *[We must show $P(k + 1)$, namely that $\sum_{i=1}^{2(k+1)+1} a_i$ is odd, or, equivalently, that $\sum_{i=1}^{2k+3} a_i$ is odd.]* But

$$\sum_{i=1}^{2k+3} a_i = \sum_{i=1}^{2k+1} a_i + (a_{2k+2} + a_{2k+3}).$$

Since the sum of any two odd integers is even, $a_{2k+2} + a_{2k+3}$ is even, and, by inductive hypothesis, $\sum_{i=1}^{2k+1} a_i$ is odd. Therefore, $\sum_{i=1}^{2k+3} a_i$ is the sum of an odd integer and an even integer, which is odd. *[This is what was to be shown.]*

32. *Hint:* Use proof by contradiction.

33. **a.** K_6:

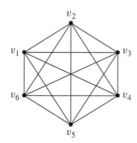

b. A proof of this fact was given in Section 5.6 using recursion. Try to find a different proof.

Hint for Proof 1: There are as many edges in K_n as there are subsets of two vertices (the endpoints) that can be chosen from a set of n vertices.

Hint for Proof 2: Use mathematical induction. A complete graph on $k + 1$ vertices can be obtained from a complete graph on k vertices by adding one vertex and connecting this vertex by k edges to each of the other vertices.

Hint for Proof 3: Use the fact that the number of edges of a graph is half the total degree. What is the degree of each vertex of K_n?

35. Suppose G is a simple graph with n vertices and $2n$ edges where n is a positive integer. By exercise 34, its number of edges cannot exceed $\frac{n(n-1)}{2}$. Thus $2n \leq \frac{n(n-1)}{2}$, or $4n \leq n^2 - n$. Equivalently, $n^2 - 5n \geq 0$, or $n(n-5) \geq 0$. This implies that $n \geq 5$ since $n > 0$. Hence a simple graph with twice as many edges as vertices must have at least five vertices. But a complete graph with five vertices has $\frac{5(5-1)}{2} = 10$ edges and $10 = 2 \cdot 5$. Consequently, the answer to the question is yes because K_5 is a graph with twice as many edges as vertices.

36. a. $K_{4,2}$:

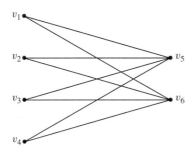

37. a. This graph is bipartite.

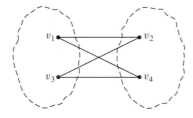

b. Suppose this graph is bipartite. Then the vertex set can be partitioned into two mutually disjoint subsets such that vertices in each subset are connected by edges only to vertices in the other subset and not to vertices in the same subset. Now v_1 is in one subset of the partition, say V_1. Since v_1 is connected by edges to v_2 and v_3, both v_2 and v_3 must be in the other subset, V_2. But v_2 and v_3 are connected by an edge to each other. This contradicts the fact that no vertices in V_2 are connected by edges to other vertices in V_2. Hence the supposition is false, and so the graph is not bipartite.

39. a.

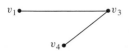

41. b.

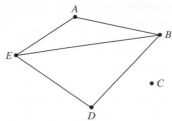

42. *Hint:* Consider the graph obtained by taking the vertices and edges of G plus all the edges of G'. Use exercise 33(b).

44. c. *Hint:* Suppose there were a simple graph with n vertices (where $n \geq 2$) each of which had a different degree. Then no vertex could have degree more than $n - 1$ (why?), so the degrees of the n vertices must be $0, 1, 2, \ldots, n - 1$ (why?). This is impossible (why?).

45. *Hint:* Use the result of exercise 44(c).

46.

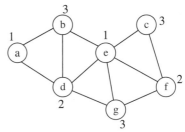

Vertex e has maximal degree, so color it with color #1. Vertex a does not share an edge with e, and so color #1 may also be used for it. From the remaining uncolored vertices, all of d, g, and f have maximal degree. Choose any one of them, say d, and use color #2 for it. Observe that vertices c and f do not share an edge with d but do share an edge with each other, which means that color #2 may be used for one but not the other. Choose to color f with color #2 because the degree of f is greater than the degree of c. The remaining uncolored vertices, b, c, and g, are unconnected, and so color #3 may be used for all three. At this point all vertices have been colored.

47. *Hint:* There are two solutions:
  (1) Time 1: hiring, library
      Time 2: personnel, undergraduate education, colloquium
      Time 3: graduate education
  (2) Time 1: hiring, library
      Time 2: graduate education, colloquium
      Time 3: personnel, undergraduate education

Section 10.2

1. a. trail (no repeated edge), not a path (repeated vertex$-v_1$), not a circuit
  b. walk, not a trail (has repeated edge$-e_9$), not a circuit
  c. closed walk (starts and ends at the same vertex), trail (no repeated edge since no edge), not a path or a circuit (since no edge)

d. circuit, not a simple circuit (repeated vertex, v_4)

e. closed walk (starts and ends at the same vertex but has repeated edges $-\{v_2, v_3\}$ and $\{v_3, v_4\}$)

f. path

3. a. No. The notation $v_1v_2v_1$ could equally well refer to $v_1e_1v_2e_2v_1$ or to $v_1e_2v_2e_1v_1$, which are different walks.

4. a. Three (There are three ways to choose the middle edge.)

b. $3! + 3 = 9$ (In addition to the three paths, there are $3!$ with vertices $v_1, v_2, v_3, v_2, v_3, v_4$. The reason is that from v_2 there are three choices of an edge to go to v_3, then two choices of different edges to go back to v_2, and then one choice of different edge to return to v_3. This makes $3!$ trails from v_2 to v_3.)

c. Infinitely many (Since a walk may have repeated edges, a walk from v_1 to v_4 may contain an arbitrarily large number of repetitions of edges joining a pair of vertices along the way.)

6. a. $\{v_1, v_3\}$, $\{v_2, v_3\}$, $\{v_4, v_3\}$, and $\{v_5, v_3\}$ are all the bridges.

8. a. Three connected components.

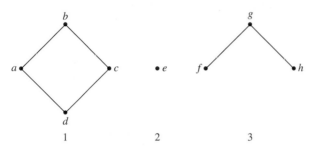

9. a. No. This graph has two vertices of odd degree, whereas all vertices of a graph with an Euler circuit have even degree.

12. One Euler circuit is $e_4e_5e_6e_3e_2e_7e_8e_1$.

14. One Euler circuit is $iabihbchgcdgfdefi$.

19. There is an Euler trail since $\deg(u)$ and $\deg(w)$ are odd, all other vertices have positive even degree, and the graph is connected. One Euler trail is $uv_1v_0v_7uv_2v_3v_4$ $v_2v_6v_4wv_5v_6w$.

23. $v_0v_7v_1v_2v_3v_4v_5v_6v_0$

25. *Hint:* See the solution to Example 10.2.8.

26. Here is one sequence of reasoning you could use: Call the given graph G, and suppose G has a Hamiltonian circuit. Then G has a subgraph H that satisfies conditions (1)–(4) of Proposition 10.2.6. Since the degree of b in G is 4 and every vertex in H has degree 2, two edges incident on b must be removed from G to create H. Edge $\{a, b\}$ cannot be removed because doing so would result in vertex d having degree less than 2 in H. Similar reasoning shows that edge $\{b, c\}$ cannot be removed either. So edges $\{b, i\}$ and $\{b, e\}$ must be removed from G to create H. Because vertex e must have degree 2 in H and because edge $\{b, e\}$ is not in H, both edges $\{e, d\}$ and $\{e, f\}$ must be in H. Similarly, since both vertices c and g must have degree 2 in H, edges

$\{c, d\}$ and $\{g, d\}$ must also be in H. But then three edges incident on d, namely $\{e, d\}$, $\{c, d\}$, and $\{g, d\}$, must be all in H, which contradicts the fact that vertex d must have degree 2 in H.

28. *Hint:* This graph does not have a Hamiltonian circuit.

32. *Partial answer:*

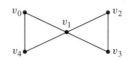

This graph has an Euler circuit $v_0v_1v_2v_3v_1v_4v_0$ but no Hamiltonian circuit.

33. *Partial answer:*

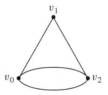

This graph has a Hamiltonian circuit $v_0v_1v_2v_0$ but no Euler circuit.

34. *Partial answer:*

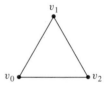

The walk $v_0v_1v_2v_0$ is both an Euler circuit and a Hamiltonian circuit for this graph.

35. *Partial answer:*

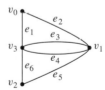

This graph has the Euler circuit $e_1e_2e_3e_4e_5e_6$ and the Hamiltonian circuit $v_0v_1v_2v_3v_0$. These are not the same.

37. a. <u>Proof:</u> Suppose G is a graph and W is a walk in G that contains a repeated edge e. Let v and w be the endpoints of e. In case $v = w$, then v is a repeated vertex of W. In case $v \neq w$, then one of the following must occur: (1) W contains two copies of vew or of wev (for instance, W might contain a section of the form $vewe'vew$, as illustrated below); (2) W contains separate sections of the form vew and wev (for instance, W might contain a section of the form $vewe'wev$, as illustrated below); or (3) W contains a section of the form $vewev$ or of the form $wevew$ (as illustrated below). In cases (1) and (2), both vertices v and w are repeated, and in case (3), one of v or w is repeated. In all cases, there is at least one vertex in W that is repeated.

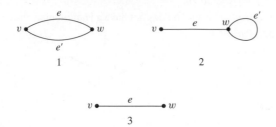

38. Proof: Suppose G is a connected graph and v and w are any particular but arbitrarily chosen vertices of G. *[We must show that u and v can be connected by a path.]* Since G is connected, there is a walk from v to w. If the walk contains a repeated vertex, then delete the portion of the walk from the first occurrence of the vertex to its next occurrence. (For example, in the walk $ve_1v_2e_5v_7e_6v_2e_3w$, the vertex v_2 occurs twice. Deleting the portion of the walk from one occurrence to the next gives $ve_1v_2e_3w$.) If the resulting walk still contains a repeated vertex, do the above deletion process another time. Then check again for a repeated vertex. Continue in this way until all repeated vertices have been deleted. (This must occur eventually, since the total number of vertices is finite.) The resulting walk connects v to w but has no repeated vertex. By exercise 37(b), it has no repeated edge either. Hence it is a path from v to w.

40. The graph to the right contains a circuit, any edge of which can be removed without disconnecting the graph. For instance, if edge e is removed, then the following walk can be used to go from v_1 to v_2: $v_1v_5v_3v_2$.

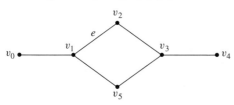

42. *Hint:* Look at the answer to exercise 40 and use the fact that all graphs have a finite number of edges.

44. Proof: Let G be a connected graph and let C be a circuit in G. Let G' be the subgraph obtained by removing all the edges of C from G and also any vertices that become isolated when the edges of C are removed. *[We must show that there exists a vertex v such that v is in both C and G'.]* Pick any vertex v of C and any vertex w of G'. Since G is connected, there is a path from v to w (by Lemma 10.2.1(a)):

$$v = v_0e_1v_1e_2v_2 \ldots v_{i-1}e_iv_ie_{i+1}v_{i+1} \ldots v_{n-1}e_nv_n = w.$$

$$\quad\uparrow \qquad\qquad\qquad \uparrow \quad\ \uparrow \qquad\qquad\qquad \uparrow$$
$$\text{in } C \qquad\qquad\qquad \text{in } C \ \text{ not in } C \qquad\qquad \text{in } G'$$

Let i be the largest subscript such that v_i is in C. If $i = n$, then $v_n = w$ is in C and also in G', and we are done. If $i < n$, then v_i is in C and v_{i+1} is not in C. This implies that e_{i+1} is not in C (for if it were, both endpoints would be in C by definition of circuit). Hence when G' is formed by removing the edges and resulting isolated vertices from G, then e_{i+1} is not removed. That means that v_i does not become an isolated vertex, so v_i is not removed either. Hence v_i is in G'. Consequently, v_i is in both C and G' *[as was to be shown]*.

45. Proof: Suppose G is a graph with an Euler circuit. If G has only one vertex, then G is automatically connected. If v and w are any two vertices of G, then v and w each appear at least once in the Euler circuit (since an Euler circuit contains every vertex of the graph). The section of the circuit between the first occurrence of one of v or w and the first occurrence of the other is a walk from one of the two vertices to the other.

Section 10.3

1. a. Math 110

2. a.

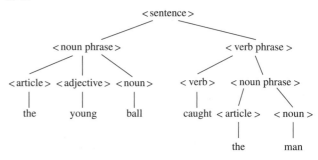

3. *Hint:* The answer is $2n - 2$. To obtain this result, use the relationship between the total degree of a graph and the number of edges of the graph.

4. a.

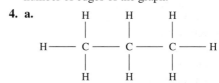

d. *Hint:* Each carbon atom in G is bonded to four other atoms in G, because otherwise an additional hydrogen atom could be bonded to it, and this would contradict the assumption that G has the maximum number of hydrogen atoms for its number of carbon atoms. Also each hydrogen atom is bonded to exactly one carbon atom in G, because otherwise G would not be connected.

5. *Hint:* Revise the algorithm given in the proof of Lemma 10.3.1 to keep track of which vertex and edge were chosen in step 1 (by, say, labeling them v_0 and e_0). Then after one vertex of degree 1 is found, return to v_0 and search for another vertex of degree 1 by moving along a path outward from v_0 starting with e_0.

7. a. Internal vertices: v_2, v_3, v_4, v_6
Terminal vertices: v_1, v_5, v_7

8. Any tree with nine vertices has eight edges, not nine. Thus there is no tree with nine vertices and nine edges.

9. One such graph is

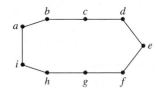

10. One such graph is

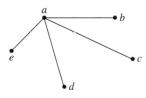

11. There is no tree with six vertices and a total degree of 14. Any tree with six vertices has five edges and hence (by Theorem 10.1.1) a total degree of 10, not 14.

12. One such tree is shown.

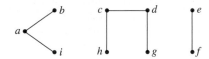

13. No such graph exists. By Theorem 10.3.4, a connected graph with six vertices and five edges is a tree. Hence such a graph cannot have a nontrivial circuit.

14.

22. Yes. Since it is connected and has 12 vertices and 11 edges, by Theorem 10.3.4 it is a tree. It follows from Lemma 10.3.1 that it has vertex of degree 1.

25. Suppose there were a connected graph with eight vertices and six edges. Either the graph itself would be a tree or edges could be eliminated from its circuits to obtain a tree. In either case, there would be a tree with eight vertices and six or fewer edges. But by Theorem 10.3.2, a tree with eight vertices has seven edges, not six or fewer. This contradiction shows that the supposition is false, so there is no connected graph with eight vertices and six edges.

26. *Hint:* See the answer to exercise 25.

27. Yes. Suppose G is a circuit-free graph with ten vertices and nine edges. Let $G_1, G_2, \ldots, G_k$ be the connected components of G *[To show that G is connected, we will show that $k = 1$.]* Each G_i is a tree since each G_i is connected and circuit-free. For each $i = 1, 2, \ldots, k$, let G_i have n_i vertices. Note that since G has ten vertices in all,

$$n_1 + n_2 + \cdots + n_k = 10.$$

By Theorem 10.3.2,

$$G_1 \text{ has } n_1 - 1 \text{ edges,}$$
$$G_2 \text{ has } n_2 - 1 \text{ edges,}$$
$$\vdots$$
$$G_k \text{ has } n_k - 1 \text{ edges.}$$

So the number of edges of G equals

$$(n_i - 1) + (n_2 - 1) + \cdots + (n_k - 1)$$
$$= (n_1 + n_2 + \cdots + n_k) - \underbrace{(1 + 1 + \cdots + 1)}_{k \text{ 1's}}$$

$$= 10 - k.$$

But we are given that G has nine edges. Hence $10 - k = 9$, and so $k = 1$. Thus G has just one connected component, G_1, and so G is connected.

28. *Hint:* See the answer to exercise 27.

Section 10.4

1. a. 3 **b.** 0 **c.** 5 **d.** u, v
e. d **f.** k, l **g.** m, s, t, x, y

3. a.

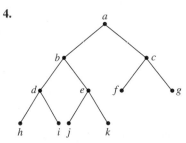

Exercises 4 and 8–10 have other answers in addition to the ones shown.

4.

5. There is no full binary tree with the given properties because any full binary tree with five internal vertices has six terminal vertices, not seven.

6. Any full binary tree with four internal vertices has five terminal vertices for a total of nine, not seven, vertices in all. Thus there is no full binary tree with the given properties.

7. There is no full binary tree with 12 vertices because any full binary tree has $2k + 1$ vertices, where k is the number of internal vertices. But $2k + 1$ is always odd, and 12 is even.

8.

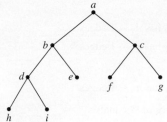

9.

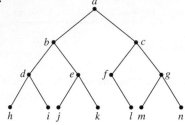

10.

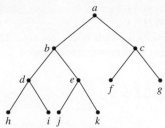

11. There is no binary tree that has height 3 and nine terminal vertices because any binary tree of height 3 has at most $2^3 = 8$ terminal vertices.

20. a. Height of tree $\geq \log_2 25 \cong 4.6$. Since the height of any tree is an integer, the height must be at least 5.

INDEX

CREDITS

This page constitutes an extension of the copyright page. We have made every effort to trace the ownership of all copyrighted material and to secure permission from copyright holders. In the event of any question arising as to the use of any material, we will be pleased to make the necessary corrections in future printings. Thanks are due to the following authors, publishers, and agents for permission to use the material indicated.

Chapter 1 **10** Problemy monthly, July 1959

Chapter 2 **23** Bettmann/CORBIS; **32** Culver Pictures; **59** Indiana University Archives

Chapter 3 **65** (top) Culver Pictures; **65** (bottom) Friedrich Schiller, Universitat Jena; **72** Public Domain; **89** Public Domain; **100** Culver Pictures

Chapter 4 **118** Courtesy of Donald Knuth; **124** (top) Bettmann/CORBIS; **124** (bottom) Andrew Wiles/Princeton University; **164** Bettmann/CORBIS; **168** (top) Courtesy Ben Joseph Green; **168** (middle) UCLA; **168** (bottom) The Art Gallery Collection/Alamy

Chapter 5 **174** CORBIS; **224** Academie Royale de Belgique; **225** (left) Courtesy of Francis Lucas; **225** (right) Courtesy of Paul Stockmeyer; **229** Bettmann/CORBIS

Chapter 6 **249** David Eugene Smith Collection, Columbia University; **253** Royal Society of London; **254** Stock Montage; **287** CORBIS; **290** Sylvia Salmi; **291** Public Domain

Chapter 7 **295** Stock Montage; **333** iStockphoto.com/Steven Wynn; **338** (top) Public Domain; **338** (bottom) Bettmann/CORBIS

Chapter 8 **363** Bettmann/CORBIS

Chapter 9 **403** Reprinted by permission of United Feature Syndicate, Inc.; **407** Bettmann/CORBIS; **464** Hulton-Deutch Collection/CORBIS

Chapter 10 **481** Wikipedia/Chris 73; **494** (top) Merian-Erben; **494** (bottom) Bettmann/CORBIS; **504** Bettmann/CORBIS; **514** (top) Courtesy of IBM Corporation; **514** (bottom) Courtesy of Peter Naur; **515** Bettmann/CORBIS

Reference Formulas

Topic	Name	Formula	Page
Logic	De Morgan's law	$\sim(p \wedge q) \equiv \sim p \vee \sim q$	32
	De Morgan's law	$\sim(p \vee q) \equiv \sim p \wedge \sim q$	32
	Negation of $\rightarrow$	$\sim(p \rightarrow q) \equiv p \wedge \sim q$	41
	Equivalence of a conditional and its contrapositive	$p \rightarrow q \equiv \sim q \rightarrow \sim p$	42
	Nonequivalence of a conditional and its converse	$p \rightarrow q \not\equiv q \rightarrow p$	43
	Nonequivalence of a conditional and its inverse	$p \rightarrow q \not\equiv \sim p \rightarrow \sim q$	43
	Negation of a universal statement	$\sim(\forall x \text{ in } D, Q(x)) \equiv \exists x \text{ in } D \text{ such that } \sim Q(x)$	76
	Negation of an existential statement	$\sim(\exists x \text{ in } D \text{ such that } Q(x)) \equiv \forall x \text{ in } D, \sim Q(x)$	76
Sums	Sum of the first n integers	$1 + 2 + \cdots + n = \dfrac{n(n+1)}{2}$	189
	Sum of powers of r	$1 + r + r^2 + \cdots + r^n = \dfrac{r^{n+1} - 1}{r - 1}$	193
Counting and Probability	Probability in the equally likely case	$P(E) = \dfrac{N(E)}{N(S)}$	405
	Number of r-permutations of a set with n elements	$P(n, r) = \dfrac{n!}{(n-r)!}$	418
	Number of elements in a union	$N(A \cup B) = N(A) + N(B) - N(A \cap B)$	427
	Number of subsets of size r of a set with n elements	$\dbinom{n}{r} = \dfrac{n!}{r!(n-r)!}$	449
	Pascal's formula	$\dbinom{n+1}{r} = \dbinom{n}{r-1} + \dbinom{n}{r}$	464
	Binomial theorem	$(a+b)^n = \displaystyle\sum_{k=0}^{n} \binom{n}{k} a^{n-k} b^k$	468
	Probability of the complement of an event	$P(A^c) = 1 - P(A)$	427